T0189713

Advances in Intelligent Systems and Computing

Volume 468

Series editor

Janusz Kacprzyk, Polish Academy of Sciences, Warsaw, Poland
e-mail: kacprzyk@ibspan.waw.pl

About this Series

The series "Advances in Intelligent Systems and Computing" contains publications on theory, applications, and design methods of Intelligent Systems and Intelligent Computing. Virtually all disciplines such as engineering, natural sciences, computer and information science, ICT, economics, business, e-commerce, environment, healthcare, life science are covered. The list of topics spans all the areas of modern intelligent systems and computing.

The publications within "Advances in Intelligent Systems and Computing" are primarily textbooks and proceedings of important conferences, symposia and congresses. They cover significant recent developments in the field, both of a foundational and applicable character. An important characteristic feature of the series is the short publication time and world-wide distribution. This permits a rapid and broad dissemination of research results.

Advisory Board

Chairman

Nikhil R. Pal, Indian Statistical Institute, Kolkata, India
e-mail: nikhil@isical.ac.in

Members

Rafael Bello, Universidad Central "Marta Abreu" de Las Villas, Santa Clara, Cuba
e-mail: rbellop@uclv.edu.cu

Emilio S. Corchado, University of Salamanca, Salamanca, Spain
e-mail: escorchado@usal.es

Hani Hagras, University of Essex, Colchester, UK
e-mail: hani@essex.ac.uk

László T. Kóczy, Széchenyi István University, Győr, Hungary
e-mail: koczy@sze.hu

Vladik Kreinovich, University of Texas at El Paso, El Paso, USA
e-mail: vladik@utep.edu

Chin-Teng Lin, National Chiao Tung University, Hsinchu, Taiwan
e-mail: ctlin@mail.nctu.edu.tw

Jie Lu, University of Technology, Sydney, Australia
e-mail: Jie.Lu@uts.edu.au

Patricia Melin, Tijuana Institute of Technology, Tijuana, Mexico
e-mail: epmelin@hafsamx.org

Nadia Nedjah, State University of Rio de Janeiro, Rio de Janeiro, Brazil
e-mail: nadia@eng.uerj.br

Ngoc Thanh Nguyen, Wroclaw University of Technology, Wroclaw, Poland
e-mail: Ngoc-Thanh.Nguyen@pwr.edu.pl

Jun Wang, The Chinese University of Hong Kong, Shatin, Hong Kong
e-mail: jwang@mae.cuhk.edu.hk

More information about this series at http://www.springer.com/series/11156

Suresh Chandra Satapathy · Vikrant Bhateja
Amit Joshi
Editors

Proceedings of the International Conference on Data Engineering and Communication Technology

ICDECT 2016, Volume 1

Editors
Suresh Chandra Satapathy
Department of CSE
Anil Neerukonda Institute of Technology and Sciences
Visakhapatnam, Andhra Pradesh
India

Vikrant Bhateja
Shri Ramswaroop Memorial Group of Professional Colleges (SRMGPC)
Lucknow, Uttar Pradesh
India

Amit Joshi
Sabar Institute of Technology
Tajpur, Sabarkantha, Gujarat
India

ISSN 2194-5357 ISSN 2194-5365 (electronic)
Advances in Intelligent Systems and Computing
ISBN 978-981-10-1674-5 ISBN 978-981-10-1675-2 (eBook)
DOI 10.1007/978-981-10-1675-2

Library of Congress Control Number: 2016944918

© Springer Science+Business Media Singapore 2017
This work is subject to copyright. All rights are reserved by the Publisher, whether the whole or part of the material is concerned, specifically the rights of translation, reprinting, reuse of illustrations, recitation, broadcasting, reproduction on microfilms or in any other physical way, and transmission or information storage and retrieval, electronic adaptation, computer software, or by similar or dissimilar methodology now known or hereafter developed.
The use of general descriptive names, registered names, trademarks, service marks, etc. in this publication does not imply, even in the absence of a specific statement, that such names are exempt from the relevant protective laws and regulations and therefore free for general use.
The publisher, the authors and the editors are safe to assume that the advice and information in this book are believed to be true and accurate at the date of publication. Neither the publisher nor the authors or the editors give a warranty, express or implied, with respect to the material contained herein or for any errors or omissions that may have been made.

Printed on acid-free paper

This Springer imprint is published by Springer Nature
The registered company is Springer Science+Business Media Singapore Pte Ltd.

Preface

The First International Conference on Data Engineering and Communication Technology (ICDECT 2016) was successfully organized by Aspire Research Foundation, Pune during March 10–11, 2016 at Lavasa City, Pune. The conference has technical collaboration with Div-V (Education and Research) of Computer Society of India. The objective of this international conference was to provide opportunities for the researchers, academicians, industry persons, and students to interact and exchange ideas, experience, and gain expertise in the current trends and strategies for information and intelligent techniques. Research submissions in various advanced technology areas were received and after a rigorous peer-review process with the help of program committee members and external reviewer, 160 papers in separate two volumes (Vol-I: 80, Vol-II: 80) were accepted. All the papers are published in Springer AISC series. The conference featured seven special sessions on various cutting-edge technologies, which were conducted by eminent professors. Many distinguished personalities as Dr. Ashok Deshpande, Founding Chair: Berkeley Initiative in Soft Computing (BISC)—UC Berkeley CA; Guest Faculty: University of California Berkeley; Visiting Professor: New South Wales University, Canberra and Indian Institute of Technology Bombay, Mumbai, India; Dr. Parag Kulkarni, Pune; Prof. Amit Joshi, Sabar Institute, Gujarat; Dr. Swagatam Das, ISI, Kolkata, graced the event and delivered talks on cutting-edge technologies.

Our sincere thanks to all Special Session Chairs (Dr. Vinayak K. Bairagi, Prof. Hardeep Singh, Dr. Divakar Yadav, Dr. V. Suma), Track Manager (Prof. Steven Lawrence Fernandes), and distinguished reviewers for their timely technical support. Thanks are due to ASP and its dynamic team members for organizing the event in a smooth manner. We are indebted to Christ Institute of Management for hosting the conference in their campus. Our entire organizing committee, staff of CIM, and student volunteers deserve a big pat for their tireless efforts to make the event a grand success. Special thanks to our Program Chairs for carrying out an immaculate job. We would like to place our special thanks here to our publication chairs doing a great job in making the conference widely visible.

Lastly, our heartfelt thanks to all authors without whom the conference would never have happened. Their technical contributions to make our proceedings rich are praiseworthy. We sincerely expect readers will find the chapters very useful and interesting.

Visakhapatnam, India
Lucknow, India
Tajpur, India

Suresh Chandra Satapathy
Vikrant Bhateja
Amit Joshi

Organizing Committee

Honorary Chair

Prof. Sanjeevi Kumar Padmanaban

Organizing Committee

Mr. Satish Jawale
Mr. Abhisehek Dhawan
Mr. Ganesh Khedkar
Mayura Kumbhar

Program Committee

Prof. Hemanth Kumbhar
Prof. Suresh Vishnudas Limkar

Publication Chair

Prof. Vikrant Bhateja, SRMGPC, Lucknow

Publication Co-Chair

Mr. Amit Joshi, CSI Udaipur Chapter

Technical Review Committee

Le Hoang Son, Vietnam National University, Hanoi, Vietnam
Nikhil Bhargava, CSI ADM, Ericsson India
Kamble Vaibhav Venkatrao, P.E.S. Polytechnic, India
Arvind Pandey, MMMUT, Gorakhpur (U.P.), India
Dac-Nhuong Le, VNU University, Hanoi, Vietnam
Fernando Bobillo Ortega, University of Zaragoza, Spain
Chirag Arora, KIET, Ghaziabad (U.P.), India
Vimal Mishra, MMMUT, Gorakhpur (U.P.), India
Steven Lawrence Fernandes, Sahyadri College of Engineering and Management
P.B. Mane, Savitribai Phule Pune University, Pune, India
Rashmi Agarwal, Manav Rachna International University, Faridabad, India
Kamal Kumar, University of Petroleum and Energy Studies, Dehradun
Hai V. Pham, Hanoi University of Science and Technology, Vietnam
S.G. Charan, Alcatel-Lucent India Limited, Bangalore
Frede Blaabjerg, Aalborg University, Denmark
Deepika Garg, Amity University, Haryana, India
Bharat Gaikawad, Vivekanand College campus, Aurangabad, India
Parama Bagchi, MCKV Institute of Engineering, Kolkata, India
Rajiv Srivastava, Scholar tech education, India
Vinayak K. Bairagi, AISSMS Institute of Information Technology, Pune, India
Rakesh Kumar Jha, Shri Mata Vaishnodevi University, Katra, India
Sergio Valcarcel, Technical University of Madrid, Spain
Pramod Kumar Jha, Centre for Advanced Systems (CAS), DRDO, India
Chung Le, Duytan University, Da Nang, Vietnam
V. Suma, Dayananda Sagar College of Engineering, Bangalore, India
Usha Batra, ITM University, Gurgaon, India
Sourav De, University Institute of Technology, Burdwan, India
Ankur Singh Bist, KIET, Ghaziabad, India
Agnieszka Boltuc, University of Bialystok, Poland
Anita Kumari, Lovely Professional University, Jalandhar, India
M.P. Vasudha, Jain University Bangalore, India
Saurabh Maheshwari, Government Women Engineering College, Ajmer, India
Dhruba Ghosh, Amity University, Noida, India
Sumit Soman, C-DAC, Noida, India
Ramakrishna Murthy, GMR Institute of Technology, A.P., India
Ramesh Nayak, Shree Devi Institute of Technology, Mangalore, India

Contents

About the Editors

Dr. Suresh Chandra Satapathy is currently working as Professor and Head, Department of Computer Science and Engineering at Anil Neerukonda Institute of Technology and Sciences (ANITS), Andhra Pradesh, India. He obtained his Ph.D. in Computer Science and Engineering from JNTU Hyderabad and M.Tech. in CSE from NIT, Rourkela, Odisha, India. He has 26 years of teaching experience. His research interests are data mining, machine intelligence, and swarm intelligence. He has acted as program chair of many international conferences and edited six volumes of proceedings from Springer LNCS and AISC series. He is currently guiding eight scholars towards Ph.D. Dr. Satapathy is also a Senior Member of IEEE.

Prof. Vikrant Bhateja is Professor, Department of Electronics and Communication Engineering, Shri Ramswaroop Memorial Group of Professional Colleges (SRMGPC), Lucknow and also the Head (Academics & Quality Control) in the same college. His area of research include digital image and video processing, computer vision, medical imaging, machine learning, pattern analysis and recognition, neural networks, soft computing, and bio-inspired computing techniques. He has more than 90 quality publications in various international journals and conference proceedings. Professor Bhateja has been on TPC and chaired various sessions from the above domain in international conferences of IEEE and Springer. He has been the track chair and served in the core-technical/editorial teams for international conferences: FICTA 2014, CSI 2014 and INDIA 2015 under Springer-ASIC Series and INDIACom-2015, ICACCI-2015 under IEEE. He is associate editor in International Journal of Convergence Computing (IJConvC) and also serving in the editorial board of International Journal of Image Mining (IJIM) under Inderscience Publishers. At present, he is guest editor for two special issues floated in International Journal of Rough Sets and Data Analysis (IJRSDA) and International Journal of System Dynamics Applications (IJSDA) under IGI Global publications.

Mr. Amit Joshi has an experience of around 6 years in academic and industry in prestigious organizations of Rajasthan and Gujarat. Currently, he is working as Assistant Professor in Department of Information Technology at Sabar Institute in Gujarat. He is an active member of ACM, CSI, AMIE, IEEE, IACSIT-Singapore, IDES, ACEEE, NPA, and many other professional societies. Currently, he is Honorary Secretary of CSI Udaipur Chapter and Honorary Secretary for ACM Udaipur Chapter. He has presented and published more than 40 papers in National and International Journals/Conferences of IEEE, Springer, and ACM. He has also edited three books on diversified subjects including Advances in Open Source Mobile Technologies, ICT for Integrated Rural Development, and ICT for Competitive Strategies. He has also organized more than 25 national and international conferences and workshops including International Conference ETNCC 2011 at Udaipur through IEEE, International Conference ICTCS 2014 at Udaipur through ACM, International Conference ICT4SD 2015 by Springer. He has also served on organizing and program committee of more than 50 conferences/seminars/ workshops throughout the world and presented six invited talks in various conferences. For his contribution towards the society, The Institution of Engineers (India), ULC, has given him Appreciation Award on the Celebration of Engineers, 2014 and by SIG-WNs Computer Society of India on ACCE, 2012.

Singular Value Decomposition and Discrete Wavelet Transform-Based Fingerprint Gender Classification

Ganesh B. Dongre and S.M. Jagade

Abstract Nowadays gender classification has immense value in technology and science because it helps to analyze the data easily. Due to uniqueness and reliability of fingerprint they can be used in civilian, industrial, commercial, unique Id of nation (AADHAR card) and even in judicial matters also. There are various methods of fingerprint based gender classification with their own limitations and strengths. In this paper, we presented Singular Value Decomposition (SVD) with Discrete Wavelet Transform (DWT) method for gender classification. SVD has good information packing characteristics and potential strengths in showing the results and DWT has good efficiency and less complexity for the gender identification using fingerprint. Fingerprint images from known gender are obtained for this study. After enhancement, features in frequency domain and spatial domain are obtained by DWT and SVD respectively. Euclidean distance is used to compare with the database. Our proposed method has success rate of around 92 %.

Keywords Gender · Fingerprint · Discrete wavelet transform · Singular value decomposition · Euclidean distance

1 Introduction

A fingerprint forms during pregnancy and remains for whole span of life with the person. If the finger of a person is injured, burned, the prints damages for little period of time. After getting well, it reconstructs as usual. It means there are no changes in the fingerprints. There is no chance to have same fingerprints of two persons even in twins also. Due to uniqueness of fingerprint it can be accepted in

G.B. Dongre (✉)
CSMSS College of Polytechnic, Aurangabad, Maharashtra, India
e-mail: ganeshbdongre@gmail.com

S.M. Jagade
College of Engineering, Osmanabad, Maharashtra, India
e-mail: smjagade@yahoo.co.in

© Springer Science+Business Media Singapore 2017

S.C. Satapathy et al. (eds.), *Proceedings of the International Conference on Data Engineering and Communication Technology*, Advances in Intelligent Systems and Computing 468, DOI 10.1007/978-981-10-1675-2_1

civilian, industrial, commercial and unique Id of nation and even in judicial matters also. Fingerprints have immense importance, with the help of which we can easily find out the various properties like age, gender etc. and which will be very much useful for the various applications.

Different biometric characteristics-based techniques has been used to differentiate the gender. In this paper, we used SVD and DWT to classify the gender from fingerprint. The 2D-Discrete Wavelet Transform is used to find out the frequency domain vector and Singular Value Decomposition (SVD) is used to find the spatial feature of the non-zero singular values.

2 Literature Review

For any fingerprint-based gender determination study, the basic step is fingerprint recognition in which feature extraction is a key point. Bana et al. in [1] proposed image segmentation method for recognition, in which specific area, i.e. RIO of the fingerprint image is obtained using block-based and morphological methods and are operated on. Park et al. [2] proposed systematic strategy of applying SIFT to fingerprint images. They extracted characteristic SIFT feature points in scale space and perform matching based on the texture information around the feature points using the SIFT operator. Using a public domain fingerprint database, they demonstrated that the proposed approach complements the minutiae-based fingerprint representation.

Gornale et al. [3] carried out gender identification using combining the features like Fast Fourier Transform, Eccentricity and Major Axis Length. From the available database, they selected left thumb impression from 450 male samples and 550 female samples. For better results, they selected a threshold for each transform. Their proposed method predicts results for male as 80 % and for female it is predicted as 78 %. Tom et al. [4] used 200 males and 200 females fingerprint samples of various age group persons for their analysis which is based on 2D discrete wavelet transforms in combination with PCA. They found that by increasing the samples of each category in the database leads to accuracy of the system.

Sudha Ponnarasi and Rajaram [5] carried out an analysis among 500 public people which includes 250 samples of males and 250 samples of females of various age groups from 1 to 90 years. They extracted features like ridge count, ridge count asymmetry, ridge thickness to valley thickness ratio, white lines count, and pattern type concordance. Support Vector Machines (SVM) were used for gender identification using the most dominant features.

Fingerprint identification systems generally work on stored database. If database is relatively small, the systems efficiency is much better than the system having heavy database. To deal with this issue, classification and indexing methods are used. Shuai et al. [6] addressed this issue using indexing method based on Scale Invariant Feature Transform local features to reduce generated features. On the cost of effectiveness they gained better efficiency in image indexing by proposed method.

According to Ritu Kaur et al. [7] ridge-based parameters were used in the traditional methods for identification of the gender. Using frequency domain analysis, relatively less work has been carried out. Therefore, they worked on frequency domain image analysis method. This technique is an efficient one which can be used for large biometric database for identification of gender. Badawi et al. [8] collected 2200 fingerprints from various persons. They extracted different ridge related features for their analysis. They used Fuzzy C Means and obtained 80.39 % result. For LDA, the result was 86.5 % and using Neural Network, 88.5 % result was obtained.

Specific region wise gender determination study based on fingerprint feature like ridge count is carried out in [9] by Kumar et al. According to their study, there is considerable difference in the ridge count between male and females in that region and is universal for other regions also.

Singular value decomposition and Scale Invariant Feature Transform based study on common fingerprint feature, i.e. Minutiae feature is carried out in [10, 11] respectively. Liu et al. [10] addressed common and weak feature descriptors issue using SVD minutiae matching method, where Zhou et al. [11] used SIFT-based minutiae descriptor which can perform feature extraction and matching between the features.

3 System Development

For the proposed system development, we are using DWT and SVD for feature extraction. The Euclidean distance is used for the calculation.

3.1 Feature Extraction

Fundamental step of any pattern recognition and machine learning problems is feature extraction. Here we used SVD and DWT for extraction of features from fingerprint images. Singular Value Decomposition is used for getting Eigen vectors and feature vectors in spatial domain, whereas the 2D-Discrete Wavelet Transform is used to determine the energy vectors and feature vectors in frequency domain. The fingerprint image is used as the input so as to obtain the feature vectors.

3.1.1 Feature Extraction Using DWT

Wavelet transform has been used mostly in image processing. 2-D Discrete Wavelet Transform will decompose an image into sub bands that are localized in frequency and orientation. This decomposition of images into different frequency subbands allows the isolation of the frequency components into certain subbands. This process is further used to isolate small changes in an image mainly in high frequency subband images. Hence 2-D Discrete Wavelet Transform is a suitable tool that can be used for designing a classification system.

3.1.2 Feature Extraction Using SVD

The Singular Value Decomposition is an algebraic technique for factoring any rectangular matrix into the product of three other matrices. Each of which has important properties.

3.2 Euclidean Distance

The Euclidean distance is used in this for the calculation of the distance between any two points p and q. It is also useful for the decision to be taken during the gender identification and classification.

3.3 Method

Fingerprint images from known gender are obtained for this proposed method. It consisted of 500 male samples and 500 female samples, of age group ranging from 16 to 50 years. Images are taken using SecuGen Hamster plus fingerprint scanner having following specifications

Model name	Hamster Plus
Optical fingerprint sensor	SDU03P
Image resolution	500 dpi
Image size	260*300 pixels
Effective sensing area	13.2 mm * 15.2 mm
Capture speed	0.2–5 s
Operating temperature	−20–65 °C
Supply voltage	+5v DC
Maximum current	140 mA

These sample images are used as it is for the experimentation. Features in frequency domain and spatial domain are obtained by DWT and SVD respectively. Euclidean distance is used to compare with the database. On the basis of minimum Euclidean distance the result are displayed. MATLAB simulation tool is used for testing the results.

3.3.1 Process

1. Select samples of fingerprint from database.
2. Extract the feature vector for all samples.
3. Select a query image and extract feature vector of 256*1 using SVD and 1*256 using DWT.
4. Combine obtained feature vectors into 512*1 vectors.
5. Compare it with database and calculate Euclidean distance and judge the result 1.
6. Compare query image with the average male and female values obtained from database to calculate D1 and D2.
7. Compare D1 and D2 and judge the result 2.
8. Compare query image with threshold Euclidean distance obtained from database and judge the result 3.

In our implementation we use singular value decomposition which extracts fingerprint features from a query fingerprint and gives 256*1 feature vector for query image (Fig. 1).

The image undergoes 2D discrete wavelet transform returns with another feature vector of size 1*256. Further we have combined these vectors to form a new vector with 512 feature values. These values are compared with the database using Euclidean distance method. On the basis of minimum Euclidean distance the result is displayed as shown in Table 1.

As per some of the papers referred, it is found that nearly about 78–80 % efficiency can be obtained using FFT, Eccentricity and major axis length [3] and using FCM, LDA and NN, the efficiency was 80–88 %, whereas the proposed method gives 90–94 % accuracy (Table 2).

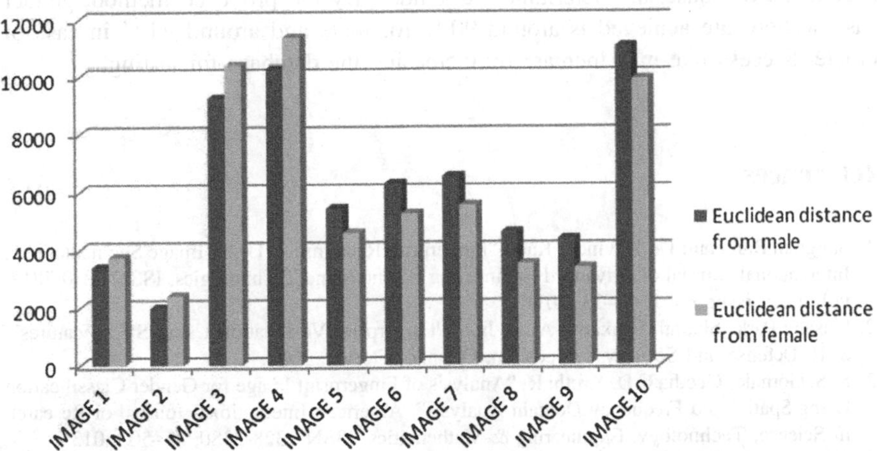

Fig. 1 Graphical representation of sample results

Table 1 Euclidean Distance (ED) between dataset of male (D1) and dataset of female (D2)

Image	Known gender	ED of male sample	ED of female sample	Results found
Image 1	Male	3391.575608	3749.530969	Male
Image 2	Male	1987.307843	2403.459813	Male
Image 3	Male	9261.837898	10400.52664	Male
Image 4	Male	10289.83464	11376.80079	Male
Image 5	Male	5429.380551	4580.434083	Female
Image 6	Female	6285.152526	5238.860764	Female
Image 7	Female	6533.97457	5541.497258	Female
Image 8	Female	4599.818816	3501.922643	Female
Image 9	Female	4369.721854	3373.649246	Female
Image 10	Female	11045.74848	9905.809451	Female

Where, D1 = Euclidian Distance from male average values
D2 = Euclidian Distance from female average values

Table 2 Confusion matrix for male–female identification based on combined results of SVD and 2-D DWT

Actual\predicted	Male	Female	Total
Male	447	53	500
Female	35	465	500
Total	482	518	1000

4 Conclusion

Frequency features from the wavelet transform and the spatial features from the singular value decomposition are used to classify gender. Euclidean distance is used to compare database and determine the gender. By the proposed method, gender classification rate achieved is around 90 % for male and around 94 % in case of female. Success rate may increase by increasing the database for testing.

References

1. Sangram Bana and Dr. Davinder Kaur, “Fingerprint Recognition Using Image Segmentation”, International Journal of Advanced Engineering Sciences and Technologies, ISSN 2230-7818, vol. no. 5, issue no. 1, 12–23, 2011.
2. Unsang Park, Sharath Pankanti, A. K. Jain, “Fingerprint Verification Using SIFT Features”, SPIE Defense and Security Symposium, Orlando, Florida, 2008.
3. S. S. Gornale, Geetha C D, Kruthi R, “Analysis of Fingerprint Image for Gender Classification Using Spatial and Frequency Domain Analysis”, American International Journal of Research In Science, Technology, Engineering & Mathematics, ISSN 2328-2380, 46–50, 2013.
4. Rijo Jackson Tom, T. Arulkumaran, “ Fingerprint Based Gender Classification Using 2D Discrete Wavelet Transforms and Principal Component Analysis”, international Journal of Engineering Trends and Technology, ISSN 2231-5381, volume 4, Issue 2, 199–203, 2013.

5. S. Sudha Ponnarasi, M. Rajaram, "Gender Classification System Derived From Fingerprint Minutiae Extraction", International Conference On Recent Trends In Computational Methods, Communication and Controls, Published in International Journal of Computer Applications, 1–6, 2012.
6. Xin Shuai, Chao Zhang and Pengwei Hao, "Fingerprint Indexing Based On Composite Set of Reduced SIFT Features" International Conference on Pattern Recognition, ISBN 978-1-4244-2175-6, 1–4, 2008.
7. Ms. Ritu Kaur, Mrs. Susmita Ghosh Mazumdar, Mr. Devanand Bhonsle, "A Study On Various Methods of Gender Identification Based On Fingerprints", International Journal of Emerging Technology and Advanced Engineering ISSN 2250-2459, Volume 2, Issue 4, 532–537, 2012.
8. Ahmed Badawi, Mohamed Mahfouz, Rimon Tadross, Richard Jantz, "Fingerprint-Based Gender Classification" International Conference on Image Processing, Computer Vision & Pattern Recognition, Volume 1, 41–46, 2006.
9. Lalit Kumar, Sandeep Agarwal, Rajesh Garg, Amit Pratap, V K Mishra, "Gender Determination Using Fingerprints In the Region of Uttarakhand", J Indian Acad Forensic Med., ISSN 0971-0973, vol. 35, no. 4, 308–311, 2013.
10. Fei Liu, Gongping Yang, Yilong Yin, Shuaiqiang Wang, " Singular Value Decomposition Based Minutiae Matching Method for Finger Vein Recognition" Neurocomputing 145, 75–89, 2014.
11. Ru Zhou, Dexing Zhong, and Jiuqiang Han, " Fingerprint identification using sift-based minutia Descriptors and improved all descriptor-pair matching", Sensors 13, ISSN 1424-8220, 3142–3156, 2013.

Isolated Word Speech Recognition System Using Deep Neural Networks

Dhavale Dhanashri and S.B. Dhonde

Abstract Speech recognition is the process of converting speech signals into words. For acoustic modeling HMM-GMM is used for many years. For GMM, it requires assumptions near the data distribution for calculating probabilities. For removing this limitation, GMM is replaced by DNN in acoustic model. Deep neural networks are the feed forward neural networks having more than one or multiple layers of hidden units. In this work, we have presented the isolated word speech recognition system using acoustic model of HMM and DNN. We are using Deep Belief Network pre-training algorithm for initializing deep neural networks. DBN is a multilayer generative probabilistic model with large number of stochastic binary units. The features used are the mel-frequency cepstrum coefficients (MFCC). Experimental results are calculated on TI digits database. Proposed system has achieved 86.06 % accuracy on TI digits database. System accuracy can be further increased by increasing the number of hidden units.

Keywords Isolated word speech recognition · Deep neural networks · Deep belief networks · Acoustic model

1 Introduction

According to definition biometric is something, you are. For every individual there are some physical characteristics that are unique. It includes mainly fingerprint, iris, speech. A combination of these characteristics can be used to build a model which will be used to identify individuals. Speech signal carries important information. It gives information about the speech and speaker identity. Speech identification systems extract information from the speech signal independent of the speaker.

Dhavale Dhanashri (✉) · S.B. Dhonde
Department of Electronics, AISSMS-IOIT, Pune, Maharashtra, India
e-mail: dhanashrimd@gmail.com

S.B. Dhonde
e-mail: dhondesomnath@gmail.com

© Springer Science+Business Media Singapore 2017

S.C. Satapathy et al. (eds.), *Proceedings of the International Conference on Data Engineering and Communication Technology*, Advances in Intelligent Systems and Computing 468, DOI 10.1007/978-981-10-1675-2_2

Since 1950 speech recognition has been in the research field for developing techniques for isolated digits or for continuous speech. Speech recognition can be considered as a process which is converting speech signals into spoken words [1]. In this paper, we are explaining about the speech recognition system for isolated words identification.

Much research has been done in the speech recognition field for isolated digits as well as for continuous speech. Many algorithms are invented like Dynamic Time Wrapping, Hidden Markov Model, and Gaussian Mixture Model [2]. Neural networks are used as a classifier for speech recognition system. There are many types of neural networks like feed forward neural networks and feedback neural networks and recurrent neural networks [3]. Many researchers came with the new learning algorithms for the neural networks. Later it came to know that more the deeper network, performance of speech recognition system is higher.

Deep neural networks are used in speech recognition for classification. Deep neural networks are the multilayer, feedforward neural networks. In this paper our approach introduces a system containing acoustic model of HMM and DNN used for identifying digits from zero to nine.

2 Literature Review

Much research has been done in the field of speech recognition technology related to the application of neural networks. Neural networks are similar to human brain. They consist of a processing element called neuron. Combination of artificial neural network and hidden markov model is used for speech recognition technology. Previously ANN is used to estimate the posterior probabilities of a continuous density HMMs' state given the acoustic features and then backpropagation algorithm is used to train that network [4]. But by using backpropagation algorithm it is found that it is difficult to train network having more than two layers.

Under this section, we are addressing the survey about research made in neural networks. In [5] they have proposed a spectral masking system to noise robust speech recognition with the help of deep neural networks [5]. A spectral masking system has been proposed by them in which power spectral domains are calculated using DNN. For further increasing performance, Lin Adaptation is applied to mask estimator and acoustic model of DNN. They have evaluated the system accuracy on Aurora2 and Aurora4 database. The system yielded word error rates of 4.6 and 11.8 %. Polur et al. in [6] implemented a system using artificial neural networks. They have calculated system accuracy for two words 'yes' and 'no.' They have used artificial neural networks in conjunction with cepstral analysis in isolated word recognition system.

In [7], they have introduced two generalized maxout units called as p-norm and soft-maxout. They evaluated their results on large vocabulary and calculated that p norm generalization works better. In [8] they have introduced the use of neural networks for speaker identification. As for neural networks it is difficult to train on large number of units. In this paper, they have proposed neural network array combining the binary partitioned approach with decision trees. This approach reduces the computational cost and classification error rate. Mohamed et al. in [9] proposed an acoustic model of HMM and DNN for speech recognition system. They have evaluated the system performance on TIMIT database. A context-dependent model for phone recognition has been presented by Dahl [4]. This model consists of hybrid architecture of HMM and DNN. They have applied this model to large vocabulary tasks. A new algorithm for learning deep neural networks is given in this paper. It is the DBN pre-training algorithm which helps in reducing recognition error. They performed experiments on business search dataset. They have compared their results with conventional HMM-GMM system. They got improvement in accuracy up to 5.8 and 9.2 %. In [10] they have proposed an acoustic model of HMM and DNN for speech recognition system. They have implemented the system using auto encoders for pre-training algorithm. Their results shows that system gives better performance for auto encoders.

In [11] they have proposed the use of deep neural networks for extracting features from mel scale filter banks and have used DBN in combination with HMM. They have evaluated their results on Tagalog database. In [3] they have make use of speech enhancement to remove noise from speech signal before extracting it. Dynamic Programming algorithm is used to calculate the similarity between templates Use of deep recurrent neural networks is presented in [12]. They have introduced end to end learning method for recurrent neural networks. They have evaluated their results on TIMIT dataset with 462 speakers. They have achieved test error of 17.7 % on TIMIT phoneme recognition benchmark.

There are many approaches of using neural networks for speech recognition. Many authors have implemented the systems using neural networks. Some of them used combination of HMM and DNN for large vocabulary speech and others have used artificial neural networks for small vocabulary. In this work instead of using that approach, we are using deeper neural network to replace traditional neural network. Also we are trying to improve the earlier hybrid approach using DBN pre-training algorithm for training DNN.

3 System Overview

In this paper, we are going to introduce a system, i.e., isolated word speech recognition system using acoustic model of HMM and DNN. Related to learning algorithms of neural networks there are many sentences made in past. Many algorithms have introduced in the literature. In this approach we are using deep belief network pre-training algorithm for the deep neural networks [4]. Many researchers proved that neural networks with pre-training achieve better response than networks without pre-training. We are using the feed forward, multilayer neural network for producing posterior probabilities of HMM states as output by using feature vectors as input [9].

It has been concluded that when using one or two hidden layers, back propagation algorithm is useful. Results are not satisfactory if the numbers of hidden layers are increased for back propagation algorithm.

In this paper, we have replaced GMM by HMM for narrating HMM states to feature vector. Here we are using multilayer, feed forward neural networks that use feature vectors as input and produces posterior probabilities of HMM states as output [9]. Previously for training neural networks backpropagation algorithm have been used. In this paper we are using pre-training algorithm to train multilayer neural networks to achieve better performance.

3.1 *Preprocessing*

The input speech signals are in the form of audible waveform, they are converted to electrical pulses. After that speech signal is preprocessed. Recognition rate can be distorted by noise and difference in amplitude of the speech signal [13]. These problems can be overcome by doing pre-processing part which consists of normalization, pre-emphasizing, windowing [14]. In order to adjust the volume of audio files to a standard level normalization is used.

In order to preprocess a signal, it should be pre-emphasized before. At the time when sound is produced, some part of speech signal is suppressed. So this suppressed part which is a high frequency part is compensated by doing pre-emphasis of speech signal.

A signal which is converted into frames is taken. It is an unstable signal. It remains stationary for very short time. In windowing method, we are applying hamming window to speech signal. It reduces the signal discontinuities mainly at the end and start of signal. Hamming window equation is given by

$$w(n) = 0.54 - 0.46\cos\left(\frac{2\pi n}{M-1}\right) \tag{1}$$

3.2 *Feature Extraction*

Mel-Frequency Cepstral Coefficients (MFCC) have been the most widely used speech features for speech recognition [13]. Feature extraction converts higher dimensional input signals into lower dimensional vectors. In the process of feature extraction first the speech signal is converted to frequency domain by doing FFT on signal. It is applied to speech signal in order to obtain magnitude frequency response of each frame. The spectrums obtained are mapped onto the mel scale using triangular overlapping window. For mel-frequency scale it is having linear frequency spacing below 1000 Hz and a logarithmic spacing above 1000 Hz [11]. Here the frequency axis in Hz is converted into mel scale using formula

$$\text{mel frequency} = 2595 * \log_{10}(1 + \frac{f}{700}) \tag{2}$$

Finally log is taken for the output and DCT, discrete cosine transform is applied to the log mel powers. The magnitudes of the resulting spectrum are the required MFCC's. The equation for DCT is given by

$$X_k = \sum_{n=0}^{N-1} x_n e^{-j2kn\pi/N}, k = 0, 1, \ldots N-1 \tag{3}$$

4 Acoustic Modeling

When applying deep neural networks as an acoustic model to speech recognition system in combination with HMM, it is used to produce posterior probabilities of HMM states as output [9]. Here we are using DBN pre-training algorithm for building deep neural networks. It is better to learn a generative model when there is large number of unlabeled data in combination with the training data [9]. Deep belief network is a multilayer generative model. Here the deep neural networks are trained layer by layer.

5 Deep Belief Networks

These are the multilayer generative probabilistic models with large number of stochastic binary units. In this the bottom layer is visible layer v, and others are the hidden layers, h. Figure 1 shows the graphical model of deep belief network. It shows the layer of hidden and visible layers. It shows, there is a undirected connection between top two layers and RBM is formed by this. And the bottom layers are in top-down directed manner [4]. By training deep neural networks for input feature vectors, probability distributions for HMM states can be obtained. Using some preexisting model, correct state can be obtained and weights of neural networks are adjusted in order to increase the probability for correct HMM state [11].

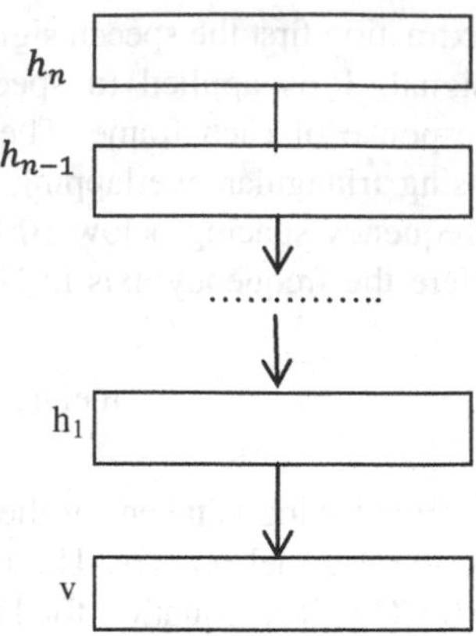

Fig. 1 Graphical model of deep belief network

6 Restricted Boltzmann Machines

Restricted Boltzmann machines (RBMs), these are the undirected graphical models consisting visible layer and hidden layer. These are not actually deep architectures [4]. All units are connected in a well manner. Also there is no connection between the units of same layer. RBM's are the building blocks of deep belief networks.

RBM's assign an energy value to each configuration so that dependencies between variables can be obtained with more probable configurations having a lower energy [9, 4]. Energy term associated with visible, v and hidden, h units is given by equation,

$$E(v,h) = -\sum_{i=1}^{V}\sum_{j=1}^{H} w_{ij}v_ih_j - \sum_{i=1}^{V} b_iv_i - \sum_{j=1}^{H} a_jh_j \tag{4}$$

where w_{ij} is the weight assigned between the connections of hidden and visible units, b_i and a_j are their bias terms. The probability of a given configuration is given by eq,

$$P(v,h) = \frac{e^{-E(v,h)}}{Z} \tag{5}$$

where z is the normalization factor, $Z = \sum_{v,h} e^{-E(v,h)}$

As RBM is an energy-based model, training of RBM is like obtaining probability distribution. In RBM it is done by decreasing the energy of most probable regions of the state space while boosting the energy of less probable regions. Contrastive divergence is used for learning RBM [4].

It uses the following rules for updating weights and biases,

$$\Delta W_{ij}^{0} \propto \langle v_i h_j \rangle_0 - \langle v_i h_j \rangle_n \tag{6}$$

$$\Delta b_i^{0} \propto \langle v_i \rangle_0 - \langle v_i \rangle_n \tag{7}$$

$$\Delta b_i^{1} \propto \langle h_i^1 \rangle_0 - \langle h_i^1 \rangle_n \tag{8}$$

where $\langle . \rangle_0$ is the average when visible units are imposed to the input values and hidden units are taken out, and $\langle . \rangle_n$ is the average after assigning the input data to visible units.

After pre-training a stack of RBMs, the weights and biases of the hidden units can be utilized to initialize the hidden layers of a deep belief network.

Once we have trained an RBM, we can use it to represent it as data. For each input vector, v we compute a hidden unit activation probabilities, h which we can be used to train next RBM. Hence RBM weights are used to extract features from pervious layer. After completing the training of RBM, we got weights of the hidden layers which will be used to build deep neural network. After that back propagation algorithm is used for fine tuning all weights in the network.

7 Experimental Results

7.1 Dataset Description

In this paper, we have evaluated all the results on TI Digits database. It is speaker independent connected digit database. The data were collected in a quiet environment and digitized at 20 kHz. The utterances collected from the speakers are digit sequences. Eleven digits were used: "zero," "one," "two," …, "nine," and "oh." The data is divided into two sets, training and testing. Each contains 55 men and 57 women while 56 men and 57 women samples.

The training data is used to train the network. Each file of speaker contains two utterances of each word. The speech samples from training data are read using wavread command. Features are extracted using MFCC. We have used voicebox speech processing toolbox for our experiment (Fig. 2).

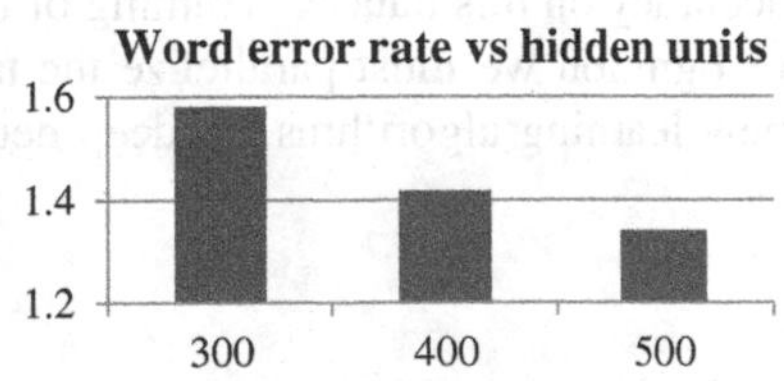

Fig. 2 Relationship between hidden units and word error rate

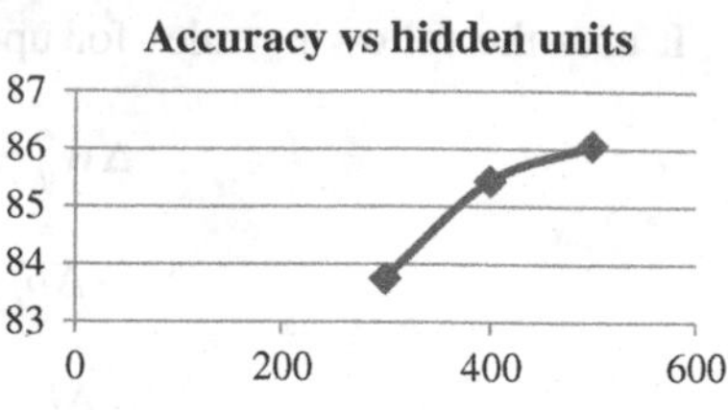

Fig. 3 Relationship between hidden units and accuracy

7.2 Results

We tested the results on dataset of 56 men and 57 women samples. In this we have calculated the system accuracy for different number of hidden layers. We have calculated it for 300, 400, 500 hidden units. Our experiment for 300 hidden units achieved 83.76 % accuracy. By increasing hidden units to 400, accuracy is increased up to 85.44 %. We have got 86.06 % accuracy for 500 hidden units. These results are shown in Fig. 3. Also the word error rate for each word is calculated. Figure 2 shows the graph of word error rate versus hidden units.

As shown in Fig. 3, accuracy increases with increase in number of hidden units. For RBM, we have used 100 epochs. In the experiment, we have calculated results for using deep neural networks only for classification. We have got results for different hidden units such as for 300 units accuracy achieved is 66.33 %.

8 Conclusion

In this paper, we have proposed an isolated word speech recognition system using acoustic model of HMM and DNN. Having large number of hidden units, deep neural networks are used as a classifier in speech recognition system. In case of GMM they are statistically incapable at modeling high-dimensional data that has any kind of componential structure. As a replacement for GMM, DNN is used in combination with HMM in acoustic modeling. In this paper, we gave a brief view about the deep neural networks and the newly invented pre-training algorithm using deep belief networks.

In [15] they have implemented the system using DTW technique. They have used Least Mean Square algorithm for removing the noise. They have got 93 % recognition rate using enhancement and 82 % without enhancement. We have evaluated our results on TI Digits dataset. The system has achieved 86.06 % accuracy on this dataset. Training of DNN requires much time. For achieving better recognition we must parallelize the training. Research is going on for discovering new learning algorithms for deep neural networks.

References

1. M.A. Anusuya, S.K. Katti.: Speech Recognition by Machine: A Review. In: International Journal of Computer Science and Information Security, Vol. 6, No. 3, pp. 181–205, (2009).
2. Khalid Saeed, Mohammad Kheir Nammous.: A Speech-and-Speaker Identification System: Feature Extraction, Description, and Classification of Speech-Signal Image, In: IEEE Transactions on Industrial Electronics, Vol. 54, No. 2, pp 887–897, (2007).
3. Veera Ala-Keturi.: Speech Recognition Based on Artificial Neural Networks. In: Helsinki hnology Institute of Technology, (2004).
4. George E. Dahl, Dong Yu, Li Deng, and Alex Acero.: Context-Dependent Pre-Trained Deep Neural Networks for Large-Vocabulary Speech Recognition. In: IEEE Transactions On Audio, Speech, And Language Processing, Vol. 20, No. 1, pp, 30–42, (2012).
5. Bo Li and Khe Chai Sim.: A Spectral Masking Approach to Noise-Robust Speech Recognition Using Deep Neural Networks. In: IEEE/ACM Transactions On Audio, Speech, And Language Processing, Vol. 22, No. 8, pp. 996–1305, (2014).
6. Prasad D Polur, Ruobing Zhou, Jun Yang, Fedra Adnani, Rosalyn S.: Hobsod Isolated Speech Recognition Using Artificial Neural Network. In: 2001 Proceedings of the 23rd Annual EMBS International Conference Istanbul, pp 1731–1734, (2001).
7. Xiaohui Zhang, Jan Trmal, Daniel Povey, Sanjeev Khudanpur.: Improving Deep Neural Network Acoustic Models Using Generalized Maxout Networks. In: IEEE International Conference On Acoustic, Speech and Signal, pp 214–219, (2014).
8. Xicai Yue, Datian Ye, Chongxun Zheng, Xiaoyu Wu.: Neural networks for improved text independent speaker identification. In: IEEE Engineering In Medicine And Biology, pp 53–58, (2002).
9. Abdel-rahman Mohamed, George E. Dahl, and Geoffrey Hinton.: Acoustic Modeling Using Deep Belief Networks. In: IEEE Transactions On Audio, Speech, And Language Processing, Vol. 20, No. 1, pp. 14–22, (2012).
10. Geoffrey Hinton, Li Deng, Dong Yu, George E. Dahl, Abdel-rahman Mohamed, Navdeep Jaitly, Andrew Senior, Vincent Vanhoucke, Patrick Nguyen, Tara N. Sainath, and Brian Kingsbury.: Deep neural networks for acoustic modeling in speech recognition. In: IEEE Signal Processing Magazine, pp 82–97, (2012).
11. Jonas Gehring, Wonkyum Lee, Kevin Kilgour, Ian Lane, Yaije Miao, Alex Waibel.: Modular Combination of Deep Neural Networks for Acoustic Modeling. In: INTERSPEECH (2013).
12. Alex Graves, Abdel-rahman Mohamed and Geoffrey Hinton.: Speech Recognition With Deep Recurrent Neural Networks. In: ICASSP, pp 6445–6449, (2013).
13. Song Yang, Meng Joo Er, and Yang Gao.: A High Performance Neural-Networks-Based Speech4 Recognition System. In: IEEE, pp 1527–1531, (2000).
14. Wouter Gevaert, Georgi Tsenov, Valeri Mladenov.: Neural Networks used for Speech Recognition. In: Journal Of Automatic Control, University Of Belgrade, Vol. 20, pp. 1–7, (2010).
15. Siva Prasad Nandyala, T. Kishore Kumar.: Real Time Isolated Word Recognition using Adaptive Algorithm. In: International Conference on Industrial and Intelligent Information, pp 163–168, (2012).

A Proposed Resource Sharing Architecture for Multitenant SaaS Applications

G.B. Pallavi and P. Jayarekha

Abstract In today's era of cloud computing technology, Software as a Service (SaaS) is one of the most widely adopted services by the customers. In a SaaS scenario, both service providers and customers yield tremendous economy of scale. Service providers deliver a single application instance among multiple organizations resulting in reduced operational and maintenance cost. They, in turn offer the service at a lower cost to customers. This has become possible by embedding a promising multitenant architecture in the development of SaaS applications, where multiple organizations referred as tenants share a common application instance and in turn a common database. However, designing such multitenant applications for its effective delivery to thousands of customer pose several challenges in the arena of resource sharing, security, scalability etc. This work in particular attempts to address the problem of fixed resource allocation through virtualized instances among tenants and propose a possible dynamic resource sharing architecture that can support in extemporizing resource usage nonetheless maintaining service level agreement requirements and isolation among tenants.

Keywords SaaS · Multitenancy · Resource sharing · FCFS

1 Introduction

Multitenancy is one of the key concerns in SaaS. It refers to a principle in software architecture which is the ability to enable SaaS applications to serve multiple client organizations using a single service instance. In a multitenant scenario, multiple tenants share the same resources while partaking the software, resulting in higher hardware utilization. These resources may include CPU, memory, secondary stor-

---

G.B. Pallavi (✉) · P. Jayarekha
BMS College of Engineering, Bangalore, India
e-mail: pallavi.cse@bmsce.ac.in

P. Jayarekha
e-mail: jayarekha.ise@bmsce.ac.in

© Springer Science+Business Media Singapore 2017

S.C. Satapathy et al. (eds.), *Proceedings of the International Conference on Data Engineering and Communication Technology*, Advances in Intelligent Systems and Computing 468, DOI 10.1007/978-981-10-1675-2_3

age etc. However, if one tenant blocks up the resources, the performance of all the other tenants may be restricted and a fair sharing of the resources may be compromised. In a virtualized instance situation this problem is solved by assigning an equal amount of resource to each tenant [1]. This solution may again lead to inefficient utilization of resources as tenants may utilize them to the maximum limit only during peak workload and at the rest of the time they will be idle resulting in underutilization. These issues limit the kind of applications that can migrate to multi-tenant cloud and leverage shared services. They also hinder cloud providers from offering differentiated service levels in which some tenants can pay for performance isolation and predictability while others choose standard "best-effort" behavior [2].

In this paper, underutilization of resources is addressed by proposing a resource sharing mechanism by considering CPU as the resource of study. In order to avoid the drawbacks of assigning equal amount of CPU resource we consider a pool of shared CPU which can be accessed by different tenants based on their requirement. In any computing scenario, scheduling is the process by which processes are given access to CPU [3]. The need for a scheduling algorithm arises from the requirement of fast computer systems in a cloud scenario to perform multi-tasking. In this work, we focus on implementation of two different CPU scheduling algorithm viz., round robin and weighted first come first served to access in order to gain access to pooled CPU.

The rest of the paper is organized as follows. In Sect. 2, the related work on multitenant applications is discussed. In Sect. 3, the architecture and design details of the proposed system and its settings are described. In Sect. 4 experimental setup of a simple multi-user, shared application scenario and results are elaborated. Finally Sect. 5 concludes the paper.

2 Related Work

Many related work has been carried out in the area of multitenancy. Research on architecture, implementation and security issues are in a move. Shue et al. [2], consider storage as a shared resource among tenants, has developed a system called "Pisces" for efficient and fair sharing of the same. However, the system does not take other resources into consideration. In Das et al. [4], have worked on CPU sharing techniques for a database as service structure where they have developed system called SQLVM which guarantees to provide a promised set of reserved key resources provided tenant has sufficient demand for the same. They suggest algorithms like largest deficit first which aims at avoiding tenants starving for the CPU in case of lower level SLA. Nevertheless, the system developed is for a static environment. Authors Cui et al. [5] have suggested a scheduling algorithm for multitenant instance intensive workflow targeting various workflow instances from multiple tenants to be scheduled which aims at improving quality of service and saving the execution cost of workflow.

TBRAM-a model on tenant based resource allocation has been developed by the authors of [6] where they investigate the influence of tenant based resource allocation model on cost effectiveness of SaaS system. SmartSLA—a cost aware resource management system developed by Xiong et al. [7] which uses system modeling module and resource allocation decision module. The system modeling module uses machine learning techniques to learn a model that describes the potential profit margin for each client under different resource allocations. Based on the learning module, the resource allocation decision module dynamically adjusts the resource allocation in order to achieve optimum profit. Sudipto et al. [8] has developed a technique called "Albatross" for live migration of tenant databases serving OLTP style workloads where the persistent database image is stored in network attached storage, ensuring minimal impact on transaction execution and guaranteeing serializability. In Yang et al. [9], propose a hybrid approach for placement of tenants with the purpose of lowering cost of ownership by high economies of scale which uses a combination of service selection with genetic algorithm, case based reasoning, resource consumption estimation model, and heuristic approach. Zhiming et al. [10] has developed a system termed "CloudScale" that employs a prediction-driven elastic resource scaling system for multi-tenant cloud computing that can meet the SLO requirements of the applications with minimum resource and energy cost. Rouven et al. [11] demonstrates resource usage control by implementing admission control mechanisms and has tested the same using a multitenant enabled TPC-W benchmark application. However research on architecture, development and engineering multitenant SaaS applications has been carried in [12–14]. Stefan et al. [15] has made comparisons of various PaaS offerings to develop SaaS applications from the perspective of portability of the application code base, available support for creating and managing multi-tenant aware applications, and quality of the tool support.

3 Architecture and Design

In this section, an architecture which avoids the need of assigning equal amount of CPU to each tenant in order to execute its workload is proposed. Rather than managing individual CPU partitions, provider can consolidate the entire available CPU as a single black box, in which aggregate CPU capacity and request rates can be elastically scaled on demand. Tenants will be contending for CPU and the degree of sharing CPU between tenants may be significantly higher and runnier than with fixed VM resource allocation. To improve usability of shared CPU with a high degree of resource sharing and contention, fair and efficient scheduling algorithms are implemented at an intermediary level between tenant's virtual machines and shared CPU. Figure 1 shows the high level architecture of shared CPU service in a multitenant scenario that provides system-wide, per-tenant fairness and isolation.

In this architecture, per-tenant fairness is provided at the scheduling level based on class of tenants. The design is restricted for two classes of tenants, GOLD and

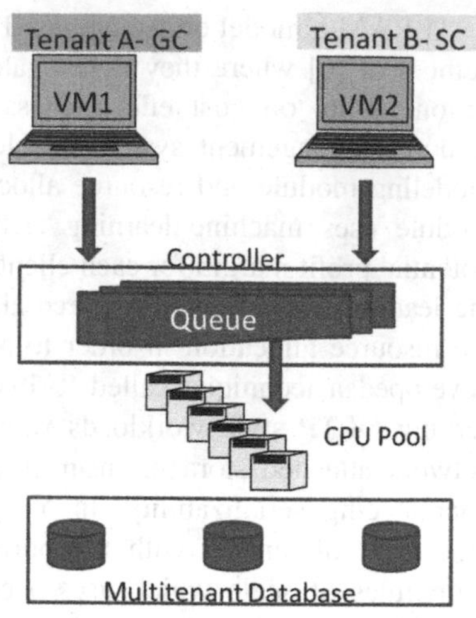

Fig. 1 Controller associated with pooled CPU

SILVER class. Two tenants say T1 and T2 one belonging to GOLD and the other to SILVER are assumed to run the shared multitenant application instances in two virtual machines V1 and V2. Followed by these machines will be the intermediary block termed as "Controller" block which embraces the required scheduling algorithm. Based upon the scheduling algorithms implemented in the block and class of tenants, the queries/tasks from any of the tenants gets queued up in this block to get access to pooled CPU. The CPU in turn performs the required processing of individual tenant by gaining access to multitenant database.

4 Experiments and Results

The proposed architecture has been implemented for a non-multitenant, multi-user application. Application is developed using Java and JSP. The web server to host the application is Apache Tomcat and Cloudsim 3.0 to simulate cloud environment. We have developed an application named Application Downloader which can be hosted as a service. Various customers can register and gain access to the service and download the required software. Different customers get registered under different classes to gain access to application. Once they access the application and pose a request to download particular software, the requests will be scheduled using scheduling algorithms like round robin, first come first serve or Weighted FCFS algorithm by the Controller module. Two customers T1, T2 are simulated where T1 belongs to gold class and T2 belong to silver class.

The following steps indicate the working of the proposed solution.

- A scenario of 2 virtual machines V1, V2 are considered.
- Two customers T1, T2 of GOLD and SILVER class respectively, run their applications on two individual virtual machines V1, V2.

- The CPU power of remaining machines is pooled as one. If there are two virtual machines V3, V4 then their CPU is merged as V3_CPU + V4_CPU
- An intermediary block called CONTROLLER where the scheduling algorithm resides is implemented between {V1, V2} and pooled CPU.
- Customers on V1 and V2 run their applications during which they submit their query for downloading software to the Controller. These queries are treated as tasks to be executed.
- These tasks will be stored in a queue in the Controller which operates as per the implemented scheduling algorithm and gains access to shared CPU.

The following three scenarios (cases) are used in the experimentation process

A. *Weighted Round Robin Algorithm*
When two customers try to download required software via Application Downloader and if round robin algorithm is applied, each customer's request will be stored in a separate queue and gain access to CPU in a context switch basis with more resources allocated to T1 per switch (As T1 belongs to gold class). This is illustrated in the following graphs. Figure 2 shows the amount of download at a given transfer rate for Gold customer and Fig. 3 shows the same for a Silver

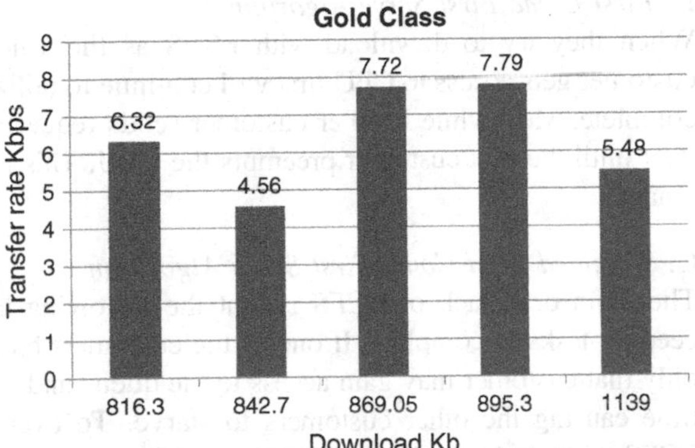

Fig. 2 Gold class downloads

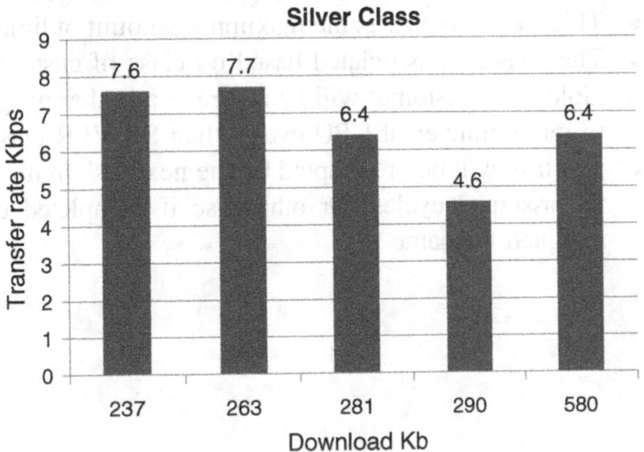

Fig. 3 Silver class downloads

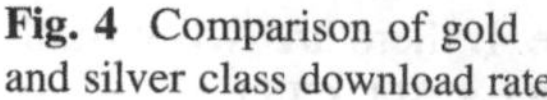
Fig. 4 Comparison of gold and silver class download rate

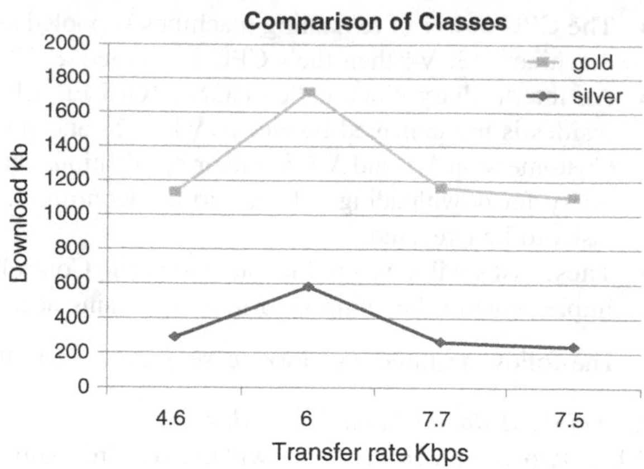

customer. Figure 4 shows the comparison of download by both classes at a given transfer rate. We can see that a Silver class customer downloads at a lower speed as compared to a Gold class as she will be given a lower priority while sharing the resource.

B. *First Come First Serve Algorithm*

When they try to download with FCFS as the scheduling algorithm whichever customer gets access to CPU first will continue to utilize the same until download is complete. Meanwhile if other customer sends request for downloading, she has to wait until the first customer preempts the CPU. This is applicable for all classes of tenants.

C. *Weighted First Come First Serve Algorithm*

The main drawback of FCFS is that the following task has to wait for the succeeding tasks to complete. If one of the customers blocks CPU over the other then only that customer may gain access to the queue and to CPU for a larger amount of time causing the other customers to starve. To overcome this drawback we use FCFS with slight modification. It is as follows

- The tasks that wait in the queue will be assigned a weight.
- This weight indicates the maximum amount of time, tasks can be assigned CPU.
- The weight is associated based on class of customer.
- Gold class customer will be assigned a higher weight indicating they get access to more number of CPU cycles than SILVER class.
- The task will be pre-empted by the next task in the queue once they finish using the assigned cycles. Or otherwise if completed earlier the next task will be assigned the same

5 Conclusions

Multitenant applications are gaining popularity day by day due to its simplicity, ease of use and cost effectiveness. Nevertheless, there are many unexplored problems which can be researched. We have tried to address a problem related to resource sharing and have identified one of the possible architecture. However, we have adopted the architecture and implemented a multi-user application which is non-multitenant and implemented using Cloudsim as a startup. We will continue our research and aim to apply the proposed architecture for a multitenant SaaS application exploring further problems in the said field and try to bring out solutions implemented in private or public clouds.

Acknowledgments The work reported in this paper is supported by the college through the TECHNICAL EDUCATION QUALITY IMPROVEMENT PROGRAMME [TEQIP-II] of the MHRD, Government of India.

References

1. Cor-Paul Bezemer, Andy Zaidman.: Multi-Tenant SaaS Applications: Mainte-nance Dream or Nightmare?", Report TUD-SERG-2010-031, Delft University of Technology Software Engineering Research Group Technical Report Series (2010)
2. Shue, David, Michael J. Freedman, and Anees Shaikh.: Performance Isolation and Fairness for Multi-Tenant Cloud Storage. In OSDI, vol. 12, pp. 349–362 (2012)
3. Gahlawat, Monica, and Priyanka Sharma.: Analysis and Performance Assessment of CPU Scheduling Algorithms in Cloud using Cloud Sim, International Journal of Applied Information Systems, FCS, Volume5-No 9 (2013)
4. Das, Sudipto, Vivek R. Narasayya, Feng Li, and Manoj Syamala.: CPU sharing techniques for performance isolation in multi-tenant relational database-as-a-service. Proceedings of the VLDB Endowment 7, no. 1, pp. 37–48 (2013)
5. Cui, L. Z., T. T. Zhang, G. Q. Xu, and Dong Yuan.: A scheduling algorithm for multi-tenants instance-intensive workflows. Appl. Math 7, no. 1L, pp. 99–105 (2013)
6. Stolarz, Wojciech, and Marek Woda.: Proposal of cost-effective tenant-based resource allocation model for a SaaS system. In New Results in Dependability and Computer Systems, pp. 409–420. Springer International Publishing (2013)
7. Xiong, Pengcheng, Yun Chi, Shenghuo Zhu, Hyun Jin Moon, Calton Pu, and Hakan Hacigümüş. "Intelligent management of virtualized resources for database systems in cloud environment." In Data Engineering (ICDE), 2011 IEEE 27th International Conference on, pp. 87–98. IEEE (2011)
8. Das, Sudipto, Shoji Nishimura, Divyakant Agrawal, and Amr El Abbadi.: Albatross: lightweight elasticity in shared storage databases for the cloud using live data migration. Proceedings of the VLDB Endowment 4, no. 8, pp. 494—505 (2011)
9. Yang, Enfeng, Yong Zhang, Lei Wu, Yulong Liu, and Shijun Liu.: A hybrid approach to placement of tenants for service-based multi-tenant saas application." In Services Computing Conference (APSCC), 2011 IEEE Asia-Pacific, pp. 124–130. IEEE (2011)
10. Shen, Zhiming, Sethuraman Subbiah, Xiaohui Gu, and John Wilkes.: Cloudscale: elastic resource scaling for multi-tenant cloud systems. In Proceedings of the 2nd ACM Symposium on Cloud Computing, p. 5. ACM (2011)

11. Krebs, Rouven, Simon Spinner, Nova Ahmed, and Samuel Kounev.: Resource usage control in multi-tenant applications. In Cluster, Cloud and Grid Computing (CCGrid), 2014 14th IEEE/ACM International Symposium on, pp. 122–131. IEEE (2014)
12. Jadhav, C. M., V. V. Bandgar, G. A. Fattepurkar, and P. S. Bhandare.: An Approach for Development of Multitenant Application as SaaS Cloud. International Journal of Computer Applications 106, no. 14 (2014)
13. Sengupta, Bikram, and Abhik Roychoudhury.: Engineering multi-tenant software-as-a-service systems. In Proceedings of the 3rd International Workshop on Principles of Engineering Service-Oriented Systems, pp. 15–21. ACM (2011)
14. Espadas, Javier, David Concha, David Romero, and Arturo Molina.: Open Architecture for Developing Multitenant Software-as-a-Service Applications. In First International Conference on Cloud Computing, GRIDs and Virtualization. Published by XPS (Xpert Publishing Services), pp. 92–97 (2010)
15. Walraven, Stefan, Eddy Truyen, and Wouter Joosen. Comparing PaaS offerings in light of SaaS development. Computing 96, no. 8, pp. 669–724 (2014)

Privacy-Preserving Profile Matching System for Trust-Aware Personalized User Recommendations in Social Networks

Vaishnavi Kulkarni and Archana S. Vaidya

Abstract Trust is one of the important points to be considered regarding the security of social networks. In the proposed system, a framework that handles trust in social network is introduced, this uses the reputation mechanism. The reputation mechanism differentiates the implicit and explicit connections that exist in the network members. The semantics and dynamics of these connections are analyzed, and personalized user recommendations to other users of network are provided. Using the semantics of trust, recommendations will be provided by the system considering both the positive trust and negative trust between users. Along with this, the proposed system matches profiles of the users under consideration. The profile matching is used in reputation ratings calculated for suggestions of friends. For computing the reputation of each member, the properties of trust such as transitivity, personalization, and context are adopted by the proposed system. In social networks, trust cannot be perfectly transitive and also it decreases along the transition path, but people can communicate the trust. The aim of this work is to design a web-based recommender system in social network that will provide suggestions to the users by analyzing the behaviour of each user in the social network as well as filtering out the similar users from the network.

Keywords Personalization · Rating · Recommender system · Social networks · Sharing · Trust

---

Vaishnavi Kulkarni (✉) · A.S. Vaidya
Department of Computer Engineering, Gokhale Education Society's
R. H. Sapat College of Engineering, Management Studies & Research,
P T A Kulkarni Vidyanagar, Nashik, Maharashtra, India
e-mail: vaishnavikulkarni1691@gmail.com

A.S. Vaidya
e-mail: archana.s.vaidya@gmail.com

Vaishnavi Kulkarni · A.S. Vaidya
Savitribai Phule Pune University, Pune, Maharashtra, India

© Springer Science+Business Media Singapore 2017
S.C. Satapathy et al. (eds.), *Proceedings of the International Conference on Data Engineering and Communication Technology*, Advances in Intelligent Systems and Computing 468, DOI 10.1007/978-981-10-1675-2_4

1 Introduction

Initial research on such recommendation systems in social networks such as online social network, the blog sphere, social bookmarking applications do not incorporate the "trust" in the suggestions. Calculation of such kind of trust worthy recommendation is somewhat difficult because trust is personalized and also it is subjective and affected by each user's personal thoughts and beliefs, as well as those of members whom the user trust and respects, and this is the bottle neck problem. The main goal and challenge of the system is recommending personalized users to another user by matching their profiles as well as considering trust between them.

The goal of this research is to design a web-based recommender system in social network that will provide suggestions to the users by analyzing the behaviour of each user in the social network as well as filtering out the similar users from the network. Literature survey is the first step towards the goal. The survey will focus on the requirements. The next step is to plan the project. To make the system platform independent, JAVA language is used. To handle a large database of social network, MySQL database engine is used. To test the system, the standard dataset of Epinions which is a large product review community site, is used. In order to give suggestions based on profile matching concept, generation of the users' profiles is done.

2 Related Work

The content and links analysis in social networks has increased the research in the related fields [1]. The largest body of work that considered positive trust and/or its propagation in the recommender systems focused mainly on item recommendations [2–5]. Walter et al. has introduced Time dynamics. The trust propagation is employed through transitivity and, similarly to this proposed recommender system, discounting takes place by multiplying trust values along paths. Making new connections, according to personalized preferences is an important service in social networking, where an initiating user can search matching users from a group of users in physical proximity of the user. According to the work in [6], FindU, a privacy-preserving profile matching scheme is proposed. In FindU, an initiating user can find the one whose profile best matches with user, from a group of users; to limit the risk of exposing privacy, only necessary and minimal information regarding the private attributes belonging to the profile of participating users is exchanged, by preserving the privacy of users. The task of providing personalized recommendations requires the ability of predicting the items. Such a prediction is typically based on (1) content—recommending items with content similar to content of items which users have already consumed; (2) social networks—providing items related to users who are related to other users either by explicit connection, or by some kind of similarity.

It has been shown that considering social network relationships and respective opinions/ratings improve the prediction and in turn the recommendation process [7, 8]. A similar type of work focused on content ranking, which is consequently employed to recommend the top-ranked items to users. A more generic model has been presented in the previous work [9], which can be applied to any social medium. To recommend users in social media, work defined local metrics and global metrics. But, in that work, the notion of negative trust among users was not incorporated. Recently, negative trust has been introduced for user recommendations in social networks [10, 11]. In this model, the trust of any user to another user considers personalized reputation rating. This rating uses information of explicit connections among users and implicit connections that are inferred from the interactions among users of the social network. Social networking sites help users to articulate their social networks by adding other users to their "friend lists" [8]. Leskovec et al. [12] tried to predict negative and positive links in social networks by the use of machine-learning framework and ideas, which are drawn from sociology, have derived opposite results. Recommendations are based on aggregated social network information from various sources across the organization [13]. Trust is nothing but the belief of user in the behaviour of other user to act reliably and honestly unlike of distrust [14]. The previous work done mainly focused on the item and user recommendation without considering the trust relationship between them. Because of this, the security of user to user connection might be disturbed. Thus, we propose a trust-aware system for providing users recommendations in order to make connections of social network trustworthy by giving positive and negative recommendations to the users while matching their profiles for strong connection.

3 Proposed System

The proposed recommender system is based on reputation mechanism which gives rating to the participants using observations, their past experiences and other user's view/opinion about them. For computing reputation of each member, the properties of trust such as transitivity, personalization and context are adopted. Also, to address the social network dynamics, the element of time has been included in the proposed system. To this direction, suggestion is given that reputation varies with time. Hence, value of the positive or negative reputation of a user tends to zero unless and until new explicit or implicit trust/distrust statements are added frequently. Finally, we assume that the trust context is same among members of the community. Specifically, after processing information posted on the network, both explicit and implicit connections are formed that bear trust statements between network members (stage 1), estimation of reputation ratings are done (stage 2) and personalized recommendations (considering both positive and negative trust) are generated (stage 3). Before providing recommendations, profile matching is done to filter out the results. (Contribution)

Stage 1: User Connection Formation

The proposed system considers the difference between explicit trust or distrust connections amongst users that have strong trust semantics and implicit trust statements that have more transient connections between users in the network. These user connection formation or trust bonds can be categorized as follows:

(A) Explicit user-to-user connection: A user can directly relate to other user by making trust or distrust connections. These connections express permanent bonds between users of a network.
(B) Explicit user-to-item connection: In this type, the user gives a like/dislike statement to a particular item posted by other user.
(C) Implicit user-to-item connection: In this each content item posted by a user has a timestamp and unique identifier. Preference to an item is given implicitly.
(D) Implicit user-to-user connection: In this connection, the information of user-to-item connection is mapped to the user-to-user level, which is aggregated for providing a single implicit connection between two users.

Stage 2: Reputation Rating Estimation

The proposed reputation rating mechanism considers the effect of time by modelling the fact that newly added trust or distrust connections and recently added like/dislike statements should have more value in the evaluation of the overall reputation rating of target user by the evaluator. So, dynamic aspect of social network is taken into consideration and is effectively addressed. Following are the reputation rating systems (A) Local Rating (B) Collaborative Rating (C) Transitivity of Trust (D) Trust aware personalized recommendations [15].

Contribution: The proposed system considers the negative trust between users to help them getting connected to another trustworthy user and to alert them from getting connected to such untrustworthy user. Before providing the list, apart from these filters, we can contribute one more step which is mode of filter, i.e. profile matching of the users. In this mode the proposed system can provide list of friends or enemies using set of privacy-preserving profile matching schemes. In this step, the initiating user can find from group of users the one, whose profile best matches with user [16]. Active user in social network will get suggestions of active users matching profile with each other.

Stage 3: Recommendations Generations

The proposed system generates personalized positive/negative user recommendations that are obtained as a result of the overall reputation ratings of the members of social network estimated by the evaluator user and the result of profile matching. Both positive and negative recommendations could help user to update his trust and distrust network connections. Figure 1 shows the overall architecture of proposed system: [17]

In order to provide recommendations to the user, following steps are executed by the system. Let us assume the presence of N users $U = \{U1, U2, \ldots, Un\}$ in a social network. Every member $Uj \epsilon U$, posts several items while in the network.

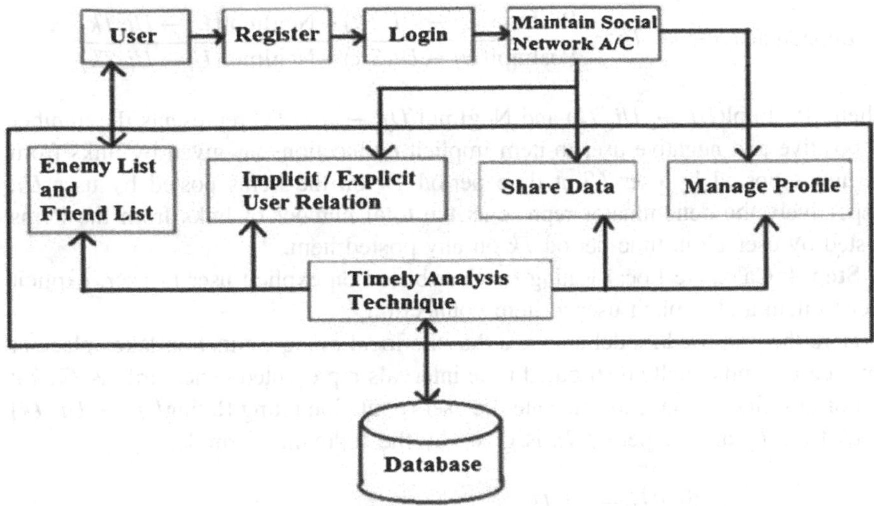

Fig. 1 Overall system architecture

Additionally, *Fr* (*Uj*) and *En* (*Uj*) represents the friend list and the enemy list of user *Uj*, respectively.

Step 1: Calculate explicit user-to-user trust/distrust, i.e.

$$\text{UserConn}(Uj \to Ui, Tk) \tag{1}$$

It is assumed that UserConn($Uj \to Ui, Tk$) lies within the [–1, 1] range, where a value close to 1(–1) indicates that the target *Ui* is a friend(enemy) of the evaluator user *Uj*.

Step 2: Calculate explicit user-to-item connection

It corresponds to the explicit user-to-item connections. This factor has been assumed to lie within the [–1, 1] range and is defined as follows.

$$\text{ExplConn}(Uj \to Ui, Tk) = \frac{\text{PosExpl}(Uj \to Ui, Tk) - \text{NegExpl}(Uj \to Ui, Tk)}{\text{PosExpl}(Uj \to Ui, Tk) + \text{NegExpl}(Uj \to Ui, Tk)} \tag{2}$$

where, PosExpl($Uj \to Ui, Tk$) and NegExpl($Uj \to Ui, Tk$) gives the number of positive and negative user-to-item explicit opinions, respectively as represented by user *Uj*, at time period Tk on the items posted by user *Ui*. The denominator represents the total number of opinions expressed by user *Uj* in time period *Tk* on any posted item.

Step 3: Calculate implicit user-to-item connection ImplConn($Uj \to Ui, Tk$)

It corresponds to the implicit user-to-item connections. This factor also lies within the [–1, 1] range and is given by the following equation

$$\text{ImplConn}(Uj \rightarrow Ui, Tk) = \frac{\text{PosImpl}(Uj \rightarrow Ui, Tk) - \text{NegImpl}(Uj \rightarrow Ui, Tk)}{\text{PosImpl}(Uj \rightarrow Ui, Tk) + \text{NegImpl}(Uj \rightarrow Ui, Tk)} \quad (3)$$

where, PosImpl($Uj \rightarrow Ui, Tk$) and NegImpl($Uj \rightarrow Ui, Tk$) represents the number of positive and negative user-to-item implicit connections, as given by links from the items posted by user *Uj* at time period *Tk* on the items posted by user *Ui*, respectively the denominator represents the total number of links from the items posted by user *Uj* in time period *Tk* on any posted item.

Step 4: Calculate Local Rating of User based on explicit user to user, explicit user to item and implicit user to item connection

Here the proposed model assumes that the local rating estimation takes place at consecutive and equally distributed time intervals represented henceforth as *Tk*, $k \in N$. For this, first we have to calculate the user reputation rating Rating$Uj \rightarrow Ui, Tk$) of *Ui* from *Uj* at time period *Tk* is given by the following formula,

$$\begin{aligned} \text{Rating}(Uj \rightarrow Ui, Tk) & \\ &= \text{Wuser.UserConn}(Uj \rightarrow Ui, Tk) \\ &+ \text{Wexpl.ExplConn}(Uj \rightarrow Ui, Tk) \\ &+ \text{Wimpl.ImplConn}(Uj \rightarrow Ui, Tk) \end{aligned} \quad (4)$$

where, Wuser + Wexpl + Wimpl = 1.

Weights Wuser; Wexpl and Wimpl provide the relative significance of the three factors user-to-user connections, user-to-item explicit connections and user-to-item implicit connections, respectively. Rating ($Uj \rightarrow Ui, Tk$) lies within the [–1, 1] range. Using Rating ($Uj \rightarrow Ui, Tk$) we can calculate LocalRating ($Uj \rightarrow Ui, Tc$). For the formation of the local user reputation rating at the current time period *Tc*, the evaluator considers only the r more recent ratings formed by the user. The value of r determines the memory of the system. The local reputation rating LocalRating ($Uj \rightarrow Ui, Tc$) of user *Ui*, as estimated by *Uj* at time period *Tc*, is defined as follows:

$$\text{LocalRating}(Uj \rightarrow Ui, Tc) = \sum_{k=c-r+1, k>0}^{c} \text{dfk.Rating}(Uj \rightarrow Ui, Tk) \quad (5)$$

Step 5: Calculate collaborative rating using Local rating

$$\begin{aligned} &\text{CollRating}(Uj \rightarrow Ui, Tc) = \text{cred}(Uj \rightarrow Uj, Tc).\text{LocalRating}(Uj \rightarrow Ui, Tc) \\ &+ \textstyle\sum_{q=1, q \neq i,j}^{Q} cred(Uj \rightarrow Uq, Tc).\text{LocalRating}(Uj \rightarrow Ui, Tc) \end{aligned} \quad (6)$$

Here, the weight cred($Uj \rightarrow Ui, Tc$) is a measure of the credibility of witness *Uq* and the respective rating of *Ui* in the eyes of the evaluator *Uj*.

4 Experimental Setup

For experimental purpose we have used systems that act as client and server. This proposed system is implemented using JAVA environment. HTML, CSS and JavaScript technologies are used for front end development. This system is client–server architecture. At the server end apache tomcat container is used. For database MySQL is used. We have setup jdk-7, apache tomcat-7 and mysql-5.3 on this system. To test our system functionality we have built experimental setup. In this we use already existing dataset as previous user input to the system. Also the dataset is modified as per the proposed system's requirement.

5 Results

Epinions is used as dataset [18]. Epinions supports various types of interactions between users, such as explicit user-to-user trust statements and product reviews written by the community members and rated by other members [18]. For Profile dataset we have used online fake profile generator [19]. Using this generator we have generated records with multiple attributes such as: <name, age, gender, city, occupation, address.>

As per the experimental setup, we have used modified dataset which consists of profiles mapped to the users in Epinions dataset. As per our setup user can become the part of our system and send friend request to each other. After establishment of relationship between them user can send comments to each other, user can explicitly like and dislike the comments. User can rate for particular comment. All these data transactions are considered to calculate the local rating; collaborative rating which forms the trust about the user and it is helpful for us to recommend the friends. As part of contribution user profiles are also matched from the list recommended on the basis of collaborative rating.

From the Table 1, it is clear that when we apply profile matching, the number of recommendations is reduced. Though the number is reduced, we get more precise and accurate recommendations.

6 Performance Evaluation

To evaluate the performance of the proposed system, we can analyze the results from the Table 1. The collaborative recommendations are generated by considering both positive and negative trust of users. Thus we get more number of recommendations. Now when we filter results by negative local rating, the users having negative local rating are removed from the list of recommendations. Again when we apply filter of profile matching, the users having zero percent profile match with the

Table 1 Comparison between collaborative recommendations, recommendations filtered by negative local rating and recommendations filtered by negative local rating and profile matching

Users	Collaborative recommendations (Positive trust + negative trust)	Recommendations filtered by negative local rating	Recommendations filtered by negative local rating and profile matching
1	20	17	15
2	10	5	2
3	30	20	10
4	20	10	5
5	40	38	35
6	35	30	25
7	50	35	15
8	5	3	2
9	45	40	37
10	10	8	5

evaluator user are removed from the list of recommendations. Thus we get more accurate and precise list of recommendations.

From the graph shown in Fig. 2, we can analyze the results. The complexity of the system is calculated based on reputation rating estimation and profile matching algorithm. Let 'n' be the number of users. To match the profile with every other member we have (n) iterations. Each user in the network is connected with his friends. The reputation rating of every user is computed with respect to ratings provided by his friends. Thus, for reputation rating estimation, the algorithm is executed $(n - 1)$ times. Hence the computation is $(n - 1) + (n) = (2 * n) - 1$. Thus the time complexity is $O(n)$.

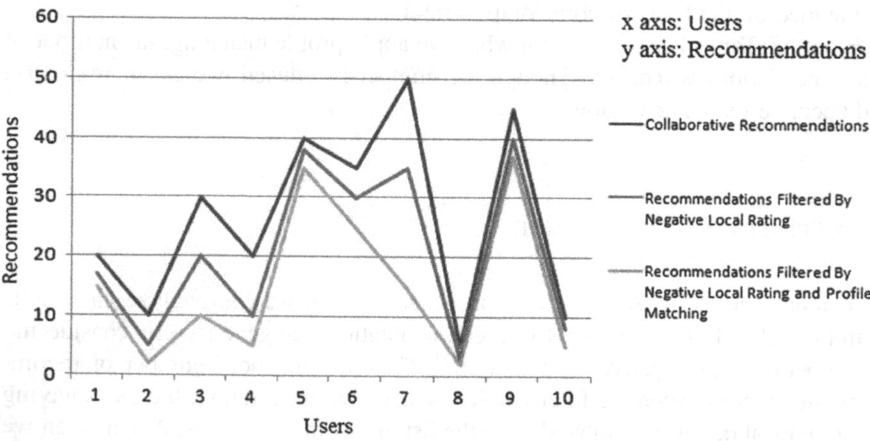

Fig. 2 Analysis of result

Table 2 Performance analysis

A		B		C	
Precision	Recall	Precision	Recall	Precision	Recall
0.9	0.99	0.93	1	0.94	1
0.8	0.9	0.83	0.94	0.9	0.97
0.96	0.97	0.96	0.983	0.97	0.992
0.95	1	0.97	1	0.978	1
0.9	1	0.92	1	0.934	1
0.97	0.987	0.973	0.99	0.973	1
0.95	0.979	0.952	0.979	0.98	0.98
0.99	0.987	0.993	0.98	0.992	0.99
0.95	0.962	0.964	0.98	0.972	0.984
0.96	0.95	0.96	0.96	0.973	0.97

The generated values are compared with real values by analyzing the dataset. After evaluation we have calculated precision and recall. The analysis is shown in Table 2. In table, 'A' is Collaborative Recommendations, 'B' is Recommendations filtered by Local Rating and 'C' is that filtered by Local rating and profile matching.

7 Conclusion and Future Scope

The previous work done mainly focused on the item and user recommendation without considering the trust relationship between them due to which the security of user connection might be disturbed. Thus, we propose a trust-aware user recommender system to make connections of social network trustworthy by giving positive and negative recommendations to the users while matching their profiles for strong connection. So that, positive recommendations will help in connecting trustworthy users while negative recommendations will alert users not to connect to the untrustworthy users and making aware of the items posted by such user. Profile matching gives more filtered results.

In future, more profile attributes can be added for matching in order to get more efficient and accurate recommendations. Also, for generating recommendations based on local rating and collaborative rating, personal chats can be semantically analyzed along with the comments and posts. The analysis of posted images can be done in order to provide personalized recommendations.

References

1. Magdalini Eirinaki, Malamati D. Louta, Member, IEEE, and Iraklis Varlamis, Member, IEEE. "A Trust-Aware System for Personalized User Recommendations in Social Networks" IEEE Trans. Systems, Man, And Cybernetics: Systems, Vol. 44, No. 4, pp. 409–421 APRIL 2014.
2. J. Golbeck, "Trust and nuanced profile similarity in online social networks," J. ACM Trans. Web, vol. 3, no. 4, pp. 1–33, 2009.
3. R. V. Guha, R. Kumar, P. Raghavan, and A. Tomkins, "Propagation of trust and distrust," in Proc. 13th Int. Conf. World Wide Web, 2004, pp. 403–412.
4. P. Massa and P. Avesani, "Trust-aware recommender systems," in Proc. ACM Conf. Recommender Syst., 2007, pp. 17–24.
5. C.-N. Ziegler, "On propagating interpersonal trust in social networks," in Computing With Social Trust, J. Golbeck, Ed. London, U.K.: Springer, 2009, ch. 6.
6. Ming Li, Member, IEEE, Shucheng Yu, Member, IEEE, Ning Cao, Student Member, IEEE, and Wenjing Lou, Senior Member, IEEE. "Privacy-Preserving Distributed Profile Matching in Proximity-based Mobile Social Networks" IEEE Trans. Wireless Communications Vol.: 12 No.: 5 pp. 2024–2033 Year 2013.
7. I. Guy, N. Zwerdling, D. Carmel, I. Ronen, E. Uziel, S. Yogev, and S. OfekKoifman, "Personalized recommendation of social software items based on social relations," in Proc. 3rd ACM Conf. Recommender Syst., 2009, pp. 53–60.
8. I. Konstas, V. Stathopoulos, and J. M. Jose, "On social networks and collaborative recommendation," in Proc. 32nd Int. ACM SIGIR Conf. Res. Develop. Inf. Retrieval, 2009, pp. 195–202.
9. I. Varlamis, M. Eirinaki, and M. D. Louta, "Application of social network metrics to a trust-aware collaborative model for generating personalized user recommendations," in The Influence of Technology on Social Network Analysis and Mining(Lecture Notes in Social Networks Series), vol. 6, T. Ozyer, J. G. Rokne, G. Wagner, and A. H. P. Reuser, Eds. Berlin, Germany: Springer, 2013, pp. 49–74.
10. J. Leskovec, D. P. Huttenlocher, and J. M. Kleinberg, "Predicting positive and negative links in online social networks," in Proc. 19th Int. Conf. World Wide Web, 2010, pp. 641–650.
11. J. Kunegis, A. Lommatzsch, and C. Bauckhage, "The slashdot zoo: Mining a social network with negative edges," in Proc. 18th Int. Conf. World Wide Web, 2009, pp. 741–750.
12. J. Chen, W. Geyer, C. Dugan, M. J. Muller, and I. Guy, "Make new friends, but keep the old: Recommending people on social networking sites," in Proc. SIGCHI Conf. Human Factors Comput. Syst., 2009, pp. 201–210.
13. I. Guy, I. Ronen, and E. Wilcox, "Do you know?: Recommending people to invite into your social network," in Proc. 14th Int. Conf. Intell. User Interfaces, 2009, pp. 77–86.
14. T. Grandison and M. Sloman, "A survey of trust in internet applications," IEEE Commun. Surveys Tuts., vol. 3, no. 4, pp. 2–16, Oct. 2000.
15. Kulkarni Vaishnavi Shripad and Prof. Archana S. Vaidya, "A Review on Trust-Aware and Privacy Preserving Profile Matching System for Personalized User Recommendations in Social networks", International Journal of Computer Applications, October 2014, Vol - 104, No. 12.
16. Elie Raad, Richard Chbeir, Albert Dipanda. "User profile matching in social networks." Network-Based Information Systems (NBiS), Sep 2010, Japan. pp. 297–304. <hal-00643509>.
17. Kulkarni Vaishnavi Shripad and Prof. Archana S. Vaidya, "Privacy Preserving Profile Matching System for Trust Aware Personalized User Recommendations in Social networks", International Journal of Computer Applications, October 2014, Vol - 122, No. 11.
18. For System Dataset - http://www.trustlet.org/wiki/Extended_Epinions_dataset.
19. For Profile Dataset - http://www.fakenamegenerator.com/order.php.

Performance Analysis of Single-Phase Bi-directional Converter for Power Factor Correction and Voltage Stabilization

Rahul Ganpat Mapari, D.G. Wakde and V.N. Patil

Abstract This paper presents the performance analysis and the comparison between two techniques implemented for bi-directional converter. The analysis is carried out in accordance with rectification and inversion process using two cases. In both cases the modulation techniques are implemented digitally. The overall analysis incorporates the parameters such as power factor, total harmonic distortions (THDs), bi-directional operation, and voltage regulation for various loads. For each load all above parameters are checked and compared between two mentioned cases. These converter topologies are evaluated on 1 KW prototype model.

Keywords Power factor correction (PFC) • AC-DC-AC converters • Voltage stabilization • THD's

1 Introduction

In most of the applications, for AC-DC power conversion it is necessary to havethe output dc voltage to be well regulated with good steady state and transient performance. The rectifier with filter capacitor is cost effective but it severely deteriorates the quality of the supply, [1] thereby disturbing the performance of other loads connected to it further causing other troubles. Power electronics engineers

R.G. Mapari (✉)
Department of Electronics Engineering, Sant Gadge Baba Amravati University, Amravati 444602, India
e-mail: rahul_mapari272153@yahoo.com

D.G. Wakde
P.R. Patil College of Engineering, Amravati, India
e-mail: dr_dgwakde@rediffmail.com

V.N. Patil
IsquareIT, Hinjawadi, Pune, India
e-mail: principal@isquareit.edu.in

© Springer Science+Business Media Singapore 2017

S.C. Satapathy et al. (eds.), *Proceedings of the International Conference on Data Engineering and Communication Technology*, Advances in Intelligent Systems and Computing 468, DOI 10.1007/978-981-10-1675-2_5

have been developing new approaches for better utility interface, to meet these imposed standards [2]. These new circuits are called as power factor correction circuits.

Three-phase AC-DC conversion of electric power is widely employed in adjustable speed drive (ASDs), uninterruptible power supplies (UPSs), HVDC systems, and utility interfaces with nonconventional energy sources such as solar photovoltaic systems (PVs) etc., [3–5] battery energy storage systems (BESSs), in process technology such as electroplating, welding units, etc., battery charging for electric vehicles and power supplies [6].

The current controller senses the input current and compares it with a sinusoidal current reference. To obtain the current reference, the phase information of the utility voltage or current is required. This information is obtained by employing a phase lock loop (PLL), which creates transients if the frequency ratio changes [7]. The single-switch rectifier has one of the simplest circuit structures. The two switch rectifier performs the same switching action as the single-switch rectifier but has the advantage of higher efficiency [8–10]. However, the scheme based on OCC exhibit instability in operation when magnitude of the load current falls below a certain level or when the converter is operating in the inverting mode of operation. To avoid it a modified OCC Bi-directional high power factor AC-to-DC converter is proposed in [11]. This scheme uses saw-tooth wave to generate PWM pulses which incorporate low frequency harmonics. OCC presents some drawbacks intrinsic with its physical realization: the controller and its parameters cannot be modified without hardware re-design; moreover they are influenced by temperature drifts, typical of analog systems. To overcome these limitations the OCC technique is implemented digitally using Field-programmable gate array (FPGA) [12]. This system uses PLL to find phase information of utility voltage and current. Another drawback of this system is that controller takes integer numbers only. The split operation is limited only to dividing number by a power of two.

A new single-phase multilevel flying capacitor active rectifier using hysteresis-based control is reported in [13], the power factor and output voltage regulation is achieved by controlling the input current. PLL is used to find the phase information of input voltage which creates the transients as the frequency ratio changes. Smart charger for electrical vehicle using three wire distribution feeders is proposed in [14]. To achieve bi-directional flow six active switches are used, which increases switching losses, hardware, and complexity. A simple control technique for a single-phase bridgeless active rectifier with high power factor and voltage stabilization using partial digital implementation is implemented in [15].

This paper not only presents the comparative study between two cases but addresses the aforementioned drawbacks. The used converter is bridgeless, transformer-less, and output DC current sensor less. All the above mentioned schemes are implemented digitally to incorporate the advantages of digital implementation. The detail simulation is carried out using MATLAB/Simulink and to validate it experimental setup is develop using DSPic33FJ64MC802 digital controller for 1 KW system.

2 System Modeling

For controlling and digital implementation of most of the part of the system author preferred DSPic33FJ64mc802 controller, which having all the features of digital signal processor (DSP). The detail simulation is carried out in Simulink of MatLab. The Sim-Power tool of MatLab is used to implement the block schematic of single-phase bi-directional converter. Figure 1 shows the single-phase representation of the converter. V_{IN} is the AC line voltage of 140 V and 50 Hz. 'L' represents the line inductor carrying current Iin (t). Controlled Switches S1–S4 are arranged in a bridge. 'C' is the DC side capacitor of 440 F and 'R' is the variable resistive load. Io and IL are output and load currents respectively.

2.1 *Digital Implementation of SPWM Technique*

For implementation of sinusoidal PWM technique SPic33FJ64MC802 digital controller is used. This controller operates on 3.3 VDC supply voltage. It has 10-bit (Analog to Digital conversion) ADC, hence its total full scale count is 210 = 1024. It means that for analog input 3.3 V to ADC controller pin gives 1024 count. For the implementation we consider this value as 2.5 V. The conversion of output AC voltage to its linear DC value is made using series connection of step-down transformer and precision rectifier shown in Fig. 2.

This sinusoidal PWM scheme, in this section is used for inversion operation; hence its output is 230 V_{AC} as per the design. To implement it digitally, an array of percentage duty cycle is developed and filled it into the look-up table. The look-up table consists of totally 200 values based on the line frequency and switching frequency.

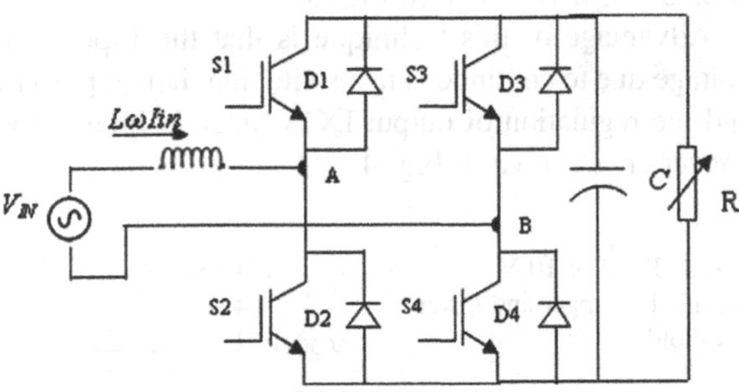

Fig. 1 Single-phase full bridge converter

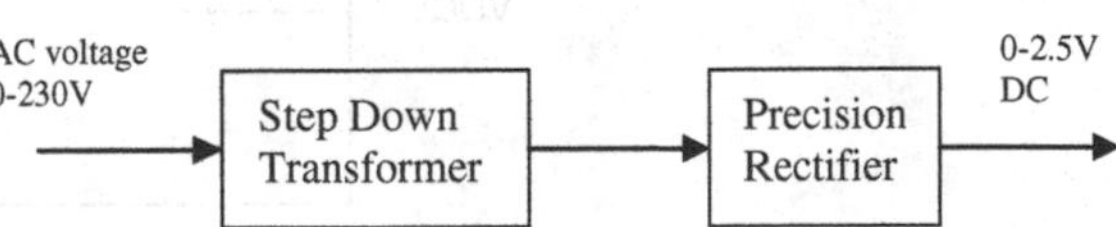

Fig. 2 Conversion of output AC voltage to its linear DC value

For implementation of sinusoidal PWM technique DSPic33FJ64MC802 digital controller is used. The ADC count value for the 2.5 V operating voltage is same as mentioned in above section. To implement it digitally, an array of percentage duty cycle is developed and filled it into the look-up table. The look-up table consists of total 200 values of duty cycles, based on the line frequency and switching frequency. When power is on, initially K_{factor} value is less for soft start mechanism. Slowly, it increases with the increment value 0.01 to achieve desired AC voltage that is 160 V. It is represented in terms of embedded C code and it is updated after every 1 ms, and duty cycle value is updated after every 50 s. For the analysis purpose, author takes 160 V as input AC RMS voltage to boost it up to 380 V_{DC} link voltages.

2.2 *Digital Implementation of SPWM Technique*

This scheme is used for the rectification operation. Soft start mechanism is used in which when power is ON duty cycle increases from 0 % till to achieve desired voltage. For implementation of continuous switching PWM technique DSPic33FJ64MC802 digital controller is used. This controller operates on 3.3 VDC supply voltage. It has 10 bit (Analog to Digital conversion) ADC, hence its total full scale count is 210 = 1024. It means that for analog input 3.3 V to ADC controller pin gives 1024 count. For the implementation we consider this value as 2.5 V. The conversion of feedback DC voltage to its equivalent step-down value is done using voltage divider circuit. The desired boosted DC voltage is 380 V for line voltage 230 VAC. For digital implementation the feedback voltage is converted to its equivalent step down voltage as described and shown in above paragraph. We convert the 380 VDCFB into 2.5 V desirable for the digital controller. The band is provided of 10 VDC as an upper threshold (390 VDC its equivalent step-down voltage is 2.58 V) and lower threshold (370 VDC its equivalent step-down voltage is 2.45 V). It is shown in Fig. 3.

Advantage of this technique is that the input current flowing similar to input voltage due to continuous fast switching, hence power factor value is nearer to unity and the regulation of output DC voltage is better. The control block for presented converter is shown in Fig. 4.

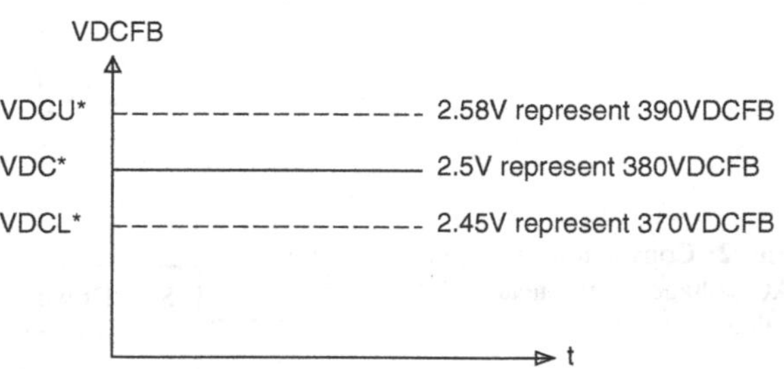

Fig. 3 Band of 10 V provided as upper and lower threshold

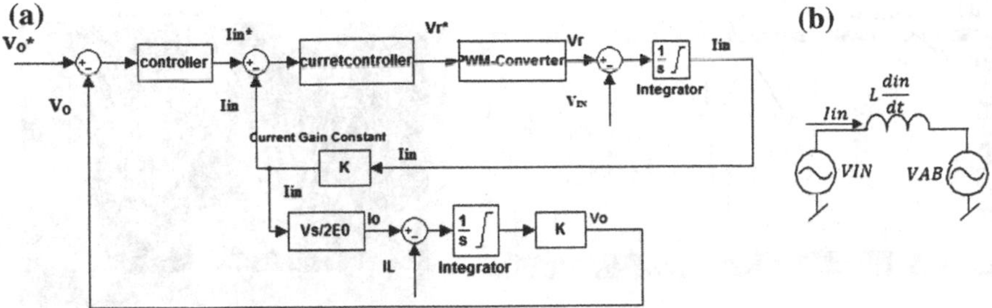

Fig. 4 **a** Control block for single-phase converter and **b** input side representation of the single-phase converter circuit

3 Simulated and Experimental Results

In order to check the performance of converter using CSPWM and SPWM techniques for rectification and inversion using SPWM technique detail simulation is carried out on MATLAB/Simulink platform and the DSPic33FJ64MC802 digital controller is used for the implementation.

3.1 Simulated and Experimental Results of Rectification Using CSPWM

Simulated and experimental result waveforms for rectifying mode of operation are shown in Fig. 5. Trace 1 of Fig. 5a shows the simulated waveform of line current following the line voltage to indicate the unity power factor. The experimental result for the same is shown in Trace 2 of Fig. 5a. This shows the exactly similar result like simulated waveform. The channel 1 is at 50 V/Div and channel 3 is at 5 A/Div.

The DC side load is changed from 5 to 80 % periodically to show the output voltage regulation which is shown in Fig. 5b. Trace 1 of Fig. 5b shows simulated result, it can be seen that as load changes from 80 to 5 % and from 5 to 80 % the DC link voltage remains regulated. Trace 2 shows experimental result for the same. It can be observed that the DC link voltage regulated within very small period which is negligible when load change occurred. Channel 2 and 4 are at 50 V/Div and channel 3 is at 2 A/Div. Trace 3 visualizes the step change and period of DC link voltage regulation after load change occurred. When load changes from 5 to 80 %, time/division knob is at 250 ms which shows the voltage regulated within the period of 40 ms, the line current increases and line voltage remains constant. The line current harmonics pattern is show in Fig. 5c. It can be observed that all the lower order harmonics and third harmonics are totally discarded while fifth and seventh harmonics are reduced up to negligible level.

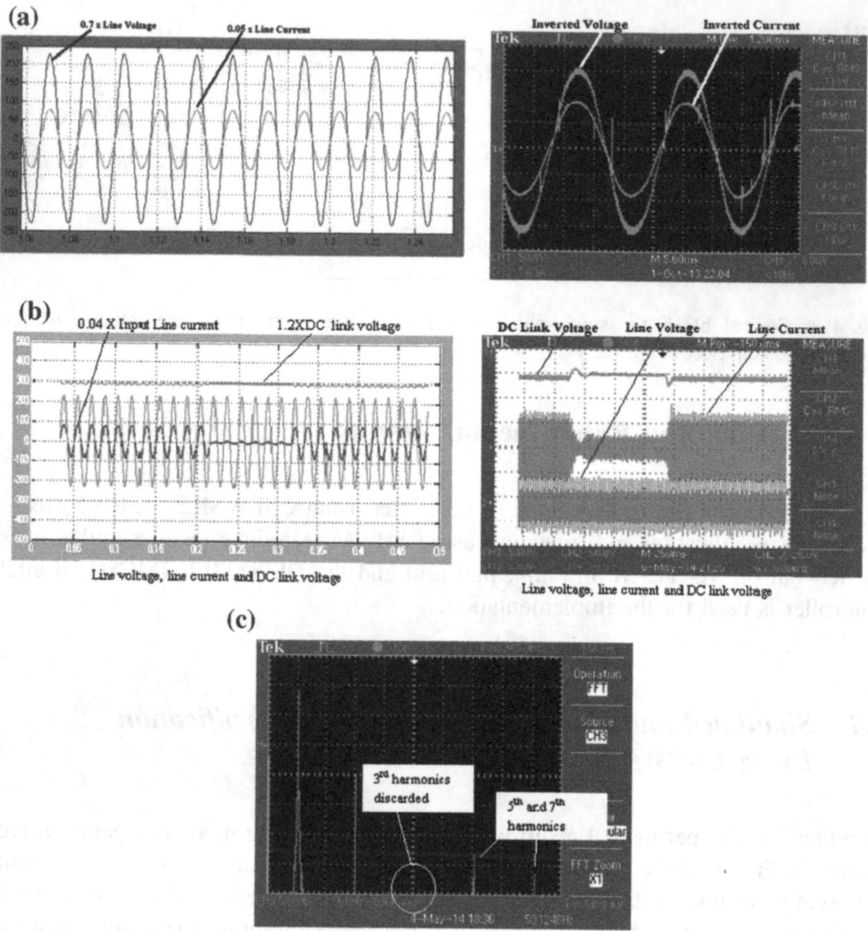

Fig. 5 **a** Simulated and experimental results of line current and line voltage to show high power factor. **b** Simulated and experimental results of line voltage, line current, and DC link voltage when load changes from 80 to 5 % periodically to show voltage stabilization period. **c** Line current harmonic pattern to show third harmonics discarded and reduction in fifth and seventh harmonics

3.2 *Simulated and Experimental Results of Rectification Using SPWM*

Simulated and experimental result waveforms for rectifying mode of operation are shown in Fig. 6. Trace 1 of Fig. 6a shows the simulated waveform of line current flowing maximum at the peak of line voltage and almost zero at the starting to indicate the degradation in power factor. The experimental result for the same is shown in Trace 2 of Fig. 6a. This shows the exactly similar result like simulated waveform. The channel 1 is at 50 V/Div and channel 3 is at 5 A/Div.

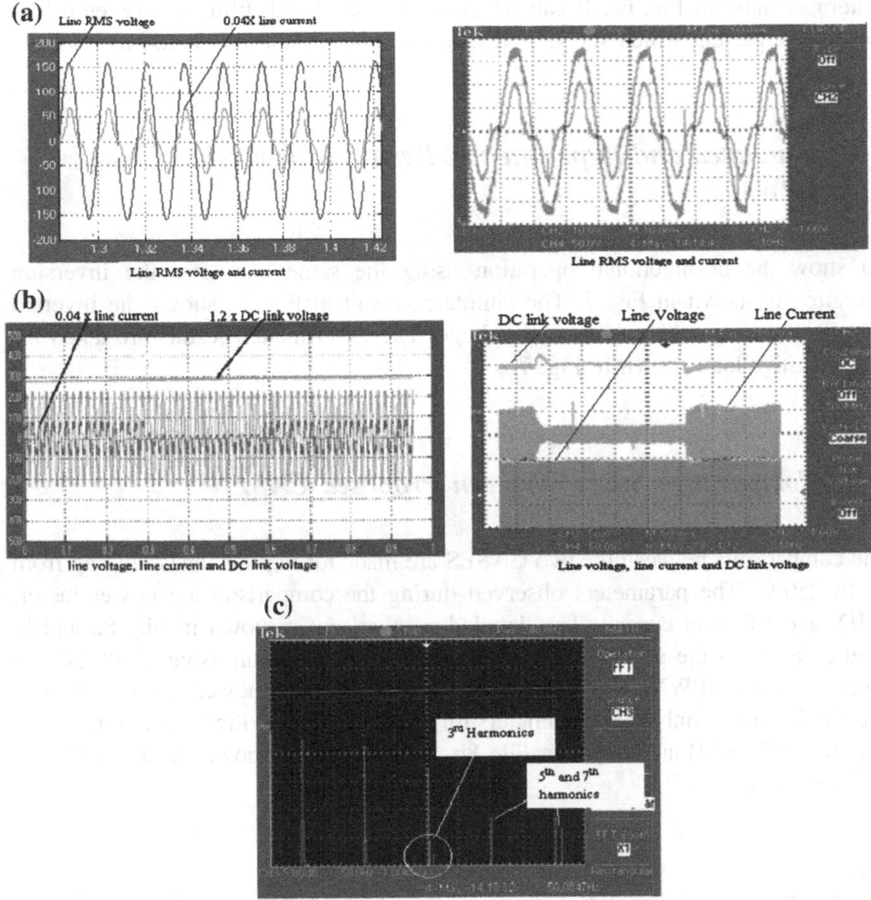

Fig. 6 **a** Simulated and experimental results of line current and line voltage to show degraded power factor. **b** Simulated and experimental results of line voltage, line current and DC link voltage when load changes from 80 to 5 % periodically to show distortions during voltage stabilization and stabilization period. **c** Line current harmonics pattern to show third, fifth, and seventh harmonics are present

The DC side load is changed from 5 to 80 % periodically to show the output voltage regulation which is shown in Fig. 6b. Trace 1 of Fig. 6b shows simulated result, it can be seen that as load changes from 80 to 5 % and from 5 to 80 % the DC link voltage becomes unregulated. Trace 2 shows experimental result for the same. It can be observed that the DC link voltage regulation take more period when load change occurred. Channel 2 and 4 are at 50 V/Div, channel 3 is at 2 A/Div and time/division knob is at 500 ms. Trace 3 visualizes the step change and period of DC link voltage regulation after load change occurred. It shows that when load changes from 5 to 80 % the voltage regulated within the period of 100 ms, also the line current and line voltage disturbance increases. The line current harmonics pattern is show in Fig. 6c. It can be observed that thirrd, fifth, and seventh harmonics are present which shows that total harmonic distortion (THD) increases.

3.3 Simulated and Experimental Results of Inversion Using SPWM

To show the bi-directional operation using the same converter the inversion operation is shown in Fig. 7. The simulated result in Fig. 7a shows the inverted current is exactly 180° of inverted voltage. The experimental results are taken for pure resistive load shown in Fig. 7b.

3.4 Comparative Study Between Proposed CASES

The comparisons between the two CASES are made for various loads ranging from 90 to 250 Ω. The parameters observed during the comparison are power factor, THD, and DC load current. The detailed comparison is shown in Fig. 8a and b. Figure 8a shows the parameters for the CASE-I (rectification using CSPWM and inversion using SPWM). It can be observed that full load power factor is 97.8 % and THD is 6.3 % only. The parameters for the CASE-II (rectification and inversion both using SPWM) are shown in Fig. 8b. It is shown that power factor is 80.3 % and THD is 19.8 %.

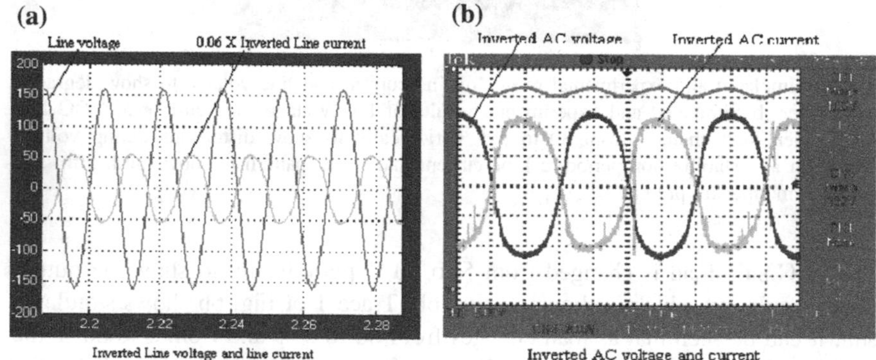

Fig. 7 **a** Simulated results of inverted line current and line voltage at exactly 180° phase shifted to show inversion action, **b** experimental results

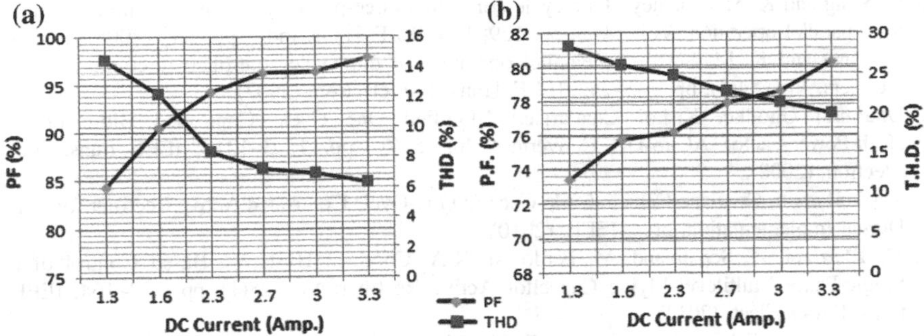

Fig. 8 **a** Variation in power factor and THD with change in load and DC load current for case-I. **b** Variation in power factor and THD with change in load and DC load current for case-II

4 Conclusion

Two schemes CSPWM and SPWM are reported for the same converter for rectification and inversion operation. The scheme using rectification with CSPWM and inversion with SPWM exhibit remarkable advantages such as high power factor, less DC link voltage stabilization period, and low THDs. Compared to this scheme second scheme (rectification and inversion both using SPWM) is less efficient. The detail simulation studies are carried out to show the comparison and effectiveness of the schemes.

References

1. A. H. Noyola, M. J. Samotyj, and W. M. Grady.: Survey of active power line conditioning methodologies., 5, pp. 1536–1542, IEEE Trans. Power Delivery, (1990).
2. Ashish Pandey, Brij N. Singh, Ambrish Chandra & Bhim Singh.: A Review of Single-Phase Improved Power Quality AC–DC Converters. 50(5), IEEE Trans. Industrial Elect, (2003).
3. J. W. Kolar, U. Drofenik, and F. C. Zach.: VIENNA rectifier II—A novel single-stage high-frequency isolated three-phase PWM rectifier system. 46(4), IEEE Trans. Ind. Electron., (1999).
4. A.D. Pathak, R.E. Locher and H.S. Mazumdar.: 3-Phase Power Factor Correction using Vienna rectifier approach and modular construction for improved overall performance, efficiency and reliability. Power electronics conf. in Long Bench, CA, (2003).
5. F. Zach, J. Kolar.: A novel three phase utility interface minimizing line current harmonics of high power telecommunication rectifier modules. 44(4), IEEE trans. Ind. Electron, (1997).
6. D. S. Lim and D. C. Lee.: AC voltage and current sensorless control of three-phase PWM rectifiers. IEEE Trans. Power Electron. 17(6), (2002).
7. Heinz Willi, Hans-Christoph, Georg Victor.: Analysis and Realization of a Pulsewidth Modulator Based on Voltage Space Vector. IEEE Trans. on Ind. Applications, 24(1), (1988).
8. S. Cuk and M. K. Smedley.: One cycle control of power converters. Vol. 10, pp. 475–481, IEEE Trans. Power Electron. Nov. (1995).

9. C. Yang and K. M. Smedley.: One-cycle-controlled three-phase grid connected inverters and their parallel operation. Vol. 44(2), pp. 129–136, IEEE Trans. Ind. Applicat., (2008).
10. C. Yang and K. M. Smedley.: Parallel operation of one-cycle controlled three-phase PFC rectifiers. Vol. 54(6), pp. 745–752, IEEE Trans. Ind. electron, (2007).
11. Dharmraj Ghodke, Kishor Chattarjee.: Modified One Cycle Controlled Bidirectional High-Power-Factor AC-to-DC Converter. Vol. 55(6), pp. 1203–1210, IEEE trans. Ind. Electron, (2008).
12. Mario Barbati, Cristiano Caluisi, Carlo Cecati.: One-Cycle Controlled Active Rectifier for Full Digital Implementation. proc. IEEE, (2010).
13. M. Khazraei, H. Sepahvand, M. Ferdowsi, K.A. Corzine.: Hysteresis-Based Control of a Single-Phase Multilevel Flying Capacitor Active Rectifier. Vol. 28(1), pp. 154–164, IEEE trans. Power Elect. (2013).
14. T. Tanaka, T. Sekiya, M. Okamoto, E. Hiraki.: Smart Charger for Electric vehicles with Power-Quality compensator on Single-Phase three wire Distribution Feeders. Vol. 49(6), pp. 2628–2635, IEEE trans. Power delivery, (2013).
15. Rahul G Mapari, D G Wakde,: A Simple Control Strategy Technique for a Single-phase Bridgeless Active Rectifier with High Power Factor and Voltage Stabilization Using Partial Digital Implementation, Vol. 324, pp. 17–26, Springer series of Advances in Intelligent Systems and Computing, (2014).

Ensemble Method Using Correlation Based Feature Selection with Stratified Sampling for Classification

Shweta B. Meshram and Sharmila M. Shinde

Abstract Ensemble methods are preferred as they represent good significance over specific predictor regarding accuracy and confidence in classification. This paper proposes here the ensemble method with multiple independent feature subsets in order to classify high-dimensional data in the area of the biomedicine using Correlation feature selection with Stratified Sampling and Radial Basis Functions Neural Network. At first, method select the feature subsets using Correlation based feature Selection with Stratified Sampling. It minimizes the redundancy in the features. After generating the feature subsets, each feature subset is trained using base classifier and then these results are combined using majority voting. The proposed method uses CFS-SS in ensemble classification method.

Keywords CFS-SS · Ensemble method · High dimensional data · RBFNN

1 Introduction

The information is constantly growing day by day. It becomes important to dig out useful information from such huge data. It also becomes necessary to sort the large information. So, there is growing interest to help people to discover the knowledge from such information. Knowledge discovery is one of the main activities of the data mining. The knowledge about procedures and rules are discovered from datasets using the data mining which is further used for classification and predicting the future events. Classification algorithm is presented with a set of records in which each record is defined by specific features. These features belong to class label and it represents its target. The classification performance can be deteriorates if it is directly applied on datasets with the high dimensional, high noise, class imbalance or small

S.B. Meshram (✉) · S.M. Shinde
Computer Engineering Department, JSCOE, Pune, Maharashtra, India
e-mail: shwet.meshram@gmail.com

S.M. Shinde
e-mail: sharmi_anant@yahoo.co.in

© Springer Science+Business Media Singapore 2017
S.C. Satapathy et al. (eds.), *Proceedings of the International Conference on Data Engineering and Communication Technology*, Advances in Intelligent Systems and Computing 468, DOI 10.1007/978-981-10-1675-2_6

sample data. We can notice that classification has wide applications in biomedicine, bioinformatics, and image processing and pattern recognition.

The data dimensionality is the growing problem in the datasets and it poses many challenges for supervised prediction. The single classification methods like decision tree, neural network, naive bayes and support vector machines cannot directly applied on high-dimensional datasets [1]. The ensemble classification methods produce better classification accuracy than that of single classifier. Hence ensemble classifier methods are used to achieve better accuracy. Individual classification method cannot handle the noise and imbalance data in the high dimensional datasets [2–4]. There are many ensemble methods proposed till now but they are not much accurate on bioinformatics dataset. Hence we are proposing here more efficient method for classification of high dimensional data.

The paper consists of mainly three parts. First part describes the literature survey. Second part discusses about the proposed ensemble method in brief and third part represents the implementation. Lastly experimental setup and result discussion is explained of the proposed method.

2 Literature Survey

One of the ensemble methods is bagging that was proposed by L. Breiman. It is more robust to noisy data and can handle the unstable procedures more efficiently [5]. Stability in the bagging decreases with the number of the variables used in the predictor decreases and results in poor performance. There was a need to improve the performance of boosting because of the two reasons [6]. They are as (i) it generates a hypothesis whose error on the training set is small by combining many hypotheses whose error may be large, (ii) variance reduction. To overcome these two problems, AdaBoost.M1 and AdaBoost.M2 are proposed. But both are highly sensitive to the noise and outlier. Breiman proposed the "random forest" [7]. The random forest improves the performance significantly when it is compared with single decision tree. But they are difficult to analyze, overfits data that are noisy and cannot predict beyond the range of training data.

S.B. Cho, H. Won proposed ensemble of neural networks for cancer classification with multiple significant gene subsets [8]. It maintains the originality of the features, performance of the classification doesn't improve much i.e. 87.9 % on colon dataset. H.I. Elshazly, A.M. Elkorany, A.E. Hassanien proposed the new multiple classifier system built with independent decision tree [9]. It achieves good accuracy on prostate cancer but with the loss of originality. Y. Piao, H.W. Park, C.H. Ji, K. Ho Ryu proposed the ensemble method that uses the Fast Correlation-Based Filter method (FCBF) to generate multiple feature subsets [1]. The computation time of ensemble method is improved, but it may create instable behavior of predictive algorithms which alternately affects the accuracy of the classification. Hence, new ensemble method is proposed here which preserve the originality of the features with improved accuracy and reduced redundancy.

3 Proposed Ensemble Method

Ensemble classifier increases the performance of classification and confidence of the results [10]. Ensemble classifier generation methods using homogeneous base classifiers can be broadly classified into five groups that are [2] that are based on (i) manipulation of the training parameters, (ii) manipulation of the error function, (iii) manipulation of the feature space, (iv) manipulation of the output labels, and (v) manipulation of the training patterns. The proposed ensemble classifier belongs to the third category. The ensemble methods using independent feature subsets are more efficient than other methods because these methods are fast because of the reduced input size and it also decreases the correlations between the classifiers.

3.1 Feature Subsets Creation

Figure 1 represents the detailed architecture of the proposed ensemble method. Feature subsets creation is the most important phase of the proposed model. The independent feature subsets are created using different methods i.e. FCBF, CFS,

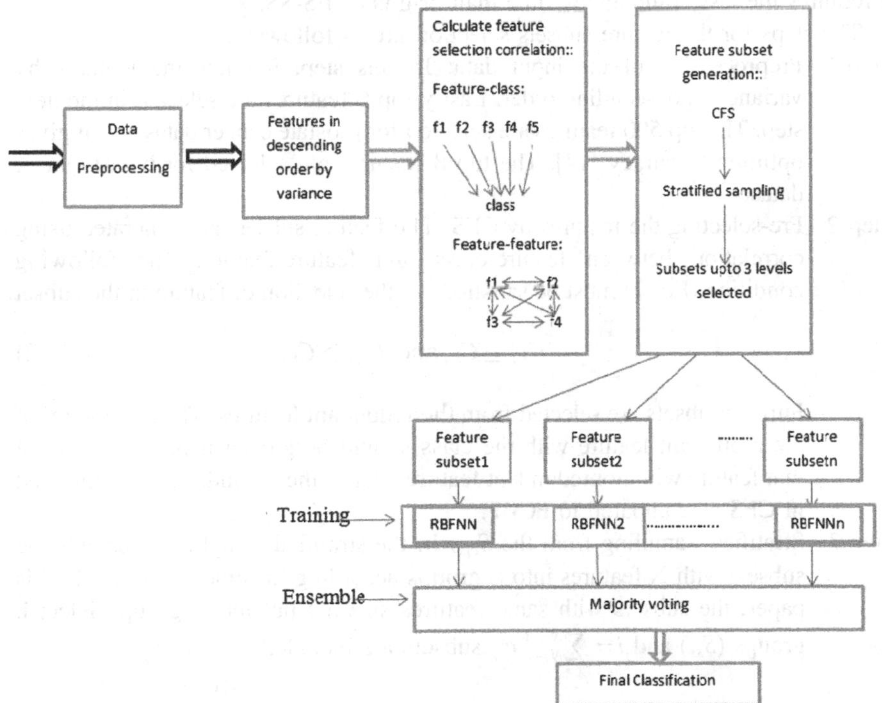

Fig. 1 Training phase of proposed method

K-S CBF, SVM RFE, RL, SBS, SFBS, CFS-SS [11–14]. FCBF is not much accurate as CFS but it requires less running time as compared to the other methods. SVM-RFE is the most accurate method but it requires the highest computational time which is not considerable when we deal with large datasets with large features. CFS requires the considerable computation time and also it is highly accurate. CFS-SS is the more accurate than CFS but requires more computation time because it searches for the best subset for learning [14]. CFS-SS is a feature selection method based on CFS. In CFS, the features are selected only by calculating the correlation between features-classes and features-features. The output of the CFS is given to the CFS-SS to select final feature subsets. The CFS ranks the feature subsets instead of individual features by merit function n. It is given as follows:

$$Merit_s = \frac{k\bar{C}_{cf}}{\sqrt{k + k(k-1)\bar{C}_{ff}}} \tag{1}$$

Equation (1) is the merit function of the s feature subset with k features. $\overline{C_{cf}}$ represents the mean class-feature correlation and $\overline{C_{ff}}$ represents the mean feature-feature correlation.

In the proposed ensemble method, the subsets selected after the stratified sampling are learned using the base classifiers and not search for the best subset. Hence it requires the less computation time than original CFS-SS.

The steps for the feature subsets selection are as follows:

Step 1 Preprocessing of the input data: In this step, it ranks the features by variance in descending order. Lastly top S features are selected in the next step. The top 500 features are selected for prostate cancer dataset as it gives optimum accuracy [14]. The top 8 features are selected for breast cancer dataset.

Step 2 Pre-selecting the features by CFS: The feature subsets are generated using correlation between feature-class and feature-feature. The following condition (Eq. 2) must be satisfied for the selection of feature in the subset

$$C_{i,c} \geq C_{i,j} \text{ and } C_{i,c} \geq C_{j,c} \tag{2}$$

Further subsets are selected from the redundant features. The correlation of the redundant feature with the class should be greater than correlation of that feature with nonredundant feature. Hence the redundancy is minimized in CFS as compared to FCBF.

Step 3 Stratified Sampling from the S_{cfs}: In the stratified sampling, separates the subsets with N features into L groups according to some strategies. In this paper, the subsets with same features size are put into a group. Select k groups (S_{ss}) and $i = \sum_{N-k}^{N-1} c_N^i$ subsets are included.

The proposed method applies CFS-SS for feature subsets in ensemble method. It selects three groups. The redundancy is reduced in the CFS as it compares the correlation of redundant feature-class with the correlation of feature-feature that is nonredundant rather than comparing it with the redundant feature.

There are many feature extraction methods. In previous ensemble methods, it considers all subsets to learn the model. But most of the subsets contain repeated features and it increases computation time. Hence the proposed method uses the stratified sampling to select subsets. So that we can focus on limited subsets while learning of the model.

3.2 Training Classifiers

SVM is adopted as the base classifier in most of the ensemble methods. It can accurately classify the data and avoids the over fitting of the data. But it requires large training and testing time. Also it is more suitable for binary classification only. It is difficult to find best kernel for given dataset in SVM classifier.

RBFNN overcome the limitations of base classifiers which are used in the earlier ensemble method. It will efficiently predict the sample class which has a low complexity than other classifier. RBFs have many good features like localization, cluster modeling, quasi orthogonality, functional approximation and interpolation [15]. These properties made them attractive in many applications.

3.3 Majority Voting

Lastly, the classification of each predictor is combined using majority voting. The majority voting classifies the features depending on maximum votes which are assigned by each classifier. The mathematical equation of the proposed method after majority voting with i classifiers can be as follows where f represents feature and c represents classifier:

$$class(z) = \arg\ \max(\sum_i f_i(z),\ c_i) \tag{3}$$

4 Experimental Setup and Result Discussion

We proposed a new ensemble method which is efficient and robust than other ensemble methods. The proposed method uses the CFS-SS which select the feature subsets for classification. Then RBFNN is used as the base classifier. That is each feature subset is learned using RBFNN. Final classification is performed using majority voting.

4.1 Datasets

To demonstrate the proposed system we have used two publicly available datasets in biomedicine field. They are breast cancer (Wisconsin) dataset and prostate cancer dataset. The prostate cancer dataset includes 102 samples and 12600 genes. Classification builds the model to differentiate between normal and tumorous prostate tissues. The breast cancer dataset includes 699 samples and 10 plus class attribute. The dataset consists of 458 benign and 241 malignant records. Classification builds the model to differentiate between benign and malignant records. Both datasets are available for classification evaluation from the UCI Machine Learning Repository. Both are binary classification problem.

4.2 Performance Evaluation

The accuracy of the proposed ensemble method is calculated using the accuracy measure. It is given as follows:

$$\text{Accuracy} = (\text{TP} + \text{TN}) / (\text{TP} + \text{TN} + \text{FP} + \text{FN}) \tag{4}$$

In the Eq. (4), TP represents the positive records that are classified positive by the classifier. TN represents the negative records that are classified as negative by the classifier. FP represents the negative records classified as positive while FN represents the positive records classified as negative. Here the positive records are malignant samples and negative records are benign samples.

The performance of the method is compared with the accuracy of the previous ensemble method using the 10 fold cross validation tool.

4.3 Results

The proposed ensemble method is compared with existing ensemble method [1] in terms of accuracy. Table 1 gives the result each fold on breast cancer and prostate cancer datasets.

The comparative chart of result is given as follows in Fig. 2.

Table 1 Comparison of classification accuracy

Sr. no.	Runs	For breast cancer dataset		For prostate cancer dataset	
		Accuracy of existing	Accuracy of proposed	Accuracy of existing	Accuracy of proposed
1	1	91.4285	96.33	95.1	97.56
2	2	93.33	93.9	95.1	96.89
3	3	95.833	96.42	95.1	97.56
4	4	96.12	93.95	94.12	97.56
5	5	97.1	97.05	94.12	97.99
6	6	94.35	97.15	93.14	95.65
7	7	95.39	97.05	95.1	97.56
8	8	95.51	94.75	96.08	98.32
9	9	95.68	97.29	95.1	97.43
10	10	95.77	96.33	95.1	97.56
Average		95.05115	96.022	94.806	97.408

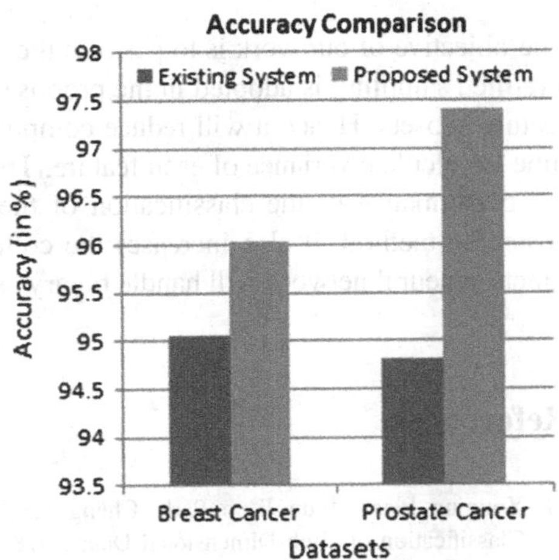

Fig. 2 Comparative chart of accuracy

The proposed ensemble method is compared with existing ensemble methods like AdaBoost M1, Random subspace and voting for both datasets. We are using weka tool to analyze the results of these classifiers. The comparison is using 10 fold cross validation tool. The analysis of results is shown in the Fig. 3.

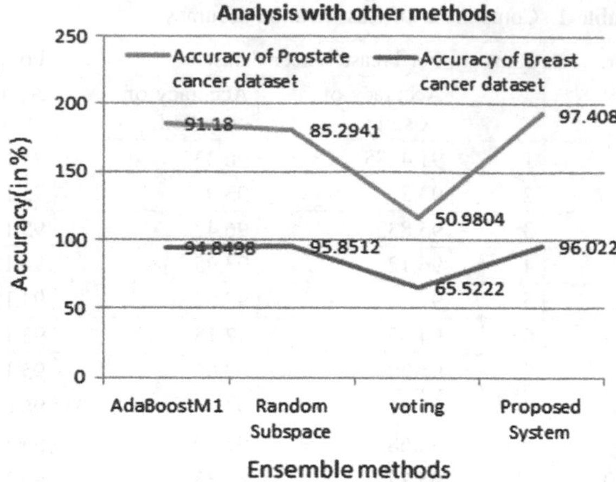

Fig. 3 Analysis with different classification methods

5 Conclusion

The objective of our work is to perform the classification task more accurately. The stratified sampling is adopted in the proposed ensemble method to focus on limited feature subsets. Hence it will reduce computation time while learning, but requires time to calculate variance of each feature. The proposed ensemble method improves the performance of the classification of the high dimensional data than previous ensemble method. It also increases the confidence in the results. The radial basis function neural network will handle binary as well as multiclass problem efficiently.

References

1. Yongjun Piao, Hyun Woo Park, Cheng Hao Ji, Keun Ho Ryu, "Ensemble Method for Classification of High-Dimensional Data", 978-1-4799-3919-0/14/ IEEE Big Comp 2014.
2. Rahman, B. Verma,"Ensemble Classifier Generation using Non uniform Layered Clustering and Genetic Algorithm", Knowledge-Based System, 2013, in press.
3. Sung-Bae Cho, Hong-HeeWon, "Cancer classification using ensemble of neural networks with multiple significant gene subsets", published online: 12 November 2006 Springer Science +Business Media, LLC 2007.
4. Yongjun Piao, Minghao Piao, Kiejung Park and Keun Ho Ryu, "An Ensemble Correlation-Based Gene Selection Algorithm for Cancer Classification with Gene Expression Data", Bioinformatics Advance Access published October 11, 2012.
5. L. Breiman, "Bagging predictors", Mach. Learning. 24, 1996, pp. 123–140.
6. Y. Freund, R.E. Schapire, "Experiments with a new boosting algorithm", In the Proceeding of the Thirteenth International Conference on Machine Learning, 1996, pp. 148–156.
7. Leo Breiman, "Random Forests", Springer Journal, Machine Learning, 45, 5–32, 2001.
8. Sung-Bae Cho, Hong-Hee Won, "Cancer classification using ensemble of neural networks with multiple significant gene subsets", published online: 12 November 2006 Springer Science & Business Media, LLC 2007.

9. Hanaa Ismail Elshazly; Abeer Mohamed Elkorany, Aboul Ella Hassanien, "Ensemble-based classifiers for prostate cancer diagnosis", 9th International Computer Engineering Conference Faculty of Engineering, Cairo University Giza, EGYPT December 28–29, 2013.
10. Josef Kittler, Mohamad Hatef, Robert P.W. Duin, and Jiri Matas, "On Combining Classifiers" IEEE Transactions on, "Pattern Analysis and Machine Intelligence", Vol. 20, NO. 3 March 1998.
11. M. A. Hall, "Correlation based Feature Selection (CFS) for Discrete and Numeric Class Machine Learning", Proceedings of the Seventeenth International Conference on Machine Learning (ICML 2000), Stanford University, Stanford, CA, USA, June 29 - July 2, 2000.
12. Jacek Biesiada, Wodzisaw Duch, "A Kolmogorov-Smirnov Correlation-Based Filter for Microarray Data", 14th International Conference, ICONIP 2007, Kitakyushu, Japan, November 13–16, 2007.
13. Guyon I, Weston J, Barnhill S, Vapnik V. "Gene selection for cancer classification using support vector machines", Machine Learning. 2002, 46:389–422.
14. Xinguo Lu, Xianghua Peng, Ping Liu, Yong Deng, Bingtao Feng, Bo Liao, "A Novel Feature Selection Method Based on CFS in Cancer Recognition", Systems Biology (ISB), 2012 IEEE 6th International Conference on 18–20 Aug. 2012, 226 – 231.
15. J. Dhande, D.R. Dandekar, S.L. Badjate, "Performance Improvement of Ann Classifiers using Pso". In Proceedings IASTED VIIP Conference, pages 733–738, 2003.

Suffix Array Blocking for Efficient Record Linkage and De-duplication in Sliding Window Fashion

Yamini Warke

Abstract Record linkage is an essential process in information mix, which is utilized as a part of combining, coordinating and copy expulsion from a few databases that allude to the same substances. De-duplication is the procedure of uprooting copy records in a solitary database. Because of multifaceted nature of today's database, coordinating records in single database is an essential one. Indexing strategies are utilized to productively actualize record linkage and De-duplication. Our additional gathering strategy with jaro-winkler similarity measure exploits the ordering used by the list to combine comparative pieces at negligible additional cost, bringing about a much higher exactness while holding the high adaptability of the base suffix array method. We complete an inside and out examination of our system what's more, show results from cxaminations using Cora, restaurant and real identity data which highlights the significance of utilizing proficient as a part of indexing and hindering in true applications where information sets contain a large number of records. This paper presents suffix array blocking for efficacious record linkage and de- duplication in sliding window fashion.

Keywords Record linkage • Blocking • Suffix array

1 Introduction

As different government offices, business, and examination tasks assemble outstandingly a lot of information, expertise that offers ascend to handling and mining of huge databases have as of late respect with both institute and industry for holding the consideration. In the period of processing information of numerous information mining activities, connecting, or coordinating records which identified with same element from more than two databases get to be grater errands. The point of such

---

Yamini Warke (✉)
Dr. D.Y. Patil School of Engineering and Technology, Savitribai
Phule Pune University, Pune, India
e-mail: warkeyamini85@gmail.com

© Springer Science+Business Media Singapore 2017
S.C. Satapathy et al. (eds.), *Proceedings of the International Conference on Data Engineering and Communication Technology*, Advances in Intelligent Systems and Computing 468, DOI 10.1007/978-981-10-1675-2_7

Table 1 Motor gas station example

Associated address	Description
SR.#23/2 Near yash Hotel, dapodi, Pune, Maharashtra	Residential location of business
Patil A Suyash 345 Hallmark avenue Ravet Road No.7	Residential location of holder of business
P A S, Inc C/o Suresh S Mahajan Ravet Road no.4 Pune	Incorporated name of business accountant does books and government form

linkages is to match and make cement of all records identifying with the same element, for example, wiped out individual, a buyer, venture, a customer item, a copyright reference.

Future utilization of existing information for new studies and the expense and decided endeavor in information securing, record linkage and de-duplication is used [1]. Evacuating copy records in a solitary database is additionally essential one. In motor gas station example, this is illustrated in Table 1 [2]. The main name alludes to name of Business and its area of residency. The second is the business holders name with his location. Third is the location of bookkeeper who does the books for the organization. The name 'P A S. Inc' is a contraction of the genuine name of the business 'Patil A Suyash' which is the holder of engine washing focus. It is potential that distinctive rundown partner with the arrangement of organizations may have passages equal to anybody of the recorded types of the element which is the engine overhauling station. For this situation there are such a variety of indistinguishable Entries discovered, that indistinguishable (duplications) are adjusted when that specific individual give back the structure. However, it is extremely monotonous errand in the event that we need [2] that data after a few years, as that individual may be not at the comparing location. Table 1 elucidate this sample. Then again, as the measure of advanced data is quickly expanding everywhere throughout the world and a large portion of the information is unstructured one, for example, picture, sound, feature, and report records. This fast development of information size causes a few issues, for example, stockpiling impediment, expanding expense. We can beat this issue by utilizing de-duplication system.

2 Related Work

Dunn [3] and Marshall [4], and Fellegi and Sunter [5] proposed a hypothesis in light of measurable grouping, in which record linkage is utilized first. Record Linkage can radically expand the data accessible for thing planned, for example, substantial medicinal wellbeing frameworks [6], business investigation, and misrepresentation identification [7] in subtle element. Indexing strategies, or blocking techniques as they are known in the connection of record linkage, were immediately perceived as a key segment for an opportunity to update or copy edit the Papers, which is not possible due to time constraints. Blocking calculations normally contain additional usefulness over standard indexing, to tackle particular record linkage issues. Blocking arrangements endeavor to diminish the quantity of applicant records for examination however much as could be expected, while as yet holding a precise result by guaranteeing that competitor records that would coordinate the question record are not left out of the hopeful set because of the blocking tenets. There are assortment of blocking routines right now utilized as a part of record linkage systems, with the most surely understood ones including customary blocking, sorted neighborhood [8], Q-gram based blocking [9], Canopy Clustering [10], string guide based blocking [11] and Suffix Array blocking [12].

Every blocking system characterize an arrangement of key fields from the information to be coordinated, that are utilized to figure out which piece (or obstructs) every record is to be put into. Huge numbers of these methodologies oblige a solitary string to be utilized as the key on which to locate the right square. In this manner, the estimations of the key fields are commonly connected together into one long string. This string is known as the blocking key value (BKV) [13]. The choice of key fields to incorporate in the BKV and additionally the requesting of these fields is vital to consider.

A suitable BKV ought to be the quality or mix of properties which are as recognizing as could be expected under the circumstances, consistently dispersed, and having low lapse likelihood. Initiate [14] thought about and assessed these blocking methods, changed two of them to make them more vigorous with respect to parameter settings, a critical thought for any calculation that is to be considered for genuine applications. The trial results demonstrated that there are expansive Contrasts in the quantity of genuine coordinated applicant record sets produced by the diverse systems, when tried utilizing the same information sets.

As different vast associations, organizations have on the whole extensive measure of information. With a specific end goal to prepare and investigate that information, coordinating of records that identify with the same substances from a few databases is important.

There are a few distinctive indexing methodologies are accessible, which includes, Conventional blocking, q-gram base indexing, covering grouping, string-guide based indexing, postfix exhibit indexing. The time intricacy of conventional blocking is O (dn log n) where n is the quantity of records in each of the two information sets that are being coordinated and d is the quantity of key fields picked [15].

Essential thought behind postfix cluster indexing is to embed the BKVs and their Additions into a postfix exhibit based modified record. In this indexing procedure, Postfixes down to a Minimum length, lm, are embedded into the addition cluster.

For instance, for a BKV "bannana" and lm = 5, the qualities 'bannana', 'annana', "nnana" will be created, and the identifiers of all records that have this BKV will be embedded into the relating four transformed Index records.

3 Methodology

3.1 Proposed System

In proposed system, Suffix array blocking in sliding window fashion is used with Grouping function. In this grouping function, suffixes are compared using different similarity measures including edit based Jaro and Jaro-Winkler [16] Similarity measures. This will give more improved results. Time complexity of algorithm is $o\ (n)^2$.

Figure 1 shows system architecture of proposed system.

In proposed system, as shown in Fig. 1, in first step one dataset is taken as input. Dataset include any Japanese and bibliographic data.

In second step, suffix array indexing is applied on that data. While applying suffix array indexing firstly blocking key value (BKV) is generated by concatenating key fields. Then suffixes are generated of that key field. All that suffixes are stored in index structure.

In third step, maximum block size is set. then for every record corresponding to that suffix is checked with block size. If no of record corresponding to that suffix is greater than maximum block size, all suffix-reference pair of that corresponding suffix are removed.

In fourth step, grouping of suffixes is done. In this for each unique suffix in inverted index comparison is done (compare *sf* to previous suffix *sg*).using comparison function jaro Winkler. Threshold is set.(if Jaro Winkler(*sf*, *sg* > *jt*)) all suffix reference pairs are grouped together corresponding to sf and sg using set join.

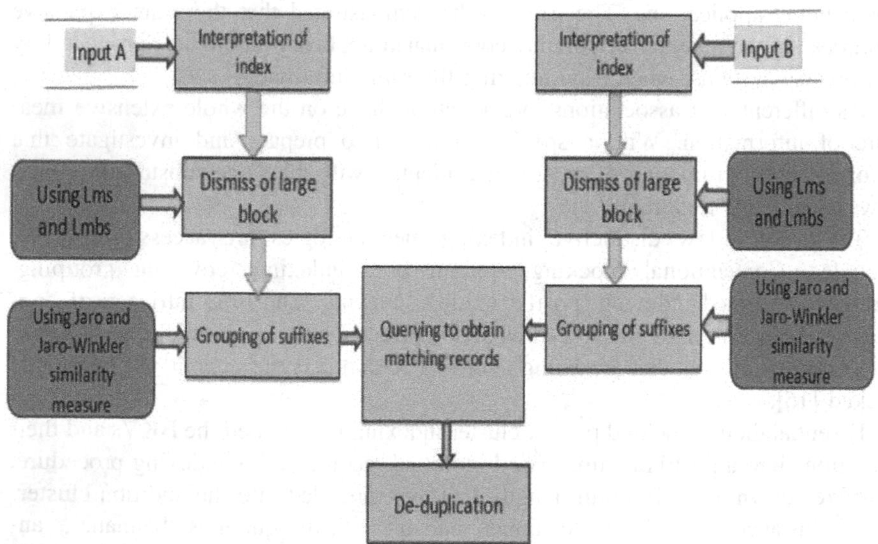

Fig. 1 System architecture

In last step, for calculating matching records, all first three steps are applied on another (Second) data set. And duplicated records are removed (De-duplication).

3.2 Algorithm Pseudo Code

Input:

1. *A* and *B*, the sets of records to find matches between
2. The suffixes comparison function similarity threshold ts
3. The minimum suffix length lms and the maximum block size lmbs

Let **I** be the inverted index structure used.
Let *Ci* be the resulting set of candidates to be used when matching with a record *A*

Interpretation of Index structure

1. For record ri1*ε A*
2. Construct BKV by concatenating Key fields
3. Generate suffixes in sliding window fashion
4. Insert suffixes and reference records to suffixes into I

Dismiss Large Block

5. For every unique suffix *Sf* in *I*
6. If the number of record reference paired with *Sf* > Lmbs
7. Remove all suffix reference pairs where the suffix is *Sf*

Grouping of suffixes

8. For each, unique suffix *Sf* in *I*
9. Compare all suffixes *Sf* with previous suffix *Sg*
10. Using chosen comparison function (e.g.,Jaro-Winkler)
11. If Jaro-Winkler (*Sf*, *Sg*) > *ts*
12. Group together the suffix reference pairs
13. Corresponding to *Sf* and *sg*.

Querying to gather candidate sets for matching:

14. For record ri1*ε B*
15. Construct BKV by concatenating key fields
16. Generate suffixes of BKV
17. Match suffixes of *A* and *B*
18. Ci resulting set of records with no duplication

4 Results and Comparison

This section briefly describes the results of the existing system, proposed system along with experiments carried out. The brief comparison of same also provided.

4.1 Description of System Execution and Results

Our experiments are designed to compare improved suffix Array blocking against proposed system primarily using the measurements of pair's completeness and pair's quality [17]. We run the experiments on Cora, Restaurant and Real identity data. As mentioned in the Sect. 2, Record linkage is done on the basis of similarity between two string records. Thus for calculating similarity between string records, some similarity measure is used like Jaro and Jaro-Winkler similarity measure [16], as described in Sect. 3.1. Following Figs. 2 and 3 shows results obtained by existing jaro and proposed jaro_winkler.In this case, all suffixes with their corresponding records are compared using Jaro and Jaro-Winkler similarity measure by observing both the Figs. 2 and 3, it can be seen that proposed (Jaro-Winkler) similarity measure gives maximum similarity than existing (jaro) similarity measure. Also time required by Jaro and Jaro-Winkler on Cora, restaurant and real identity dataset is shown in Figs. 4, 5, and6, respectively. All experiment results shown use Jaro-Winkler for the grouping similarity function, and the threshold for determining Jaro-Winkler similarity between two strings is set at 0.85 for all experiments.

Suffix1	Record1	Suffix2	Record2	Jaro Max-Si..
a-1993	R22	a-1994	R55	0.81
a-1993	R22	a-1994	R55	0.81
a-1993	R22	a-1994	R55	0.81
a-1993	R22	a-1997	R75	0.81
a-1993	R22	a-1994	R55	0.81
a-1993	R22	a-1994	R55	0.81
a-1993	R22	a-1994	R55	0.81
a-1993	R22	a-1997	R75	0.81
a-1993	R22	a-1994	R55	0.81
a-1993	R22	a-1994	R55	0.81
a-1993	R22	a-1994	R55	0.81
a-1993	R22	a-1997	R75	0.81
a-1994	R55	a-1997	R75	0.81
a-1994	R55	a-1997	R75	0.81
a-1994	R55	a-1997	R75	0.81
cesa-1993	R21	cesa-1994	R56	0.88
cesa-1993	R21	cesa-1994	R56	0.88
cesa-1993	R21	cesa-1994	R56	0.88
cesa-1993	R21	cesa-1997	R75	0.88
cesa-1993	R21	cesa-1994	R50	0.88
cesa-1993	R21	cesa-1994	R56	0.88
cesa-1993	R21	cesa-1994	R56	0.88

Fig. 2 Suffixes with Jaro similarity (Existing system)

Suffix1	Record1	Suffix2	Record2	Jaro-Winkl...
a-1993	R22	a-1994	R55	0.8480000...
a-1993	R22	a-1994	R55	0.8480000...
a-1993	R22	a-1994	R55	0.8480000...
a-1993	R22	a-1997	R75	0.8480000...
a-1993	R22	a-1994	R55	0.8480000...
a-1993	R22	a-1994	R55	0.8480000...
a-1993	R22	a-1994	R55	0.8480000...
a-1993	R22	a-1997	R75	0.8480000...
a-1993	R22	a-1994	R55	0.8480000...
a-1993	R22	a-1994	R55	0.8480000...
a-1993	R22	a-1994	R55	0.8480000...
a-1993	R22	a-1997	R75	0.8480000...
a-1994	R55	a-1997	R75	0.8480000...
a-1994	R55	a-1997	R75	0.8480000...
a-1994	R55	a-1997	R75	0.8480000...
cesa-1993	R21	cesa-1994	R56	0.904
cesa-1993	R21	cesa-1994	R56	0.904
cesa-1993	R21	cesa-1994	R56	0.904
cesa-1993	R21	cesa-1997	R75	0.904
cesa-1993	R21	cesa-1994	R56	0.904
cesa-1993	R21	cesa-1994	R56	0.904
cesa-1993	R21	cesa-1994	R56	0.904

Fig. 3 Suffixes with Jaro-Winkler similarity (Proposed system)

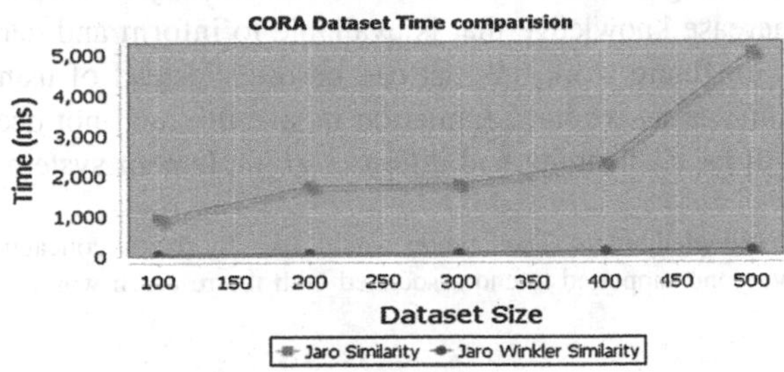

Fig. 4 Time obtained by Jaro and Jaro-winkler similarity on Cora dataset

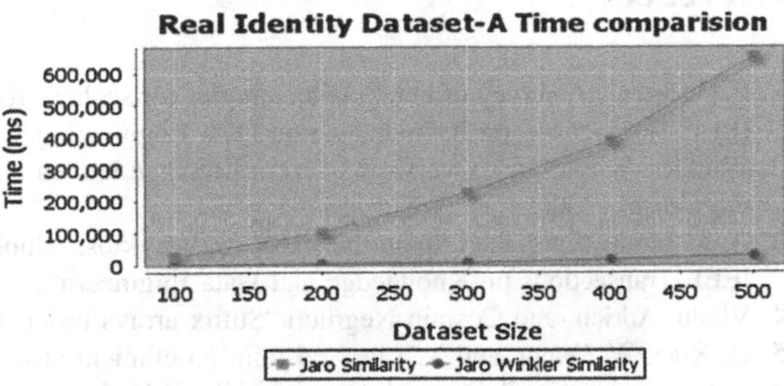

Fig. 5 Time obtained by Jaro and Jaro-winkler similarity on real identity dataset

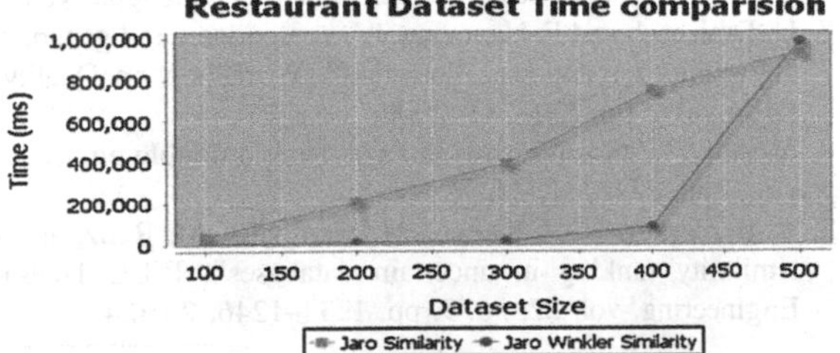

Fig. 6 Time obtained by Jaro and Jaro-winkler similarity on restaurant dataset

5 Conclusion and Future Scope

Suffix array blocking in sliding window fashion is highly capable and relevant to outperform traditional methods in scalability, at the cost of indicative amount of accuracy, depending on the attributes of the data used. Proposed improvement derives these qualities, but significantly improves the accuracy at the cost of very small amount of extra processing. The qualities of suffix array blocking in sliding window fashion make it well suited for large-scale applications of record linkage. We have also shown that the accuracy or pair completeness of proposed Suffix Array blocking is much higher than improved Suffix Array blocking for the data sets we used in our experiments. For example, identity matching of Rina and Tina, proposed approach gives more accuracy as compared to improved suffix array blocking. Because proposed approach generates suffixes in sliding window fashion. As in many industries; it is common situation that many large data sets exist including archival and current. It is necessary to keep that data together, in order to increase knowledge that is available to inform and derive decisions.

In future work link list can be used instead of using suffix array. As by using suffix array we have limitation in size this will not occur in case of link list. So it will be challenging and different to implement system by using link list.

Acknowledgmens The author would like to thank colleagues, friends, all researchers and everyone supported to and associated with the research work.

References

1. P. Christen. "A survey of indexing techniques for scalable record linkage and de- duplication", IEEE Transactions on Knowledge and Data Engineering, Vol. 24.9, pp. 1537–1555, 2012.
2. Winkler, William E. "Overview of record linkage and current research directions." *Bureau of the Census*. 2006.
3. A. K. Elmagarmid, P. G. Ipeirotis, and V. S. Verykios. "Duplicate record detection: A survey", IEEE Transactions on Knowledge and Data Engineering, vol. 19, no. 1, pp. 116, 2007.
4. Vladu, Adrian, and Cosmin Negrueri."Suffix arrays programming contest approach",2005.
5. C. Xiao, W. Wang, and X. Lin. "Ed-join: an efficient algorithm for similarity joins with edit distance constraints", Proceedings of the VLDB Endowment, vol. 1, no. 1, pp. 933–944, 2008.
6. A. Behm, S. Ji, C. Li, and J. Lu. "Space-constrained gram-based indexing for efficient approximate string search", IEEE ICDE09, Shanghai vol. 2,pp. 604–615, 2009.
7. U. Draisbach and F. Naumann. "A comparison and generalization of blocking and windowing algorithms for duplicate detection", Workshop on Quality in Databases, held at VLDB09, Lyon vol. 3,pp. 274–283, 2009.
8. N. Adly. "Efficient record linkage using a double embedding scheme", DMIN09, Las Vegas vol. 2,pp. 274–281, 2009.
9. T. Bernecker, H.-P. Kriegel, N. Mamoulis, M. Renz, and A. Zuefle. "Scalable probabilistic similarity ranking in uncertain databases", IEEE Transactions on Knowledge and Data Engineering, vol. 22, no. 9, pp. 1234–1246, 2010.4.

10. Gog, Simon, Alistair Moffat, J. Culpepper, Andrew Turpin, and Anthony Wirth. “Largescale pattern search using reduced-space on-disk suffix arrays”, IEEE Transactions on Knowledge and Data Engineering, VOL. 26, NO. 8, AUGUST 2014.
11. Winkler, William E.. “Overview of record linkage and current research directions”, US Bureau of the Census., Tech. Rep. vol. 2, 2006.
12. M. Weis, F. Naumann, U. Jehle, J. Lufter, and H. Schuster. “Industry-scale duplicate detection ”, Proceedings of the VLDB En- dowment, vol. 1, no. 2, pp. 1253–1264, 2008.
13. G. V. Moustakides and V. S. Verykios. “Optimal stopping: A record-linkage approach”, Journal Data and Information Quality vol. 1, pp. 9:19:34, 2009.
14. P. Christen and A. Pudjijono “Accurate synthetic generation of realistic personal information”, IEEE Transactions on Knowledge and Data Engineering, vol. 5476, pp. 507–514,20095.
15. P. Christen. “Automatic record linkage using seeded nearest neighbour and support vector machine classification”, ACM SIGKDD08, Las Vegas, pp. 151–159, 2008.
16. van der Loo, M., van der Laan, J., Team, R. C. & Logan, N, “Package stringdist”, 2013.
17. T. de Vries, H. Ke, S. Chawla, and P. Christen, “Robust record linkage blocking using suffix arrays,” *ACM CIKM’09*, pp. 305–314, 2009.

Robust Dynamic Sliding Mode Control for a Class of Uncertain Multi-variable Process

B.J. Parvat and B.M. Patre

Abstract This paper deals with a design of dynamic sliding mode tracking control for a class of uncertain multiple input multiple output (MIMO) process. A dynamic sliding mode control (DSMC) gives more accuracy with reduced/removed chattering resulting from high frequency switching control input. To demonstrate the effectiveness of the proposed DSMC, multi variable coupled tank process is simulated. From simulation results it has been found to be satisfactory.

Keywords Robust control · Siding mode control (SMC) · Dynamic sliding mode control (DSMC) · Multi-variable processes · MIMO process and uncertain system

1 Introduction

During the past several decades, a lots of robust control strategy have been developed for controlling of a processes with parametric uncertainties and external disturbances [1–4]. The SMC approach has been recognized as a one of the efficient control strategy to design robust controllers for nonlinear dynamic systems operating under uncertainty and external disturbance conditions. The main advantage of SMC is, it has low sensitivity to plant parameter variations and disturbances so that it minimizes the requirement of accurate process model for designing of controller. However, the drawback of SMC is, it has been restricted in application due to an undesirable chattering occurred in control input. Because chattering is not acceptable by physical final control elements or actuators [5–8].

In the literature of SMC, different solutions are proposed to reduce the chattering effect. In process control, such a chattering can lead to high losses in terms of

B.J. Parvat (✉) · B.M. Patre
Department of Instrumentation Engineering,
S.G.G.S Institute of Engineering and Technology, Vishnupuri, Nanded 431606, India
e-mail: pbhagsen@yahoo.com

B.M. Patre
e-mail: bmpatre@yahoo.com

© Springer Science+Business Media Singapore 2017
S.C. Satapathy et al. (eds.), *Proceedings of the International Conference on Data Engineering and Communication Technology*, Advances in Intelligent Systems and Computing 468, DOI 10.1007/978-981-10-1675-2_8

damaging of actuators introduced down time and lost the production. Hence, it is necessary to reduce chattering by maintaining and control the switching frequency in to a required and acceptable level. For the chattering free control DSMC [9–11], adaptive fuzzy sliding mode observer [12], adaptive terminal sliding mode control [13] and higher order sliding mode control [14–17] are reported in the literature. A DSMC adds additional dynamics, which can be considered as compensator. The SMC with compensator is an improved system. Hence DSMC gives improved stability and performance of system. It also yield smooth and chattering free control from the high frequency switching control [3, 9, 18]. In this paper for showing the effectiveness of DSMC, it has been demonstrated for both uncertain nonlinear single input single output (SISO) and MIMO processes. For MIMO processes decentralized control structure is used. It uses decoupler designed from plant model to minimize the interactions among the multiple control loops [19–22].

The rest of the paper is organized as follows. Section 2 briefly describes about design of DSMC. A DSMC for SISO system is designed in Sect. 3. In Sect. 4, application of DSMC is carried out for MIMO system. The effectiveness of designed controller is illustrated by results obtained from simulation study in Sect. 5 with conclusions in Sect. 6.

2 Dynamic Sliding Mode Control (DSMC)

In DSMC a new constructed switching function from normal switching function of classical SMC gives relative first order or high order derivatives in the control input. Accordingly, a continuous DSMC is obtained and the chattering phenomenon can be reduced sufficiently [10, 18].

2.1 Problem Formulation

Consider a uncertain nonlinear system described as follows:

$$\begin{aligned} \dot{x}_1 &= x_2 \\ \dot{x}_2 &= f(x,t) + g(x,t)u + \triangle f(x,t) + \triangle g(x,t)u + \delta(x,t) \\ y &= x_1 \end{aligned} \tag{1}$$

where x_1, x_2 are state variables, x is state vector, y is system output, u is control input, $f(x,t)$, $g(x,t)$ are known system functions, $\triangle f(x,t)$, $\triangle g(x,t)$ are uncertainties in $f(x,t)$, $g(x,t)$ and $\delta(x,t)$ is unmeasurable external disturbance.

Assumptions System uncertainties $\triangle f(x,t)$, $\triangle g(x,t)$ and disturbance $\delta(x,t)$ satisfy the matching conditions given by [4, 23].

$$\triangle f(x,t) + \triangle g(x,t)u + \delta(x,t) = g(x,t)d(t) \tag{2}$$

where d(t) indicates the uncertain part of the system Eq. (1). Therefore, system Eq. (1) can be written as

$$\begin{aligned} \dot{x}_1 &= x_2 \\ \dot{x}_2 &= f(x,t) + g(x,t)u + d(x,t) \\ y &= x_1 \end{aligned} \tag{3}$$

The main aim of the proposed control method is to design a robust DSMC for the uncertain system Eq. (3) with minimum control input efforts to stabilize it.

2.2 Design of Controller

For designing of DSMC let us define the tracking error $e(t)$ and the switching function $s(t)$ as below

$$e(t) = y(t) - y_d(t) \tag{4}$$

where $y_d(t)$ is desired output for the system.

$$s(t) = ce(t) + \dot{e}(t) \tag{5}$$

where c is positive and real. Therefore,

$$\dot{s}(t) = f(x,t) + g(x,t)u + d(x,t) - \ddot{y}_d(t) + c\dot{e}(t) \tag{6}$$

Now constructing a new dynamic switching function as

$$\sigma(t) = \lambda s(t) + \dot{s}(t) \tag{7}$$

where λ is positive and real. If $\sigma(t) = 0$, is asymptotically stable and $e(t)$, $\dot{e}(t)$ tends to zero.

From Eqs. (6) and (7), we get

$$\sigma(t) = f(x,t) + g(x,t)u + d(x,t) + c\dot{e}(t) + \lambda s(t) - \ddot{y}_d \tag{8}$$

Hence

$$\begin{aligned} \dot{\sigma}(t) = \dot{f}(x,t) + (c+\lambda)f(x,t) + (\dot{g}(x,t) + \lambda g(x,t))u \\ + g(x,t)\dot{u} + \dot{d}(x,t) + (c+\lambda)d(x,t) + \lambda c\dot{e}(t) \\ -(c+\lambda)\ddot{y}_d(t) - \dddot{y}_d(t) \end{aligned} \tag{9}$$

Let the dynamic control u be expressed as

$$\dot{u} = \frac{1}{g(x,t)}(-\dot{f}(x,t) - (c+\lambda)f(x,t) - (\dot{g}(x,t) + cg(x,t) + \lambda g(x,t))u - \lambda c\dot{e}(t) + (c+\lambda)\ddot{y}_d(t) + \dddot{y}_d(t) - \rho sgn(\sigma)) \tag{10}$$

where ρ positive and real.

2.3 Stability Condition

For stability analysis of σ a Lyapunov function is introduced in the form as

$$V = \frac{1}{2}\sigma^2 \tag{11}$$

Assumptions $V > 0$
Therefore, the control u computed in Eq. (10) will derive the sliding surface variable σ to zero in finite time and will keep it zero thereafter. The derivative of V is computed as

$$\dot{V} = \sigma\dot{\sigma} \tag{12}$$

From Eqs. (9) and (10), we can write

$$\dot{\sigma}(t) = \dot{d}(x,t) + (c+\lambda)d(x,t) - \rho sgn(\sigma) \tag{13}$$

and Eqs. (12) and (13) gives

$$\sigma\dot{\sigma} = \sigma(\dot{d}(x,t) + (c+\lambda)d(x,t)) - \rho|\sigma|$$

for stability condition $\sigma\dot{\sigma}$ should be negative definite and it gives condition for selection of ρ is as

$$(\dot{d}(x,t) + (c+\lambda)d(x,t))\sigma - \rho|\sigma| < 0 \tag{14}$$

3 DSMC Design for SISO Process

To demonstrate the effectiveness of DSMC, simulation is conducted for first order plus delay time (FOPDT) model of SISO process. For illustrative example following process model is considered as

$$G(s) = \frac{0.75e^{-14s}}{137s+1} \tag{15}$$

For designing of DSMC plant model is transformed in to linear state space model as below

$$\begin{aligned} \dot{x}_1 &= x_2 \\ \dot{x}_2 &= -0.0787x_1 - 0.0005x_2 + u \\ y &= x_1 \end{aligned} \tag{16}$$

Let us consider uncertain part in the system is $d(x, t)$. then we can write Eq. (15) as

$$\begin{aligned} \dot{x}_1 &= x_2 \\ \dot{x}_2 &= -0.0787x_1 - 0.0005x_2 + u + d(x, t) \\ y &= x_1 \end{aligned} \tag{17}$$

From Eqs. (10) and (15) final control $\dot{u}$ is obtained and studied in simulation. For simulation $f(x, t) = -0.0787x_1 - 0.0005x_2$, $g(x, t) = 1$, $\lambda = 15$, $c = 5$, $\rho = 5$, $y_d(t) = 2$(step input with magnitude 2), $d(x, t) = sin(t)$ and 1 for $t \geq 250$ are taken. The simulated results for without disturbance and with disturbance are presented in Fig. 3a–c.

4 DSMC for MIMO Process

In this section, design of DSMC is extended from SISO to MIMO process. To design of controller for MIMO process, decentralized control structure with decoupler is used. For application demonstration, process model of coupled tank process is taken from [24]. The objective of decentralized controller is to control two interacting tank levels of coupled tank process. The process flow diagram as shown in Fig. 1 describes the coupled tank process with two inputs and two outputs. This process model is found in open loop by scaling of all process variables in to 0 to 5 V electrical signal. Therefore model is taken as

$$G_p(s) = \begin{bmatrix} \frac{0.43e^{-5s}}{29s+1} & \frac{0.172e^{-10s}}{35s+1} \\ \frac{0.145e^{-10s}}{40s+1} & \frac{0.37e^{-5s}}{27s+1} \end{bmatrix} \tag{18}$$

For decentralized control structure two DSMC controllers $GC1(s)$ and $GC2(s)$ are decoupled with plant model $G_p(s)$ through decoupler block D(s), so that the controller manipulates the variable $m1$ and $m2$ instead of the u_1 and u_2, as shown in Fig. 2. In this sp_1, sp_2 and y_1, y_2 are set points and plant outputs respectively. This configuration of the control system gives a set of two independent controller with the minimum interactions.

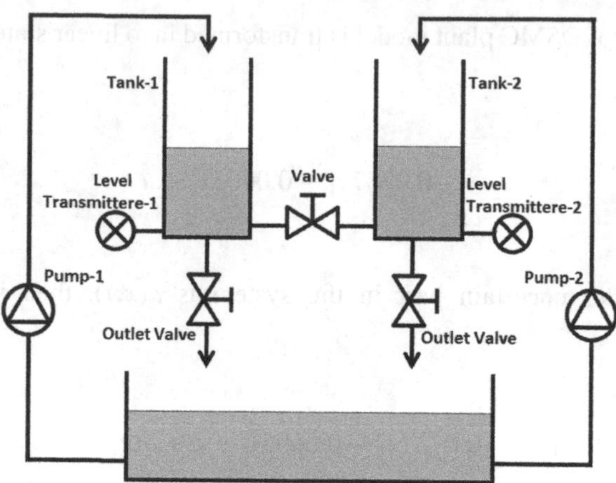

Fig. 1 Multi-variable coupled tank process

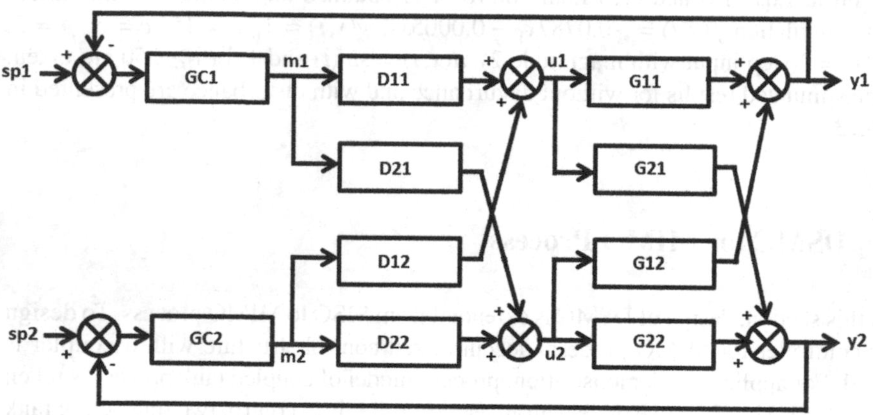

Fig. 2 General 2×2 system with decouplers and single-loop controllers

The decoupler $D(s)$ is obtained from process model $G_p(s)$ Eq. (18) is

$$D(s) = \begin{bmatrix} \frac{2.55}{27s+1} & \frac{-e^{-5s}}{35s+1} \\ \frac{-e^{-5s}}{40s+1} & \frac{2.5}{29s+1} \end{bmatrix} \tag{19}$$

for design of DSMC matrix Q(s) is needed and can be found as

$$Q(s) = G_p(s)D(s) \tag{20}$$

where

$$Q(s) = \begin{bmatrix} Q_{11}(s) \; Q_{12}(s) \\ \\ Q_{21}(s) \; Q_{22}(s) \end{bmatrix}$$

After solving Eq. (20) two diagonal elements $Q_{11}(s)$ and $Q_{22}(s)$ are resulted and $Q_{12}(s)$ and $Q_{21}(s)$ are cancelled. The determined diagonal elements $Q_{11}(s)$ and $Q_{22}(s)$ are of higher order and can be reduced in to FOPDT model [19, 25] as:

$$Q_{11}(s) \simeq \frac{0.9245e^{-12.7176s}}{113.178s+1} \tag{21}$$

$$Q_{22}(s) \simeq \frac{0.78e^{-12.7178s}}{113.186s+1} \tag{22}$$

The controller $GC1$ and $GC2$ in Fig. 2 are designed separately from $Q_{11}(s)$ and $Q_{22}(s)$ respectively. Therefore $Q_{11}(s)$ and $Q_{22}(s)$ are transformed in to linear state space models as

$$\begin{aligned} \ddot{y}_1(t) &= -0.087466\dot{y}_1(t) - 0.0006947y_1(t) \\ &\quad +0.0006423u_1(t) + d_1(x,t) \end{aligned} \tag{23}$$

$$\begin{aligned} \ddot{y}_2(t) &= -0.087460\dot{y}_2(t) - 0.0006947y_2(t) \\ &\quad +0.0005418u_2(t) + d_1(x,t) \end{aligned} \tag{24}$$

where $d_1(x,t)$ and $d_2(x,t)$ are considered as uncertain and external disturbance elements in the process. Therefore from Eq. (10) $\dot{u}_1$ and $\dot{u}_2$ are computed from Eqs. (23) and (24) as

$$\begin{aligned} \dot{u}_1 = &\frac{1}{g_1(x,t)}(-\dot{f}_1(x,t) - (c_1+\lambda_1)f_1(x,t) - (\dot{g}_1(x,t) \\ &+c_1g_1(x,t) + \lambda_1 g_1(x,t))u_1 - \lambda_1 c_1 \dot{e}_1(t) \\ &+(c_1+\lambda_1)\ddot{y}_{d1}(t) + \dddot{y}_{d_1}(t) - \rho_1 sgn(\sigma_1)) \end{aligned} \tag{25}$$

where $f_1(x,t) = -0.087466\dot{y}_1(t) - 0.0006947y_1(t)$, $g_1(x,t) = 0.0006423$, $e_1(t) = y_1 - y_{d1}$, $s_1 = c_1e_1 + \dot{e}_1$ and $\sigma_1 = \lambda_1 s_1 + \dot{s}_2$.
and

$$\begin{aligned} \dot{u}_2 = &\frac{1}{g_2(x,t)}(-\dot{f}_2(x,t) - (c_2+\lambda_2)f_2(x,t) - (\dot{g}_2(x,t) \\ &+c_2g_2(x,t) + \lambda_2 g_2(x,t))u_2 - \lambda_2 c_2 \dot{e}_2(t) \\ &+(c_2+\lambda_2)\ddot{y}_{d2}(t) + \dddot{y}_{d_2}(t) - \rho_2 sgn(\sigma_2)) \end{aligned} \tag{26}$$

where $f_2(x,t) = -0.087460\dot{y}_2(t) - 0.0006947y_2(t)$, $g_2(x,t) = 0.0005418$, $e_2(t) = y_2 - y_{d2}$, $s_2 = c_2e_2 + \dot{e}_2$ and $\sigma_2 = \lambda_2 s_2 + \dot{s}_2$.

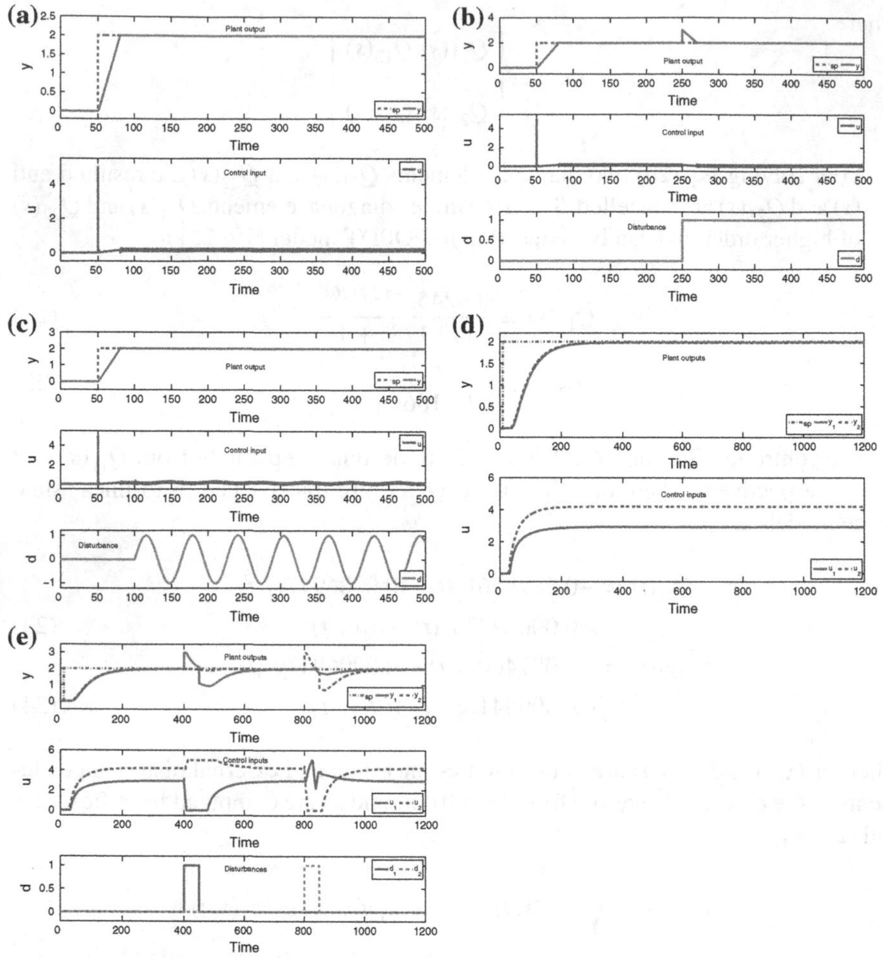

Fig. 3 Simulation results for SISO and MIMO systems. **a** SISO system-without disturbance. **b** SISO system-with disturbance. **c** SISO system-with disturbance. **d** MIMO system-without disturbance. **e** SISO system-with disturbance

The above obtained DSMC controls in Eqs. (23) and (24) are studied in simulation and its results are presented in Fig. 3d and e for without and with disturbances respectively. For simulation study $\lambda_1, \lambda_2 = 15$, $c_1, c_2 = 15$, $\rho_1 = 35$, $\rho_2 = 41$, $d_1(x, t) = 1$ for $400 \leq t \leq 450$ and $d_2(x, t) = 1$ for $800 \leq t \leq 850$ are selected.

5 Simulation Results

A simulation study is carried out for SISO as well as MIMO uncertain process models. The closed loop performance of SISO process is depicted in Fig. 3a–c. In this results y and u are output and control input of SISO process. Similarly in Fig. 3d and e y_1, y_2 and u_1, u_2 are two output interacting tank levels and two control inputs to coupled tank process. For SISO process uncertain and external disturbance element $d = sin(t)$ and 1 for $t \leq 250$ and for coupled tank process uncertain and external disturbance elements $d_1 = 1$ for $400 \leq t \leq 450$ and $d_2 = 1$ for $800 \leq t \leq 850$ are applied. From this results it is clear that designed DSMC is successfully able to handle uncertainties and external disturbances in both SISO and MIMO processes.

6 Conclusion

In this paper a robust dynamic sliding mode control have been studied. It is technique for improving and achieving of desired behaviour of system, by adding extra dynamics. This paper has represented its applications for both SISO and MIMO process. From simulation results it proved that DSMC gives continuous and chattering free control input. Also it is able to compensate uncertainties and external disturbances significantly. Hence it is robust and accurate control method.

References

1. M. Khan, and S. Spurgeon, Robust MIMO water level control in interconnected twin-tanks using second order sliding mode control, Control Engineering Practice 14, 375–386 (2006).
2. S. Mondal, C. Mahanta, A fast converging robust controller using adaptive second order sliding mode, ISA Transactions, 51, pp. 713–721 (2012).
3. X. Yan, S. Spurgeon, and C.Edwards, Dynamic Sliding Mode Control for a Class of Systems with Mismatched Uncertainty, European Journal of Control, Volume 11, Issue 1, Pages 110 (2005).
4. P.V. Suryawanshi, P. D. Shendge, and S. B. Phadke, Robust sliding mode control for a class of nonlinear systems using inertial delay control, Nonlinear Dynamics, Springer, Vol. 78, Number 3, pp. 1921–1932 (2014).
5. Utkin V. Variable structure systems with sliding modes. IEEE Transactions on Automatic Control, 22(2), 212–222 (1977).
6. C. Edwards, and S. Spurgeon, Sliding Mode Control: Theory and Applications, Taylor & Francis, 1998.
7. A. Levant, Sliding order and sliding accuracy in sliding mode control, International Journal of Control, 1993, Vol. 58, Issue 6, 1247–1263 (1993).
8. D.M. Tuan, Z. Man, C. Zhang, J. Jin and H. Wang, Robust sliding mode learning control for uncertain discrete-time multi-input multi-output systems, in Control Theory & Applications, IET, Vol. 8, No. 12, pp. 1045–1053, (2014).
9. A.J. Koshkouei, K.J. Burnham and A.S. Zinobar, Dynamic sliding mode control design, IEEE proceeding, Control Theory Applications, Vol. 152, No. 4 (2005).

10. H. Ramirez, and O. Santiogo, Adaptive dynamical sliding mode control via backstepping, Proceedings of the 32th conference on Decision and Control, San Antonia, Texas, 1422–1427 (1992).
11. S. Mobayen, An adaptive chattering-free PID sliding mode control based on dynamic sliding manifolds for a class of uncertain nonlinear systems, Nonlinear Dynamic, Springer (2015).
12. A.Gholami, A.H.D. Markazi, A new adaptive fuzzy sliding mode observer for a class of MIMO nonlinear systems, Nonlinear Dynamic, Springer, 70, pp. 2095–2105 (2012).
13. L. Fang, T. Li, Z. Li and R. Li, Adaptive terminal sliding mode control for anti-synchronization of uncertain chaotic systems, Nonlinear Dynamic, Springer, 74, pp. 991–1002 (2013).
14. Levant A. Chattering Analysis, IEEE Transactions on Automatic Control, 55(6), 1380–1389 (2010).
15. L. Fridman, J. Moreno, and R. Iriarte, Sliding Modes after the First Decade of the 21st Century, Lecture notes in Control and Information Sciences 412, Springer, Mexico City (2010).
16. I. Eker, Second-order sliding mode control with experimental application, ISA Transactions, Volume 49, Number 394–405 (2010).
17. H. Joe, M. Kim, and S. Yu, Second-order sliding-mode controller for autonomous underwater vehicle in the presence of unknown disturbances, Nonlinear Dynamic, Springer, 78, pp. 183–196 (2014).
18. L. Liu, J. Pu, X. Song, Z. Fu, and X. Wang, Adaptive sliding mode control of uncertain chaotic systems with input nonlinearity, Nonlinear Dynamic, Springer, 76, pp. 1857–1865 (2014).
19. S.Tavakoli, I. Griffin, and Fleming P. J.,Tuning of decentralised PI (PID) controller for TITO processes, Control Engineering Practice, Vol. 14, pp. 1069–1080, (2006).
20. P. Nordfeldt, and T. Hagglund, Decoupler and PID controller design of TITO systems. Journal of Process Control, 16(9), pp. 923–936, (2006).
21. A.A. Khandekar, and B. M. Patre, Advances and applications in sliding mode control system, book chapter, Springer, pp. 225–277, (2015).
22. B.B. Musmade, and B. M. Patre, Design decentralized sliding mode controllers with dynamic decouplers for TITO processes, Variable Structure Systems (VSS), 12th International Workshop, IEEE, (2012).
23. M. Das, and C. Mahanta, Optimal second order sliding mode control for linear uncertain systems, ISA Transactions, 53, 1807–1815, (2014).
24. B. J. Parvat, and B.M. Patre, Design of SMC with decoupler for multi-variable coupled tank process, India Conference (INDICON), 2014 Annual IEEE, vol., no., pp. 1, 5, 11–13, (2014).
25. Q. G. Wang, Lee T. H., Fung H. W., Bi Q. and Zhang Y., PID Tuning for Improved Performance, IEEE Trans. on Control System Technology, Vol. 7, No. 4, pp. 457–465 (1999).

Fault Tolerance Communication in Mobile Distributed Networks

D. Bhuvana Suganth and R. Manjunath

Abstract In mobile distributed networks, there may be possibility of unreliable communications and lack of flexibility in storing the data. In this paper, authors have proposed a distributed-based caching technique and error-free communication in mobile distributed networks. For every mobile host of the multicast tree, a mobile is placed to monitor the failure and to initiate failure recovery technique. Then the data storage is performed by selecting the caching policy based on game theory. The simulation result shows that this proposed technique enhances the reliability and reduces the communication failures.

Keywords Mobile distributed networks · Content distribution · Mobile host · Reliability

1 Introduction

Routing in mobile networks generally involves multiple hops, whereas in distributed networks, its a challenging task [1]. Distributed computing systems can provide several advantages, such as scalability, fault tolerance, and load balancing. Dealing with distributed systems and data storage together, introduces several challenges and difficulties [2], such as dynamic load balancing, unstable connections, communication failures, lack of flexibility in storing data, lack of auto-reconfigurations, limited radio range.

D. Bhuvana Suganth (✉) · R. Manjunath
AMC Engineering College, Bangalore, Karnataka, India
e-mail: bhuvanasuganthi@gmail.com

R. Manjunath
e-mail: manju_r99@yahoo.com

© Springer Science+Business Media Singapore 2017

S.C. Satapathy et al. (eds.), *Proceedings of the International Conference on Data Engineering and Communication Technology*, Advances in Intelligent Systems and Computing 468, DOI 10.1007/978-981-10-1675-2_9

1.1 Reliability in Mobile Distributed Networks

A network's reliability has always been a major concern. Among different other factors such as software extensibility, maintainability, and usability, etc., Reliability have greater impact on software's life, because it can make the running application out of order [1, 2]. Reliability is a very broad term and any software application running in distributed environment can have various definitions.

A Software application is reliable if it can

- Performs well in specified time *t*.
- Do exactly the way it is designed as per requirements.
- Resist various failures and recover in case of any failure that occurs during system execution without producing any incorrect result.

The major key issues for ensuring reliability [3] is considered as

- errors
- inconsistency
- denial of services.

1.2 Problem Identification

The Failure detection is not as much efficient from [4, 5]. Failure recovery is not well organized from [5, 6]. The protocol used for multicasting is not having a considerable reliability from [7]. By considering all these problems, authors propose a technique called distributed caching and fault-tolerant communication in mobile distributed networks.

2 Methodology

A distributed mobile agent is deployed in each MH of the multicast tree [7] which performs the following tasks:

- Monitoring the failures of multicast tree members [4]
- Recovering from failures [4]
- Selecting the caching policy for content distribution based on Game theory.

In multicast network, a distributed mobile agent (MA) is deployed in each mobile host (MH) of the multicast tree. The main role of MA is to either broadcast the messages to all MHs in its cell or to a specific MH in its cell. However, MH can only transmit the messages to MA of the cell in which it is located.

The parameters used in this network model is taken as

- Mobile Host Identifier (ID_{MH}): Identity the mobile host
- First Level Node Identifier (ID_{FN}): Identity the immediate higher level nodes next to MH.
- Second Level Node Identifier (ID_{FN}): Identity the immediate higher level nodes next to first level nodes.
- Root Node Identifier (ID_{RN}): Identity the root node of the tree.
- Timer (*t*): Timer interval to detect whether the lower level nodes are in active or sleep mode.
- MA after identifying the neighbor list, it performs, monitoring the failures of multicast tree members.

2.1 Monitoring the Failures of Multicast Tree Members

The procedure for monitoring the failures of multicast tree members is mentioned below

1. Each Mobile Agent (MA_i) periodically transmits the 'Active' (AC) message only to its Mobile Host (MH_i.)

   $$\mathrm{MA_i} \xrightarrow{AC} MH_i$$

   This causes MA to be aware of active or failed MH among its immediate lower level nodes.
   Note: Also MA decrements t for its immediate lower level nodes by one for every pre-defined time interval.
2. Upon receiving AC message from MA_i, MH_i sends active acknowledgement (AACK) message to MA_i.

   $$\mathrm{MH_i} \xrightarrow{AACK} MA_i$$

3. If MA_i does not receive any AACK message from MH_i until timer expiry, it concludes that MH_i is subjected to failure.
   This technique of monitoring failure nodes results in low-failure free overhead.
4. If MH_i wishes to have a new mobile agent to perform effective monitoring of the immediate next level nodes, it generates the new mobile agent MA_{new}.
5. MA_{new} initiates the monitoring task and informs its location to all the nodes existing in between the MH_i and itself.
6. When MH_i decides that it is not efficient to guarantee the required monitoring performance, it forcefully shifts its MA_i to other trusted MH_j, where MA_i begins its monitoring task. If the MH detects that its immediate level nodes have failed, it initiates failure recovery technique

2.2 *Failure Recovery Technique*

If the MH detects that its immediate level nodes have failed, it initiates failure recovery technique. The recovery is based on the availability and capability of nodes. It involves following three scenarios:

Scenario 1

When MH identifies that its immediate low level nodes have failed, then

- Finds a new node as alternate node N_{new}
- Generates the new MA (MA_{new})
- Transfers the MA_{new} to N_{new}
- MA_{new} performs the similar task as it was performing earlier.

Scenario 2

When MH identifies that its immediate lower level node has failed and there is availability of node for replacing the failed one, then

- It verifies its lower level nodes whether there is any suitable node for replacement.
- If node exists, then it is allowed to take over the role of failed MH
- The new node informs its location and its replacement to all the nodes in the path between the MH and failed node.

Scenario 3

When there is no new node or lower level nodes available for failed node replacement, then

- MH recommends the immediate higher level node to take the role of failed node.
- The new node informs its location and its replacement to all the nodes in the path.

2.3 *Selecting the Caching Policy for Content Distribution Based on Game Theory Model*

The caching policy is selected for content distribution based on game theory model which is explained below: When the mobile nodes receive the data from the Internet, they prefetch it, store, and share the content [8]. Also there is a possibility that the nodes may be subjected to dynamic content more than once.

Let Q be the set of all websites that node have possibility to download

Let CP be the caching policy.

CP: $\{n_i, n_j, \sigma, \tau\}$

R represents the rule that selects four-tuple $\{n_i, n_j, \sigma, \tau\}$ to transmit at time slot t.

where $\{\sigma, \tau\}$ = time slot

$$\{n_i,\ n_j\} = \text{neighboring nodes.}$$

The steps involved in this approach are as follows:

1. Each node n_i maintains the public cache where it stores the most downloaded x_i websites, where
   $x_i \leq |Q|$.
2. The nodes make the contents of the cache to be available for other encountered nodes.
   i.e. If n_i visits n_j, any website stored in n_i will be made available for n_j and vice versa.

$$\forall q \in Q,\ \text{let}\ c_{ni,q} = \begin{cases} 1, & \text{if } n_i \text{ stores } q \\ 0, & \text{otherwise} \end{cases}$$

   Here, 0–1 variable indicates that if the node $n_i \in N$ stores websites q.
3. The caching policy (CP) of n_i can be selected using the utility function (UF_{ni}) at time t.

$$UF_{ni}(CP,\ t) = \begin{cases} n_i(q(t)), & \text{if } n_i \text{stores } q \\ 0, & \text{otherwise} \end{cases}$$

4. For all $q \in Q$, $v_{i,q} = \sum_{n \in T_i} c_{ni,q}/|T_i|$ be the fraction of nodes in tree T_i storing website q
   This reveals that $v_{i,q}$ is v's replication ratio in tree T_i.

$$v_i = [v_i, q]\ q \in Q,\ i \in 1, \ldots, S$$

5. If the request is raised by the node for a website during time slot $(1-\sigma)$ {0 $\leq \sigma \leq$ 1 of time slot}, a feasibility condition is defined as follows:

$$0 < (1-\sigma)\,(n_i q(t))$$

   The strict inequality assures that the request by the node is satisfied.
   i.e. If $CP_{n,\ q} = 1$

   Then

   Request is satisfied

   Else

   Request is backlogged

   End if

   Here $CP_{n,q} = 1$ reveals that the node stores the requested website in its public cache

The above condition will be subsequently satisfied during following scenarios:

- When n_i meets n_j, such that $c_{nj,q} = 1$
- n_i accesses the infrastructure

6. The optimal value of τ is estimated using following equation

$$\tau^* = \frac{(1-\sigma)CP_{ninj}^{nj}}{\sigma CP_{nj}^{ni}}$$

If $\tau < \frac{(1-\sigma)CP_{ninj}^{nj}}{\sigma CP_{nj}^{ni}}$

Then

  y does not have sufficient cache to store the website.

End if

If $\tau > \frac{(1-\sigma)CP_{ninj}^{nj}}{\sigma CP_{nj}^{ni}}$

Then

  It is concluded that there is sufficient cache available.

End if

Thus the notation $R = \{n_i, n_j, \sigma\}$ is used for decision of caching policy.

7. In order to maximize the total expected utility of nodes while selecting caching policy, we perform the following

$$O(R,t) = \sum_{ni}^{N}\sum_{nj}^{N} UF_{ninj}^{Ni}(t) + UF_{ninj}^{Nj}(t)$$

Note:

- A request from q can be satisfied when any copy of the website is retrieved.
- The above process becomes realistic during following scenarios:
  - When the websites are updated infrequently.
  - Content is diffused quickly among users that cache it.

3 Simulation Results

3.1 Simulation Parameters

Authors used NS2 [9] to simulate our proposed Distributed Caching and Fault-tolerant Communication (DCFC) method. Authors used the IEEE 802.11 for Mobile Distributed Network as the MAC layer protocol. It has the functionality to notify the network layer about link breakage. In simulation, the packet sending rate

Table 1 Simulation parameters

No. of mobile nodes	4, 6, 8, 10 and 12
Arca	1250 × 1250
MAC	802.11
Simulation time	50 s
Traffic source	CBR
Rate	150 Kb
Propagation	TwoRayGround
Antenna	OmniAntenna

150 Kb. The area size is 1250 m × 1250 m square region for 50 s simulation time. The simulated traffic is constant bit rate (CBR) and exponential (Exp).

Our simulation settings and parameters are summarized in Table 1

3.2 *Performance Metrics*

Authors evaluate performance of the new protocol mainly according to the following parameters:

Average Packet Delivery Ratio It is the ratio of the number of packets received successfully and the total Number of packets transmitted

Average end-to-end delay The end-to-end-delay is averaged over all surviving data packets from the Sources to the destinations

Throughput The throughput is the amount of data that can be sent from the sources to the destination

Packet Drop It is the number of packets dropped during the data transmission.

3.3 *Results and Analysis*

3.3.1 Based on Mobile Nodes

In first experiment by varying the number of mobile nodes as 4, 6, 8, 10, and 12, the various factors are analysed and the simulation results are given below (Fig. 1).

Figures 2, 3, 4 and 5 show the results of delay, delivery ratio, packet drop, and throughput by varying the mobile nodes from 4 to 12 for the CBR traffic in DCFC and RMP protocols. When comparing the performance of the two protocols, we infer that DCFC outperforms RMP by 29 % in terms of delay, 42 % in terms of delivery ratio, 24 % in terms of packet drop, and 47 % in terms of throughput.

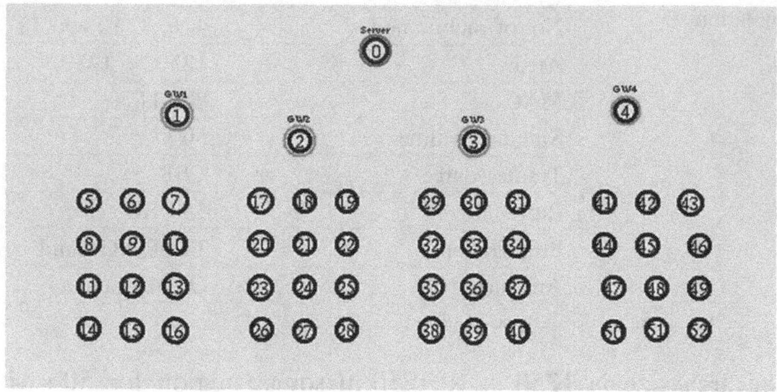

Fig. 1 Simulation topology

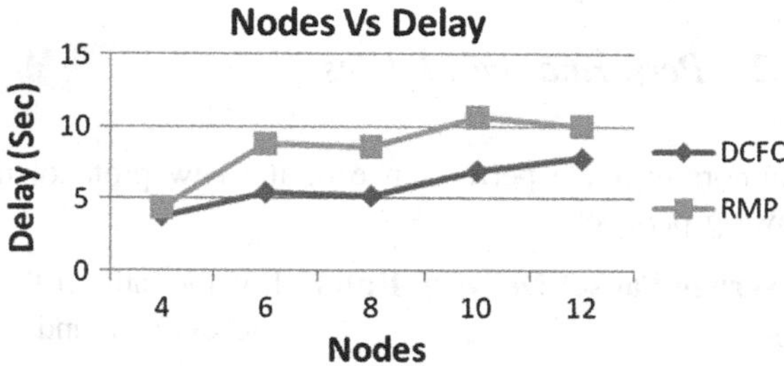

Fig. 2 Nodes versus delay

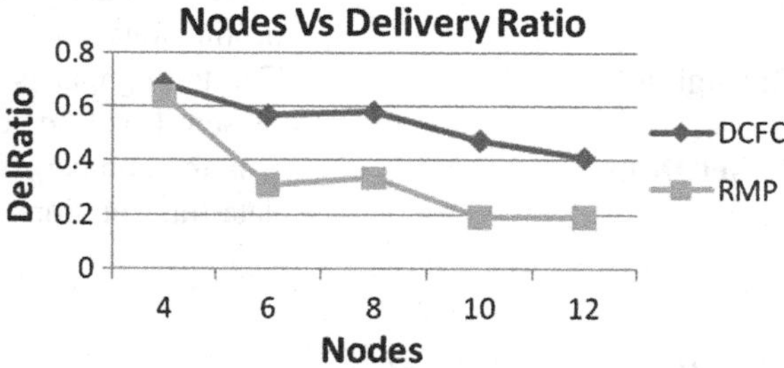

Fig. 3 Nodes versus delivery ratio

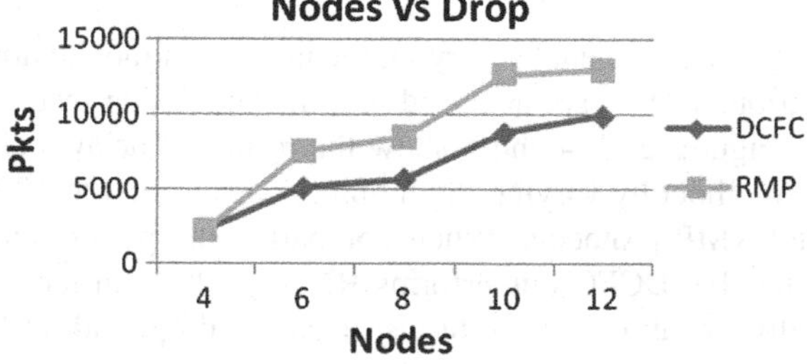

Fig. 4 Nodes versus drop

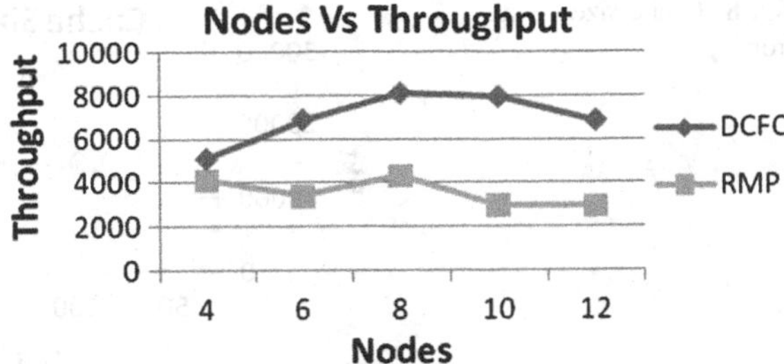

Fig. 5 Nodes versus throughput

3.3.2 Based on Cache Size

In second experiment by varying the cache size as 50, 100, 150, 200, and 250, the various factors are analysed and the simulation results are given below.

Figures 6, 7, 8, and 9 show the results of delay, delivery ratio, packet drop, and throughput by varying the Cache Size from 50 to 250 for the CBR traffic in DCFC and RMP protocols. When comparing the performance of the two protocols, we infer that DCFC outperforms RMP by 18 % in terms of delay, 50 % in terms of delivery ratio, 16 % in terms of packet drop, and 53 % in terms of throughput.

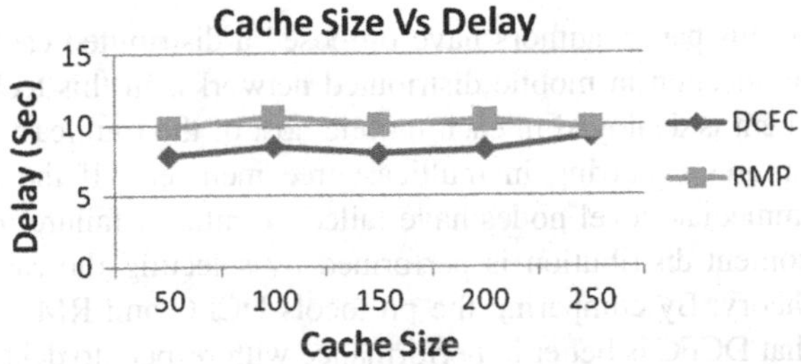

Fig. 6 Cache size versus delay

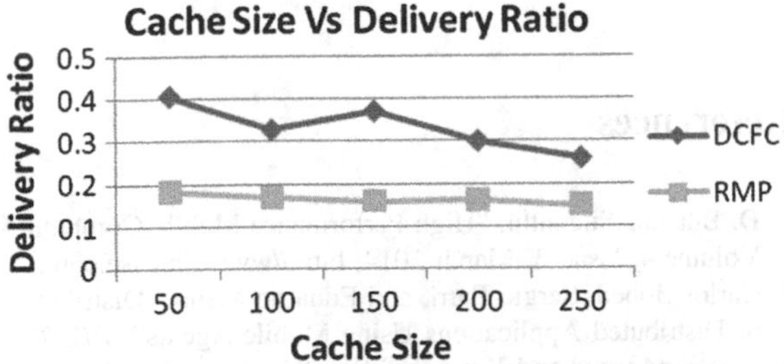

Fig. 7 Cache size versus delivery ratio

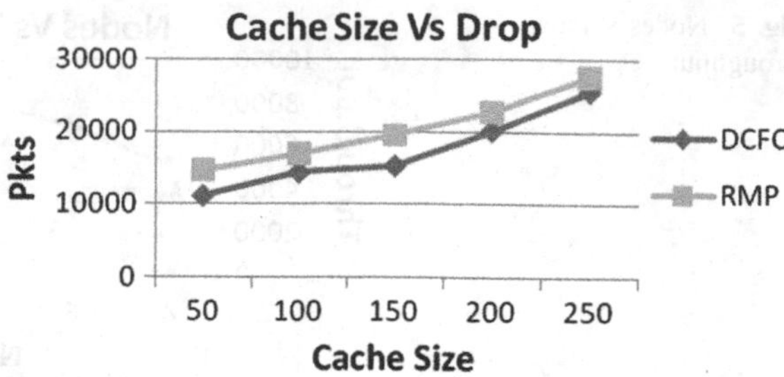

Fig. 8 Cache size versus drop

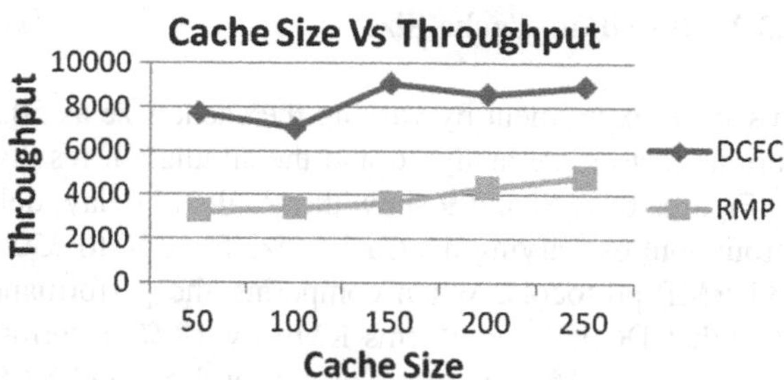

Fig. 9 Cache size versus throughput

4 Conclusions

In this paper, authors have proposed a distributed caching and fault tolerant communication in mobile distributed networks. In this technique, a distributed mobile agent is deployed in each mobile host of the multicast tree. It initially monitors the failures occurring in multicast tree members. If the mobile host detects that its immediate level nodes have failed, it initiates failure recovery technique. Then the content distribution is performed by selecting the caching policy based on game theory. By comparing the protocols DCFC and RMP, the simulation results show that DCFC is better in performance with respect to delay, delivery ratio, throughput, and drop based on both the cache size and number of mobile nodes. Hence the proposed technique DCFC enhances the reliability and reduces the communication failures.

References

1. D. BhuvanaSuganthi, "High Performance Mobile Computing Nodes in Distributed Networks", Volume 4, Issue 3, March 2014, http://www.ijarcsse.com.
2. Carlos Bobed, Sergio Ilarri, and Eduardo Mena, "Distributed Mobile Computing Development of Distributed Applications Using Mobile Agents", *PDPTA*. 2010.
3. WaseemAhmed and Yong Wei Wu, "A survey on reliability in distributed systems", Journal of Computer and System Sciences 79 (2013), Elsevier Inc. Pages 1243–1255.

4. Jinho Ahn, "Fault-tolerant Mobile Agent-based Monitoring Mechanism for Highly Dynamic Distributed Networks", IJCSI International Journal of Computer Science Issues, Vol. 7, Issue 3, No 3, May 2010.
5. Stratis Ioannidis, Laurent Massoulie and Augustin Chaintreau, "Distributed Caching over Heterogeneous Mobile Networks", ACM SIGMETRICS Performance Evaluation Review. Vol. 38. No. 1. ACM, 2010.
6. Chien-Lung Hsu and Yu-Li Lin, "Improved migration for mobile computing in distributed networks", Computer Standards and Interfaces, Volume 36. Issue 3, March 2014: Pages 577–584.
7. Giuseppe Anastasi, Alberto Bartoli and Francesco Spadoni, "A Reliable Multicast Protocol for Distributed Mobile Systems: Design and Evaluation", Parallel and Distributed Systems, IEEE Transactions, 2001. Pages 1009–1022.
8. Huayong Wang and Li-ShiuanPeh, "MobiStreams: A Reliable Distributed Stream Processing System for Mobile Devices", Parallel and Distributed Processing Symposium, IEEE 2014. Pages 51–60.
9. Network Simulator: http://www.isi.edu/nsnam/ns.

Analysis of Bathymetry Data for Shape Prediction of a Reservoir

Sushama Shelke and Selva Balan

Abstract The collection of bathymetry data remotely and its analysis is a challenging task due to the complexity of the data acquisition devices like echo sounders and the large amount of data. The data obtained is also affected by noise and needs to be processed to predict the bottom surface of the water body. This paper presents a novel approach for analysis of bathymetry data of a reservoir. The proposed system involves multipath noise removal and interpolation of the data to find the volume of water in the reservoir. The Wavelet packet decomposition technique is used for removing noise in the data. Further, nearest neighborhood and TIN interpolation techniques are used to obtain a 3D model of the reservoir. This approach is used to calculate the volume of water in the reservoir.

Keywords Bathymetry • Wavelet packet decomposition • Interpolation • TIN

1 Introduction

Recently, major part of the world is facing scarcity of water. The amount of potable water present in reservoirs is very less compared to the population. Apart from drinking, water is essential for various other purposes. This demands a need to find the amount of water in the reservoirs accurately. Finding water volume in reservoirs is challenging due to various reasons like availability of acquisition devices to measure the depth of the reservoir, sedimentation in the reservoir etc. So far, the bathymetry data is measured using echo sounders. The echo sounders, either single-beam or multi-beam, transmit and receive the sonar waves. The time duration between transmission and reception of the sonar waves is equivalent to the depth of

Sushama Shelke (✉)
NBN Sinhgad School of Engineering, Pune, India
e-mail: shelke.sushama@gmail.com

Selva Balan
Central Water and Power Research Station, Pune, India
e-mail: instcwprs@gmail.com

© Springer Science+Business Media Singapore 2017

S.C. Satapathy et al. (eds.), *Proceedings of the International Conference on Data Engineering and Communication Technology*, Advances in Intelligent Systems and Computing 468, DOI 10.1007/978-981-10-1675-2_10

the reservoir at that location. Finding the depth of the reservoir using acoustic waves is also affected by the accumulation of sediments in it. Sedimentation is hazardous to the aquatic environment and causes pollution of water. The depth calculation in case of multi beam echo sounder adds multipath noise due to reflection from several layers of sediments in the bottom of the reservoir. This demands removal of multipath noise and use of interpolation techniques [1] to predict the bottom of the reservoir. The bathymetry data is three dimensional consisting of values that are spaced in irregular manner. Thus these values fail to obtain a regular grid map. To achieve this, a gridding method is implemented that converts irregular data map into equally spaced grid map. The gridding technique generates values which are not present to obtain a complete map.

Data mapping can be done using various gridding methods. The fast gridding method is Inverse to Power gridding method [2]. It implements a weighted average interpolation technique for obtaining the new values. In Kriging [3], irregularly spaced data are used to obtain extrapolated values. Another gridding technique uses minimum curvature method [4]. Gridding is also performed by modified Shepard's algorithm that uses an inverse distance weighted least squares [5]. The Nearest Neighbor [6] and Polynomial Regression [7] are popular gridding techniques. The Radial basis function [8] and Triangulation with Linear Interpolation gridding [9] methods are popular and accurate gridding techniques. For large data sets, a Moving Average gridding [10] technique is commonly used and for smooth data sets, the Local Polynomial gridding technique [11] is implemented. Various interpolation techniques are studied by researchers in [12].

The rest of the paper is organized as follows. Section 2 describes the proposed system design. Section 3 presents the results and discussion and Sect. 4 includes the conclusion of the proposed system design.

2 Proposed System Design

The steps involved in the proposed system design for the analysis of bathymetry data of the reservoir is indicated in Fig. 1. First, the bathymetry data is collected by the data acquisition system. The collected data contains noise, which needs to be removed in the preprocessing stage. Then the data is interpolated to predict the bottom surface of the reservoir. In the post-processing stage, the interpolated data can be used for different purposes. The details of each step are explained further.

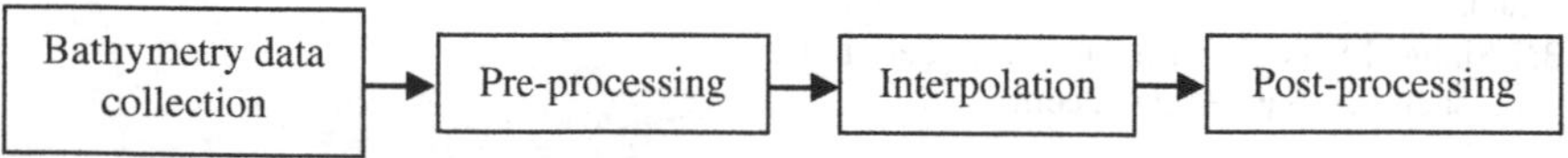

Fig. 1 Steps in proposed system design

2.1 Bathymetry Data Collection

The bathymetry data of a reservoir is collected using a multibeam echosounder. Here a reservoir X is selected for data collection. The data obtained is three dimensional, in (x, y, z) form. The (x, y) coordinates of the data correspond to the GPS location of the position at which the depth values are found using the echo sounder. And, the z coordinates correspond to the depth of the reservoir at (x, y) location. Around 7000 data points are collected from the given reservoir. A sample of the collected three-dimensional data is given in Table 1.

2.2 Preprocessing

The collection of bathymetry data using multi beam echo sounder induces multipath noise. The multipath noise occurs due to reflection and refraction of the acoustic signal in water. The reflection occurs due to the rocks, sediments in water and also due to the bottom surface of the reservoir. The multipath noise in the data set is removed by applying Wavelet Decomposition [13]. The approximation coefficients obtained after decomposition are further used for calculating the interpolated points in the data set.

2.3 Interpolation

The bathymetry data obtained after noise removal is passed to the nearest neighbor interpolation technique. This data consists of the depth of the sample points. The height of the other points has to be approximated at the other points. This interpolation is done by the triangulated irregular network model, i.e., TIN algorithm. The bathymetry data is a set of x and y coordinates with their depth or elevation values, the z coordinates. In TIN algorithm, a point lies on a vertex, an edge or in a triangle. The TIN algorithm generates a three-dimensionally connected triangular structure [14].

Table 1 Sample bathymetry data of the reservoir

x	y	z
689505.8	1254012	−4.03
689506.6	1254013	−4.46
689506.9	1254014	−4.23
689507.6	1254015	−4.55
689508	1254015	−4.51
689508.4	1254016	−4.5

2.4 Post-processing

The post-processing of the interpolated data may consist of sediment classification, calculation of volume of water, and finding aquatic life condition. The proposed system calculates the volume of the water in the reservoir.

3 Results and Discussion

The proposed system is developed in MATLAB. The gridded bathymetry data collected using echo sounder and GPS is large in size. Figure 2 shows the sample data plot of 500 points out of 7000 points. It shows sharp variations in the depth values. The negative values indicate the height below the surface of water.

The data is then passed to Wavelet packet decomposition for the removal of multipath noise in the data. Figure 3 shows the result after decomposition. It shows the irregularity in the data points.

The approximation coefficients of the decomposed data are then passed through the nearest neighbor interpolation technique to give a better approximation of the bottom of the reservoir. The data after nearest neighbor interpolation is plotted in Fig. 4.

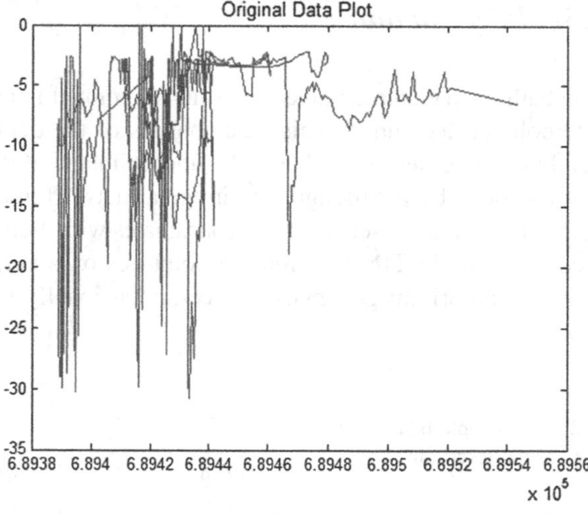

Fig. 2 Sample data plot

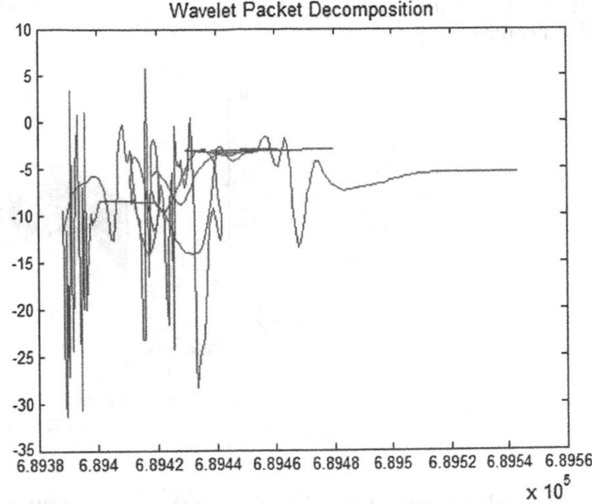

Fig. 3 Wavelet packet decomposition

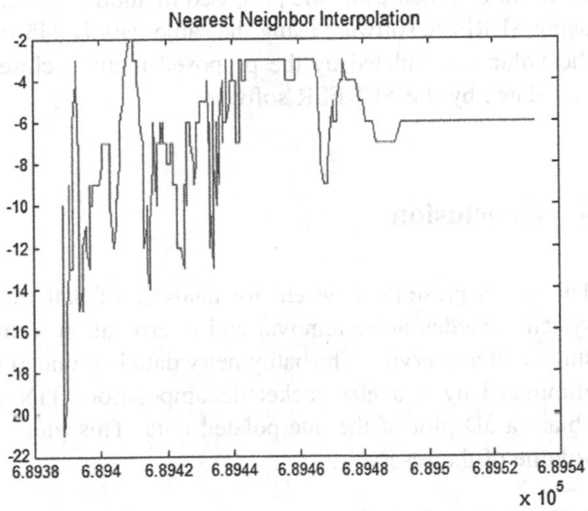

Fig. 4 Nearest neighbor interpolation

The data is finally converted into a 3D model by TIN algorithm. Here Delaunay's TIN algorithm is implemented for generating the plot. The 3D plot for sample data is shown in Fig. 5.

The surface area of the 3D plot is obtained using convex hull to calculate the volume of the plot. Another parameter required for volume calculation is the average depth. Thus the volume of the plot is calculated by,

$$\text{Volume} = \text{surface area} \times \text{average depth}. \tag{1}$$

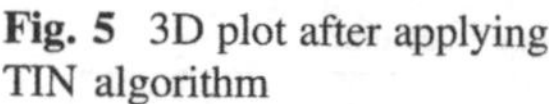

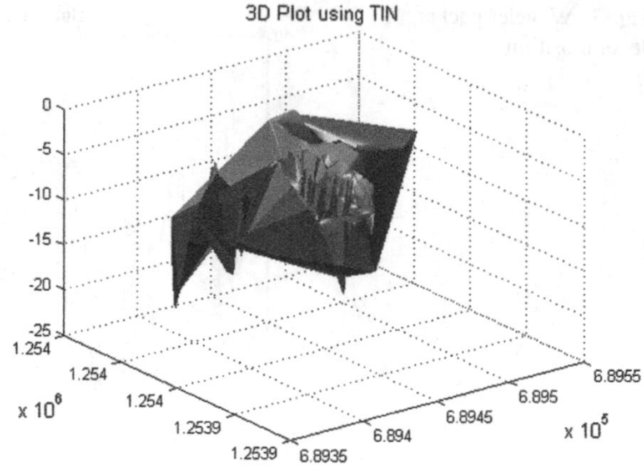

Fig. 5 3D plot after applying TIN algorithm

The volume calculated for the sample plot of 500 data values is 31 879 m^3. The volume of the reservoir is further obtained for 7000 data points. The volume of the reservoir obtained using the proposed method is 239 375 m^3. The volume obtained using SURFER software using the same data is 245 003.15 m^3. This indicates that the volume calculated by the proposed method closely approximates the volume calculated by the SURFER software.

4 Conclusion

This paper presents a system for analysis of bathymetry data of a reservoir. The system includes noise removal and interpolation techniques to predict the bottom surface of a reservoir. The bathymetry data is prone to multipath noise. This noise is eliminated by Wavelet packet decomposition. TIN algorithm is further used to obtain a 3D plot of the interpolated data. This plot is finally used for finding the volume of the reservoir.

Acknowledgments The authors are grateful to BCUD, Savitribai Phule Pune University for funding this research work. The authors are also thankful to the anonymous reviewers and the journal editors.

References

1. EL-Hattab, A.: Single Beam Bathymetric Data Modeling Techniques for Accurate Maintenance Dredging. Egypt. J. Remote Sens. Space Sci. 17, 189–195 (2014)
2. Tomczak, M.: Spatial Interpolation and its Uncertainty Using Automated Anisotropic Inverse Distance Weighting (IDW)-Cross Validation/Jackknife Approach. J. Geogr. Inform. Decis. Anal. 2, 18–33 (1998)

3. Lam, N.S.N.: Spatial Interpolation Methods: A Review. Am. Cartographer. 10, 129–149 (1983)
4. Kay, M., Dimitrakopoulos, R.: Integrated Interpolation Methods for Geophysical Data: Applications to Mineral Exploration. Nat. Resour. Res. 9, 53–64 (2000)
5. Thacker, W.I., Zhang, J., Watson, L.T., Birch, J.B., Iyer, M.A., Berry, M.W.: Algorithm XXX: SHEPPACK: Modified Shepard Algorithm for Interpolation of Scattered Multivariate Data. Technical Report, Computer Sci., Verginia Tech (2009)
6. Sibson, R.: A Brief Description Of Nearest Neighbor Interpolation Interpolating Multivariate Data. John Wiley & Sons, New York (1981)
7. Eberly, S., Swall, J., Holland, D., Cox, B., Baldridge, E.: Developing Spatially Interpolated Surfaces and Estimating Uncertainty. Technical Report, U.S. Environmental Protection Agency (2004)
8. Buhmann, M.: Radial Basis Functions: Theory and Implementations. Cambridge University Press, United Kingdom (2003)
9. El-Sheimy, N., Valeo, C., Habib, A., Valeo, C.: Digital Terrain Modeling: Acquisition, Manipulation And Applications. Artech House, United Kingdom (2005)
10. Li, Z., Zhu, Q., Gold, C.: Digital Terrain Modeling: Principles and Methodology. CRC Press, Boca Raton (2005)
11. Rawlings, J.O., Pantula, S.G., Dickey, D.A.: Applied Regression Analysis: A Research Tool. Springer, New York (1998)
12. Yang, C.S., Kao, S.P., Lee, F.B., Hung, P.S.: Twelve Different Interpolation Methods: A Case Study of Surfer 8.0. In: Proceedings of the XXth ISPRS Congress, pp. 778–785. ISPRS, Istanbul (2004)
13. Sun, F., Zhang, X., Wang, G.: An Approach for Underwater Image Denoising via Wavelet Decomposition and High-pass Filter. In: 4th IEEE International Conference on Intelligent Computation Technology and Automation, pp. 417–420. IEEE Press, Guangdong (2011)
14. Kreveld, M.: Digital Elevation Models: Overview and Selected TIN algorithms. In: Course Notes of the CISM Advanced School on Algorithmic Foundations of Geographical Information Systems, pp. 1–43. Netherlands (1996)

An Extensive Survey on Diagnosis of Diabetes Mellitus in Healthcare

Suvarna Pawar and Smita Sikchi

Abstract Healthcare is such a wide and extensive area of research wherein diabetes is such a deadly disease which hampers a common man life at the extreme end. Under healthcare, there is always a chance for uncertainty and imprecision under various aspects of medical diagnosis process. This paper reviews various existing methodologies and techniques used for early diagnosis of diabetes mellitus which works like risk alarms to save human life. This survey gives us current state of research in diagnosing disease like diabetes and helps us to find out difficulties with the existing systems. In many of the expert systems different soft computing and data mining techniques are used along with usage of real time dataset from hospitals or readily available datasets like PIMA Indian dataset, colic dataset to diagnose the disease.

Keywords Fuzzy cognitive maps (FCM) · Fuzzy diabetic ontology (FDO) · K-NN · Support vector machine (SVM)

1 Introduction

Diabetes is a metabolic disorder in which body is unable to handle sugar characterized by hyperglycemia i.e. high blood glucose levels resulting from defects in insulin secretion. It is a silent killer. The chronic hyperglycemia of diabetes is associated with disturbances of carbohydrate, fat and protein metabolism which in return results in long-term damage, dysfunction and failure of different organs, especially the eyes, kidneys, nerves, heart and blood vessels. The body needs insulin to use sugar, fat and protein from the diet for day to day activities [1]. In spite of so much medical progress still many of diabetes mellitus are unaware of their disease as many of the symptoms are common with other diseases. So it takes long duration for diagnosis of such a diabetes disease [2]. Hence there is a need to

Suvarna Pawar (✉) · Smita Sikchi
Pune, India
e-mail: pawar.suvarna@gmail.com

© Springer Science+Business Media Singapore 2017

S.C. Satapathy et al. (eds.), *Proceedings of the International Conference on Data Engineering and Communication Technology*, Advances in Intelligent Systems and Computing 468, DOI 10.1007/978-981-10-1675-2_11

develop such an expert system which not only alarms the risk of disease but also helps in finding out the solutions with expertise. In this review paper we focus on existing methods for diabetes detection so as to know the current developments in the field of diabetes under healthcare.

2 Commonly Used Approaches for Diagnosis of Diabetes

2.1 Fuzzy Approach

Bhatia and Kumar proposed an FCM (fuzzy cognitive maps) approach to model knowledge-based systems for diagnosing thyroid diagnosis. They used temporal medical data. Proposed framework adaptive algorithm used for learning FCM which works at three levels to record symptoms. Software tool is used under soft computing which tests for 50 cases and got accuracy of 96 % [3, 4].

Thirugnanam proposed to improve diabetes diagnosis using three combined approaches of fuzzy, Neural network and case based reasoning. All three are dependent to each other and which gives accuracy in prediction rate for occurrence of diabetes mellitus [1].

Lee defined a fuzzy diabetic ontology (FDO) to model the diabetes knowledge. Instances of FDO are generated by the fuzzy ontology generating mechanism. It simulates fuzzy decision making applications [5, 6].

Grant proposed an approach for controlling of diabetes through fuzzy logic and insulin pump which is small wearable device for short acting insulin. When sensor senses the deficiency then optimal insulin infusion rate has been calculated and through fuzzy logic patient has been operated through a pump. It avoids multiple daily injections [7].

Along with fuzzy logic faith-Michael Emeka Uzoka, Okure Obot, Ken Barker introduced AHP which is analytical hierarchy process. It discovered from the fuzzy logic diagnosis covary a little bit more strongly to the conventional diagnosis results than AHP. Fuzzification and inference engine is used [8].

Microalbuminuria is predictor of diabetes and other diseases like cardiovascular. Hamid Marateb, Marjan, Elham Faghihimani, Masoud, Dario Farina proposed expert based fuzzy MA classifier in which with rule induction particle sworm optimization is performed. After classifying variety of classifiers based on different diabetic parameters like BMI, HBA1c, age etc. performance is calculated on 10 folds cross validation. Hence parameters accessed are Sensitivity, Specificity, precision and accuracy for values 95, 85, 84 and 92 % respectively [9].

2.2 Classification and Decision Tree

Seera and Lim proposed hybrid intelligent system with classification and regression tree proves with experimental tasks for effective medical data classification. This article posses three important characteristics likely as online learning where it performs incremental learning from data samples with one pass. Secondly high performance whereby RF method is applied to form an ensemble of CART models. It also handles the Outliers or noise to handle robustness n accuracy and third is rule extraction whereby CART model is employed for knowledge capturing under FMM into decision tree format [10, 11].

Using the same technique of classification and clustering S Peter presented analytical study of various algorithms like decision tree, neural network etc. [12]. In continuation to same strategies B.L. Shivkumar and S. Alby focused more on association rules in comparison to clustering and classification. Also How J48 decision tree algorithm helps in identifying the disease using PIMA Indian dataset [13].

Khanna and Agarwal introduced kappa statistics along with data mining techniques where after classification on dataset D, we need to find Clusters among it by averaging the values. By calculating classifier weight of the range of the attribute and repeat the same till higher priority attribute. Finally using kappa, find out risk of diabetes as low, medium or high. SGPGI dataset is used for the same [14].

Morten Jensen, Toke Folke, Lise Tarnow proposed an study to evaluate a new pattern classification approach by comparing the performance with PCGM calibration algorithm based on samples and events w.r.t. sensitivity and specificity. This enables the clinician together with patient to adjust insulin and carbohydrates intake and thereby avoid exertions [15].

Sankaranrayan and Pramananda uses predictive approach for DMD through different data mining techniques such as rule based classification and decision tree algorithm like C4.5 and ID3 to mine the data. Using UCL ML data repository information HBA1c and micro albuminuria values are determined so that diabetes will be classified with IDDM diabetes [16].

Velu and Kashwan proposed visual data mining techniques for classification problem. It uses EM algorithm for sampling on PIMA diabetes dataset. It also makes use of H-means clustering to form the clusters having similar symptoms and genetic algorithm to crossover genetics related result analysis. Weka is effectively used for analyzing the data and got 35 % of total 768 samples with positive with diabetes [17].

2.3 Soft Computing Techniques

Karan et al. proposed an methodology to detect diabetes with neural network on mobile devices for client as well as on server i.e. two tier pervasive architecture which is based on ANN computations. Wireless communications between client and server optimized for real time use of it in healthcare services [18].

Manza worked on diabetes expert system and make use of PIMA Indian dataset. With neuro-fuzzy approach, ANFIS and PCA, most commonly used methodology for diagnosis as well as prognosis of diabetes [19].

Using the same ANN and PCA Rakesh Motka, Viral Parmer implemented the framework using mat lab [20].

Vijayan proposed prediction and diagnosis of diabetes along with KNN and ANFIS algorithm. For diagnosis of diabetes there is always some uncertainty exists for physical as well as chemical tests hence proposed algorithm of KNN and ANFIS gives 80 % accuracy than existing other one [21].

Staroula, Evangelos Bozas proposed a framework for effective communication of patient and medical health management system using different monitoring and diagnosis management approach. Here two different units are maintained and when measuring patient's details has been gathered then same records will be making available to PMU through WIFI or GPRS. This in turn proposes a home care management for diabetes sufferers [22].

Ntaganda proposes fuzzy logic for solving optimal glucose control problem and insulin for diabetes mellitus using fuzzy as well as direct approach. It also emphasis more on physical activity and healthy diet chart plan. It works on approximation method [23].

Hasslacher et al. highlighted how glycated albumin and HbA1c works in correlation with each other. Also GA shows strong positive dependence of one another by considering patients features like gender, age, renal function and anemia etc. They also emphasis on different CKD 1 to 3 stages [24].

2.4 Support Vector Machine

Abdullah, Mohammed concentrates upon predictive analysis of diabetes treatment using ODM (Oracle data miner) and SVM is used for experimental analysis. It is based on statistical learning theory. Hence instead of going for mathematical programming, kernel functions are used to choose hyper plane i.e. optimal separating interface. The factor P(o) and P(y) comparisons with soft computing tool of ODM and effectiveness of drug treatment can be found out [25].

Mansour et al. introduce SVM as popular machine learning tasks involving classification for diagnosis of diabetes. They proposed detection of hard exudates from diabetic retinal images as diabetic retinopathy which is one of earliest sign of diabetes. Discrete cosine transform and SVM makes use of color information to perform the classification. By considering 1200 images, system can achieve a diagnostic accuracy with 97 % and 98 % specificity for the same [26].

2.5 *Artificial Intelligence*

Fiuzy, Haddadnia and Kazem recommends diagnosis of diabetes by using combined approach of artificial intelligence and search algorithm to select best features through data mining method to classify and categorization of patients records using neural network modeling. Author proposed eight different factors by WHO to test and achieves 94.03 % precision in diabetic patient identification. Using Mat lab implementation user get 171 from 174 patients identified as diabetic [27].

2.6 *Traditional Approach*

Cui et al. proposed an traditional chinese medicine to cure diabetes mellitus under TCM theory with the help of data mining. They suggested about different diet therapy with herbs as part one of it and secondly treatment principle and acupoints to work on it. TCM research exploits different treatment rules, principles and herbs through network [28].

3 Analysis for Existing Literature Work

After analyzing all the approaches used for diagnosis of diabetes disease, we found still some difficulty in diagnosing the disease with higher accuracy. Herewith we try to present current scenario of some aspects of comparison of existing methodologies. This study is performed mostly on PIMA Indian dataset or SGPGI dataset. Even their exists other measuring parameters like specificity, Sensitivity along with accuracy which emphasis statistical analysis of resultant output values as shown in Fig. 1.

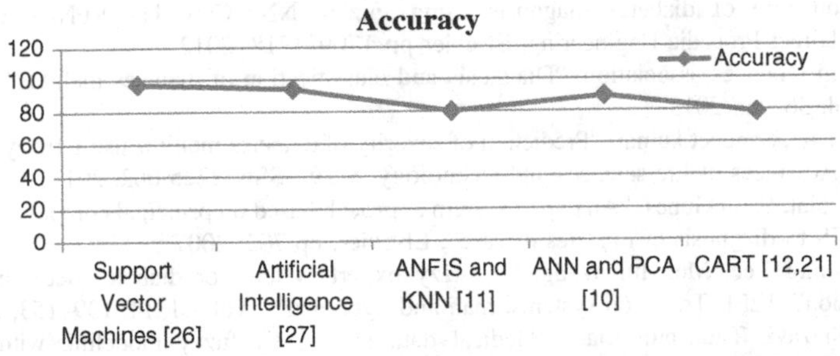

Fig. 1 Shows accuracy with respect to all approaches

Table 1 Comparison of existing approaches

Sr. no	Existing approach	% Accuracy
1	Support vector machines [26]	97
2	Artificial intelligence [27]	94.03
3	ANFIS and KNN [21]	80
4	ANN and PCA [20]	89.2
5	CART [10, 11]	78.39

Specificity = (TP)/[(TP) + (FN)] and Sensitivity = (TN)/[(TN) + (FP)] where TP = True Positives, TN = True Negatives, FP = False Positives, FN = False Negatives (Table 1).

4 Conclusion

Extensive literature of diagnosis of diabetes mellitus using various techniques for detection and classification of diabetes manifest focuses that day by day diabetes found to be very common in alternate home and widely spread over the nation like any virus. It is becoming socio-critical issue to handle irrespective of parameters like age, blood pressure etc. To deal with such annoying problem, some requisite and imperative solution should be provided which will save precious human life with early detection of diabetes and play vital role as lifesaver. So there exists a key to develop a framework which not only show risk of occurrence of diabetes disease but also provide a kind of solution for various physic-chemical tests available with affordable life style.

References

1. Mythili Thirugnanam, Dr Praveen Kumar, S Vignesh Srivatsan, Nerlesh C R, “Improving the prediction rate of diabetes diagnosis using Fuzzy, NN, Case based(FNC) approach”, ScienceDirect Procedia Engineering, Elsevier pp 1709–1718, 2012
2. American Diabetes association, “Diagnosis and classification of diabetes mellitus”, Diabetes care, vol. 36, Jan 2013
3. Nitin Bhatia, Sangeet kumar, “Prediction of severity of diabetes mellitus using fuzzy cognitive maps”, Advances of life sciences and technology, v 29, ISSN 2225-062, 2015
4. Kemal Polat, Salih Gunes, “An expert system approach based on principal component analysis & ANFIS to diagnosis of diabetes disease”, Elsevier, pp 702, 2007
5. Chang-shing Lee, Mei-Hui Wang, “A fuzzy expert system for diabetes decision support application”, IEEE Trans. on systems, man and cybernetics, vol. 41, PP 139–153, 2011
6. Sean Ghazavi, Thunshun Liao, “Medical data mining by fuzzy modeling with selected features,” Elsevier in artificial intelligence in medicine, pp 195–206, 2008
7. Paul Grant, “A new approach to diabetic control: Fuzzy logic and insulin pump technology”, Elsevier in medical Engineering and physics, 29, pp 824–827, 2007

8. Faith-Michael Emeka Uzoka, Okure Obot, Ken Barker, "An experimental comparison of fuzzy logic and Analytic hierarchy process for medical decision support system", Elsevier Computer methods in biomed- cine, pp 10–27, 2012
9. Hamid Marateb, Marjan, Elham Faghihimani, Masoud, Dario Farina, "A hybrid intelligent system for Diagnosing microalbuminiuria in type 2 diabetes patients without having to measure urinary albumin", Elsevier computers in biology and medicine, pp 34–42, 2014
10. Manjeevan Seera, Chee Peng Lim, "A hybrid intelligent system for medical data classification", Science Direct Elsevier, pp 2239–2249, 2014
11. K Lokanayaki, A Malathi, "Exploring on various prediction model in data mining techniques for disease Diagnosis", International journal of CA, Vol 77, pp 27, 2013
12. S. Peter, "An analytical study on early diagnosis and classification of diabetes mellitus", Bonfring international journal of data mining, pp 7–11, ISSN 2277-5048, 2014
13. Dr.B.L. Shivkumar, S. Alby, "A Survey on data mining technologies for prediction and diagnosis of diabetes", IEEE international conference on intelligent computing applications, pp 167–172, 2014
14. Subham Khanna, Sonali agarwal, "An integrated approach towards the prediction of likelihood of diabetes", IEEE international conference on machine intelligence and research advancement", pp 294–298, 2013
15. Morten Hasselstrom Jensen, Zeinab Mahmaudi, Toke Folke, Edmund Seto, Ole Kristain, "Evaluation of an algorithm for retrospective hypoglycemia detection using professional continuous glucose monitoring data", Journal of diabetes science and technology, vol. 8, pp 117–122, 2014
16. S Sankaranarayanan and Dr. Pramananda T., "A predictive approach for diabetes mellitus disease through data mining technologies", IEEE world congress on computing and communication technologies, 2014
17. C M Velu, K R Kashwan, "Visual data mining techniques for classification of diabetic patients", IEEE Trans., pp 1070–1075, 2012
18. Oguz Karan, Canan, Haluk and Bekir, "Diagnosing diabetes using neural networks on small mobile device Elsevier, pp 54–60, 2012
19. Ambiwade, R R Manza, "Medical expert systems for diabetes diagnosis: A survey", international journal of advanced research in computer science and software engineering, pp 647–651, Nov 2014
20. Rakesh Motka, Viral Parmar, A.R. Verma, "Diabetes mellitus forecast using different data mining tech." IEEE 4th international conference on computer and communication technology," pp 99–103, 2013
21. V Vijayan, Aswathy, "Study of data mining algorithms for prediction and diagnosis of diabetes mellitus", International journal of computer applications, vol 95, June 2014
22. Staroula Mougiakakou, Nikos, Dimitra, "SmartDiab: A Communication and Information Technology Approach for the intelligent monitoring, management and follow-up of Type-1 diabetes patients", IEEE Trans. on information technology in Biomedicine, vol. 14, May 2010
23. Jean Mare Ntaganda, "Fuzzy logic strategy for solving an optimal control of problem of glucose and insulin in diabetic human", Scientific research open journal of applied science, 3, pp 421–429, 2013
24. Christoph Hasslacher, Felix Kulozik, Isabel and Justo Bermejo, "Glycated albumin and HBA1c predictors of mortality and vascular complications in type 2 diabetes patients with normal and moderately impaired renal function: 5-year results from 380 patient", Journal of diabetes research and clinical metabolism", Herbert Open access journals, 2014
25. Abdullah, Mohammed Gulam, Mohammad Khubeb, "Applications of data mining: diabetes health care in young and old patients", Journal of King Saud University comp and information sciences, pp 127–136 Elsevier 2013
26. R.F. Mansour, E. Md. Addelrahim, Amna, "Identification of diabetic retinal exudates in digital images using SVM", Scientific research, Journal of intelligent learning and applications, pp 135–142, 2013

27. Mahammad Fiuzy, Javad Haddania, Nasrin Mollsnis, Maryam Hashemian, Kazem Pour, "An intelligent system for diabet diagnosis on combined intelligent algorithm and risk factor in patients", Life science journals, 10, pp 380–386, 2013
28. Zhaoli Cui, Dan He, Miao Jiang, Yaoxian Wang, Guang Zheng, "To discover the traditional medicine techniques applied in diabetes mellitus through data mining", IEEE 9th international conference on computer science and education, Aug 22–24, Canada, 2014

Convolution Encoder for BER Analysis Using Doppler Frequency Shift in the Packet Erasure Network

Arjun Ghule and Prabhu Benakop

Abstract Trustworthy communication with higher data rates at affordable cost and optimum performance are the requirements for any wireless communication system; system design for such network should consider the challenges of narrow spectrum and losses in wireless network components. This paper compares the performance of the model for Bit Error Ratio (BER) analysis using a Doppler frequency shift in the packet erasure network with various modes of operation of the encoder. In last decade many researchers have shown interest in developing the models for BER, while in the paper we have made an analysis and comparison of the model developed for the optimizing the error in the speech signal. Doppler frequency variation method plays a vital role in optimizing the errors of the packet erasure network. The implemented Rayleigh channel model is a generalized model over which experimentation was carried out to conclude with the performance for minimum BER.

Keywords Bit error ratio · Doppler frequency shift · Packet erasure network · Rayleigh channel model

1 Introduction

A wide range of experimentation and research is carried out in wireless communication technology over the past decade. The areas of challenges are still present in this field. The reliability of any wireless network depends upon the bit error ratio (BER) at the receivers which are mainly affected by the signal-to-noise ratio (SNR) at the receiving end [1]. The SNR is the function of the distance between the

Arjun Ghule (✉)
Department of ECE, Jawaharlal Nehru Technological University, Hyderabad, India
e-mail: arjunghule@mmcoe.edu.in

Prabhu Benakop
INDUR Institute of Engineering and Technology, Siddipet, A.P., India
e-mail: pgbenakop@rediffmail.com

© Springer Science+Business Media Singapore 2017
S.C. Satapathy et al. (eds.), *Proceedings of the International Conference on Data Engineering and Communication Technology*, Advances in Intelligent Systems and Computing 468, DOI 10.1007/978-981-10-1675-2_12

receiver and sender [2]. The Packet erasure network is basically a model where sequential packets have been transmitted or received. It is closely related to binary erasure model which uses a bit sequence instead of large data packet. When the distance between transmitter and receiver increases the frequency decreases and vice versa, this phenomenon is called Doppler Effect or Doppler shift invented by Austrian physicist Christian Doppler, who proposed it in 1842 in Prague [2, 3]. In wireless channels, several models have been developed, we are using the Rayleigh model in this paper with Doppler frequency shift. In the implemented model, we are comparing the errors in various operating modes of the encoder for convolution codes. The convolution code improves the performance of the network as operates on serial data [4].

2 Rayleigh Model

It is a statistical model for the effect of transmission surrounding on the signal, the distortion is generally caused due to multipath reception [5]. This affects the signal most when there is no line of sight between the transmitter and receiver [6].

The general representation of Rayleigh model is shown in Fig. 1, where the noise signal minimizes between transmitter and receiver. The Transmitter consists of the encoder and modulator. The encoder basically uses Linear Block Codes and Convolution Codes. We are implementing the model with convolution codes. In the model, the signal transmitted through a channel, during this travel of a signal the errors are calculated and BER value also calculated and displayed with Doppler shift.

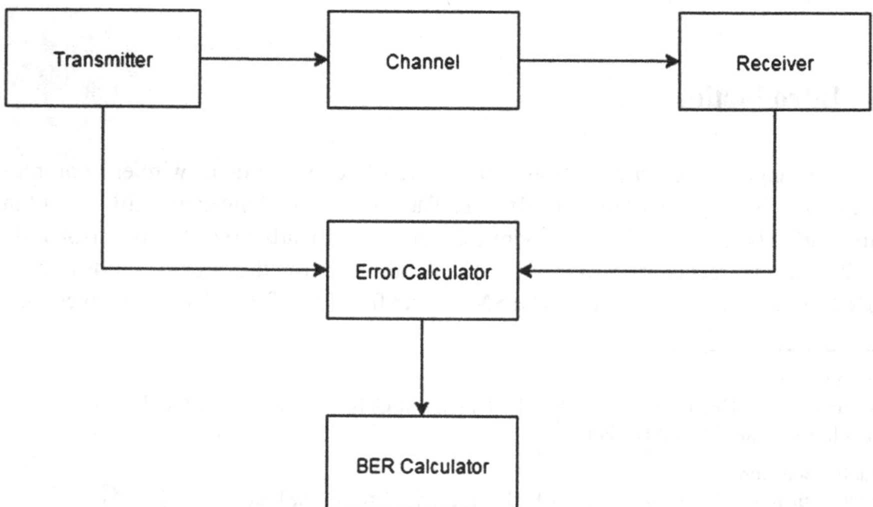

Fig. 1 General representation of Rayleigh model

3 Implemented Model

We have demonstrated the concept of BER analysis using Doppler frequency shift in the packet erasure network. It is really important to understand, what happened to a signal between transmitters to receiver stage. Bit Error Rate (BER) has been considered as the performance indicator in all models for analysis [7, 8]. The implemented model for Rayleigh fading and Doppler frequency shift channel with Additive White Gaussian Noise (AWGN) is shown in Fig. 2. The technical specifications of the model are listed in Table 1.

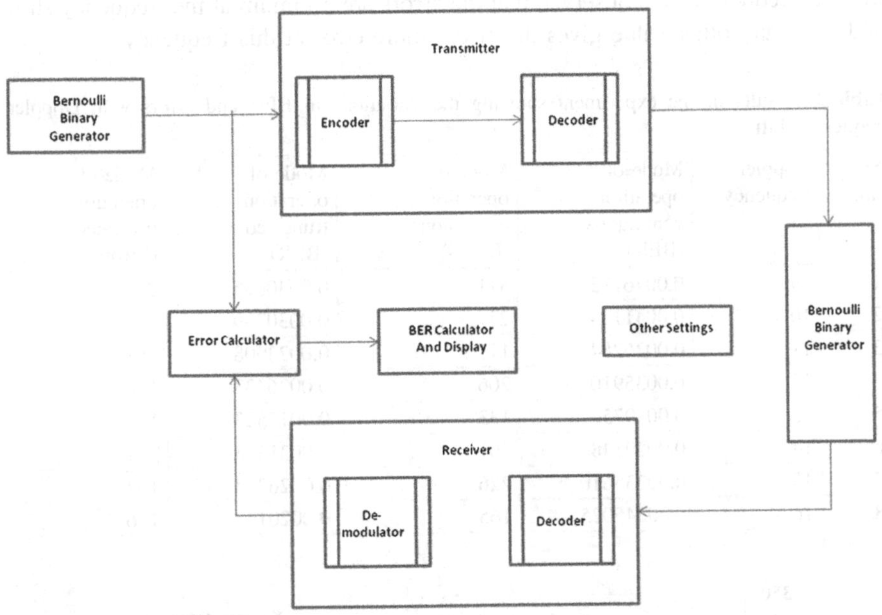

Fig. 2 Implemented Model of Rayleigh fading and Doppler frequency shift channel

Table 1 Technical details of implemented model

Sr. no.	Details	Specification
1	Models used	Poly2trellis (7, [133, 171])
2	Initial speed of AWGN	12345 bps
3	Signal to noise ratio (SNR)	10 DB
4	Probability of zero bit	0.5
5	Data rate	1.2 Kbps

4 Results and Discussions

The experiment carried out for shift in Doppler frequency, the transmitter of the model implemented consists of trellis encoder and modulator. The convolution encoding with continuous or terminate mode and truncated mode was taken to accomplish the results. The error and the Bit Error Ratio (BER) in the signal were observed with the shift in Doppler frequency, from the readings we can conclude that with increase in Doppler frequency the errors are varying, the minimum errors observed are at Doppler frequency shift of 13 Hz. Any value of Doppler frequency more or less than this will give the errors more than the value at this frequency as shown in Table 2. below.

The graph drawn below in Fig. 3 shows the variation of errors with shift in Doppler frequency, it is observed that the errors are optimum at the frequency shift of 13 Hz; any other value gives the errors more error at this frequency.

Table 2 Results of the experiment showing the variations in BER and errors with Doppler frequency shift

Sr. no.	Doppler frequency	Mode of operation continuous (BER)	Mode of operation continuous (Errors)	Mode of operation truncated (BER)	Mode of operation truncated (Errors)
1	9	0.0046183	311	0.0040095	270
2	10	0.0041134	277	0.0030739	207
3	11	0.0026284	117	0.0023908	161
4	12	0.0035910	206	0.0026433	178
5	13	0.0019750	133	0.0017523	118
6	14	0.0029848	201	0.0031333	211
7	15	0.00335610	226	0.0026284	177
8	16	0.00245025	165	0.0020196	136

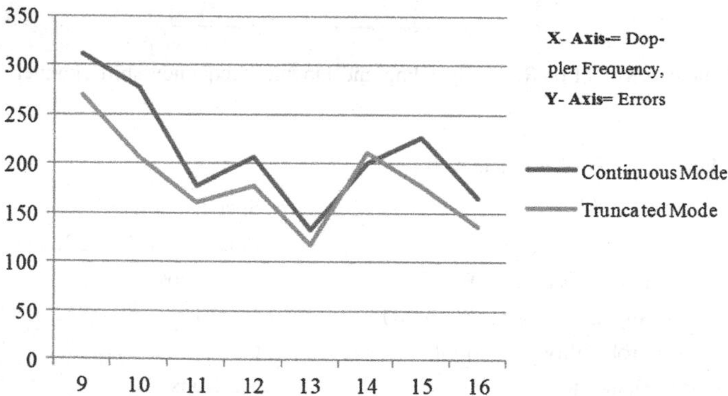

Fig. 3 Graph showing comparison of errors with Doppler frequency shift for various modes of operation

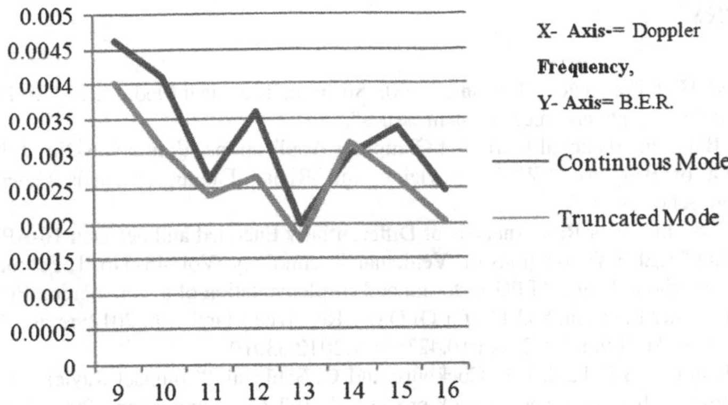

Fig. 4 Graph showing comparison of BER with Doppler frequency shift for various modes of operation

The other variation, i.e., Doppler frequency versus the Bit Error ratio (BER) is shown below in Fig. 4. we can observe that BER is also minimum at the Doppler frequency shift of 13 Hz. As BER has direct relation with SNR, it is really essential to curtail the value of BER.

5 Conclusion

The paper implemented a model to study BER analysis using Doppler frequency shift in the packet erasure network. For carrying out this, Rayleigh model is used. We have conducted the set of readings showing the variation in error and BER with Doppler frequency shift for two modes of operation of convolution encoder. From the results it was observed that the errors were minimized at the Doppler frequency shift of 13 Hz and Channel errors are increasing if Doppler frequency shift increases or decreases. Any shift in the Doppler frequency, will increase the value of BER and errors. As we know, BER value plays a vital role in the performance of any model in wireless communication, and the implemented model is used here to optimize the value of BER by Doppler frequency shift method.

Acknowledgments This paper is the part of research work carried out by Mr. Arjun Ghule for his PhD work. We would like to thank the reviewers for their valuable comments for improvement of this paper.

References

1. Gary Breed, High Frequency Electronics, 2003 Summit, Technical Media LLC "Bit Error Rate: Fundamental Concepts and measurement issues".
2. A. Sudhir Babu, International Journal of Computer Applications, Volume 26–No. 9, July 2011, "Evaluation of BER for AWGN, Rayleigh and Rician Fading Channels under Various Modulation Schemes".
3. Fumiyaki Adachi, "error Rate Analysis of Differentially Encoded and detected 16-APSK under Rician fading", IEEE Transactions on Vehicular Technology, Vol. 45, No. 1, February 1996.
4. Yan Sun1, Zhizhong Ding2, "FPGA Design and Implementation of a Convolution Encoder and a Viterbi Decoder Based on 802.11a for OFDM", Received March 5th, 2012; revised April 1st, 2012; accepted April 9th, 2012, doi:10.4236/wet.2012.33019.
5. A. Alimohammad, S.F. Fard, B.F. Cockburn and C. Schlegal, "Compact Rayleigh and Rician fading simulation based on random walk processes", IET Communications, 2009, Vol. 3, Issue 8, pp 1333–1342.
6. Yahong Rosa Zheng and Chengshan Xiao, "Simulation models with correct statistical properties for Rayleigh fading channels", IEEE Transactions on communications, Vol. 51, No. 6, June 2003.
7. Sanjay Kumar Khadagade, "Comparison of BER of OFDM System using QPSK and 16QAM over Multipath Rayleigh Fading Channel using Pilot-Based Channel Estimation", International Journal of Engineering and Advanced Technology (IJEAT) ISSN: 2249–8958, Volume-2, Issue-3, February 2013.
8. A. R. S. Bahai AND B. R. Saltzberg, "Multi-Carrier digital communications theory and applications of OFDM", Kluwer Academic/Plenum, 1999.

Study of Variation in Ambient Noise with Fluctuations of Surface Parameters for the Indian Ocean Region

Piyush Asolkar, Arnab Das, Suhas Gajre and Yashwant Joshi

Abstract Ambient noise variability is a critical challenge encountered by multiple stakeholders, including sonar designers and operators. Among the sources of ambient noise in the ocean, wind related noise has significant impact on sonar performance. The tropical waters in the Indian Ocean Region (IOR), present random fluctuations in the surface parameters, namely the wind speed, surface temperature, wave height, etc. resulting in variations in the ambient noise characteristics. The site-specific surface fluctuations in the tropical regions restrict the possibility of generalized algorithm design to mitigate the ambient noise impact. The work attempts to study the variations in the ambient noise levels corresponding to the fluctuations in the surface parameters. The site-specific behavior of the tropical IOR is demonstrated using surface data available from moored buoy at three distinct locations of the IOR. The analysis methodology can be used to characterize, predict and improve sonar performance, particularly in severe conditions of the tropical IOR.

Keywords Ambient noise characterization · Wind noise · Tropical littoral waters · Indian ocean region · Surface fluctuations · Sonar performance

---

Piyush Asolkar (✉) · Suhas Gajre · Yashwant Joshi
Department of Electronics and Telecommunication Engineering, SGGSIE&T, Nanded Maharashtra, India
e-mail: asolkarpiyush@sggs.ac.in

Suhas Gajre
e-mail: ssgajre@sggs.ac.in

Yashwant Joshi
e-mail: yvjoshi@sggs.ac.in

Arnab Das
Acoustic Research Lab, Tropical Marine Science Institute, National University of Singapore, Singapore, Singapore
e-mail: arnabdas1972@hotmail.com

© Springer Science+Business Media Singapore 2017

S.C. Satapathy et al. (eds.), *Proceedings of the International Conference on Data Engineering and Communication Technology*, Advances in Intelligent Systems and Computing 468, DOI 10.1007/978-981-10-1675-2_13

1 Introduction

The ambient noise in the oceans has been on the rise and the acoustic systems present sub-optimal performance due to poor Signal to Noise Ratio (SNR) at the deployment location. Modern underwater acoustic systems are largely passive due to their obvious advantages for both military (related to stealth) and also non-military (related to power budgeting in offshore deployments) applications [1]. Hence, the ambient noise becomes a key limitation in improving sonar performance. The ambient noise in the ocean has a multiple source components manifesting at varying frequency bands. Among these ambient sources, the wind noise by far has the maximum impact on the sonar performance due to its overlapping frequency spectrums ranging from 3 kHz up to 5 kHz [2–4].

The different regions, namely the tropical, temperate, and polar present diverse environmental behavior that has direct ramifications on the sea surface parameters and the related wind fluctuations. Further, among them the tropical region displays significant high random diurnal, seasonal and site-specific surface fluctuations namely sea surface temperature, wind speed, wind direction, wave height, current speed and current direction [5]. Thus, the tropical littoral waters of the Indian Ocean Region (IOR) are highly sensitive to the surface disturbances that translate to wind noise [6–10].

The classical SNR enhancement techniques fail to mitigate the noise impact due to random behavior of the ambient noise. However, the understanding of the relation between the measurable surface parameters and the ambient noise can potentially improve our ability to estimate the ambient noise characteristics. Such estimation will facilitate improved mitigation of the wind related ambient noise on sonar performance [11]. Ambient noise measurement and characterization at certain sites have demonstrated high correlation with the surface parameters like wind speed, wave height and sea surface temperature [12–15]. Pioneering work by Knudsen et al. [2] in 1948 had investigated ambient noise variations based on surface parameters and found that ambient noise increases with increasing wind speed and wave height. However, the tropical water of IOR has not been studied that extensively. There are no studies representing site-specific comparison of ambient noise which is the key feature of IOR and is critical for any ambient noise estimation and subsequent SNR enhancement efforts.

The work attempts to study the variations in the ambient noise levels corresponding to the fluctuations in the surface parameters that can be easily measured using Commercially-Off-The-Shelf (COTS) equipment based on available models from the open source literature. Section 2 introduces measurements and ambient noise dependency on surface parameters. Results and site-specific behavior in IOR is demonstrated in Sect. 3.

2 Measurement and Processing Methodology

The data was collected in Arabian Sea off the cost of Goa. Measurements were done periodically eight times a day over a period of 2 month with moderate environmental condition which gives 118 data samples. ITC 8264 omnidirectional hydrophones were used with sensitivity −175 dB re 1 μPa and bandwidth 10 Hz−100 kHz. The data were acquired at rate of 256 kHz filtered and digitized to 16 bits resolution [9]. The site-specific surface data was made available from moored buoy owned by Indian National Centre for Ocean Information Services (INCOISE) at three distinct location of the IOR, namely Arabian Sea, Bay of Bengal and Indian Ocean.

The spectral analysis was carried out using Welch power spectral density [16]. Ambient noise in dB was plotted for different wind speed and wave heights for a frequency range of 10–5000 Hz. Ambient Noise level at different frequencies have been observed for varying wind speed and wave heights. A mathematical model has been designed based on sea surface temperature, wave height, and wind speed. The time series variations of wind speed and wave heights from three distinct sites from IOR namely Arabian Sea, Indian Ocean and Bay of Bengal are analyzed for a yearly data. Wind–wave relationship of the three sites was observed by plotting wind speed against wave height. Ambient noise levels at three sites were compared based on modeled data.

3 Results and Discussions

The results of ambient noise variation with wind speed and wave height are presented in Figs. 1 and 2. Measurements showing dominant evidence of noise from ship (46 samples) were not included in analysis. Figure 1 shows ambient noise variation with respect to variation of wind speed over the spectrum of 10–5000 Hz. It is observed that noise level increases with the increase in wind speed. The frequency spectrum up to 500 Hz shows less variation with respect to wind speed.

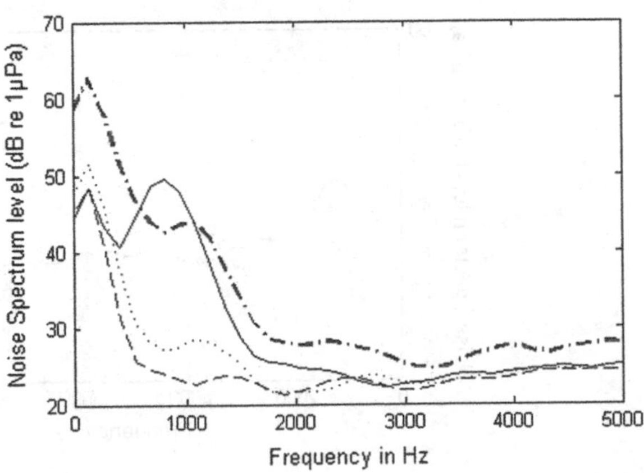

Fig. 1 Ambient noise spectrum for wind speed of 1 m/s (*dotted line*), 2 m/s (*dashed line*), 3 m/s (*solid line*) and 4 m/s (*dash-dot line*)

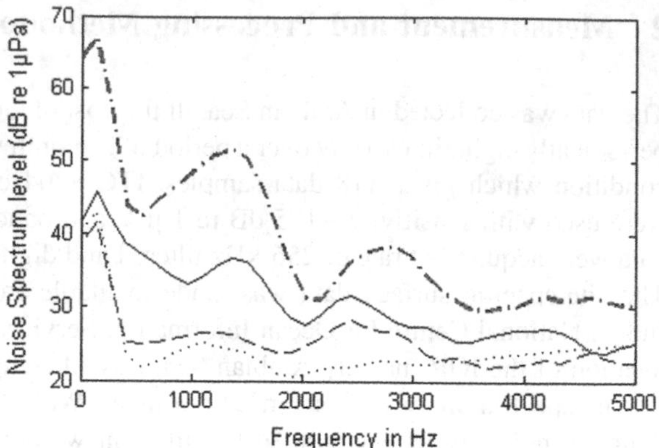

Fig. 2 Ambient noise spectrum for 0.5 m (*dotted line*), 1 m (*dash line*), 1.5 m (*solid line*) and 2.1 m (*dash-dot line*) wave heights

However, spectrum of 500–4500 Hz shows higher dependency on wind speed. Spectrum follows Knudsen [2] spectra with a decreasing slope of 6 dB per octave for a frequency range of 100–5000 Hz. It shows higher noise level and slope up to 1 kHz, however, slope decreases with increase in frequency representing weak dependency of ambient noise due to wind at higher frequency. Figure 2 shows ambient noise variation with respect to wave height over spectrum of 10–5000 Hz. A peak is observed at a low frequency of 70–100 Hz which is due to turbulence. Wide variations between 1000–5000 Hz are mainly due to bubbles and spray. Noise due to bubbles show higher variations with increase in wave height [17, 18]. It shows a decreasing slope of 11 dB/octave for frequency band of 10–5000 Hz. Higher slope at frequency of 1–3 kHz clearly represents the higher variation of noise level due to wave height, bubbles and spray [17].

This analysis represents high correlation of ambient noise with wind speed and wave height. Ambient noise model have been designed based on nonlinear regression analysis given by Eq. 1.

$$NL = a + b * \log_{10}(W_s) + c * \log_{10}(\mathrm{SST}) + d * W_h^2 + e * W_h \tag{1}$$

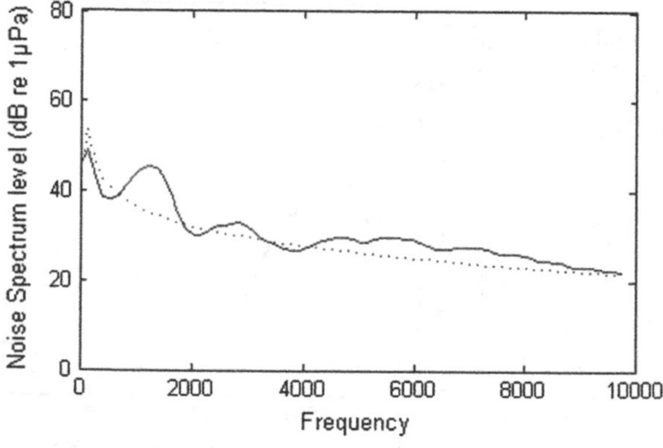

Fig. 3 Comparison of field (*solid line*) and modeled (*dotted line*) spectra

where, *NL* is noise level, W_s is wind speed, W_h is wave height and *a–e* are the regression coefficients.

Figure 3 shows comparison of field measurements with modeled results. It shows that predicted values by the model are close to the measured ambient noise value.

3.1 Comparison of Field Environments

Indian Ocean Region is well-known for the random fluctuation of the environment; hence generalized mitigation algorithm fails in IOR. In order to design efficient mitigation methods it is necessary to analyze dependency of physical parameters and their effect on ambient noise. Sea surface data from moored buoy (owned by INCOISE) at three distinct sites namely Arabian Sea, Indian Ocean, and Bay of Bengal is used for analysis of the relation between surface parameters in IOR. Figure 4 shows time series variation of sea surface temperature, wind speed and

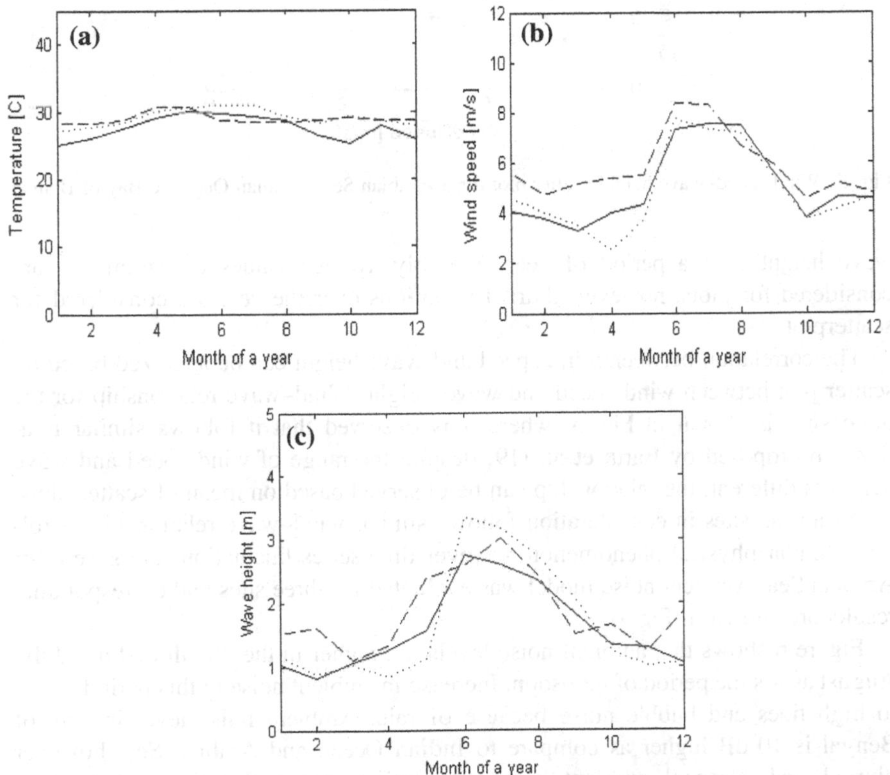

Fig. 4 Time series plot of **a** sea surface temperature, **b** wind speed and **c** wave height for Arabian Sea (*dotted line*), Indian Ocean (*solid line*) and Bay of Bengal (*dash line*)

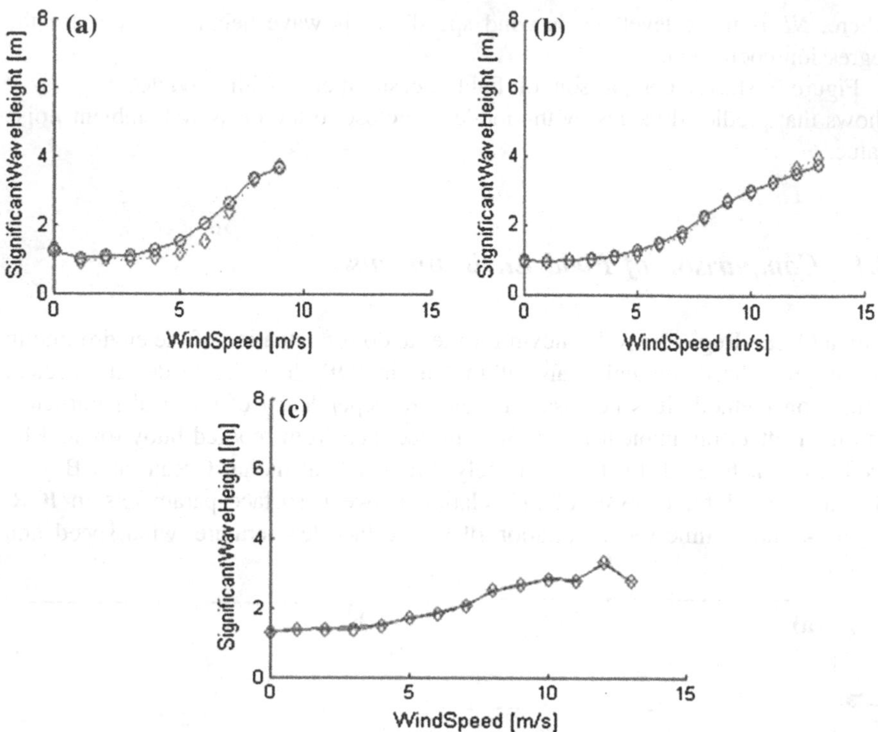

Fig. 5 Wind speed–wave height scatter plot for **a** Arabian Sea, **b** Indian Ocean, **c** Bay of Bengal

wave height over a period of year. Monthly average values of parameters are considered for plots, however, diurnal variations over the year are considered for scatterplot.

The correlation between wind speed and wave height can be analyzed based on scatter plot between wind speed and wave height. Wind–wave relationship for the three sites is shown in Fig. 5, where it is observed that it follows similar relationship proposed by Barth et al. [19] despite the range of wind speed and wave height is different, the relationship can be observed based on mean of scatter plots.

As all the sites in consideration follows similar wind–wave relationship, it follows similar physical phenomenon however time series fluctuations are severe for Arabian Sea. Ambient noise model was evaluated for three sites and corresponding results are shown in Fig. 6.

Figure 6 shows that ambient noise levels are higher in the month of June, July, August as it's the period of monsoon. Increase in ambient noise in this period is due to high tides and bubble noise because of rain. Ambient noise level in Bay of Bengal is 10 dB higher as compare to Indian Ocean and Arabian Sea, however diurnal and seasonal ambient noise fluctuations are higher in Arabian Sea. Figure 7a–c shows relationship of normalized variance of sea surface temperature,

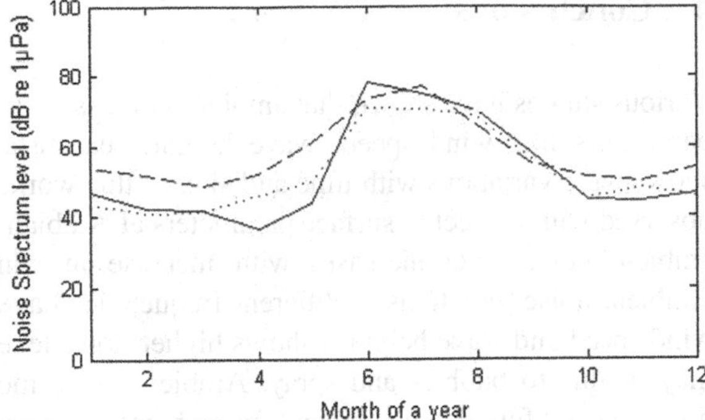

Fig. 6 Modeled ambient noise for Arabian Sea (*solid line*), Indian Ocean (*dotted line*) and Bay of Bengal (*dash line*)

variance of wave height, and variance of wind speed for IOR. Figure 7d compares the variance of ambient noise for IOR.

Variance of ambient noise with respect to surface parameters at different sites of IOR can be observed in Fig. 7. Where it is observed that ambient noise level is highly correlated to wind speed and wave height, however variance of surface parameters is site-specific, which results in site-specific variance of Noise level (Fig. 7d).

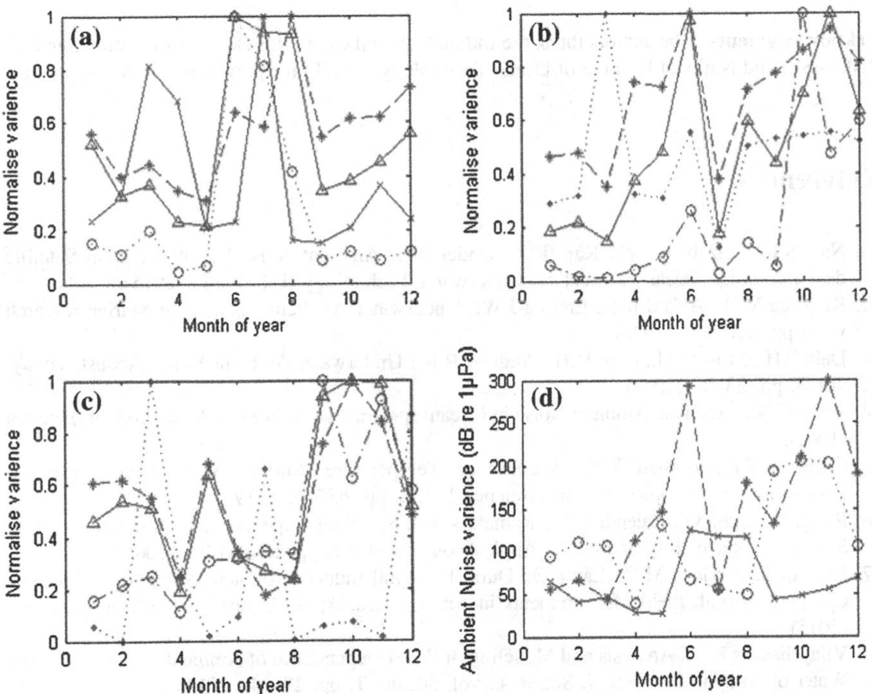

Fig. 7 Comparison of variance of SST (*dot line*), wind speed (*dash line*), wave height (*dash dot line*) and ambient noise (*solid line*) for **a** Arabian Sea, **b** IOR, **c** Bay of Bengal. **d** Comparison of variance of ambient noise for Arabian Sea, (*solid line*), IOR (*dash line*), Bay of Bengal (*dotted line*)

4 Conclusions

Various studies have shown that ambient noise is highly correlated with sea surface parameters like wind speed, wave height, sea surface temperature; however, it shows wide variations with time and site. In this work, ambient noise variations are observed with respect to surface parameters at Arabian Sea. The analysis shows that ambient noise level increases with increase in wind speed and wave height. Ambient noise variations at different frequencies have been observed for range of wind speed and wave height it shows higher noise levels at 1 kHz frequency which may be due to bubbles and spray. Ambient noise model have been verified with field data and found to be efficient for ambient noise analysis. Sea surface parameter fluctuations at three sites of IOR have been analyzed and ambient noise variations are plotted based on monthly average sea surface parameters for three sites which shows seasonal variation of ambient noise in IOR. It shows remarkable fluctuations in ambient noise level. This is the first such effort to compare seasonal variation of ambient noise in different regions of IOR. However it can be further improved by improving ambient noise model using real-time data with wide variation of sea surface parameter. The analysis methodology can be used to characterize, predict and improve sonar performance, particularly in severe conditions of the surface fluctuations in the tropical IOR.

Acknowledgments The authors thank the Indian National Centre for Ocean Information Services (INCOISE) and National Institute of Ocean Technology (NIOT) for providing moored buoy data.

References

1. Nair N.R., Elizabeth S.N., Raju R.P.: Underwater Ambient Noise Variability from Satellite data–An Indian Ocean perspective. Underwater Technology (UT) IEEE, (2015).
2. Knudsen V.O., Alford R.S., Emling J.W.: Underwater Ambient Noise. J. of Marine research vol 7 pp. 410–429 (1948).
3. Dahl P.H., Miller J.H., Cato D.H., Andrew R.K.: Underwater Ambient Noise. Acoust. Today, vol. 3, pp. 23–33 (2007).
4. Wenz G.M.: Acoustic Ambient Noise in Ocean:Spectra and Sources. J. Acou. Soc. of America (1962).
5. Graham N.E., Barnett T.P.: Sea-Surface Temperature, Surface Wind Divergence and Convection over Tropical Oceans. Science J. 238, pp. 657–659 (1987).
6. Ramji S., Latha G., Rajendran V., Ramakrishnan S.: Wind Dependence of Ambient Noise in Shallow Water of Bay of Bengal. Appl. Acoust., vol. 69, pp. 1294–1298 (2008).
7. Najeem S., Sanjana M.C., Latha G., Durai P.: Wind Induced Ambient Noise Modelling and Comparison with Field Measurements in Arabian Sea. Appl. Acoust., vol. 89, pp. 101–106 (2015).
8. Vijayabaskar R.V.: Analysis and Modeling of Wind Dependence of Ambient Noise in Shallow Water of Arabian Sea. Eur. J. Sci. Res., vol. 50, no. 1, pp. 28–34, (2011).
9. Das A.: Shallow Ambient Noise Variability due to Distant Shipping Noise and Tide. Appl. Acoust. vol. 72, pp. 660–664, (2011).

10. Murugan S.S., Natarajan V., Kumar R.R.: Noise Model Analysis and Estimation of Effect due to Wind Driven Ambient Noise in Shallow Water. Int. J. Oceanogr., vol. 2011, pp. 1–4, (2011).
11. Peyvandi H., Farrokhrooz M., Roufarshbaf H., Park S.J.: SONAR Systems and Underwater Signal Processing: Classic and Modern Approaches. Intech Open access publisher, (2011).
12. Roth E.H., Hildebrand J.A., Wiggins S.M., Ross D.: Underwater Ambient Noise on the Chukchi Sea continental slope from 2006–2009. J. Acoust. Soc. Am., vol. 131, no.1, pp. 104–108, (2012).
13. Jung S.K., Choi B.K., Kim B.C., Kim B.N, Kim S.H., Park Y., Lee Y.K.: Seawater Temperature and Wind Speed Dependences and Diurnal Variation of Ambient Noise at the Snapping Shrimp Colony in Shallow Water of Southern Sea of Korea J. J. Appl. Phys., vol. 51, pp. 12–14, (2012),
14. Kim B., Hahn J., Choi B.K.: Wind Dependence of Oceanic Ambient Noise Measured at the Ieado Ocean Research Station Proceeging Symp. Ultrason. Electron., vol. 31, pp. 411–412, (2010).
15. Poikonen A.: Measurements, Analysis and Modeling of Wind Driven Ambient Noise in Shallow Brackish Water (2012).
16. Welch P.D.: The Use of Fast Fourier Transform for the Estimation of Power Spectra: A Method based on Time Averaging over Short, Modified Periodograms. IEEE Trans. Audio and electroacoust. Au-15 pp 70–73, (1967).
17. Harland E.J., Jones S.A., Clarke T.: SEA 6 Technical Report: Underwater Ambient Noise. Report by QINETIQ (2005).
18. Harland E.J., Richards S.D.: SEA 7 Technical Report: Underwater Ambient Noise. Report by QINETIQ (2006).
19. Barth S., Eecen P.J.: Description of the Relation of Wind, Wave and Current Characteristics at the Offshore Wind Farm Egmond aan Zee (OWEZ) Location in 2006. Report by ECN (2012).

ANIDS: Anomaly Network Intrusion Detection System Using Hierarchical Clustering Technique

Sunil M. Sangve and Ravindra C. Thool

Abstract The Intrusion detection system (IDS) is an important tool to detect the unauthorized use of computer network and to provide the security for information. The IDS consists of two types signature-based (S-IDS) and anomaly-based (A-IDS) detection system. S-IDS detect only known attacks whereas A-IDSs are capable to detect unknown attacks. In this paper, we are focusing on A-IDS. The proposed system is Anomaly network intrusion detection system (ANIDS). The ANIDS is implemented using metaheuristic method, genetic algorithm and clustering techniques. The two different clustering techniques are used i.e. *K*-mean clustering and hierarchical clustering to check the performance of system in terms of false positive rate (FPR) and detector generation time (DGT). The system includes modules like input dataset, preprocessing on input dataset, clustering and selection of sample training dataset, testing dataset, and performance analysis using training and testing dataset. The experimental results are calculated based on large–scale dataset, i.e., NSL-KDD for detector generation time and false positive rate (FPR). Our proposed technique gives better result for false positive rate and detector generation time as compared to *K*-means clustering technique.

Keywords Intrusion detection system (IDS) · Anomaly network intrusion detection system (ANIDS) · NSL-KDD

S.M. Sangve (✉)
ZCOER, Computer Engineering, SP Pune University, Pune, Maharashtara, India
e-mail: sunilsangve@gmail.com

R.C. Thool
SGGIET Computer Science and Engineering, Nanded, Maharashtara, India
e-mail: rcthool@yahoo.com

© Springer Science+Business Media Singapore 2017
S.C. Satapathy et al. (eds.), *Proceedings of the International Conference on Data Engineering and Communication Technology*, Advances in Intelligent Systems and Computing 468, DOI 10.1007/978-981-10-1675-2_14

1 Introduction

Now-a-days, dependency on networked computers increased and is still increasing, along with this, growing expertise in such networked system requires brilliant and adaptive threat detection. Because of this computer network security becomes a major issue. Thus, in computer security, confidentiality, integrity, and availability (CIA) plays a very vital role [1]. To identify improper or unauthorized modifications, the integrity mechanism has divided into two classes: prevention and detection [2]. Therefore, for detection of attacks, intrusion detection system (IDS) is used and for prevention of unauthorized user, intrusion detection and prevention system (IDPS) is used.

The problem of identifying the intrusion in the system is resolved by checking violation of privilege levels in the system, misuse of the system and unauthorized use. The heterogeneous computer network gives additional burden to detect the intrusions [3]. Originally, the concept of intrusion detection was proposed by Anderson in 1980 [4]. Basically, there are two types of IDS Host-based Intrusion Detection System (H-IDS) and Network-based Intrusion Detection System (N-IDS). The H-IDS detects the intrusions on the single system but N-IDS detect attacks on multiple systems by connecting the systems with each other by a network. In this paper, we are focusing on network-based intrusion detection system. The N-IDS consists of two types signature-based N-IDS and anomaly-based N-IDS whereas Signature-based N-IDS used to detect only known attack and anomaly-based N-IDS used to detect unknown attack [5].

This paper represents the network anomaly detection using metaheuristic method including genetic algorithm and clustering techniques. The metaheuristic method defined by Osman and Laporte in 1996, is an iterative generation process which gives guidance to subordinate heuristic by combining concepts for exploring and exploiting the search space, as well as learning strategies are used to find efficiently near optimal solutions [6]. Blum and Roli [7], gives fundamental properties of metaheuristic: (1) to guide search process, metaheuristic strategies are used. (2) Explore the search space to find near optimal solution.

2 Related Work

The anomaly detection methods are classified into several types. One of the methods among them is statistic-based method. It identifies the intrusion by using the predefined threshold, standard deviation, mean, and the probabilities [8]. Another category is rule-based methods. It uses the if-then and if-else rules, in order to construct the model of detection for some previously known intrusions [9]. Additionally, the State-Based approach is also there. It makes the use of Finite state machine, which is derived from the network topologies to determine the attacks [10]. Negative selection algorithm (NSA) is one of the artificial immune system (AIS) algorithms which motivated by immune system microorganism development and tolerance toward oneself in human immune system [11]. It builds a model of non-self information by producing examples that didn't match existing ordinary (self) designs, then utilizing this model to match non-ordinary examples to recognize anomalies. NSA detectors are structured with different geometric shapes, for example, hyper-rectangles, hyper-circles, hyper-ellipsoids or various hyper-shapes. Anna Sperotto et al. [12], proposed the automatic approach for anomaly network intrusion detection using SSH (secure shell is the encrypted protocol which allows to operate remotely over an unsecured network) traffic. They suggested the procedure which selects the system parameter automatically and increases the system performance. Alexander et al. [13], have considered the problem of online anomaly detection in computer network traffic effectively. This is done using changepoint detection method.

3 Implementation Details

The main idea is based on combination of multi-start metaheuristic algorithm, genetic algorithm, and hierarchical clustering technique. The number of detectors is very important to detect anomaly. In ANIDS, we are using hierarchical clustering technique to reduce FPR and DGT and compare the results with existing k-mean clustering. The clustering techniques are used to select multiple initial points using multi-start method. Using multi-start method, the radius of hyper sphere detector is obtained. This radius is optimized using genetic algorithm. The rule reduction is used to remove redundant detectors to reduce detector generation time. The detector generation process is repeated to increase the detection quality. As shown in Fig. 1, we use the hierarchical clustering and K-mean clustering to divide the large training dataset into number of clusters. The Anomaly Network Intrusion Detection System consists of following modules:

3.1 Input Dataset

The input dataset is NSL-KDD dataset [14]. It contains Normal, Probe, U2R, R2L, and DoS attacks. The total 41 columns headers are added that contain information such as duration, protocol type, service, src_bytes, dst_bytes, flag, land, wrong fragment, etc.

3.2 Data Preprocessing

Preprocessing is applied on input dataset (I). To remove unnecessary data or words which are not useful for extracting the features, data preprocessing is used. The

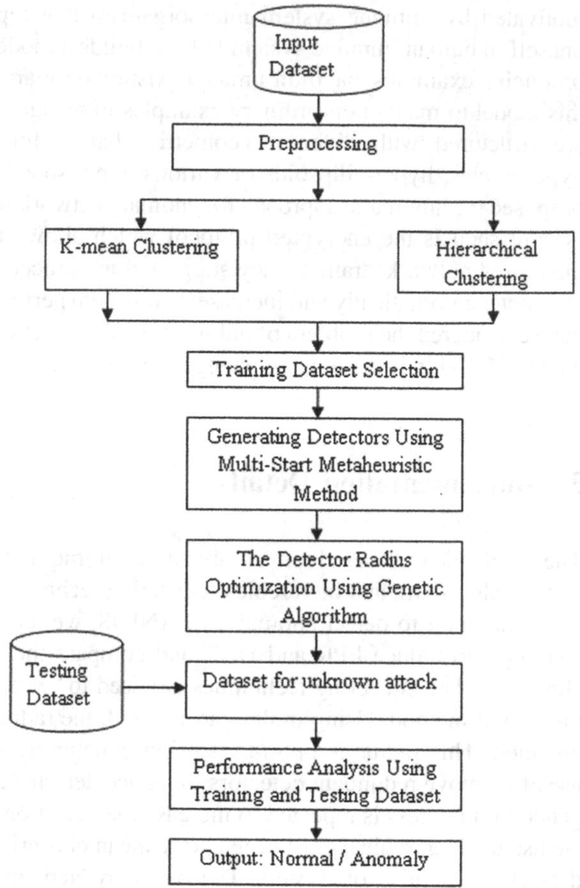

Fig. 1 Anomaly network intrusion detection system (ANIDS)

main benefit of data preprocessing is that the time required for processing will also decrease. The following example describes how preprocessing applied on input dataset:

Let us consider the one sample from I,

{0, tcp, ftp_data, SF, 491, 0, 0, 0, 0, 0, 0, 0, 0, 0, 0, 0, 0, 0, 0, 0, 0, 2, 2, 0.00, 0.00, 0.00, 0.00, 1.00, 0.00, 0.00, 150, 25, 0.17, 0.03, 0.17, 0.00, 0.00, 0.00, 0.05, 0.00, normal}

When we apply the preprocessing on the above single sample, the words like tcp, ftp_data, SF (start flag) are removed to decrease the processing time. The preprocessed sample consists of numeric. The last word in sample denotes the class normal or anomaly. Therefore, the obtained vector contains two important features, i.e., pattern in numeric form and class name 'normal'.

{491, 0, 0, 0, 0, 0, 0, 0, 0, 0, 0, 0, 0, 0, 0, 0, 0, 2, 2, 0.00, 0.00, 0.00, 0.00, 1.00, 0.00, 0.00, 150, 25, 0.17, 0.03, 0.17, 0.00, 0.00, 0.00, 0.05, 0.00}

3.3 Clustering

3.3.1 Hierarchical Clustering Algorithm

Given a set of N items to be clustered, and an $N * N$ matrix (distance or similarity), the main idea about hierarchical clustering is defined by Johnson [15].

3.3.2 Pseudo Code

The pseudocode for hierarchical clustering is given in [16] which are given below: First, we compute the $N * N$ similarity matrix. The algorithm executes $N - 1$ steps to merge the most similar cluster.

Hierarchical Clustering (d_1, d_2, d_3, d_4,….d_N)

```
For n ← 1 to N
Do for I ← 1 to N
Do C[n] [i] ← SIM (d_n, d_i)
I[n] ← 1 (used to keeps track of active clusters)
A ← [·] (assembles clustering as a sequence of merges
For k ← 1 to N − 1
Do (I, m) ← arg max{<i, m>: i ≠ m ∧ I [i] = 1 ∧ I [m] = 1} C[i] [m]
A. Append (<i, m>) (store merge)
For j ← 1 to N
Do C[i] [j] ← SIM (i, m, j)
C[j] [m] ← SIM (i, m, j)
I [m] ← 0 (Deactivate cluster)
Return A;
```

SIM (i, m, j)—used to compute similarity of cluster j with i and m cluster which are merged together. It is function of $C[i]$ $[j]$ and $C[j]$ $[m]$. The time complexity of Hierarchical Clustering is $\Theta(N3)$, here we scan $N * N$ matrix C with largest similarity in each of $N - 1$.

3.4 Selection of Dataset and Detector Generation Using Multi-start Metaheuristic Method

After applying the clustering algorithm, we select some training dataset samples from given dataset. The multiple initial start points are selected from clustering as input to generate the detectors. The detector shape is hyper sphere. Thus, we calculate the radius of hyper sphere to identify the anomaly by using following rules [17]:

The detector radius $R = \{r \in R | \ 0 < r \leq \text{hpu}\}$ where hpu is the hyper-sphere radius upper bound. Thus,

$U_j = \max(x_{ij})$ where $i = 1, 2, 3\ldots m$, $L_j = \min(x_{ij})$ where $i = 1, 2, 3\ldots, m$

UB—upper bound and LB—lower bound are used for solution space.

UB = $(u_1, u_2, u_3\ldots,u_n, \text{hpu})$, LB = $(l_1, l_2, l_3\ldots 0)$, the detectors $D = \{d_1, d_2, d_3\ldots d_{isp}\}$

The solution space obtained by multi-start framework is calculated as: $D_i = (u_{i1}, u_{i2}, u_{i3}\ldots u_{in}, r_i)$ where hyper-sphere center is at $D_{center} = (u_{i1}, u_{i2}, u_{i3}\ldots u_{in})$ and hyper sphere radius is r_i.

The objective function to control the detector generation process is:

$$F(D_i) = N_{\text{abnormal}}(d_i) - N_{\text{normal}}(d_i) \quad (1)$$

where, N_{abnormal} is the number of abnormal samples covered by detector d_i and N_{normal} is the number of normal samples covered by detector d_i.

Anomaly Detection is done from the generated detectors and rule is, If (dist $(D_{center}, x) \leq r$) then {normal} else {abnormal} where r is the detector hyper-sphere radius and (dist (D_{center}, x)) is the Euclidean distance between detector hyper sphere center D_{center} and test samples x.

3.5 Testing Dataset

The testing dataset is the additional dataset to detect unknown pattern. It is used to test training patterns in the training dataset, If the system was trained successfully, outputs produced by the system would be similar the actual outputs.

3.6 Performance Analysis Using Training and Testing Dataset

Finally performance of the system is analyzed using training and testing dataset for false positive rate, detector generation time.

4 Experimental Results

The experimental results are calculated based on NSL-KDD dataset. For experimental set up, we use Windows 7 operating system, Intel i5 processor, 4 GB RAM, 80 GB Hard disk, Net Beans IDE 8 + JDK tool. The False positive rate is calculated using formula:

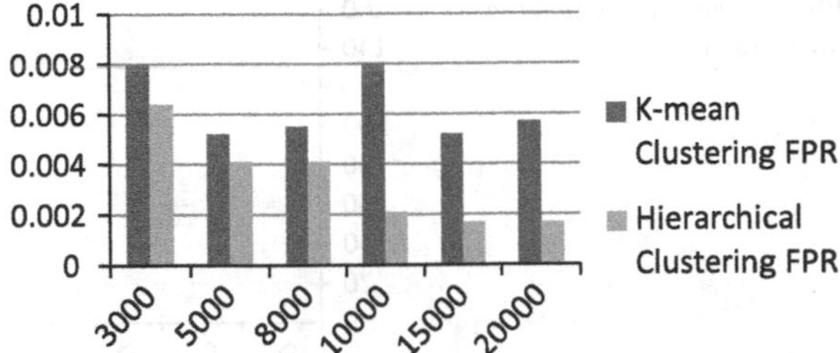

Fig. 2 False positive rate graph

The False positive rate (FPR):

$$\text{FPR} = \frac{\text{False Positive}}{\text{FP} + \text{FN}} \tag{2}$$

where, FP-False Positive, FN-false Negative.

4.1 False Positive Rate (FPR)

The minimum false positive rate is obtained for the anomaly network intrusion detection system using hierarchical clustering technique. Therefore, the results show that, Hierarchical clustering gives minimum FPR than the *K*-mean clustering. The minimum FPR is 0.0017 obtained at training dataset 15000. Fig. 2 shows comparison of false positive rate using *K*-mean and hierarchical clustering approach (Table 1).

4.2 Detector Generation

Fig. 3 shows the comparison of DGT by using *K*-mean and Hierarchical-clustering algorithm. The time required for generation of detector is less by Hierarchical-clustering. The observation is that detector generation time increases with the increase in dataset size (Table 2).

Table 1 False positive rate (FPR)

Dataset size	FPR using *K*-mean	FPR using hierarchical clustering
3000	0.008	0.0064
5000	0.0052	0.0041
8000	0.0055	0.0041
10000	0.008	0.0021
15000	0.0052	0.0017
20000	0.0057	0.0017

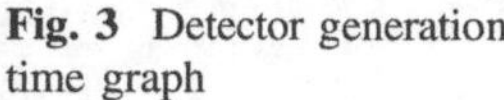
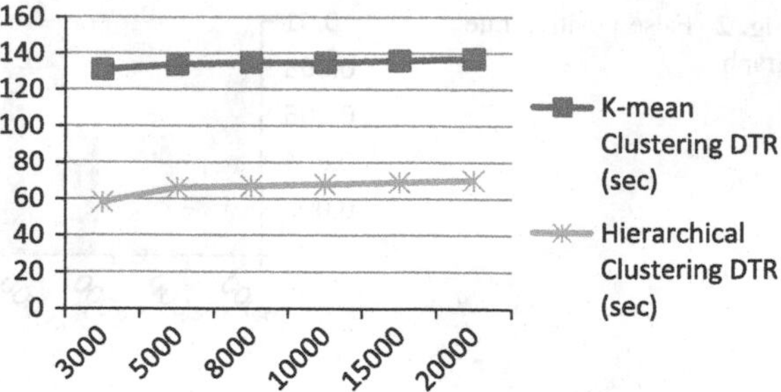

Fig. 3 Detector generation time graph

Table 2 Detector generation time

Dataset size	FPR using *K*-mean (s)	FPR using hierarchical clustering (s)
3000	131	58
5000	134	66
8000	135	67
10000	135	68
15000	136	69
20000	134	70

5 Conclusion

The anomaly-based network intrusion detection system is implemented using metaheuristic method, clustering techniques and Genetic algorithm. The two algorithms *K*-mean clustering and hierarchical clustering are used to check the performance for detection accuracy and false positive rate. The proposed approach use hierarchical clustering to reduce false positive rate. The clustering techniques are used to divide training dataset to reduce time and processing complexity. The metaheuristic method with evolutionary algorithm, i.e., genetic algorithm plays an important role to select multiple initial start points, to generate number of detectors, to calculate the radius limit of hypersphere detector and to remove redundant detectors to give final output, i.e., anomaly or normal. The experimental results are calculated using NSL-KDD dataset. Using hierarchical clustering we have obtained false positive rate and detector generation time 0.0017 and 69 s, respectively, for training dataset of size 15000. The benefit of hierarchical clustering is that it gives minimum false positive rate and detector generation time as compared to *k*-mean clustering. In future, the results will be calculated on other dataset like online to reduce false positive rate and detector generation time.

References

1. Morteza Amini, Rasool Jalili, Hamid Reza Shahriari. RT-UNNID: A practical solution to real-time network-based intrusion detection using unsupervised neural networks. Computers & security 25 (2006) 459–468.
2. Bishop M. Computer security, art and science. Addison-Wesley; 2003
3. James Brentano, Steven R Snapp et al. Architecture for Distributed Intrusion Detection. Division of computer science, University of California, 1991.
4. J.P. Anderson. Computer security threat monitoring and surveillance.Technical Report, James P. Anderson Co., Fort Washington, PA, April 1980
5. Tamer F. Ghanem, Wail S. Elkilani, Hatem. A hybrid approach for efficient anomaly detection using metaheuristic methods. Journal of advanced research, volume 6,issue 4 (2014) 609–619.
6. Osman, I.H., and Laporte, G. Metaheuristics bibliography. *Ann. Oper. Res. 63*, 513–623, 1996.
7. Blum, C., and Andrea R. Metaheuristics in Combinatorial Optimization: Overview and Conceptual Comparison. *ACM Computing Surveys, 35(3),* 268–308, 2003.
8. Xu X. Sequential anomaly detection based on temporal difference learning: principles models and case studies. Applied Soft Computing 2010.
9. Kartit A, Saidi A, Bezzazi F, El Marraki M, Radi A. A new approach to intrusion detection system. JATIT 2012.
10. Garcia-Teodoro P, Diaz-Verdejo J, Macia -Fernandez G, Vazquez E. Anomaly-based network intrusion detection: techniques, systems and challenges. Computer Security, volume 24, Issue 1–2, (2009) 18–28.
11. Forrest S, Perelson AS, Allen L, Cherukuri R. Self- NonSelf discrimination in a computer. In: Proceedings of the 1994 IEEE symposium on security and privacy; Oakland, USA: IEEE Computer Society; 1994.
12. Anna Sperotto, Michel Mandjes, RaminSadre, Pieter-Tjerk de Boer, and AikoPras. Autonomic Parameter Tuning of Anomaly-Based IDSs: an SSH Case Study. IEEE Transactions On Network And Service Management, Vol. 9, No. 2, June 2012.
13. Alexander G. Tartakovsky, Senior Member, IEEE, Aleksey S. Polunchenko, and Grigory Sokolov. Efficient Computer Network Anomaly Detection by Changepoint Detection Methods. IEEE Journal Of Selected Topics In Signal Processing, Vol. 7, No. 1, February 2013.
14. The NSL-KDD dataset. The available World Wide Web is http://nsl.cs.unb.ca/NSL-KDD/
15. S. C. Johnson (1967). Hierarchical Clustering Schemes. *Psychometrika*, 2:241–254
16. Chapter 17, Hierarchical Clustering, DRAFT!© April 1, 2009 Cambridge University Press
17. Tamer F.Ghanem,Wail S. Elkilani, Hatem. A hybrid approach for efficient anomaly detection using metaheuristic methods. Journal of advanced research, 2014.

Semantic-Based Service Recommendation Method on MapReduce Using User-Generated Feedback

Ruchita Tatiya and Archana Vaidya

Abstract Service recommender systems provide same recommendations to different users based on ratings and rankings only, without considering the preference of an individual user. These ratings are based on the single criteria of a service ignoring its multiple aspects. Big data also affects these recommender systems with issues like scalability and inefficiency. Proposed system enhances existing recommendations systems and generates recommendations based on the categorical preferences of the present user by matching them with the feedback/comments of the past users. System semantically analyzes the users feedback and distinguishes it into positive and negative preferences to eliminate the unnecessary reviews of the users which boosts the system accuracy. Approximate and exact similarity between the preferences of present and past users is computed and thus the recommendations are generated using SBSR algorithm. To improve the performance, i.e., scalability and efficiency in big data environment, SBSR is ported on distributed computing platform, Hadoop.

Keywords Service recommender systems · Big data · Semantic analysis · Jaccard co-efficient · Cosine similarity · Hadoop · MapReduce

1 Introduction

The overabundance of data on the web drives away the focus of users, landing them to surf for the data that they were not searching for initially. Information filtering systems are used to overcome these problems and to eliminate the unnecessary information before presenting it to the user. The subclass of these systems, called as

Ruchita Tatiya (✉) · Archana Vaidya
Gokhale Education Society's R. H. Sapat College of Engineering, Management Studies and Research, P T A Kulkarni Vidyanagar, Nashik, Maharashtra, India
e-mail: ruchitatatiya@gmail.com

Archana Vaidya
e-mail: archana.s.vaidya@gmail.com

© Springer Science+Business Media Singapore 2017
S.C. Satapathy et al. (eds.), *Proceedings of the International Conference on Data Engineering and Communication Technology*, Advances in Intelligent Systems and Computing 468, DOI 10.1007/978-981-10-1675-2_15

recommendation systems assist by predicting the services or items that the user would like. Service recommender systems [1] provide appropriate recommendations and have become popular in variety of practical applications like recommending the users about hotels, books, movies, music, travel, etc. [2, 3]. The enlarged number of Internet users is contributing to immense amount of data everyday [4]. Such immense data, known as Big data, is not only difficult to capture and store but also managing, processing, and analysing such data with the available current technology within the tolerable speed and time is a difficult task.

1.1 Motivation

The service recommender systems present the same ratings and rankings of the services to the different users and also provide the same recommendations to them without considering the user's personal likings and taste [1]. Also many recommendation systems provides single-criteria ratings i.e. just the overall rating of any service is being considered which makes them less accurate [5]. Due to the ever increasing amount of data, the Big-data management poses a heavy impact on service recommender systems with issues like scalability and inefficiency. The proposed system, Semantic-Based Service Recommendation (SBSR), considers the issues and drawbacks of the existing system and contributes to generate recommendations more accurate according to the user likings and the categorical preferences by considering the multiple aspects of the service and also tries to improve the efficiency and performance of the system in the big data environment.

2 Literature Survey

There are various recommendation methods based on the information or knowledge source they use for making the apt recommendations. Reference [6] describes various methods to generate recommendations and also focuses on the algorithmic methods like memory-based and model-based algorithms. Author Hiralall [7] has compared various methodologies which can be used to generate recommendations. Different pros and cons are also stated which helps the user to select the apt approach according to his/her application. Adomavicius and Tuzhilin [5], has described the current generation of the recommendation methods and have stated that they are based on rating and rankings only, without considering the taste and choices on an individual and just provide the recommendation based on the single criteria ratings. To overcome the drawback of single-criteria ratings, authors Adomavicius and Kwon [8] incorporated and leveraged multi-criteria rating which improved the accuracy of the system as compared with single-rating recommendations. The problem faced by many recommendation algorithms is its scalability, i.e. when the volume of the dataset is very large, the computation cost would be very high.

The development of cloud computing software tools such as Apache Hadoop, MapReduce, and Mahout, made possible to design and implement scalable recommender systems in Big data environment. Reference [9] implemented the collaborative filtering algorithm on the cloud computing platform, Hadoop which solves the scalability problem for large scale data by dividing the dataset. Meng et al. [1] proposed a keyword aware service recommendation method, which utilizes the reviews of previous users to get both, user preferences and the quality of multiple criteria of candidate services, and computes similarity with the preferences of active user which in turn makes the recommendations more accurate. Moreover, they implemented their approach on MapReduce which showed favorable scalability and efficiency. Turney [10] presented an algorithm for classifying the reviews as recommended (thumbs up) or not recommended (thumbs down). The classification of a review is predicted by the average semantic orientation of the phrases in the review that contain adjectives or adverbs. The algorithm presented has three steps: extract phrases containing adjectives or adverbs, estimate the semantic orientation of each phrase, and classify the review based on the average semantic orientation of the phrases.

3 Proposed System

3.1 Architecture

The proposed recommender system is specially designed for the large scale data processing. While recommending particular service, the system mainly considers the user preferences and uses the previous users' comments/reviews which accounts to the immense data on the web. In the following, Fig. 1 shows the architecture of the proposed system SBSR, which is specifically the information filtering architecture that uses the distributed computing platform to reduce the processing time. Here, the system needs to filter the previous users' comments according to the active user preferences and semantically analyze them for removing the negative reviews, to present a personalized service recommendation list. Hence, the system manages to deal with large scale data with the help of Hadoop (a distributed computing platform using the MapReduce parallel processing paradigm for big data). The processing of data can be distributed across various nodes by splitting the input into multiple Map() and Reduce() phases and the response time of the system can be decreased. To test the working of the system, test dataset regarding hotels is used that helps us to analyze the throughput of the system. Later a more generalized form of this system can be developed using precision of experiments.

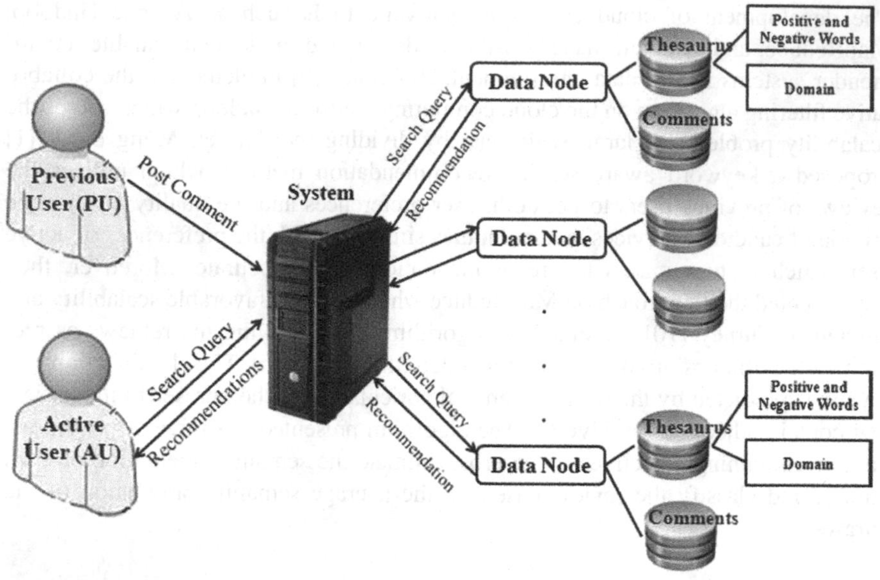

Fig. 1 System architecture

3.2 System Flow

Before mentioning the system flow, following are the descriptions of terminologies used.

1. Aspect keyword list (AKL): It is a keyword set related to the users' preferences searching for a particular aspect and also multiple criteria regarding that service are mentioned into it. For example, if the service is recommending the hotels then the aspect keyword list will contain all the main aspect keywords regarding the hotels like cleanliness, food, value, location, etc. [1].
2. Thesaurus: A thesaurus is the group of words collected according to their similarity of the meaning. Basically a domain thesaurus is associated with the aspect included in the AKL then all the related words of food like breakfast, tea, lunch, etc., are included in the thesaurus. Also the positive and negative words thesaurus consists of all the positive and negative words, phrases which are used in common natural language that are used for the semantic analysis.
3. Preference Weight Vector: The preference keyword sets of the active and previous users will be transformed into *n*-dimensional weight vectors, respectively, denoted as $W = [w1; w2;\ldots; wn]$ where '*n*' is the number of keywords and '*wi*' is the weight of the keyword *Ki* in the AKL [1].

Figure 2 depicts the flow of the system diagrammatically and is explained below. The proposed system, SBSR, is divided into two processes as swing application and web application, respectively. The swing application mainly deals

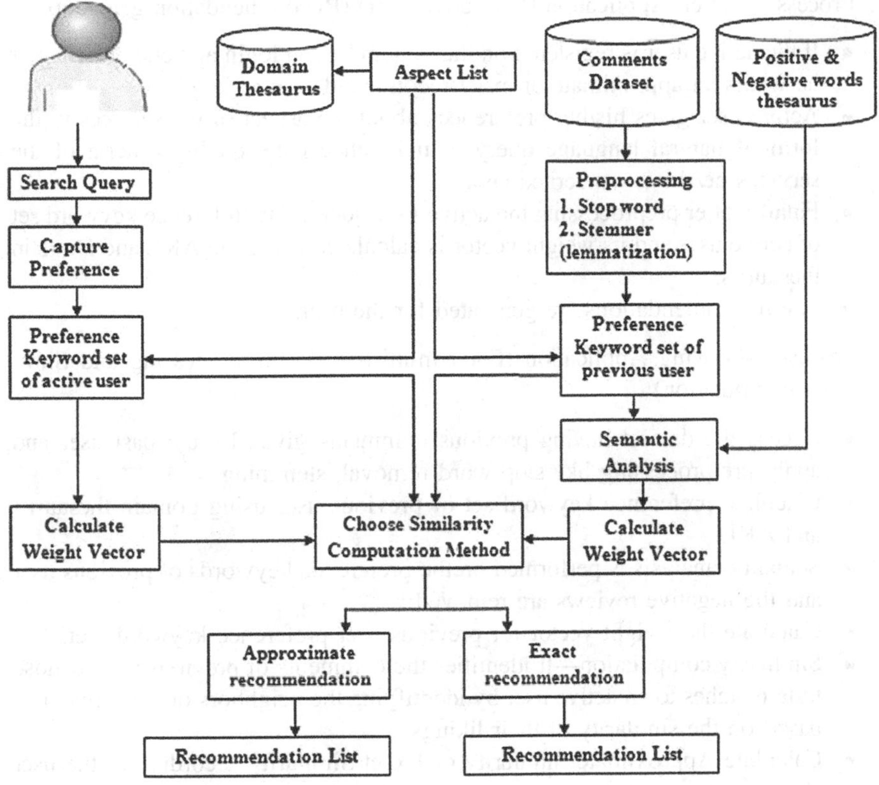

Fig. 2 System flow

with preprocessing the dataset which consists of the previous user comments. The administrator handles this swing application in which, the raw comments are pre-processed by applying the stop word removal and stemming algorithm and stored in the database. Then semantic analysis is performed on these processed comments and the positive and negative keywords in the comments are identified and categorized with respect to the aspects presented in the AKL. Later, the reviews for a particular hotel are amalgamated and the weight vector is calculated for every aspect and cached into the database. As the dataset is large and this processing is vast, the system is ported on hadoop which reduces the processing time. The web application is the recommendation generation system for the active users who receive the recommendations according to the personal likings and taste. In this, the active user can register and login to the system and choose approximate and exact recommendations of hotels according to his desires. The active user can search for hotels by using a natural language search query specifying his requirements. Following is the system flow which is divided into two parallel executable processes.

1. Process 1—Web Application (For active user) (Recommendation generation)
    - If the active user is registered on the system he can login and choose whether he wants an approximate or exact recommendation of hotels.
    - Active user gives his/her preferences about the aspect of the services in the form of natural language query, which reflects the quality criteria of the services he/she is concerned about.
    - Finally, after preprocessing the active user query, the preference keyword set of active user and its weight vector is calculated using the AKL and Domain thesaurus.
    - The recommendations are generated for the user.
2. Process 2—Swing Application (For administrator) (Pre-processing and Similarity computation)
    - Access the dataset having previous comments given by the past user and apply pre-processing like stop word removal, stemming.
    - Calculate preference keyword set of previous user using domain thesaurus and AKL.
    - Semantic analysis is performed on the preference keywords of previous user and the negative reviews are removed.
    - Calculate the weight vector for previous user preference keywords set.
    - Similarity computation—It identifies the comments of previous users whose taste matches to an active user by identifying the neighbors of the active user based on the similarity of their likings.
    - Calculate Approximate Similarity or Exact Similarity according to the user choice.

4 Implementation Details

4.1 Environment

The proposed system is designed for open source operating system Linux—Ubuntu 14.04. The implementation of this system is based on Java jdk-7 and Hadoop 2.3 platform using the MapReduce framework. MySQL 5.5.41 database is used for storing the datasets by configuring the LAMP server in Ubuntu. Also the configuration of phpMyAdmin in Ubuntu helps to perform various tasks such as creating, modifying or deleting databases with the use of a web browser. Eclipse (Luna) environment is being used for the system development. Initially for the testing purpose a single node Hadoop framework is being established. Also the Hadoop is configured with Eclipse to execute the hadoop programs in Eclipse environment.

4.2 Dataset

For the previous user comments or reviews regarding hotels, entity-ranking-dataset [11] is being used which is in the text format and contains: Full reviews of hotels in 10 different cities and there are about 80–700 hotels in each city which accounts to ~259,000 total number of reviews. The review format is: Date1 <tab> Full review1. For creating Domain Thesaurus related to aspect keyword list, the use of FeatureWords is done, downloaded from the Tripadvisor (http://www.tripadvisor.com) site and was in the text format having the following form of: #cat = <category or aspect>. For semantic analysis of comments there is a need of positive and negative words. It has been downloaded from [12]. These lists of words were downloaded in the .xls format.

4.2.1 Conversion of Raw Dataset

The datasets used for the system are in the raw format and immense in nature which therefore requires huge processing to convert it in the usable format. The text and .xls files of domain thesaurus and positive/negative words were converted into .csv format which were then imported into the MySQL database for further processing.

5 Results

5.1 Pre-processing and Semantic Analysis of Previous User Comments

As the comments dataset is huge in size and in the raw format, the pre-processing of it is done on the Hadoop platform. The mapper() class of the hadoop is responsible for applying the stemming and stop word removal algorithms on the comments dataset & the intermediate pre-processed data (comment id, file name, date, original review, stemmed review, stop-word removed review, country and city) is stored in the database. Then the semantic analysis of the pre-processed comments is done with the help of positive and negative thesaurus. The domain of the preference word is found and the result is stored in the database (total domains occurring in the comment, negative words, domain related to negative words and domains related to positive words). After applying the semantic analysis phase the reducer() class of hadoop is responsible for the calculation of weight vector related to every aspect for a particular hotel, by considering all the reviews related to that one hotel. The positive & negative count for every aspect of that hotel is cached into the database so that it can be used while generating recommendation. If the positive count is high then it means that, the hotel is good for that particular aspect and vice versa. The time required for pre-processing the data on hadoop is also noted. Likewise data is pre-processed & semantically analyzed for all the hotels in each city. After training raw data country, by country, completes the processing stage and the processed data can then be used by the web application while generating the recommendations.

5.2 Recommendations Generation

Using the web application, the active user can register into the system for approximate or exact recommendation generation. The active user queries the system regarding the hotels in natural language format, which generates the recommendation list of hotels for them according to their query and wish. A sample query fired for both approximate and exact recommendations was "Food should be tasty. Wifi should be there." (Aspects mentioned in the query are food and business service) and the results generated are shown in the figures below. Figure 3 shows the approximate recommendations for the active user query using the Jaccard co-efficient. Even if any of the aspect mentioned in active user query is matched, the hotel is included in the approximate recommendation list. Figure 4 shows the exact recommendations using the cosine similarity function. If all the aspects mentioned in the active user query are matched then only that hotel is included in the recommendation list.

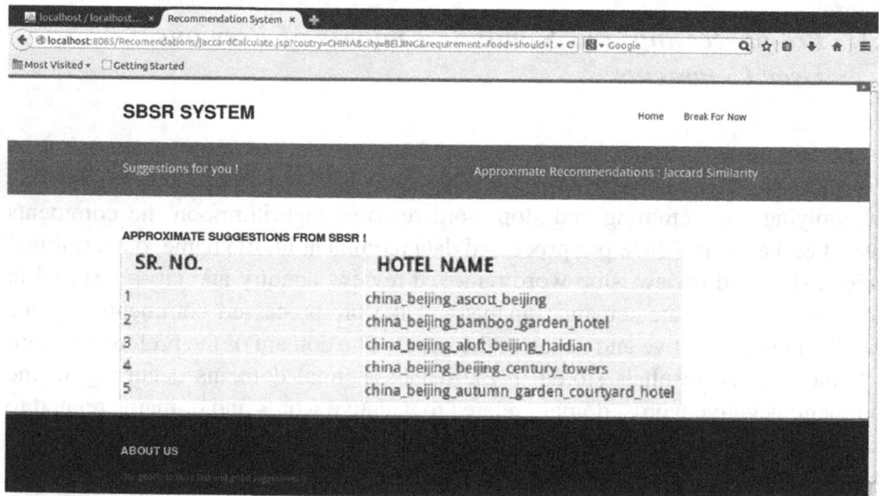

Fig. 3 Approximate recommendation

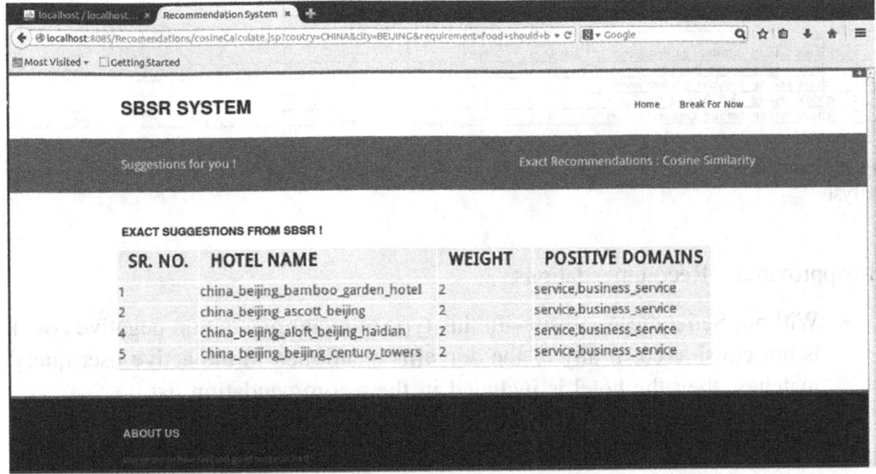

Fig. 4 Exact recommendation

6 Performance Evaluation

6.1 Comparison of the Recommendation System with and Without Semantic Analysis

For this comparison purpose, five hotels in the Beijing city and for each hotel 4 different aspects like rooms, service, location and business service were considered. Fig. 5 shows the weight vector table generated after processing the comments for these 5 hotels. For testing purpose the active user query was, "Rooms should be clean. Food should be tasty. Location should be pleasant and wifi should be there always." After pre-processing this query the domains recognized were rooms, service, location and business-service. Based on this query and above weight vectors, the comparison results for recommendation generation with and without semantic analysis were noted as shown in Fig. 6. From the above drawn results following is the conclusion and description about how the recommendations vary from each other.

Sr. No.	Hotel Name	Categories/Aspects							
		Rooms		Services		Location		Business Service	
		Pcount	Ncount	Pcount	Ncount	Pcount	Ncount	Pcount	Ncount
1	china_beijing_autumn_garden_courtyard_hotel	3	5	1	2	0	3	3	4
2	china_beijing_ascott_beijing	28	15	9	11	26	10	15	6
3	china_beijing_bamboo_garden_hotel	42	20	27	10	42	23	22	3
4	china_beijing_aloft_beijing_haidian	6	1	0	0	2	3	4	0
5	china_beijing_beijing_century_towers	2	1	2	1	2	0	0	0

Fig. 5 Weight vector table for testing

Sr. No.	Hotel Name	Approximate Recommendation		Exact Recommendation	
		Without Semantic Analysis	With Semantic Analysis	Without Semantic Analysis	With Semantic Analysis
1	china_beijing_autumn_garden_courtyard_hotel	Yes	No	Yes	No
2	china_beijing_ascott_beijing	Yes	Yes	Yes	No
3	china_beijing_bamboo_garden_hotel	Yes	Yes	Yes	Yes
4	china_beijing_aloft_beijing_haidian	Yes	Yes	No	No
5	china_beijing_beijing_century_towers	Yes	Yes	No	No

Fig. 6 Comparison of approximate and exact recommendation with and without using semantic analysis

1. Approximate Recommendations
   - Without Semantic Analysis—In this type of recommendation negative count is not considered. If any of the domains mentioned in the active user query matches, then the hotel is included in the recommendation list.
   - With Semantic Analysis—Negative count about the hotel services is considered. If any of the domains mentioned in the active user query matches, then the hotel is included in the recommendation list. But if the positive count of that hotel in respective domain is less then that hotel is not included.
2. Exact Recommendations
   - Without Semantic Analysis—In this type of recommendation negative count is not considered. The hotel is included in the recommendation list only if all the domains mentioned in the active user query are talked about in a particular hotel.
   - With Semantic Analysis—Negative count about the hotel services is considered. If all of the domains mentioned in the active user query matches and if the positive count of all the domains is greater, then only the hotel is included in the recommendation list. But if the positive count of that hotel in respective domain is less then that hotel is not included in the recommendation list.

6.2 Comparison of Pre-processing Time with and Without Using Hadoop

To evaluate the system functioning, processing was carried out on the experimental dataset to test the working of the system on hadoop platform and without using it. The testing was carried out for five cities. Each city consisted of 5 hotels and multiple comments inside it. The processing of comments was done using hadoop platform and also without using hadoop platform and time required to complete the processing was noted down and accordingly the graph was plotted measuring the time in seconds on Y-axis and number of comments on X-axis as shown in Fig. 7. Also the speed-up of the system is calculated using the same results of processing time and it was concluded that if the processing is done on the Hadoop platform, the

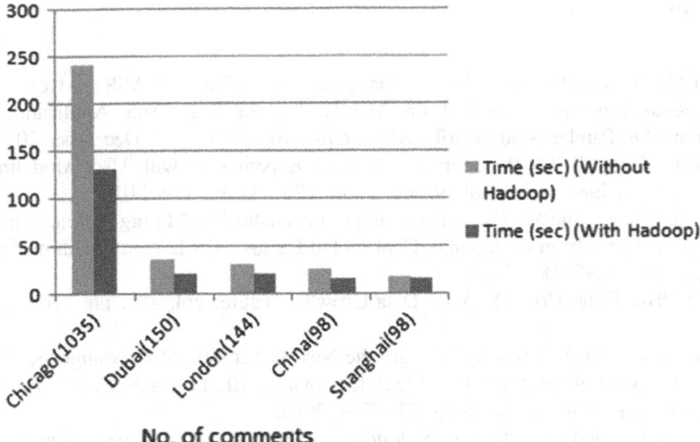

Fig. 7 Processing time required with and without using Hadoop

data processing is faster. The average speed-up of the system that was observed is 34 %. Also it is noticed that if the data size is larger, then the percentage speed-up was more, marking a favorable difference between processing on Hadoop platform and without it. The task of recommendation can also be divided among multiple nodes, which can decrease the processing and response time of the system and hence the efficiency of the system would increase. Also as the implementation of this system is on Hadoop platform which distributes its task across many map() and reduce() phases, the scalability of the system increases. This can lead to an increase in the overall performance of the system.

7 Conclusion

The SBSR system deals with generating the recommendations according to the personalized likings and taste of the users and by considering the multiple aspects of the service. The incorporation of semantic analysis of the previous user comments distinguishes the positive and negative preferences and avoids the negative comments to increase the accuracy of the recommendations. A comparative study is done to mark the difference between approximate and exact recommendation generation strategies with and without using semantic analysis. Thus, the results depict that the recommendations generated using semantic analysis are more accurate than without using it. As this accounts a large dataset, it is affected by the factors like scalability and inefficiency which is improved by 34 % by implementing the system in distributed platform known as Hadoop which uses MapReduce framework and can manage large amount of data in these service recommendation systems. The SBSR system shows a good accuracy, efficiency, and scalability when compared to other systems.

References

1. Shunmei Meng, Wanchun Dou, Xuyun Zhang and Jinjun Chen, "KASR: A Keyword Aware Service Recommendation Method on MapReduce for Big Data Applications", IEEE Transactions On Parallel And Distributed Systems, vol. 25, no. 12, December 2014.
2. M. Bjelica, "Towards TV Recommender System Experiments with User Modeling", IEEE Trans. Consumer Electronics, vol. 56, no. 3, pp. 1763–1769, Aug. 2010.
3. Y. Chen, A. Cheng, and W. Hsu, "Travel Recommendation by Mining People Attributes and Travel Group Types from Community Contributed Photos", IEEE Trans. Multimedia, vol. 25, no. 6, pp. 1283–1295, Oct. 2013.
4. C. Lynch, "Big Data: How Do Your Data Grow?", Nature, vol. 455, no. 7209, pp. 28–29, 2008.
5. G. Adomavicius and A. Tuzhilin, "Toward the Next Generation of Recommender Systems A Survey of the State of the Art & Possible Extensions", IEEE Transactions on Knowledge and Data Engineering, Vol. 17, No. 6 pp. 734–749, 2005.
6. Ruchita V. Tatiya and Prof. Archana S. Vaidya, "A Survey of Recommendation Algorithms", International Organization of Scientific Research Journal of Computer Engineering (IOSR-JCE), Volume 16, Issue 6, Ver. V, PP 16–19, Nov–Dec. 2014.
7. ManishaHiralall, "Recommender systems for e-shops", VrijeUniversiteit, 2011.
8. G. Adomavicius and Y. Kwon, "New Recommendation Techniques for Multicriteria Rating Systems", IEEE Intelligent Systems, vol. 22, no. 3, pp. 48–55, May/June 2007.
9. Z. D. Zhao, and M. S. Shang, "User-Based Collaborative-Filtering Recommendation Algorithms on Hadoop", In the third International Workshop on Knowledge Discovery and Data Mining, pp. 478–481, 2010.
10. Peter Turney, "Thumbs Up or Thumbs Down? Semantic Orientation Applied to Unsupervised Classification of Reviews".
11. Kavita Ganesan and Cheng Xiang Zhai, "Opinion-Based Entity Ranking", Information Retrieval, 2011. Comments dataset: http://www.kavita-ganesan.com/entity-ranking-data
12. For semantic analysis: http://mpqa.cs.pitt.edu/lexicons/effectlexicon/

Implementation of DSTC and PSO Algorithm Using CLD to Improve the Performance of MANET

Sheetal Bhale, Abhilasha Mishra and Mazher Khan

Abstract Throughput is a valuable parameter for mobile ad hoc network (MANET), it is defined as a number of successfully received packets in a unit interval of time. In wireless communication for the extension of coverage as well as capacity MANET is used. In this paper cross-layer is designed between physical and network layer using received signal strength (RSS) as a cross-layer interaction parameter. The given technique proposes strong route formation in ad hoc on-demand distance vector (AODV) through cross-layer design (CLD), distributed space-time coding (DSTC), and particle swarm optimization (PSO) in MANET. The simulation results show the improvement in performance of MANET.

Keywords Mobile ad hoc network (MANET) · Ad hoc on-demand distance vector (AODV) · Cross layer design (CLD) · Distributed space-time coding (DSTC) · Particle swarm optimization (PSO)

1 Introduction

As wireless LAN has flexible as well as simple architectures, it has been utilized in many applications of wireless networks. However, the technical growth in wireless networks relied on infrastructures like an access point and a router. So, many researchers proposed ad hoc networks with no need of infrastructures and still the research is continued. Ad hoc networks are made up of mobile nodes with router functions as well as wireless media (Figs. 1 and 2).

Sheetal Bhale (✉) · Abhilasha Mishra · Mazher Khan
G. S. Mandal's Marathwada Institute of Technology, Aurangabad, Maharashtra, India
e-mail: sheetalbhale@ymail.com

Abhilasha Mishra
e-mail: abbhilasha@gmail.com

Mazher Khan
e-mail: mazher.engg@gmail.com

© Springer Science+Business Media Singapore 2017

S.C. Satapathy et al. (eds.), *Proceedings of the International Conference on Data Engineering and Communication Technology*, Advances in Intelligent Systems and Computing 468, DOI 10.1007/978-981-10-1675-2_16

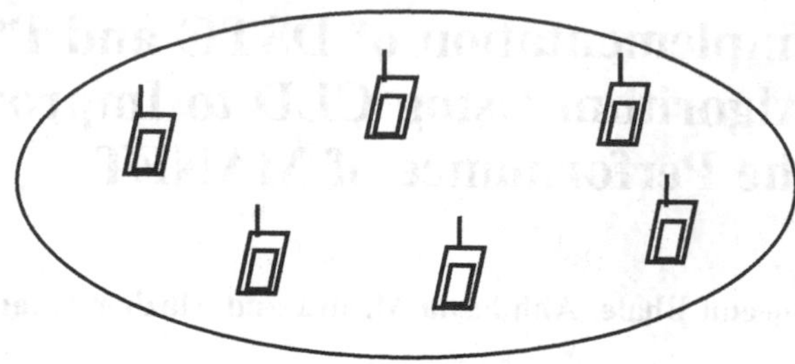

Fig. 1 Ad hoc network [3]

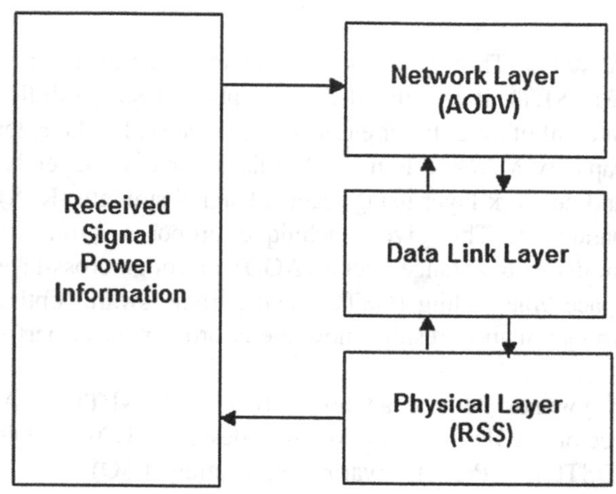

Fig. 2 Cross-layer design for information sharing

2 Theory

2.1 MANET

Wireless mobile ad hoc networks include a collection of mobile nodes which forms a network to communicate with each other without the help from stationary infrastructure like access points. So as to forward packets nodes communicate directly through wireless links when in radio range but nodes which are not in each other's radio range utilize other nodes as intermediate routers via multiple-hop routing. The nodes that communicate directly are said to be neighboring nodes. Furthermore, due to the movement of nodes, the network topology changes rapidly. Therefore, an efficient routing protocol is needed in order to do better communication between the nodes in ad hoc network [1]. Numerous advantages coming out of wireless technology are not the physical setup for data transfer. They are lower installation and maintenance costs [2]. Mobile ad hoc network has many real life applications such as in business meetings, outside the offices, in Bluetooth, Wi-Fi Protocols, etc. [3].

2.2 AODV

This topic tells about AODV, which is the routing protocol of on-demand types under study by MANET. Here, every node consists of the routing table, and the newly produced routes use sequence number with every routing information. At the reception of control packet which occurs in on-demand by each node the routing table updation is done dependent upon the sequence number or the number of hops. Route discovery phase establishes the route while route maintenance phase maintains the route (Figs. 3 and 4).

AODV is based on two steps

1. Route Discovery: It has RREQ (Route Request) and RREP (Route Reply)
2. Route Maintenance: It has RERR (Route Error) and Hello message [4].

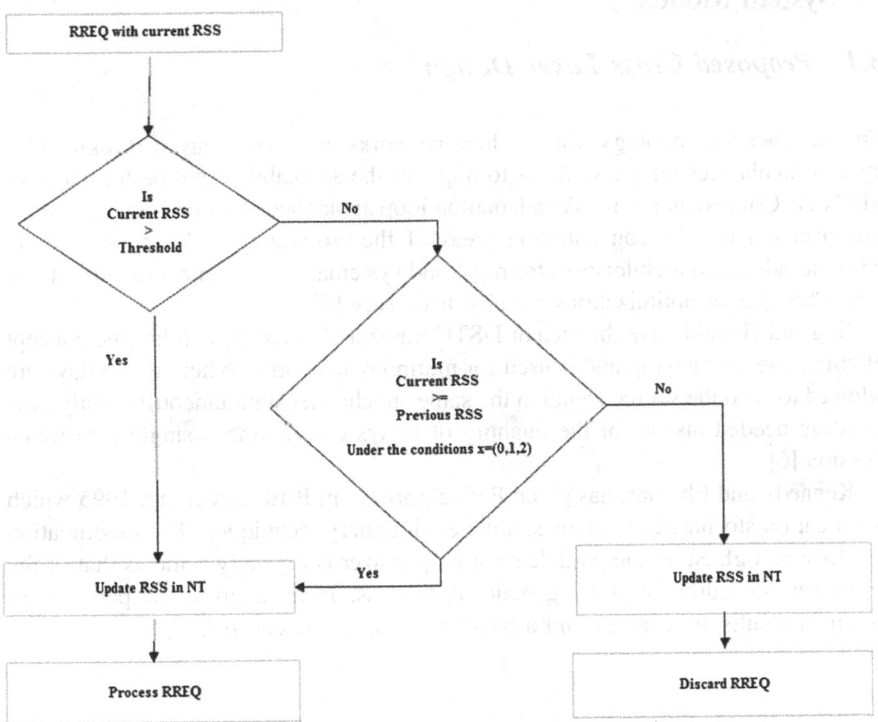

Fig. 3 Flow diagram of the cross-layer protocol with different values of x

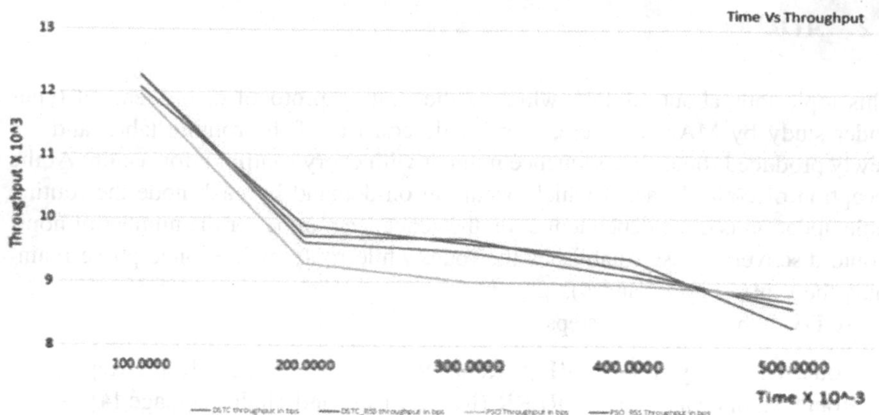

Fig. 4 Time versus throughput

3 System Modeling

3.1 *Proposed Cross Layer Design*

The advanced technology for ad hoc networks is a cross-layer design. This approach enhances the possibilities to improve the adaptability and performance of MANET. Coincident points like adaptation loop avoidance as well as protocol stack improvement are the concentrating areas of the cross-layer architectures. Along with this advanced architectures for rapid and systematic deployment of current and new Cross Layer optimizations are also necessary [5].

Jing and Hassibi gave the idea of DSTC transmission strategy. It has the concept of simultaneous relaying and is used for mutirelay networks. Where total relays are allowed to send the source signal in the same subchannel simultaneously. Only time slots are needed instead of the quantity of relays considering a single data transmission [6].

Kennedy and Eberhart has given PSO algorithm in IEEE Evocomp, 1995 which is reliant on stochastic techniques, alike evolutionary techniques. The modification of pBest and gBest by the particle swarm optimizer is logically same as that of the crossover procedure used by genetic algorithms. Here, a group of particles as potential results are used to find a better solution at the end [6].

3.2 *Proposed System Flow Chart*

Proposed system for improving performance of MANET based on AODV protocol by using RSS as an interaction parameter for implementing CLD. Strong route will be selected in AODV as RSS values are being used and updated in NT (Neighbor

Table) [7]. This will result in improving throughput of system. This can be understood by following flow chart under different values of x, as below

1. $x = 0$ (Using only RSS values),
2. $x = 1$ (Using RSS values with DSTC),
3. $x = 2$ (Using RSS values with PSO)

Now to improve route formation in AODV, DSTC and PSO algorithms are implemented.

3.3 Steps of DSTC

1. DSTC technique has two stages which utilizes a concept of listen and transmit protocol.
2. Relays carry the vector of symbols transmitted by the source node, which is the first stage of transmission.
3. Second stage performs multiplication operation of the distributed relays with received vectors with a quite arbitrary matrix. Further transmission of the obtained vectors toward the destination is done.
4. In the received signal of destination, i.e., final stage of DSTC, a linear space–time codeword is generated [6].

3.4 Steps of PSO

1. A group of particles as potential solutions is taken in search space for few iterations. We have to initialize pBest and gBest where pBest represents best location having best fitness value which the particle has individually inspected since the first stage and gBest denotes the location of best fitness experienced.
2. Initialization of random position and velocity of particles is done, i.e., $x_p{}^n(t)$ and $v_p{}^n(t)$ for a particle $p(1 \leq p \leq N_{\text{Particle}})$.
3. For tth iteration $(1 \leq t \leq N_{\text{Iteration}})$ in the nth dimension of search space $(1 \leq n \leq N_{\text{Dimension}})$ the position of particle is modified as

$$x_p{}^n(t+1) = x_p{}^n(t) + v_p{}^n(t+1) \tag{1}$$

4. As per the upgradation velocity of pth particle is

$$v_p{}^n(t+1) = \omega.\, v_p{}^n(t) + c_1.\, \text{rand}_{1.}\left(\text{pBest}_p{}^n - x_p{}^n(t)\right) + c_2.\, \text{rand}_{2.}\left(\text{gBest}_p{}^n - x_p{}^n(t)\right) \tag{2}$$

Factor ω is a constant in between the scale 0 and 1 which denotes the velocity rising range towards gBest and pBest. The cognitive and social rates are scaled in respective to acceleration factors, i.e., constants c_1, c_2. Factors rand_1 and rand_2 return uniformly distributed random numerals in the range 0 and 1 [6].

4 Implementation of Proposed System

For formation of strong route in AODV following steps are implemented.

(1) RSS values are used for selecting route.
(2) DSTC algorithm is applied in AODV for selecting route.
(3) DSTC algorithm is applied in AODV for selecting route by using RSS as an interaction parameter in CLD.
(4) Power is allocated to all nodes and PSO algorithm is used to select strong route.
(5) Power is allocated to all nodes and PSO algorithm is used to select strong route by using RSS as an interaction parameter in CLD.

5 Simulation Scenario

Simulation setup can be summarized in following table:

Parameter	Meaning	Value
x * y	Size of the scenario	1000 * 1000 (m)
Nn	Number of nodes	50
Initial energy	Initial energy in Joules	100
Relay	Relay	35
Simulator		ns-2.32
Routing protocol		AODV
Traffic		Constant Bit Ratio (CBR)
Mac layer		IEEE 802.11
Antenna type		Omni antenna
Total energy consumption		25.9315

5.1 *Simulation Steps*

(i) Creation of wireless network topology for DSTC algorithm only.
(ii) Creation of wireless network topology for DSTC algorithm along with RSS values (CLD).
(iii) Creation of wireless network topology for PSO algorithm only.
(iv) Creation of wireless network topology for PSO algorithm along with RSS values(CLD).

6 Results

6.1 *Throughput Graph for RSS, DSTC and PSO*

See (Fig. 5).

6.2 *Packet Delivery Ratio Calculation for RSS, DSTC, and PSO*

Above graph shows the improvement of throughput and PDR for CLD with DSTC and CLD with PSO.

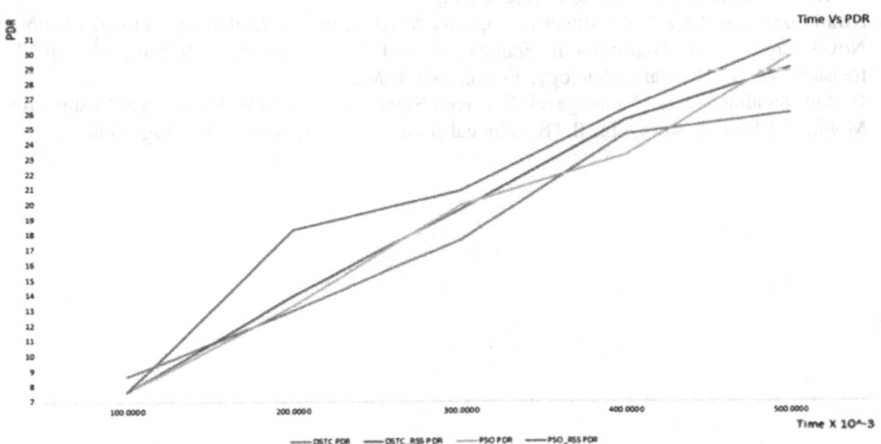

Fig. 5 Time versus packet delivery ratio

7 Conclusion

Mobile ad hoc networks serve as a promising technology to increase coverage area and capacity of wireless networks with no infrastructure. With the use of received signal strength as an interaction parameter for a cross-layer design for AODV-based MANET with the help of DSTC and PSO algorithms performance of the system improves. The proposed system for formation of best route in AODV by using RSS, DSTC, and PSO algorithm shows better results for wireless mobile network. Proposed system will provide reliable network for wireless communication. As future scope, we can implement energy-based sleep scheduling algorithm for AODV protocol and compare its performance with the proposed technique.

References

1. Manoj Kumar Singh, Brajesh Kumar, Chiranjeev Kumar, Manish Gupta: Preemptive Multipath-Adhoc On Demand Distance Vector Routing Protocol. In: MIT International Journal of Computer Science & Information Technology Vol. 1 No. 1 Jan 2011 pp 36–40
2. Waqas Ikram & Nina F. Thornhill: Wireless Communication in Process Automation: A Survey of Opportunities, Requirements, Concerns and Challenges. In: Presented at Control 2010, Coventry, UK, September 2010
3. Asma Tuteja, Rajneesh Gujral, Sunil Thalia: Comparative Performance Analysis of DSDV, AODV and DSR Routing Protocols in MANET using NS2. In: 2010 International Conference on Advances in Computer Engineering
4. Masayuki Tauchi, Tetsuo Ideguchi, Takashi Okuda: Ad-hoc Routing Protocol Avoiding Route Breaks Based on AODV. In: Proceedings of the 38th Hawaii International Conference on System Sciences -2005
5. Zhijiang Chang, Georgi Gaydadjiev, Stamatis Vassiliadis: Infrastructure for Cross-Layer Designs Interaction. In: IEEE 2007 Engineering
6. Sara Efazati and Paeiz Azmi: Effective Capacity Maximization in Multirelay Networks With a Novel Cross-Layer Transmission Framework and Power-Allocation Scheme. In: IEEE transactions on vehicular technology, Vol 63, No 4, May 2014
7. B. Ramchandran and S. Shanmugavel: Received Signal Strength-based Cross-layer Designs for Mobile Ad Hoc Networks. In: IETE technical review, Vol 25, Issue 4, Jul-Aug 2008

Progressive Review Towards Deep Learning Techniques

Poonam Chaudhari and Himanshu Agarwal

Abstract Deep learning is a thing of tomorrow which is causing a complete drift from shallow architecture to deep architecture and an estimate shows that by 2017 about 10 % of computers will be learning rather than processing. Deep learning has fast growing effects in the area of pattern recognition, computer vision, speech recognition, feature extraction, language processing, bioinformatics, and statistical classification. To make a system learn, deep learning makes use of a wide horizon of machine learning algorithms. Gene expression data is uncertain and imprecise. In this paper, we discuss supervised and unsupervised algorithms applied to gene expression dataset. There are intermediate algorithms classified as semi-supervised and self taught which also play an important role to improve the prediction accuracy in diagnosis of cancer. We discuss deep learning algorithms which provide better analysis of hidden patterns in the dataset, thus improving the prediction accuracy.

Keywords Deep learning · Supervised and unsupervised algorithm · Semi-supervised and self-taught algorithm · Gene expression data

1 Introduction

Soft Computing is a branch of Computer Science wherein we try to find out tractable solutions from data which is uncertain and imprecise. These algorithms are inspired by human brain. Dealing with data that is partially true and difficult to predict, leads to development of machine learning approaches. Research shows the involvement of evolutionary algorithms with bio-inspired algorithms, to solve NP hard problems. Soft computing includes the study of evolutionary algorithms, neural networks, genetic programming, Fuzzy sets, and chaos theory. These

Poonam Chaudhari (✉) · Himanshu Agarwal
Symbiosis Institute of Technology, Pune, India
e-mail: poonam.chaudhari@ges-coengg.org

Himanshu Agarwal
e-mail: himanshu.agarwal@sitpune.edu.in

© Springer Science+Business Media Singapore 2017

S.C. Satapathy et al. (eds.), *Proceedings of the International Conference on Data Engineering and Communication Technology*, Advances in Intelligent Systems and Computing 468, DOI 10.1007/978-981-10-1675-2_17

algorithms help in prediction and forecasting. The most important application area is biomedical.

Cancer is a ruthless disease claiming many lives. According to the survey, the percentage of cancer patients has increased by 37 % in urban population. This calls for immediate attention to the detection of patients at primary stage. Affymetrix dataset of a patient can be considered to help us predict whether the cancer is benign or malignant. Machine learning techniques, also called as predictive modeling, enable to learn from the available data and predict the output based on the given input. These techniques are superficially segregated as supervised and unsupervised learning techniques.

2 Machine Learning Algorithm

2.1 Supervised Learning

Supervised learning is a technique where the training and trained data deal with labeled instances. The model is trained using known data and known instances. This model then helps predict the class of the new instance. Supervised methods include various algorithms like support vector machine, naïve Bayesian classifier, decision tree, etc. However, all these algorithms are not suitable when we consider unknown instances of data.

2.2 Unsupervised Learning

Unsupervised learning algorithms include clustering, hidden markov model, relevance vector machine, etc. When we try to deduce data output from the induced data, we find that data is hidden in between intermediate layers. This gives rise to the need of **deep learning algorithms**. We try to go deep within the available data, to discover hidden patterns/data to enhance prediction accuracy.

2.3 Deep Belief Network

Deep belief network is an amalgamation of neural network and deep learning techniques. DBN is stacked auto-encoder over restricted Boltzmann machine. Issues like complementary priors are handled using wake and sleep algorithm.

This paper presents a progressive review through the inception of neural networks till deep learning algorithms.

3 Supervised Learning Algorithm

Classification is broadly classified as discriminative, generative, and probabilistic. Depending upon the prior conditions and evidences, the likelihood is considered which gives us the probability of the posterior (next).

Bayesian classification gives the basic probability theorem. Learning joint probability is better justified using naive Bayesian classifier (Fig. 1).

This classifier assumes that all the input features are conditionally independent of each other [1]. However, when we consider genetic data, the features are greatly dependent on each other. Linear classification requires mainly largest classification interval, i.e., maximum margin hyper plane [2].

Support vector machine is primarily designed for classification which yields two groups. Problems, which have a nonlinear boundary decision, can be solved by transforming data into other plane, using kernel function [3]. However, tuning SVM's is the critical task as selecting a specific kernel and parameters is done on a trial and error basis.

Decision trees are easy to implement as rules can be applied for decision making. Various algorithms like ID3 and C4.5 have been applied for effective prediction on decision trees. Data uncertainty arises naturally in many applications due to various reasons: measurement errors, data staleness, repeated measurements, limitations of the data collection process, etc. [4]. The gene expression information is extremely vague in nature and it is difficult to predict the boundaries between the groups. An evolutionary algorithm is used for decision tree induction that can be applied to different classification problems (a hyper heuristic approach) at generic level [5].

4 Semi-Supervised Learning Algorithm

Supervised learning deals with labeled instances. However, labeling instances requires human expertise and thus, is time consuming. Unlabeled data is relatively easy to obtain. Semi-supervised learning makes use of labeled and unlabelled data to build better classifiers. Thus, reducing human effort and giving higher accuracy [6].

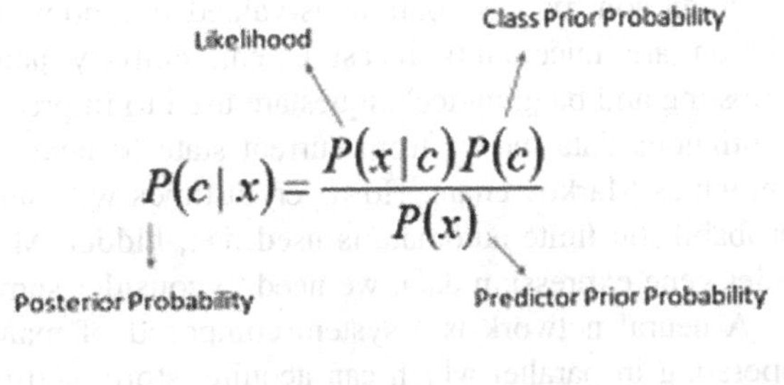

Fig. 1 Naïve Bayes theorem, calculating probability considering the likelihood and prior probability. *Source* Wikipedia

Likelihood
Class Prior Probability

$$P(c \mid x) = \frac{P(x \mid c)P(c)}{P(x)}$$

Posterior Probability
Predictor Prior Probability

$$P(c \mid X) = P(x_1 \mid c) \times P(x_2 \mid c) \times \cdots \times P(x_n \mid c) \times P(c)$$

Semi-supervised learning techniques are further bifurcated as generative and discriminative algorithms. A straightforward, generative semi-supervised method is the expectation maximization (EM) algorithm. The EM approach for naive Bayes text classification models is discussed by Nigam [6]. The author assigned weights to unlabelled data. The strengths of unlabelled data were dynamically adjusted with the help of a weighing factor. A many-to-one correspondence was calculated using multiple mixture components [7].

Rie Kubota [8] worked with a generative model and discovered effective parametric feature representations. The two-view model created a small set of features from unlabeled data. The prediction was optimized and represented as a linear combination of those features. However, data redundancy was the biggest drawback.

5 Unsupervised Learning Algorithms

The data provided by the user is unlabelled and no class values of the data instances are given. These techniques explore the data to find some intrinsic structures or hidden pattern within them.

Unsupervised learning is very commonly addressed as clustering. K-means is the radical algorithm of clustering. K-means forms clusters based on the distance function (similarity or dissimilarity). However, K-means faces the problem of an outlier [9]. Attributes of medical data are continuous with high dimension [10]. A hybrid method which combines K-means and k-mode algorithm was introduced [10].

Hierarchical clustering generates a sequence of clusters known as dendograms. Perceptron learning works on the principle that a new (unseen) input pattern that is similar to an old (seen) input pattern is likely to be classified correctly. Gallant [1] in his work came up with a pocket algorithm which made perceptron learning better behaved in spite of noisy and non separable data. The algorithm considers all existing labeled data patterns and checks their classification with a current weight vector [1]. If labels are unknown a quantity is added to the weights which are proportional to the product of the input pattern with the desired output Z (1 or −1). However, this algorithm applies to only linearly separable data and has limited functionality of hyper plane [1].

Prediction models continuous-valued functions. Major factors influencing prediction are uncertainty measurement, entropy analysis, expert judgments, etc. Boosting and bagging techniques are used to improve the prediction accuracy. This continuous data moves from current state to next state (transition). This chain is known as Markov chain. However, at times we cannot observe few states. Hence, probabilistic finite automata is used, i.e., hidden Markov model [11]. As we consider gene expression data, we need to consider summing up of exponential paths.

A neural network is a system composed of many simple processing elements operating in parallel which can acquire, store, and utilize experiential knowledge.

However, lot of information is stored at the middle layer. These intermediate calculations and data are stored within hidden units. The network then is known as artificial neural network. If data moves only in a linear fashion in one direction, the network is called as feedforward network [12]. If data moves in arbitrary paths, the network is called as recurrent neural nets.

For nonlinear learning task, a single layer feed forward network does not suffice, hence, the need of multi-layer feedforward network. Back propagation method is used to obtain a stable network. The important network parameter is how many layers to consider. Less number of hidden layers does not impart any knowledge and too many layers make it difficult to generalize the system.

6 Deep Learning

Deep learning techniques state few methods to decide the number of hidden layers according to the application domain. Auto-encoder is a new way to train multilayer neural networks. Each of the (nonoutput) layers is trained to be an auto-encoder. It is forced to learn good features that describe what comes from the previous layer. An auto-encoder is trained with an absolutely standard weight adjustment algorithm to reproduce the input. By making this happen with fewer units than the inputs, this forces the 'hidden layer' units to become good feature detector. Intermediate layers are each trained to be auto encoders. Final layer is trained to predict class based on outputs from previous layers (Fig. 2).

Instead of deterministic approach, restricted Boltzmann machine [13] uses stochastic units with particular distribution (usually Gaussian). Intuition behind RBMs is that there are some visible random variables and some hidden variables [14]. The task of training is to find out how these two sets of variables are actually linked to each other.

Convolutional neural network [14] is somewhat similar to these two, but instead of learning single global weight matrix between two layers, they aim to find a set of locally connected neurons. The idea is the same as that of autoencoders and RBMs—translate many low-level features to the compressed high-level representation—but weights are learned only from neurons that are spatially close to each other.

Deep Sigmoid Belief Network is a bunch of stacked restricted Boltzmann machines. RBM is a bunch of stacked autoencoders. DBN's at the top layer are

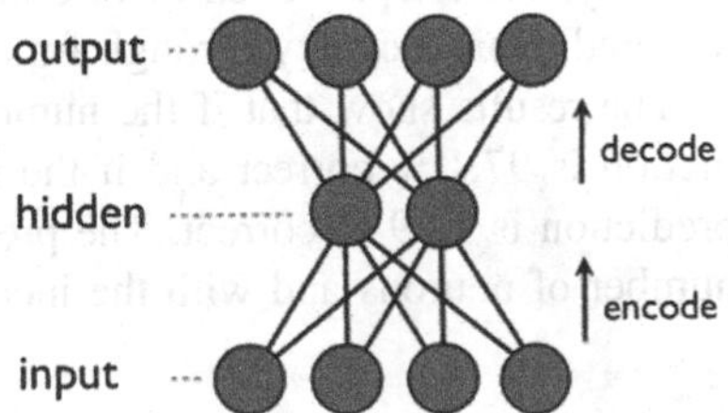

Fig. 2 Auto-encoder. *Source* Wikipedia

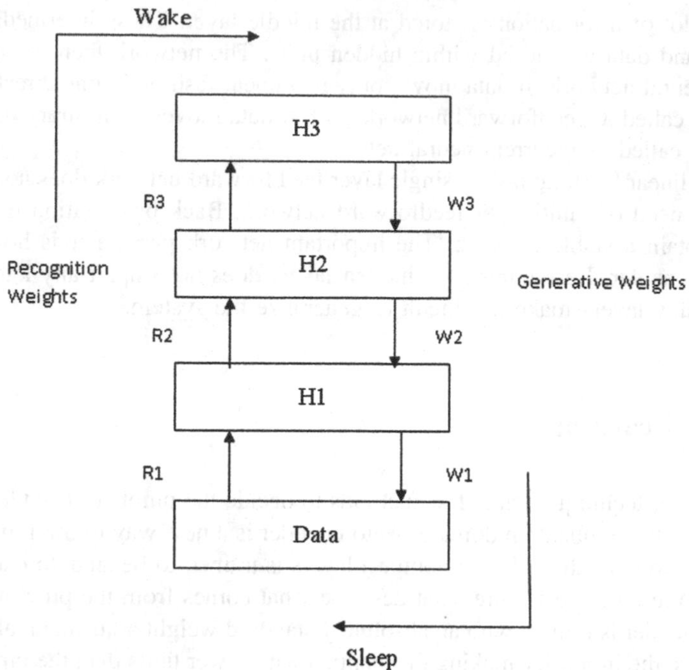

Fig. 3 Wake and Sleep algorithm

undirected with untied weights, computed to form Deep Sigmoid Belief Nets (directed) [14]. This network is fine tuned using wake–sleep algorithm. This algorithm has a recognition phase and a generative phase. It averages the weights between the intermediate nodes, to enhance better output value (Fig. 3).

7 Experiments

An experiment was performed considering the dataset for breast cancer. Data source is www.ncbi.nlm.nih.gov. Input data was preprocessed and feature selection was done. The dataset consisted of 37 parameters. Seven features were selected as input parameters, i.e., age, weight, lump size, lump position, hormone induction, heredity, and relapse. Feedforward neural network algorithm is applied to calculate the prediction accuracy through f-measure.

The results show that if the number of neurons selected is 10, the overall prediction is 97.1 % correct and if the number of neurons selected is 12 the overall prediction is 97.9 % correct. The prediction rate increases with the increase in the number of neurons and with the increase in the size of dataset (Fig. 4).

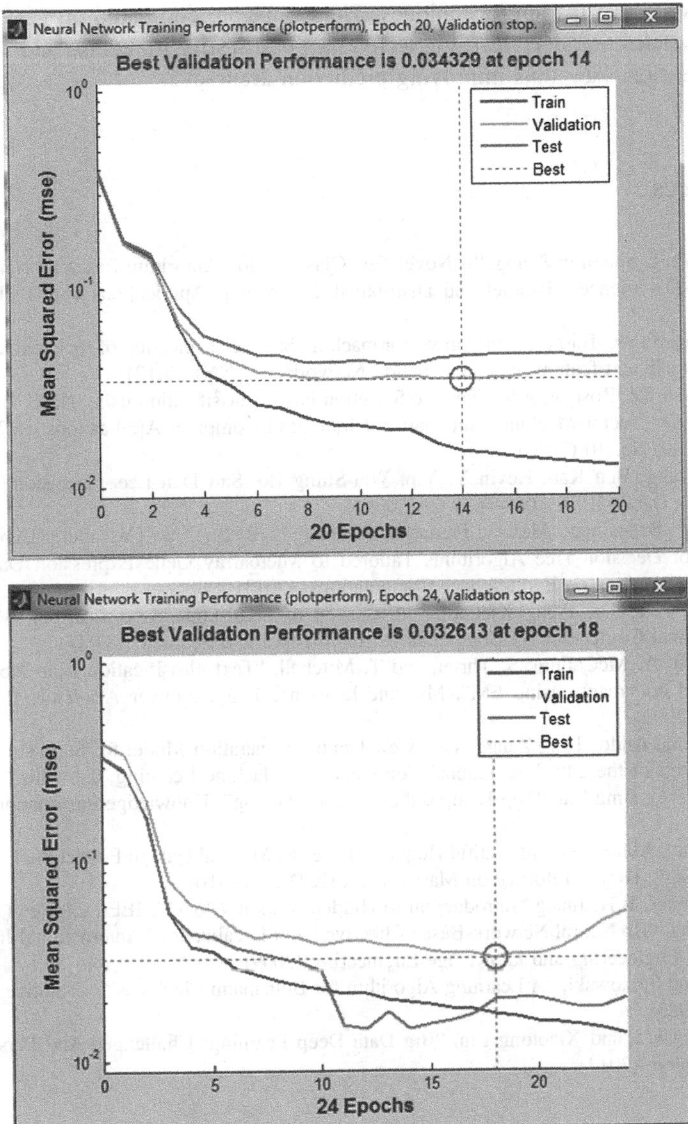

Fig. 4 Validation performances

8 Conclusions

This paper provides a detailed discussion of machine learning algorithms spanning from supervised learning techniques to semi-supervised to unsupervised learning techniques. It also throws light to the latest upcoming algorithms of deep belief networks. The study provides a smooth inclination from shallow learning techniques to deep learning techniques. The results are computed on genetic data. The paper states that deep learning techniques best justify imprecise and uncertain gene expression data, thus improving prediction accuracy.

References

1. Baoyi Wang, Shaomin Zhang, "A Novel Text Classification Algorithm Based on Naïve Bayes and KL-Divergence", Parallel and Distributed Computing, Applications and Technologies (2005)
2. Changjing Shang, Barnes, "Support vector machine-based classification of rock texture images aided by efficient feature selection" Neural Networks (IJCNN) (2012)
3. Dipali Bhosale, Roshani Ade, "Feature Selection based Classification using Naive Bayes, J48 and Support Vector Machine", International Journal of Computer Applications (0975–8887) Volume 99, No. 16 (2014)
4. Smith Tsang, Ben Kao, Kevin Y. Yip, Wai-Shing Ho, Sau Dan Lee, "Decision Trees for Uncertain Data", IEEE 1084-4627/09 (2009)
5. Márcio P. Basgalupp, Alex A. Freitas, and André C. P. L. F. de Carvalho, "Evolutionary Design of Decision-Tree Algorithms Tailored to Microarray Gene Expression Data Sets", IEEE Transactions on Evolutionary Computational (2013)
6. Nitin N. Pise, Dr. Parag Kulkarni, "A Survey of Semi-Supervised Learning Methods", International Conference on Computational Intelligence and Security (2008)
7. K. Nigam, A. McCallum, S. Thrun, and T. Mitchell, "Text classification from labeled and unlabeled documents using EM", Machine Learning, 1–34, Kluwer Academic Publishers, Boston (2000)
8. Rie Kubota, Ando, Tong Zang, "Two-view Feature Generation Model for Semi-supervised", Proceedings of the 24th International Conference on Machine Learning, Corvallis (2007)
9. Xindong Wu, Bing Liu, "Top 10 algorithms in Data Mining", Knowledge Information System (2007)
10. Razan Paul, Abu Sayed Md. Latiful Hoque "Clustering Medical Data to Predict the Likelihood of Diseases", Digital Information Management (ICDIM) (2010)
11. L.R. Rabiner, B.H. Juang "Introduction to Hidden Markov Model", IEEE explore (1996)
12. Eric Wong, "Bp Neural Network-Based Effective Fault Localization", International Journal of Software Engineering and knowledge Engineering (2009)
13. Hinton and Sejnowski, "A Learning Algorithm for Boltzmann Machines", Cognitive Science, Vol 9 (1983)
14. Xue-wen Chen and Xiaotong Lin, "Big Data Deep Learning: Challenges And Perspective" IEEE explore (2014)

Design and Simulation of Two Stage Operational Amplifier with Miller Compensation in Nano Regime

Rohini A. Sarode and Sanjay S. Chopade

Abstract The paper represents a design procedure of basic two stage CMOS operational amplifier using Miller compensation technique. The LtSpice simulation tool is used to present system result at low capacitive load with different characteristics. The Miller capacitor creates an undesirable right-half-plane (RHP) zero due a non inverting feedforward signal path is induced in the input of the second stage towards its output, which can be eliminated by using voltage buffer. My work shows the two stage amplifier with Miller compensation techniques, simulated using LtSpice simulation tool for 180, 130 and 90 nm CMOS technology process. When a 10-pF capacitive load is drive, the amplifier achieves voltage gain approximate 20 % more with exactly double gain bandwidth (GBW) which shows phase margin of 44.8°, 49.06°, and 53.70°, slew rate of 44.48, 10.29, and 9.77 V/μs, with dissipating power value of 830 μW at 2.5 V, 504.06 μW at 1.5 V, 486 μW at 1.2 V supply voltage for 180 nm, 130 nm, 90 nm CMOS technology, respectively.

Keywords CMOS · Compensation · Gain bandwidth · LtSpice · Miller compensation · Output swing · Poles · RHP · Slew rate · Two stage op amp · Voltage gain · Zero

1 Introduction

The two stage CMOS operational amplifier is mostly used due its simple structure and robustness [1, 2], it also provides high dc gain and large output voltage swing [3]. In designing of an op amp with numerous electrical characteristics [4] have to be taken into account such as UGB, SR, ICMR, etc. [1, 5–8].

R.A. Sarode (✉) · S.S. Chopade
Sandip Foundations, Sandip Institute of Technology and Research Center (SITRC), Nashik, Maharastra, India
e-mail: rohini_sarode@rediffmail.com

S.S. Chopade
e-mail: sanjay.chopade@sitrc.org

© Springer Science+Business Media Singapore 2017

S.C. Satapathy et al. (eds.), *Proceedings of the International Conference on Data Engineering and Communication Technology*, Advances in Intelligent Systems and Computing 468, DOI 10.1007/978-981-10-1675-2_18

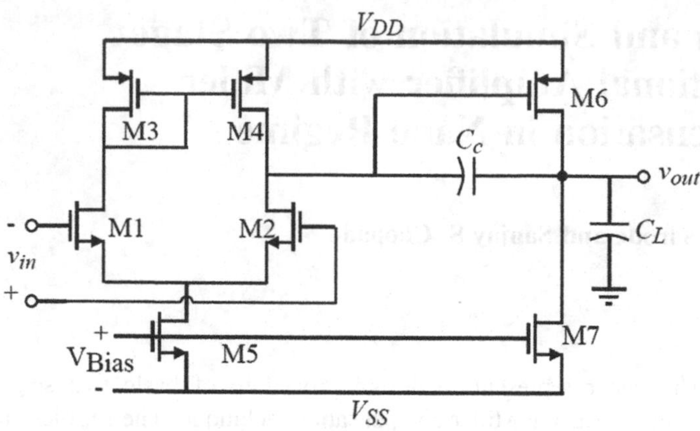

Fig. 1 Schematic for two stage operational amplifier

As shown in Fig. 1 [3, 9–14], the two stage op amp configuration. The first stage includes differential amplifier form by M1 and M2, which converts differential voltage to current. These differential current further sends to current mirror circuit formed by M3 and M4, which recover differential stage voltage. Here we get first stage op amp output which is nearly same as the differential voltage amplifier. The M5 supplies the bias current I_{B1} to differential pair. The second stage composed by a common source MOSFET amplifier M6 which converts input voltage of second stage into current. The CS transistor is actively loaded with current sink load M7, which converts current again into voltage at output. A M7 does not provide biasing for M6 and it is biased from the gate side. Therefore second stage is like a current–sink inverter.

As two stage operational amplifiers (op amps) are mostly used to achieve high dc gain and large output voltage swing, which is achieved by frequency compensation [1, 3, 4]. In general, a frequency compensation is essential for closed-loop stability as op amp work with negative-feedback connection. The stability is generally shown by phase margin [11, 15]. The simplest frequency compensation technique here is Miller effect; considers a connection of a compensation capacitor between the high-gain stages.

The conventional Miller capacitor is connected between input of second stage from output of first [3, 9, 12, 14, 16]. It has two poles connected with internal and output node of second stage which gives dominant and nondominant poles are typically widely spaced. The Miller capacitor provides a feedforward signal path that introduces a RH plane zero in the transfer function of op amp [11, 16–18].

2 Miller Compensation

In two stage op amps analog systems design simple frequency compensation and relaxed stability criterions that use dominant topologies [19]. It conventionally compensated the so-called *Miller compensation* or *Direct Compensation* technique [1, 3, 15, 20].

In Miller Compensation, the compensation capacitor (Cc) is an establish connection between first stage outputs to second stage input, which separates input and output poles by acquiring dominant and nondominant poles which are placed away from each other [15]. It also produces the feedforward current from output of first stage to output of op amp to bring RH plane zero [10].

3 Design of Two Stages Operational Amplifier

The design procedure begins by selecting a device length parameter which is used throughout the circuit. This value will determine the value of the **channel length modulation parameter λ** which is required for calculating amplifier gain, because transistor modeling varies strongly with channel length [3]. The selected device length used in the design allows for more accurate simulation models. The minimum valued compensation capacitor Cc established by placing the output pole P_2, is 2.2 times greater than the unity gain (GB) permitted with 60° phase margin (by assuming that the RHP zero Z_1 is placed at or beyond ten times the value of GB) [1, 3, 9, 19].

As in Fig. 2 [3, 9], the equivalent small signal circuit of two stage operational amplifier with Miller capacitance C_C which connects the two stages of op amp.

There are two results due to compensation capacitor C_C, first is the effective capacitance shunts R_1. It increases by the additive amount of $g_{mII}(R_{II})(C_C)$, which moves p_1 near to complex frequency plane origin by a significant amount (by assuming the second stage gain is large). Second is the p_2 that is moved far away from origin of the complex frequency plane, obtaind from negative feedback. It reduces the second stage output resistance [21].

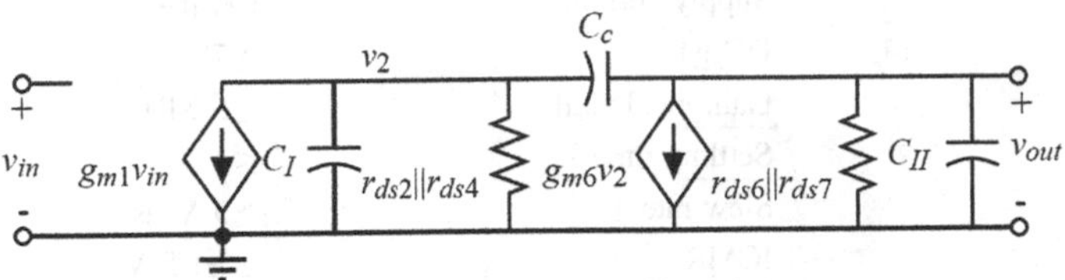

Fig. 2 Small signal equivalent circuit of operational amplifier

The overall transfer function for circuit with C_C is [3, 9],

$$\frac{V_{O(S)}}{V_{IN(S)}} = \frac{(g_{mII})(R_I)(R_{II})(1 - sC_C/g_{mII})}{1 + s[R_I(C_I + C_C) + R_{II}(C_{II} + C_C) + R_I R_{II} C_C)] + s^2 R_I R_{II}[C_I C_{II} + C_C C_I + C_C C_{II}} \tag{1}$$

The design procedure for two stage operational amplifier assumes following parameter specifications as mentioned in Table 1 [3, 5, 22].

By considering equation given below, we can calculate parameters for op amp [3, 9],

1. Gain at dc, $A_V(0)$.

$$A_V = \frac{2(g_{m2})(g_{m6})}{I_5(\lambda_2 + \lambda_4)I_6(\lambda_6 + \lambda_7)} \tag{2}$$

2. Gain bandwidth, GB.

$$\text{GB} \cong \frac{g_{mI}}{C_C} = \frac{g_{mII}}{C_C} \tag{3}$$

3. Load capacitance, C_L.

$$Cc > \frac{2.2C_L}{10} = 0.22C_L \tag{4}$$

4. Slew Rate, SR.

$$\text{SR} = \frac{I_5}{C_C} \tag{5}$$

5. Power dissipation, P_{diss}.

$$P_{\text{diss}} = (I_5 + I_6)(V_{DD} + |V_{SS}|) \tag{6}$$

Table 1 Typical CMOS operational amplifiers parameters specifications [3]

Boundary conditions	Requirement
Supply voltage	±2.5 V ± 10 %
Supply current	100 μA
DC gain	≥ 70 dB
Gain bandwidth	≥ 5 MHz
Settling time	≤ 1 μs
Slew rate	≥ 5 V/μs
ICMR	≥ ±1.5 V
CMRR	≥ 60 dB
PSRR	≥ 60 dB
Output swing	≥ ±1.5 V

4 Simulation Result

The Parameters and the size of CMOS can be calculated by considering the above equations and by assuming parameter value given in table we find the values given in Table 2 is as follows. The resulting simulation graph for considered value for CMOS is given thereafter.

4.1 *Designed CMOS Size*

See Table 2.

4.2 *Simulated Output Graphs*

See Figs. 3, 4, 5, 6, 7, and 8.

Table 2 Calculated CMOS operational amplifiers size

Technology (W/L) (nm)	S1	S2	S3	S4	S5	S6	S7	S8
180	12	12	15	15	4.45	138	20.47	4.45
130	8.45	8.45	13.78	13.7	5.09	126.79	23.42	5.09
90	1.39	1.39	5.09	5.09	0.47	58.54	2.7	0.47

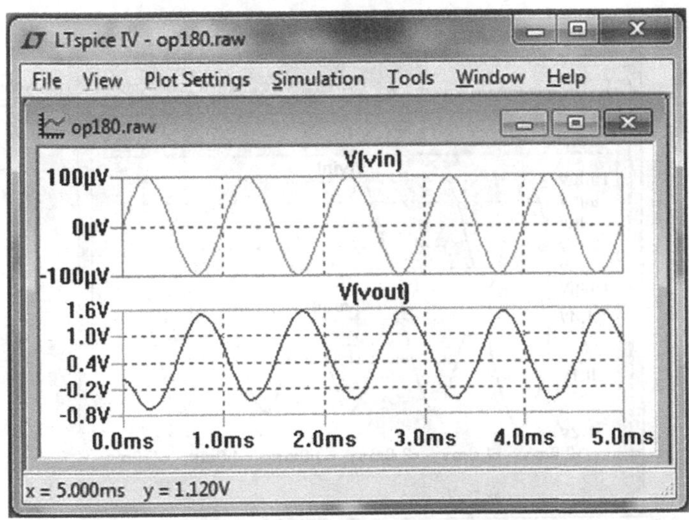

Fig. 3 Input/Output waveform for 180 nm process

4.2.1 For 180 nm Technology

See Figs. 3 and 4.

4.2.2 For 130 nm Technology

See Figs. 5 and 6.

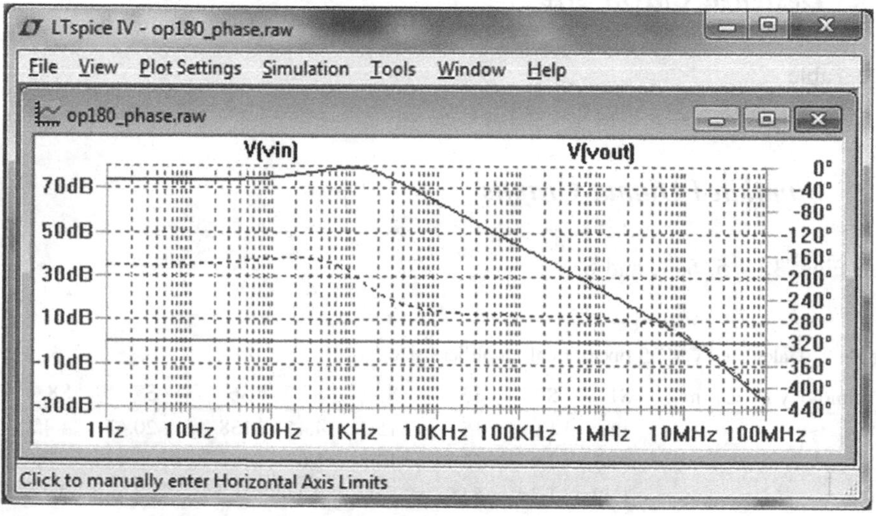

Fig. 4 Gain margin/phase margin plot for 180 nm process

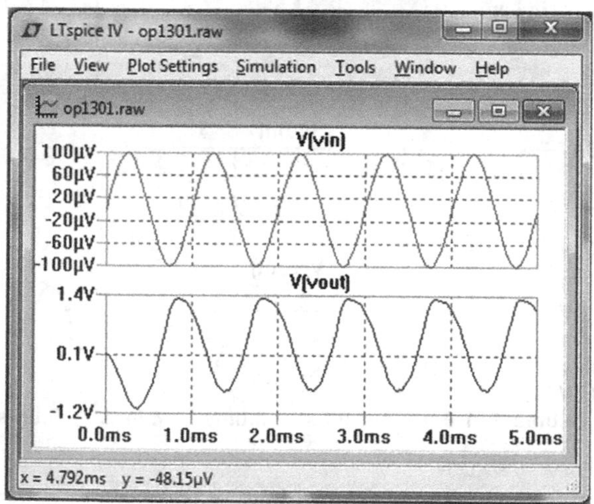

Fig. 5 Input/Output waveform for 130 nm process

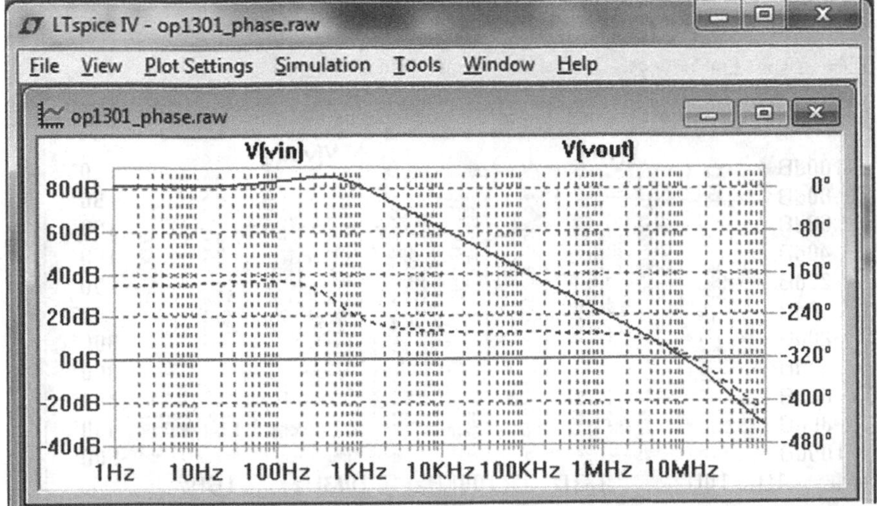

Fig. 6 Gain Margin/Phase margin plot for 130 nm process

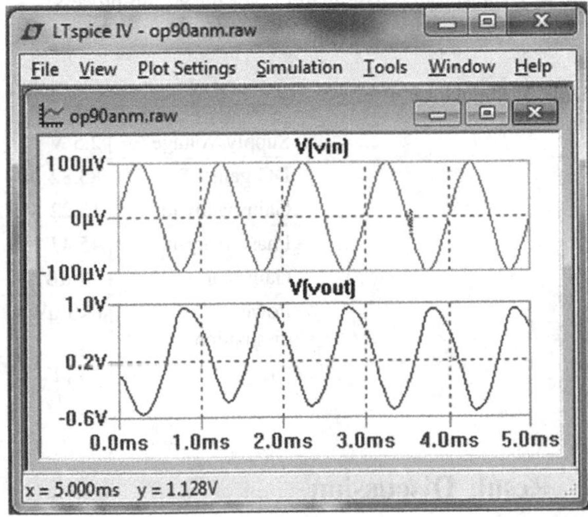

Fig. 7 Input/Output waveform for 90 nm process

4.2.3 For 90 nm Technology

See Fig. 7 and 8.

5 Result Summary

The Table 3 gives summary of parameters from above simulation result.

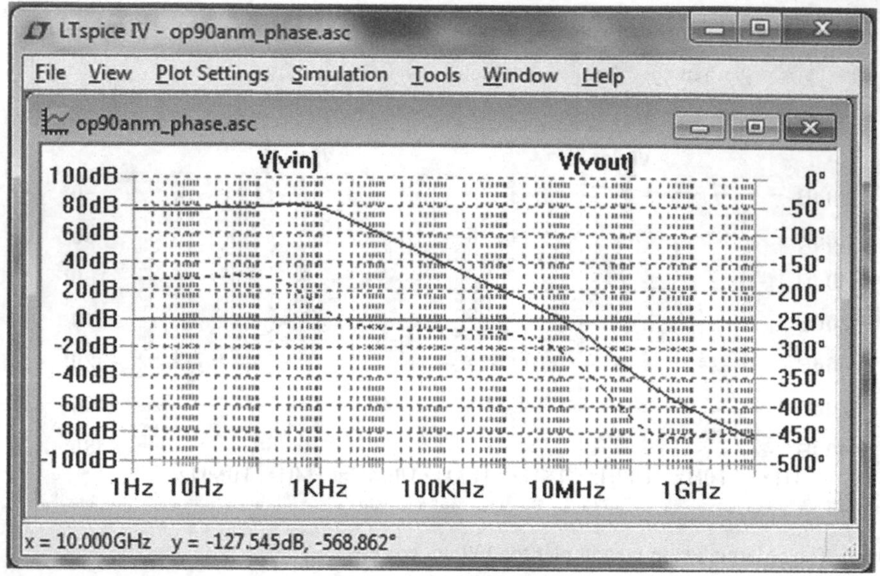

Fig. 8 Gain Margin/Phase margin plot for 90 nm process

Table 3 Simulation results parameter

Parameters	180 nm	130 nm	90 nm
Supply voltage	2.5 V	1.5 V	1.2 V
DC gain	83.82 dB	83.81 dB	81.87 dB
Gain bandwidth	13.22 MHz	10.13 MHz	8.11 MHz
Phase margin	45.47°	49.31°	53.23°
Gain margin	79 dB	84.78 dB	80.28 dB
Power dissipation	840 μW	504.06 μW	486.1 μW
CL	10 pF	10 pF	10 pF

6 Result Discussion

The simulated parameter value is approximately same as calculated parameter value. In case of all **180, 130, and 90 nm** technology the **achievable gain is increased by approximately 15 %,** with less power consumption at low supply voltage and the better closed-loop stability we get at low capacitive load connection with greater bandwidth.

7 Conclusion

In this paper, the design procedure is followed and simulated using Ltspice for two stage operational amplifier with Miller compensation technique in 180, 130, and 90 nm technology. As in Miller compensation technique allows the relatively low valued compensation capacitor and increases the capable value of bandwidth. As given result is compared with standard two stage amplifier parameters, the proposed design parameters provides improved GBW with given slew rate and reduces on chip area with less power dissipation. From result summary the given design achieve better performance for low dc power supply and low output load value.

References

1. P. R. Gray, P. J. Hurst, S. H. Lewis, and R. G. Meyer, "Analysis and Design of Analog Integrated Circuits", 4th edition, New York: Wiley, pp. 644–645, (2001).
2. Gaetano Palumbo, and Salvatore Pennisi, "Design Methodology and Advances in Nested-Miller Compensation", IEEE TRANSACTIONS ON CIRCUITS AND SYSTEMS—I: FUNDAMENTAL THEORY AND APPLICATIONS, VOL. 49, NO. 7, pp. 893–903, (JULY 2002).
3. Philip E. Allen & Douglas R. Holberg, "CMOS Analog Circuit Design", Second Edition, Oxford University Press, pp. 269–278, (2002).
4. Hoi Lee and Philip K. T. Mok, "Active-Feedback Frequency-Compensation Technique for Low-Power Multistage Amplifiers", IEEE JOURNAL OF SOLID-STATE CIRCUITS, VOL. 38, NO. 3, pp. 511–520, (MARCH 2003).
5. Mohammad Yavari, Student Member, IEEE, Nima Maghari, and Omid Shoaei, "An Accurate Analysis of Slew Rate for Two-Stage CMOS Opamps", IEEE TRANSACTIONS ON CIRCUITS AND SYSTEMS—II: EXPRESS BRIEFS, VOL. 52, NO. 3, pp. 164–167, (MARCH 2005).
6. Alfio Dario Grasso, Gaetano Palumbo and Salvatore Pennisi, "Advances in Reversed Nested Miller Compensation", IEEE TRANSACTIONS ON CIRCUITS AND SYSTEMS—I: REGULAR PAPERS, VOL. 54, NO. 7, pp. 1459−1470, (JULY 2007).
7. Aldo Pena Perez, Nithin Kumar Y.B., Edoardo Bonizzoni, and Franco Maloberti, "Slew-Rate and Gain Enhancement in Two Stage Operational Amplifiers", IEEE, pp. 2485–2488, (2009).
8. Wei Wang, Zushu Yan, Pui-In Mak, Man-Kay Law and Rui P. Martins, "Micropower Two-Stage Amplifier Employing Recycling Current-Buffer Miller Compensation", IEEE, pp. 1889–1892, (2014).
9. Jirayuth Mahattanakul and Jamorn Chutichatuporn, "Design Procedure for Two-Stage CMOS Opamp With Flexible Noise-Power Balancing Scheme", IEEE TRANSACTIONS ON CIRCUITS AND SYSTEMS—I: REGULAR PAPERS, VOL. 52, NO. 8, pp. 1508–1514, (AUGUST 2005).
10. R. Jacob Baker, "CMOS Circuit Design, Layout & Simulation", Third Edition, IEEE Press Editorial Board, A JOHN WILEY & SONS, INC., PUBLICATION, (2010).
11. Zushu Yan, Pui-In Mak, and Rui P. Martins, "Two-Stage Operational Amplifiers: Power-and-Area-Efficient Frequency Compensation for Driving a Wide Range of Capacitive Load", IEEE CIRCUITS AND SYSTEMS MAGAZINE, pp. 26–42, (FIRST QUARTER 2011).
12. D.S. Shylu, D. Jackuline Moni and Benazir Kooran, " DESIGN AND ANALYSIS OF A TWO STAGE MILLER COMPENSATED OP-AMP SUITABLE FOR ADC

APPLICATIONS", IJRET: International Journal of Research in Engineering and Technology eissn: 2319-1163 | pissn: 2321-7308, Volume: 03 Special Issue: 07 |, pp. 249–256, (May-2014).
13. M. I. Idris, N. Yusop, S. A. M. Chachuli, M.M. Ismail, Faiz Arith & A. M. Darsono, "Low Power Operational Amplifier in 0.13um Technology", Modern Applied Science; Vol. 9, No. 1; ISSN 1913-1844 E-ISSN 1913-1852, Published by Canadian Center of Science and Education, pp. 34–44, (2015).
14. Mohd Haidar Hamzah, Asral Bahari Jambek and Uda Hashim, "Design and Analysis of a Two-stage CMOS Op-amp using Silterra's 0.13 μm Technology", 2014 IEEE Symposium on Computer applications & Industrial Electronics (ISCAIE 2014), Penang, Malaysia, (April 7–8, 2014).
15. Vishal Saxena, "INDIRECT FEEDBACK COMPENSATION TECHNIQUES FOR MULTI-STAGE OPERATIONAL AMPLIFIERS", A thesis submitted in partial fulfillment Of the requirements for the degree of Masters of Science in Electrical Engineering Boise State University, (October 2007).
16. Paul J. Hurst, Stephen H. Lewis, John P. Keane, Farbod Aram and Kenneth C. Dyer, "Miller Compensation Using Current Buffers in Fully Differential CMOS Two-Stage Operational Amplifiers", IEEE TRANSACTIONS ON CIRCUITS AND SYSTEMS—I: REGULAR PAPERS, VOL. 51, NO. 2, pp. 275–285, (FEBRUARY 2004).
17. Feng Zhu, Shouli Yan, Jingyu Hu, and Edgar Sánchez-Sinencio, "Feedforward Reversed Nested Miller Compensation Techniques for Three-Stage Amplifiers", IEEE, pp. 2575–2578, (2005).
18. Annajirao Garimella, M. Wasequr Rashid and Paul M. Furth, "Nested Miller Compensation Using Current Buffers for Multi-stage Amplifiers", IEEE (2011).
19. Kin-Pui Ho, Cheong-Fat Chan, Chiu-Sing Choy, and Kong-Pang Pun, "Reversed Nested Miller Compensation With Voltage Buffer and Nulling Resistor", IEEE JOURNAL OF SOLID-STATE CIRCUITS, VOL. 38, NO. 10, pp. 1735–1738, (OCTOBER 2003).
20. Xiaohong Peng, Willy Sansen, Ligang Hou, Jinhui Wang, and Wuchen Wu, "Impedance Adapting Compensation for Low-Power Multistage Amplifiers", IEEE JOURNAL OF SOLID-STATE CIRCUITS, VOL. 46, NO. 2, pp. 445–451, (FEBRUARY 2011).
21. Hassan Khameh, Hossein Mirzaie and Hossein Shamsi, "A New Two-Stage Op-Amp Using Hybrid Cascode Compensation, Bulk-Driven, and Positive Feedback Techniques", IEEE, pp. 109–112, (2010).
22. Xiaohua Fan, Chinmaya Mishra and Edgar Sánchez-Sinencio, "Single Miller Capacitor Frequency Compensation Technique for Low-Power Multistage Amplifiers", IEEE JOURNAL OF SOLID-STATE CIRCUITS, VOL. 40, NO. 3, pp. 584–592, (MARCH 2005).

Mobile Apps Classification with Risk Score by Exploiting the Enriched Information of App Context

Prajakta P. Lokhande and Shivaji R. Lahane

Abstract With the use of mobile devices increasing rapidly day by day, huge numbers of mobile apps are coming into the market, many of which provide same functionality; as a result, having a proper classification of these apps can be useful for various purposes like making it time efficient and easy to the user for selecting the required app, understanding the user preferences which can motivate the intelligent personalized services, etc. But for having proper mobile app usage analysis, effective classification of apps is required for which detailed information about the apps is needed. However this is a nontrivial task as limited contextual information is available. As the information available about the apps is short and sparse, the classification of these apps can also be considered as coming in the category of classification of short and spares text. To classify these short and spares text, various methods are present that can be used to classify the mobile apps. In this paper, we have presented a method in which we extract the information about the apps from the sources like information from the labels (app name), information from the web search engine (snippets), contextual usage logs of users and the permissions the app requests before installation. This gives an effective and secure classification of the apps as source for most of these apps are some unknown vendors and so they are having the higher possibility of being malicious. With the contextual information collected the designed system is able to recommend the apps to the user based on their preferences.

Keywords Classification · Mobile apps · Security · Risk score · Recommendation · Enriched information

---

P.P. Lokhande (✉) · S.R. Lahane
Department of Computer Engineering, Gokhale Education Society's R.H. Sapat College of Engineering, Management Studies & Research, Nashik-5, Maharashtra, India
e-mail: prajaktalokhande7@gmail.com

S.R. Lahane
e-mail: shivajilahane@gmail.com

© Springer Science+Business Media Singapore 2017
S.C. Satapathy et al. (eds.), *Proceedings of the International Conference on Data Engineering and Communication Technology*, Advances in Intelligent Systems and Computing 468, DOI 10.1007/978-981-10-1675-2_19

1 Introduction

The booming market of android has increased the demand of android apps. Today the numbers of android apps being designed are more than the number of software's being designed for the computers. Every vendor today is trying to make use of mobile apps for providing their services and as a result, large numbers of mobile apps having similar functionality are coming into the market. Studying the usage of these apps can be used for various intelligent services like recommender system, user segmentation and target advertising [1]. The key step for mobile usage analysis is to have the apps classified into some predefined category. However it is a nontrivial task as limited contextual information about the apps is available. Then, again the large number of mobile apps coming from the unknown third-party vendors has increased the security concerns, as the devices provide access to personal and sensitive information such as phone number, relocation, messages, etc. if the apps are malicious.

Our goal is to design a classification system that gives a detailed analysis of the mobile apps by effectively classifying them into their appropriate category and identifying the risk factor of that app depending upon the different information collected, i.e. the information obtained from the name of the app, snippets obtained from the web search engine, usage records of the app collected from different people and permissions of the app that are extracted from their manifest file. We have used multiple algorithms to achieve our goal. To design our system, we have used divide and conquer strategy. We have developed and tested each algorithm separately. We then merged these algorithms and tested the final outcome for around 150 mobile apps. The main objectives of the proposed system are

- To find the different ways through which the information about the app can be obtained.
- To define new taxonomy, i.e. a new set of category list, which will give a more appropriate understanding of what operation a particular app performs.
- To obtain analysis of the mobile app by categorizing the app into its appropriate new defined category based on the explicit and implicit information obtained.
- Improve the security mechanism of android by providing risk score of the app to the user in an easy-to-understand manner. This is obtained by extracting the permissions the app requests.
- Recommend the apps based on user preferences or the profile they have created.

2 Related Work

Proper classification requires to have detailed information on the topic, but this becomes a nontrivial task if the topic is short and sparse [1]. As the apps name are small consisting of 2–3 words, they can be considered as short and sparse text.

Different solutions have been proposed with the aim to classify these short and sparse texts by considering different parameters, like information obtained from the web search engine is being used to classify the short and sparse text [2, 3]. The traditional methodology to measure the similarity between the short and sparse text is improved using the relevance weight inner product of the term occurrence and using the machine learning technology improving the earlier web kernel-based method [4].

X.-H. Phan et al. presented a framework for classifying the short and sparse text based on synonym and hyponym. They have also considered the semantics between them. It gave us the idea of classification based on title and description [5]. H. Ma et al., proposed an approach which leverages search snippets to build vector space for both categories and app usage; and classifies the app usage records using the cosine space distance [6]. H. Zhu et al., proposed an approach to automatically classify the mobile apps using the web-based features and explicit contextual features [7].

Permissions of the apps are considered as an important factor with respect to the security parameter. Studies have shown how the permissions can be used to get access to user's private data and information, which threatens the security of the user. The defence mechanism provided by android is to warn the user about the permissions the app will request before installation, trusting that the user will make the right decision [8]. W. Enck et al., proposed a real-time privacy monitoring tracking system on smart phones. Here it informs the user when the application may be trying to send sensitive data off the phone [9].

A.P. Felt et al. used a static analysis technique to determine if an android app is over privileged, i.e. if the app is requesting for the permissions it never used. It helps us to develop a module that calculates the risk factor of particular app [10]. E. Chin et al., studied the permissions requested by the app belonging to a particular category and checked it with the permissions requested by the other apps belonging to that category. For that they studied the apps belonging to two datasets; one includes 158,062 Android apps from the Android Market, and the other of 121 malicious apps [11]. B.P. Sarma et al. studied the attitude of the user with respect to the privacy and security of their device which can be useful for designing a secure and effective mobile ecosystem. Authors conducted a survey in which they studied the user's insights of user's perception towards the security and installation habits of around 60 smartphone users in two stages, in the first stage they interviewed the users for their willingness to try some tasks on their device, which people normally avoid due to the privacy and security issues. Then in the second stage, they studied the habits of users for selecting the applications, where they came to the conclusion that users mainly focus on the ratings and comments, and sometimes the total downloads of the application and overlook the permissions requested by the applications [12].

3 Proposed System

The proposed system uses the web-based and context-based information to classify the mobile apps; to make this system a more effective analysis, the permissions requested by the app to be classified is considered. Based on the analysis of these permissions, the system provides a risk score to the app. A mobile application is designed which gives the user a detailed analysis of the app along with the risk score in an easy-to-understand manner, and based on the preferences selected user gets the recommendations of the apps from the system. Specifically, the working of the system is divided into different phases; Fig. 1 shows the overall architecture of the system.

Phase 1: Web-based feature extraction
Here the feature extraction takes place with the information obtained from the web search engine. Both the explicit and implicit features are obtained as explained below.

(i) Explicit web-based feature extraction
Here the explicit features of the app are extracted from web search engine. In simple words, after giving the name of the app to the system, it extracts the top results from the search engine to place the app in the appropriate category. To achieve this goal, vector space model is being used., which consists of the three steps, according to which first, one category profile dc will be built by integrating all the M snippets retrieved for some app. Here, the stop words will be removed and the verbs and

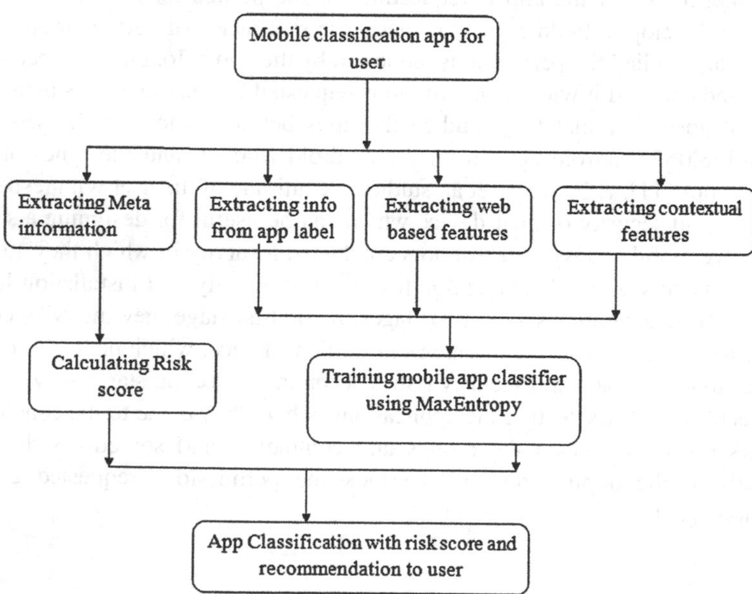

Fig. 1 Overall system architecture

adjectives will be normalized Then in the second step, normalized word vector will be built for each app category $\overrightarrow{w_c} = \dim[n]$ using the following formula [1]:

$$\dim[i] = \frac{freq_{i,c}}{\sum_i freq_{i,c}} (1 \le i \le n) \tag{1}$$

where $freq_{i,c}$ is the frequency of ith word in the category profile. Similarly, in the last step for each snippet s retrieved for the app a; we will build word vector $\overrightarrow{w_{a,s}}$ and calculate the cosine distance between $\overrightarrow{w_c}$ and $\overrightarrow{w_{a,s}}$. From the result obtained, we will consider the max similarity result and assume that category to be the appropriate category for app a.

$$c^* = \arg\max \text{Similarity}\,(\overrightarrow{w_{a,s}}, \overrightarrow{w_c}) \tag{2}$$

Then, to confirm our result, we calculate the general label confidence score by [1]:

$$\text{GConf}(c, a) = \frac{M_{c,a}}{M} \tag{3}$$

where, $M_{c,a}$ is the number of returned-related search snippets of app a, whose category labels are c after mapping.

(ii) Implicit web-based features extraction
Here the latent semantic meaning of the snippets retrieved will be considered; as a result this will give us a more refined result. This is because, in the explicit feature selection the latent semantic meaning of the words is not considered, i.e. for example, the words like "play", "game" and "fun" are considered totally different while calculating the distance between the word vectors. But after considering their latent semantic, they can be placed in the same semantic topic of "entertainment". To understand the latent semantic meaning, we will use the latent dirichlet allocation (LDA) model [13]. Using this model, a category profile dc is built. Then for given app a, top M result will be retrieved and the words which do not match with category profile will be removed. Then *KL* Divergent will be used to map each snippet with the category. The formula for *KL* Divergent is given by [1]

$$D_{KL}(P(z|s)||P(z|c)) = \sum_k P(z_k|s) ln \frac{P(z_k|s)}{P(z_k|c)} \tag{4}$$

where, z is the latent topic of the category. The category with the smallest *KL* divergence is selected. To confirm the obtained result, the topic confidence score for the given category is then calculated as follows [1]:

$$\text{TConf}(a, c) = \frac{T_{a,c}}{M} \tag{5}$$

where, $T_{a,c}$ is the number of returned snippets of app *a* with respect to category label *c*. The topic confidence score gives the confidence that app *a* is labelled as category *c* with respect to the latent semantic topic.

Phase 2: Context-based feature extraction
In this phase, the contextual features of the app from real-world context log are considered in terms of two factors, i.e. explicit features and implicit features.

(i) Explicit contextual feature
Here, the real-world context information are collected from different user's logs in the terms of feature-value pair. We have used App Usage Tracker application to trace the user logs. User logs contains: <application name, usage date time, usage times-pan>. This information is exported to .xls file using tracker. This information is further passed to the system with user details like: <user age, gender, occupation>. The context logs of user are preprocessed and create context profile R_a, for each app. Similarly, a context profile is built for each category R_c, by combining the context profiles of the preselected apps labelled with *c*. At last, the cosine distance calculated is arranged in descending order and the one with max similarity is assumed as the category. The result is then confirmed by calculating category rank distance given by [1]:

$$\text{CRDistance}(a, c) = Rk(c) - 1 \tag{6}$$

where, *Rk*(*c*) is the rank of the category *c*, which is obtained by comparing the vector distance to app *a*. Here, the smaller the distance, more accurate is our selected category.

(ii) Implicit contextual feature
Here, the semantic meaning of the contextual feature-value pairs will be obtained from the explicit contextual features available. Like for example, the feature-value pairs like "Time-span: 2 h", "age: 15–20" are grouped together under the topic "trendy app". To achieve this goal, here the category profile *Rc* is built using Latent Dirichlet Allocation on context model. Accordingly, for a given app *a*, category profile *Ra* is built using the historic logs database. Then, *KL* Divergence is calculated between each category and the app using the given formula [1].

$$D_{KL}(P(z|Ra)||P(z|Rc)) = \sum_k P(z_k|Ra)\ln\frac{P(z_k|Ra)}{P(Z_k|Rc}\tag{7}$$

From this, the smallest KL-Distance is considered as the category for the app and finally, the topical Rank distance is calculated to confirm the result. The topical rank distance is calculated by

$$\text{TRDistance}(a, c) = Rk(c) - 1 \tag{8}$$

Phase 3: Classification
After the extraction of the features, a classifier is trained to classify the app into its appropriate category by combing these different features. From the different classifier present, we have considered the MaxEntropy classifier [14] as it gives best results when the information available is sparse and it is easy to incorporate different features using this classifier.

Phase 4: Rerank classification using risk score
The earlier stage calculates the app category; the risk score of an application is calculated based on the permission dataset using the naïve byes classifier [15]. Finally in this phase app classification based on risk score is obtained, i.e. apps in each category are ranked in decreasing order as per the risk score.

Phase 5: Generate Recommendations
Using an application created for the end user; the user can get classified results of the app. For recommendation, user has to register first. Using his/her contextual information profiling applications are suggested to the user.

4 Implementation

The system proposes an effective to way to get the detailed analysis of any app being requested based on its information obtained from the web search, context usage records, and the permissions requested by the app. A client server model is designed, where all the classification is done on the server side and the analysis of the requested app is displayed to the user through the mobile app developed for client. Different web services are created to carry out the server side processing. The server side consists of three web services namely web-based information, context-based information and respond to user request. These web services are created using java and apache server. For obtaining the context logs of the volunteer users we have used a app called “App Usage Tracker”, and the information obtained is exported in xls file. Through the client side app developed, the user can view the details of the desired app, search for the requested app and get recommendations based on his preferences. This app is developed using android sdk in eclipse and it communicates with the web service using http protocol. All the information obtained is stored in the database, for which we have used the wamp server.

Multiple datasets are used for the experiment, dataset for App Category consist of two levels of categories, level one consists of basic category and level two includes its sub categories, for example, if level 1 category is “Business”, then level 2 category for it consists of “Office Tools, Job, Security, etc.”. 44 such categories are manually created by studying the categories and different apps present on the Play Store. Second is the Permission Dataset, this data set contains level-2 category-wise permission details, in which the commonly used permissions for a particular category are stored. The stored permissions are given a weight value depending upon the category they are present, with the help of permission information present on the android development site [16]. Third is the List of Category keyword. This list contains category level-2 specific keyword list which are created by analyzing the snippets obtained from the Google search engine. We have created this dataset as ready-made dataset for the words describing the app is not available. This data is stored in the format of <catid, keyword list, wt>, which includes around 700 hundred keywords. Last is the Context log dataset, it is created with a help of the information collected from around 30 to 40 users using a app called "App usage Tracker", which they installed it on their device and their app usage information is collected. Then a dataset is created containing the user specific app access information in the form <uid, app name, access details> and the dataset of the users in the format <name, age, occupation, gender>.

5 Results

The system is tested for around 150 mobile apps belonging to different category. User gets the proper classification of the app and its risks core in easy-to-understand manner. Based on the profile created or the preferences selected by the user, he can

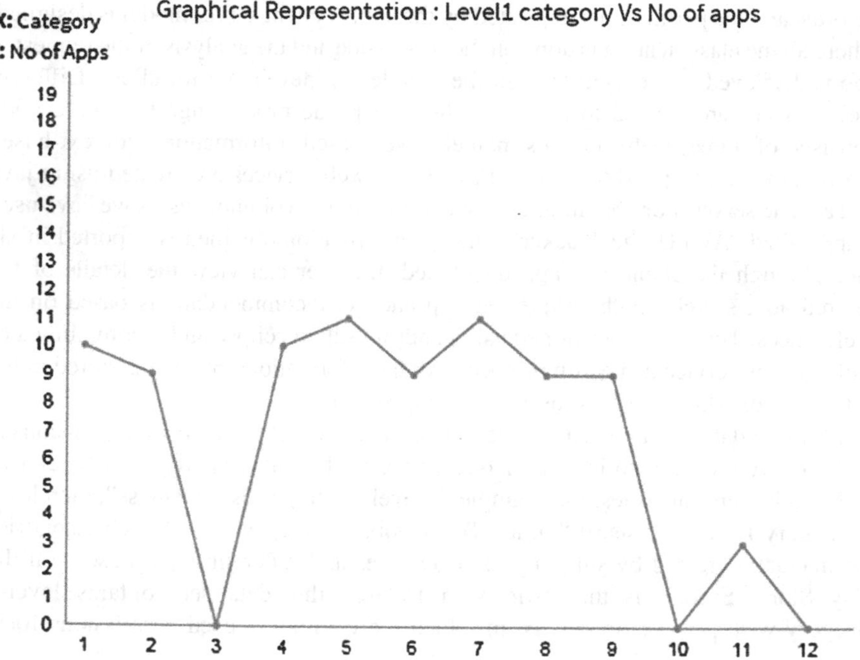

Fig. 2 Analysis of result

get the recommendation of the apps belonging to different category. The system is tested for different application with the help of the datasets mentioned above. Figure 2 shows the graph of server side result, displaying the number of applications accurately matched with the level 1 category for the tested mobile apps.

Figure 3 shows the snapshot of the mobile app developed, in which for the "Audio" category is selected by the user; here, a list of apps present in that category

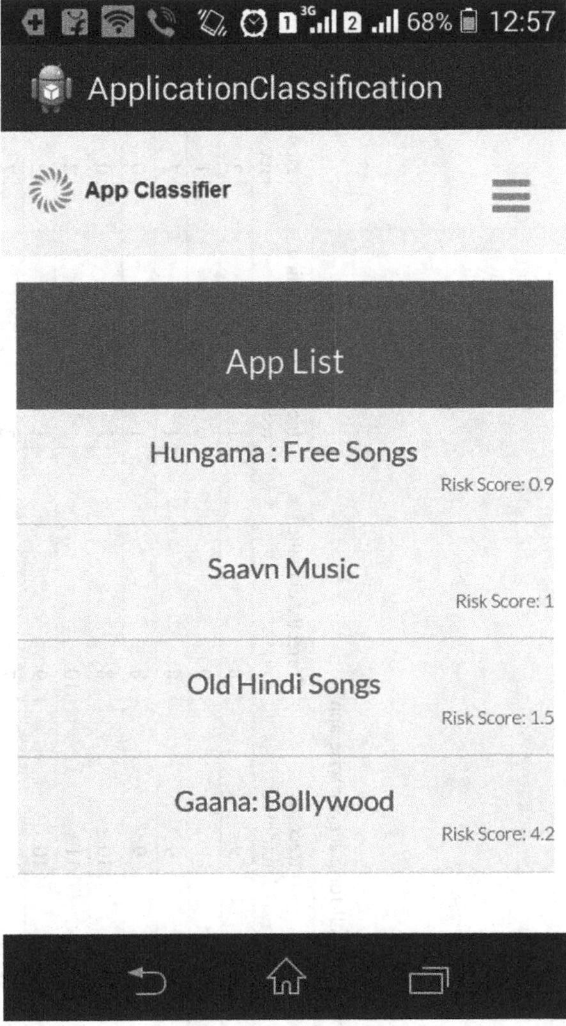

Fig. 3 Snapshot of mobile app developed

Table 1 Precision and recall for category wise app

Category	Total apps	Records retrieved	Correctly classified records	Incorrectly classified records	Non-classified records	Precision %	Recall %
Games	10	9	8	1	2	88.88	80
Business	10	9	9	0	1	100	90
Multimedia	10	9	8	1	2	88.88	80
Communication	8	9	8	1	0	88.88	100
System	8	10	8	2	0	80	100
Entertainment	12	11	10	1	2	90	83.33
Education	11	10	9	1	2	90	81.81
Medical	9	9	7	2	2	77.77	77.77
Average						88.05	86.61

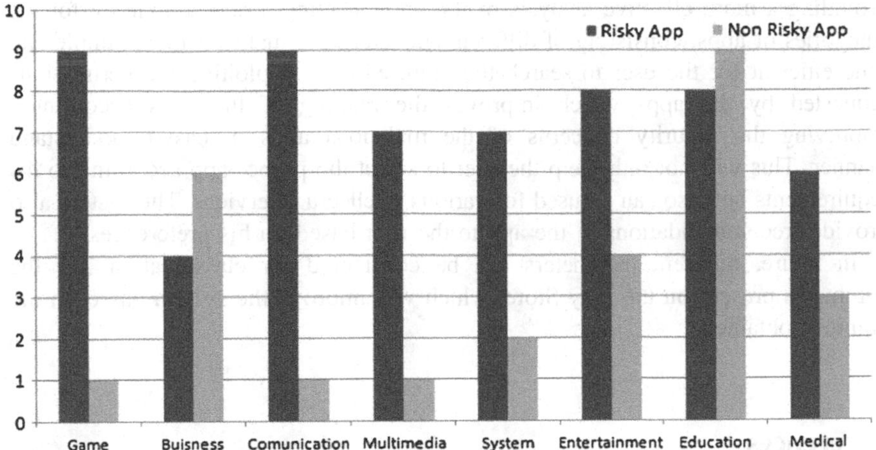

Fig. 4 Risk score-wise category app distribution

along with their risk score is being displayed to the user. As shown, all the apps are being arranged in the ascending order of their risk score which makes it easy for the user while deciding which app to use.

6 Performance Evaluation

For evaluating the performance of the system, precision and recall is calculated for the tested applications belonging to different categories. Here 8–10 applications are considered for different categories. The results obtained are shown in Table 1, according to which for the eight categories considered, the average precision is 88.05 % and recall is 86.61 % and for all the categories performance obtained is in the range of around 80–100 % in terms of precision and recall.

We have then estimated the number of risky and non-risky applications for the same applications and the results are shown in the Fig. 4 which shows the category-wise number of risky and non-risky applications.

7 Conclusion and Future Scope

As everyday there are number of similar kind of apps coming in the market, effective classification of the mobile apps is important. The previous work done considers single factor for classifying the mobile apps thus not giving the effective results. The proposed system effectively classifies the mobile apps with the help of extracted information from multiple sources which improve the classification providing a more effective analysis of the apps. Having a new taxonomy for the categories of apps, consisting of different subcategories makes it more simple and time efficient for the user to search the required app. Exploiting the permissions requested by the apps which improves the ranking of the apps accordingly improving the security concerns of the malicious apps in easy-to-understand manner. This will not only help the user to select the proper app according to his requirements but also can be used for various intellectual services. The system also provides recommendations of the app to the user based on his preferences.

In future, different parameters can be considered for classification like the comments present on the Play Store, which will improve the system based on the opinions obtained.

References

1. H. Zhu, E. Chen, H. Xiong and H. Cao, "Mobile App Classification with Enriched Contextual Information," IEEE Transactions on mobile computing, (Volume:13, Issue:07), 7 July 2014.
2. M. Sahami and T.D. Heilman, "A web-based kernel function for measuring the similarity of short text snippets," in Proc. WWW, Edinburgh, U.K., pp. 377–386, 2006.
3. Z. Broder et al., "Robust classification of rare queries using web knowledge," in Proc. SIGIR, Amsterdam, Netherlands, pp. 231–238, 2007.
4. Wen-tau and C. Meek, "Improving Similarity Measures for Short Segments of Text", in Proc. 22nd Nat. Conf. Artif. Intell., vol. 2. 2007, pp. 1489–1494.
5. X.-H. Phan et al., "A hidden topic-based framework toward building applications with short web documents," IEEE Trans. Knowl. Data Eng., vol. 23, no. 7, pp. 961–976, Jul. 2010.
6. H. Ma, H. Cao, Q. Yang, E. Chen, and J. Tian, "A habit mining approach for discovering similar mobile users," in Proc. WWW, Lyon, France, pp. 231–240, 2012.
7. H. Zhu, H. Cao, E. Chen, H. Xiong, and J. Tian, "Exploiting enriched contextual information for mobile app classification," in Proc CIKM, Maui, HI, USA, pp. 1617–1621, 2012.
8. Christopher S. Gates, et al., "Generating Summary Risk Scores For Mobile Applications", IEEE Transactions on dependable and secure computing, (Volume:11, Issue:03), May–June 2014.
9. W. Enck, P. Gilbert, B. Chun, L.P. Cox, J. Jung, P. McDaniel, and A.N Sheth, "TaintDroid: An Information-Flow Tracking System for Realtime Privacy Monitoring on Smartphones," Proc. Ninth USENIX Conf. Operating Systems Design and Implementation, article 1–6, 2010.
10. A.P. Felt, E. Chin, S. Hanna, D. Song, and D. Wagner, "Android Permissions Demystified," Proc. 18th ACM Conf. Computer and Comm. Security, pp. 627–638, 2011.
11. E. Chin, A.P. Felt, V. Sekar, and D. Wagner, "Measuring User Confidence in Smartphone Security and Privacy," Proc. Eighth Symp. Usable Privacy and Security, (SOUPS'12), article 1, 2012.
12. B.P. Sarma et al., "Android Permissions: A Perspective Combining Risks and Benefits," Proc. 17th ACM Symp. Access Control Models and Technologies (SACMAT12), 2012.
13. William M. Darling, "A theoritical and practical implementation Tutorial on Topic Modeling and Gibbs Sampling" December 1, 2011.
14. K. Nigam et al., "Using Maximum Entropy for Text Classification," IJCAI Workshop Machine Learning for Information Filtering 1999, pp. 61–67.
15. NaïveBayesClassification, http://www.saedsayad.com/naive_bayesian.html.
16. Android Manifest Permission Details, http://developer.android.com/reference/android/Manifest.permission.html.

Ultra-Wideband Differential Fed Antenna with Improved Radiation Pattern

Shukla Aditi and Medhane Dipak

Abstract The main objective of this work is to design compact size Ultra-wideband antenna with improved radiation pattern. In this project, the effect of change in antenna dimension has been studied also technique like Beveling, DGS, and Differential feed has been implemented and the radiation pattern is enhanced. The antenna has simulated and evaluated in Finite Element Method based ANSOFT-HFSS v11.0 simulation software using FR4_epoxy substrate with dielectric constant of 4.4, loss tangent of 0.02. The Proposed antenna shows return loss of S11 < −10 dB (3.1–10.6 GHz), 85–90 % radiation efficiency within operating frequency band and 10–15 dB cross polarization has been reduced. The differential feeding technique shows an improved radiation pattern.

Keywords Radiation pattern · Microstrip Antenna (MSA) · Ultra-wideband antenna · Defected Ground Structure (DGS)

1 Introduction

Ultra-wideband (UWB) systems due to their advantages like high data rate, wide bandwidth, and short-range characteristic are paid more attention since the Federal Commercial Commission (FCC) issued the frequency band 3.1–10.6 GHz for commercial UWB systems. A vital component of Ultra-wideband system, UWB antenna is required to have features such as compact, easy integration with the radio-frequency front end circuits, low-cost, stable radiation pattern, and constant gain in the required direction [1, 2].

Shukla Aditi (✉)
Sandip Institute of Technology & Research Center, Nashik, Maharashtra, India
e-mail: aditi.shukla020@gmail.com

Medhane Dipak
Late G. N. Sapkal College of Engineering, Nashik, Maharashtra, India
e-mail: medhanedipak@gmail.com

© Springer Science+Business Media Singapore 2017

S.C. Satapathy et al. (eds.), *Proceedings of the International Conference on Data Engineering and Communication Technology*, Advances in Intelligent Systems and Computing 468, DOI 10.1007/978-981-10-1675-2_20

Various types of antenna such as the circular monopole antennas [3], rectangular aperture antenna [4], fractal bow tie antenna [5], open slot antenna [6], multimode slot line antenna [7, 8], and self-complementary antenna [9] has been introduced. But then, the radiation pattern of these antennas was not stable across the whole frequency band, especially above 9 GHz.

To reduce this problem and improve the radiation pattern of the UWB antenna several techniques have been proposed. In [10], the (EBG) mushroom-like electromagnetic band-gap structures are implemented for intensification of the gains of the antenna at high frequencies. To stabilize the radiation pattern across the whole operating frequency band in [11, 12] the strip-loaded wide slot antenna (SWSA) is used. However, though at the high frequency radiation pattern is improved to some extent, the cross polarization of the antennas is at a high level, thus the polarization purity still needs to be improved.

In order to reduce the cross polarization, which is caused by higher order mode, especially higher-odd-mode, can be suppressed when the antenna is symmetrically driven using differential feeding systems [13, 14]. Therefore, in this study, a compact differential-fed microstrip antenna is presented and analyzed for UWB applications. The proposed antenna exhibits S11 < 10 dB, stable omnidirectional radiation patterns in H-plane, with bidirectional radiation patterns in E-plane. By implementing differential feeding systems, the cross-polarization level is kept low across the whole operating frequencies, which results in high polarization purity of the antenna.

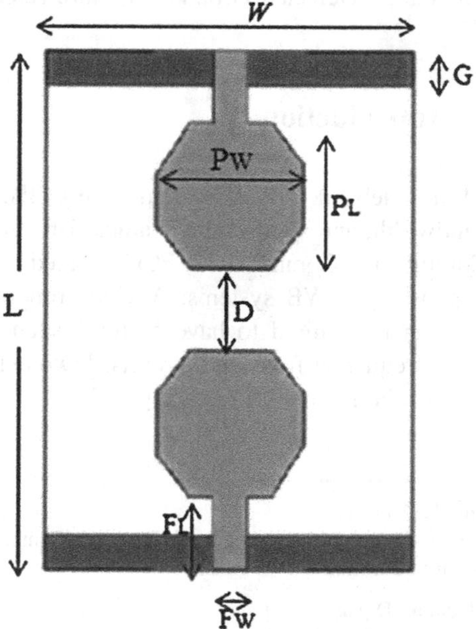

Fig. 1 Geometries of proposed antenna ($L = 40$ mm, $W = 26$ mm, $G = 3.7$ mm, $Pw = 16$ mm, $PL = 12.25$ mm, $D = 5.75$ mm, $FL = 4.87$ mm, $FW = 2.55$ mm)

2 Antenna Fabrication

The differential-fed antenna with compact size is designed. It has improved the radiation pattern with maintaining high polarization purity. Also, radiation efficiency has enhanced by the unique structure shown below.

Figure 1 shows the configuration of the proposed antenna. Here, an FR4_epoxy substrate is used, that has a thickness of 1.6 mm, relative dielectric constant of 4.4, and loss tangent of 0.02. The octagon patch along with the differential microstrip feeding lines is etched on the top of the substrate. The width of the microstrip feeding lines is chosen as 2.55 mm to achieve the characteristic impedance of 50 Ω. The partial ground plane is on the bottom side of the substrate. The proposed differential-fed antenna has a compact size of 26 mm × 40 mm.

3 Performance Improvement Stages

3.1 Beveling Technique

The rectangular monopole antenna (RMA) is chosen as a basic structure and it is optimized to the octagonal-patch-shaped antenna. Beveling the edges of rectangular radiator has been demonstrated and it is found that any reshaping of radiating area, strongly affects the current path. Here the angle of beveling is 45°.

The simulated results before and after beveling are depicted in Fig. 2. It shows that, before the beveling, the current flowing at the corner will cause the main beam tilting away and increase in cross polarization. Also, it affects return loss, depicted in Fig. 3.

With the decrease in return loss and VSWR, the bandwidth and directivity increases. The return loss has been reduced as depicted in Fig. 3 with octagon radiator. The E-plane shown in Fig. 4 is unidirectional and it has become more

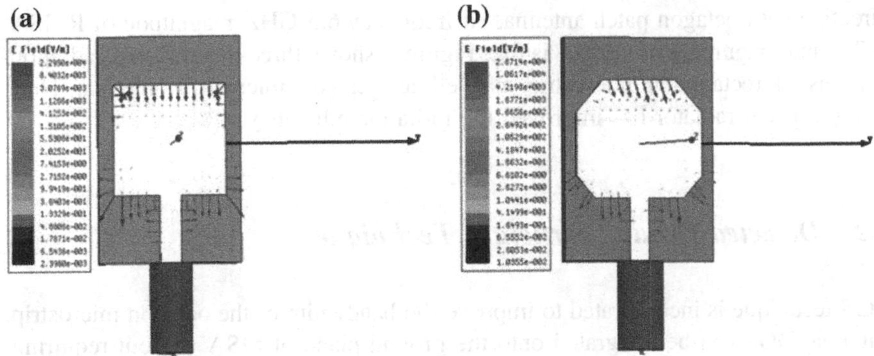

Fig. 2 Current distribution of **a** Rectangular MSA and **b** Octagonal MSA

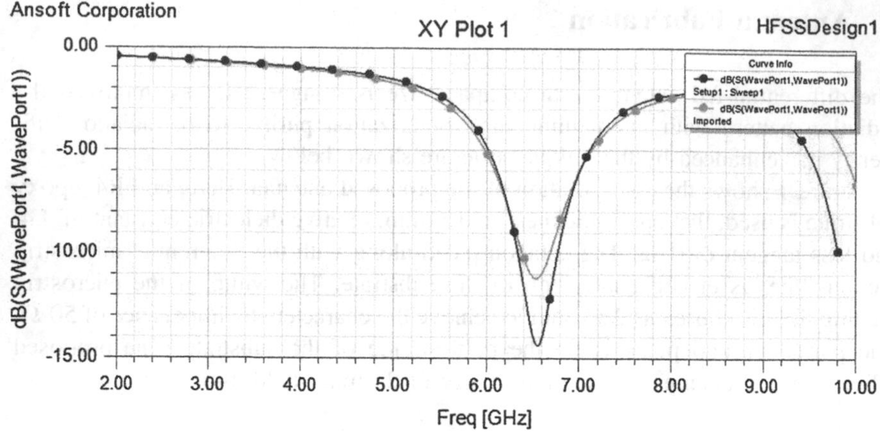

Fig. 3 Return loss verses Frequency curve: Rectangle patch (*green*), Octagon patch (*Red*)

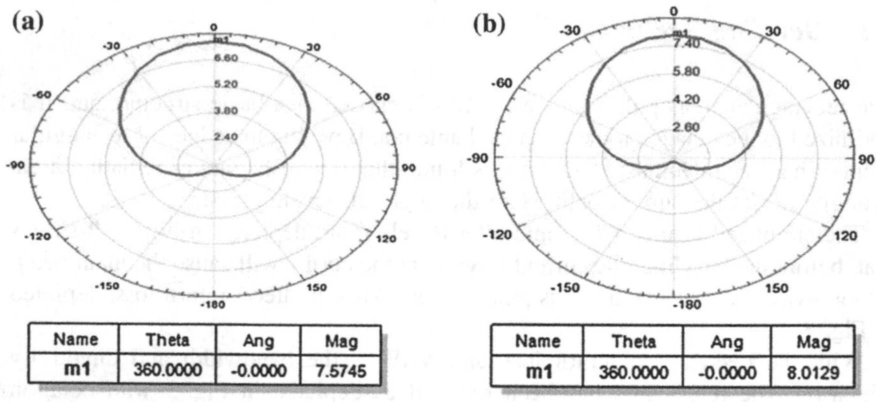

Name	Theta	Ang	Mag
m1	360.0000	-0.0000	7.5745

Name	Theta	Ang	Mag
m1	360.0000	-0.0000	8.0129

Fig. 4 Radiation pattern of **a** Rectangular MSA, **b** Octagon MSA

directive with octagon patch antenna. At frequency 6.5 GHz, magnitude of RMSA is 7.6 and magnitude of OMSA is 8.2. Figure 5 shows three-dimensional radiation patterns of rectangular microstrip as well as octagon microstrip antenna. Thus octagon patch radiator has improved the radiation efficiency.

3.2 *Defected Ground Structure Technique*

DGS technique is incorporated to improve the bandwidth of the octagon microstrip antenna. DGS can be integrated onto the ground plane of MSA without requiring any additional circuit for implementation. Introduction of DGS will etch the ground

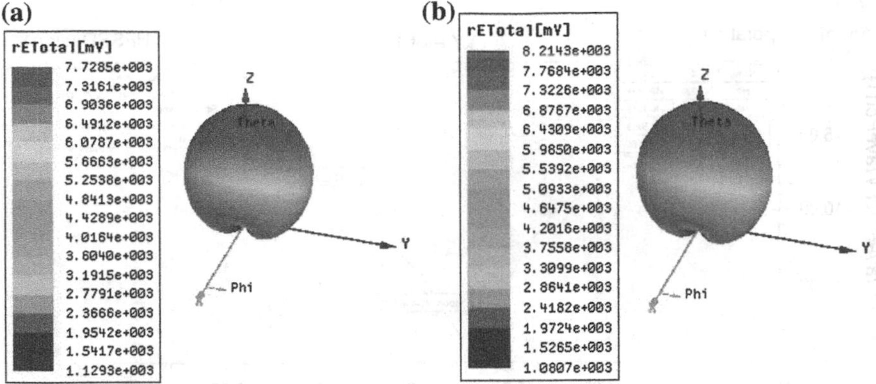

Fig. 5 Three-dimensional radiation pattern of **a** Rectangular MSA **b** Octagon MSA

plane and will thus disturb the shielded current distribution depending on shape and dimension of the defect. The excitation and electromagnetic propagation through the substrate layer can also be controlled by DGS.

DGS influences the radiation parameter. Figure 6a depicts the return loss versus frequency. The octagonal patch with rectangular DGS has the lowest return loss. By implementing DGS, the bandwidth has been greatly enhanced. However, bandwidth and directivity are inversely proportional to each other and hence with increase in bandwidth the directivity has decreased. Here we can observe this effect in Fig. 6b. Also, we have achieved the bidirectional radiation pattern, as the slot of rectangular shape is made in the ground plane, the E- and H-plane with cross polarization are shown in Fig. 7.

3.3 *Differential Feed Technique*

The bandwidth is improved with DGS technique, but the directivity has decreased as noted Sect. 3.1. And now to improve the directivity and polarization purity, differential-fed antenna is being introduced.

Differential-fed microstrip antenna is differentially excited by two probes to suppress the unwanted radiation, with the signals of the two probes being equal in magnitude but 180° out of phase. The differential feed antenna depicted in Fig. 1, has small size of 40 × 26 mm^2.

Figure 8 illustrates the simulated result DFA, $|S_{11}| < -10$ dB (3.1–10.6 GHz). Isolation and Cross polarization radiation are the important electrical parameters. Antenna to antenna mutual coupling describes energy absorbed by one antenna receiver when another nearby antenna is operating. That is, mutual coupling is typically undesirable because energy that should be radiated away is absorbed by a

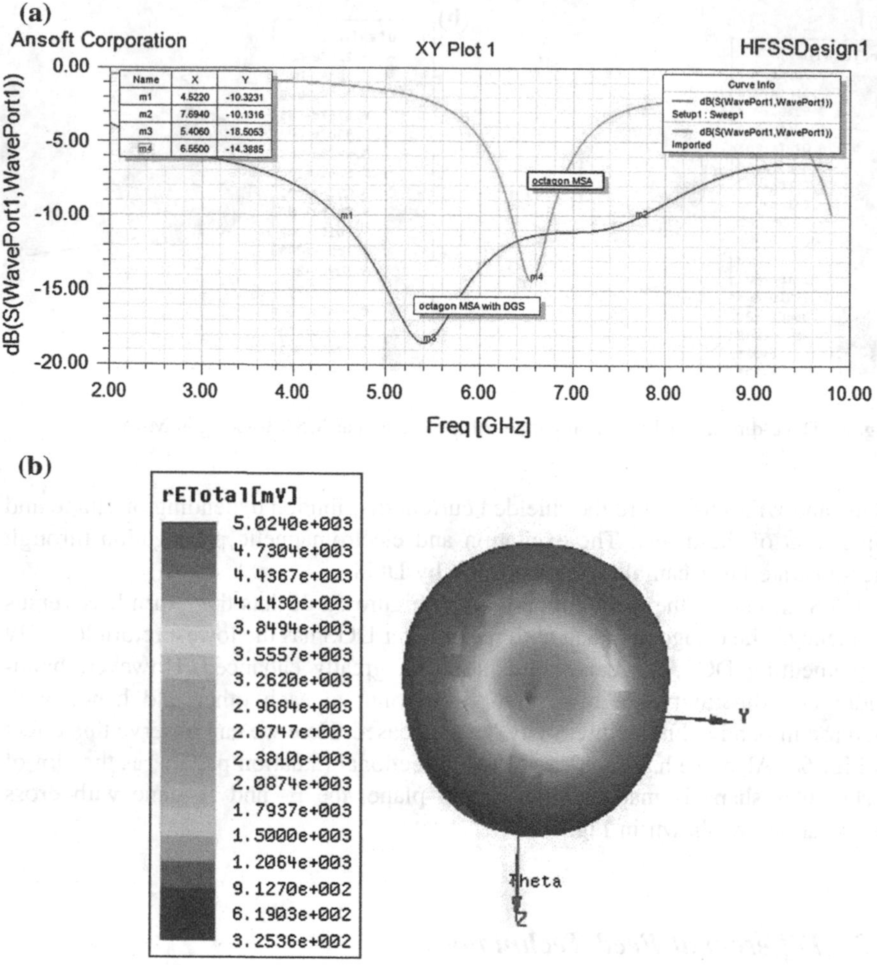

Fig. 6 **a** |S11| showing the effect of DGS, **b** Three-dimensional radiation pattern

nearby antenna; more than 85–90 % radiation efficiency is achieved within the operating frequency band with 9 DB gain.

In comparison to the radiation pattern in Fig. 7, the radiation pattern in Fig. 9 has 10–15 dB reduced cross-polarization level. Thus, using differential feed system, we have improved the radiation pattern of UWB antenna and made antenna suitable for various wireless applications.

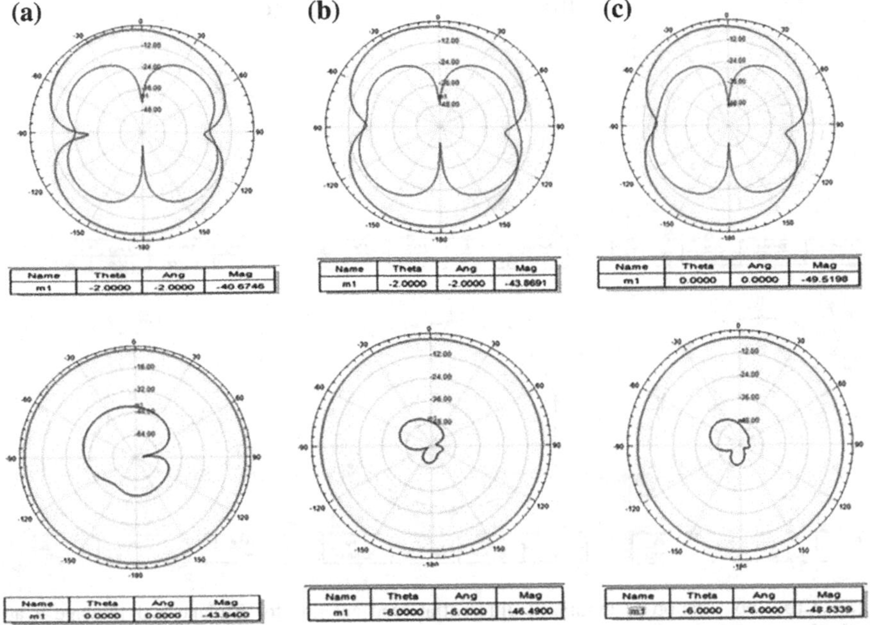

Fig. 7 Cross polarization of Octagon MSA with Rectangular DGS for E-plane and H-plane **a** at 5 MHz **b** at 6 MHz **c** at 7 MHz

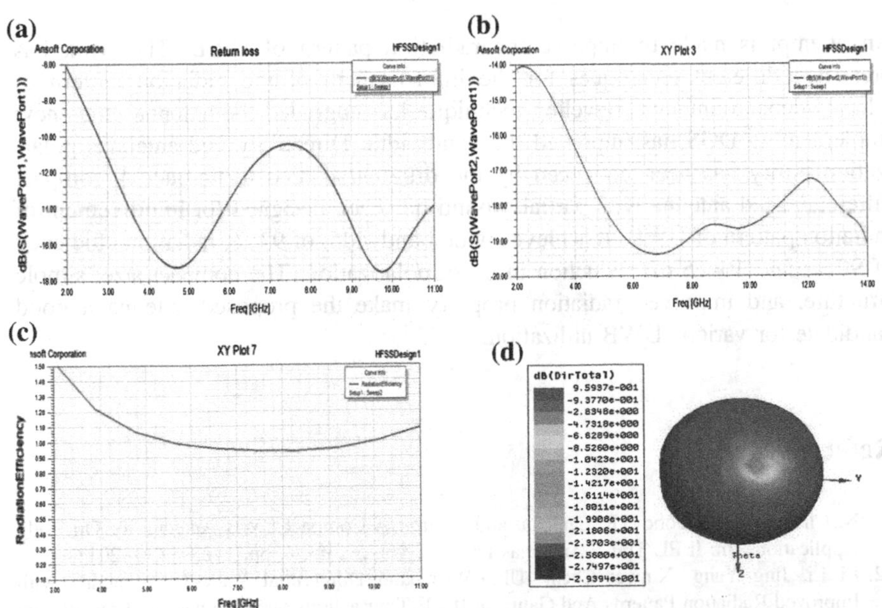

Fig. 8 **a** Return loss |S11|, **b** Mutual coupling |S21|, **c** Radiation efficiency versus frequency, **d** three-dimensional gain

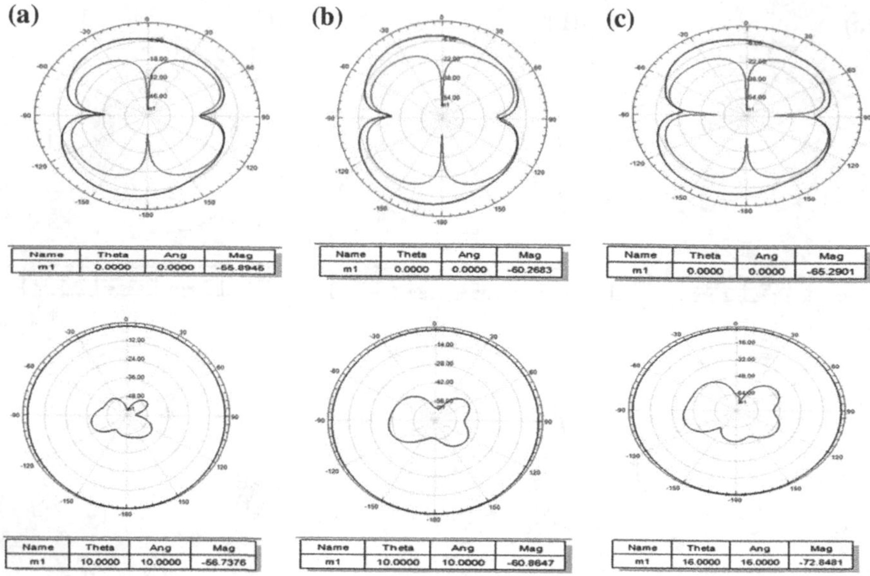

Fig. 9 Cross polarization of optimized differential feed antenna for E-plane and H-plane **a** at 5 MHz **b** at 6 MHz **c** at 7 MHz

4 Conclusion

An attempt is made to improve the radiation pattern of UWB. The work has presented different techniques for the improvement of the radiation pattern of Ultra-wideband antenna. Beveling technique has upgraded the antenna efficiency; Implementing DGS has improved the bandwidth; Directivity and antennas polarization purity has been enhanced by the differential feed technique. A compact differential-fed antenna with partial ground has been designed for improvement of radiation pattern of UWB. It achieves wide bandwidth of 93 %, radiation efficiency of 90 %, and 10–15 dB reduction in cross polarization. The compact size, simple structure, and improved radiation property make the proposed antenna a good candidate for various UWB utilization.

References

1. N. Chahat, M. Zhadobov, R. Sauleau, and K. Ito.: A Compact UWB Antenna for On-Body Applications. In: IEEE Trans. Antennas Propag., vol. 59, no. 4, pp. 1123–1130 (2011).
2. Li Li, Jing Yang, Xinwei Chen.: Ultra-Wideband Differential Wide-Slot Antenna with Improved Radiation Patterns And Gain. In: IEEE Transactions on Antennas and Propagation, vol. 60, no. 12, December (2012).

3. Liang J. X., C. C. Chiau, X. D. Chen, and C. G. Parini.: Study of a Printed Circular Disc Monopole Antenna For UWB Systems. In: IEEE Trans. Antennas Propag., vol. 53, No. 11, pp. 3500–3504 (2005).
4. Lin, Y. C. and K. J. Hung.: Compact Ultra Wideband Rectangular Aperture Antenna and Band- Notched Designs. In: IEEE Trans. Antennas Propag., vol. 54, no. 11, pp. 3075–3081 (2006).
5. Li, D. T. and J. F. Mao.: A Koch-Like Sided Fractal Bow-Tie Dipole Antenna. In: IEEE Trans. Antennas Propag., vol. 60, no. 5, pp. 2242–2251 (2012).
6. Sim, C.Y., W. T. Chung, and C. H. Lee.: Compact Slot Antenna for UWB Applications. In: IEEE Antennas Wireless Propag. Lett., vol. 9, pp. 63–66 (2010).
7. Q. Wu, R. H. Jin, J. P. Geng, and M. Ding.: Printed Omni-Directional UWB Monopole Antenna with Very Compact Size. In: IEEE Trans. Antennas Propag., vol. 56, pp. 896–899 (2008).
8. Huang, X. D., C. H. Cheng, and L. Zhu.: An Ultra wideband (UWB) Slotline Antenna under Multiple-Mode Resonance. In: IEEE Trans. Antennas Propag., vol. 60, No. 1, pp. 385–389 (2012).
9. Sayidmarie, K. H. and Y. A. Fadhel.: A Planar Self-Complementary Bow-Tie Antenna for UWB Applications. In: Progress In Electromagnetics Research, vol. 35, pp. 253–267 (2013).
10. Fereidoony, F., S. Chamaani, and S. A. Mirtaheri.: UWB Monopole Antenna with Stable Radiation Pattern and Low Transient Distortion. In: IEEE Antennas Wireless Propag. Lett., vol. 10, pp. 302–305 (2011).
11. Qu, Shi, J. Li, J. Chen, and Q. Xue.: Ultra Wideband Strip-Loaded Circular Slot Antenna with Improved Radiation Patterns. In: IEEE Trans. Antennas Propag., vol. 55, no. 11, pp. 3348–3353 (2007).
12. C. A. Balanis.: Antenna Theory and Design. John Wiley and Sons, New York (1997).
13. Zhang, Y. P. and J. J. Wang.: Theory and Analysis of Differentially-Driven Microstrip Antennas. In: IEEE Trans. Antennas Propag., vol. 54, no. 4, pp. 1092–1099 (2006).
14. Tong, Z., A. Stelzer, and W. Menzel.: Improved Expressions for Calculating the Impedance of Differential Feed Rectangular Microstrip Patch Antennas. In: IEEE Microwave. Wireless Compon. Lett., vol. 22, no. 9, pp. 441–443 (2012).

An Automatic Approach to Segment Retinal Blood Vessels and Its Separation into Arteries/Veins

Medhane Dipak and Shukla Aditi

Abstract The retinal fundus image consists of blood vessels which are further classified as arteries and veins. The measurement of retinal microvasculature changes by classifying arteries and veins using image processing opens window to find biomarkers and gives signs related to diabetic retinopathy, hypertensive retinopathy, hyperglycemia and blood pressure, etc. The purpose of this paper is to find major vessels in retinal image and automatically distinguish them into artery and vein. This paper gives an automated approach for artery–vein classification by analyzing graphical vasculature tree extracted from retinal image. Here, the proposed method distinguish the graphical retinal network by classifying each graphical node as end point, intersection point, and separate point node furthermore labeling each graphical links as artery or vein. Finally, artery–vein classification is performed on the basis of structural as well as intensity-based features. We have tested results of this method on publically available DRIVE database.

Keywords Artery • Artery–vein classification • Morphology • Retinal images • Retinal vascular graph • Vascular segmentation

1 Introduction

Blood vessels are classified into two types, veins and arteries; both transport blood in body organs [1, 2]. Segmentation and classification of arteries and veins are a primary constituent of automated retinal disease showing system. Retinal blood vessels are disturbed by many general diseases like hypertension, diabetic retinopathy, and high blood pressure [3, 4].

Medhane Dipak (✉)
Late G. N. Sapkal College of Engineering, Nashik, Maharashtra, India
e-mail: medhanedipak@gmail.com

Shukla Aditi
Sandip Institute of Technology & Research Center, Nashik, Maharashtra, India
e-mail: aditi.shukla020@gmail.com

© Springer Science+Business Media Singapore 2017

S.C. Satapathy et al. (eds.), *Proceedings of the International Conference on Data Engineering and Communication Technology*, Advances in Intelligent Systems and Computing 468, DOI 10.1007/978-981-10-1675-2_21

Both having various features with the differences between artery and vein are [5, 6].

- Veins are darker as it transports blood at low oxygen content
- Arteries are brighter as it transports blood which having high oxygen content
- Veins are thicker than arteries
- Central reflex is present in arteries which are larger and in veins it is minor

Contracting artery in retinal image is one of sign-related hypertension [7]. Veins are typically wide in diabetic patients [8] One of the signs of high blood pressure which is found on retinal image is thickened arteries, this blood pressure is typically stated by Artery to Venular ratio (AVR) [9, 10]. This AVR also indicates signs related to hypertensive retinopathy and diabetic retinopathy [11]. To find artery—vein ratio(AVR) it is needed to successfully segment the retinal blood vessel tree from input retinal color image, effective measurement of vessel width, and classification of blood vessels in arteries and veins [12, 13].

In this paper, we present an effective automated method for classifying blood vessels into artery and veins in retinal fundal images along with retinal vessel segmentation algorithm. It consists of three autonomous processing stages. First, retinal color image is preprocessed; and then morphological operations are performed on it to get segmented vessel centerlines; finally, we assign labels to each pixels for segmented retinal vessels which are then automatically classified into arteries and veins with the help of classifier.

2 Related Work

By calculating visual and geometrical feature, one can distinguish between artery and vein [14]. Based on this, various techniques were proposed in last several years to discriminate between artery and veins.

Martinez-Parez et al. [15] developed a semi-automated method for evaluating retinal vascular network. In this, they used multiscale feature extraction based on first and second spatial derivatives of images which gives topology of a retinal image. For tracking, particular tree segment input from user is needed and then it decides which vessel is artery and which is vein.

Grisan and Ruggeri [16] described a divide et impera method for automatic differentiation of retinal blood vessels into arteries and veins. Here they subdivided a concentric zone of retina around optic disk into four quadrants and performed local classification. One drawback of this system is that it is not planned for operate on whole vascular vessel zone, but it divides image in four quadrants.

Kondermann et al. [17] work on two different feature extraction methods; first is to extract vessel profile-based features and other is region of interest-based features of vessel centerline pixels. After extracting features, classification is done with help of two classifier support vector machine (SVM)-based and neural network-based.

Ouyang and Shao [18] proposed an optical coherence tomography (OCT)-based method to distinguish between arteries and veins. It mainly deals with geometrical properties of blood vessels, that is, vessel diameter can be calculated with help of OCT.

Rothaus et al. [19] described a semi-automatic rule-based technique with the help of presegmented and some hand-labeled retinal images. This rule-based system propagates the retinal blood vessel class (artery–vein) in whole vascular tree. Relan and Trucco [20] gives a Gaussian mixture modeled method which is based on color features of an image to distinguish between arteries and veins. Sivakami and Ahamed Gani [21] presented a SVM and graph-based technique to artery–vein classification in retinal images.

All above-mentioned methods used intensity-based features to classify blood vessels in artery and vein (A/V). Hence, the performance of above methods gets affected by the input image which consists of nonlinear luminosity, variation in contrast, image blurring, etc. For this reason, we proposed an automated method for A/V classification based on both structural information and visual information with the representation of vascular graph.

3 Proposed Method for Artery–Vein Classification

The proposed approach for A/V classification follows graph-based method; in this we concentrated on structural and visual characteristics of the retinal vessels. With vessel structural point of view, we have defined three different types of nodes (points) those are end nodes, splitting nodes, and intersection nodes. An end node is the vessel point where particular vessel gets terminated. Splitting node is a point where a vessel gets split to narrower parts. Intersection node is the point in vascular graph where two types of vessels intersect with each other.

Figure 1 shows the proposed method works in three phases (1) Morphological Segmentation (2) Nodal Analysis (3) A/V Classification. In morphological image segmentation, we extract retinal vascular graph from input image, and after that nodal analysis is performed on vascular graph to decide the type of nodes as mentioned. Based on nodal analysis, by extracting feature sets and using classifier, we assign labels to vessels whether it is artery or vein.

3.1 Morphological Segmentation

Segmentation divides an image into constitutes parts. It partitions an image into regions and extracts individual word or characters from the background. Vascular segmentation applies on contrast present between blood vessels and neighboring retinal background.

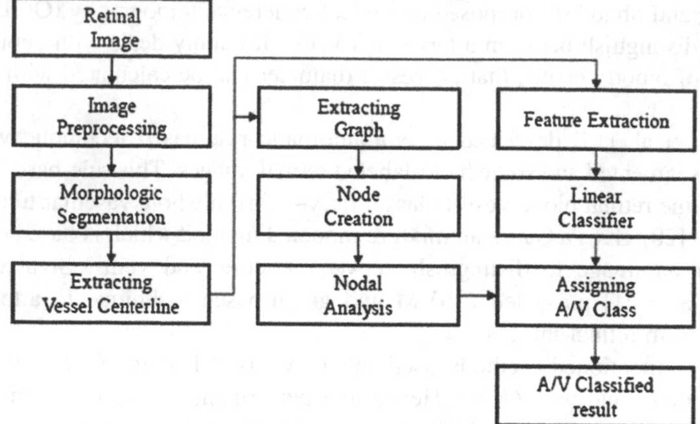

Fig. 1 Block diagram of proposed method of Artery–Vein Classification

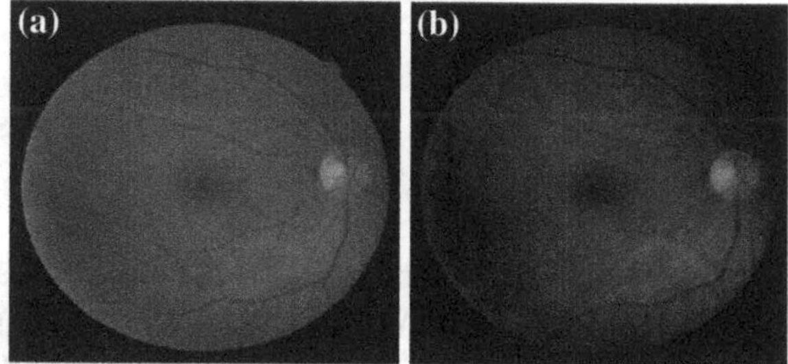

Fig. 2 Image preprocessing **a** Input image; **b** Green channel image

Input retinal image as in Fig. 2a is captured from digital fundal camera which is in tree channels red, green, and blue (RGB) form. These images require three-dimensional array in which, first plane presents intensities of red pixels, while second plane presents intensities of green pixels, and third plane presents intensities of blue pixels.

In preprocessing of retinal fundal image, we first divide image into gray level and extracts red, green, and blue channels Fig. 2b. We used green channel as it enhances contrast between retinal blood vessels and retinal background, such as hemorrhages. Similarly, red and blue channels provide too bright or dark retinal images respectively. In proposed work, we have performed retinal segmentation operation on monochromatic images in which blood vessels show high intensities than that of background.

In morphological segmentation, multiscale morphological vessel enhancement is performed using the bottom-hat transformation which uses morphological closing operation with large linear structuring elements of different sizes and in different directions. As in Fig. 3b, ball shaped with radius and height of ball, disk shaped with radius of disk, square shaped with size of square in pixels.

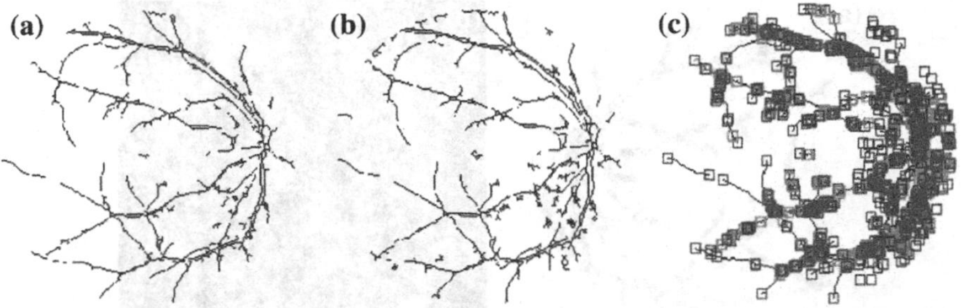

Fig. 3 Morphological vessel segmentation and nodal analysis **a** Global structuring element; **b** Ball-shaped structuring element (radius = 3) **c** Nodal analysis of segmented image (*Red* end nodes, *Green* splitting nodes, *Blue* intersection nodes) (color figure online)

In bottom-hat transform original binary image is get subtracted from morphologically closed (dilation followed by erosion) image resulting in enhanced blood vessels.

Here, sensitivity to noise is one limitation of bottom-hat transform, as pixel values in closed image operation are always greater than or equal to original image. To overcome this problem, we followed two steps; first we take mean of given image and performed morphological opening operation which removes all object pixels which are having edges and peaks and smoothen the vessels. In the next step, perimeter determination of pixels is carried out with eight connected neighborhoods in binary image. Here, vessel tracking is performed with morphological filtering operators.

3.2 *Nodal Analysis*

In nodal analysis, graph points are taken out from pixel centerline image by finding the connecting components in vessels. Output of nodal analysis is the decision of type of connecting points that are end nodes, intersection nodes, and splitting nodes. The output of this is represented in Fig. 3c as end node with red color, splitting node with green color, intersection node with blue color.

3.3 *Artery–Vein Classification*

A/V classification method consists of features, extracting and classifier stage. In nodal analysis, we have performed node classification and extract structural information, while in next stage we have to extract intensity-based features of retinal image for final differentiation among artery–vein classes.

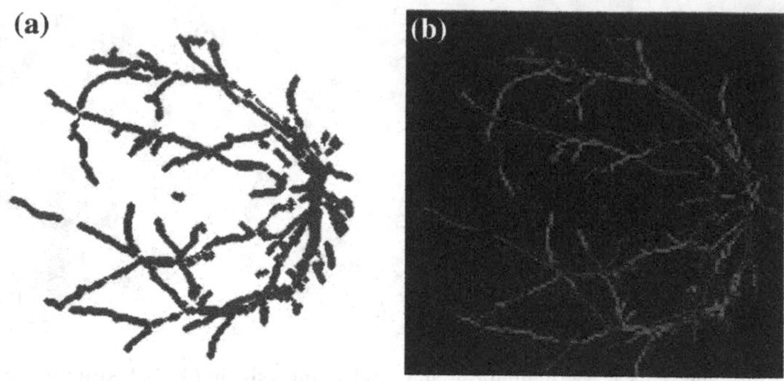

Fig. 4 A/V classification **a** Centerline pixel output; **b** Artery–vein classification result (*Red* classified artery, *Blue* classified vein) (color figure online)

We have enumerated some of following features for each centerline pixel-color intensity, hue, saturation, standard deviation of intensities of centerline pixels in vessel. We have supervised most frequently used k- Nearest Neighbor (k-NN) classifier on the DRIVE dataset. It estimates possibility of test data for its correctness to a specific class and depending on the class of training data samples in its nearest neighbor assign classes to vessel centerline pixels as shown in Fig. 4b where artery is represented by red color, while veins are represented by blue color.

4 Experimental Results and Discussion

The automated approach present in the previous section was tested on retinal images of Digital Retinal Images for Vessel Extraction (DRIVE) dataset, which contains 40 images. Images in DRIVE database were captured using Canon CR5 non-mydriatic 3CCD camera with 768 × 584 pixels along with 8 bits per color plane [22]. We have carried out results using MATLAB R2012b Software.

In the DRIVE dataset, for successfully classifying segments of vessels, we have used structural features along with k-NN classifier. The outcomes of proposed method for both (1) Automatic segmentation of retinal blood vessels (2) Automated artery–vein classification of blood vessels are compared with previous existing methods and we found that our proposed method of A/V classification outperforms by achieving segmentation accuracy of almost 94 % and A/V classification accuracy of 92.3 % for major retinal vessels as in Fig. 5 and in Fig. 6.

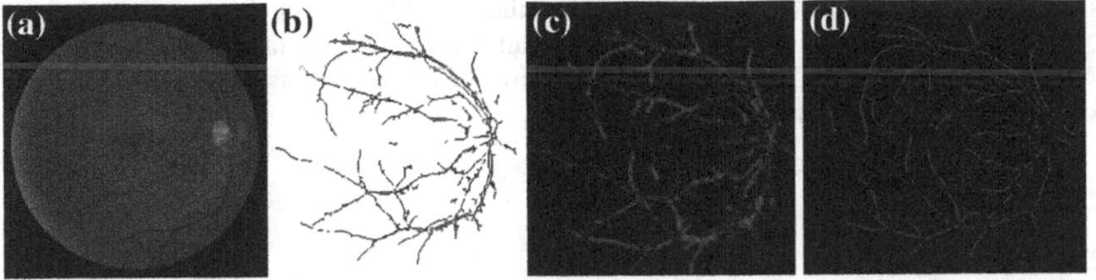

Fig. 5 DRIVE dataset best results **a** Original Image; **b** Morphologically segmented image; **c** Artery–Vein classification result; **d** Manually labeled image

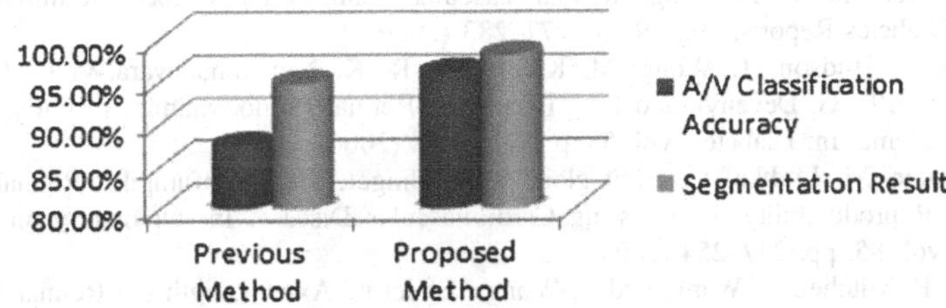

Fig. 6 Comparison of proposed and existing methods for Morphological segmentation and AV classification result

5 Conclusion

We have presented automatic approach for differentiating arteries and veins which is a necessary stage in automated diagnosis of retinal vascular changes. In this we have proposed a new different approach compared to that of all existing methods for classifying arteries and veins. One major advantage is that our method generates A/V classification result of whole retinal image instead of specific regions of retinal image.

For blood vessel segmentation purpose, most of the existing methods follows methods like matched filtering, supervised methods, vessel profile models; our method uses morphological vascular segmentation with the help of bottom-hat transformation. Also we have classified retinal segmented nodes in end nodes, intersection nodes, and splitting nodes. Proposed method for A/V classification uses structural as well as intensity-based features to distinguish between arteries and veins.

Furthermore, we compared achieved results of vascular segmentation and A/V classification with existing methods. We have found that our proposed method achieves high accuracy for major retinal vessels compared with all existing methods. Supplementary research is planned for automated retinal image analysis package; this will automatically give AVR, branching angles, vessel tortuosity, and fractal dimensions which will be helpful in early diagnosis of various systemic diseases like diabetes, hypertension, and various vascular disorders.

Acknowledgments The author would like to thank the Signal Processing department for pursuing this work; we have received help and support from all corners; and my family and friends for the encouragement and support. We convey our sincere thanks to authors of DRIVE dataset for making retinal image dataset open source.

References

1. N. Patton, T. M. Aslam, T. Mac Gillivray, I. J. Deary, B. Dhillon, R. H. Eikelboom, K. Yoge-sa, and I. J. Constable.: Retinal Image Analysis: Concepts, Applications and Potential. In: Progress in Retinal and Eye Research, vol. 25, pp. 99–127 (2006)
2. T. T. Nguyen and T. Y. Wong.: Retinal Vascular Changes and Diabetic Retinopathy. In: Current Diabetes Reports, vol. 09, pp. 277–283 (2009)
3. K. Guan, C. Hudson, T. Wong, M. Kisilevsky, R. K. Nrusimhadevara,W. C. Lam, M. Man-delcorn, R. G. Devenyi, and J. G. Flanagan.: Retinal Hemodynamics in Early Diabetic Macular Edema. In: Diabetes, vol. 55, pp. 813–818 (2006)
4. S. Neubauer, M. Ludtke, C. Haritoglou, S. Priglinger, and A. Kampik.: Retinal Vessel Analysis Reproducibility in Assessing Cardiovascular Disease. In: Optometry and Vision Science, vol. 85, pp. 247–254 (2008)
5. G. Liew, P. Mitchell, J. Wang, and T. Wong.: Effect of Axial Length On Retinal Vascular Network Geometry. In: American Journal of Ophtholmology, pp. 648–653 (2005)
6. S. R. Lesage, T. H. Mosley, T. Y. Wong, M. Szklo, D. Knopman, D. J. Catellier, S. R. Cole, R. Klein, J. Coresh, L. H. Coker, and A. R. Sharrett.: Retinal Microvascular Abnormalities and Cognitive Decline-The ARIC 14-year follow-up study. In: Neurology, vol. 73, pp. 862–868 (2009)
7. J. Leandro, R. Cesar, and H. Jelinek.: Blood Vessels Segmentation in Retina: Preliminary Assessment of the Mathematical Morphology and of The Wavelet Transform Technique. In: XIV Brazilian Symposium on Computer Graphics and Image Processing, pp. 84–90 (2009)
8. L. Gang, O. Chutatape, and S. Krishnan.: Detection and Measurement of Retinal Vessels in Fundus Images Using Amplitude Modified Second-Order Gaussian Filter. In: IEEE Transactions on Biomedical Engineering, vol. 49, pp. 168–172 (2002)
9. Hoover, V. Kouznetsova, and M. Goldbaum.: Locating Blood Vessels in Retinal Images by Piecewise Threshold Probing of a Matched Filter Response. In: IEEE Transactions on Medical Imaging, pp. 203–210 (2000)
10. T.T. Nguyen, J.J. Wang, and T.Y. Wong.: Retinal Vascular Changes In Prediabetes And Prehypertension: New Findings And Their Research And Clinical Implications. In: Diabetes Care, pp. 2708–2715 (2007)
11. Ciulla TA, Amador AG, Zinman B.: Diabetic Retinopathy and Diabetic Macular Edema: Pathophysiology, Screening, And Novel Therapies. In: Diabetes Care, pp. 2653–2664 (2003)
12. Leung H, Wang JJ, Rochtchina E, Tan AG, Wong TY, Klein R, Hubbard LD, Mitchell P.: Relationships Between Age, Blood Pressure, and Retinal Vessel Diameters in an Older Population. In: Investigative Ophthalmology & Visual Science PUBLICATION, vol. 44, pp. 2900–2904 (2004)
13. Wong TY, Islam FM, Klein R, Klein BE, Cotch MF, Castro C, Sharrett AR, Shahar E.: Retinal Vascular Caliber, Cardiovascular Risk Factors, and Inflammation the Multi-Ethnic Study of Atherosclerosis (MESA). In: Investigative Ophthalmology & Visual Science PUBLICATION, vol. 47, pp. 2341–2350 (2006)
14. Mitchell P, Cheung N, de Haseth K, Taylor B, Rochtchina E, Islam FMA, Wang JJ, Saw SM, Wong TY.: Blood Pressure and Retinal Arteriolar Narrowing in Children. In: Hypertension, pp. 1156–1162 (2007)

15. M. E. Martinez-Perez, A. D. Hughes, A. V. Stanton, S. a Thom, N. Chapman, A. a Bharath, and K. H. Parker.: Retinal Vascular Tree Morphology: A Semi-Automatic Quantification. In: IEEE Transactions On Biomedical Engineering, pp. 912–917 (2002)
16. E. Grisan and A. Ruggeri.: A Divide Et Impera Strategy for Automatic Classification of Retinal Vessels into Arteries and Veins. In: Proceedings of the 25th Annual International Conference of the IEEE Engineering in Medicine and Biology Society, pp. 890–893 (2003)
17. Kondermann, D. Kondermann, and M. Yan.: Blood Vessel Classification into Arteries and Veins in Retinal Images. In: Proceedings of SPIE 6512, Medical Imaging: Image Processing (2007)
18. Yanling Ouyang, Qing Shao.: An Easy Method To Differentiate Retinal Arteries From Veins By Spectral Domain Optical Coherence Tomography: Retrospective, Observational Case Series. In: BMC Ophthalmology, pp. 1471–2415 (2014)
19. K. Rothaus, X. Jiang, and P. Rhiem.: Separation of The Retinal Vascular Graph in Arteries and Veins Based Upon Structural Knowledge. In: Image & Vision Computing, vol. 27, pp. 864–875 (2009)
20. Relan, E. Trucco.: Retinal Vessel Classification: Sorting Arteries and Veins. In: IEEE International Conference Proceeding engineering in medicine & Biology Society, pp. 7396–7396 (2013)
21. S. Sivakami, V.K.U. Ahamed Gani.: SVM & Graph Based Artery/Vein Classification in Retinal Images. In: IJIRCCE, pp. 291–295 (2015)
22. DRIVE: Digital Retinal Images for Vessel Extraction, http://www.isi.uu.nl/Research/Databases/DRIVE/

Design of Microstrip Antenna with Improved Bandwidth for Biomedical Application

K. RamaDevi, A. Jhansi Rani and A. Mallikarjuna Prasad

Abstract In this paper, a single element microstrip antenna will be designed at 6 GHz in rectangular shape for biomedical application to detect breast cancer tissues. The antenna will be designed in HFSS-Ansoft version 13 and simulated and analyzed in terms of return loss and gain. This work further extended to enhance bandwidth applying some techniques like air-gap, capacitive feed, and slot on patch with different structures.

Keywords Rectangular patch · Air-gap · Capacitive feed · Slot on patch

1 Introduction

The opt antenna for biomedical application for detecting breast cancer tissues [1, 2] at microwave range is a microstrip antenna which is also called as patch antenna [3–5]. Microstrip patch antenna consists of a patch of metal which acts as radiating element on top of the grounded dielectric substrate of thickness h in mm, with relative permittivity and permeability ε_r and $\mu_r = 1$ respectively as shown in Fig. 1. The metallic patch may be of various shapes with rectangular, circular, and triangular, hexagon, etc.

Generally microstrip antennas are excited with different feed techniques like **coaxial feed**, inset feed, coplanar feed, aperture coupling, etc. The simplest feed technique is a coaxial feed at 50 Ω. The schematic diagram is shown in Fig. 2a. One of the limitations of single layer antenna is low bandwidth. To enhance the

K. RamaDevi (✉) · A.M. Prasad
JNTU College of Engineering, Kakinada, Andhra Pradesh, India
e-mail: kolisettyramadevi@gmail.com

A.M. Prasad
e-mail: a_malli65@yahoo.com

A.J. Rani
V.R. Siddhartha Engineering College, Vijayawada, Andhra Pradesh, India
e-mail: jhansi9rani@gmail.com

© Springer Science+Business Media Singapore 2017

S.C. Satapathy et al. (eds.), *Proceedings of the International Conference on Data Engineering and Communication Technology*, Advances in Intelligent Systems and Computing 468, DOI 10.1007/978-981-10-1675-2_22

Fig. 1 Rectangular patch microstrip antenna

Fig. 2 **a** Schematic diagram of coaxial feed microstrip antenna. **b** Schematic diagram of capacitive feed microstrip antenna

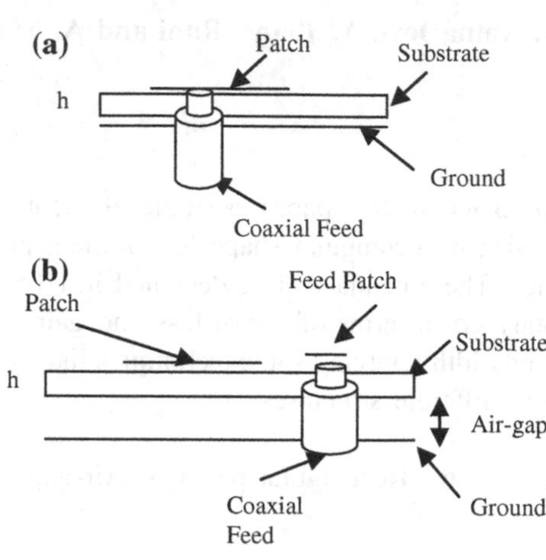

bandwidth, coaxial feed is connected to small patch (capacitive feed) through air-gap and substrate shown in Fig. 2b.

The performance of an antenna depends on the shape of the radiator and generally measured in terms of return loss S_{11} in dB, radiation pattern, gain, directivity, frequency bandwidth from return loss of below 10 dB, impedance, etc.

Capacitive-feed technique [2] and their response are observed and further extended with air-gap microstrip antenna for breast cancer detection applications with improved band width and gain [6]. Pentagon-shaped radiator for wireless communications that makes a slot to the patch was observed [7–9] for high gain and improvement in impedance bandwidth. Before designing, the practical antenna can be simulated using software like **ANASOFT-HFSS**^{TM} [10]. By simulation, antenna characteristics can be analyzed and synthesized.

2 Design Procedure

In this paper, an antenna is designed at 6.0 GHz microstrip antenna with rectangular-shaped patch (Fig. 3) having linear polarization with coaxial feed. Further, work is extended to different models of patch antennas to enhance the bandwidth and gain. Here, rectangular patch dimensions 'W' and 'L' are calculated from Eq. (1). This antenna will be designed and simulated in HFSS-Ansoft designer

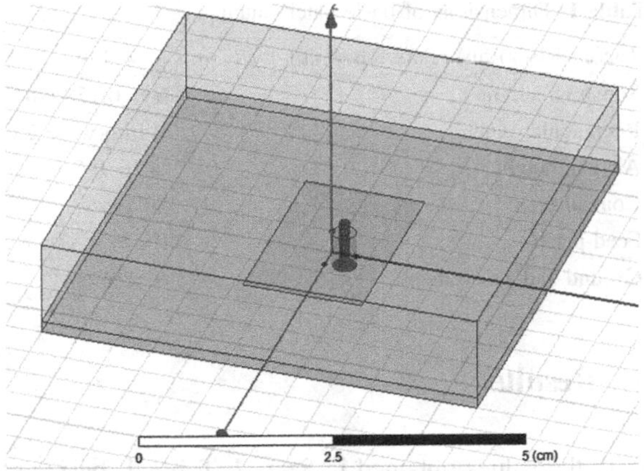

Fig. 3 Rectangular microstrip antenna in Ansoft-HFSS

version 13 which is working in finite discrete time domain (FDTD). Then bandwidth will be calculated from return loss (dB) below 10 dB and radiation pattern in terms of gain (dB).

After simulating antenna, air-gap (h_a) in 'mm' is applied between ground and substrate (h_{d1}) [8]. The average permittivity (ε_{av}) is calculated from Eq. (2). Later, the work is extended to design capacitive feed antenna instead of direct coax feed antenna. Later, slots are designed on patch and simulated. Simulated results will be discussed in Sect. 3.

Equations for Rectangular Patch Antenna:

$$\text{Width of the patch } W = \left[\frac{C}{2f_r} * \left[\frac{2}{\in_r + 1}\right]^{1/2}\right] \tag{1}$$

where 'C' is velocity of light in free space

f_r is a resonant frequency or operating frequency.

Length of the patch 'L' is generally **λo/3 < L < λo/2**

$$\in_{av} = \left[\frac{\in_{rhd1} + \in_{rha}}{\frac{h_t}{2}}\right] \tag{2}$$

where $h_t = h_{d1} + h_a$.

The antenna structure shown in Figs. 1 and 2 is designed with TaconicRF-30 (tm) material $\varepsilon_r = 3.0$ and loss Tangent 0.0013 as substrate and calculated values are given in Table 1.

Table 1 Dimensions of basic patch antenna

TaconicRF-30(tm) relative permittivity	$\varepsilon_r = 3.0$
Substrate height	h = 0.158 cm
Rectangular patch	Width(w) = 2.65 cm, Length(L) = 1.65 cm
Air-gap height	'a' = 0.6 cm
Coaxial feed	50 Ω
Feed-patch	Fw = 0.37 cm and Fl = 0.12 cm
Ground and substrate size	$6 \times 6\ \text{cm}^2$

3 Results

With the values obtained in Sect. 2, Table 1, basic antenna has been designed and simulated in HFSS-Ansoft. Results have been determined in terms of return loss curve and gain pattern. The results are as follows:

- The rectangular patch antenna at 6 GHz with coaxial feed is shown in Fig. 3 and simulated in ANSOFT-HFSS. The return loss in dB as shown in Fig. 4 shows the antenna resonation at two frequencies $f_r1 = 6.27$ and $f_r2 = 9.94$ GHz. The bandwidth range at fr2 calculated from return loss below 10 dB is 9.44 GHz to 10.44 GHz and bandwidth is 1 GHz. The antenna is also called dual-frequency antenna.
  Figure 5 shows the gain response of Fig. 3 at first resonance that is 5.89 dB maximum along 30° end-fire direction and maximum gain is 9.17 dB obtained at fr2, i.e., 9.94 GHz along 30° (red line at Φ = 0°) as shown in Fig. 6.

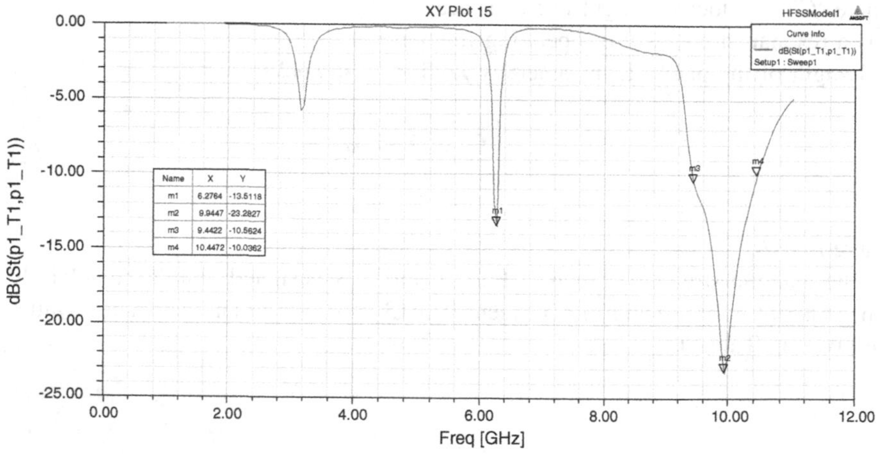

Fig. 4 Return loss S_{11} in dB of basic antenna

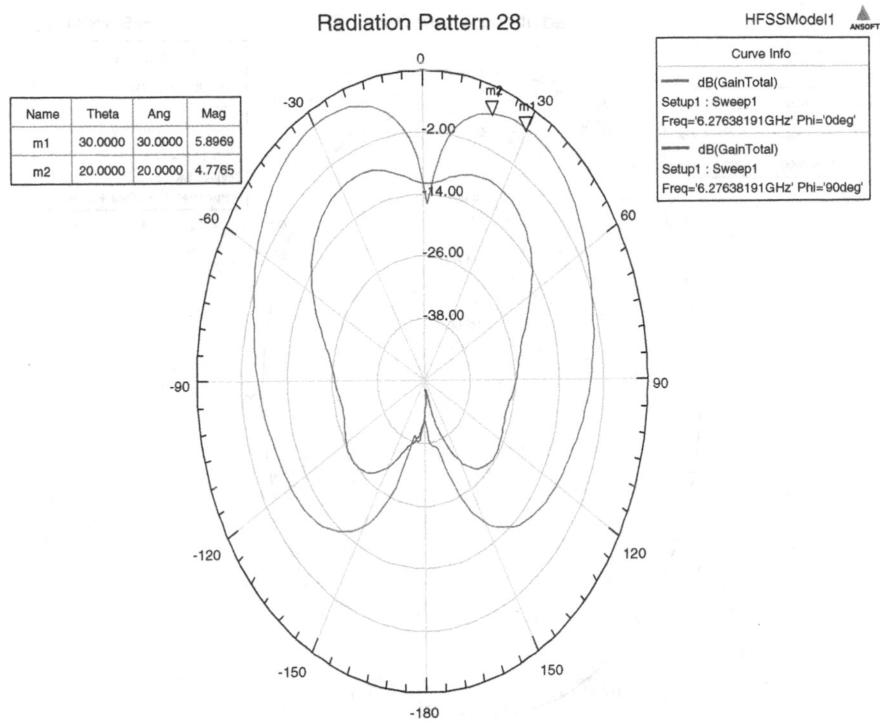

Fig. 5 Radiation pattern, Gain in dB at 6.27 GHz

- Figure 3 is modified by including an air-gap of 4 mm between ground and substrate and simulated and observed the response. An air-gap antenna without lateral dimension change is shown in Fig. 7.
  Return loss curve (Fig. 8) shows that the antenna has resonated at 3.86 GHz and with gain 5.49 dB along 0° direction (Fig. 9) is also known as miniature antenna.
- Air-gap size was adjusted from 4 mm to 6 mm for shifting the resonant frequency towards operating frequency, and feed is shifted to capacitive feed and observed. The response in terms of return loss and gain in dB is shown in Figs. 10, 11 and 12.
- Return loss Fig. 11 shows that antenna is resonated at 5.42 GHz with return loss 7.88 dB and gain is 4.71 dB along 30° direction.
- To enhance the bandwidth of an above antenna, a slot is made here on the patch. Rectangular slot will make on a patch parallel to feed. Structure and response of an antenna shown in Figs. 13, 14 and 15.

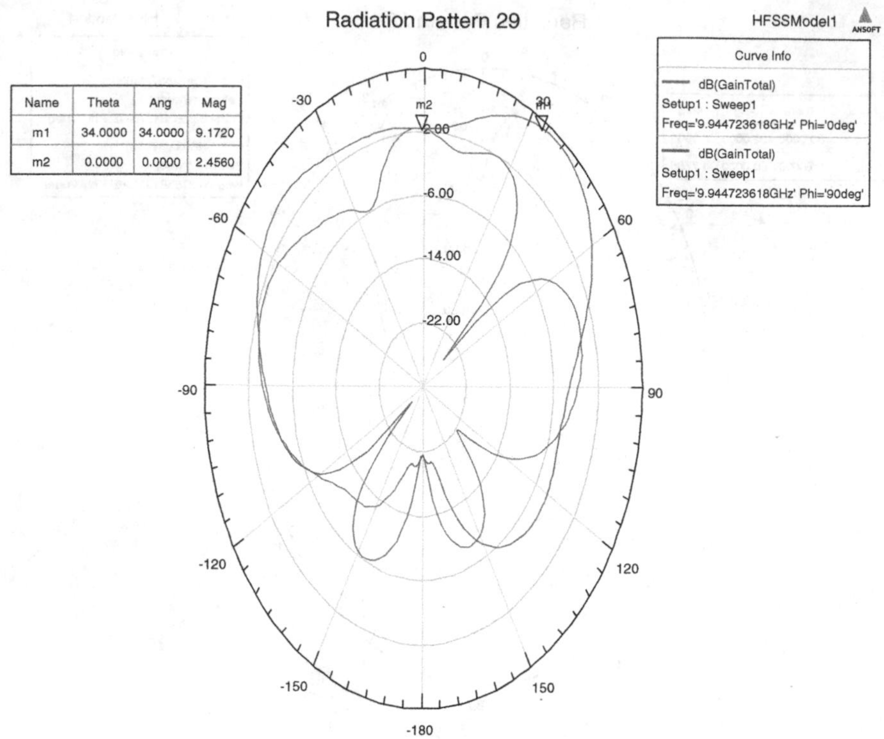

Fig. 6 Radiation pattern, gain in dB at 9.94 GHz

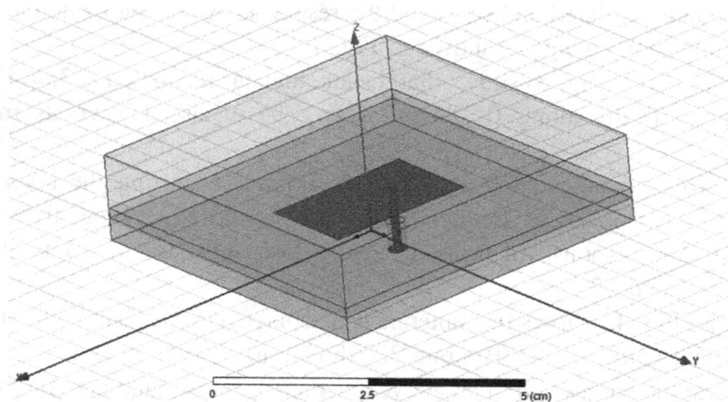

Fig. 7 Air-gap microstrip antenna in Ansoft-HFSS

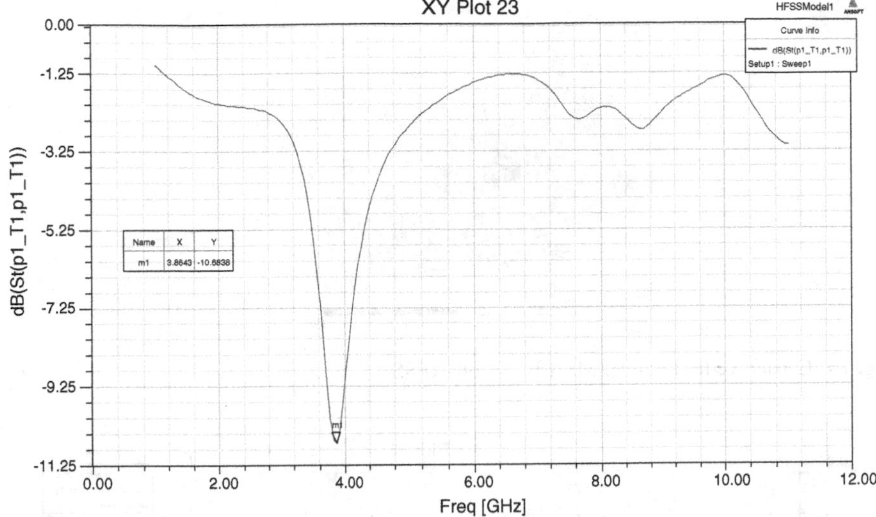

Fig. 8 Return loss S_{11} in dB of Air-gap MSA

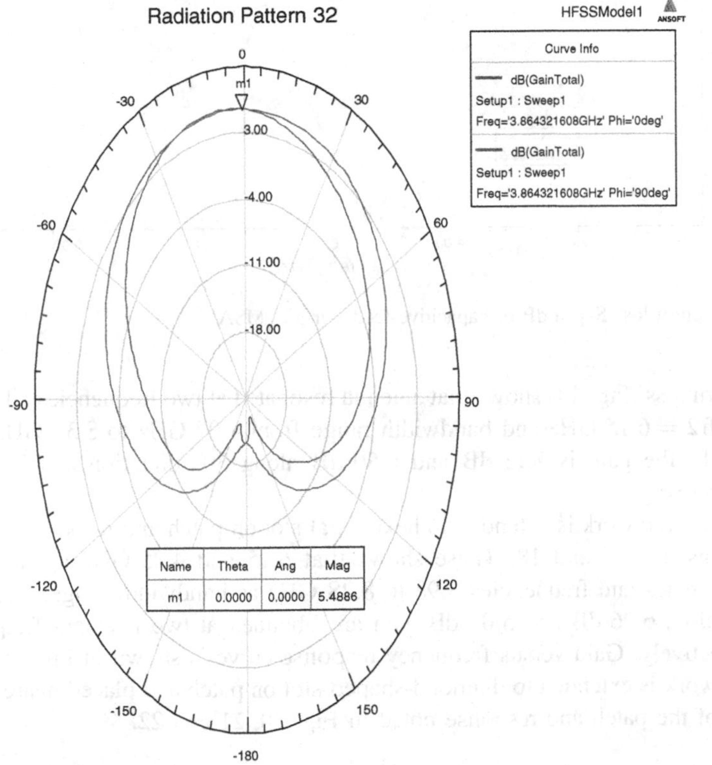

Fig. 9 Radiation pattern, gain in dB at 3.86 GHz

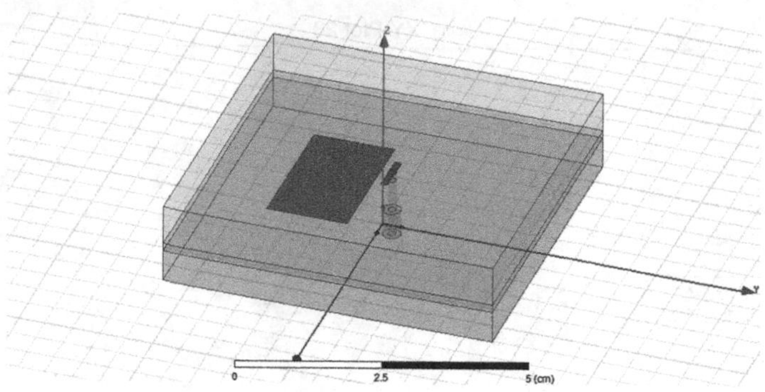

Fig. 10 Capacitive-feed air-gap MSA in Ansoft-HFSS

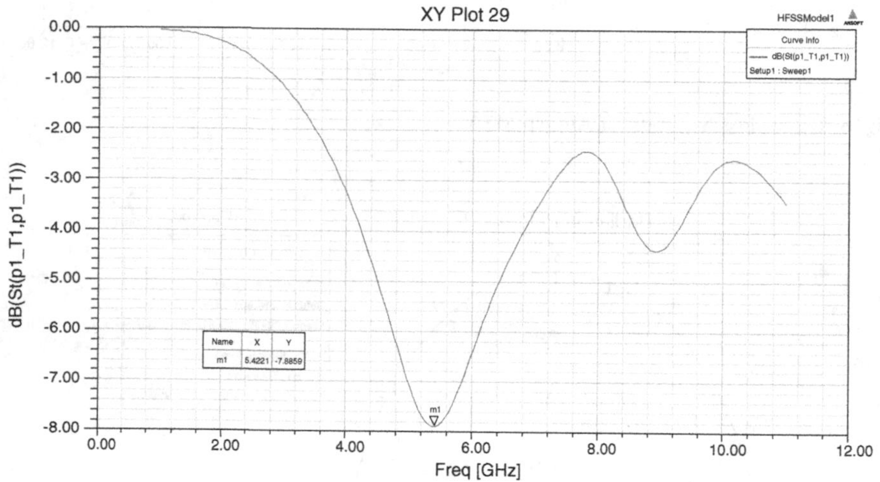

Fig. 11 Return loss S_{11} in dB of capacitive-feed Air-gap MSA

Return loss (Fig. 14) shows that antenna resonated at two frequencies $f_r1 = 6.87$ and fr2 = 6.15 GHz and bandwidth range from 4.07 GHz to 8.37 GHz. From Fig. 15 the gain is 7.13 dB and 6.50 dB along 30° direction at f_r1 and f_r2, respectively.

- Further, the work is extended to hexagonal slot on patch and response is shown in Figs. 16, 17 and 18. These shows that 6.15 and 4.25 GHz are resonance frequencies and frequencies 3.98 to 8.28 GHz as bandwidth range. Along 30° direction, 6.96 dB and 5.05 dB gain are obtained at two resonant frequencies respectively. Gain versus frequency response curve is shown in Fig. 19.
- The work is extended to diamond-shaped slot on patch and placed nearer to one end of the patch and response noted in Fig. 20, 21 and 22.

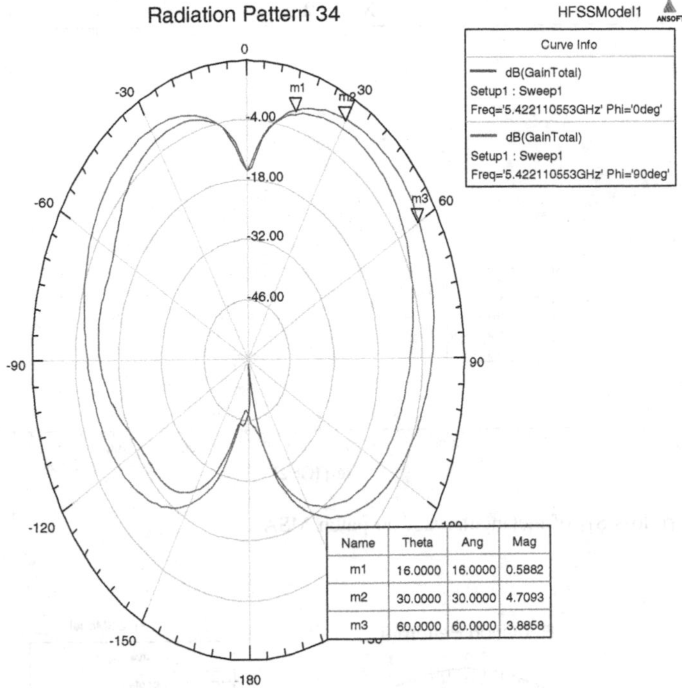

Name	Theta	Ang	Mag
m1	16.0000	16.0000	0.5882
m2	30.0000	30.0000	4.7093
m3	60.0000	60.0000	3.8858

Fig. 12 Radiation pattern, gain in dB at 5.42 GHz

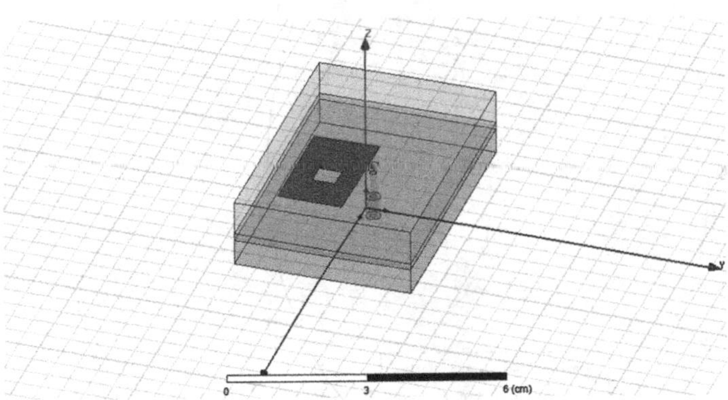

Fig. 13 Rectangular slot on patch MSA

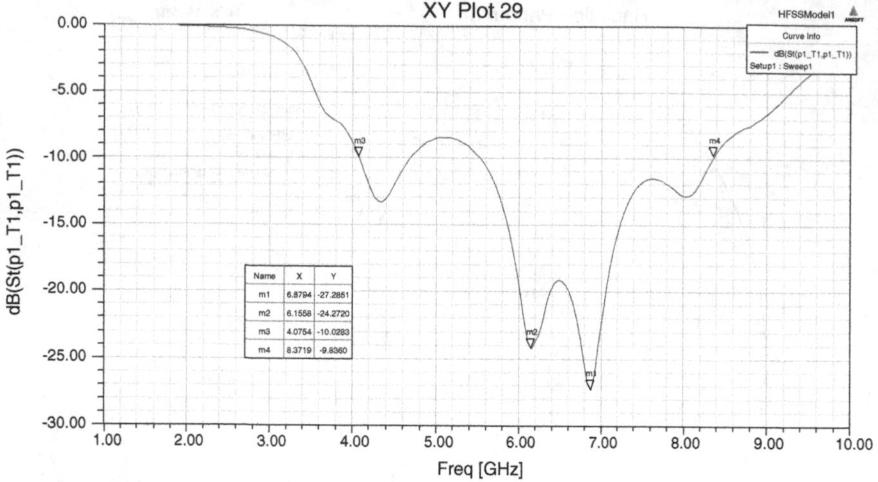

Name	X	Y
m1	6.8794	-27.2851
m2	6.1558	-24.2720
m3	4.0754	-10.0283
m4	8.3719	-9.8360

Fig. 14 Return loss S_{11} of rectangular slot on patch MSA

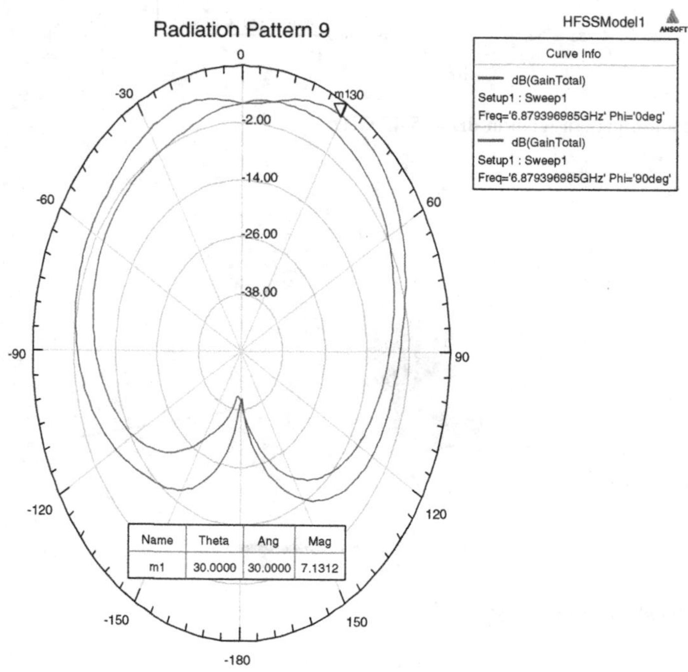

Name	Theta	Ang	Mag
m1	30.0000	30.0000	7.1312

Fig. 15 Gain in dB at 6.87 GHz

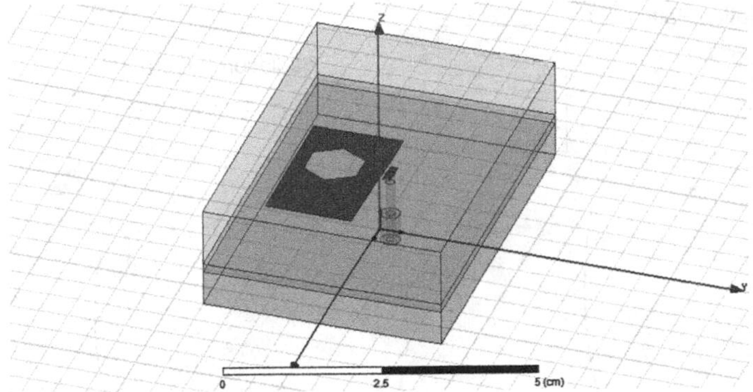

Fig. 16 Hexagonal slot on patch MSA

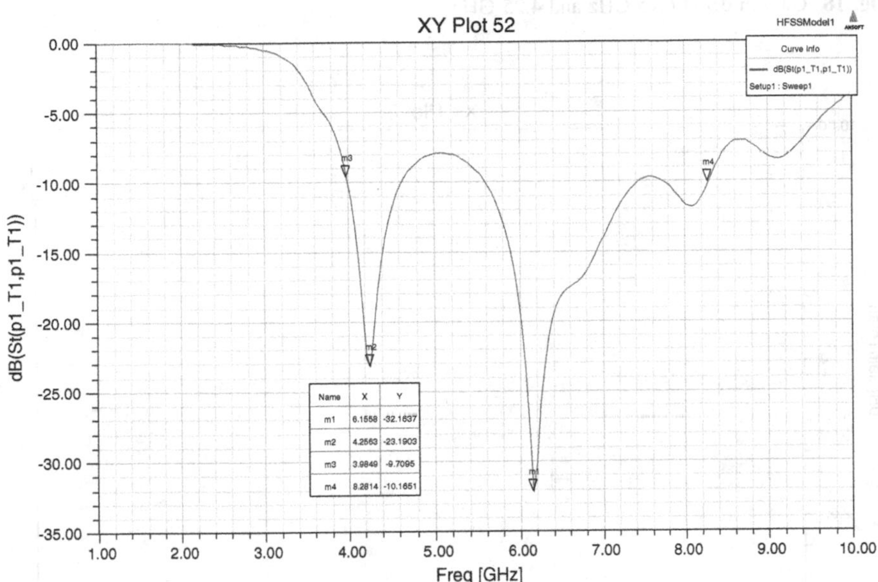

Fig. 17 Return loss S_{11} of hexagonal slot on patch MSA

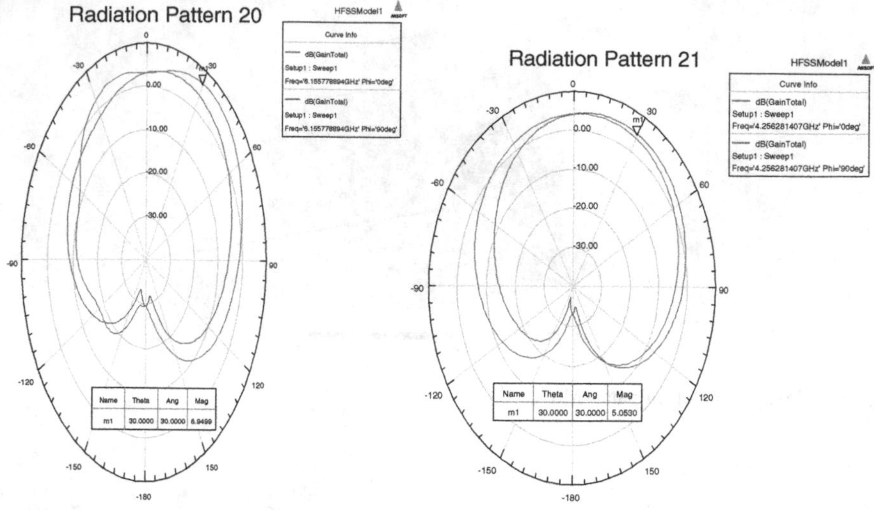

Fig. 18 Gain in dB at 6.15 GHz and 4.25 GHz

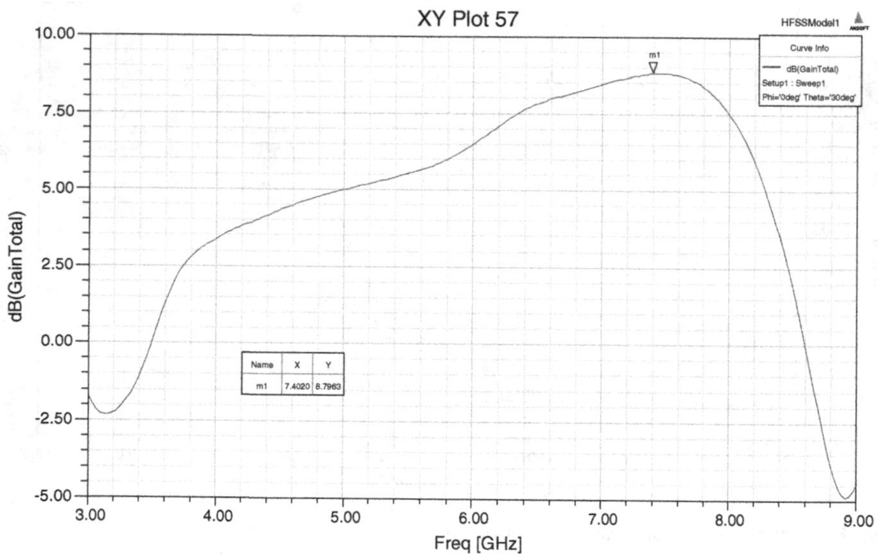

Fig. 19 Gain(dB) versus frequency response

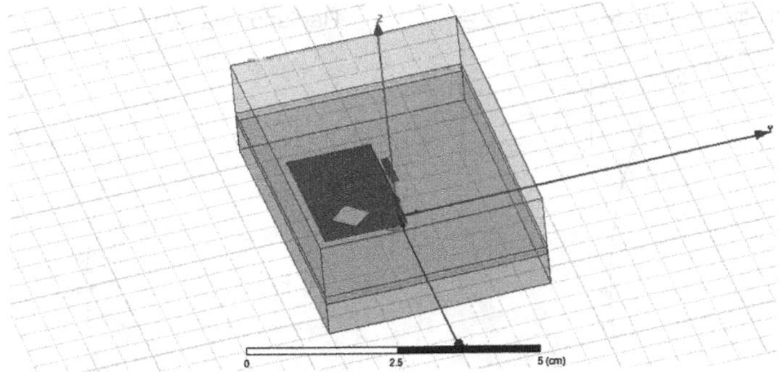

Fig. 20 Diamond-Shaped Slot on patch MSA

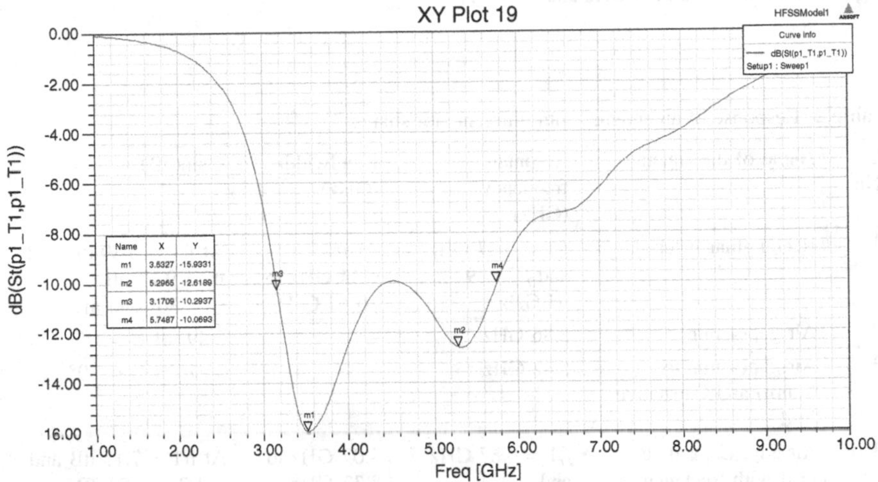

Fig. 21 Return loss S_{11} of diamond slot on patch MSA

The antenna resonated at 3.53 GHz and 5.34 GHz with 3 dB and 5.28 dB gain along 15° direction respectively. The bandwidth of an antenna is 3.17 GHz to 5.75 GHz. The performance of an antenna with different slot shapes is tabulated in Table 2.

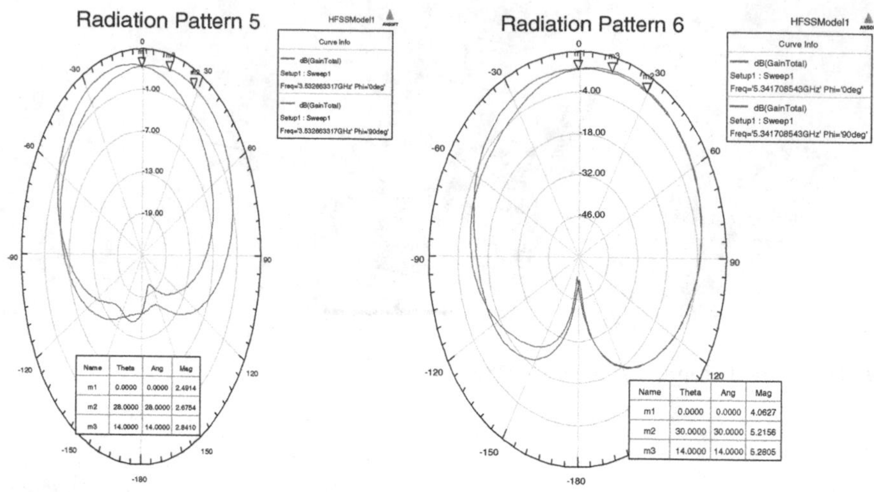

Fig. 22 Gain in dB at 3.53 GHz and 5.34 GHz

Table 2 Response of an antenna with different slot shapes

S. No	Name of the antenna	Resonant frequency (GHz)	Bandwidth (GHz)	Gain(dB)
1	Rectangular patch	f_r1 = 6.27 GHz and f_r2 = 9. 94 GHz	At fr2 9.44 GHz to 10.44 GHz	At fr1 = 5.89 dB and fr2= 9.17 dB Θ = 30° end-fire
2	Air-gap(4 mm)	3.86 GHz	–	5.49 dB Θ = 0°
3	Air-gap(4 mm to 6 mm) and Capacitive feed	5.42 GHz	–	4.71 dB Θ = 30°
4	Air-gap, Capacitive feed with Rectangular slot on patch	f_r1 = 6.87 GHz and fr2 = 6.15 GHz	4.07 GHz to 8.37 GHz	At fr1 = 7.13 dB and at fr2 = 6.50 dB Θ = 30°
5	Air-gap, capacitive feed with hexagonal slot on patch	f_r1 = 6.15 GHz and f_r2 = 4.25 GH	3.98 GHz to 8.28 GHz	At fr1 = 6.96 dB and fr2 = 5.05 dB Θ = 30°
6	Air-gap, capacitive feed with diamond slot on patch	f_r1 = 3.53 GHz and f_r2 = 5.34 GHz	3.17 GHz to 5.75 GHz	At fr1 = 3 dB and fr2 = 5.28 dB Θ = 15°

4 Conclusion

Primary structured microstrip antennas are suitable at microwave frequencies because of low weight. But they suffered low bandwidth. Some of the techniques like air-gap placed between substrate and ground, capacitive feed and slot on patch are enhanced the bandwidth. In this paper, the basic antenna dimensions are calculated from Eq. (1) and simulated in HFSS-Ansoft version 13. Figure 4 shows that antenna has been resonated at two frequencies with less bandwidth. The work has been extended step by step and observed. And the results show noticeable enhancement in bandwidth with small change in gain. The basic antenna radiation pattern shows that maximum gain has radiated in 30° direction.

References

1. Xiaolu Guo.: Simulation and design of an UWB imaging system for breast cancer detection: INTEGRATION: pp. 1–12. the VLSI journal, 2014
2. Mahdi Ali, Abdennacer Kachouri and Mounir Samet: Novel method for planar microstrip antenna matching impedance: Volume 2, Issue 2, May 2010, pp. 131–138, Journal of Telecommunications
3. J. L. Volakis: Antenna Engineering Handbook: 4th ed. New York: McGraw Hill, 2007
4. Balanis C.A: Antenna Theory: Analysis and Design: 2nd Edition, John Wiley and Sons, NewYork
5. Ramesh Garg, Prakash Bhartia, Inder Bhal, Apisak Ittipiboon: Microstrip Antenna Design Handbook: Inc., 2001, Artech House
6. Sangam Kumar Singh and Arun Kumar Singh: UWB Rectangular Ring Microstrip Antenna with Simple Capacitive Feed for Breast Cancer Detection: March 23–27, 2009, pp. 1639–1642. Progress In Electro magnetics Research Symposium, Beijing, China
7. Sunil Kumar, Rajgopal and Satish Kumar Sharma: Investigation on UWB Pentagon Shape Microstrip Slot Antenna for Wireless Communications: Vol. 57, No.5, May 2009. IEEE Transactions on Antenna And Propagation
8. RSA Raja Abdullah, D Yoharaaj & A Ismail: Bandwidth Enhancement for Microstrip Antenna in Wireless Applications: Vol2.No.6, 2008, pp. 179–187. Modern Applied Science
9. K. Rama Devi, A. Mallikarjuna Prasad, A. Jhansi Rani: Design of A Pentagon Microstrip Antenna for Altimeter Application: Vol.3,No-4, October2012, ISSN 0976-2280, PP. 31–42 International journal of Web &Semantic Technology(IJWesT)
10. Ansoft Corporations, Designer and High Frequency Structure Simulator (HFSS) [Online]. Available: www.ansoft.com

Extended ExOR Opportunistic Routing Algorithm Based on Broadcasting

Smita Shukla Patel and A.B. Bagwan

Abstract Extensive research has been done to use broadcasting nature of wireless links, instead of making wireless links as good as wired one. ExOR is the milestone piece of work in this area. Routing protocols based on the inherent nature of wireless link (i.e., broadcast) trigger the opportunity to extensively used network layer to maintain the correct path for Mac layer. This research work proposed broadcasting nature of wireless network to implement opportunistic routing algorithm. This is the extension of ExOR routing protocol which is based on opportunistic in on demand routing creation and topology changes propagate to all nodes.

Keywords ExOR · Opportunistic routing · Ad hoc network

1 Introduction

There are two areas of work in opportunistic routing in ad hoc network (Table 1):

1. Proactive opportunistic routing algorithm: In this area, nodes try to maintain the routing paths for all destination any time.
2. Reactive (On Demand) opportunistic routing algorithm: In this area, nodes try to identify the routing path for required destination, and at the same time try to incorporate topology changes

Proactive routing is more popular and realistic as topology changes can be identified in advance. Also lightweight proactive routing protocol is used to gather the neighboring node information to maintain the route paths [1, 2].

S.S. Patel (✉)
Karpagam University, Coimbatore, India
e-mail: Smitapatel7122006@gmail.com

A.B. Bagwan
J.S.P.M College of Engineering, Pune, India
e-mail: Aliakbar.bagwan@gmail.com

© Springer Science+Business Media Singapore 2017

S.C. Satapathy et al. (eds.), *Proceedings of the International Conference on Data Engineering and Communication Technology*, Advances in Intelligent Systems and Computing 468, DOI 10.1007/978-981-10-1675-2_23

Table 1 HELLO message

Maximum	Number of period of ICDECT 2016
ALLOWED_HELLO_LOSS. HELLO_INTERVAL	HELLO_INTERVAL to wait without receiving a HELLO message
NETWORK_ADDRESSES	At least one network address of each MANET interface of the router
NODE_CAPACITY	**BATTERY_RANK, PROCESSING_POWER, FORWARDING_CAPACITY**
CONNECTED_LINKS	ROUTE_TABLE for connected links ROUTE_TABLE NETWORK_ADDRESS and latest HELLO message received from it contains

HELLO message computing: HELLO messages received by a router are used to update the Interface Information Base for the MANET interface over which that HELLO message was received. HELLO message propagation makes sure that each node has up to date table for all connected node. This up to date information will be used to perform route identification for packets destination

Some works have been done in proactive opportunistic routing based on clustering the nodes. These are based on identifying the most suitable node as cluster head. Cluster head is responsible to maintain the routing table for all cluster members including distance, cost, and energy level. It has extra overhead to perform clustering routine.

Few other works proposed probabilistic routing protocol, which is based on the history of past encounter and transitivity to estimate node's delivery capability. It is not suitable for highly mobile ad hoc network, and delivery ratio is very low. Link-State protocol is useful to estimate the weight of each path between source and destination [3, 4].

Based on the existing opportunistic routing protocols survey, following are the major points I am considering in my work.

2 Proposed Work

2.1 Neighborhood Discovery

Maintaining complete path for any destination on each node is cumbersome and most of the time not fully utilized. However, bare minimum topology information is required on each node to perform opportunistic routing. I am using controlled HELLO messaging, which is a modified version of traditional MANET HELLO message protocol (RFC 6130) to contain node's resources capacity. Node resources capacity is based on self-rating mechanism. Each HELLO message generator marks them from 1 to 10 to provide resources capacity information. '1' is lowest and '10' is highest capacity [2–5].

2.2 Capacity Calculation

BATTERY_RANK = Abs ((REMAINING_BATERRY_PERCENTAGE/MEMORY_CONSUMPTION_PERCENTAGE) * TOLERANCE_FACTOR)
PROCESSING_POWER = Abs((AVAILABLE_STORAGE + VIRTUAL_MEMORY)/MEMORY_CONSUMPTION_PERCENTAGE) * TOLERANCE_FACTOR)

** The node based on available MEMORY CAPACITY decides TOLERANCE_FACTOR. For 150 MB memory capacity, consider '10' as TOLERANCE_FACTOR.

2.3 FORWARDING_CAPACITY

Is based on the available route table and other capacity available. It is a table containing FORWARDING_CAPACITY to all other connected nodes.

E.g.

- N1 is connected with N2 and N3.
- FORWARDING_CAPACITY will contain entry for N2 and N3.
- Formula to calculate FORWARDING_CAPACITY for any connected node is

FORWARDING_CAPACITY = [Average (N1's BATTERY_RANK, N1's PROCESSING_POWER, FORWARDING_CAPACITY) of {N2| N3}] TOLERANCE_FACTOR.

3 HELLO Message

- HELLO message contains the following:

4 Routing Algorithm

Traditional Opportunistic broadcast routing is used. Following are the routine steps every forwarder will follow:

1. Route Candidate Selection: Source use the ***route table*** to identify the candidate for packets to be sent. Candidate selection will be based on **NODE_CAPACITY.** Selected candidates will be embedded in packet's header as proposed forwarder list. Nodes in the list are ordered according to the **FORWARDING_CAPACITY.**

2. Packet Batch creation: Packets are sent as batches. Batch creation is based on the NODE_CAPACITY of selected *mid level forwarder.* Reason for not creating the batch based on the highest priority node is to reduce the probability of packet retransmission in next forwarder.
3. Broadcast the packets with forwarder list

- Receiver priority in 'routing path list' is based on link strength (closer one has high priority) [ExOR]. Traditional opportunistic protocol does not consider the resources availability in prioritized receiver node.
- Batch packet sending is efficient idea, but it can be further optimize in following way: o In case of retransmission due to bath delivery failure, batch recreation can be an efficient way to save resources.
- Broadcast the acknowledgment to confirm as sender. Based on priority forwarder. Every other forwarder holds on, and waits for ***CONFIRM_SENT*** signal. Confirm Sent signal will be sent by the ultimate sender of the packet to next hop. Ultimate sender rerun the routine protocol, i.e., update the forwarder list, recreate the batch (if required).
- Once MAC transmits the batch packets, it will broadcast the ***CONFIRM_SENT*** signal. ***CONFIRM_SENT*** signal contains the range of packets successfully sent to next hop [range contain the packet_ids). Other nodes drop the transmitted packet range from the cache.
- If receiver is the ultimate destination, it will broadcast the **REMAINING_PACKET_REQUEST** or PACKET_**RECEIVED_MESSAGE**

5 Conclusion

There are so many algorithms that are available for opportunistic routing. This paper proposed routing algorithm which is an extension of ExOR algorithm.

In this paper, we discussed routing algorithm which is based on broadcast nature of opportunistic network.

Proposed algorithm is advantageous over previous algorithm on various parameters like reliability and packet loss.

References

1. Network Protocols, 2008, ICNP, IEEE International Conference, 2008.
2. Dimitrios Koutsonikolas, Chih-Chun Wang, Y. Charlie Hu, 'CCACK Efficient Network Coding Based Opportunistic Routing Through Cumulative Coded Acknowledgments', INFOCOM, Proceedings IEEE, 2010.
3. Kurth, M., Humboldt-Univ. zu Berlin, Berlin, Zubow, A., Redlich, J.P., 'Cooperative Opportunistic Routing Using Transmit Diversity in Wireless Mesh Networks', INFOCOM, The 27th Conference on Computer Communications, IEEE, 2008.

4. Szymon Chachulski, Michael Jennings, Sachin Katti and Dina katabi, 'MORE A Network coding approach to Opportunistic routing', ACM Sigcomm, 2007.
5. Yunfeng Lin, Baochun Li, Ben Liang, 'CodeOR Opportunistic Routing in Wireless Mesh Networks with Segmented Network Coding'.
6. Xinyu Zhang, Baochun Li, 'Optimized Multipath Network Coding in Lossy Wireless Networks', IEEE journal on selected areas in communications, vol. 27, no. 5, June 2009.
7. Dimitrios Koutsonikolas, Y. Charlie Hu and Chih-Chun Wang, 'XCOR Synergistic Interflow Network Coding and Opportunistic Routing', In the ACM International Conference on Mobile Computing and Networking, San Francisco, CA, September 14–19, 2008.
8. Che-Jung Hsu, Huey-Ing Liu, Winston Seah, 'Economy a duplicate free Opportunistic Routing', Mobility '09 Proceedings of the 6th International Conference on Mobile Technology, Application and Systems, ACM New York, USA, 2009.

Factors Influencing Group Member Satisfaction in the Software Industry

Thamatam Teja Goud, V. Smrithirekha and G. Sangeetha

Abstract Software Development is a group process involving several stakeholders making decisions based on a number of factors. Research in Group Decision-Making reveals that satisfaction of group members plays a significant role in the performance of the group and the outcomes. In this paper, we present the details of our study on the factors influencing the satisfaction of group members in software projects. While there has been good amount of work on satisfaction of group members through experiments, very few of them talk about satisfaction levels of current practitioners. We surveyed 80 software professionals from different companies in India and abroad at various levels of hierarchy. We found that participation of leader and other group members in the group decision-making process has significant impact on the satisfaction with the decision outcome. While currently the software industry is focusing on technology to make better decisions, our survey gives key insights on how social factors influence the decision outcomes and the role of a leader in enhancing the satisfaction of group members.

Keywords Group decision-making · Software projects · Member satisfaction

T.T. Goud (✉) · V. Smrithirekha · G. Sangeetha
Amrita School of Business, Amrita Vishwa Vidyapeetham, Coimbatore, India
e-mail: ttejagoud@gmail.com

V. Smrithirekha
e-mail: v_smrithirekha@cb.amrita.edu

G. Sangeetha
e-mail: g_sangeetha@cb.amrita.edu

V. Smrithirekha
Center for Research in Advanced Technologies for Education, Amrita Vishwa Vidyapeetham, Kollam, India

© Springer Science+Business Media Singapore 2017

S.C. Satapathy et al. (eds.), *Proceedings of the International Conference on Data Engineering and Communication Technology*, Advances in Intelligent Systems and Computing 468, DOI 10.1007/978-981-10-1675-2_24

1 Introduction

Major organizational decisions are often made in groups that involve people from diverse experience and expertise. The stakeholders in these groups share their ideas, opinions, and preferences on the issues at hand and make optimal decisions that are satisfactory to most stakeholders. Since groups are very valuable in an organization, lot of efforts are taken in keeping the members together, resolving conflicts, formalizing the meetings, developing a shared mental model, and maintaining the heterogeneity of the group. Such efforts to enhance group decision-making are supported by robust theories in the literature and experiments conducted with real-world groups. In the management literature, research in Group Decision-Making (GDM) has been carried out along multiple dimensions including group characteristics, the decision-making process, the challenges faced, group member attitude, conflict resolution, member satisfaction, and the outcome of the decision-making process [1–5]. Several researchers and professionals acknowledge that the satisfaction of group members influences the effectiveness of the group and thereby on the decisions made [6, 7].

Software projects involve a group of stakeholders with a variety of expertise, including customers, codesigning, and developing software. Software teams could be colocated or globally distributed or virtually connected. Similar to groups in other industries, software groups follow an organization-specific method of group decision-making where issues are identified, potential alternate solutions are discussed and based on the preference of stakeholders a decision is made that satisfies stakeholder concerns and quality criteria. In order for this process to be carried out effectively, the satisfaction of the group members is important. In this paper, we present our study on the factors that influence the satisfaction of software project group members. A survey of 80 software professionals was conducted. The respondents to the survey were asked to score their satisfaction levels on various aspects of GDM in their organizations. The key research questions that drive our research are (a) On what aspects of GDM are Software Group members having a higher satisfaction level? (b) Satisfaction levels on which of the GDM aspects enhances the likelihood of satisfaction with the final decision? The results of our study provide interesting insights into how leadership influences the decision outcomes.

2 Background and Related Work

Group member satisfaction is known to have significant impact on the success and performance of a team [7]. The effectiveness of group decision-making process has also been observed to be influenced by the satisfaction of group members. Hence, satisfaction of members has become a key factor in increasing the probability of a solution being adopted. Team satisfaction gains more importance especially when there is low agreeableness among team members as observed by Kong et al. [8]. This satisfaction in turn is impacted by a number of organizational, social, and individual factors. Keyton has classified the research works on group satisfaction into (a) group context (b) interpersonal context and (c) individual context [9]. Researchers have identified several factors including organizational, technical, and social factors that determine the satisfaction of group members. Of these, leadership has a great influence on member satisfaction [10]. In a study by Zeitun et al., it was observed that the correlation between team performance and performance was very high. They also observed that, while demographics played only a partial role, cultural differences influenced team satisfaction [11].

In our study, we have identified the aspects of organizational GDM from various sources in the literature. A key reference that has guided our survey is the seminal on the Generic Group Problem Solving model proposed by Aldag and Fuller that has various elements of the group decision-making process [12].

A Generic Group Decision-Making Model: Aldag and Fuller have expanded the Groupthink model to make it value neutral and proposed the Generic Group Solving Model. Janis defined Groupthink as *"mode of thinking that people engage in when they are deeply involved in a cohesive in-group, when the members' strivings for unanimity override their motivation to realistically appraise alternative courses of action.... Groupthink refers to a deterioration of mental efficiency, reality testing, and moral judgment that results from in-group pressures, (p. 9)"*. Aldag and Fuller mention that Groupthink research has often focused on defective decisions and negative outcomes. Such negative association has biased research on groupthink. Hence they felt the need to enhance the mode by making it free from negative antecedents, symptoms, or outcomes. They proposed the General Group Problem Solving Model (GGPS). They challenge certain assumptions of the groupthink model by referring to experimental studies and other works in the literature. The value-neutral GGPS model has antecedents for decision-making that includes (a) decision characteristics (b) group structure and (c) decision-making context. The emergent group characteristics include the behavioral aspects of the group and its members. The decision process characterize includes the various steps involved in decision-making, i.e., survey of objectives, problem definition quantity and quality of alternatives, group decision rule, selection space, solution space, and other related information. These lead to the final outcome that is the acceptance and implementation of the decision and the satisfaction of the group members with the leader, the process, and decision made. In our previous work, we used the GGPS model as a basis for evaluating the decision-making methods in Software Architecture [13]. We found that very few of these methods supported GDM. Inspite of recognizing that software development is inherently a group process, the rich knowledge available in the GDM literature is yet to be harnessed by the software community.

3 Methodology

The research questions that guide our study are the following:

RQ1: On what aspects of GDM do Software Group members have a higher satisfaction level?
RQ2: Satisfaction levels on which of the GDM aspects enhance the likelihood of satisfaction with the final decision?

We designed a questionnaire based on the literature. The questionnaire had 36 questions of which six questions were related to their personal and professional details, 27 were likert scale questions where participants had to indicate their satisfaction levels on various aspects of GDM on a scale of 1–5. The remaining three questions were related to the usage of tools to support GDM in the organization, the satisfaction levels with the tool and the overall satisfaction of the group members.

The questionnaire was circulated through emails to personally known contacts in the software industry and through online forums like Google groups, Linkedin, and Facebook.

Inorder to answer the RQ1, we ran descriptive statistics tests. The response frequencies were calculated.

Inorder to answer RQ2, we used ordinal logistic regression. The work experience of respondents was classified into four categories and included in the regression model. The workplace location was classified into two categories: India and Abroad.

The Y and X variables are both satisfaction based likert scale questions. The model has 1 Y variable and 22 X variables. The model is as follows:

$$Y_i = \alpha_0 + \beta' X_i + U_i$$

Y = Chosen Alternative/Decision Taken.

X1 = Medium of Communication, X2 = Problem Definition, X3 = Understanding of Risks, X4 = Group Composition-Member Expertise, X5 = Group Composition-Member Hierarchy, X6 = Group Discussion time, X7 = Leader Participation, X8 = Discussion Facilitation by Leader, X9 = Chance given to Members to express opinion, X10 = Alternative Quantity, X11 = Preference Process, X12 = Conflict Management, X13 = Member attention, X14 = Process time provided, X15 = Group Identity, X16 = Group Member Tenure, X17 = Member Dominance, X18 = Participation of Others, X19 = Customer Participation, X20 = Depth of Alternative Exploration, X21 = Experience Category, X22 = Location.

4 Analysis

4.1 Participant Profile

We had 80 responses to the survey. All respondents were from the software Industry. They had a wide range of work experience. The experience levels of respondents are shown in Fig. 1.

84 % of the respondents are from India and the remaining 16 % are from other countries. Respondents had indicated 43 unique designations (work profile) of which 18 % of them were software engineers.

4.2 Descriptive Statistics

Based on the frequency count of responses, we observed that among all the other aspects of GDM, the following were rated as *"Highly satisfied"* by more number of respondents compared to the other factors: (a) Leader Participation (27 % of the respondents) (b) Chance given to Members to express opinion (20 %) (c) Medium of Communication (18.75 %). Overall, a large proportion of the respondents have rated *"Satisfied"* for almost all the aspects of GDM and very few of them rated *"Highly Dissatisfied"*. The descriptive statistics only gives a broad picture of the satisfaction/dissatisfaction of respondents. A deeper analysis of impact of these aspects on the *decision taken* may be required to get a clearer understanding. Only 20 % of the respondents have mentioned that they use a tool for arriving at a consensus. Only 6 % of those who use tools are *"Highly Satisfied"* with the same.

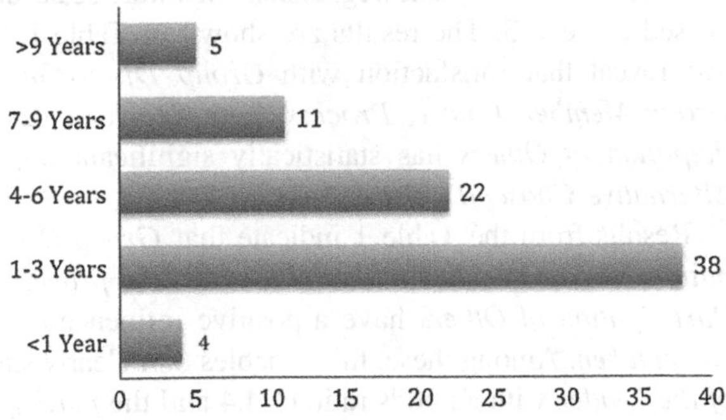

Fig. 1 Number of respondents in various categories of job experience

GDM aspect	Mean rating
Medium of Communication	4.0135
Problem Definition	3.9595
Understanding of Risks	3.7162
Group Composition-Member Expertise	3.595
Group Composition-Member Hierarchy	3.581
Group Discussion time	3.568
Leader Participation	4.027
Alternative Quantity	3.486
Preference Process	3.284
Conflict Management	3.581
Member attention	3.608
Group Identity	3.6216
Group Member Tenure	3.8243
Member Dominance	3.554
Minority Opinion	3.176
Customer Participation	3.61
Process time provided	3.419
Chance given to Members to express opinion	3.797
Discussion Facilitation by Leader	3.8649
Participation of Others	3.541

4.3 Regression Analysis

We ran Ordinal Logistic regression on likert scale data based on the model discussed in Sect. 3. The results are shown in Table 1. The results of the regression test reveal that satisfaction with *Group Discussion time, Leader Participation, Group Member Tenure, Process Time Provided, Experience Category and Participation of Others* has statistically significant impact on the satisfaction with *Alternative Chosen/Decision Taken.*

Results from the Table 1 indicate that *Group Discussion time, Leader Participation, Group member tenure, Process time provided, Experience Category and Participation of Others* have a positive influence on the satisfaction with the *decision taken.* Among these, the variables that clearly stands out are the *Participation of the Leader* with an odds ratio of 4.4 and the *participation of others* with an odds ratio of 3.9 indicating that these are the two most influential variables. This means that respondents who are satisfied with the participation of the leader and the other group members in GDM tend to be four times more satisfied with the decision taken. Studies reveal that enhancing the participation of members in the decision-making process enhances their satisfaction levels with the decision and overall with the organization as enhancing participation increases the flow of

Table 1 Results of ordinal logistic regression

Decision satisfaction	Odds ratio
Medium of Communication	1.666328
Problem Definition	1.026611
Understanding of Risks	0.7240656
Group Composition-Member Expertise	0.9850087
Group Composition-Member Hierarchy	1.001214
Group Discussion time	2.101216
Leader Participation	4.424488
Alternative Quantity	1.787605
Preference Process	0.9246735
Conflict Management	0.8331165
Member attention	1.138051
Group Identity	1.708863
Group Member Tenure	2.880494
Member Dominance	0.5548677
Minority Opinion	1.200755
Customer Participation	1.562849
Process time provided	2.352457
Experience Category	2.225577
Location	1.154761
Chance given to Members to express opinion	0.3196512
Discussion Facilitation by Leader	0.1500695
Participation of Others	3.920567

information within the group. This increases the availability of information to make key decisions. Affective models of participative effects indicate that allowing members to freely participate enhances their self-esteem, respect, independence, and equality that in turn influence the satisfaction levels [14]. Our study also indicates that *Discussion Facilitation by Leader* and *Chance given to Members to express opinion* have less favorable influence on the satisfaction with the decision taken. These aspects have to explored further to ascertain their negative influence on the outcome.

5 Conclusion

In this paper, we have presented our study on the satisfaction levels of software project members on the various aspects of group decision-making and the factors that influence the satisfaction with the decision taken. We observed that (a) Leader Participation (b) Chance given to Members to express opinion and (c) Medium of Communication are the aspects that most respondents are highly satisfied with. *Leader participation* and *participation of other members* in the group significantly impact the satisfaction with the decision taken. This study gives us only a glimpse of the satisfaction of group members in the industry and the influencing aspects. A more detailed study on the GDM processes used, the satisfaction with those processes and how it impacts the performance of group members in the software industry will be of interest to us in the future.

References

1. Ambrus A, Greiner B, Pathak P.: Group Versus Individual Decision-Making: Is There a Shift. Economics Working Papers. 2009 May; 91.
2. Brodbeck FC, Kerschreiter R, Mojzisch A, Schulz-Hardt S.: Group Decision Making Under Conditions Of Distributed Knowledge: The Information Asymmetries Model. Academy of Management Review. 2007 Apr 1; 32(2):459–79.
3. Ingham AG, Levinger G, Graves J, Peckham V.: The Ringelmann Effect: Studies Of Group Size And Group Performance. Journal of Experimental Social Psychology. 1974 Jul 31; 10 (4):371–84.
4. Mullen B, Copper C.: The Relation Between Group Cohesiveness and Performance: An Integration. Psychological Bulletin. 1994 Mar; 115(2):210.
5. Sims RR, Sauser WI.: Toward a Better Understanding of the Relationships among Received Wisdom, Groupthink, and Organizational Ethical Culture. Journal of Management Policy and Practice. 2013 Aug 1; 14(4):75–90.
6. Gladstein DL. Groups in context: A Model of Task Group Effectiveness. Administrative Science Quarterly. 1984 Dec 1:499–517.
7. Shaw ME, Blum JM.: Group Performance as Function Of Task Difficulty and The Group's Awareness of Member Satisfaction. Journal of Applied Psychology. 1965 Jun; 49(3):151.
8. Kong DT, Konczak LJ, Bottom WP.: Team Performance as a Joint Function of Team Member Satisfaction and Agreeableness. Small Group Research. 2015 Apr 1; 46(2):160–78.
9. Keyton J.: Evaluating Individual Group Member Satisfaction as a Situational Variable. Small Group Research. 1991 May 1; 22(2):200–19.
10. Lee P, Gillespie N, Mann L, Wearing A.: Leadership and Trust: Their Effect on Knowledge Sharing and Team Performance. Management Learning. 2010 Jun 28.
11. Zeitun RM, Abdulqader KS, Alshare KA.: Team Satisfaction and Student Group Performance: A Cross-Cultural Study. Journal of Education for Business. 2013 Jan 1; 88(5):286–93.
12. Aldag RJ, Fuller SR.: Beyond fiasco: A Reappraisal of The Groupthink Phenomenon and a New Model of Group Decision Processes. Psychological Bulletin. 1993 May; 113(3):533.
13. Miller KI, Monge PR.: Participation, Satisfaction, and Productivity: A Meta-Analytic Review. Academy of management Journal. 1986 Dec 1; 29(4):727–53.
14. Rekha S, Muccini H.: Suitability Of Software Architecture Decision Making Methods for Group Decisions. In Software Architecture 2014 Jan 1 (pp. 17–32). Springer International Publishing.

PI-PD Smith Predictor Based Cascade Controller Designing

Snehal A. Wankhade, Chandrakant B. Kadu and Bhagsen J. Parvat

Abstract Cascade Control (CC) scheme is one of those most successful control strategies that can be used in process control applications where time delays are long and disturbances are strong. Cascade control strategy combined with Smith Predictor (SP) scheme proves to be more beneficial in improving the control performance as compared to conventional cascade scheme. This paper presents a control strategy, using a PI-PD SP structure for outer loop and results in significant improvement in control. PI-PD controller is designed in such a way that the delay free part of output of system will follow the response of outer loop, assuming that the model and actual plant has a perfect match. Plant models are obtained using Process Reaction Curve Method. Simulation results show comparative analysis of conventional cascade, conventional SP, single loop and PI-PD SP approach.

Keywords Cascade control · Smith predictor · Process reaction curve

1 Introduction

The variables that cannot be handled by the control systems are known as disturbances. To compensate and prevent the propagation of undesired effects caused by these disturbances and generate a stable process response an easy and inexpensive conventional single loop feedback strategy can be implemented. However, the dis-

S.A. Wankhade (✉) · C.B. Kadu · B.J. Parvat
Padmashree Dr. Vithalrao Vikhe Patil College of Engineering,
Ahmednagar, Maharashtra, India
e-mail: snehiwankhade07@gmail.com

C.B. Kadu
e-mail: chandrakant_kadu@yahoo.com

B.J. Parvat
e-mail: pbhagsen@yahoo.com

S.A. Wankhade · C.B. Kadu · B.J. Parvat
Pravara Rural Engineering College, Loni, Maharashtra, India

© Springer Science+Business Media Singapore 2017

S.C. Satapathy et al. (eds.), *Proceedings of the International Conference on Data Engineering and Communication Technology*, Advances in Intelligent Systems and Computing 468, DOI 10.1007/978-981-10-1675-2_25

turbances in single loop feedback control strategy, has to propagate through the system before the controller recognises its effect and starts compensating action for them. Hence it becomes difficult to use conventional single loop in case of strong disturbances and long time delays. In such cases, CC strategy proves to be beneficial for improving the control performance, as it uses a secondary controller in cascade with primary controller, where primary controller output will act as a set point input for secondary controller. Cascade strategy which was introduced many years ago by Franks and Worley, is one of those complex multi loop control systems which are favoured extensively in process industries to have better control over process variables and aids in improving the system performance mainly in presence of disturbances [1]. If various disturbances are expected in conventional single loop, application of CC strategy proves to be beneficial [2].

Dead time is present in almost all industrial processes and may arise due to actual process transport lag or computational time, measurement lag and controller scan interval. This dead time makes improved control difficult [3, 4]. In such cases, the closed loop performance of system can be improved by using SP. However, in presence of model uncertainties SP may not give satisfactory performance. Dead time processes, thus are very strenuous to control using PID [5]. A method of finding an ideal controller giving desired closed loop response and then approximating PID using Maclaurin series is described by Park and Lee [6]. Author proposed improved cascade control (ICC) that uses SP structure, but the uncertain system proved unsatisfactory in giving desired responses [7].

Chen and Seborg presents a controller based on Direct Synthesis (DS) method for disturbance rejections. The proposed method requires approximation of higher order model to a lower order model or the higher order controller can be approximated either using a frequency domain approximation or series expansion [8]. Modified PI-PD SP leading to significant improvement for large time constant, unstable plant for variations in reference input and disturbances is proposed in [9]. The closed loop performance of the system can be improved using traditional PI or PID in unison with SP [10, 11]. However, this method results in instability for unstable and delay systems and does not take into consideration model uncertainties [12].

An approach that is based on relay auto-tuning to find controller parameters using SP is stated in [13]. CC however may not give satisfactory closed loop responses for set point changes if long time delay is present in the primary loop and standard forms were used to obtain PI-PD SP controller and involves certain trade-off difficulty between selected values of proportional gain and integral time of PI controller [14]. Experimental validation of a simple CC system is done using PLC, SCADA, OPC and internet [15]. On-line tuning methods available for under performing controllers, always does not provide a satisfactory response for set point and disturbance variations [16].

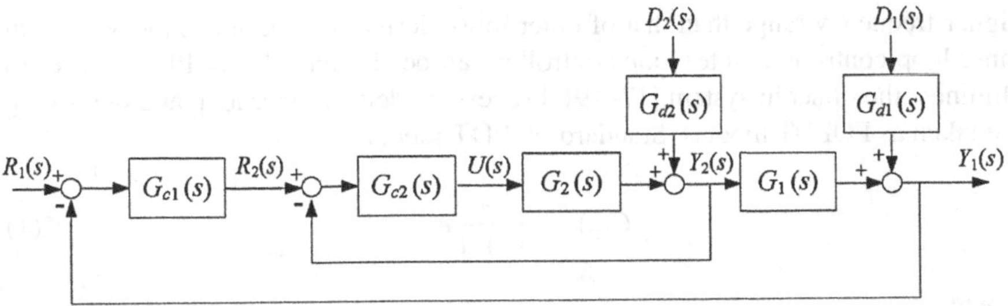

Fig. 1 Cascade control system block diagram

1.1 Cascade Control Structure

Cascade structure shown in Fig. 1, possess the ability to handle multiple disturbances and results in improved performance for set point changes. This scheme employs two sensors, two controllers and one final control element. Inner secondary or slave loop has a conventional feedback control structure and is nested within the outer primary or master loop. Output of primary controller, rather than going directly to final control element, will act as a secondary controller set point. Inner loop settling time should be remarkably faster than the outer loop settling time in order to attenuate disturbance at a faster rate and to minimize the effects of disturbances before they affects the output of primary controller. Since this control strategy consists of two controllers, tuning of controller however is a strenuous task [17–19]. PID controllers are the most preferred controller in industries for its effectiveness and simplicity in control performance.

$R_1(s)$ and $R_2(s)$ is primary and secondary set point respectively, $G_{c1}(s)$ is the master controller, $G_{c2}(s)$ is slave controller, $G_1(s)$ is primary level process, $G_2(s)$ is secondary flow process, $D_1(s)$ is the disturbance in outer loop and $D_2(s)$ is the disturbance in inner loop, $Y_2(s)$ is the output of inner process and $Y_1(s)$ is the final controlled variable of interest.

1.2 Process Models for Inner and Outer Loop

Process reaction curve method can be used for obtaining primary and secondary plant model transfer functions and first order plus delay time (FOPDT) process parameters are obtained accordingly. Modelling is done by collecting set of about 1770 readings from the process, which consists information like Process variables, Set points to both processes, output of primary and secondary controller. Thus from the experimental analysis, the required process models are found by providing the step input to the system. Once the plant models are available, the parameters for controllers can easily be obtained. Inner secondary controller is usually chosen as P or PI controller. Since the disturbance in inner process is quite noisy and also have higher frequency range than that of outer loop, derivative action cannot be used in inner loop controller. Outer loop controller can be chosen as PI or PID in order to eliminate the offset in system [17–19]. Process models of inner loop and outer loop are taken as FOPDT models. Standard FOPDT model is

$$G(s) = \frac{K_p}{\tau s + 1} e^{-\theta s} \tag{1}$$

where
K_p = process gain
θ = time delay
τ = process time constant

Therefore secondary process model is

$$G_2(s) = \frac{K_{p2}}{\tau_2 s + 1} e^{-\theta_2 s} = \frac{2.2}{4.2s + 1} e^{-3s} \tag{2}$$

and primary process model is

$$G_1(s) = \frac{K_{p1}}{\tau_1 s + 1} e^{-\theta_1 s} = \frac{1.7}{140s + 1} e^{-9s} \tag{3}$$

Process models (2) and (3) are used to tune the controllers using Direct Synthesis (DS) and Internal Model Control (IMC) approach.

2 Controller Design

PID controllers are used extensively and perform well in diverse control industrial applications. CC strategy is implemented for improving the system response to disturbances. However, CC strategy alone may not be able to deal with long time delays that prevail within the outer loop of system and hence to deal with such difficulties SP time delay compensation scheme can be employed that gives satisfactory performance for set point changes. Thus, uniting SP and conventional CC scheme, improved results can be obtained when the time delays are long and disturbances are strong. Earlier many research articles has shown that a SP employed in unison with PI-PD controller in outer loop of CC system gives superior closed loop responses in case of large time constants [9–11, 13, 14]. However, simple tuning formulas that have physically meaningful parameters like natural frequency ω_o and damping ratio ζ can be used to obtain the PI-PD controllers tuning parameters for outer loop of CC system. Desired overshoot and rise time can be used as a basis for finding the values of natural frequency ω_o and damping ratio ζ.

2.1 Controller Tuning

The procedure of finding controller parameter is called tuning. The control loop performs well, if controllers P/PI/PID parameters are chosen properly else if not chosen appropriately the controller performs poorly resulting in unstable system. In process control applications the most common controller structure is PID and its variations as P, PI, PD or PID structure.

2.2 Inner Loop Controller Tuning

As per the guidelines stated for cascade controller tuning, inner controller needs to be tuned first that serves the purpose of deviation reduction and aids in good set point tracking capability [17–19]. DS or IMC designing approach, both based on process models can be used to tune the slave controller. DS design method takes into concern desirable closed loop transfer function and helps to understand the relationship between process models and resulting controlling action. Secondary controller is expected to reject the disturbances $D_2(s)$ in inner loop as quickly and as stably as possible and hence secondary variable should follow its set point, with little overshoot and oscillations, as quickly as possible and as stably as possible. To satisfy this requirement slave controller should be so designed such that $\frac{Y_2(s)}{R_2(s)}$ gives stable over damped response and hence IMC controller designing approach proves to be more beneficial as compared to DS approach since it considers the model uncertainties and make tradeoffs between robustness and control system performance easier. Therefore, consider a stable FOPDT process model of inner loop which can be factorised as follows

$$\tilde{G_2}(s) = \tilde{G_2+}(s)\tilde{G_2-}(s) \tag{4}$$

where

$$\tilde{G_2-}(s) = \frac{K_{p2}}{\tau_2 s + 1} \tag{5}$$

and

$$\tilde{G_2+}(s) = e^{-\theta_2 s} \tag{6}$$

Forming idealized IMC controller which is inverse of $\tilde{G\_}(s)$

$$\tilde{G_{imc}}(s) = \tilde{G_2-}^{-1}(s) \tag{7}$$

A filter is added to make the controller proper or realizable.

$$G_{imc}(s) = \tilde{G_{imc}}(s)f(s) = \tilde{G_2-}^{-1}(s)f(s) \tag{8}$$

Filter transfer function usually possess the form

$$f(s) = \frac{1}{(\lambda s + 1)^n} \tag{9}$$

Thus λ is the only filter tuning parameter, which is adjusted in order to change the speed of closed loop response and adjust the robustness of system. If model is assumed to be perfect then,

$$\frac{Y_2(s)}{R_2(s)} = \frac{e^{-\theta_2 s}}{\lambda_2 s + 1} \tag{10}$$

Standard feedback controller is a function of internal model $G_2(s)$ and IMC $G_{imc}(s)$ or $q_2(s)$. Therefore we get

$$G_{c2}(s) = \frac{q_2(s)}{1 - G_2(s)q_2(s)} \tag{11}$$

Thus

$$q_2(s) = \frac{G_{c2}(s)}{1 + G_2(s)G_{c2}(s)} \tag{12}$$

Substituting

$$q_2(s) = \frac{\tau_2 s + 1}{K_2(\lambda_2 s + 1)} \tag{13}$$

and

$$G_2(s) = \frac{K_{p2}}{\tau_2 s + 1} e^{-\theta_2 s} \tag{14}$$

into (11) we get

$$G_{c2}(s) = \frac{\tau_2 s + 1}{K_2(\lambda s + 1 - e^{-\theta_2 s})} \tag{15}$$

Controller $G_{c2}(s)$ can be expressed as

$$G_{c2}(s) \equiv \frac{f(s)}{s} \tag{16}$$

Above ideal controller can be approximated to PID controller using Maclaurin series in s, which gives

$$K_{c2} = \frac{\tau_I}{K_2(\lambda_2 + \theta_2)} \tag{17}$$

where

$$\tau_I = \tau_2 + \frac{\theta_2^2}{2(\lambda_2 + \theta_2)} \tag{18}$$

and the overall outer loop plant transfer function will be

$$G(s) = G_1(s)G_{c2}(s)G_2(s) = G_1(s)H_{r2}(s) = \frac{K_{p1}}{\tau_1 s + 1}e^{-\theta_1 s}\frac{e^{-\theta s}}{\lambda_2 s + 1} \quad (19)$$

For FOPDT, the overall outer loop plant transfer function from above (19)

$$G(s) = G_1(s)H_{r2}(s) = \frac{K_1 e^{-(\theta 1+\theta 2)s}}{(\tau_1 s + 1)(\lambda_2 s + 1)} \quad (20)$$

Thus considering stable second order plus delay time (SOPDT) process model as shown in (21) below PI-PD SP controller can be designed for the outer loop master controller.

$$\tilde{G}(s) = \frac{K_1 e^{-(\theta 1+\theta 2)s}}{(\tau_1 s + 1)(\tau_2 s + 1)} \quad (21)$$

2.3 PI-PD Smith Predictor Controller for Outer Loop

Figure 2 above shows the PI-PD SP configuration for primary loop of cascade control system. An additional PD controller in outer loop modifies the locations of pole for a stable processes whereas PI controller is meant for taking care of servo tracking. PI controller in outer loop is considered as $G_{c1}(s)$ and PD controller as $G_{c3}(s)$. When $G_{c3}(s)$ is equal to zero, standard structure of SP can be obtained. Assuming a perfect match between $G_{pm}(s)$ and $G_1(s)$ set point response can be given as

$$T_r(s) = \frac{G_{pm}(s)G_{c1}(s)e^{-\theta_m s}}{1 + G_{pm}(s)[G_{c1}(s) + G_{c3}(s)]} \quad (22)$$

Equation (22) reveals that the delay free part i.e. $G_{pm}(s)$ can be used to obtain the parameters of controllers $G_{c1}(s)$ and $G_{c3}(s)$. Delay free part of SOPDT is

$$G_{pm}(s) = \frac{K_m}{(\tau_{1m}s + 1)(\tau_{2m}s + 1)} \quad (23)$$

$$G_{c1}(s) = K_p(1 + \frac{1}{T_i s}), G_{c3}(s) = K_f(1 + T_f s) \quad (24)$$

Substituting (23) and (24) in (22) and solving for $T_r(s)$ yields,

$$T_r(s) = \frac{\frac{K_m K_p}{\tau_{1m}\tau_{2m}}}{s^2 + \frac{1}{\tau_{2m}}(1 + K_m K_f)s + \frac{K_m K_p}{\tau_1 m \tau_2 m}} \quad (25)$$

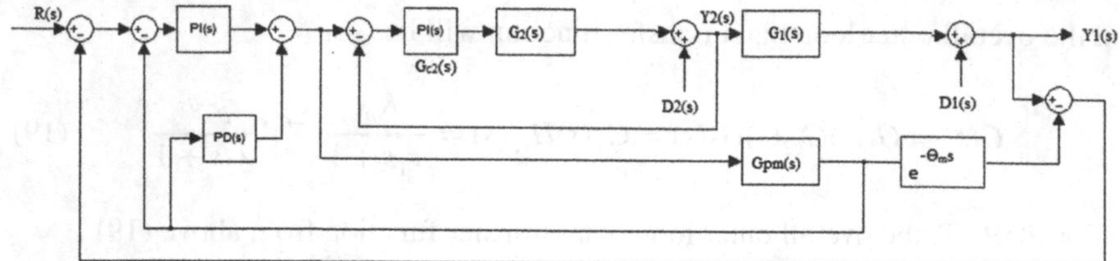

Fig. 2 PI-PD Smith predictor based cascade control system

which can be equated to

$$T_r(s) = \frac{\omega_o^2}{s^2 + 2\zeta\omega_o s + \omega_o^2} \tag{26}$$

where ω_o and ζ are natural frequency and damping ratio respectively. Comparing R.H.S of (25) and (26) we get

$$K_p = \frac{\omega_o^2 \tau_{1m}\tau_{2m}}{K_m} \tag{27}$$

$$K_f = \frac{2\zeta\omega_o\tau_{2m} - 1}{K_m} \tag{28}$$

and

$$T_i = \tau_{1m}, T_f = \tau_{1m} \tag{29}$$

Finding values of ω_o and ζ completes the designing of PI-PD controller and to find these values time domain characteristics like maximum overshoot (M) and rise time (Tr) is used which states the relationship between M, Tr, ω_o and ζ as

$$M = e^{-\frac{\zeta\Pi}{\sqrt{1-\zeta^2}}} \tag{30}$$

$$Tr = \frac{1 - 0.416\zeta + 2.917\zeta^2}{\omega_o} \tag{31}$$

from (30) and (31) we get

$$\zeta = \sqrt{\frac{(lnM)^2}{\pi^2 + (lnM)^2}} \tag{32}$$

$$\omega_o = \frac{1 - 0.416\zeta + 2.917\zeta^2}{Tr} \tag{33}$$

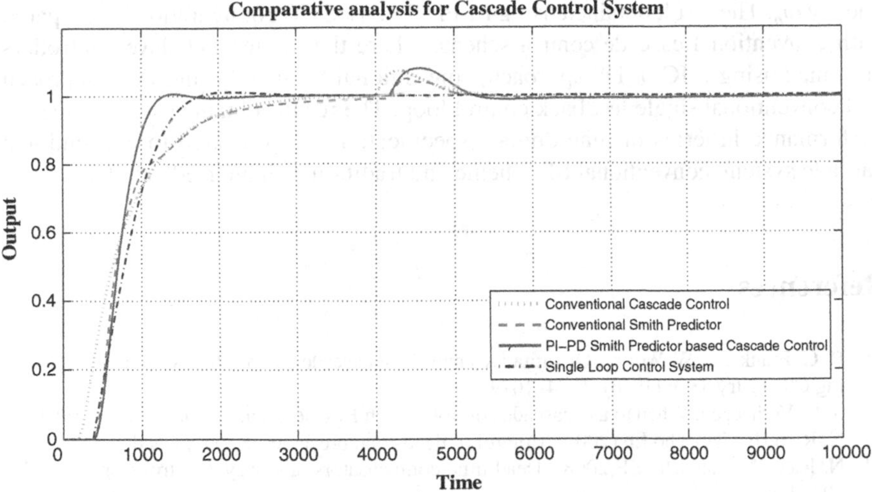

Fig. 3 Comparative analysis of cascade control system

Therefore, once the values of maximum overshoot and rise time are specified, ζ and ω_o can be obtained from (32) and (33) respectively and hence values of K_p and K_f can be calculated.

3 Simulation Results

To elicit the performance of the proposed controller designing and tuning method a laboratory level-flow cascade control system is used. Primary and secondary loop process transfer functions for the same are derived using Process-Reaction Curve Method. Simulation result shown in Fig. 3 shows a comparative analysis of conventional cascade control system, conventional SP system, the proposed PI-PD SP scheme and conventional single loop configuration. Figure illustrates the process response when a disturbance of 1 % is simulated. The PI-PD based SP presents a better performance than other strategies. The method gives better response for rise time, settling time and peak overshoot. The controller also results in improved set point responses and disturbance rejection as compared to other strategies.

4 Conclusion

Outer loop model of cascade control scheme is taken as lower second order with delay time process model and hence tuning of master controller becomes very easy. Primary loop PI-PD controller can be tuned using simple formulae. Desired overshoot and rise time is used to obtain the values of damping ratio ζ and natural frequency ω_o. The result obtained using PI-PD SP based CC configuration is compared with conventional cascade control scheme where the master and slave controllers are tuned using IMC or DS approach, conventional SP tuned using IMC approach and conventional single feedback control loop. The results obtained shows superior performance in terms of time domain specifications as compared to conventional cascade system, conventional SP scheme and traditional single feedback loop.

References

1. R. G. Franks, C. W. Worley, Quantitative analysis of cascade control. Industrial and Engineering Chemistry 48, 6 (1956), 1074–1079
2. S. C. Verhaegen, When to use cascade control. Intech Engineer's notebook, October 1991
3. N. Rico JE, Camacho EF (2007), Control of dead-time processes, Spring, London
4. N. Rico JE, Camacho EF(2008), Dead-time compensators: a survey, Control Eng. Pract 16, 407–428
5. DK Lee, MY Lee, SW Snng, IB Lee, Robust PID tuning for Smith Predictor in the presence of model uncertainty. Journal of Process Control, 9 (1999), 79–85
6. Yongho Lee and Sunwon Park, PID Controller Tuning To Obtain Desired Closed Loop Responses for Cascade Control Systems. Ind. Eng. Chem. Res. 37 (1998), 1859–1865
7. Ibrahim Kaya, Improving performance using cascade control and a Smith Predictor. ISA Transaction 40 (2001), 223–234
8. Dan Chen and Dale E. Seborg, PI/PID Controller Design Based on Direct Synthesis and Disturbance Rejection. Ind. Eng. Chem. Res. 41 (2002), 4807–4822
9. Ibrahim Kaya, Autotuning of a new PI-PD Smith Predictor based on time domain specifications. ISA Transaction 42 (2003), 559–575
10. Ibrahim Kaya, Obtaining controller parameters for a new PI-PD Smith Predictor using autotuning. Journal of Process Control, 13 (2003), 465–472
11. Ibrahim Kaya, A new Smith Predictor and controller for control of processes with long deadtime. ISA tans., 42 (2003), 101–110
12. A. Nortciff, J. Lore, Varying time-delay Smith Predictor process control, ISA Trans., 43 (2004), 41–71
13. Ibrahim Kaya, IMC based automatic tuning method for PID controllers in a Smith Predictor configuration. Science Direct Computers & Chemical Engineering 28 (2004) 281–290
14. Ibrahim Kaya, Nusret Tan, Derek Atherton, Improved cascade control structure for enhanced performance. Science Direct Journal of Process Control 17 (2007) 3–16
15. A. Lakshmi Sangeetha, B. Naveenkumar, A. Balaji Ganesh and N. Bharathi, Experimental validation of PID based cascade control system through SCADA-PLC-OPC and internet architectures. Measurement (ELSEVIER) 45 (2012), 643–649
16. J. C. Jeng, Simultaneous closed-loop tuning of cascade controllers based directly on set-point step response data, Journal of Process Control 24 (2014), 652–662
17. Thomas E. Marlin, Control Systems for Dynamic Performance, Tata McGraw-Hill
18. B. Wayne Bequette, Process Control Modelling, Design & Simulation, Prentice Hall, New Jersey, 2003
19. Dale E. Seborg, Thomas F. Edgar, Duncan A. Mellichamp, Process Dynamic and Control, Wiley India Pvt. Ltd. Second Edition

Speech Enhancement Using Transform Domain Techniques

Pradnyesh Deshmukh and D.G. Bhalke

Abstract The techniques of speech enhancement are used to minimize the undesirable external noises; the speech signal get degraded due to these external noises and these noises are background noise, wind noise, and other noises. These noises are generated due to environmental sources such as vehicular noise, street noise, babble noise, etc. Therefore, the speech enhancement plays an important role to reduce unwanted noise from the speech signal. This paper shows the single channel, the speech enhancement techniques based on a spectral subtraction method (SSM) and wavelet packets method (WPM) have been used for removing the noises. The performance of enhanced speech signal has been evaluated by measuring the speech quality parameters such as cross-correlation, average absolute distortion (AAD), peak signal-to-noise ratio (PSNR), and mean square error (MSE). The speech signal recorded in male voice and the Noizeus Speech database with different speech signals has been used to test the performance of this system. The experimental result shows that the WPM is more suitable for speech enhancement.

Keywords Wavelet Packet Method (WPM) · Signal-to-Noise Ratio (SNR) · Minimum Mean Square Error (MMSE) · Spectral Subtraction Method (SSM) · Speech enhancement · Average Absolute Distortion (AAD) · Peak Signal-to-Noise Ratio (PSNR) · Mean Square Error (MSE)

Pradnyesh Deshmukh (✉) · D.G. Bhalke
Department of Electronics and Telecommunication Engineering, JSPMs, Rajarshi Shahu College of Engineering, Pune, India
e-mail: pradnyesh.deshmukh@gmail.com

D.G. Bhalke
e-mail: bhalkedg2000@yahoo.co.in

© Springer Science+Business Media Singapore 2017

S.C. Satapathy et al. (eds.), *Proceedings of the International Conference on Data Engineering and Communication Technology*, Advances in Intelligent Systems and Computing 468, DOI 10.1007/978-981-10-1675-2_26

1 Introduction

The usage of speech communication and the related products like voice over internet protocol (VoIP), teleconferencing, audio video calls, hearing aids, etc. has been increased due to the revolution in mobile phone technology and its different applications. Today, most of the speech communications applications are working fine in a noise-free environment whereas the speech communication quality deteriorates in an external noisy condition. The speech enhancement method is used for the improvement of quality of noisy speech signal so that there will be less noise present in the speech signal. A noise may be generated by different sources of environment conditions such as vehicular noise, street noise, babble noise, exhibition noise, etc. It is very difficult to estimate various types of noises and their time variations in speech communication systems. Therefore, the complete noise elimination is also not feasible. From the past three decades, researchers are trying to develop new techniques to make better the performance of speech enhancement methods held at different noisy environments. In this paper, there is an overview of speech enhancement techniques by using spectral subtraction method (SSM) and wavelet packets method (WPM). These methods are used to minimize the noise present in the speech signal [1].

In Fig. 1, we can see the block diagram of the proposed organization. First of all, take clean speech signals take which are in .wav format. Then add the different noises into this signal to get the noisy speech signal. There are six different noises added in the speech signal. To enhance this signal remove this noise by using one of the methods such as SSM and WPM. Then we get the enhanced speech signal. Lastly, the different parameters of the speech signal are measured.

The organization of the paper is as follows. Section 2 describes of the proposed method, SSM. Section 3 describes the proposed WPM. Section 4 evaluates the measures of performance parameters of the proposed system. Section 5 shows the results of the proposed system, and the last Sect. 6 shows the conclusion of this paper.

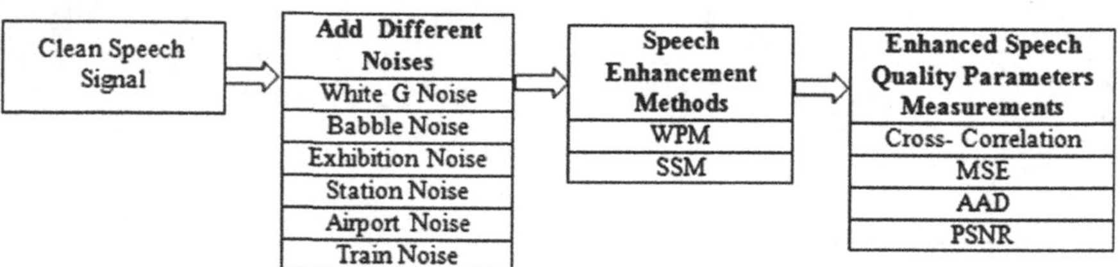

Fig. 1 Proposed block diagram

2 Spectral Subtraction Method (SSM)

SSM is one of the methods which is used over a large area for speech enhancement because it has a lower computational load. The input given to this method is noisy speech.

The windowing technique that has been used for taking the unchanging number of specimen of the speech signal method is without a break in nature. The Fourier transform has been applied for converting the signal from frequency domain to time domain, In this paper, fast Fourier transform and a Hamming window is used for windowing and Fourier transform.

For estimating the noise minimum mean square error (MMSE) estimator has been used. An estimated noise spectrum has been deducted from the noisy speech input for obtaining clean speech. A magnitude spectrum of a speech signal can be easily restored with the help of this technique [2–4]. Figure 2 shows the spectral subtraction technique. For the implementation of the MMSE estimator, the probability distribution of the noise Fourier and Speech Fourier expansion coefficients should be known, this is calculated by using MMSE filter. To obtain the signal in its time domain form inverse Fourier transform of the enhanced speech is taken. At this stage the phase of the signal, in its original form, is added to the magnitude. In this way, an enhanced version of the noisy speech signal is obtained as output.

3 The Wavelet Packet Method (WPM)

The wavelets theory is used in various fields like, statistics, physics, medicine, biology, etc. In this paper wavelet packet, Coiflet 5 is used for speech enhancement. The wavelet packet decomposition (WPD) is a wavelet transform in which the discrete-time signal passes through more filters in comparison with a discrete wavelet transform (DWT) [5–7]. A wavelet packet de-noising is carried out in four steps, as given below.

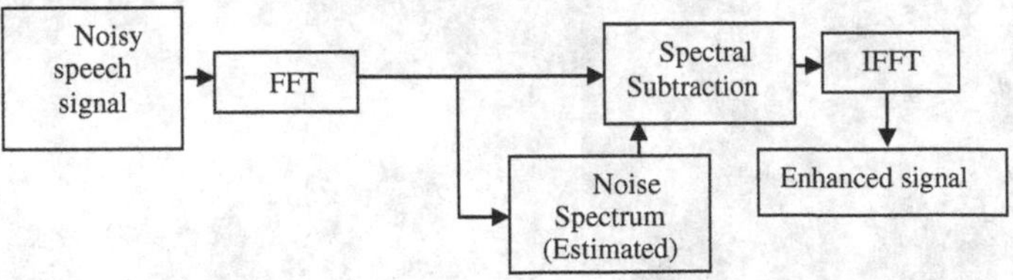

Fig. 2 Spectral subtraction

3.1 Wavelet Packet Decomposition

In decomposition process, signal S has been decomposed up to N levels. If the human voice is considered and if more-frequency components are taken off from the speech signal, then the voice sounds are different. Whereas the words present in it are clearly recognized and still audible by the person. But, if enough less frequency components are taken off, then resulting audio sign is not audible. Figure 1 determines the stage for signal decomposition. In this, the signal is decomposed over three levels to the high-pass filter and low-pass filter for calculating diagonal and approximation components from the signal. Where g[n] are coefficients of low-pass approximation and the h[n] are the high-pass detail coefficients. The multiple-level decomposition process has been calculated by using the successive approximations. In this, one signal is broken down into a many smaller resolution components which are called wavelet decomposition tree as shown in Fig. 2.

3.2 Entropy

It is used to calculate optimal trees i.e., based on nature of the signal a particular number of stages are selected. The entropy is one measure of knowledge present in speech signal regularity. There are various types of entropies such as Shannon entropy, *P*-order standard entropy, log energy entropy, entropy thresholding, and SURE entropy. In this paperwork, the Shannon entropy is applied for detecting the best tree (Fig. 3).

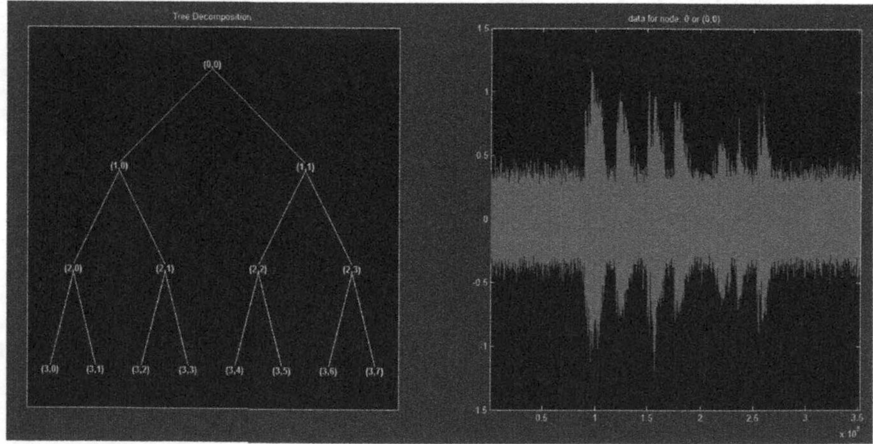

Fig. 3 Wavelet decomposition **a** Tree **b** Signal

3.3 Wavelet Thresholding

It is used for the calculation of the threshold values present in a speech signal. There are two types of thresholding namely, hard thresholding and soft thresholding. In the hard thresholding, data less than or equal to the threshold value becomes zero, otherwise data is unchanged. In this paper, soft thresholding is used. In the soft thresholding, the absolute value of the signal is compared with the threshold value of the signal. When the absolute value data is less than or equal to the threshold value, it will reduce to zero. If the data is greater than the threshold value, the data becomes the difference between the data and the threshold value.

(1) The hard threshold function is defined as

$$\begin{aligned} w_j, k &= w_j, k \ldots \left|w_j, k\right| \geq \lambda \\ &= 0 \ldots \left|w_j, k\right| < \lambda \end{aligned} \tag{1}$$

(2) The soft threshold function can be expressed as

$$\begin{aligned} w_j, k &= \left[\operatorname{sgn}\left(w_j, k\right)\right]\left(w_j, k - \lambda\right) \ldots \left|w_j, k\right| > \lambda \\ &= 0 \cdots \left|w_j, k\right| < \lambda \end{aligned} \tag{2}$$

where, $w_{j,k}$ is the wavelet packet coefficients, $w_{j,k}$ is the estimated wavelet packet coefficients, and λ is a threshold.

3.4 Wavelet Reconstruction

The next step is to reconstruct the wavelet packet. In this method, the *N*th-high-frequency coefficients and *N*th-low-frequency wavelet coefficients are calculated to reconstruct the wavelet packets. Out of these four steps, the most difficult step is selecting a threshold value and how to quantify this threshold value.

4 Measures of Performance Parameters

This section shows the different parameters measured from the enhanced speech to measure the quality of enhanced speech w.r.t. noisy speech signal.

4.1 Mean Square Error (MSE)

The MSE is used to show the true values of the quantity being estimated from the signal and determines the difference between the values implied by an estimator. The MSE shows the wanted value of a squared error loss. This method helps to measure the average of squares of the errors and also helps to differentiate between the values implied by estimator and the amount of quantity that is to be estimated which gives us an error value of a signal. A MSE can be represented by using

$$\mathrm{MSE} = \frac{\sum(\mathrm{Enhanced} - \mathrm{Input})^2}{\mathrm{Length}(\mathrm{Input})} \tag{3}$$

4.2 Average Absolute Distortion (AAD)

It is the AAD which is the warping of the original shape from the signal. Distortion is always not wanted in a signal and it is often tried to reduce it as much as possible [6]. In the paper, we use the parameter average absolute distortion to measure the distortion in addition to the speech signal in comparison with the original signal.

$$\mathrm{AAD} = \frac{\sum(\mathrm{Enhanced} - \mathrm{Input})}{\mathrm{Length}(\mathrm{Input})} \tag{4}$$

4.3 Cross-Correlation

The co-relation between the two signals is the parameter to compare enhanced speech signal quality with clean speech signal. It shows relation between the two signals. If cross-correlation is one then it means both signals are matching 100 %.

4.4 Peak Signal-to-Noise Ratio (PSNR)

In case of an image compression, for an example, a signal is an original image and the noise is the error between them introduced due to the compression. PSNR is nearly correct value for the human perception of the good quality of reconstruction of speech signal [7]. A higher PSNR shows that the reconstructed signal has the better quality, but the reverse may be true in case of some applications. PSNR can be calculated by

$$\mathrm{PSNR} = 10 \times \log_{10} \frac{\mathrm{Length} \times \max[\mathrm{Input}^2]}{\mathrm{Input}^2 - \mathrm{Enhanced}^2} \tag{5}$$

5 Results

This is the section which reports experimental results of considered speech adjustment using transform domain techniques. The techniques as mentioned above are used to remove white Gaussian noise and output speech signals were evaluated as indicated below.

Figure 4a determines the original speech signal that is also in .wav format, Figure 4b determines the noisy speech signal in which the white Gaussian noise is added.

Figure 4c determines the enhanced speech signal by using spectral subtraction method.

Figure 5 determines the experimental results of the speech signal. First graph determines the original speech signal found in .wav format; the second graph determines the noisy speech signal in which the white Gaussian noise is added, and a third graph shows the enhanced speech signal.

Table 1 shows experimental results of enhanced speech signal using SSM and WPM method.

Figure 6a shows the comparison of cross-correlation for spectral subtraction and wavelet packets. Figure 6b shows the comparison of AAD for spectral subtraction and wavelet packets. Figure 7a shows the comparison of MSE for spectral subtraction and wavelet packets. Figure 6b shows the comparison of PSNR for spectral subtraction and wavelet packets.

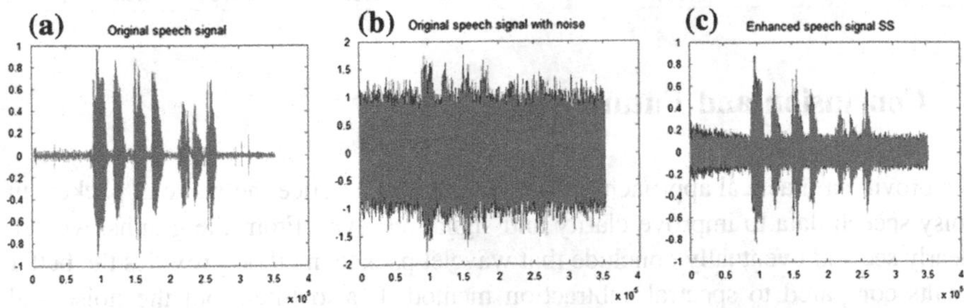

Fig. 4 **a** Original speech signal, **b** Original speech signal with noise, **c** Enhanced speech signal using spectral subtraction

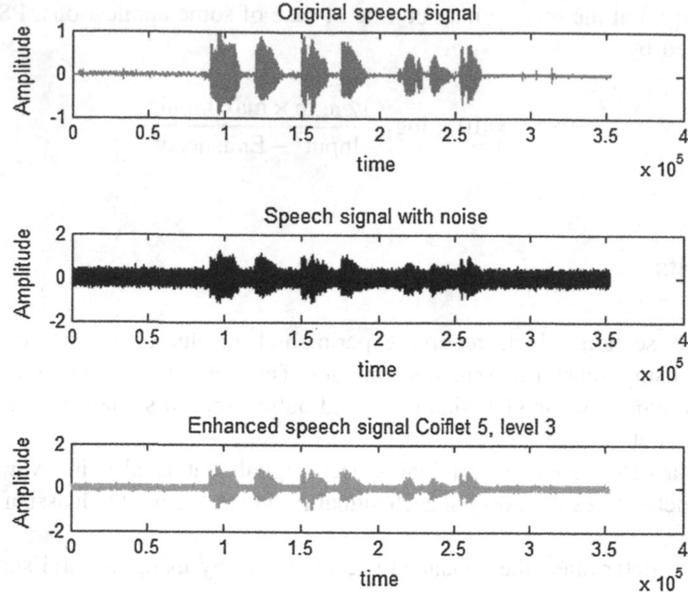

Fig. 5 Results of wavelet packet method

Table 1 Results of enhanced speech signal quality measurement

Results for cross-correlation			Results for AAD		Results for MSE (db)		Results for PSNR (db)	
Input-SNR (db)	SSM	WPM	SSM	WPM	SSM	WPM	SSM	WPM
0	0.2703	0.9375	0.00044	0.00002	0.0298	0.0019	14.95	27.00
2	0.3765	0.9572	0.00024	0.00012	0.0192	0.001200	16.87	28.79
5	0.5604	0.9753	0.00034	0.00013	0.0113	0.000693	19.16	31.28
10	0.8081	0.9885	0.00029	0.00012	0.0048	0.000318	22.89	34.67
15	0.9267	0.9927	0.00003	0.00001	0.002	0.000201	26.66	36.67
20	0.9694	0.994	0.00001	0.00001	0.00093	0.000164	30.03	37.54

6 Conclusion and Future Scope

We provide a practical approach on how to put into practice the wavelet packets in noisy speech data to improve clarity and signal retrieval. From the graphs, we can clearly see and eventually conclude that wavelet packets method provides the better results compared to spectral subtraction method. It also filters out the noise and renders us with clean speech in a more flat line peak SNR value, with less distortion in a varying loudness of the speech.

In this area of research still there is wide scope for improvement, till date no one was able to detect and remove 100 % of background noise.

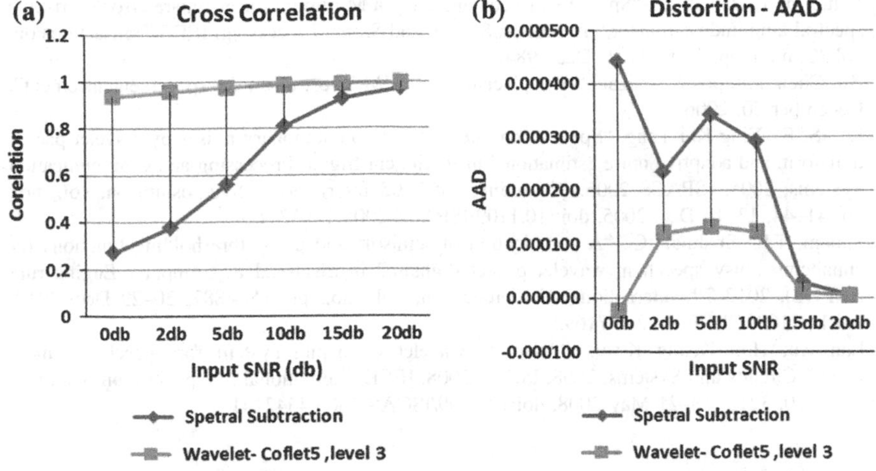

Fig. 6 **a** Comparison of cross-correlation for spectral subtraction and wavelet packets, **b** Comparison of AAD for spectral subtraction and wavelet packets

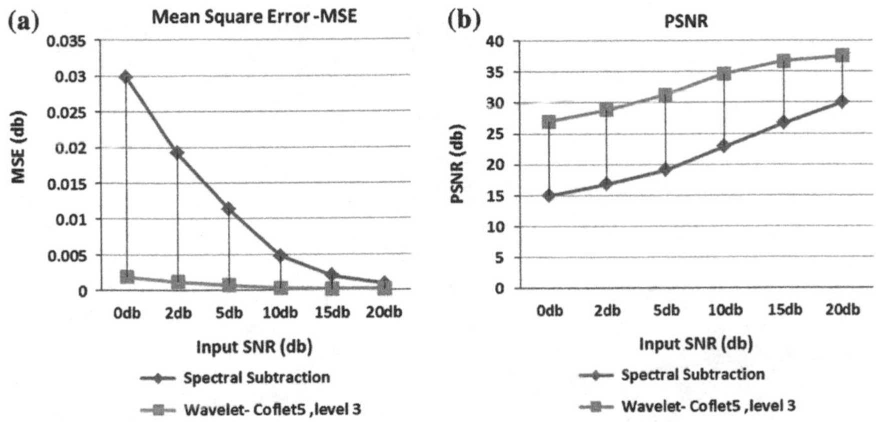

Fig. 7 **a** Comparison of cross MSE wavelet packets, **b** Comparison of PSNR for spectral subtraction and wavelet packets

References

1. Islam, M.T.; Shahnaz, C.; Fattah, S.A., "Speech enhancement based on a modified spectral subtraction method," Circuits and Systems (MWSCAS), 2014 IEEE 57th International Midwest Symposium on, vol., no., pp. 1085, 1088, 3–6 Aug. 2014, doi:10.1109/MWSCAS.2014.6908607.
2. Gupta, V.K.; Bhowmick, A.; Chandra, M.; Sharan, S.N., "Speech Enhancement Using MMSE Estimation and Spectral Subtraction Methods," Devices and Communications (ICDeCom), 2011 International Conference on, vol., no., pp. 1, 5, 24–25 Feb. 2011, doi:10.1109/ICDECOM.2011.5738532.

3. Ephraim, Y., Malah D., "Speech enhancement using a Minimum mean square error short-time spectral amplitude estimator," Acoustics, Speech and Signal Processing, IEEE Transactions on, vol.32, no.6, pp. 1109–1121, Dec 1984.
4. Bin Chen, Philipos C. Loizou, "A Laplacian based MMSE estimator for speech enhancement", December 20, 2006.
5. Lei, S.-F.; Ying-Kai Tung, "Speech enhancement for non-stationary noises by wavelet packet transform and adaptive noise estimation," in Intelligent Signal Processing and Communication Systems, 2005. ISPACS 2005. Proceedings of 2005 International Symposium on, vol., no., pp. 41–44, 13–16 Dec. 2005, doi: 10.1109/ISPACS.2005.1595341.
6. Sanam, T.F.; Shahnaz, C., "A combination of semisoft and μ-law thresholding functions for enhancing noisy speech in wavelet packet domain," in Electrical & Computer Engineering (ICECE), 2012 7th International Conference on, vol., no., pp. 884–887, 20–22 Dec. 2012, doi:10.1109/ICECE.2012.6471692.
7. Lan Xu; Hon Keung Kwan, "Adaptive wavelet denoising system for speech enhancement," Circuits and Systems, 2008. ISCAS 2008. IEEE International Symposium on, vol., no., pp. 3210, 3213, 18–21 May 2008, doi:10.1109/ISCAS.2008.4542141.

Indian Election Using Twitter

Gayatri P. Wani and Nilesh V. Alone

Abstract Today's world is ruled by Social Media. It is giving user, an access to real information without space and time constraints. Along with this information this gives a great opportunity to users to share their views freely on open platform. This current happening with people's opinion generates huge database. This generated database can be studied by scientist to analyze the generated data. The proposed system studies user's opinion on Indian election. System studies user opinion to understand if there exist any similarities in user's views on election. The system tries to identify the feasibility of classification model. The system studies the political orientation of Twitter users by analyzing the tweet content and other user's Twitter features. There are systems like voting advice applications (VAAs), online tools which are popularly used during an election in countries like Greece, Cyprus for getting real opinion on parties/Candidates during election but still in India there is no such application which can guide user. The proposed system plans to develop an application for recommending and comparing user's political opinions.

Keywords Social media · Election campaign · Indian politics · Tweets · Twitter · Facebook and Hadoop

---

G.P. Wani (✉) · N.V. Alone
Department of Computer Engineering, Gokhale Education Society's
R. H. Sapat College of Engineering, Nashik, Maharashtra, India
e-mail: gayatriwani@gmail.com

N.V. Alone
e-mail: nilesh.alone@gmail.com

G.P. Wani · N.V. Alone
Management Studies and Research, P T A Kulkarni Vidyanagar, Nashik
Maharashtra, India

G.P. Wani · N.V. Alone
Savitribai Phule Pune University, Pune, Maharashtra, India

© Springer Science+Business Media Singapore 2017

S.C. Satapathy et al. (eds.), *Proceedings of the International Conference on Data Engineering and Communication Technology*, Advances in Intelligent Systems and Computing 468, DOI 10.1007/978-981-10-1675-2_27

1 Introduction

We call today's world a connected world. The best way of connectivity is Internet. For every new term or question; we first use Google search to find out relevant information. India is at third position in using Internet [1]. In India people, especially, youth spend lot of their time on sites like Facebook, Twitter, etc. Every current topic is getting discussed on social media generating huge public's reviews which can make a clear picture of people's views on that particular topic to world. The main challenges with social media are privacy and especially accuracy or reality of the opinion of the user [2]; but still it has given new opportunities for collaboration, sharing, and engagement of users. It provides the platform to various subjects and one of them is politics. Social media like Twitter, Facebook, and YouTube motivate people to get involved in all the political activities by sharing their view about party and candidates. Elections are having a great impact of social media [3]. It is specially used by both general public for getting updates on political events and parties/Candidate to update their strategies and policies based on this social media's valuable feedback [4]. This paper is organized as follows. Section 2 reviews existing system used for election-based analysis. In Sect. 3, we present our proposed system. In Sect. 4, it provides the implementation details. Section 5 evaluates the dataset and results.

1.1 Problem Statement

Using the real-time data (tweets) collected over election period, the system analyzes the collection of tweets to draw meaningful inferences. The given system analyzes the tweets with the perspective of volume analysis, trend analysis, and sentiment analysis. Using volume analysis, system studies the popularity of candidate/Party, terms, active users during election period. In sentiment analysis, the system studies the sentiments of the user's tweets in terms of positive, negative, and neutral tweets for identifying user's orientation toward political party/candidate. In trend analysis, the system studies the popularity by area and tag wise. The input for the system is set of hashtags which download relevant tweet set which becomes the input for further analysis.

2 Related Work

Social media is a very important part of today's modern civilization. The social media influence is generating huge information which challenges analyst to make use of this generated information for various purposes starting from health to market needs, finance, global cliental, etc. These users' opinions are also largely generated when it comes to any political activity such as election; which is very crucial for any country. Most of the countries like Greece, United States are studied by analyst from social media like Twitter and Facebook point of view. They have adopted broad-brush technologies and limited analysis possibilities.

Author Song et al. [5, 6] have studied a election relevant to Twitter dataset for 2012 Korean election by storing real-time tweets by using tweet4j api of Twitter. They developed a system to analyze 2012 Korean presidential election, by carrying out Dirichlet-multinomial regression (DMR) topic modeling, network analysis using mention-based Twitter network, and term co-occurrence analysis relevant to presidential issues embedded in Twitter data.

Indian general election 2014; was studied by Author Bhola [7] using Twitter. The system studies user's orientation toward parties and candidates. The dataset used consists of 17.60 million of tweets; which were analyzed for popular user, candidate, and party; based on location, topic, and peak of time. A sentiment analysis was performed with the help of human annotators using classification algorithms.

Countries like Greece use online voting advice application called voting advice applications (VAAs) [8]. It is quite popular and users are getting benefit from it by choosing which party/candidate to vote. VAAS proposed variety of approaches for community-based recommendation systems which are evaluated on five different datasets. First it was evaluated on VAA dataset for identifying prediction accuracy of the system.

Kaczmirek and his team from GESIS [9] have compared Facebook and Twitter dataset for German election with a local survey for giving new insight on importance of social media during election period. This research is part of research project called German Longitudinal Election Study (GLES). This research will examine the German federal elections for the years 2009, 2013, and 2017.

3 Proposed System

The main goals of the system are analyzing Indian election Twitter data for studying the impact of twitter on Indian election and Maharashtra state. The system uses Twitter streaming and rest API for tweet collection. Our system also uses data collected for Indian election by Mr. Shreyansh Bhatt from Knoesisn University for Twitris project [10]. For collecting tweets for Maharashtra, hashtags like BJP, Namo, UthaMaharashtra, etc. are used as filters. Twitter gives platform for discussion which is quite popularly used by academicians, Journalist, and Politicians. Twitter has potential political value. Many politicians express their agendas and strategies with common man using Twitter. These tweets with geolocation and date, time, follow, and follower along with retweet facility provide quite popularity of party. This analysis can be used by politicians for defining their winning strategies. The system mainly covers volume analysis to identify popular days, topics, parties, and candidates along with active users during election period. Sentiment analysis is performed in system in order to understand positive and negative popularity of parties/Candidates. Trend Analysis covers geolocation and word-based clustering to analyze popular parties and candidates. The system is developed in three phases starting from tweet download followed by the analysis of collected data and lastly visualizes the results discussed in Fig. 1. The tweets are preprocessed before actually being used in the system.

The algorithms used in the system are discussed below

Algorithm 1: Volume Analysis

Input: - Preprocessed tweets Output: - Top hashtags, Top active users, Top trends Begin

1. Transfer all tweets from local file system to HDFS.
2. In map phase, all tweets are split into words with whitespace as a separator and mapped to one.
3. In reduce, all similar words are counted and written to HDFS.
4. Transfer all output tweets to local file system.
5. Sort all tweets as per the count and filter all with hashtags, users, and trends.

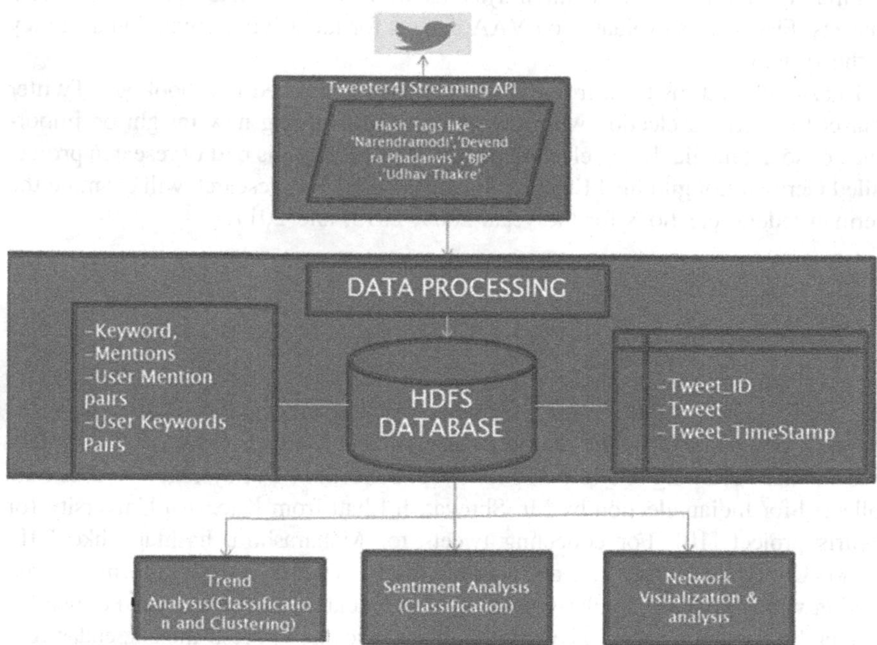

Fig. 1 System architecture

6. Put all top hashtags, users, and trends.
7. Visualize results for better perspective.

End

Algorithm 2: K-Means Clustering Algorithm
Input: Preprocessed tweets Output: Clusters formed

***K*-Means Clustering Algorithm Using Geolocation**
Input: Set of geolocation points where $X = \{X1, X2 \ldots Xn\}$, Output: Clusters Begin

1. Provide the number of tweets to be used for clustering [11–14].
2. Provide numbers of clusters to be determined
3. Transfer vectors and clusters from local file system on HDFS
4. For initial centroid of the clusters; randomly choose the geolocation value
5. Repeat
   a. In map phase, calculate closeness in terms of longitude and latitude of geolocation of each vector with every cluster centroid.
   b. In reduce phase, assign vector to cluster which is the most close to centroid.
6. Until
   a. No more changes in the clusters center OR
   b. Object's clusters are not changed further
7. Classify each tweet from cluster on provided party type.

End

***K*-Means Clustering Algorithm Using TF-IDF**
Input: Set of TF-IDF of tweets where $X = \{X1, X2 \ldots Xn\}$ Output: Clusters Begin

1. Find out unique words and TF-IDF of each unique [11–14].
2. Create set of tweet vectors using dictionary of unique words.
3. Provide the numbers of clusters to be determined.
4. For initial centroid of the clusters, randomly choose/provide the TF-IDF
5. Transfer vectors and clusters from local file system on HDFS
6. Repeat
   a. In map phase, calculate closeness in terms of TF-IDF of each vector with every cluster centroid.
   b. In reduce phase, assign vector to cluster which is the most close to centroid.
7. Until
   a. No more changes in the clusters center OR
   b. Object's clusters are not changed further

End

Algorithm 3: Naive Bayes Classifier
Political orientation of users toward party, topics can be analyzed from tweets. Map Reduce version of Naive Bayes algorithm is implemented to classify tweets into positive, negative, and neutral classes.

Input: Preprocessed tweets, Positive and Negative dictionary Output: Classified Tweets.

Begin

1: Create a data for the classifier

- 1.1: Create a list of positive words
- 1.2: Create a list of negative words
- 1.3: Provide a tweet file which needs to be analyzed

2: Design a classifier

- 2.1: Extract the word feature list from the list with its frequency count
- 2.2: Using this words list, create feature extractor which contains the words which will match with a dictionary created by us indicating what words are contained in the input passed

3: Training the classifier using training dataset

- 3.1: Generate Label Positive Probability which contains total number of positive words in input file
- 3.2: Generate Label Negative Probability which contains total number of Negative words in input file

4: Calculate the probability score for the positive and negative word for individual Tweet

- 4.1: Calculate positive score by total number of positive words in the tweet divided by Positive Probability
- 4.2: Calculate negative score by total number of negative words in the tweet divided by Negative Probability

5: Compare this probability to identify the tweet category as positive, negative, or neutral

4 Results

The system uses two datasets for election-related analysis. The first dataset is about India's 16th general election conducted in 2014 and the other database is for Maharashtra state assembly election for year 2014.

4.1 Volume Analysis

The system using the importance of hashtag analyzes the election for finding important messages or themes of communication during election period. Similarly trending users and trending topics are identified by the system.

Figures 2 and 3 show top hashtags of the Maharashtra and Indian Election.

4.2 Sentiment Analysis

Sentiment analysis was performed to understand user's orientation toward any party or candidate. During Maharashtra election, it was observed that people were having more orientation toward Shivsena party and Mr. Modi. Whereas for Indian election, the results were quite in favor of BJP due to the popularity of Mr. Narendra Modi.

The pie chart in Fig. 4 below describes the popularity of BJP, Shivsena, Congress, Mr. Arvind Kejriwal, and Mr. Narendra Modi.

4.3 Trend Analysis

The trend analysis provides the popular trends like popular personality or topics and geolocation-wise popular parties.

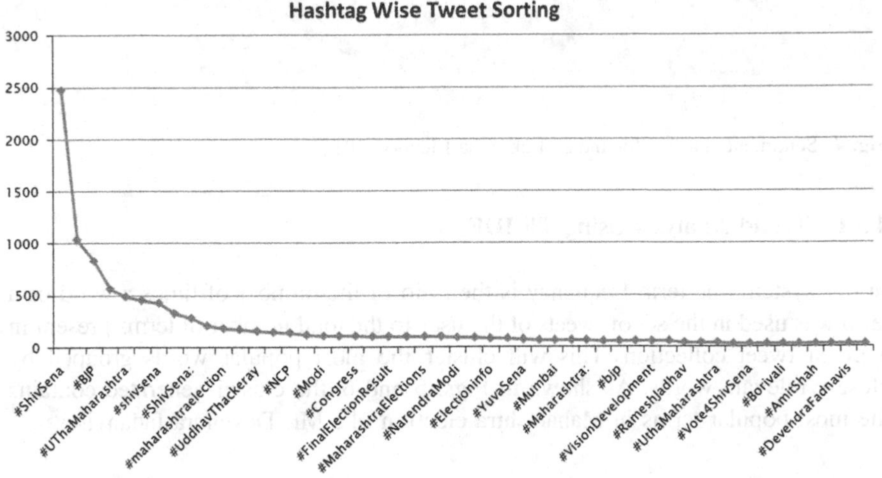

Fig. 2 Top Hashtag for Maharashtra State Assembly Election-2014

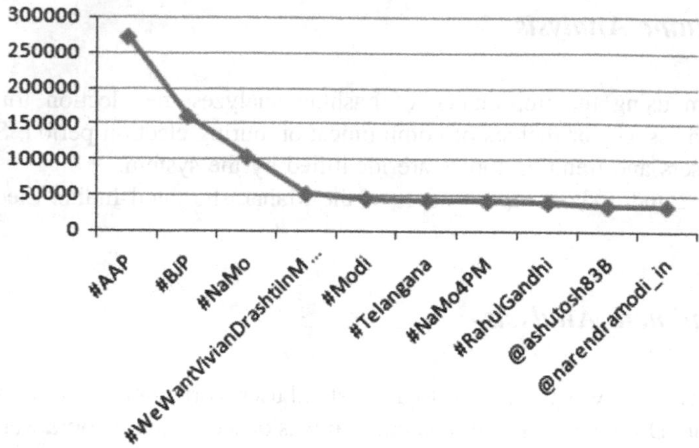

Fig. 3 Top Hashtag for India's 16th Loksabha Election-2014

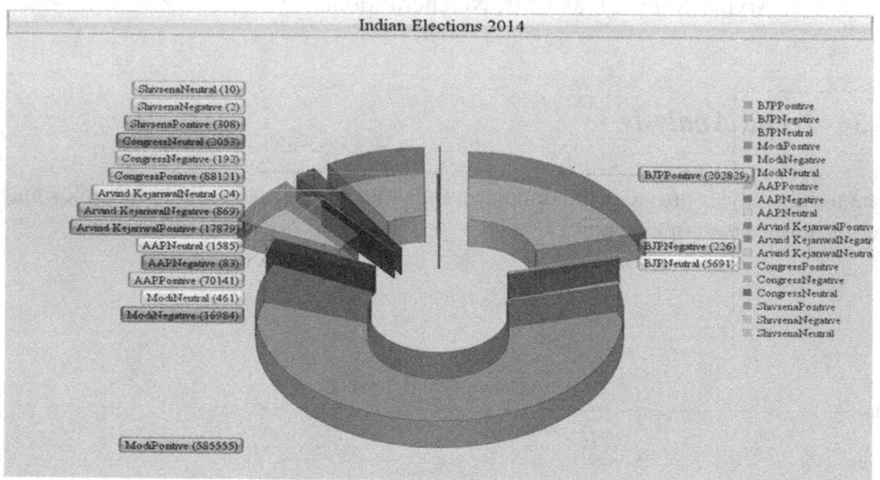

Fig. 4 Sentiment analysis for Indian Loksabha Election-2014

4.3.1 Trend Analysis Using TF-IDF

In our system, the term frequency is the ratio of the number of times a word or a term was used in the set of tweets of the user to the total number of terms present in a given tweet collection. This will cluster the most popular words grouped by closely relevant words. As shown in Figs. 5 and 6, the cluster generated contains the most popular terms in Maharashtra election like Mr. Devendra fadanvis.

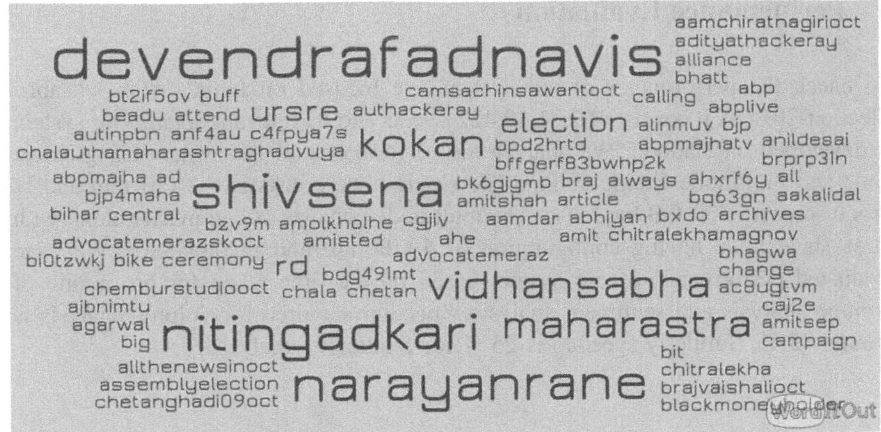

Fig. 5 Word cloud of cluster0 for Maharashtra State Assembly Election-2014

Fig. 6 Party-wise region-wise geolocation-based tweet Clustering (Color figure online)

4.3.2 Trend Analysis Using Geolocation

The trend analysis of the system is performed using K-Means clustering. The K-Means clustering algorithm works on the principal of distance-based clustering of the given point. For India's 16th Loksabha election the clustering is performed using geolocation coordinates. These tweets were classified on party basis for identifying which plotted geolocation belongs to which party.

5 Performance Evaluation

To check the performance of our system, we focused on India's 16th loksabha Election-2014 as it consist of 2,71,49,422 tweets and its relevant details.Our system is developed on Hadoop. We analyze the performance of the system using single node and multinode Configuration. The single node consists of 8 GB ram and I7 Processor with 500 GB hard disc. The multinode consists of one master along with three slave node having configuration of 4 GB ram and i3 processor. We have evaluated performance of our system based on the time taken for processing of volume analysis and sentiment analysis. Speedup acquired for volume analysis is around 45 %. Similarly speedup is 25 % for Sentiment analysis.

6 Conclusion and Future Scope

The experimental results show that the algorithm used in designing system is implemented successfully and also capable of eliminating the drawback of existing system. As per result of 16th Loksabha election it is verified that the predictions computed using our system are correct. The speedup obtained by using distributed system for volume analysis is 45 % and for sentiment analysis is 25 %. As per our analysis of Twitter data, maximum people were pro-BJP/Modi. In Loksabha election, it could be observed that people from maximum states of India elected BJP to form the government of India. In Maharashtra state, assembly election, we analyzed that Shivsena, Mr. Udhav Thakrey, and B.J.P. were quite popular even though this election was just after the Loksabha election. People were having mixed feeling for both the parties.

References

1. http://www.thehindu.com/sci-tech/technology/internet/india-is-now-worlds-third-largest-internet-user-after-uschina/article5053115.ece
2. https://www.academia.edu/7486078/Use_of_New_Media_in_Election_Campaigning_Lok_Sabha_Elections_2014
3. http://www.bbc.com/news/world-asia-india762391?OCID=fbbbcindia
4. http://www.slideshare.net/RaviTondak/social-media-for-political-campaign
5. http://www.ijcsit.com/docs/Volume%205/vol5issue06/ijcsit20140506100.pdf
6. Min Song, Meen Chul Kim; Yoo Kyung Jeong, Analyzing the Political Landscape of 2012 Korean Presidential Election in Twitter 1541-1672/14/ Published by the IEEE Computer Society
7. Abhishek Bhola "Twitter and Polls: Analyzing and estimating political orientation of Twitter users in India General Elections 2014" arXiv:1406.5059 [cs.SI]
8. Ioannis Katakis, Nicolas Tsapatsoulis, Fernando Mendez, Vasiliki Triga, and Constantinos Djouvas "Social Voting Advice Applications - Definitions, Challenges, Datasets and Evaluation" IEEE TRANSACTION CYBERNETICS, VOl. 44 No. 7

9. Lars Kaczmirek, Philipp Mayr, Ravi Vatrapu, Arnim Bleier, Social Media Monitoring of the Campaigns for the 2013 German Bundestag Elections on Facebook and Twitter
10. http://twitris.knoesis.org/
11. Aibek Makazhanov, Davood Rafiei: Predicting Political Preference of Twitter Users, ASONAM'13 Proceedings of the 2013 IEEE/ACM International Conference on Advances in Social Networks Analysis and Mining Pages 298–305
12. Use and Rise of Social media as Election Campaign medium in India, Narasimhamurthy N, (IJIMS), 2014, Vol 1, No. 8, 202–209
13. http://en.wikipedia.org/wiki/Indian_general_election,_2014
14. http://www.indiaonlinepages.com/population/india-current-population.html

Adaptive Data Transmission in WSN Using Enhanced Path Assured Transmission Protocol

Avinash Devare, G.K. Mohan and Hruturaj B. Nikam

Abstract Wireless sensor network is a highly dynamic network environment and a class of wireless network specially designed for monitoring. Large number of applications such as military systems, object tracking, monitoring, disaster reporting are designed on top of the WSN. These are real-time applications and generate very sensitive data or urgent data. This sensitive information needs to communicate reliably. Accurate delivery of sensitive information has direct impact on the overall performance of the system. Achieving reliability and congestion-free communication is important for the WSN. Some urgent data transmission protocol in WSN mainly focuses on transmission of sensitive data only at the same time it neglects the normal data traffic. Motivated by these challenges, we propose a EPAT system which is autonomous and distributed. This system achieves reliable and congestion-free urgent data as well as normal data communication at the same time.

Keywords WSN · EPAT · Congestion · Urgent data

1 Introduction

Wireless sensor networks are specially designed networks that comprise autonomous system placed in distributed fashion, this autonomous devices like sensor can monitor as well as gather the context. Context may be either location, noise level, traffic condition, vibration, pressure, sound, motion, temperature, weather condition, etc., these autonomous systems have limited computation power battery

Avinash Devare (✉) · G.K. Mohan · H.B. Nikam
Department of Computer Engineering KLU, Vijaywada, India
e-mail: devarea9@gmail.com

G.K. Mohan
e-mail: gvlkm@kluniversity.in

H.B. Nikam
e-mail: hrutu4nike@gmail.com

Avinash Devare · G.K. Mohan · H.B. Nikam
Department of Computer Engineering and Information Technology,
VJTI, Mumbai, India

© Springer Science+Business Media Singapore 2017

S.C. Satapathy et al. (eds.), *Proceedings of the International Conference on Data Engineering and Communication Technology*, Advances in Intelligent Systems and Computing 468, DOI 10.1007/978-981-10-1675-2_28

powered, less memory, and limited bandwidth. Wireless sensor network is a wide research area where hardware designs software developer user application domain expert cooperatively design efficient system autonomous device also called as node, which has its own sensing computation and wireless communication capability. Quality of service is the main design issue in WSN to achieve the QOS and efficient network performance, many efficient routing protocols, power management technique, and data dissemination protocol are specially designed. [21] wireless sensor network is self-configuration network where every sensor node communicates with each other using radio signal, it is deployed in quality to sense monitored and understand the physical world contextual info is the data sensed by sensor. Data aggregation is the process of collection of useful contextual information in wireless sensor network. It is an efficient and effective way to save limited resources data receiving from number of sensor node is aggregate as is they are about the sane attribute of the phenomenon when they reach the same routing node on the way back to the sink while collecting info by reducing the number of transmission and network. computation at intermediate node can substantially increase network efficiency on the other end, it also increases the amount of info contained in single packet and makes the system vulnerable to packet loss.

Rather than retransmitting dropped packet that causes additional delay wireless broadcasting is an effective strategy to improve delay performance data confidentiality integrity and security issue [22]. In data aggregation process is crucial when the sensor network is deployed in adverse environment, because of high packet loss rate in WSN more reliability in data transmission is desirable because of WSN application require various level of communication reliability. We proposed urgent data transmission scheme where sensitive data is transmitted on priority basis or via dedicated path at the same time normal data packet transmission is also takes place using efficient mechanism [1].

2 Transport Layer Protocol Characteristics

In wireless sensor network, there are two major functionality of transport layer protocols that are congestion control and reliability. Transport layer in WSN supports reliable message delivery, congestion control mechanism, and efficient energy management.

Packet loss in the WSN is occurred due to congestion, wireless channel quality and sensor failure, memory full, bad radio communication, and packet collision. Proper mechanism needs to apply to recover the lost packet and in multi-hop WSN, packet should reach the destination to achieve high reliability in data delivery. Most of the applications require each packet should send correctly, i.e., packet level reliability is required. In WSN, most of the protocol provide unidirectional reliable message delivery but some application require bidirectional reliability. Reliability can be measured in terms of hope-by-hope reliability, end-to-end reliability, sensor to sink (upstream reliability), and sink to sensor (Downstream reliability).

Data reliability can be defined in terms of packet reliability and event reliability. Packet reliability refers to the successful transmission of of all the packet to the destination. To achieve the packet reliability, it is required that all packet from sensor nodes must reach to the sink node but due to some interference or noisy communication channel packet are lost, retransmission of that lost packet is required which result in wastage of sensor node energy.

Event reliability refers to the successful detection of the event. It is required that event data from each sensor region needs to reach successfully, but loss of packet can be tolerated as long as sink node receive at least one packet from sensor node [2].

Reliability can be measured as per the direction of flow, i.e., classified as upstream reliability, downstream reliability, and bidirectional reliability. Upstream reliability refers to the unicast or converge cast communication between sensor node and sink node. Downstream reliability refers to the broadcasting communication between sink node and sensor node. Data transmission is broadcasting rather than unicasting, because there is only one sender node, i.e., sink node. Bidirectional reliability refers to the mechanism of two-way transmission of data, one is from sensor to sink node and other is sink to the sensor node. Instead of using two unidirectional protocol, it is important to use single bidirectional protocol because it reduces the heterogeneity of the network as well as complexity. Also it reduces the consumption of energy and one more important use is piggybacking.

For improving the reliability, it is important to identify the loss of data, i.e., loss detection technique needs to apply and the result of this technique need to intimate or notify to take appropriate action to recover from the loss. To perform this task of intimation, various methods are used. In explicit intimation technique, when packet are received at node, it send acknowledgment to the sink node which indicates, when and which packet is received correctly. On the other hand in implicit intimation technique, node understand the successful delivery of packet when it overhear the transmission of packet from neighbor node which node has sent recently. Also node explicitly intimate to the sender about the incorrectly received packet so that sender can retransmit the packet.

Error recovery mechanism in wireless sensor network is classified as end-to-end error recovery and hope-by-hope error recovery. Packet in the wireless sensor network is reached to destination via a number of intermediate node. In end-to-end recovery technique some protocol like TCP, when packet received at intermediate node, without verifying or detecting the loss and requesting for retransmission if lost, packets are sent as it is to the destination. But after receiving at destination, accuracy of packet is checked and if loss is occurred, retransmission request is send, i.e., final destination is responsible for error recovery mechanism. This approach will cause the large delay and low throughput. Some protocols like STCP, ART, RCRT, CTCP, and CRRT offer end-to-end error recovery. In hope-by-hope recovery scheme, loss detection and recovery mechanism are performed at all the intermediate node rather than just the final node. Hope-by-hope loss detection scheme is more energy-efficient scheme than end-to-end approach.

In wireless sensor network root causes of congestions are data rate of sender that is higher than receiver, interference between concurrent data transmission, collision in the physical channel, addition or removal of sensor node in the node, many-to-one network topology. Traditional protocols such as UDP do not provide the functionality of congestion control mechanism, on the other hand protocol like TCP provide sliding window-based approach to avoid congestion. Main cause of congestion is packet service rate that is greater than the packet arrival rate. And this scenario usually found at sensor node closer to the sink node. Due to congestion in WSN, packet get dropped or delayed because of large and filled queue. Dropping of the packet results in wastage of energy due to retransmission.

Some congestion detection protocols are existing that provide the mechanism of finding the location of congestion, i.e., this mechanism identify whether congestion occurred or not and at which location. As mentioned above in the causes of congestion, lack of memory at sensor node is also cause of congestion. Sensor nodes have a memory limitation, i.e., size of buffer is less and when node receives excessive incoming traffic, buffer cannot hold the excess packet, which result in drop. In WSN, buffer size controls the contention level because when number of source nodes increased, contention level also gets increased.

Packet rate is also one cause of congestion, packet rate defines the number packet sent or received within a specific time interval. At the particular node, the rate of receiving packet is higher than sending packet rate, buffer overflow is possible because the node has limited memory capacity.

After congestion detected at the node, the congested node conveys the congestion occurrence information to the neighbor node, i.e., node in the vicinity of congested node. This information is conveyed using control packet, i.e., it sets congestion notification field (CN Bit) in the header field of control packet. Also it may specify that packet may contain the allowable data rate. The way of intimating the congestion occurrence information is explicit or implicit. In implicit intimation, congestion information is piggybacks in the normal data packet, whereas in explicit intimation, explicit control message can be sent to notify the congestion.

After congestion detected and notified node needs to avoid the congestion. Congestion avoidance can be done by simply stoping the sending packet in the network or send the packet at lower rate. Also dynamic rate adjustment policy is applied where congested node can make the rate adjustment decision and notify to all its neighbor. Rate adjustment can be decide by two ways, centralize or distributed. In centralize policy, decision of rate adjustment is done at the sink node and in distributed it is done at each hope of network. Another method for congestion avoidance is traffic redirection where excessive traffic is redirected via high-speed network or network has long communication range.

3 Literature Review

There are a number of transport layer protocols that are designed for WSN. These protocols are characterized into three different categories.

1. Protocols which provide only reliability
2. Protocols which provide only congestion control mechanism
3. Protocols which provide both reliability and congestion control mechanism.

3.1 Reliability Guarantee Protocols

PSFQ: Pump Slowly Fetch Quickly (PSFQ) protocol.
This protocol works in downstream direction, i.e., from sink node to the sensor node. It uses hope-by-hope error recovery mechanism. The key features of this protocols are low signaling overhead. High error tolerance, scalable, and efficient, finally it is a customizable protocol. It operates in three-step pump operations, fetch operations, and report operations. Working of this protocol is it distributes the data slowly, i.e., pump the packet into network slowly. As it provides hope-by-hope recovery, node can recover quickly from errors. This protocol performs multimodal operation where packet is sent by sink at very low data rate on the other hand intermediate node store and forward at very high data rate. Fetch mode starts when it found any missing sequence number of packet, for that it send NACK. NACK is not propagated beyond 1 hope. It can also send cancel NACK if node overhears same NACK or repair request from other node. At the end sensor report hope-by-hope the delivery acknowledgment to the sink node [3].

RMST: Reliable Multi Segment Transport Protocol
RMST protocol is transport layer protocol and specially designed to provide the reliability in upstream direction, i.e., from sensor to sink node. RMST provides end-to-end data packet transfer reliability. This protocol consists of two mode: caching and noncaching mode. In caching mode, The nodes are assigned as RMST node if they are being used to transfer the data to the sink node. It is selective NACK-based protocol and it is configured for network caching and repair. RMST node caches the fragment and WatchDog timer is maintained. If the fragment is not received before timer expires negative acknowledgment is sent backward in the reinforced path. RMST depends on the directed diffusion scheme for recovery. RMST provides guaranteed end-to-end delivery of all the fragments, but it does not guarantee in-order delivery [4].

3.2 Congestion Control Protocols

CODA: Congestion Detection and Avoidance
It is one of the best algorithm that directly pointing to congestion control and avoidance in wireless sensor network. In CODA three different mechanisms are involved such as congestion detection, open-loop hope-by-hope backpressure notification, and closed-loop multisource traffic control to reduce traffic congestion. In CODA, congestion is detected based upon current buffer occupancy as well as present and past

channel loading. Continuous listening to the channel causes energy consumption so in CODA instead of continuous listening to the channel, it provide sampling scheme which listen the channel when required. When congestion is detected, the node notifies about congestion to its neighbor via backpressure mechanism. This backpressure control packet is propagated toward the source and the nodes, which receives this packet will take action (rate reduction or packet drop) based upon their local congestion status. Also in converge network, i.e., multiple sources are sending data to the single sink node, this protocol employ closed-loop multisource traffic regulation. In this traffic regulation, when source do not get the acknowledgment of data it automatically reduces the transmission rate. Though it is energy-efficient mechanism due to backpressure and ACK scheme, extra energy will be wasted also reduction in transmission rate may result in the quality of service (QOS). Specially in long period of congestion, latency and high data error rate is occurred because of mitigation in the response [5].

3.3 Reliability and Congestion Guarantee Protocols

There are number of protocols in Wireless Sensor Networks for achieving both reliability and congestion control.

[6] **ESRT**: Event to Sink Reliable transport protocol
This protocol offers the reliability at the application level also provides reliable delivery of the packet from sensor node to the sink node, i.e., in upstream direction. ESRT applies regulation on sensor report frequency to achieve desired reliability. This protocol runs on the sink node, because sensor nodes have limited resources, e.g., Power. ESRT is a dynamic state changing protocol which takes action depending upon the current state of network. First, it computes reliability from successfully received packet within a time interval. Based on the reliability, it computes required sensor report frequency. This sensor report frequency is informed to the all sensor nodes. If the reliability at node is less than the required reliability, ESRT increases the sensor report frequency to achieve target reliability also if the reliability is greater than required reliability it reduces the sensor report frequency. This dynamic nature of ESRT is required in random and dynamic topologies of Wireless Sensor Networks.

Drawback of ESRT is it treats all node equally so if congestion in one area of the network, all other nodes have to reduce their data rate as per sensor report frequency message which ultimately affect the network throughput.

[9] **ATP**: Ad hoc Transport Protocol
The core functionality if ATP protocol is rate-based transmission, i.e., it transmits the fixed size of data in each time interval. Unlike TCP it uses the timer to clock the new data, not window-based transmission. For controlling the congestion in the network source node uses the data packet (feedback) from intermediate node to adjust transmission rate. This feedback is piggybacked from receiving node to the sender node. Feedback sends periodically as well as in case of path failure and queuing delay.

If feedback packet is lost, sender waits for certain time period, if it is not received within a epoch time then it multiplicatively reduces the transmission rate also till the end of third epoch if feedback packet is not received, sender sends the probe packet to receiver. Also for achieving the desired reliability, receiver uses selective ACK to report any new holes in the data stream. Based on SACK, sender mark the packet for retransmission and provide high priority than new data packets. advantages of ATP is (1) sending rate estimation is accurate (2) it lowers the data traffic on reverse path (3) Recovery of lost packet at single time. Some issues of this protocols are (1) It need assistance and coordination from the intermediate node (2) It has detect and recover the lost packet very fast.

[10] **STCP**: Sensor Transmission Control Protocol
Sensor transmission control protocol (STCP) is reliable transport layer protocol. This Protocol offers congestion detection and control mechanism as well as Upstream reliability. In SCTP, before actual transmission of data, Sensor node transmit data packet called session initiation packet which contain the Transmission Rate, Type of data, Expected reliability and number of currently running transmission and receiving activity. When this packet is received at receiver (sink) end, Receiver (sink) send the acknowledgement (ACK) to the source sensor node. Then source node start transmitting the data. As the sink node knows the transmission rate from source node, it estimate the arrival time of data packets. If packet do not received within an estimated time it sends negative acknowledgement. Also as expected Reliability is provided to the sink in session initiation packet, Sink node measure the reliability in terms of number of packet received successfully. It offers end-to-end reliability. SCTP has a functionality of congestion detection by analyzing the size of buffer. To avoid the congestion at sink node it provide mechanism of setting the congestion bit in ACK packet. If sink node face the problem of buffer overflow it intimate to the source node by sending ACK packet in which congestion bit is set. To Reduce the congestion it offers the traffic redirection scheme as well as end-to-end rate adjustment mechanism.

[12] **CRRT**: Congestion-Aware and Rate-Controlled Reliable Transport
CRRT is reliable and congestion control protocol for WSN. It offers the hope-by-hope as well as end-to-end reliability in upstream direction, i.e., from sensor to the Sink node. In CRRT, for increasing the hope-by-hope reliability it provided the efficient retransmission mechanism. For retransmission of dropped data it employ reservation based mechanism where sender reserve the transmission medium. It uses the PACK and NACK to achieve the desired reliability. Congestion detection is simply carried out by analyzing the buffer size as well as the transmission rate of the node. For reducing the congestion, congested node send the Control packet containing the information (Buffer Memory occupancy) about congestion so that source node automatically reduce their data transfer rate.

[13] **CTCP**: Collaborative Transport control Packet
CTCP is a transport layer protocol which offers both the functionalities, reliable of data transmission and congestion detection and avoidance mechanism. CTCP provide end-to-end reliability in upstream direction, i.e., from sensor node to the sink

node. Efficiency of this protocol is measured in terms of the packet delivery ratio and energy consumption of the network. Data delivery ratio is fraction of number of packet received to the number of packet sent and Energy consumption of Network is the sum of energy consumption of individual sensor node. Congestion detection scheme is based on analysis of buffer occupancy and error rate. To reduce the congestion, sink node sends explicit acknowledgement for reducing the transfer rate. The key features of protocol is reliable delivery of all transmitted packet source to sink node and energy efficiency, it is also capable to identify error loss.

3.4 Congestion Control with Decentralized Parameters

[14] **PORT**: Price-Oriented Reliable Transport protocol
This transport layer protocol offers the Reliability in upstream direction, energy efficiency and congestion control mechanism. In PORT, It defines the price of node which is measured as total number of transmission attempt between sensor node and sink in the network. It avoid the links which have high communication cost which causes reduction in energy consumption. PORT provide application based optimization approach to the sink node so that sink can send the optimal reporting rate of each source and energy consumption of communication (sensor to sink) as a feedback. Also for informing congestion and increase the node cost it provide optimal routing scheme based on feedbacks from source node to the sink node. Above mentioned two offers will reduce the consumption of energy across the network. Congestion detection scheme is based on two parameters, node price and link loss rate. To reduce the congestion it uses the traffic redirection scheme as well as rate adjustment mechanism.

[15] **ART**: Asymmetric and Reliable Transport
ART is transport layer protocol which offers upstream end-to-end event reliability, downstream end-to-end query reliability and upstream congestion control. The major functionalities of ART are reliable event and query transfer and distributed congestion control. From the network, ART select nodes either essential node (E node)or nonessential node (N node). Essential nodes are those who take part in reliable data transfer between sink node and sensor node in both direction, i.e., upstream and downstream. To achieve the reliable data transfer in both direction it uses ACK and NACK mechanism. It has four characteristic (1) nonessential do not take part in end-to-end communication (2) congestion method is decentralized to regulate the data traffic effectively. (3) Only few nodes can participate in the recovery from lost message. (4) It uses distributed energy aware congestion control. Drawback of ART protocol is. It provide reliability guarantee for essential node only and not for nonessential node.

[16] **RCRT**: Rate-Controlled Reliable Transport
It is Transport layer protocol which offers the upstream reliability, i.e., it provide reliable transmission of sensor data from sensor node to the sink node. It Guaran-

tee about reliable transmission of data based on NACK scheme. In NACK scheme, sink request for missing packet to the sensor node so whenever data will be reached it is 100 % correct. RCRT implemented three regulation at sink node, (1) Congestion detection scheme (2) Rate Adaptation (3) Rate Allocation. Congestion detection scheme is based on the Round Trip Time (RTT). In Rate Adaptation regulation, Every node can dynamically change the data transfer rate, i.e., if congestion is in the network it will decrease the transfer rate and increase when congestion is not there. In Rate Allocation regulation, Depending on the application requirement data transfer rate will be allocated. This protocol provide end-to-end loss recovery. Main goal of this protocols are (1) Reliable end-to-end data transmission (2) Avoid the collapse of the network due to congestion (3) Application oriented data transfer rate allocation scheme (4) Support for the dynamic environment.

[17] **RTMC**: Reliable transport with memory consideration
It is reliable transport layer protocol which offers hope-by-hope data transmission policy and congestion control mechanism. Rate adjustment scheme is also used to reduce the congestion but it is possibility of drop of control packet which contain the information of rate adjustment. This protocol efficiently reduce the congestion by sending the memory occupancy information. This protocol include memory information field in the header of the packet. This causes all sender get the information of memory occupancy of receiver so accordingly sender adjust their transmission speed or rate. This protocol improve the throughput of the network as well as energy efficient, i.e., reduce the consumption of energy.

3.5 Congestion Elimination Protocols for Urgent Data

Majorly Buffer occupancy notification and rate adjustment scheme are used for the congestion control. But it may be possible that In case of emergency situation large amount of traffic is injected into the network in very short amount of time and it is required to get the information of the event quickly. There are number of protocols are designed for communication and avoid the congestion control in wireless sensor network but there are few of them which offers the transmission of urgent data.

[18] **RETP-UI**: Reliable Transmission Protocol for Urgent information
This protocol categorized traffic into three different types and for each kind it maintain different queue. Congestion in the network is predicted accurately based upon the queue length and its fluctuation. This protocol provide high throughput, less delay and probability of loss of packet is very less.

[23] **Fast and Reliable Transmission mechanism for urgent information in sensor network**
This is reliable transport layer protocol for urgent data transmission. In case of any emergency data transmission, node will establish the alive connection between source node to the sink node. All nodes which are part of live connection keep awake all the time for transmission of the data as well as it refrain from emission of normal

Table 1 Comparative analysis of congestion elimination protocols for urgent data in wsn

Protocol name	Congestion detection	Congestion avoidances	Reliability level	Reliability type	Reliability direction	Acknowledge
RETP-UI	Queue occupancy fluctuation	Multistage rate adjustment	Event	H-B-H	Upstream	ACK
CP EDCA	Emergency detection	Normal data preemption	Event	H-B-H	Upstream	ACK
ADMQOS	Event detection	Priority wise categorization	Event	H-B-H	Upstream	ACK
OD AODV	Event classification	Priority wise shortest path	Event	H-B-H	Upstream	ACK
FMUMUWSN	Event classification	Multistage path	Event	H-B-H	Upstream	ACK
PAT	Urgent event	Blocking of normal data	Event	H-B-H	Upstream	ACK

packet, i.e., it neglect the transmission of normal packet until all emergency packet are not delivered. This protocol also added retransmission mechanism of lost emergency packet on priority basis. In implementation they showed the alive connection can be established very quickly and emergency data is transferred with accuracy. This protocol achieve efficient Urgent data transfer to the sink node with 92 % of packet delivery ratio and less than 92 ms delay. Congestion is eliminated by reducing or neglecting the data transmission of normal packet.

[20] **CP-EDCA**: Channel Preemptive EDCA
This reliable transport layer protocol is designed for the transmission of the urgent data. Working of this protocol is urgent data traffic is preempt the services of regular data routine traffic. This protocol guarantee the QoS of the emergency data. Simulation result of emergency data transmission shows that 50–60 % decrement in the MAC layer delays. This advantage Preemption strategy it to expand from 802.11e standard to distributed emergency reporting.

[24] **PAT**: Path Assured Data Transfer Protocol
This protocol operates three steps. In the first step emergency data node initiate the blocking request to the other node to block their normal data transmission. Due to blocking mechanism path will get cleared. In the second step, Urgent data is transferred to the sink or master node and acknowledgement is received for the same. When data transmission of urgent data completed,Sink node or Master node will send the release message. This dedicated path for a moment will guarantee collision free data transmission and reduce the delays due to retransmission of data (Tables 1 and 2).

Table 2 Comparative analysis of congestion detection and avoiding protocols

Protocol name	Congestion detection	Congestion mitigation	Reliability level	Reliability direction	Reliability type	Acknowledge	Delay	Congestion notification
PSFQ	Packet level rate	–	Packet level	Downstream	H-B-H	NACK	Medium	Implicit
RMST	Packet level rate	–	Packet level	Upstream	H-B-H	NACK	–	Implicit
CODA	Packet level rate, Queue occupancy	Rate adjustment	–	Downstream	H-B-H	–	Small	Implicit
ESRT	Queue occupancy	Rate adjustment	Event Level	Upstream	Evt to sink	–	Large	Implicit
ATP	Queue occupancy	Rate adjustment	Packet level	Upstream	EtoE	NACK	Medium	Explicit
STCP	Queue occupancy	Rate adjustment, traffic redirection	Packet level	Upstream	EtoE	NACK, Eack	Large	Implicit
CRRT	memory overflow	Rate adjustment	Packet level	Upstream	H-B-H	–	Large	Implicit
CTCP	queue transmission error loss	Rate adjustment	Packet level	Upstream	E to E	eack, Double ack	Medium	Explicit
PORT	Node price, link loss rate	Rate adjustment	Event Level	Upstream	–	–	–	Implicit
ART	ack to core node, link loss reduce	Reduce traffic non core node	Packet level	Both	E toE	NACK	Small	Implicit

(continued)

Table 2 (continued)

Protocol name	Congestion detection	Congestion mitigation	Reliability level	Reliability direction	Reliability type	Acknowledge	Delay	Congestion notification
RCRT	Time to receive loss	Rate adjustment	Packet level	Upstream	E to E	NACK	–	Implicit
RTMC	Memory overflow	Header memory info	Packet level	Upstream	H-B-H	–	Large	Implicit
FLUSH	Ack to core node, link loss reduce	Rate adjustment	Packet level	Upstream	E to E	NACK	Small	Implicit

4 Proposed System

For Urgent data transmission there are number of protocol are designed. In Path assured data transmission protocol, Sensor node send request for transmission of urgent data to the sink node. Sink node block all other transmission of normal packet by sending the blocking request. Existing systems for Urgent data transmission mainly focused on the only transmission of urgent data. Proposed scheme aims not only transmission of urgent data packet but also the normal data packet at the same time. In existing PAT scheme, while transmission of urgent data (sensitive) from sink node to sensor node, all the normal data traffic get blocked to avoid the congestion and provide 100 % reliability for urgent data but at the same time normal data is generated at other sensor node and due to blocking request from sink node, sensor node do not send the data in the network even if the urgent data transmission is not in the vicinity. And which result in normal data packet generated at sensor node could not store at node due to Less memory. In the proposed scheme, this issue is resolved by using intelligence (Fig. 1).

It works in three phases, when sensor node has urgent data information, it send UREQ (urgent data request) to the sink node. This request will reach to the sink node via number of hops. The intermediates node will add their ID information to the request packet. When sink node receive the request packet, It immediately broadcast the blocking request which contain the ID information of intermediate node which

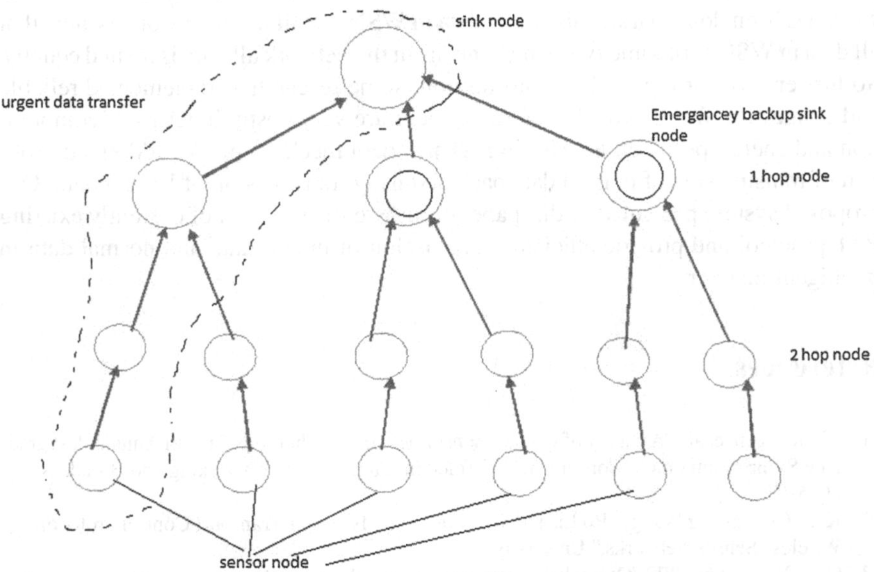

Fig. 1 Proposed system

received in request packet. When all other nodes will receive the blocking request it compare its all neighbor ID and all ID contained in blocking request. If it found any neighbor ID in the blocking request ID list then it will block the normal data traffic and if not then it will forward the traffic toward the sink node. Also when Sink node broadcast the blocking request, immediately its one hope neighbor node send the status information, i.e., currently available power, buffer occupancy to the sink node. When this information contained packet received at sink node, it will select one of them as a backup sink node and broadcast (BUPSINK) request so that all other normal data generator node transfer data to the backup sink node and urgent data is transmitted to the original sink node. Finally when urgent data transmission is completed then original sink node will broadcast the block release (BRELEASE) request. When this request is received at backup sink node, it will start sending normal data which is aggregated from normal data generator node.

5 Conclusion

In this paper, we described comparative study of many existing reliable, congestion detection and avoidance and urgent data transmission protocols in wireless sensor network. This paper describe the problems and limitations of existing transport layer protocols. We have studied and examined various requirements and some design issues of transport layer protocols. This survey of existing protocol directed us to problem in Urgent or Sensitive Data transmission protocols. Though lot of research work has been done in transmission of data in WSN but all research work assume that all data in WSN is of same type which means in the network all data is treated equally. So for sensitive or urgent data transmission, some researcher implemented reliable and congestion-free protocols. These protocol are very costly in terms of computation and energy perspective. Also in PAT protocol mechanism, We addressed problem of transmission of normal data packed during transmission of Urgent data. Our proposed system presented in this paper eliminate the problem of currently existing PAT protocol and provide efficient transmission of urgent data and normal data in intelligent manner.

References

1. Khemapech, et al., "A survey of wireless sensor networks technology," in 6th Annual Post graduate Symposium on the Convergence of Telecommunications, Networking and Broadcasting, 2005.
2. K. S. Chonggang Wang1, Bo Li, and Weiwen Tang,"Issues of Transport Control Protocols for Wireless Sensor Networks," University.
3. C.-Y. Wan, et al., "PSFQ: a reliable transport protocol for wireless sensor networks," in Proceedings of the 1st ACM international workshop on Wireless sensor networks and applications, 2002, pp. 1–11.
4. F. Stann and J. Heidemann, "RMST: reliable data transport in sensor networks," in Sensor Network Protocols and Applications, 2003. Proceedings of the First IEEE. 2003 IEEE International Workshop on, 2003 pp. 102–112.

5. S. B. E. a. A. T. C. C.-Y. Wan, "CODA: Congestion detection and avoidance in sensor networks," in Proceedings of ACM Sensys03, November 5–7, 2003.
6. K. S. C. Wang, and B. Li, "SenTCP: A hop-by-hop congestion control protocol for wireless sensor networks," in Proceedings of IEEE INFOCOM2005 (PosterPaper), Mar. 2005.
7. K. S. C. Wang, V. Lawrence, B. Li, and Y. Hu,"Priority-based congestion control in wireless sensor networks," in Proc. IEEE International Conference on Sensor Networks, Ubiquitous, and Trustworthy Computing(SUTC06), pp. 2231.
8. O. B. A. Y. Sankarasubramaniam, and I. F. Akyidiz, "ESRT: Event-to-sink reliable transport in wireless sensor networks," in Proceedings of ACMMobihoc03, June 1–3, 2003.
9. V. A. K. Sundaresan, H. Y. Hseeh, and R. Sivakumar, "ATP: a reliable transport protocol for ad-hoc networks," in Proceedings of the ACM Symposium on Mobile Ad Hoc Networking and Computing (MobiHoc03), pp. 6475.
10. Y. G. Iyer, et al., "STCP: a generic transport layer protocol for wireless sensor networks," in Computer Communications and Networks, 2005. ICCCN 2005. Proceedings. 14th International Conference on, 2005, pp. 449–454.
11. S. Kim, et al., "Flush: a reliable bulk transport protocol for multihop wireless networks," in Proceedings of the 5th international conference on Embedded networked sensor systems, Sydney, Australia, 2007, pp. 351–365.
12. M. M. A. a. C. S. Hong, "CRRT: congestion-aware and rate-controlled reliable transport in wireless sensor networks," in IEICE Transactions on Communications, pp.184–199.
13. F. J. E. Giancoli, and A. Pedroza, "CTCP: reliable transport control protocol for Sensor networks," in Proceedings of the International Conference on Intelligent Sensors, Sensor Networks and Information Processing (ISSNIP 08), pp. 493–498, December 2008.
14. Y. Z. a. M. R. Lyu, "PORT: a price-oriented reliable transport protocol for wireless sensor network," in Proceedings of 16th IEEE International Symposium on Software Reliability Engineering, pp. 117–126, 2005.
15. N. T. a. W. Wang, "ART: an asymmetric and reliable transport mechanism for wireless sensor networks," International Journal of Sensor Networks, vol. 2, pp. 188–200, 2007.
16. J. P. a. R. Govindan, "RCRT: rate-controlled reliable transport for wireless sensor networks," in Proceedings of the 5th International Conference on Embedded Networked Sensor Systems, pp. 305–319, 2007.
17. X. G. H. Zhou, and C.Wu, "Reliable transport with memory consideration in wireless sensor networks," in Proceedings of the IEEE International Conference on Communications (ICC 08), pp. 2819–2824, May 2008.
18. L. Lulu, et al., "A Novel Reliable Transmission Protocol for Urgent Information in Wireless Sensor Networks," in Global Telecommunications Conference (GLOBECOM 2010), 2010 IEEE, 2010, pp. 1–6.
19. T. Kawai, et al., "A fast and reliable transmission mechanism of urgent information in sensor networks," Proceedings of the 3rd International Conference on Networked Sensing Systems (INSS 2006), 2006.
20. M. Balakrishnan, et al., "Service preemptions for guaranteed emergency medium access in Wireless Sensor Networks," in Military Communications Conference, 2008. MILCOM 2008. IEEE, 2008, pp. 1–7.
21. R. Haji, et al., "Towards an adaptive QoS-oriented and secure framework for wireless sensor networks in emergency situations," in Multimedia Computing and Systems (ICMCS), 2012 International Conference on, 2012, pp. 1007–1011.
22. S. S. a. D. Kumar, "An approach to optimize adaptive Routing Framework to provide QOS in Wireless Sensor Networks," in proceeding of International Journal of wireless Networks and Communication, vol. 1(1), pp. 55–69 2009.
23. K. Ishibashi and M. Yano, "A Proposal of Forwarding Method for Urgent Messages on an Ubiquitous Wireless Sensor Network,"in Information and Telecommunication Technologies, 2005. APSITT 2005 Proceedings. 6th Asia-Pacific Symposium on, 2005, pp. 293–298.
24. A. W. R. A D Karanjawane, S D Mali, A A Agarkar,"Designing Path Assured Data Transfer Protocol for Wireless Sensor Network,"In proceeding of International Journal of Engineering Research and Technology (IJERT), vol. 2, pp. 1151–1160, 2013.

Cohort Intelligence and Genetic Algorithm Along with Modified Analytical Hierarchy Process to Recommend an Ice Cream to Diabetic Patient

Suhas Machhindra Gaikwad, Rahul Raghvendra Joshi and Anand J. Kulkarni

Abstract Proposed new genetic algorithm (AHP-GA) and cohort intelligence algorithm (AHP-CI). Results are obtained from both the algorithm (AHP-GA) and (AHP-CI) and compared with M-AHP obtained results. The purpose of using M-AHP (Modified Analytical Hierarchy Process) as a mathematical tool is to structure a multiple criterion problem by feedback connection control loop in order to recommend a particular ice cream to the patient suffering from diabetes. Here, results of M-AHP are verified by considering different weights, ratios and by using MATLAB for the problem under consideration. Here, results of M-AHP are verified by considering different weights and ratios by using MATLAB for the problem under consideration.

Keywords M-AHP · Feedback loop AHP model · Genetic Algorithm (GA) · Cohort Intelligence (CI) · AHP-GA and AHP-CI

1 Introduction

The modified analytical hierarchy process (M-AHP) is used as a technique for organizing and analyzing complex problems by using mathematics and psychology. It can use both prejudiced individual judgments and objective assessment just by

S.M. Gaikwad (✉) · R.R. Joshi · A.J. Kulkarni
Symbiosis Institute of Technology (SIT),
Symbiosis International University (SIU), Pune, Maharastra, India
e-mail: suhasg2009@gmail.com

R.R. Joshi
e-mail: rahulj@sitpune.edu.in

A.J. Kulkarni
e-mail: anand.kulkarni@sitpune.edu.in; kulk0003@uwindsor.ca

A.J. Kulkarni
Odette School of Business, University of Windsor, 401 Sunset Avenue, Windsor, ON N9B 3P4, Canada

© Springer Science+Business Media Singapore 2017

S.C. Satapathy et al. (eds.), *Proceedings of the International Conference on Data Engineering and Communication Technology*, Advances in Intelligent Systems and Computing 468, DOI 10.1007/978-981-10-1675-2_29

eigenvector and examining the reliability of the assessment by eigenvalue. The combinations of individual performance indicator with one of key performance indicator are done in order to assign a different weight to each criterion or attribute. The process of (M-AHP) is mainly used to calculate weights. It considers ratios for paired comparison. The inputs for (M-AHP) are alternatives and criterions [1, 2]. In this paper, three different types of ice cream, viz., Breyers Homemade vanilla, Breyers vanilla and Ben and jerry butter pecan are considered as three different alternatives. The criterions considered for ice cream are sugar, proteins, cholesterol and dietary fiber, which are attributes of an ice cream [3–5]. This paper proposes a method for recommending a particular ice cream to a diabetic patient by developing a model based on M-AHP [6–8]. The organization of this paper is as follows, Section 1 gives introductory details, Sect. 2 gives details about the used methodology for analyzing problem under consideration, Sect. 3 shows results for proposed analysis, Sect. 4 contains conclusions, Sect. 5 throws light on future work of this analysis and at the last in Sect. 6 references used in this paper are listed.

2 Methodology Used for Analyzing Problem Under Consideration

The first step while developing a M-AHP model for a particular problem is to arrange its details in hierarchy while in Fig. 1, level 0 shows objective of this analysis [9–12], which is to recommend a particular ice cream to a diabetic patient. The next level, i.e., level 1 shows different criterions cum attributes of an ice cream, which are sugar, cholesterol, calories, and proteins. The last level, i.e., level 2

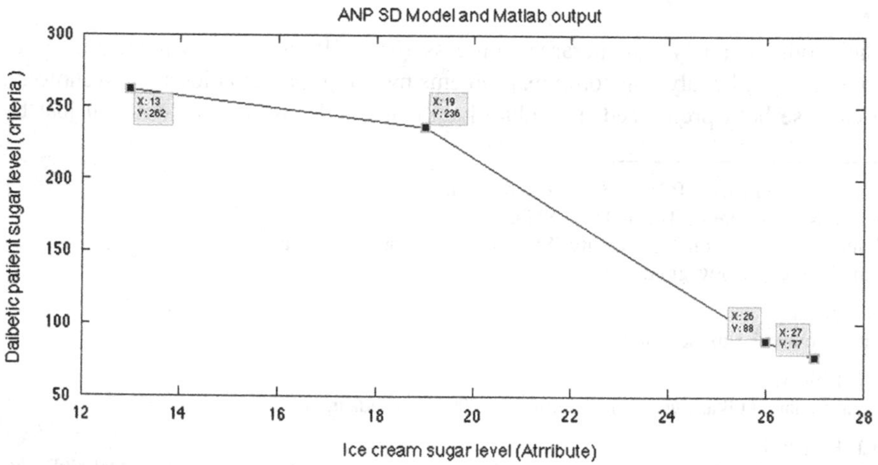

Fig. 1 M-AHP Model and MATLAB results for mapping of a diabetic patient's sugar level with sugar in the ice cream

shows different types of ice cream like Breyers butter almond (B.b), Breyers peanut butter and fudge (B.p) followed by Ben and jerry coffee (B.j), and lastly Breyers extra creamy chocolate. Here, patient details are not considered as a criterion because by separating those from details of ice cream analysis among them can be carried out easily.

2.1 Methodology for Criterion

The parameters of M-AHP like weight, Consistency Index (CI), and ratio are calculated and these calculated values are verified by using different ways, which are explained, in the subsequent sections of this paper [13]. Initially, criteria for the ice cream are considered.

The above-obtained values show weights for sugar, calories, cholesterol, and proteins which (to make it more rational) are shown in Table 1.

2.2 Methodology for Alternative (Level 2)

The parameters of M-AHP like weight, Consistency Index (CI), and ratio are calculated and these calculated values are verified by using different ways, which are explained, in the subsequent sections of this paper [2, 3]. Initially, alternatives for the ice cream are considered. Alternatives are considered consisting of the following names or four different types of ice cream are considered in the alternatives.

(1) Brayers butter almond
(2) Brayers peanut butter
(3) Ben and jerry coffee
(4) Brayers extra creamy chocolate

The cluster nodes and their arrangement on another. Hierarchy level depends on alternatives that is different ice cream will affect their weight on criteria and vice versa shown in Fig. 2.

Table 1 Matrix for calculating weight for each criterion of an ice cream

Criteria	Sugar	Calories	Cholesterol	Proteins
Sugar	1	3	5	7
Calories	1/3	1	3	5
Cholesterol	1/5	1/3	1	3
Proteins	1/7	1/5	1/3	1

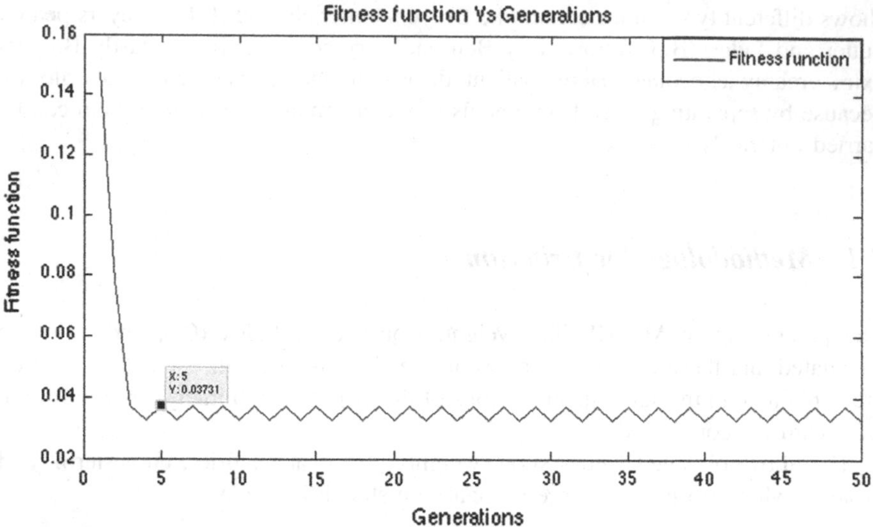

Fig. 2 Plot of fitness function values against number of generations

3 Results of M-AHP

The Feedback loop Working of M-AHP

Different weights or eigenvector of the alternative are obtained by considering one criterion at a time and assigning weight to each alternative with respect to the criteria so four different criteria will assign different weights to each alternatives and same procedure is followed for the criteria (Table 2).

Each criteria will assign weight to each alternative so four different criteria will assign four different weights to each alternatives. Resulting in Table 3 matrix.

Highest eigenvector or weight is for criteria is sugar 27 % then next is calories is 13.66 % flowed by cholesterol and finally protein with weight is 6.09 %. Now similarly for the alternatives highest eigenvector or eigenvalue is for Breyers butter

Table 2 Matrix for calculating weight for each criterion of an ice cream

Alternatives	Brayers butter almond	Brayers peanut butter	Ben and jerry coffee	Brayers extra creamy chocolate
Brayers butter almond	1	5	7	9
Brayers peanut butter	1/5	1	3	7
Ben and jerry coffee	1/7	1/5	1	5
Brayers extra creamy chocolate	1/9	1/7	1/5	1

Table 3 M-AHP feedback loop limit matrix

Cluster node label		Alternatives				Criteria				Goal
Alternatives		B.j.	B.b	B.e.	B.p.	Cl	Ch	Pr	Sg	I.c. for D.P.
	B.j.	13.10	13.10	13.10	13.10	13.10	13.10	13.10	13.10	13.10
	B.b.	28.24	28.24	28.24	28.24	28.24	28.24	28.24	28.24	28.24
	B.e	2.76	2.76	2.76	2.76	2.76	2.76	2.76	2.76	2.76
	B.p.	5.88	5.88	5.88	5.88	5.88	5.88	5.88	5.88	5.88
Criteria	Cl	13.66	13.66	13.66	13.66	13.66	13.66	13.66	13.66	13.66
	Ch	2.84	2.84	2.84	2.84	2.84	2.84	2.84	2.84	2.84
	Pr	6.09	6.09	6.09	6.09	6.09	6.09	6.09	6.09	6.09
	Sg	27.89	27.89	27.89	27.89	27.89	27.89	27.89	27.89	27.89

almond (B.b) 28.24 % then next one Breyers peanut butter and fudge (B.p) 13.66 % followed by Ben and jerry coffee (B.j) 5.88 %, and lastly Breyers extra creamy chocolate (B.e.) 2.76 % of eigenvalue with respect to goal according to assumption made in Sect. 2, a graph for diabetic patient as a function for given ice cream can be drawn and their results for the same are shown in Fig. 1.

Diabetic patient has sugar level 262 mg/dl, flowed by 236 mg/dl and 88 mg/dl lastly 77 mg/dl have their weight are 55.78, 26.335, 22.75, and 12.18 %. Mapping between ice cream and diabetic patient is shown in Fig. 1.

The diabetic patient whose sugar level 262 mg/dl, flowed by 236 mg/dl and 88 mg/dl lastly 77 mg/dl have their weight are 55.78, 26.335, 22.75 and 12.18 and these values are mapped with the considered ice creams. Breyers butter almond (B.b) 28.24 % then Breyers peanut butter and fudge (B.p) 13.66 % followed by Ben and jerry coffee (B.j) 5.88 % and lastly Breyers extra creamy chocolate (B.e.) 2.76 % of eigenvalue. In this way, mapping of ice cream with diabetic patient is done and results of their mapping are shown as an output through following Fig. 3 which obtained through MATLAB by considering M-AHP model.

3.1 AHP-GA and Algorithm Proposed Set of Equation for Different Variables

Moreover for AHP matrix number of pair comparison is limited. Pair comparison is done with criterion to criterion or attribute to attribute. In practice the comparison is limited to only n = 7. As formula indicated n (n − 1)/2. Also AHP is positive matrix [7–10].

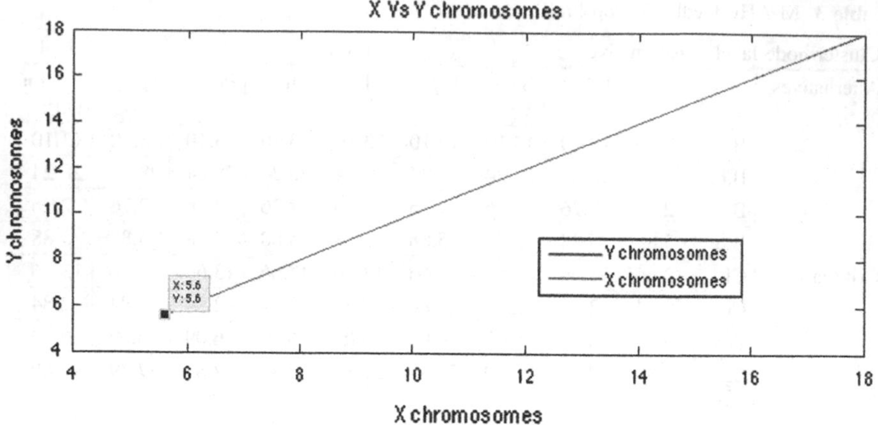

Fig. 3 Plot for new values of criterion chromosomes value

As number of paired comparison is limited to n = 7 only, so start with seven variable equations. The value of criterion or alternatives limited to seven can use these equations.

Above equation uses 7 variables and 7 equations.

$$7a + 6b + 5c + 4d + 3e + 2f + g = 0 \tag{1}$$

$$6b + 5c + 4d + 3e + 2f + g = 0 \tag{2}$$

$$5c + 4d + 3e + 2f + g = 0 \tag{3}$$

$$4d + 3e + 2f + g = 0 \tag{4}$$

$$3e + 2f + g = 0 \tag{5}$$

Consider the simple pair comparison for 3 variables and 3 equations. For equation number 5

$$4d_{11} + 3e_{12} + 2f_{13} + g_{14} = 0 \tag{6}$$

$$4d_{21} + 3e_{22} + 2f_{23} + g_{24} = 0 \tag{7}$$

$$4d_{41} + 3e_{42} + 2f_{43} + g_{44} = 0 \tag{8}$$

$$4d_{41} + 3e_{42} + 2f_{43} + g_{44} = 0 \tag{9}$$

3.2 Fitness Function and Recommending an Ice Cream to Diabetic Patients

The chromosomes that are having higher fittest probability value are to be considered for next generation. The obtained fitness probability value shows fitness for respective chromosome. The fitness value for each chromosome can be calculated as follows:

$$\text{Fitness}\,[1] = 1/(4d_{11} + 3e_{12} + 2f_{13} + g_{14} = 0) \tag{10}$$

$$\text{Fitness}\,[n] = 1/(4d_{n*n} + 3e_{n*n} + 2f_{n*n} + g_{n*n} = 0) \tag{11}$$

$$T = F[1] + \ldots\ldots. + F[\text{n}] \tag{12}$$

$$P[n] = p[n]/T \tag{13}$$

The above-mentioned formulas (6)–(13) obtain the said results. In the Fig. 2, fifty generations are considered; initially value of the fitness function is at 0.17 but finally remains between 0.03 and 0.02 which is almost equal to the zero so, obtained fitness function is good enough.

The proposed algorithm is applied on criterion of AHP for the 50 generation. The obtained results are shown in Fig. 3. It starts with value of 5.6 which conforms with the obtained values of AHP and gradually it increases up to 18.

3.3 AHP-CI Algorithm

Cohort intelligence is a group of candidates competing with one another for common goal, inherently common to all candidates. The following steps for CI are considered with respect to AHP.

Step1: Calculate probability for each candidate

$$\text{Fitness}\,[1] = 1/(4d_{11} + 3e_{12} + 2f_{13} + g_{14} = 0) \tag{14}$$

$$\text{Fitness}\,[n] = 1/(4d_{n*n} + 3e_{n*n} + 2f_{n*n} + g_{n*n} = 0) \tag{15}$$

$$T = F[1] + \cdots\cdots + F[n] \tag{16}$$

$$P[n] = p[n]/T \tag{17}$$

Step 2: Roulette wheel approach
R [t] = p (1) + p (2) + p (3) + p (4)

Step 3: Replacement of candidates
C1 = C2, C2 = C3, C3 = C4, C4 = C1;

Table 4 Time complexities of AHP-CI, AHP-GA, and respective values with M-AHP

Criterion	Sugar	Calories	Cholesterol	Proteins
M-AHP value	27.89	13.66	2.84	6.09
AHP-GA values	30	29.33	17.99	15.99
50 iterations	0.409 s	0.409 s	0.409 s	0.409 s
200 iterations	1.366 s	1.366 s	1.366 s	1.366 s
Alternatives	Brayers butter almond	Brayers peanut butter	Ben and jerry coffee	Brayers extra creamy chocolate
M-AHP value	28.24	5.88	13.10	2.76
AHP-CI values	30	8.22	19.99	3.07
50 generation	0.403 s	0.403 s	0.403 s	0.403 s
200 generation	1.381 s	1.381 s	1.381 s	1.381 s

Here, candidate C1 is represented by the Eq. (6), candidate C2 by Eq. (7) and candidate C3 by Eq. (8) mentioned earlier in this paper.

The CI gives the value or behavior of candidates within the range of 0.0625 for 50 generations. The values obtained from the GA are in the range of 0.03–0.02 and value of CI is 0.03, so values in case of both of these algorithms are nearly same. In this way, CI also satisfies the values obtained through GA for the concept under consideration.

Almost same values are obtained as compared with M-AHP to AHP-GA, and similarly for M-AHP to AHP-CI. Values obtained from M-AHP, AHP-GA, and AHP-CI are analogous to each other from the above-mentioned Table 4.

4 Conclusions

The obtained results show a proper usage of M-AHP in order to recommend a particular ice cream to the diabetic patient. The results are obtained through AHP. These results also are verified through MATLAB which shows that patient having a high sugar level of 262 mg/dl can consume an ice cream having lower sugar content like Breyers butter almond, also patient with low sugar level of 77 mg/dl can consume high sugar content ice cream like Breyers extra creamy chocolate In this paper, verification of results for proposed analysis is carried out through M-AHP model and through MATLAB and they are found to be analogous to each other.

However, proposed AHP-GA gives the same results as those obtained from M-AHP. So, M-AHP results for recommending an ice cream to diabetic patient are verified through AHP-GA. It also shows that new chromosomes produced in the upcoming generation will increase the average of M-AHP results. However, with this increased average of ice cream and diabetic patient, the same kind of recommendation can also be achieved.

Similar kinds of results are also obtained through AHP-CI. So, proposed AHP-GA results are validated and verified with respect to AHP-CI also.

5 Future Work

The obtained results from M-AHP model for the problem under consideration can also be compared with clusters of ice cream and with clusters for diabetic patients in order to achieve precise validations and verifications to predict suitability of a particular ice cream for a diabetic patient.

References

1. S. M. Gaikwad: *Cohort Intelligence and Genetic Algorithm along with AHP to recommend an Ice Cream to a Diabetic Patient.* Lecture Notes In Computer Science, 12/2015: chapter Cohort Intelligence and Genetic Algorithm along with AHP to recommend an Ice Cream to a Diabetic Patient: pages 1–9; SEMCCO 2015.
2. S. M. Gaikwad, R. R. Joshi, and Mulay, P., "Cluster Mapping with the help of New Extended MCF Algorithm and MCF Algorithm to Recommend an Ice Cream to the Diabetic Patient," METHODOLOGY, vol. 1, p. 7.
3. S. M. Gaikwad, R. R. Joshi, Mulay, P. (2015). Analytical Network Process (ANP) to Recommend an Ice Cream to a Diabetic Patient. IJCA Online 121 (12), 49–52.
4. S. M. Gaikwad, R. R. Joshi, and Mulay, P., "System Dynamics Modeling for Analyzing Recovery Rate of Diabetic Patients by Mapping Sugar Content in Ice Cream and Sugar Intake for the Day," in Proceedings of the Second International Conference on Computer and Communication Technologies, ed: Springer India, 2016, pp. 743–749.
5. S. M. Gaikwad, Mulay, P., and R. R. Joshi, "Mapping with the help of new Proposed Algorithm and Modified Cluster Formation Algorithm to recommend an Ice Cream to the Diabetic Patient based on Sugar Conatin in it," International Journal of Students' Research in Technology & Management, vol. 3, pp. 410–412, 2015.
6. S. M. Gaikwad, R. R. Joshi, Mulay, P., "Attribute visualization and cluster mapping with the help of new proposed algorithm and modified cluster formation algorithm to recommend an ice cream to the diabetic patient based on sugar contain in it," vol. 10, pp. 1–6, 2015.
7. S. M. Gaikwad, "Cluster mapping with the help of new proposed algorithm and MCF algorithm to recommend an ice cream to the diabetic Patient," International Journal of Applied Engineering Research, vol. 10, pp. 21259–21266, 2015.
8. S. M. Gaikwad, R. R. Joshi, Mulay, P. "Modified Analytical Hierarchy Process to Recommend an Ice Cream to a Diabetic Patient," pp. 1–6, 2015.

9. Kulkarni, A. J., Durugkar, I. P., & Kumar, M. (2013). Cohort Intelligence: A self supervised learning behavior. In Proceedings of IEEE International Conference on Systems, Man and Cybernetics (1396–1400).
10. Krishnasamy, G., Kulkarni, A. J., & Paramesaran, R. (2014). A hybrid approach for data clustering based on modified cohort intelligence and K-means. Expert Systems with Applications, 41(13), 6009–6016.
11. Kulkarni, A. J., & Shabir, H. (2014). Solving 0–1 Knapsack problem using cohort intelligence algorithm. International Journal of Machine Learning and Cybernetics. doi:10.1007/s13042-014-0272-y.
12. Kulkarni, A. J, Application of the cohort-intelligence optimization method to three selected combinatorial optimization problems, European Journal of Operational Research (2015), http://dx.doi.org/10.1016/j.ejor.2015.10.008.
13. S. M. Gaikwad, P. Mulay, R.R. Joshi: Analytical Hierarchy Process to Recommend an Ice Cream to a Diabetic Patient Based on Sugar Content in it. Procedia Computer Science 12/2015; 50. DOI:10.1016/j.procs.2015.04.062.

Theoretical Framework for Privacy in Interpersonal Information Communication

T. Shanmughapriya and S. Swamynathan

Abstract In the world of Internet, the number of users and applications has been increasing unprecedentedly, as well as the complexity of interactions. Users communicate directly as well as through applications and the services that each of them in turn invoke. In this fast increasing complexity, users not only share their personal information intentionally to gain a benefit, but does inadvertently end up giving away much more information than what they will willingly give, compromising privacy requirement of not only themselves but that of their contacts as well. The outcomes of earlier studies swing from too simple postulation of tradeoff between secrecy and benefit of sharing, to too complex definitions, both of which makes privacy enforceability practically hard in the current and future scenarios, while preserving universality. This paper provides a way for theoretically capturing the complex relationship of interpersonal information communication by providing a framework within which any systems interpersonal information and its communication can be dynamically modeled but yet its applicability is not restricted to only a subset of application contexts or locations or time. This model provides the necessary theoretical foundation for analyzing the different user requirement attributes like sharing and secrecy, whose complex interplay makes up for the notion of Privacy. From simple clear evidence of presenting the relationship between sharing and secrecy, to the ability of this framework to capture dynamic user and group behavior within a system is presented. This paper also lays the road ahead of how this model can be used to predict user aspirations and arrive at privacy policy specifications for any given system.

Keywords Online social network · Privacy · Interpersonal information communication

T. Shanmughapriya (✉) · S. Swamynathan
Department of Information Science and Technology, Anna University, Chennai, India
e-mail: priyanethiran@gmail.com

S. Swamynathan
e-mail: swamyns@annauniv.edu

© Springer Science+Business Media Singapore 2017

S.C. Satapathy et al. (eds.), *Proceedings of the International Conference on Data Engineering and Communication Technology*, Advances in Intelligent Systems and Computing 468, DOI 10.1007/978-981-10-1675-2_30

1 Introduction

In the current scenario, users play the role of both data publishers and service providers interchangeably, as Lev Grossman [1] Times Person of the Year says, The new web is a very different thing. It is a tool for bringing together the small contributions of millions of people and making them matter. Today, online world works on the concept of sharing and every one contributes some information to the online world either intentionally or unintentionally. In intentional sharing, the user shares whatever he knows and feels appropriate to be shared with a group in a particular context. In unintentional sharing the user behavior captured from online interactions are aggregated and correlated with other available information leading to user profiling. Sharing is a highly encouraged feature in today's web owing to which the users are subjected to a wide spectrum of privacy breach. The notion of Privacy is without a clear universal definition, notwithstanding intense debates and having seen many widely varying laws seeking to protect it across the globe. Hence, it is not surprising that the applications and services that are built on such loosely understood notion provide for vulnerabilities exploiting which, privacy can be compromised. Therefore, the focus has to be to evolve a comprehensive framework that captures interpersonal information communication, which will hold variations with respect to space (geography, application domains etc.) and time (present and future). Over this one could analyze the information secrecy and sharing dynamics, and hence provide a framework for studying and assuring privacy of systems. The rest of the paper is to address this requirement and has been organized as follows. Section 2 surveys the related work in the literature. Section 3 presents the model. Section 3 presents different hypothesis arrived at from the model. Section 4 ends with Conclusion of the paper.

2 Related Work

Voluminous work has been done in the area of engineering privacy. Still the notion of privacy is not with a clear definition. Westins [2] theory of privacy addresses how people protect themselves by temporarily limiting access of others to themselves. Privacy is the claim of individuals, groups, or institutions to determine for themselves when, how, and to what extent information about them is communicated to others.

Emphasis on the constant dialectic and dynamic regulation of privacy coined by Altman [3] stresses that there is no paramount condition of privacy that one should generally try to attain. On the contrary, desired levels of privacy fluctuate depending on the specific situation and the interaction that takes place. In order to achieve a desired level of privacy, people constantly adjust interpersonal boundaries. Petronio [4] in CPM theory coins that privacy boundaries can range from complete openness to complete closeness or secrecy. An open boundary reflects willingness to grant access to private information through disclosure or giving permission to view that information, thus representing a process of revealing. On the other hand, a closed boundary represents information that is private and not necessarily accessible, thus characterizing a process of concealing and protecting.

Attitude behavior gap proposed by Ajzen and Fishbein's [5] addresses the link between attitudes and behavior is not very strong. There is strong evidence in privacy research that such a gap exists for privacy as well. Ajzen and Fishbein provide remedy to this problem by introducing the mediating concept of behavioral intentions, but they note that even these intentions are not always perfectly correlated with actual. Olson 2005 [6] demonstrated that people vary in their overall level of comfort in sharing. Key classes of recipients and information was identified and such abstractions highlight the need for developing expressive controls for sharing and privacy. The proposal by Livingstone et al. [7] reflected that risks go always hand in hand with opportunities. Risk in sharing attributes may be minimized by limiting the sharing, which minimizes the opportunities. Catering to the raising levels of user interaction in current internet-based application scenario, solely protection is not the goal. Decisions have to be made by analyzing the tradeoff between the benefit and the risk associated in sharing attributes.

Having become a legal requirement, there are serious efforts by NIST to come out with engineering frameworks to assure privacy of systems [8]. Efforts to have a comprehensive working definition for engineering privacy of systems have led to the following complex definition from NIST draft workshop [9]: Privacy engineering is a collection of methods to support the mitigation of risks to individuals of loss of self-determination, loss of trust, discrimination and economic loss by providing predictability, manageability, and confidentiality of personal information within information systems. A draft formula for calculating system privacy risk is as follows:

System Privacy risk = Personal Information Collected or Generated + Data Actions Performed on that Information + Context.

The privacy literature surveyed so far, of which some representative samples have been provided above, swings between simple tradeoffs between information sharing and privacy to too complex notions of privacy that are hardly practically enforceable. Another part of the literature, takes to user opinion based survey about privacy. The samples chosen and specific application contexts in which they have been studied raises questions of appropriateness of application of the outcome of such studies, in other application contexts, at other locations and at other time.

3 Model for Interpersonal Information Communication and Privacy

The generalized model for information sharing presented in Fig. 1 depicts that user X shares information with users in the community Y. (This model can also alternately be simplified with Y for others that includes all users in all communities). The user

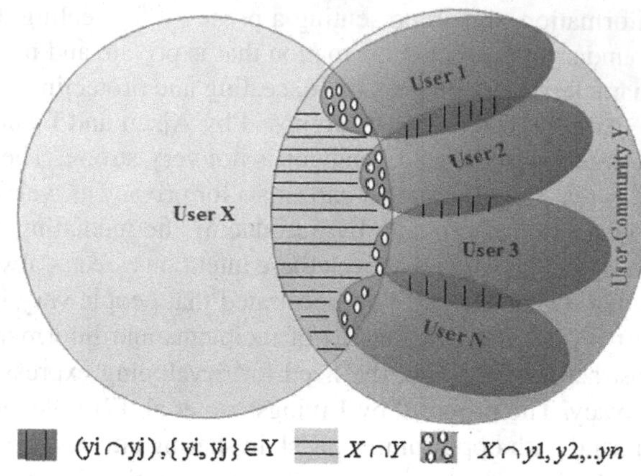

Fig. 1 Model for interpersonal information communication

X is free to communicate with any number of communities. In the model shown in Fig. 1, $\{y_1, y_2, y_3 \ldots y_n \subseteq Y\}$ where $y_1, y_2, y_3 \ldots y_n$ denote the users in the community Y with whom user X shares information. The shared region (X ∩ Y) represents the attributes shared by user X with the community Y. The shared region can also dynamically grow or shrink depending on various facts like the current mood of different users, the attributes the user shares, the context in which it is shared, to whom its shared and for what purpose it is shared, sought or collected. $(X \cap y_1, y_2, y_3 \ldots y_n)$ region represents the data shared by user X with users in community Y.

There is a region in (X ∩ Y) where the data is shared by the user X to community Y, but not all the attributes are shared with every user in the community, the relationship is represented in Eq. 1.

$$[(X \cap Y) - ((X \cap y_1) \cup (X \cap y_2) \cup (X \cap y_n))] \tag{1}$$

The area representing attributes not shared with community Y is secrecy area of X represented by Eq. 2.

$$[X - (X \cap Y)] \tag{2}$$

The attributes of X on the whole is represented as the sum of data shared by the user and the data retained by the user in secrecy.

$$X - (X \cap Y) + (X \cap Y) = X \tag{3}$$

From Eq. 3 it is understandable that by combining users shared attributes and the secret data leads to the data owned by user x.

H1: The shared data and secret data on its entirety comprise the whole of the user data.

The model validates this hypothesis (H1). From Eq. 3 it is clear that that by combining users shared attributes and the secret data leads to the data owned by user x on the whole. Figure 1 showcases a scenario of user x sharing data with community Y. It can be extended to share information with any number of communities, irrespective of the number of communities, the information shared with all the communities and the data retained in secrecy sums up to the data owned by the particular user.

H2: *The sharing decision area grows or shrinks depending on the context, what is shared and with whom its shared.*

Olson et al. [6] in his study reveal that peoples willingness to share depends on what information they are sharing and with whom they share. The study involved querying people about what information they are willing to share with whom. They were able to identify key classes of recipients and information. The context also plays an important role in sharing. People reveal their health conditions to practitioners and may not be willing to share the same information in a different context.

H3: *The data once shared between two users. The data sharer has very less control over the data recipient. The recipient becomes the co-owner of the information and gets control over the information same as that of the owner.*

The ownership of the data gets expanded once the data is shared with external participants. Once the data is shared, the control of the data is in the hands of both the data owner and the recipient with whom the data is shared. The concept of Collective privacy boundary coined by Sandra Petronio [4] regards that a person can't decide on own whether to conceal or reveal. The theory also states that the act of disclosing private information creates a confidant and draws that person into a collective privacy boundary. The information once disclosed seldom shrinks back to being solely personal.

H4: *The data shared in one context to a recipient could be revealed to other recipient in some other context.*

The data shared by user X to users in the community Y need not be same. Considering the case of data shared with y1 and y2 are different, y1 and y2 become the co-owner of the information. The collective privacy boundary is gained by y1 and y2. y1 and y2 could share the information they know about X. The user may not prefer this because the information sharing happens out of the context in which it was originally shared.

H5: *The dynamic changes in decision have minimal impact on the previously shared data.*

The data once shared in online world remain and cannot be withdrawn completely without any traces. The user boundaries are subject to dynamic expansion and contraction/shrinking. In which case the expansion leads to sharing of more information and shrinking leads to sharing of less information. The shrink after an expansion is not very effective in online environment. Even though the user might decide to reduce the level of disclosure, it cannot be certainly claimed that the attributes the user wishes not to share any more are not being used outside the context. In the digital world, the data once crosses the boundary;the visibility of the data is not fully in the hands of the data owner.

H6: *Depending on the secrecy and sharing preferences (mood) of the user at a particular place at a particular context and at a particular time the venn bubble of the user can expand or contract.*

Considering an instance when say User X either is depressed or wants to introspect, and hence chooses to not engage in interpersonal communication, the region of User X can contract to a very small volume. Same holds good for each of the other users in Y.

4 Conclusion

It is not hard within this framework, to build a model to capture any kind of dynamics of the interpersonal information sharing. Say for example if suppose one wants to assess the career/social/psychological aspiration of an individual, one could capture with respect to time, the size and movement of the individual bubble in the venn diagram, the nature and movement of the community or Other bubble and the nature of the interactions between them. By tracking these aspects it is feasible to come out with verifiable predictions on different dimensions of aspiration of an individual. Hence it is feasible to overlay attributes like secrecy and privacy preservation on this framework. Once this can be done, one could reverse engineer the same and specify the secrecy and privacy preferences as input. As an output from this model, one could obtain the restriction conditions on the individual user or group behavior. The latter conditions can go as policy configurations of the application that the user or the group uses for information sharing, making privacy practically enforceable thus assuring the privacy of the system. Further details and implications of this are part of the ongoing and future work.

Acknowledgments We thank Mr. U. Thiruvaazhi (Isha Higher Education, Coimbatore, India) toward his valuable suggestions for the paper.

References

1. Warren S, Brandeis L. The right to privacy. Harvard LawReview (1890).
2. Ds Westin A. Privacy and freedom. New York: Atheneum (1967).
3. Altman, I., The Environment and Social Behaviour: Privacy, Personal Space, Territory, Crowding, Monterey. CA (1975).
4. Child, Jeffrey T., and Sandra Petronio. Unpacking the paradoxes of privacy in CMC relationships: The challenges of blogging and relational communication on the internet, Computer-mediated communication in personal relationships (2011).
5. Ajzen, I., Fishbein, M., Attitudebehavior relations: A Theoretical Analysis and Review Of Empirical Research, Psychological Bulletin (1975).
6. Judith S. Olson, J. G., Eric Horvitz. "A Study of Preferences for Sharing and Privacy." Human Factors in Computing Systems: 1985–1988 (2005).

7. Livingstone, S., lafsson, K., Staksrud, Risky social networking practices among under age users: Lessons for evidence based policy. Journal for Computer Mediated Communication, 18(3), 303–32 (2013).
8. NIST, NIST Privacy Engineering Objectives and Risk Model Discussion Draft-http://www.nist.gov/itl/csd/upload/nist_privacy_engr_objectives_risk_model_discussion_draft.pdf (2014).
9. NIST, NIST Privacy Engineering Workshop Discussion eckhttp://www.nist.gov/itl/csd/upload/nist_privacy_engr_objectives_risk_model_discussion_deck.pdf (2014).

Gist, HOG, and DWT-Based Content-Based Image Retrieval for Facial Images

Rupali Desai and Bhakti Sonawane

Abstract In today's world there is a wide range of digitization. Everyone is living in a world of digital through text, images, videos, and many more. This raises need for development of latest technologies for retrieval of digital data of the pictorial form. Content-based image retrieval is an efficient method which automates retrieval of images with respect to its salient features. This paper defines approach to have CBIR on facial images with three different feature extraction methodologies Gist, HOG, and DWT. Each feature extraction method will extract facial features for given query image. These facial features are stored in multidimensional feature vector form then used for classifying given query image using knn algorithm. In this approach, we observe performance of every feature extraction technology with its impact on CBIR using precision and recall values.

Keywords Histogram of gradients (HOG) · Discrete wavelet transform (DWT) · Content-based image retrieval (CBIR) · Knn

1 Introduction

Today's world is digital which consists of data in the form of text, images, videos, and many more. Wide usage of Internet leads to growth of digital data. Browsing of Internet also allows users to search relevant images or videos from remote sites. One of the active research areas is the image retrieval, by which user should get relevant digital images when he browse through the search engine [1]. Image retrieval can be text based or content based. Text-based image retrieval is retrieval of images using textual information. In the text-based image retrieval, images are annotated using textual information this leads to limitation of this method. The first

Rupali Desai (✉) · Bhakti Sonawane
Shah & Anchor Kutchhi Engineering College, Govandi(E), Mumbai, India
e-mail: rupadesai92@gmail.com

Bhakti Sonawane
e-mail: bhaktisonawane@rediffmail.com

© Springer Science+Business Media Singapore 2017
S.C. Satapathy et al. (eds.), *Proceedings of the International Conference on Data Engineering and Communication Technology*, Advances in Intelligent Systems and Computing 468, DOI 10.1007/978-981-10-1675-2_31

limitation is, textual information by which images are annotated can be subjective. Second limitation is that we cannot always annotate the image with proper set of words which will completely describe content of the image. Third limitation is that if the image database is very large then manual annotation process becomes cumbersome. In content-based image retrieval, retrieval of images will be based on content of images. The image contents can be of low-level features which can be visual features or it can be high-level features. Thus in CBIR, images would be indexed by their visual features content and the desired images are retrieved from a large collection, on the basis of features that can be automatically extracted from the images itself (Figs. 1 and 2).

As shown in block diagram the facial CBIR consists of two phases; learning phase and retrieval phase. In learning phase, the images are processed to extract feature vectors based on facial features that represent numeric values of the interest

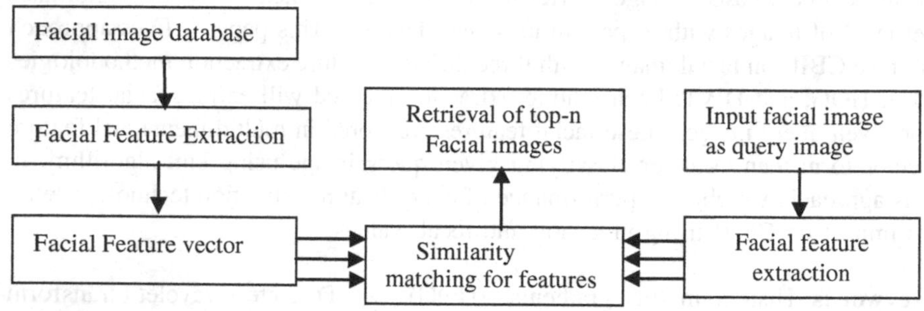

Fig. 1 Block diagram of facial CBIR

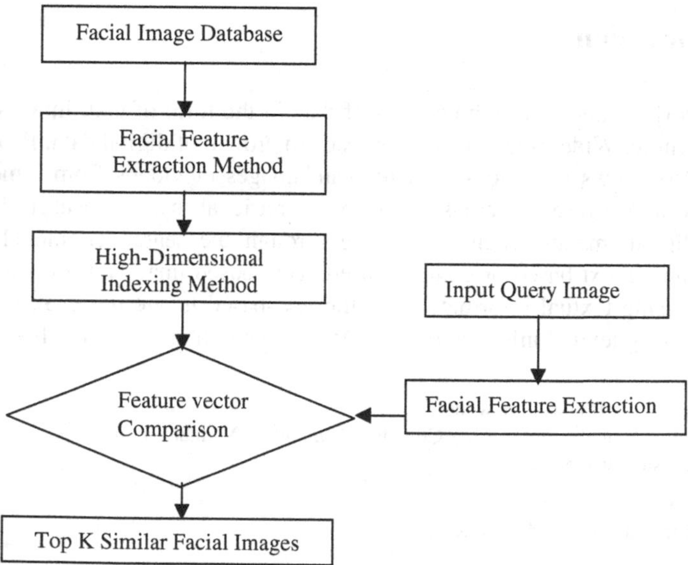

Fig. 2 Proposed CBIR system

points of facial region. In the retrieval phase, the user enters a query image whose facial feature vector is calculated and then compared to those of the candidate images in the database. Using similarity distance, top-n facial images with most similar features are retrieved. In content-based image retrieval, knowledge of different areas such as pattern recognition, object matching, machine learning, etc., is used. More efficiency means always relevant images should be retrieved by system and less computational time means retrieval of the relevant images should be in minimum time.

2 Background

Every time a user inputs query facial image, its facial features are extracted and used to find images from the database with most similar features. Various techniques may be used to find different facial features that may be based on a particular characteristic of facial images like pose variance, color, shape, etc. To have proper face alignment in uniform, author used DLK algorithm for facial images in unsupervised learning [2]. HOG feature has been introduced by Navneed Dalal and Bill Triggs [2] who have developed and tested several variants of HOG descriptors, with differing spatial organization, gradient computation, and normalization methods. In [3] HOGs which count occurrences of edge orientations in a local neighborhood of facial image are used. Here facial images are first divided into small connected regions, called cells, and for each cell a histogram of edge orientations is computed. In facial images, there are many interest points which can be referred as salient points also [1, 4]. In this paper, author proposed an improved image retrieval system using wavelet transform to extract salient points more accurately. In this novel approach, discrete wavelet decomposition [2] of an image is carried out to locate salient points. These points are tracked back to the original image to locate their positions in spatial domain. Various feature extraction methods like color distribution, texture features using Gabor coefficients, discrete cosine transform, and color moments are computed around the neighborhood of these salient points that form feature vectors. In [2], author used texture and color features are extracted through wavelet transformation and color histogram. In [5], author described basic context-based scene recognition algorithm using a multiscale set of early-visual features, which capture the “Gist” of the scene into a low-dimensional signature vector. In [6], author mentioned about Gist feature as a collection of Gabor filter response of image and it can represented as region boundary of an object or shape of a scene in an image (Fig. 3).

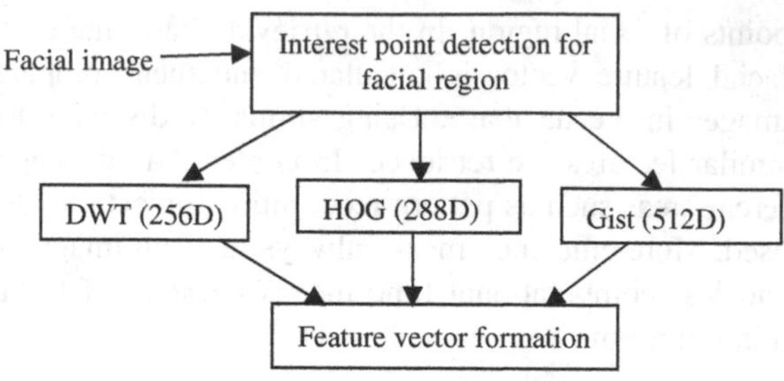

Fig. 3 Feature vector generation [2]

3 Proposed Framework

The proposed framework uses gist, HOG, and DWT features as facial features. These features are high-dimensional features. The block diagram of the proposed approach is shown as follows.

3.1 Facial Image Database

The weak labeled database named as WDB-040 K [7, 8] is used in proposed framework, which is having 400 person facial images with approx. 125–130 images/person. The database contains total 53,448 facial images with different resolutions. To have uniformity in resolution of database images, DLK algorithm is used by author in [7, 8]. The DLK aligned images are having resolution of 138 × 166 with jpg type.

3.2 Facial Feature Extraction Method

An image undergoes for facial feature extraction via GIST, HOG, and DWT features as shown in figure.

Now we describe about facial features used in proposed framework.

3.2.1 DWT Transform

DWT decomposes images in various resolutions with the help of multi-resolution property. The four subbands which are approximation, horizontal, vertical details, and diagonal details of a first-level wavelet decomposition of image are shown in Fig. 4

LL	LH
HL	HH

Fig. 4 Discrete wavelet decomposition first-level decomposition [2, 4]

At each level (scale), the image is decomposed into four frequencies. Subbands LL, LH, HL, and HH where L denotes low frequency and H denotes high frequency. When DWT [2, 4] is applied to 32 × 32 image, we get 16 × 16 LL, LH, HL, and HH bands.

3.2.2 Histogram of Oriented Gradients

Here image is first divided into cells, i.e., small connected regions and for each cells a histogram of edge orientations is computed [3]. For query image, HOG descriptor is computed as follows:

(1) Centered horizontal and vertical gradients.
(2) Compute Gradient orientation and magnitudes by using centered filter masks in x and y directions.
(3) Divide 64 × 128 image into small connected regions 16 × 16 called blocks, with 50 % overlap. Each block should consist of 2 × 2 cells with size 8 × 8.
(4) Quantize the gradient orientation into 9 bins (directions). Normalized group of histograms represents the block histogram. The set of these block histograms of 288 dimensional represents the descriptor.

3.2.3 Gist

Gist uses 8 orientations of parameters per scale in 4 different blocks [5, 6]. While computing image similarities, normalize image size before computing gist descriptor (Figs. 5 and 6).

Fig. 5 Input query image

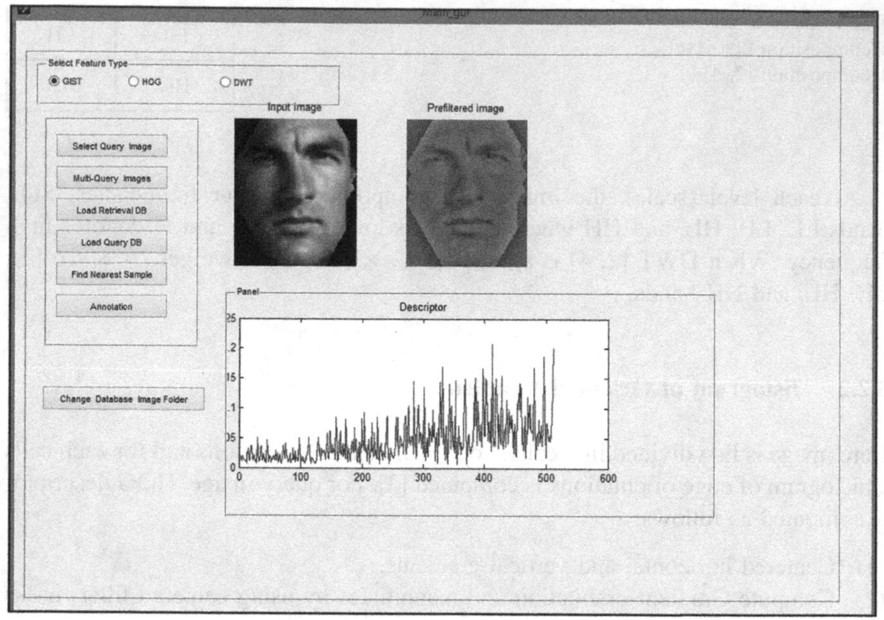

Fig. 6 gist descriptor of input image

For given an input image, a Gist descriptor [5, 6] is computed by following:

1. Convolve the image with 32 Gabor filter at 4 scales, 8 orientations, producing 32 feature maps of the same size of input image.
2. Divide each feature map into 16 regions (by 4×4 grid), and then average the feature values within each region.
3. Concatenate the 16 averaged values of all 32 features maps resulting in a $16 \times 32 = 512$ Gist descriptor.

3.2.4 High-Dimensional Indexing Method

There are different high-dimensional indexing techniques given in [9, 10]. Locally Sensitive Hashing [7, 8, 11] is one of efficient searching technique used for multidimensional data. It takes concept of hash function and hashing process (Figs. 7 and 8).

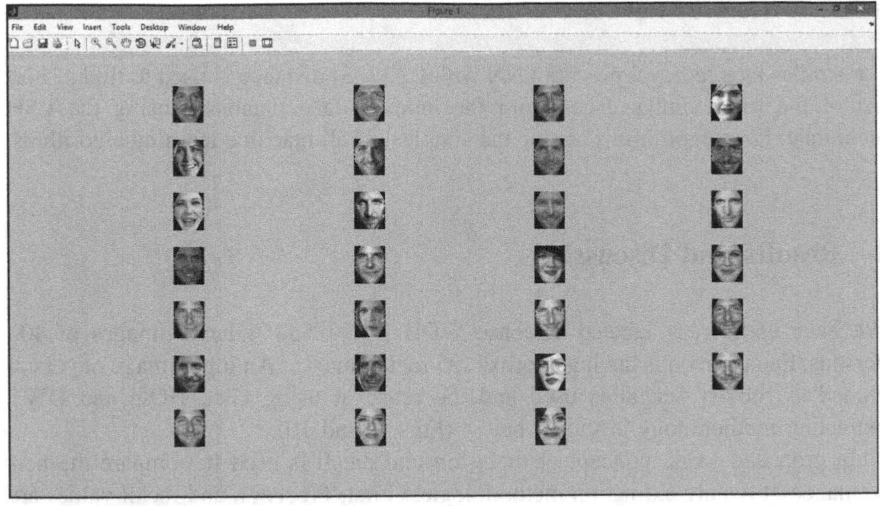

Fig. 7 Retrieval result using gist descriptor

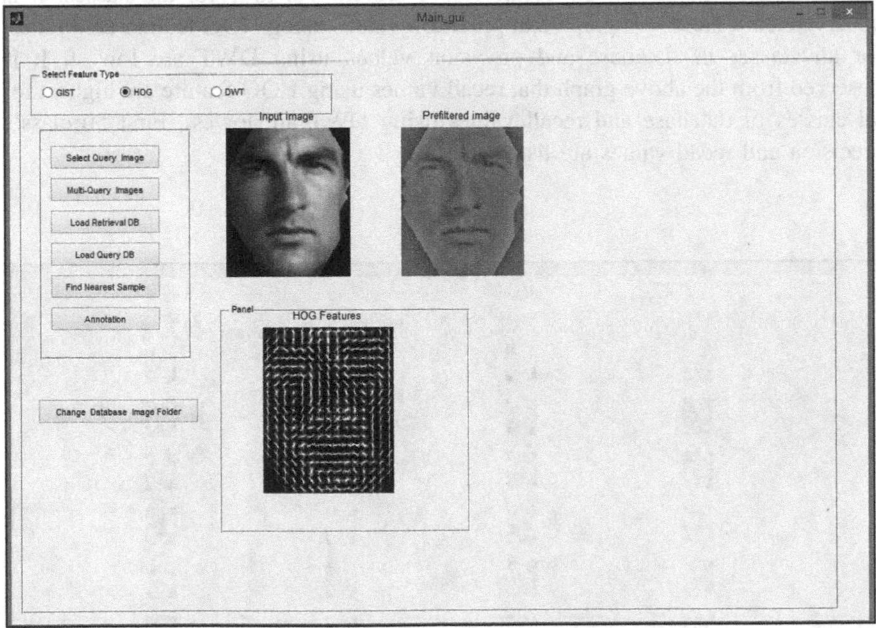

Fig. 8 HOG descriptor of input facial image

3.2.5 K-Nearest-Neighbor Classification

For similar face retrieval process kNN with L2 [7, 8] distance is used to find a short list of the most similar faces from the indexed face databases using the LSH technique. Knn algorithm is among the simplest of all machine learning algorithms.

4 Results and Discussions

We have used weak labeled database WDB [7] of 53448 facial images of 400 persons. Each person is having approx. 125 facial images. An input image of person named as Steven Seagal is used and the retrieval using Gist, HOG and DWT extraction methodology is shown below (Figs. 9 and 10).

In proposed work, concept of precision and recall is used to compare the performance of feature extraction methodologies. First, Precision and recall values are computed for all database images then class-wise aggregate value is derived.

Remark: Above Fig. 11a, b shows precision and recall values of 400 classes of weak labeled facial database for Gist, HOG, and DWT feature extraction methodologies. Precision and recall values are from 0 to 1 for all classes. It is observed from the above graph that precision values using HOG feature are highest for all classes of database and precision values using DWT are lowest. It is observed from the above graph that recall values using HOG feature are highest for all classes of database and recall values using DWT are lowest. First 50 class's precision and recall values are listed in Table 1.

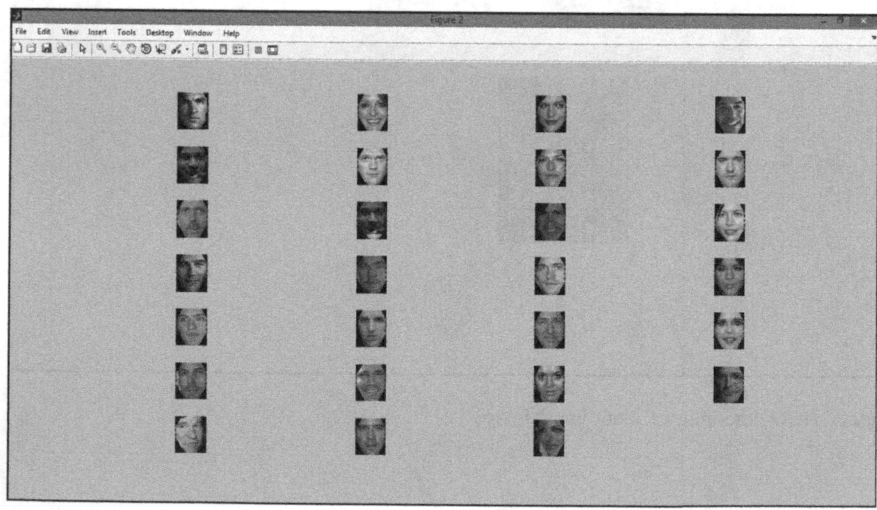

Fig. 9 Retrieval result using HOG feature

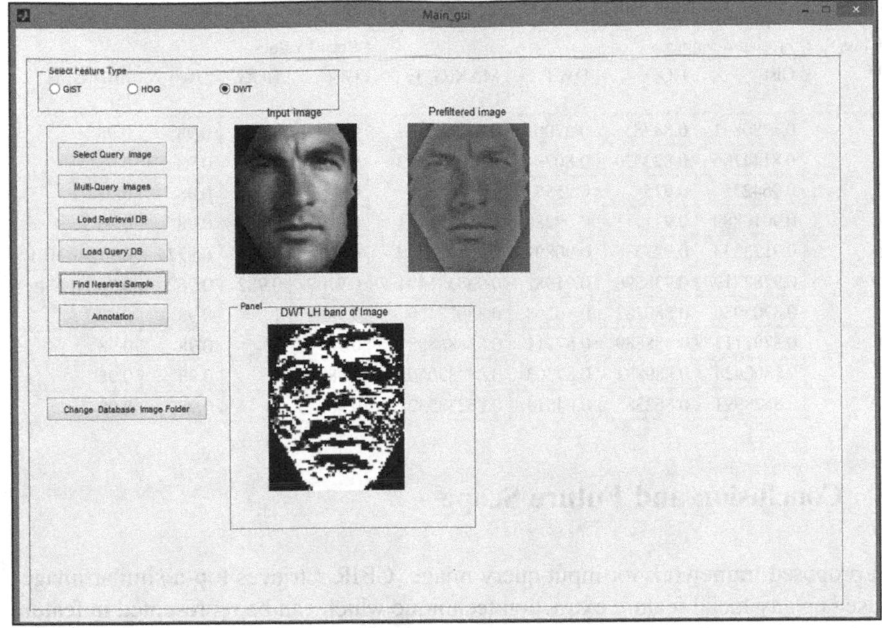

Fig. 10 DWT descriptor of input facial image

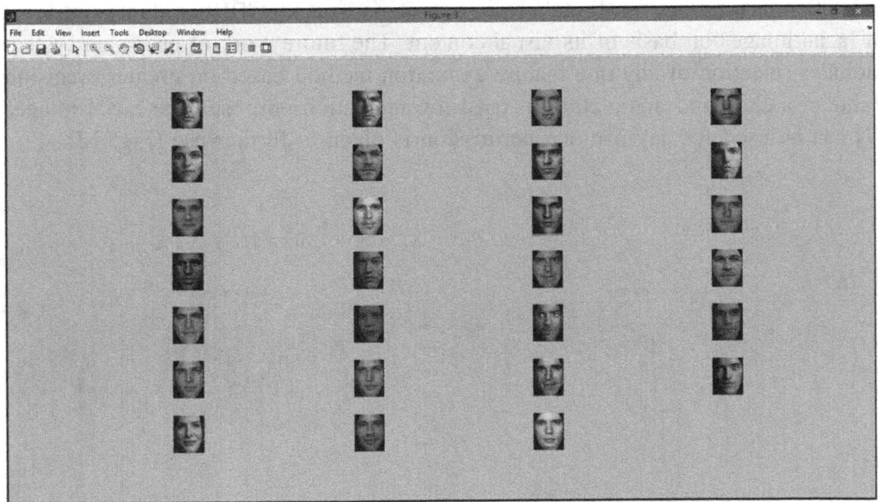

Fig. 11 Retrieval result of DWT feature

Table 1 Shows precision values of first 50 classes of database

Class no	Precision values				Recall values			
	Gist	HOG	DWT	MAX(G, H, D)	Gist	HOG	DWT	MIN(G, H, D)
1	0.8780841	0.88785	0.87009	0.887850467	0.989	1	0.98	0.98
2	0.8144706	0.823529	0.80706	0.823529412	0.989	1	0.98	0.98
3	0.964275	0.975	0.9555	0.975	0.989	1	0.98	0.98
4	0.9010889	0.911111	0.89289	0.911111111	0.989	1	0.98	0.98
5	0.9173333	0.927536	0.90899	0.927536232	0.981	0.992	0.9724	0.972403101
6	0.9282719	0.938596	0.91982	0.938596491	0.9709	0.982	0.962	0.962018349
7	0.8705986	0.880282	0.86268	0.88028169	0.989	1	0.98	0.98
8	0.8791111	0.888889	0.87111	0.888888889	0.989	1	0.98	0.98
9	0.8800424	0.889831	0.87203	0.889830508	0.989	1	0.98	0.98
10	0.8528991	0.862385	0.84514	0.862385321	0.989	1	0.98	0.98

5 Conclusion and Future Scope

In proposed framework for input query image, CBIR retrieves top-n similar images based on any facial feature extraction technique which can be represented in feature vector format. The proposed system uses three different feature extraction methodologies which are Gist, HOG, and DWT. These features vectors are indexed by using LSH high-dimensional indexing technique. For comparison of feature extraction techniques, recall and precision criteria is used. CBIR evaluates retrieval of facial images on basis of its visual content. The future work of proposed system includes selection of any one feature extraction method based on greater precision value of a class and this method is used for annotation purpose. The facial images [7] can be used for any use and permission is given to further use (Fig. 12).

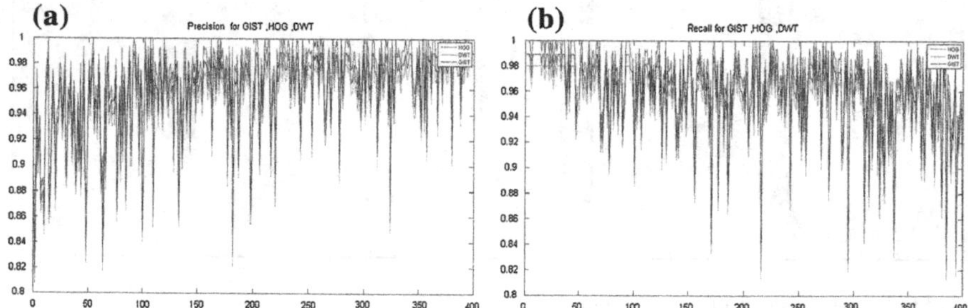

Fig. 12 **a** Precision graph for HOG, DWT and Gist of all facial database images of class in which *red line* shows precision for HOG, *green line* shows precision for DWT and *blue line* shows precision for Gist. **b** Recall graph for HOG, DWT and Gist of all facial database images of class in which *red line* shows precision for HOG, *green line* shows precision for DWT and *blue line* shows precision for Gist

References

1. Y. Tian, W. Liu, A Face Annotation Framework with Partial Clustering and Interactive Labeling Proc. IEEE Conf. Computer Vision and Pattern Recognition (CVPR), 2007.
2. Manimala Singha and K. Hemachandran Content Based Image Retrieval using Color and Texture, Signal & Image Processing: An International Journal (SIPIJ) Vol. 3, No. 1, February 2012.
3. O. Déniz, G. Bueno, J. Salido, F. De la Torre, Face recognition using Histograms of Oriented Gradients Pattern Recognition Letters 32 (2011) 1598–1603.
4. Franchesca Fernandes, Satishkumar Chavan, Content Based Image Retrieval using Localized Colour and Texture Features at International Conference on Signal, Image and Video Processing (ICSIVP) 2012.
5. http://ilab.usc.edu/siagian/Research/Gist/Gist.html.
6. http://www.researchgate.net/publication/262917557_LargeScale_Scene_Classification_Using_Gist_Feature.
7. D. Wang, S.C.H. Hoi, Retrieval-Based Face Annotation by Weak Label Regularized Local Coordinate Coding, Proc. 19th ACM Int'l Conf. Multimedia, pp. 353–362, 2011.
8. D. Wang, S. Hoi, Mining Weakly-Labeled Web Facial Images for Search-Based Face Annotation, IEEE Trans. Knowledge and Data Eng., vol. 99, no. Preprints, pp. 1–14, 2012.
9. Haiying Shen, Felix Ching, Ting Li, Ze Li, An Intelligent Locally Sensitive Hashing Based Algorithm for Data Searching Department of Computer Science and Computer Engineering.
10. W. Dong, Z. Wang, W. Josephson, M. Charikar, and K. Li, Modeling LSH for Performance Tuning Proc. 17th ACM Conf. Information and Knowledge Management (CIKM), pp. 669–678, 2008.
11. M. Guillaumin, Face Recognition from Caption-Based Supervision, Int'l J. Computer Vision, vol. 96, no. 1, pp. 64–82, Jan. 2011.

Low Cost Intelligent Irrigation System

Abhishek Kathpal, Nikhil Chawla, Nikhil Tyagi and Rakshit Yadav

Abstract Agriculture is one of the most crucial factor in sustaining human life on Earth. Also, it is a source of occupation for majority of people in many countries. Various technologies have been developed to process the raw product of agriculture for business practices, but comparably very less advancements have been made to aid the farming process. This paper focuses on a system that automates the irrigation process of fields in a precise and efficient manner, keeping its cost in the budget of farmers. This system uses the moisture content in soil, location of farm on Earth, type of crop, level of standing water in fields, and temperature of air to calculate the water requirement of crops based on empirical data prestored in a database. Then appropriate amount of water is pumped to the fields, and the procedure repeats itself periodically.

Keywords Irrigation · Autonomous · Moisture sensor · Wireless · Agriculture

1 Introduction

Agriculture is a necessary activity because it is a source of food for the people. A good yield provides surplus food that makes the population healthy. But a poor yield leads to inflation in prices and even famines.

Abhishek Kathpal (✉) · Nikhil Chawla · Nikhil Tyagi · Rakshit Yadav
National Institute of Technology Kurukshetra, Kurukshetra, India
e-mail: kathpal.abhishek@gmail.com

Nikhil Chawla
e-mail: nikhil21chawla@gmail.com

Nikhil Tyagi
e-mail: tyaginikhil94@gmail.com

Rakshit Yadav
e-mail: rrakshityadav@gmail.com

© Springer Science+Business Media Singapore 2017
S.C. Satapathy et al. (eds.), *Proceedings of the International Conference on Data Engineering and Communication Technology*, Advances in Intelligent Systems and Computing 468, DOI 10.1007/978-981-10-1675-2_32

The output of farming depends on soil, water, seed quality, sunlight, climate, minerals, pests, and diseases. Out of these, water plays a vital role in governing the quality and quantity of the produce. It is required in every stage of plant development, from seed germination till harvesting. Also, a crop with high water sensitivity has affected the most when its requirement is not met. Additionally, a crop requires more water in a hot climate than in a cooler one.

But having ample water supply does not guarantee a worthy yield. A crop may get damaged if extra amount of water is given to it due to growth of fungus and spreading of diseases in standing water. Also, overirrigation limits the air from reaching the plant's root and stem which leads to decay of the plant.

Water should be supplied critically by keeping in mind the requirement of crop at the time, then only a perfect yield can be attained. Also, water being a limited natural resource should be used responsibly. But unfortunately the farmers responsible for irrigation are not much educated and follow traditional irrigation methods. This leads to inappropriate water supply and crop degradation.

This system is a solution to above-mentioned problems. It autonomously irrigates the fields based on the database of water requirements of crops in every growing period, and in every region on Earth. This database is based on experimentally tested values and is also verified by various research organizations around the world.

This system (shown in Fig. 1) comprises of a soil moisture sensor, temperature sensor, and standing water level detector. All the three sensors provide relevant data which is transmitted wirelessly to a main controlling hub. The hub comprises of an

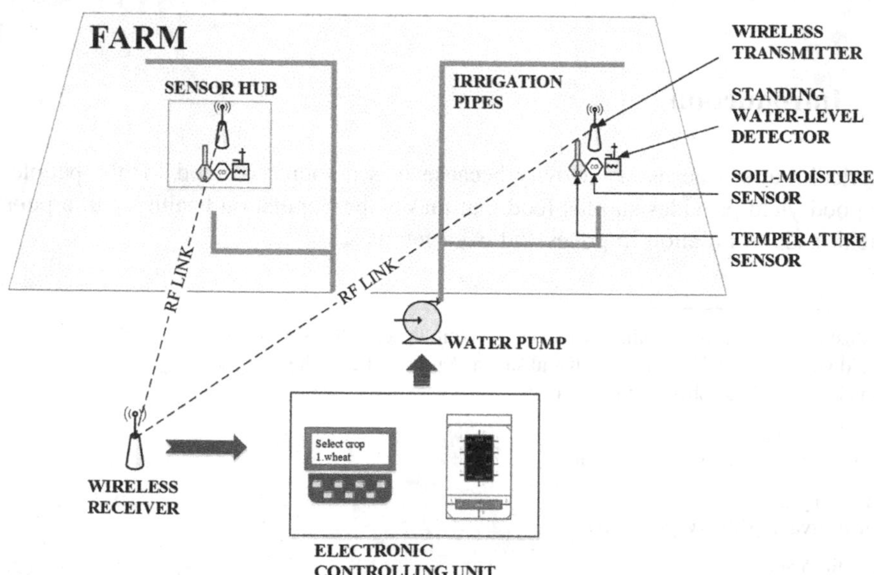

Fig. 1 Schematic of low cost intelligent irrigation system

electronic controlling unit along with a water pump. The received information along with location data is processed to calculate the amount of water required by the soil. Then the water pump is used to supply water to the fields until the correct soil moisture level is reached. This procedure repeats itself periodically five times a day to ensure proper irrigation. During rest of the time it goes into power saving sleep mode. It uses a real-time clock to schedule the irrigation cycles accurately.

2 Sensor Hub

Moisture in the soil, temperature of air, and standing water level is required to calculate the water requirement of the crop. Data collected by this sensor hub is encoded and then transmitted to the main controlling unit, where irrigation needs of the field are calculated accordingly as shown in Fig. 2.

2.1 Moisture Sensor

Amount of moisture content present in the soil is detected by simply using two metal probes. They are of cylindrical shape, with length 10 cm and diameter 7 mm. This is a resistive-type sensor. Electrical resistance between these two metal probes is inversely proportional to the moisture present in the soil. Potential difference between the two probes is sent to the microcontroller for analog-to-digital conversion (ADC) (Table 1).

$$\text{Height of water} = (\text{Volume of water})/(\text{Area of field}) \quad (1)$$

Area of observation field used for calculations is 30 cm^2.

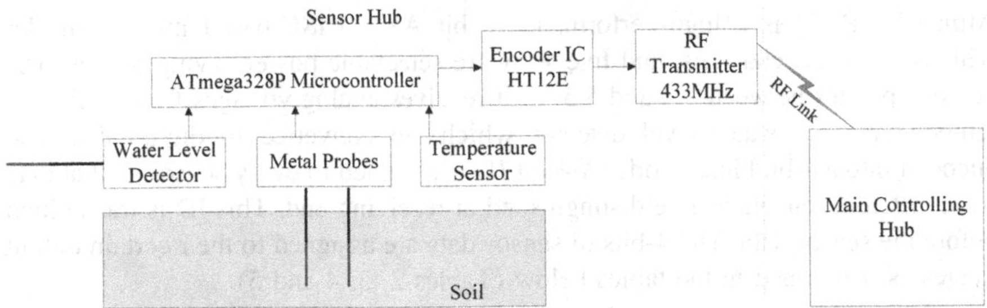

Fig. 2 Block diagram of sensor hub

Table 1 Output voltage observations for moisture sensor

Volume of water (ml)	Height of water (mm)	Voltage across metal probes (V)
0	0	4.93
160	5.3	4.92
320	10.63	4.57
480	15.69	4.97
640	21.26	3.65
800	26.57	3.18

2.2 *Standing Water Level Detector*

This sensor is used to detect standing water level. It is very helpful in flooding situations like heavy rainfall and for crops which need standing water such as rice. Two metal pins are installed at a certain height above soil. One of them is connected to a positive voltage supply and other to negative. When water reaches their height, current flows between, decreasing the potential difference. Voltage between the two metal pins is read by the microcontroller through ADC and is further processed.

2.3 *Temperature Sensor*

IC-LM35 from Texas instrument [1] is used to monitor the current temperature of field. The operating temperature range is from −55 to 150 °C, and output voltage varies by 10 mV in response to every °C rise/fall in ambient temperature, i.e., its scale factor is 0.01 V/°C. Its output pin is connected to microcontroller for ADC.

2.4 *Microcontroller Atmega328P*

Atmega328P [2] is a high-performance 8-bit AVR RISC-based microcontroller with 10-bit ADC channels, and five software selectable power saving modes. The device operates between 1.8 and 5.5 V. It receives analog voltages from moisture, temperature, and water level detector which are converted to digital data and encoded into a 4-bit binary code. A 4-bit ID is assigned to every sensor so that data from different sensors can be distinguished at receiving end. This ID is transmitted before the sensor data. The 4-bits of sensor data are assigned to their certain output ranges as mentioned in the tables below (Tables 2, 3, 4 and 5).

Table 2 4-bit sensors ID

Name of sensor	4-bit data ID
Temperature sensor	1111
Moisture detector	1110
Standing water level detector	1101
Empty	0000

Table 3 4-bit temperature data ID

Temperature range (°C)	4-bit temperature data
0–4	0001
4–8	0010
8–12	0011
12–16	0100
16–20	0101
20–24	0110
24–28	0111
28–32	1000
32–36	1001
36–40	1010
40–44	1011
44–48	1100

Table 4 4-bit standing water level data ID

Voltage of metal pins (V)	4-bit standing water level data
Less than 4 (yes)	0010
More than 4 (No)	0111

Table 5 4-bit moisture data ID

Voltage of metal probes (V)	4-bit moisture data
3.000–3.165	0001
3.165–3.332	0010
3.332–3.498	0011
3.498–3.665	0100
3.665–3.832	0101
3.832–3.998	0110
3.998–4.165	0111
4.165–4.332	1000
4.332–4.499	1001
4.499–4.665	1010
4.665–4.833	1011

2.5 *Encoder IC HT12E and Radio Frequency Transmitter*

The encoder IC [3] takes a 4-bit parallel data from the microcontroller and converts it into serial data which is transmitted to main controlling hub at 433 MHz frequency.

3 User Interface

The user interface of the system shown in Fig. 3 is very user friendly. It consists of a total of 15 push buttons for various operations and one liquid crystal display (LCD).

When the system starts two options are displayed on LCD for selecting either automatic or manual mode of operation. In automatic mode, the user has to select crop and input location details. Afterward, the system controls irrigation for the complete crop period. Whereas in manual mode, after selecting the crop, the number of days for irrigation and amount of water is input.

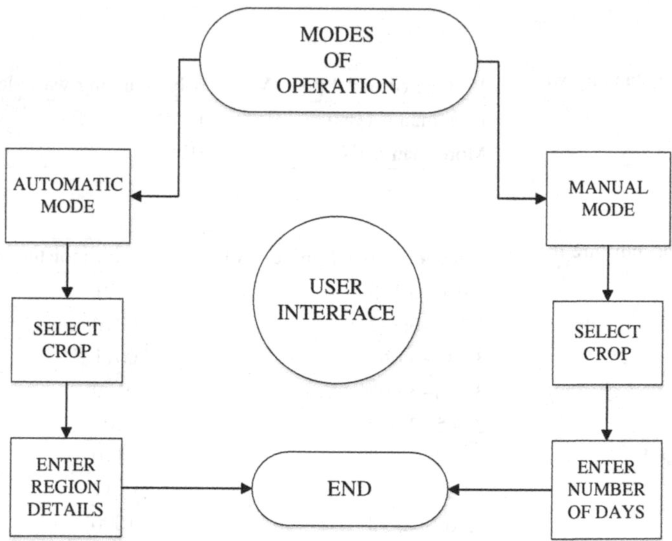

Fig. 3 Flow diagram of user interface

4 Calculations and Database

Calculations for water requirement are based on Blaney-Criddle method [4] and cross-checked with help of moisture sensor data. This method gives the amount of water that is to be delivered to the field. It does not consider the present weather conditions, that is why it calculates the same water requirement on a rainy, windy, cloudy, or a sunny day. To overcome this problem soil moisture sensor data is used to enhance the water requirement calculations. One important calculation is,

$$ET_{\mathrm{o}} = p(0.46\,T_{\mathrm{mean}} + 8) \tag{2}$$

ET_{o} average reference crop evapotranspiration (mm/day)
T_{mean} mean daily temperature (°C)
p mean daily percentage of annual daytime hours (its value varies with location)

Latitude is entered by farmer and T_{mean} is calculated from temperature sensor data received by hub. Water requirement (ET_{crop}) can be calculated as,

$$ET_{\mathrm{crop}} = ET_o \times K_c \tag{3}$$

ET_{crop} crop evapotranspiration or crop water need (mm/day)
K_{c} crop factor
ET_{o} reference evapotranspiration (mm/day)

Crop factor depends on the type of crop which is grown by farmer and development stage of crop which is monitored by RTC (real-time clock).

Water content present already in soil is subtracted from ET_{crop} to get ET_{new} which gives the current water requirement value.

Suppose, the water pump supplies X ml/hr, and T be the time for which pump is to be switched on by system. The compensating factor (M) of water supply loss is,

$$M = 3\%\ \text{of}\ (ET_{\mathrm{new}} * A) \tag{4}$$

$$T = X * ((ET_{\mathrm{new}} * A) + M)\ \ (\text{in hrs}) \tag{5}$$

The water pump will automatically turn on for T hours everyday. And if T comes out to be negative, it means sufficient water is already present in the field.

5 Circuit Diagrams

The circuit used for detection of moisture is shown in Fig. 4.

The complete circuit diagram of the transmitting unit is summarized in a schematic diagram given in Fig. 5, the water pump connections are shown in Fig. 6, and the connections of receiving unit [5] are shown in Fig. 7.

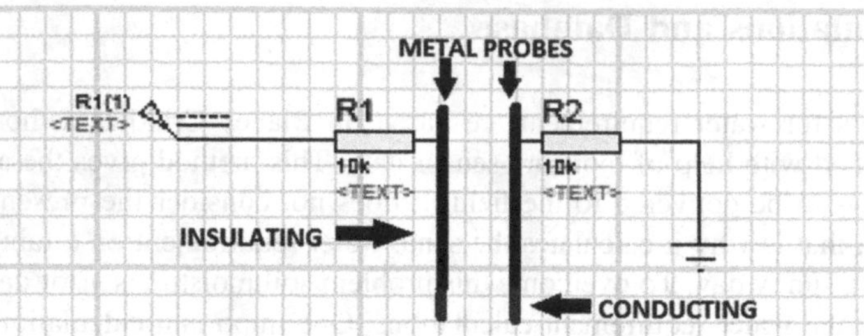

Fig. 4 Moisture detecting probe circuit

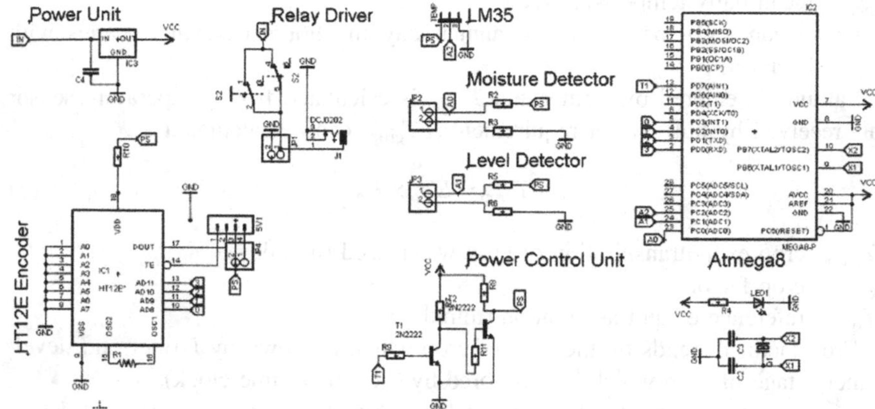

Fig. 5 Transmitting unit circuit diagram

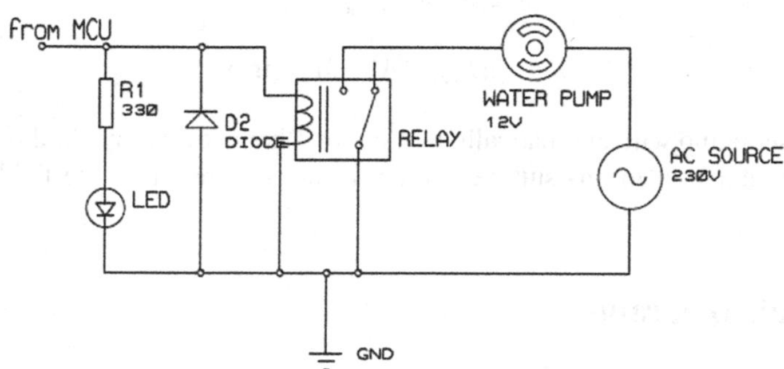

Fig. 6 Water pump driver circuit using electromechanical relay

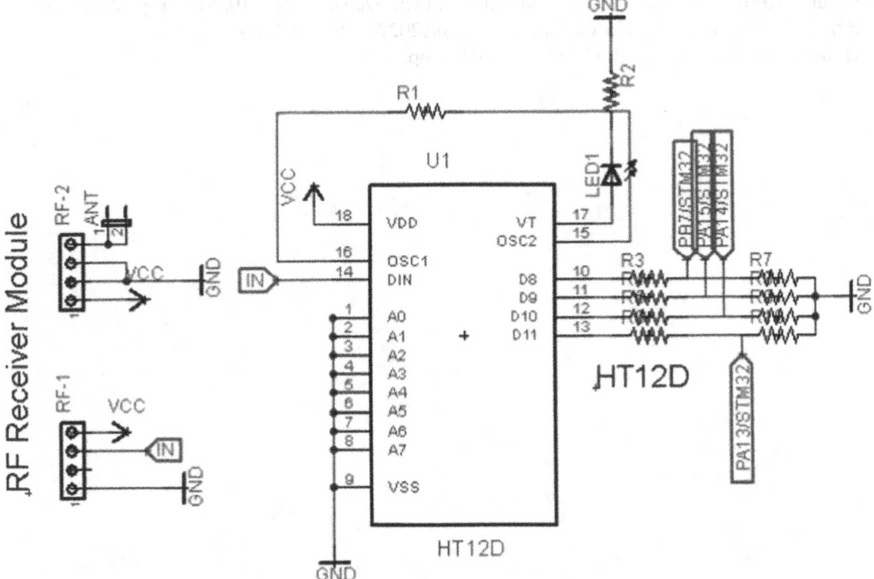

Fig. 7 Receiving unit decoder circuit diagram

6 Market Analysis

This system is designed by keeping in mind the budget of farmers. It is made using very basic inexpensive and easily available components. Also, the concept of making moisture sensor just with two metal probes obviates the need of factory-made sensors. This complete system costs 4 to 5 times less than the currently available alternatives, which makes it very inexpensive system with good technology.

This system is kept in a dust and water proof case and thus it does not require any significant maintenance. Also, being a fully automatic system, it does not needs to be configured, operated, or checked regularly. Hence, this system also mitigates the need of labor for irrigation. When the field does not require water, the system operates in power saving sleep modes thus saving electricity and money. This system has the potential to revolutionize the agricultural sector.

References

1. Texas Instruments. "LM35 Temperature Sensor." SNIS159E (January, 2015), pp. 3–4.
2. Atmel. (2009, Oct.). "8-bit AVR Microcontroller with 4/8/16/32 K Bytes In-System Programmable Flash" [Online]. Available: http://www.atmel.com/images/doc8161.pdf.
3. Holtek. "Encoder IC HT12E." Rev. 1.20 (February, 2009), pp. 1–6.

4. Natural Resources Management and Environment Department. (1986). Crop Water Needs [Online]. Available: http://www.fao.org/docrep/s2022e/s2022e07.htm.
5. Holtek. "Decoder IC HT12D." (March, 1996), pp. 1–5.

IVIFS and Decision-Making

R.K. Mohanty, T.R. Sooraj and B.K. Tripathy

Abstract Many models handle uncertainty problems. Fuzzy set (FS) is one of them. The next one is intuitionistic fuzzy set (IFS). Further generalization is interval-valued fuzzy set (IVFS). But all those models had some difficulty due to lack of parameterization tool, which motivated mathematician Molodtsov to introduce soft set model in 1999. Hybrid models of these models are more efficient. Interval-valued intuitionistic fuzzy soft set (IVIFSS) introduced by Jiyang. Following their characteristic function approach Tripathy et al. introduced fuzzy soft set as a hybrid model in 2015. Here, we continue this further to define IVI fuzzy soft sets (IVIFSS). Many related concepts like complement, null, and absolute IVIFSS are introduced and operations like intersection and union of IVIFSSs are also redefined. Recently, soft set is applied in various forms to derive decision-making (DM) by Tripathy et al. We extend it further by proposing an algorithm which uses IVIFSS in order to achieve DM. These algorithms are much improved and applicable than that of Jiyang. Also, it generalizes all the previous algorithms in this direction.

Keywords Soft sets · Interval-valued fuzzy set · IVIFSS · Decision-making

1 Introduction

Fuzzy set, introduced in 1965 [1], is being used extensively in real-life applications. As it lacks parametrization, Molodtsov [2] led to inception of soft set (SS). A SS is a parametrized family of subsets and each subset is associated with one or more

R.K. Mohanty (✉) · T.R. Sooraj · B.K. Tripathy
SCSE, VIT University, Vellore, Tamil Nadu, India
e-mail: rknmohanty@gmail.com

T.R. Sooraj
e-mail: soorajtr19@gmail.com

B.K. Tripathy
e-mail: tripathybk@vit.ac.in

© Springer Science+Business Media Singapore 2017

S.C. Satapathy et al. (eds.), *Proceedings of the International Conference on Data Engineering and Communication Technology*, Advances in Intelligent Systems and Computing 468, DOI 10.1007/978-981-10-1675-2_33

parameters. As has been observed by Dubois and Prade hybrid models are more often than not better than the individual constituents and have led to the development of many such models. Recently, in [3] the definition of SS is given using characteristic function approach. It has been proved to be very efficient in defining the basic operations on SS. In [4] the concept of fuzzy soft set (FSS) is introduced. The approach of Tripathy et al. [3] is used to redefine FSS and many operations on them in [5]. Following this, we introduce the membership functions for IVIFSS here. Decision-making is studied in the context of fuzzy soft sets by Tripathy et al. in [3]. Here, problems in the procedure in [6] are pointed out and rectified. We proceeded further by using IFSS instead of FSS. Using this, we proposed an algorithm in handling the problem of multi-criteria decision-making. Jiyang [7] proposed IVIFSS. This combines IVIFS and SS models. Applications of many soft set hybrid models are discussed in [8–14].

It has been observed that IFS proposed in [15] is a better model than FS. This is because of the existence of the notion of nonmembership function as an entity independent of the membership function. The hesitation function generated as a consequence is what real-life situations demand. In case of fuzzy sets the hesitation component is zero. Here, we follow the definition of SS from [3] to propose IVIFSS. It may be noted that following the same approach we have extended the definition in [5] to the context of FSS. The important aspect here is a proposal of a DM procedure using IVIFSS. Areal life application of this is provided. It extends the algorithm proposed in [16].

2 Definitions and Notions

Let *U* be a universe and *E* be a set of parameters. Elements of *U* are denoted by x and those of *E* are denoted by e. By *P(U)* and *I(U)* we denote the set of all subsets of *U* and the set of all fuzzy subsets of *U* respectively.

Definition 2.1 (*Soft Set*) We denote a soft set over (*U*, *E*) by (*F*, *E*), where

$$F: E \rightarrow P(U). \tag{2.1}$$

Definition 2.2 (*FSS*) We denote a FSS over (*U*, *E*) by (*F*, *E*) where

$$F: E \rightarrow I(U). \tag{2.2}$$

Definition 2.3 (*Intuitionistic Fuzzy set*) An intuitionistic fuzzy set *A* over *U* is associated with a pair of functions $\mu_A, \nu_A: U \rightarrow [0,1]$ such that for any $x \in U, 0 \leq \mu_A(x) + \nu_A(x) \leq 1$.

The hesitation function π_A is defined as $\pi_A(x) = 1 - \mu_A(x) - \nu_A(x)$, $\forall x \in U$.

Definition 2.4 (*Intuitionistic Fuzzy Soft Set*) Let *U* is the universal set and *E* be a set of parameters. Let *IF(U)* denotes the set of all intuitionistic fuzzy subsets of

U. The pair *(F, E)* is called as intuitionistic fuzzy soft set over *U*, where *F* is a mapping given by

$$F: E \rightarrow IF(U) \quad (2.3)$$

2.1 IVIFSS

Let *U* be the universal set and *E* be a set of parameters. Let *IVIF(U)* denote the set of all interval-valued intuitionistic fuzzy subsets of *U*. The membership and non-membership of IVFSS is defined as follows:
$\mu^{e}_{(F,E)} = \alpha$, $\alpha \in [0, 1]$, $\nu^{e}_{(F,E)} = \beta$, $\beta \in [0, 1]$, where $0 \leq \alpha + \beta \leq 1$

Definition 2.1.1 A pair (*F*, *E*) is called an IVIFSS over (*U*, *E*), where *F* is a mapping given by

$$F: E \rightarrow IVIF(U) \quad (2.1.1)$$

Definition 2.1.2 Let (*F*, *E*) and (*G*, *B*) are two IVIFSS. (*G*, *B*) is said to be a subset of (*F*, *E*) denoted by $(G,B) \subseteq (F,E)$ iff $B \subseteq E$, $\mu^{a}_{(G,B)}(x) \leq \mu^{a}_{(F,E)}(x)$, $\nu^{a}_{(G,B)}(x) \geq \nu^{a}_{(F,E)}(x)$ $\forall x \in U$.

Where $\mu^{a}_{(F,E)}(x)$, $\nu^{a}_{(F,E)}(x)$, $\mu^{a}_{(G,B)}(x)$, $\nu^{a}_{(G,B)}(x)$ are the membership and non-membership values of the element *x* for the parameter *a* in the soft sets (*F*, *E*) and (*G*, *B*) respectively.

Definition 2.1.3 Two IVIFSS (*F*, *E*) and (*G*, *B*) are said to be equal denoted by $(F,E) \doteq (G,B)$ iff, $\mu^{a}_{(F,E)}(x) = \mu^{a}_{(G,B)}(x)$, $\nu^{a}_{(F,E)}(x) = \nu^{a}_{(G,B)}(x)$ and $E = B$.

3 Application of IVIFSS in DM

In our previous paper [16] the decision-making example given was dependent on the normal intuitionistic fuzzy values. But, this paper gives example of decision-making using IVIFSS. Sometimes it is very difficult to predict a membership value directly. So we can predict an interval to handle complex real-life uncertainty-based problems. The concept of priority and negative parameters is carried out from [5].

$$\text{Score} = \begin{cases} -m(1+h) \text{ if } m<0 \text{ and } h<-1; \\ m(1+h) \quad \text{otherwise.} \end{cases} \tag{3.1}$$

(3.1) degenerates to membership score in case of a FSS. In the application, we have taken the summation of membership values, so it may not lie in [0, 1].

The Table 2 shows the IVIFSS which represents the selection of cars. For IVFSSs, we need to consider the pessimistic, optimistic and neutral values, where the neutral value is obtained by the following:

$$\text{neutral value} = \frac{\text{pesimistic} + \text{optimistic}}{2}$$

3.1 Algorithm

1. Input the parameter data table by ranking according to the absolute value of parameter priorities. If the priority for any parameter has not given, then take the value as 0 and that column can be opt out from further computations. The boundary condition for a positive parameter is [0, 1] and for a negative parameter is [−1, 0].
2. Input the IVIFSS table.
3. Construct the optimistic value table by taking the maximum range membership values, minimum range non-membership values from IVIFSS table and compute the hesitation values accordingly.
4. Construct the pessimistic value table by taking the minimum range membership values, maximum range non-membership values from IVIFSS table and compute the hesitation values accordingly.
5. Construct the neutral value table by taking the average of the membership ranges for membership values and average of the non-membership ranges for non-membership values from IVIFSS table and compute the hesitation values accordingly.
6. Function Deci_make(IFSS table)
    a. Multiply the priority values with the corresponding parameter values to get the priority table.
    b. Construct the comparison table by finding the entries as differences of each row sum in priority table with those of all other rows taking membership, non-membership and hesitation separately.
    c. Construct the decision table by taking the sums of membership values, non-membership values and hesitation values separately for each row in the comparison table. Compute the score for each candidate using the formula (3.1).
    d. Assign rankings to each candidate based upon the score obtained.

i. If there is more than one having same score than who has more score in a higher ranked parameter will get higher rank and the process will continue until each entry has a distinct rank.
Return (Decision Table)

7. Construct the decision tables for optimistic, pessimistic and neutral values by using the function Deci-make.
8. Calculate the normalized score of each candidate in the rank table by using the following equation.

$$\text{Normalized Score} = \frac{2\left(|c| \times |k| \times |j| - \sum_{i=1,\, k=\{o,p,n\}}^{i=|j|} RC_{ik}\right)}{|k| \times |j| \times |c| \times (|c| - 1)}$$

where $|c|$ is the number of candidates, $|j|$ is the number of judges and $|k|$ is the number of cases (pessimistic, optimistic and neutral). Here, $k = 3$.

9. The candidate with higher normalized score value will get the higher rank. In case of more than one same normalized score, resolve the conflict by the process used in the step 6.d.i.

The Real Life Application:

The example is to take a decision to purchase a car from the available cars. Let U be a set of cars U and E be the parameters. $U = \{c_1, c_2, c_3, c_4, c_5, c_6\}$ and $E = \{e_1, e_2, e_3, e_4, e_5, e_6\}$. Here, e_1 represents mileage, e_2 represents design, e_3 represents Price, e_4 represents company, e_5 represents color and e6 represents gear system.

The parameter having highest absolute value as priority will be assigned highest rank and so on. All the data about parameters are listed in the Table 1.

The selection of car can be represented as IVIFSS. In IVIFSS there are three different approaches, such as optimistic, pessimistic and neutral (Table 2). Tables 3, 4 and 5 represent the optimistic, pessimistic and neutral values respectively.

The priority tables for each approach can be obtained by multiplying the values in the tables with respective parameter priority values fixed by the authority. Note that, parameter 'price' is having negative priority, shows that, it is a negative parameter.

In the same way priority table can be constructed for pessimistic and neutral values (Table 6).

Table 1 Parameter data table

U	e_1	e_2	e_3	e_4	e_5	e_6
Priority	0.3	0.2	−0.1	0.15	0.35	0
Parameter rank	2	3	5	4	1	6

Table 2 IVIFSS

	e_1		e_2		e_3		e_4		e_5		e_6	
	μ	ν	μ	ν	μ	ν	μ	ν	μ	ν	μ	ν
C_1	0.1–0.5	0.3–0.5	0.2–0.4	0.6–0.8	0.3–0.5	0.4–0.6	0.8–0.9	0.0–0.1	0.4–0.7	0.1–0.3	0.6–0.9	0.1–0.2
C_2	0.8–1.0	0.0–0.2	0.4–0.8	0.2–0.6	0.6–0.9	0.1–0.3	0.2–0.5	0.1–0.4	0.7–1.0	0.0–0.1	0.5–0.6	0.2–0.3
C_3	0.1–0.5	0.3–0.5	0.5–0.8	0.2–0.5	0.7–0.9	0.1–0.2	0.7–0.8	0.0–0.2	0.8–1.0	0.0–0.1	0.5–0.7	0.3–0.4
C_4	0.5–0.9	0.1–0.4	0.6–0.8	0.2–0.4	0.5–0.9	0.1–0.5	0.8–1.0	0.0–0.2	0.5–0.9	0.1–0.3	0.7–0.8	0.0–0.1
C_5	0.1–0.2	0.4–0.7	0.1–0.4	0.5–0.9	0.9–1	0.0–0.1	0.3–0.6	0.3–0.5	0.1–0.5	0.2–0.4	0.8–1.0	0.0–0.1
C_6	0.9–1.0	0.0–0.1	0.7–0.9	0.0–0.3	0.5–0.7	0.3–0.4	0.1–0.3	0.4–0.7	0.2–0.4	0.1–0.4	0.3–0.7	0.2–0.3

Table 3 Optimistic values

	e_1			e_2			e_3			e_4			e_5		
	μ	ν	h	μ	ν	h	μ	ν	h	μ	ν	h	μ	ν	h
C_1	0.5	0.3	0.2	0.4	0.6	0	0.5	0.4	0.1	0.9	0	0.1	0.7	0.1	0.2
C_2	1	0	0	0.8	0.2	0	0.9	0.1	0	0.5	0.1	0.4	1	0	0
C_3	0.5	0.3	0.2	0.8	0.2	0	0.9	0.1	0	0.8	0	0.2	1	0	0
C_4	0.9	0.1	0	0.8	0.2	0	0.9	0.1	0	1	0	0	0.9	0.1	0
C_5	0.2	0.4	0.4	0.4	0.5	0.1	1	0	0	0.6	0.3	0.1	0.5	0.2	0.3
C_6	1	0	0	0.9	0	0.1	0.7	0.3	0	0.3	0.4	0.3	0.4	0.1	0.5

Table 4 Pessimistic values

	e_1			e_2			e_3			e_4			e_5		
	μ	ν	h	μ	ν	h	μ	ν	h	μ	ν	h	μ	ν	h
C_1	0.1	0.5	0.4	0.2	0.8	0	0.3	0.6	0.1	0.8	0.1	0.1	0.4	0.3	0.3
C_2	0.8	0.2	0	0.4	0.6	0	0.6	0.3	0.1	0.2	0.4	0.4	0.7	0.1	0.2
C_3	0.1	0.5	0.4	0.5	0.5	0	0.7	0.2	0.1	0.7	0.2	0.1	0.8	0.1	0.1
C_4	0.5	0.4	0.1	0.6	0.4	0	0.5	0.5	0	0.8	0.2	0	0.5	0.3	0.2
C_5	0.1	0.7	0.2	0.1	0.9	0	0.9	0.1	0	0.3	0.5	0.2	0.1	0.4	0.5
C_6	0.9	0.1	0	0.7	0.3	0	0.5	0.4	0.1	0.1	0.7	0.2	0.2	0.4	0.4

The respective comparison tables are constructed by finding the entries as differences of each row sum in priority tables with those of all other rows and compute row sum in each of the table.

Table 7 is the comparison table for the optimistic values. In the same way comparison tables for pessimistic and neutral values can be constructed.

By using the formula (3.1), the decision table can be formulated and a rank will be given to each car. By this way all the decision tables for all approaches can be constructed (Tables 8, 9 and 10).

The rank Table 11 can be constructed by using the formula "Neutral value = Optimizti + pessimistic/2". If same rank sum obtained by more than one car, then the conflict can be resolved by the same way as in decision table formation.

Decision-Making: The car having highest rank is the best. The next subsequent rank holders can be chosen in the same way. In the example discussed above, the score is same for C_4 and C_6. So, we need to calculate the sum of the scores securing by all approaches using only e_5 (Highest Ranked Parameter) values using the same formula 3.1. After computation we can find the values C_4 has good score than C_6 in all pessimistic, optimistic, and neutral approaches. Hence the ranks were given.

Table 5 Neutral values

	e_1			e_2			e_3			e_4			e_5		
	μ	ν	h	μ	ν	h	μ	ν	h	μ	ν	h	μ	ν	h
C_1	0.3	0.4	0.3	0.3	0.7	0	0.4	0.5	0.1	0.85	0.05	0.1	0.55	0.2	0.25
C_2	0.9	0.1	0	0.6	0.4	0	0.75	0.2	0.05	0.35	0.25	0.4	0.85	0.05	0.1
C_3	0.3	0.4	0.3	0.65	0.35	0	0.8	0.15	0.05	0.75	0.1	0.15	0.9	0.05	0.05
C_4	0.7	0.25	0.05	0.7	0.3	0	0.7	0.3	0	0.9	0.1	0	0.7	0.2	0.1
C_5	0.15	0.55	0.3	0.25	0.7	0.05	0.95	0.05	0	0.45	0.4	0.15	0.3	0.3	0.4
C_6	0.95	0.05	0	0.8	0.15	0.05	0.6	0.35	0.05	0.2	0.55	0.25	0.3	0.25	0.45

Table 6 Priority table for optimistic values

	e_1			e_2			e_3			e_4			e_5					
	μ	ν	h	μ	ν	h	μ	ν	h	μ	ν	h	μ	ν	h	$\sum\mu$	$\sum\nu$	$\sum h$
C_1	0.15	0.09	0.06	0.08	0.12	0	−0.05	−0.04	−0.01	0.135	0	0.015	0.245	0.035	0.07	0.56	0.205	0.135
C_2	0.3	0	0	0.16	0.04	0	−0.09	−0.01	0	0.075	0.015	0.06	0.35	0	0	0.795	0.045	0.06
C_3	0.15	0.09	0.06	0.16	0.04	0	−0.09	−0.01	0	0.12	0	0.03	0.35	0	0	0.69	0.12	0.09
C_4	0.27	0.03	0	0.16	0.04	0	−0.09	−0.01	0	0.15	0	0	0.315	0.035	0	0.805	0.095	0
C_5	0.06	0.12	0.12	0.08	0.1	0.02	−0.1	0	0	0.09	0.045	0.015	0.175	0.07	0.105	0.305	0.335	0.26
C_6	0.3	0	0	0.18	0	0.02	−0.07	−0.03	0	0.045	0.06	0.045	0.14	0.035	0.175	0.595	0.065	0.24

Table 7 Comparison table for optimistic values

	c1			c2			c3			c4			c5			c6		
	μ	ν	h	μ	ν	h	μ	ν	h	μ	ν	h	μ	ν	h	μ	ν	h
C_1	0	0	0	−0.235	0.16	0.075	−0.13	0.085	0.045	−0.245	0.11	0.135	0.255	−0.13	−0.125	−0.035	0.14	−0.105
C_2	0.235	−0.16	−0.075	0	0	0	0.105	−0.075	−0.03	−0.01	−0.05	0.06	0.49	−0.29	−0.2	0.2	−0.02	−0.18
C_3	0.13	−0.085	−0.045	−0.105	0.075	0.03	0	0	0	−0.115	0.025	0.09	0.385	−0.215	−0.17	0.095	0.055	−0.15
C_4	0.245	−0.11	−0.135	0.01	0.05	−0.06	0.115	−0.025	−0.09	0	0	0	0.5	−0.24	−0.26	0.21	0.03	−0.24
C_5	−0.255	0.13	0.125	−0.49	0.29	0.2	−0.385	0.215	0.17	−0.5	0.24	0.26	0	0	0	−0.29	0.27	0.02
C_6	0.035	−0.14	0.105	−0.2	0.02	0.18	−0.095	−0.055	0.15	−0.21	−0.03	0.24	0.29	−0.27	−0.02	0	0	0

Table 8 Decision table by optimistic values

	μ	ν	h	Score	Rank
C_1	−0.39	0.365	0.025	0.135	5
C_2	1.02	−0.595	−0.425	1.095	1
C_3	0.39	−0.145	−0.245	0.645	4
C_4	1.08	−0.295	−0.785	0.795	3
C_5	−1.92	1.145	0.775	−0.645	6
C_6	−0.18	−0.475	0.655	0.975	2

Table 9 Decision table by pessimistic values

	μ	ν	h	Score	Rank
C_1	−0.48	0.135	0.345	0.365	5
C_2	0.93	−0.615	−0.315	1.115	1
C_3	0.39	−0.315	−0.075	0.815	3
C_4	0.81	−0.375	−0.435	0.875	2
C_5	−2.04	1.485	0.555	−0.985	6
C_6	0.39	−0.315	−0.075	0.815	4

Table 10 Decision table by neutral values

	μ	ν	h	Score	Rank
C_1	−0.435	0.25	0.185	0.25	5
C_2	0.975	−0.605	−0.37	1.105	1
C_3	0.39	−0.23	−0.16	0.73	4
C_4	0.945	−0.335	−0.61	0.835	3
C_5	−1.98	1.315	0.665	−0.815	6
C_6	0.105	−0.395	−0.29	0.895	2

Table 11 Rank table

	Optimistic	Pessimistic	Neutral	Normalized score	Final Rank
C_1	5	5	5	0.06666667	5
C_2	1	1	1	0.33333333	1
C_3	4	3	4	0.15555556	4
C_4	3	2	3	0.22222222	3
C_5	6	6	6	0	6
C_6	2	4	2	0.22222222	2

4 Conclusions

In this article, we introduced a new definition of IVIFSS which uses the more authentic characteristic function approach for defining soft sets provided in [3]. This provides several authentic definitions of operations on IVIFSS. Earlier fuzzy soft sets were used for decision-making in [17]. Here, the flaws in [17] were traced and provided solutions to rectify them in [5]. This made decision-making more efficient and realistic. Here, we proposed an algorithm for decision-making using IVIFSS. Also, an application of this algorithm in solving a real-life problem is demonstrated.

References

1. Zadeh L.A.: Fuzzy sets, Information and Control, 8 (1965) pp. 338–353.
2. Molodtsov D.: Soft Set Theory - First Results, Computers and Mathematics with Applications, 37 (1999) pp. 19–31.
3. Tripathy, B.K and Arun,K.R., “A New Approach to Soft Sets, Soft Multisets and Their Properties”, International Journal of Reasoning-based Intelligent Systems, Vol. 7, no. 3/4, 2015, pp. 244–253.
4. Maji P.K., Biswas R., Roy A.R.: Fuzzy Soft Sets, Journal of Fuzzy Mathematics, 9(3), (2001) pp. 589–602.
5. Tripathy B.K., Sooraj, T.R, Mohanty, R.K., A new approach to fuzzy soft set and its application in decision making, Computational Intelligence in Data Mining-2, Advances in intelligent systems and computing, vol. 11, pp. 307–315.
6. Maji P.K, Biswas R, Roy A.R: An Application of Soft Sets in a Decision Making Problem, Computers and Mathematics with Applications, 44(2002) pp. 1007–1083.
7. Jiang, Y, Tang, Y, Chen, Q, H. Liu, Tang, J, “Interval-valued intuitionistic fuzzy soft sets and their properties”, Computers and Mathematics with Applications, vol. 60, 2010, pp. 906–918.
8. Tripathy, B.K., Sooraj, T.R., Mohanty, RK, “A New Approach to Interval-valued Fuzzy Soft Sets and its Application in Decision Making” in the proceedings of ICCI-2015, Ranchi.
9. Tripathy, B.K., Sooraj, T.R.,Mohanty, RK, “A New Approach to Interval-valued Fuzzy Soft Sets and its Application in Group Decision Making” in the proceedings of CDCS-2015, Kochi.
10. Tripathy, RK Mohanty, T.R. Sooraj, “ On Intuitionistic Fuzzy Soft Sets and Their Application in Decision Making”, accepted in ICSNCS-2016, NewDelhi.
11. Sooraj, T.R, Mohanty, RK, Tripathy, B.K., “Fuzzy soft set theory and its application in group decision making, in the proceedings of ICACCT-2015, Panipat, Springer publications, (2015).
12. Tripathy, B.K., Mohanty, RK, Sooraj, T.R, “On Intuitionistic Fuzzy Soft Set and its Application in Group Decision Making” accepted for presentation ICETETS-2016, Thanjavur.
13. Tripathy, B.K., Mohanty, RK, Sooraj, T.R, Tripathy, A, “A New Approach to Intuitionistic Fuzzy Soft Set Theory and its Application in Group Decision Making” in the proceedings of ICTIS-2015, Ahemadabad, Springer publications, (2015).
14. Mohanty, RK, Sooraj, T.R, Tripathy, B.K, “An application of IVIFSS in medical diagnosis decision making, in the proceedings of ICASME-2015, Chennai.
15. Atanassov K.T., “Intuitionistic Fuzzy Sets”, Fuzzy Set Systems, vol. 20, 1986, pp. 87–96.
16. Tripathy B.K, T.R. Sooraj, RK Mohanty and K.R. Arun, “A New Approach to Intuitionistic Fuzzy Soft Set Theory and its Application in Decision Making” in the proceedings of ICICT2015.
17. Maji P.K., Biswas R. Roy A.R.: Soft Set Theory, Computers and Mathematics with Applications, 45(2003) pp. 555–562.

Comparative Analysis of Integer and Fractional Order Controller for Time Delay System

Ravishankar Desai

Abstract It is very crucial to control time-varying parameters and uncertainty present in chemical processes. In this paper, proposed work represents the comparative analysis of integer and fractional order controller for time-delay system. To minimize the performance criteria (IAE, ITAE, ISE) with a restraint on the maximum sensitivity, fractional and integer order IMC tuning rule have been compared which is applied to the experimental model of the level control system. The achieved experimental model is used to assess the achieved performance with controller performance. For process model stability, the robustness of the system is tested for parameter variation.

Keywords Integer order (IO) controller · Proportional integral (PI) · Robust control · Fractional order (FO) controller · Internal model controller (IMC)

1 Introduction

In industrial control environment classical PID controller has a wide variety of applications because of ease, reliable, and simple control configuration. Hence, it is prime important in controller tuning and it is used in controlling the various processes. The objective of tuning is to keep a set point. Ideally, a suitably tuned control loop will work within safe restrictions of the process, increase working yield and to avoid extreme control action. The numerous researchers [1–6] propose various tuning rules in order to improve the closed-loop performance. The "Internal Model Control" (IMC) is model-based controller and it is based on design algorithm presented by Garcia and Morari [7, 8]. To combine process model and exterior signal dynamics which is dependent upon the internal model principle. Controller tuning is the process of gaining the controller parameters to meet agreed performance specifica-

Ravishankar Desai (✉)
Department of Instrumentation Engineering, Padmabhushan Vasantraodada Patil Institute of Technology, Budhgaon, Sangli 416304, India
e-mail: rpdesai@pvpitsangli.edu.in; ravishankardesai@gmail.com

© Springer Science+Business Media Singapore 2017

S.C. Satapathy et al. (eds.), *Proceedings of the International Conference on Data Engineering and Communication Technology*, Advances in Intelligent Systems and Computing 468, DOI 10.1007/978-981-10-1675-2_34

tions. Fractional order controller has two extra variables to tune the fractional order system. It offers extra degree of freedom to the dynamic properties and less sensitive than integer order controller [9, 10]. For accepting the contentment of robust enactment in control system, a broad view of fractional order of outmoded control systems transforms more adaptable time domain and frequency domain responses and extra tuning parameters [11, 12]. The paper is ordered as follows. The Sect. 2 introduce the system description and mathematical modeling, Sect. 3 represents the system identification, validation and analysis, Sect. 4 is representation of IOPI, FOPI controller and Sect. 5 representing the design of controller tuning method. In Sect. 6 simulation results are reported and robustness analysis is obtained in Sect. 7. Finally, Sect. 8 presents conclusion.

2 System Description and Mathematical Modelling

Figure 1 shows the schematic of an implemented control system to control water level of a tank. Computer, NI 6024E PCI DAQ card, variable speed positive displacement pump, tank, V to I converter, level transmitter, I to V converter are parts of it. Design of controller are applied to a computer which is interfaced with the hardware system through the NI 6024E PCI DAQ card. This card receives input signal as a voltage and gives output signal as a voltage both ranging from 0 to 5 V. Variable speed pump is used to pump the water into the tank. Level measurement is done by the level transmitter. Here, level acts as a process variable (PV). Output of the transmitter is a current of 4–20 mA. It is further converted into a voltage by using I to V converter so that it can be feed to the controller through the NI 6024E PCI DAQ card. Controller output is taken as an analog output from NI 6024E PCI DAQ card. Its a voltage signal ranging from 0 to 5 V. This voltage is converted into a current by using V to I

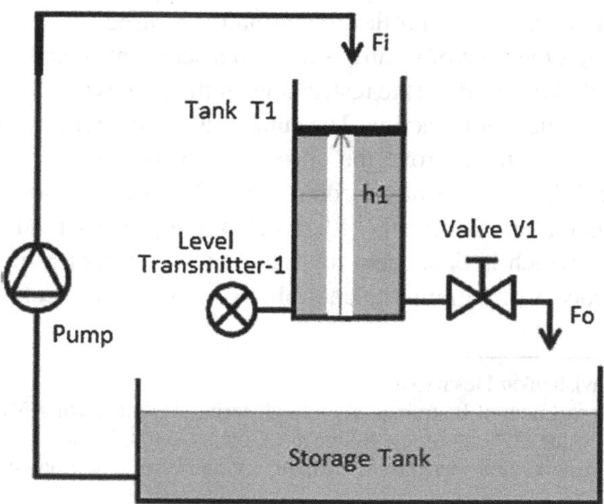

Fig. 1 Liquid level control system

converter to feed it to the pump. Water level within a tank is controlled by controlling the voltage given to the variable frequency drive which drives the pump. In order to increase the level of a tank, a high voltage is given to the pump. To decrease level within a tank, a low voltage is given to the pump. It drains some amount of water from the tank.

3 System Identification, Validation, and Analysis

3.1 Identification

To determine the system model, empirical method is used. The change of level is accounted with time by giving step of 20 %. From experimental data time constant, steady state gain, dead time is the system parameter are obtained. Then the obtained transfer function is,

$$G(s) = \frac{0.7e^{-10s}}{76.089s + 1} \tag{1}$$

3.2 Validation

From experimental data, the model (1) is identified using identification tool, two-point method, and maximum slope method. The best fit plot is considered which gives better step response. The obtained model is then validated and its best fit 96.80. The Fig. 2 shows the validated result of (1).

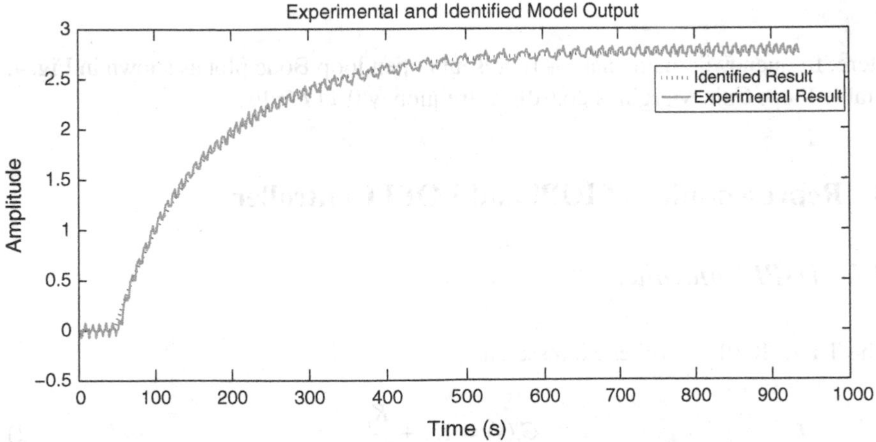

Fig. 2 Result of validation

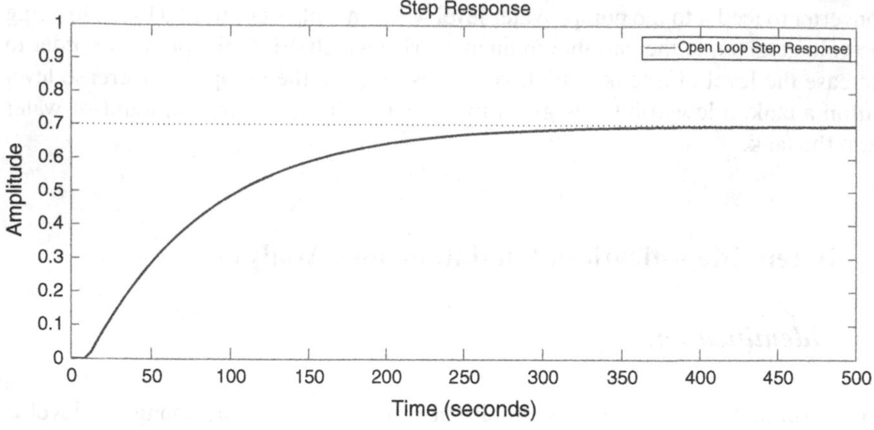

Fig. 3 Open-loop system result

Table 1 Open-loop system performance parameter for level control system

Performance parameter	Rise time	Settling time	Overshoot	Peak time
Open loop	167.183	307.669	00.0000	634.6519

3.3 *Analysis*

Here, time domain analysis represents open-loop step response as shown in Fig. 3. Table 1 illustrated the open loop system performance.

3.4 *Bode Plot*

Here, frequency domain analysis represents open loop Bode plot as shown in Fig. 4. Stable loop; Gain Margin is 26.6 dB at frequency 0.117 rad/s.

4 Representation of IOPI and FOPI Controller

4.1 *IOPI Controller*

The T.F of IOPI controller expressed as,

$$C_i(s) = K_p + \frac{K_i}{s} \tag{2}$$

Therefore, the controller contributions are dependent on: gain K_p, K_i

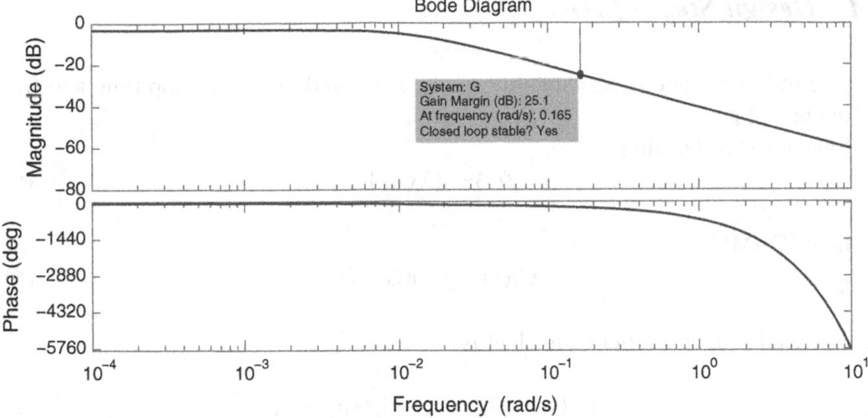

Fig. 4 Open loop bode plot result for level control system

4.2 *FOPI Controller*

The T.F. Of FOPI controller expressed as,

$$C_f(s) = K_p + \frac{K_i}{s^{\lambda}} \tag{3}$$

Therefore, the controller contributions are dependent on: gain K_p, K_i, fractional order λ. Where, λ is positive real number.

The relative dead time τ is the necessary and highly impelling controller parameter given as,

$$\tau = \frac{L}{L+T} \tag{4}$$

where L is dead time and T is lag time.

Process can be classified as lag dominated when $\tau < 0.5$, delay dominated when $\tau > 0.5$ and balanced system when $\tau = 0.5$. Thus identified process model (1) is lag dominated.

5 Design of Controller

The FOMCON toolbox of MATLAB [9, 13] had used in present work to design integer [8] and fractional order controllers [9–12] for (1).

5.1 Design Steps of IO-IMC

1. Use fotf command to test stability of (1). Also find out open loop time and frequency plot.
2. Select first order filter.
$$f(s) = (\lambda s + 1)^{-1} \tag{5}$$
then IO-IMC,
$$G_I(s) = f(s)(G\_(s))^{-1} \tag{6}$$
3. The feedback controller equivalent is
$$G_c(s) = G_I(s)(1 - G_I(s)G_m(s))^{-1} \tag{7}$$
4. Calculate gain of the controller and filter λ.
5. Use fpid_optim and simulate it.

For (1) the corresponding CLTF is,

$$G_c(s) = \frac{(76.089s + 1)(5s + 1)}{(\lambda + 7)0.7s} \tag{8}$$

5.2 Design Steps of FO-IMC

1. Use fotf command to test stability of (1). Also find out.
2. The frequency plot provide GM, PM, and ω_c.
3. Calculate λ and τ_c from step 2. Where
$$\lambda = \frac{\Pi - \psi_m}{\frac{\Pi}{2}}, \tau_c = \frac{1}{\omega_c^{\lambda+1}} \tag{9}$$
4. Design of fractional filter.
$$F(s) = \frac{S^{\alpha-1}}{1 + (\frac{\tau_c}{\theta})s^{\lambda}} \tag{10}$$
5. The FO-PI controller is:
$$C_{IMC}(s) = (1 + \frac{1}{\tau s^{\alpha}})\frac{\tau}{K\theta} \tag{11}$$
6. The feedback controller T.F. is

$$C(s) = \frac{\tau}{K\theta}(1 + \frac{1}{\tau s^{\alpha}})\frac{s^{\alpha-1}}{1 + (\frac{T_c}{\theta})s^{\lambda}} \tag{12}$$

7. Using fpid_optim command.

$$G_{cl} = \frac{e^{-\theta s}}{T_c s^{\alpha+1} + \theta s + e^{\theta s}} \tag{13}$$

8. Simulate the fpid_optim GUI.

For (1) The corresponding CLTF is,

$$G_{cl} = \frac{e^{-10s}}{21.5218 s^{1.3333} + 10s + e^{10s}} \tag{14}$$

6 Results

This section reported a result of comparative analysis of IO-PI and FO-PI for (1).

6.1 Simulation Result

Table 2 represents the controller gain and Table 3 represents performance parameter of IO-PI controller. Table 4 represents the controller gain, filter and Table 5 represents performance parameter of FO-PI controller. Figure 5 represents the comparison of IO-PI and FO-PI controller for IMC tuning method. From this result FOC gives improved result.

Table 2 Controller gain of IO-PI

Controller parameter	IMC
Kp	07.1428
Ti	00.0131

Table 3 Performance parameter of IOPI

Performance parameter	Rise time	Settling time	Overshoot	Peak time	ISE	IAE	ITAE
IO-IMC	17.8306	111.385	00.0000	131.0000	364.300	222.000	234.700

Table 4 Controller gain of FO-PI

Controller parameter	Kp	Ti	λ	Filter
FO-IMC	05.4300	00.0356	–	*f

Table 5 Performance parameter of FOPI

Performance parameter	Rise time	Settling time	Overshoot	Peak time	ISE	IAE	ITAE
IO-IMC	16.1388	93.6220	04.6202	86.0000	206.2000	139.4000	143.9000

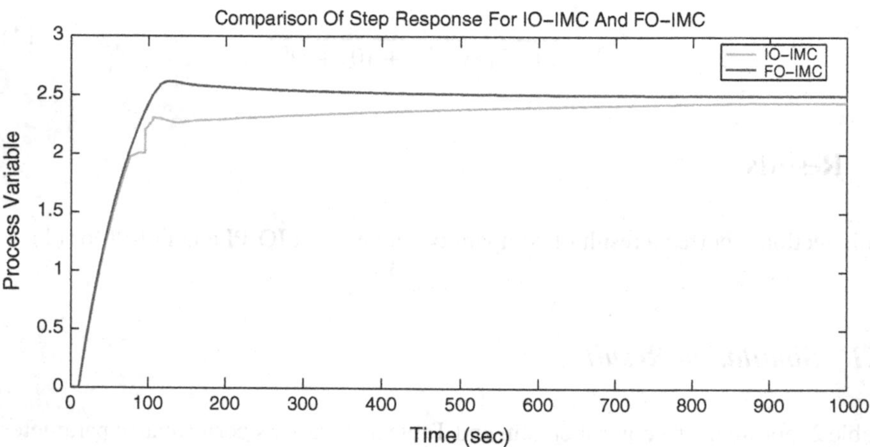

Fig. 5 IMC tuning method For IO-PI and FO-PI

6.2 Experimental Result

Figure 6 shows the experimental result of FO-IMC with control law. To check the process uncertainty, disturbance is applied to system at 400 s with a 10 % step to tank T1.

7 Robustness Analysis

The evaluation of robustness of controller is,

- Introducing a perturbation ambiguity of −10 % in time delay and +10 % in time constant into (1), the resultant transfer function is given by:

$$\frac{0.7e^{-9.0s}}{83.69s + 1} \tag{15}$$

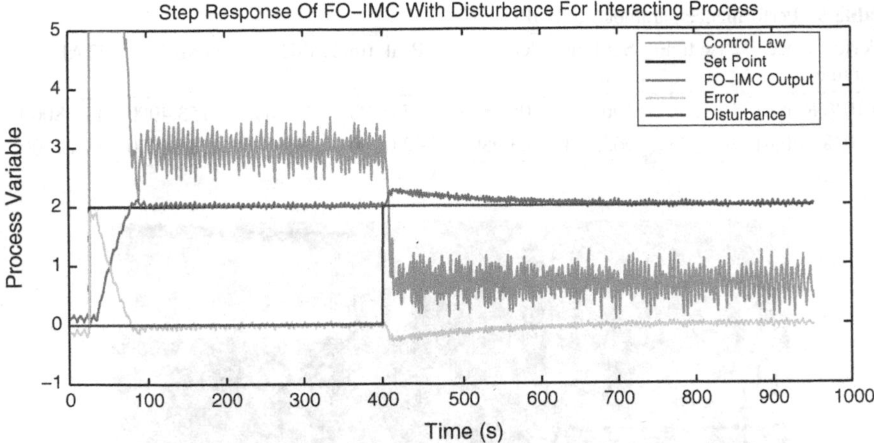

Fig. 6 Step response with disturbance of FO-IMC

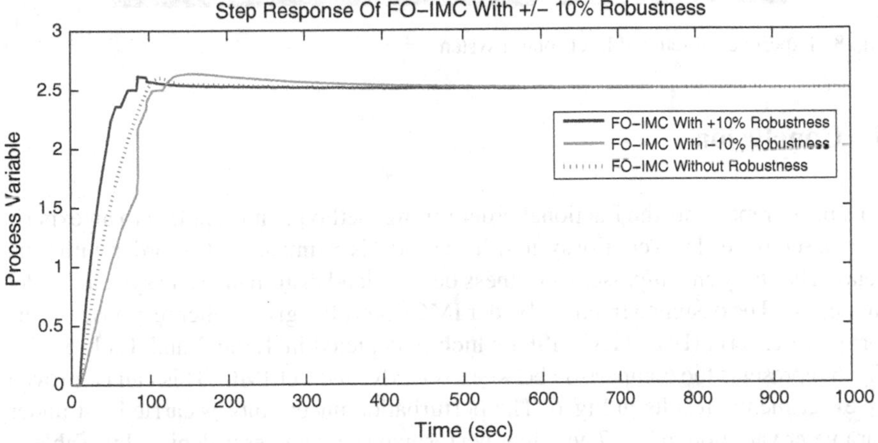

Fig. 7 FO-IMC for perturbation uncertainty

- Introducing a perturbation ambiguity of +10 % in time delay and −10 % in time constant into (1), the resultant transfer function is given by:

$$\frac{0.7e^{-11.0s}}{68.48s+1} \tag{16}$$

where as the controller settings are those provided for the nominal process. Figure 7 represents the perturbation uncertainty of ±10 % variation and their performance parameter depicted in Table 6.

Table 6 Performance parameter of FOPI

Performance parameter	Rise time	Settling time	Overshoot	Peak time	ISE	IAE	ITAE
+10 % Robust	16.1356	106.499	05.4853	97.0000	215.6000	153.4000	171.8000
−10 % Robust	16.1335	96.2524	03.9803	92.0000	198.0000	125.5000	116.7000

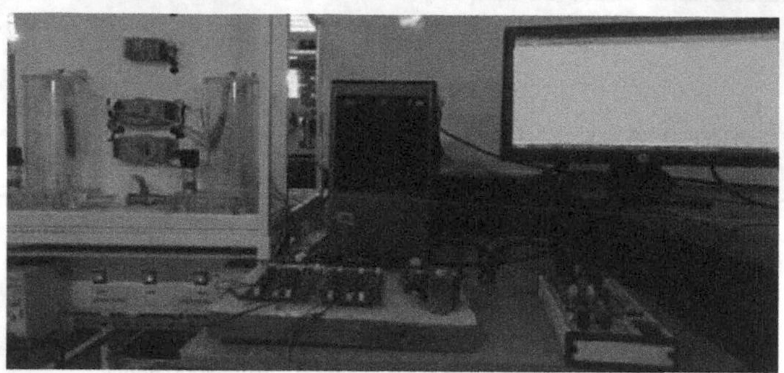

Fig. 8 Experimental setup of level control system

8 Conclusion

This paper represents the fractional order tuning method that is applied to an experimental setup, i.e., level control system. The model is estimated as first order with time delay. The study encompasses robustness besides load disturbance and system model steadiness. The designed fractional order IMC controller gives better response to performance criteria (IAE, ITAE, ISE) which is depicted in Table 3 and Table 5 with Fig. 5 represents the comparison between IO-IMC and FO-IMC. This can be proved by experimental results in Fig. 6. The perturbation uncertainty is carried out under parameter variation in Fig. 7 and their performance parameter is depicted in Table 6. As compared to IO-IMC, the FO-IMC shows best performance against steadiness and process uncertainties. The experimental setup is as shown in Fig. 8.

References

1. Astrom, K.J., Hagglund, T.: PID Controller Theory, Design and Tuning. Research Triangle Park, NC: Instrument Society of America. (1995)
2. Astrom, K.J., Hagglund, T.: Advanced PID Control, Instrument System and Automation Society (ISA). 158 (2006)
3. Hagglund, T., Astrom, K.J.: Revisiting the Ziegler-Nichols Tuning Rules for PI Control. Asian Journal of Control. vol. 4., no. 4., 364–380 (2002)
4. Vilanova, R.: IMC based Robust PID design: Tuning guidelines and automatic tuning. Journal of Process Control. 18, 61–70 (2008)

5. Marlin, T.E.: Process Control, Designing Processes and Control Systems for Dynamic Performance. Morgan Kaufmann, McGraw Hill
6. Luyben, W.L.: Process Modelling:Simulation and Control Systems for Chemical Engineers. McGraw Hill
7. Morari M., Zafiriou E.: Robust Process Control. Englewood Cliffs: Prentice-Hall. (1991)
8. Rivera D.E., Morari M., Skogested S.: Internal Model Control 4. PID Controller Design. Ind. Eng. Chem. Process Des. Dev. 25, 252–265 (1986)
9. Tepljakov, A., Petlenkov, E., Belikov, J., Halas M.:Design and Implementation of Fractional-order PID Controllers for a Fluid Tank System. American Control Conference (ACC) Washington. DC. USA. 17–19 (2013)
10. Kadu, C.B., Patil, C.Y.: Performance Assesment of IOPI and FOPI Controller for FOPDT System. IEEE Conference ICIC (2015)
11. Tajjudin, M., Rahiman, M.H.F., Arshad, N.M., Adnan R.,: Robust Fractional-Order PI Controller with Ziegler-Nichols Rules. World Academy of Science, Engineering and Technology. vol. 7. 07–29 (2013)
12. Maamar, B., Rachid M.: IMC-PID Fractional Order Filter Controllers Design for Integer Order Systems. ISA Transactions. 53 (2014)
13. Tepljakov, A., Petlenkov, E., Belikov, J.: FOMCON: a MATLAB Toolbox for Fractional-order System Identification and Control. International Journal of Microelectronics and Computer Science. vol. 2., no. 2., 5162 (2011)

A Secure Authentication Scheme in Multi-operator Domain (SAMD) for Wireless Mesh Network

Ninni Singh, Gunjan Chhabra, Kamal Preet Singh and Hemraj Saini

Abstract Wireless mesh network (WMN) is considered to be an evolving technique because of self-configuration and adaptive features, it supports large-scale network especially in an organization and academics. As with any network, communication among nodes plays an important role, when two nodes in a network communicate with each other via the Internet, secure authentication is an imperative challenge. In literature, there are many approaches that have been suggested to deliver a secure authentication between nodes in wireless mesh network (WMN), however, all these outlines contain some disadvantages, i.e. management cost of the public key and system complexity. Our suggested proposed approach is dealing with one of the wireless mesh network challenges, i.e. Mutual authentication. Here in this we have considered the three authentication techniques, i.e. Inter-, Intra- and Inter-Operator domain authentication. SAMD (**Secure Authentication Scheme in Multi-Operator Domain**) hides the location, communication path and network access and apart from that it will resist to the adversary, key forgery and modification attack. As a result, we have assessed the performance of the proposed design in terms of Authentication cost, Encryption cost, Key validation, Key Generation, Throughput and System Delay, which indicates our scheme more efficient than other schemes. Results show that it will efficiently work in real-time traffic.

Ninni Singh (✉) · Gunjan Chhabra · K.P. Singh
Department of Centre of Information Technology, University of Petroleum and Energy Studies, Dehradun, Uttrakhand, India
e-mail: ninnisingh1991@gmail.com

Gunjan Chhabra
e-mail: g_chhabra@yahoo.com

K.P. Singh
e-mail: kamalpreet1010@gmail.com

Hemraj Saini
Department of Computer Science and Engineering, Jaypee University of Information Technology, Waknaghat, India
e-mail: hemraj1977@yahoo.co.in

© Springer Science+Business Media Singapore 2017

S.C. Satapathy et al. (eds.), *Proceedings of the International Conference on Data Engineering and Communication Technology*, Advances in Intelligent Systems and Computing 468, DOI 10.1007/978-981-10-1675-2_35

Keywords Multi-operator • Inter-Domain • Intra-Domain • Inter-Operator • Wireless mesh network

1 Introduction

The Wireless Mesh Network is an emerging technology, its fast, inexpensive network deployment, easy Internet connectivity features makes it a popular choice for Wireless ISP (Internet Service Provider). WMN represents the combination of wide area cellular network and high-speed Wi-Fi networks. Nevertheless, without any security in WMN, it is impossible to securely exchange any information [1, 2]. Various research work is in progress. At present there are no formal methods to authenticate the network in WMN. Security is an open challenge in WMN. In recent times lot of research work is in progress. (Santhanam) [3] Proposed an authentication scheme grounded on Merkle tree. There whole consideration is to authenticate the client irrespective of the entire security architecture and mesh client roaming. (Fu et al.) [4] Proposed an authentication scheme in which he integrate various existing techniques, i.e. Virtual certificate authority, zone-based hierarchical structure and multi-signature scheme. (Zhang et al.) [5] Proposed an architecture, in which, if mesh client wishes to roam to another network, then it requires a pass from trusted third party. In this paper, we have proposed a novel secure authentication scheme for multi-operator domain. This is the extended version of [6] paper. The proposed technique is a broker-based three-tier hierarchical architectures. The broker is a reliable third party which lives in the first tier. Broker consist of a private key generator, whose function is to generate a private key. Gateway lives in the second tier and router lives in the third tier. Both gateways as well as router are considered as a trustworthy node because of their less mobility. Any node willing to enter into the network, it has to submit its own identity (identity of any node act as the public key of that node) to the broker and broker hand over private key (by giving identity as an input to the private key generator) to that node. If the newly entered node is the router or gateway, instead of private key it can also send a ticket and its own signing rights to them. Now onwards both gateway and router possess the same functionality of the broker. This technique is formally verified on AVISPA SPAN, which shows that there is no attack is possible and private key is not forgeable. SAWMN (Secure Authentication in Wireless Mesh Network) reduces the overall system complexity by not explicitly managing the public keys and it also efficiently works in real-time environment. Our suggested proposed approach extend above discussed technique to accomplish the authentication procedure in Inter-domain and Intra-domain. As a result, we have assessed the performance of the proposed design in terms of security analysis, which indicates our scheme more efficient than other schemes. The rest of the paper is structured into the following categories: in section second we discuss the related works. In section third we have elaborated the proposed technique (SAWMN) and Inter- and

Intra-domain Authentication. In section four we have shown the simulated results and finally in section five we conclude our paper.

2 Related Work

For secure communication among nodes, there will be a demand of some authentication mechanism in a wireless network. Many researchers are working on this area, few of them, we will discuss in this section. Authentication among nodes can be attained, when any two nodes present in the network, Inter-domain (mesh clients roam from one operator domain to other operator domain), Intra-domain (mesh clients roam within a same domain operator) and Intra operator (mesh client roams from one domain to another domain within same operator). (Summit R.) [1] Proposed a token-based authentication scheme, in which token is utilized for verification purpose. Token works same as that of digital signatures by integrating public key with subjects ID and it also verifies the authenticity of subject ID in the issuer realm. This protocol reduces the time required for authentication and also somehow restricts the communication between the home network to the roam or foreign network, but it requires a roaming credential that will be shared among servers and this incurs some cost for supervision. (Ford) [2] Proposed a key agreement protocol based in identity-based encryption technique. This scheme overcomes the above discussed problems, i.e. the administration cost. One of the biggest drawbacks of this scheme is that it cannot guard the user's privacy. (Yeh and Sun) [3] Proposed a four-party password-based authentication technique and key establishment protocol. To accomplish all this feature, there will be a requirement of public key infrastructure for the distribution and confirmation of server's public key to the clients. But the problem with this approach is that it is not well suited for lightweight computing domain. (Ren-Junn) [4] Proposed an authentication schemes, which utilizes symmetric encryption technique and hash function.

(Hung-Yu Chien) [5] Proposed an authentication scheme, which utilizes a public key encryption technique. Instead of using certificates, they utilize hash function, which decreases the management cost of certificates. To accomplish this feature additional server is required, which somehow increases the time delay. All the above discussed schemes incur some drawbacks either in terms of time delay in computation, authentication cost, having high storage cost and system complexity. In our proposed authentication technique, we work on these drawbacks. Proposed scheme reduces the authentication cost, system complexity, power consumption and timing cost, and also shows the comparative study with respect to the traditional authentication technique used in wireless mesh network.

3 Proposed Technique

In the former section we have discussed various research works of different authors. Now in this section we have elaborated the extended version of [6]. In this [6] paper we have discussed the key management in the proposed architecture. The proposed technique is a broker-based three-tier hierarchical architectures. The broker is a reliable third party which lives in the first tier. Broker consist of a private key generator, whose function is to generate a private key. Gateway lives in the second tier and router lives in the third tier. Both gateways as well as router are considered as a trustworthy node because of their less mobility. Any node willing to enter into the network, it has to submit its Own identity [7] (identity of any node act as the public key of that node) to the broker and broker hand over private key (By giving identity as an input to the private key generator) to that node. If the newly entered node is the router or gateway, instead of private key it can also send a ticket and its own signing rights to them. Now onwards both gateway and router possess the same functionality of the broker [8, 9]. This technique is formally verified on AVISPA SPAN, which shows that there is no attack is possible and private key is not forgeable. SAWMN reduces the overall system complexity by not explicitly managing the public keys and it also efficiently works in real-time environment (Fig. 1).

Our suggested proposed approach extend above discussed technique in order to accomplish the authentication procedure in Inter-domain [10] and Intra-domain. Here we used one additional server named as Main server, which contain the all related information like IP address of the sub-module, roaming information, etc.

Main server performs various functionality, like if any mesh client roams from one domain to some other domain, then this activity is first noticed by main server, which internal hand over the mesh client IP address of the foreign domain. After this foreign domain performs some authentication process between mesh clients and mesh router.

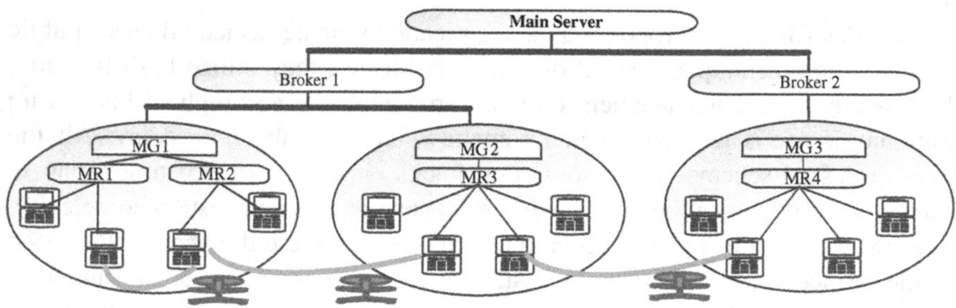

Fig. 1 Block diagram of authentication network

3.1 Inter-Domain Authentication

When mesh client roams from broker 2 domains to broker 1, Inter-domain authentication has been taking place between the mesh router and the mesh client. Following an authentication process will be followed by MC and MR_1.

(1) $MR_1 \rightarrow MC\colon = TK_{MR_1}^{B_1 MG_1} = \{\text{Exp}, ID_{B1}, t_{B1}, ID_{MG1}, ID_{MR1}, \text{Sig(gateway 1)}\}.$

(2) $MC \rightarrow MR_1\colon = TK_{MC}^{B2} = \{\text{Exp}^!, ID_{B2}, t_{B2}, ID_{MC}, \text{Sig(broker } B2)\}.$

(3) $MR_2 \rightarrow MC\colon = TK_{MC}^{MR_1 MG_1} = \{\text{Exp}^!, ID_{MG1}, ID_{MR1}, ID_{MC}, \text{Sig(Mesh router1)}t_1, N_1\}.$

(4) $MC \rightarrow MR_1\colon = \{t_2, N_2\}.$

The mesh router periodically broadcasts message 1 to its coverage area. When mesh client roams from broker 2 to broker1 called inter-domain. After receiving message 1 following operations are performed (Fig. 2).

1. It first checks the freshness of the Expiration or validity of the ticket.
2. Retrieve broker 1 public key and from broker's public key, it verifies the signature of gateway 1.
3. After verification, it computes the shared key $K_{MC--MR1} = \text{e}\,(ID_{MC}, ID_{MR1})$, where $= ID_{MR1} = \text{H1(Exp, IDMG1)}$.

Mesh client now sends a message (2) to mesh router 1. After receiving the message (2) it performs following tasks.

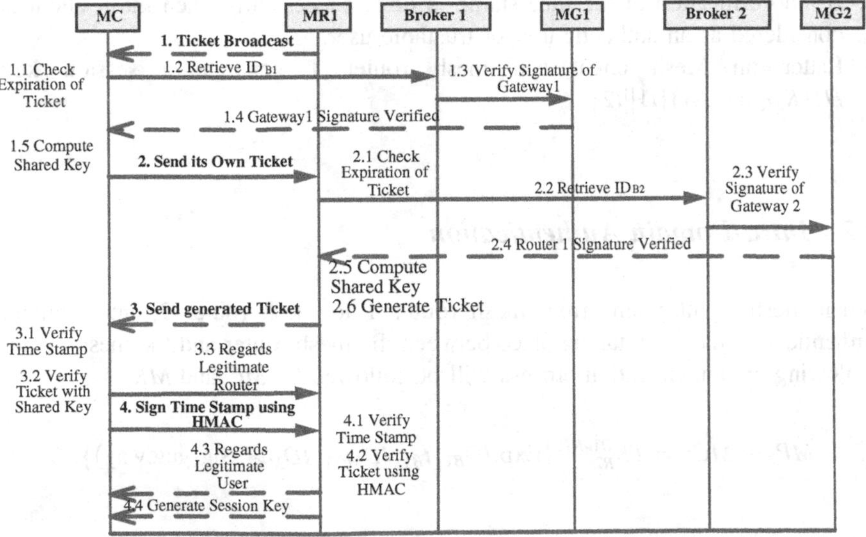

Fig. 2 Inter-Domain authentication

1. Check for the expiry date on the Client ticket and make certain that it is not expired.
2. Retrieve broker 2 public key and from broker's public key, it verifies the signature of gateway 2.
3. After verification, it computes the shared key $K_{MR1--MC} = e\ (ID_{MR1}, ID_{MC})$.
4. Mesh router 1 generates tickets for newly entered nodes, i.e. Mesh client.

$$TK_{MC}^{MR_1MG_1} = \{\text{Exp}', ID_{MG1}, ID_{MR1}, ID_{MC}, \text{Sig(Mesh router1)}t_1, N_1\}.$$

5. Before sending to the mesh client, mesh router 1 sign the ticket with HMAC, $1 = TK_{MC}^{MR_1MG_1}$.

Mesh router 1 now sends a message (3) to mesh client. After receiving the message (3) it performs following operations.

1. Check the newness of timestamp and the expiry of the ticket.
2. Verify the ticket $TK_{MC}^{MR_1MG_1}$ Using shared key $K_{MC--MR1}$ (Computed by mesh client).
3. If the verification of ticket is done successfully, then mesh router is considered as an authentic router or trustable router.
4. Generate a timestamp and create a signature on it, by signing it with the shared key $(K_{MC--MR1})$.

Mesh client now sends a message (4) to mesh router 1, after receiving a message (4) it perform the following operations.

1. Check the newness of timestamp and the expiry of the ticket.
2. Verify the timestamp using a shared key $K_{MR1--MC}$ (Computed by mesh router 1).
3. If the verification of the time stamp is done successfully, then mesh client is considered as an authentic user or trustable user.
4. Latter on Mesh client and mesh router 1 generate a session key $H1(K_{MC--MR1})\{t1||t2\}$.

3.2 Intra-Domain Authentication

When mesh client roams from mesh router 1 to mesh router 2, Intra-domain authentication has been taking place between the mesh router and the mesh client. Following an authentication process will be followed by MC and MR_2.

1. $MR_2 \to MC{:} = TK_{R2}^{B1MG1}\{\text{Exp}.ID_{B1}, t_{B1}, ID_{MG1}, ID_{MR2}, \text{Sig(gateway)}\}$.

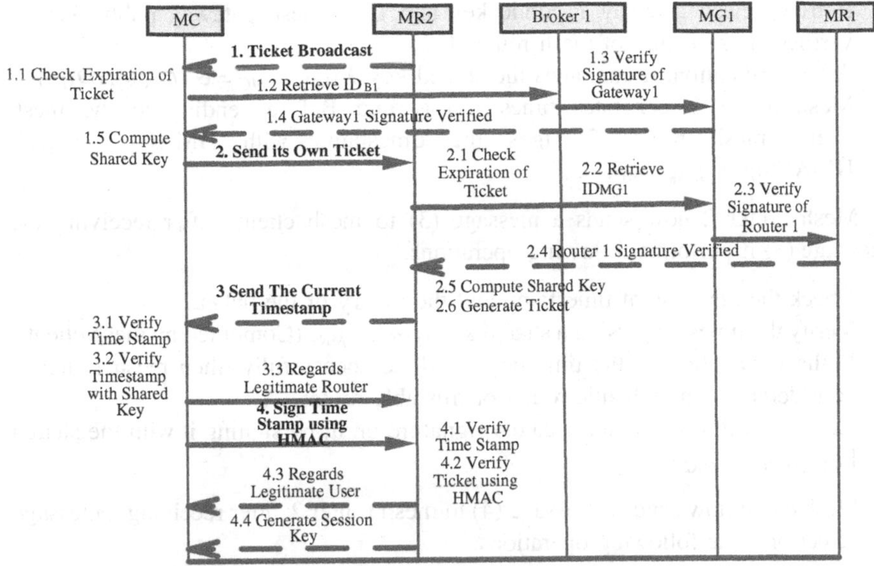

Fig. 3 Intra-Domain authentication

2. $MC \rightarrow MR_2 := TK_{MC}^{MG1MR1}\{\text{Exp}^!, ID_{MG1}, ID_{MR1}, ID_{MC}, \text{Sig (mesh router 1)}\}$.

3. $MR_2 \rightarrow MC: = \{t_3, N_3\}$.

4. $MC \rightarrow MR_2: = \{t_4, N_4\}$.

Mesh router 2 periodically broadcast message 1 to its coverage area. When mesh client roams from mesh router 1 to mesh router 2 called intra-domain. After receiving message 1 following operations are performed (Fig. 3).

1. It first checks the freshness of the Expiration or validity of the ticket.
2. Retrieve broker 1 public key and from broker's public key, it verifies the signature of gateway 1.
3. After verification, it computes the shared key $K_{MC--MR2}\Im \leftarrow e \leftarrow ID_{MC}$, $ID_{MR2} \leftarrow$, where

$$ID_{MR2} = \text{H1}(\text{Exp}, ID_{MG1}, ID_{MR2}).$$

Mesh client now sends a message (2) to mesh router 2. After receiving the message (2) it performs following tasks.

1. Check for the expiry date on the Client ticket and make certain that it is not expired.

2. Retrieve mesh gateway 1 public key and from mesh gateway public key, it verifies the signature of mesh router 1.
3. After verification, it computes the shared key $K_{MR2--MC}$ = e (ID_{MR2}, ID_{MC}).
4. Mesh router 2 generates timestamp t_3 and Before sending to the mesh client, mesh router 2 signs the timestamp with HMAC $N_3 = \{t_3\}$ HMACSig_$K_{MR2--MC}$.

Mesh router 2 now sends a message (3) to mesh client. After receiving the message (3) it performs following operations.

1. Check the newness of timestamp and the expiry of the ticket.
2. Verify the timestamp using a shared key $K_{MC--MR2}$ (Computed by mesh client).
3. If the verification of the timestamp is done successfully, then mesh router is considered as an authentic router or trustable router.
4. Generate a timestamp and create a signature on it, by signing it with the shared key $(K_{MC--MR2})$.

Mesh client now sends a message (4) to mesh router 2, after receiving a message (4) it perform the following operations.

1. Check the newness of timestamp and the expiry of the ticket.
2. Verify the timestamp using a shared key $K_{MR2--MC}$ (Computed by mesh router 1).
3. If the verification of the timestamp is done successfully, then mesh client is considered as an authentic user or trustable user.
4. Latter on Mesh client and mesh router 2 generate a session key $H1(K_{MC--MR2})\{t3||t4\}$.

3.3 *Inter-Operator Authentication*

When mesh client roams from mesh router 1 in One domain to mesh router 3 in other domain, Inter-Operator authentication has been taking place between the mesh router 3 and the mesh client. Following an authentication process will be followed by MC and MR_3.

1. $MR_3 \rightarrow MC: = TK_{R3}^{B1MG2}\{\text{Exp}.ID_{B1}, t_{B1}, ID_{MG1}, ID_{MR3}, ID_{MG2}, \text{Sig}(\text{gateway 2})\}.$

2. $MC \rightarrow MR_3: = TK_{MC}^{MG1MR1}\{\text{Exp}^{!}, ID_{MG1}, ID_{MR1}, ID_{MC}, \text{Sig}(\text{mesh router 1})\}.$

3. $MR_3 \rightarrow MC: = TK_{MC}^{MR_3MG_2} = \{\text{Exp}^{!}, ID_{MG2}, ID_{MR3}, ID_{MC}, \text{Sig}(\text{Mesh router3})t_5, N_5\}.$

4. $MC \rightarrow MR_3: = \{t_6, N_6\}.$

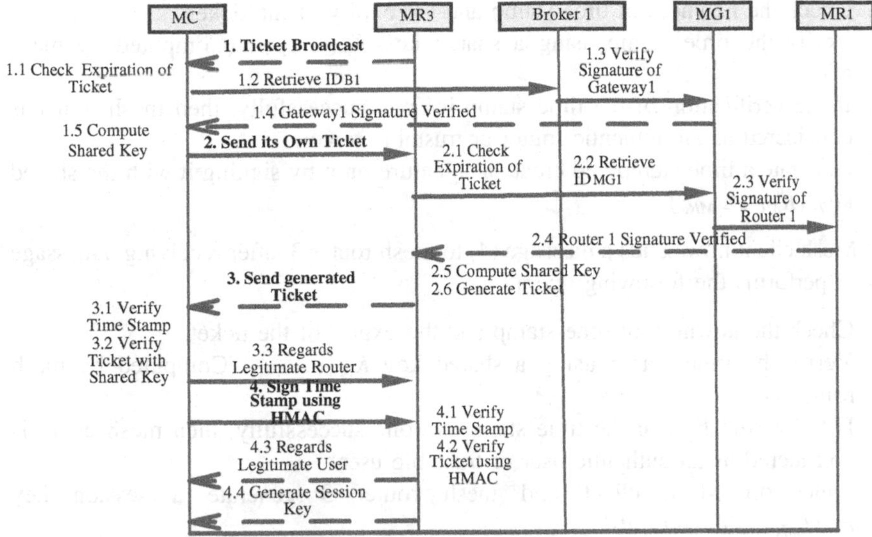

Fig. 4 Inter-Operator domain

Mesh router 3 periodically broadcast message 1 to its coverage area. When mesh client roams from mesh router 1 to mesh router 3 called an inter-Operator domain. After receiving message 1 following operations are performed (Fig. 4).

1. It first checks the freshness of the Expiration or validity of the ticket.
2. Retrieve broker 1 public key and from broker's public key, it verifies the signature of gateway 1.
3. After verification, it computes the shared key $K_{MC--MR2}$ = e (ID_{MC}, ID_{MR2}), where

4. $ID_{MR3} = \mathrm{H1}(\mathrm{Exp}, ID_{MG1}, ID_{MR3})$.

Mesh client now sends a message (2) to mesh router 3. After receiving the message (2) it performs the following tasks:

1. Check for the expiry date on the Client ticket and make certain that it is not expired.
2. Retrieve mesh gateway 1 public key and from mesh gateway public key, it verifies the signature of mesh router 1.
3. After verification, it computes the shared key $K_{MR3--MC}$ = e (ID_{MR3}, ID_{MC}).
4. Mesh router 3 generates time stamp t_5 and before sending to the mesh client, mesh router 3 signs the time stamp with HMAC $N_5 = \{t_5\}$HMACSig_ $K_{MR3--MC}$.

Mesh router 3 now sends a message (3) to mesh client. After receiving the message (3) it performs the following operations:

1. Check the newness of time stamp and the expiry of the ticket.
2. Verify the time stamp using a shared key $K_{MC--MR3}$ (Computed by mesh client).
3. If the verification of the time stamp is done successfully, then mesh router is considered as an authentic router or trustable router.
4. Generate a time stamp and create a signature on it by signing it with the shared key $(K_{MC--MR3})$.

Mesh client now sends a message (4) to mesh router 3, after receiving a message (4) it performs the following operations:

1. Check the newness of time stamp and the expiry of the ticket.
2. Verify the time stamp using a shared key $K_{MR3--MC}$ (Computed by mesh router 1).
3. If the verification of the time stamp is done successfully, then mesh client is considered as an authentic user or trustable user.
4. Latter on Mesh client and mesh router 3 generate a session key $H1(K_{MC--MR3})\{t5||t6\}$.

4 Results

In this section we discourses the simulation of the projected algorithm. It also describes the framework for simulation for the proposed scheme.

System Configuration

For the simulation, we required the following system configuration:

- One GB RAM
- Core to duo Processor
- Windows or Linux Operating System
- Net bean Framework
- Mysql Database
- Net bean 7.4

This section gives the detail discussion about the results. In which we have focused on some parameters like Encryption Cost, Authentication Cost, Throughput, Key Generation, Key validation and System Delay. Now in this section I would like to discuss the various computation results in the form of graphs.

Encryption is the technique in which, any message is converted into an unreadable format, i.e. Ciphertext. This will be very helpful if any entity wishes to send some confidential information to any other party, i.e. before sending data it has to encrypt the message so that only the intended user can able to read the message. A lot of research work is going on this field. Figure 5 shows that the different encryption cost of different file size.

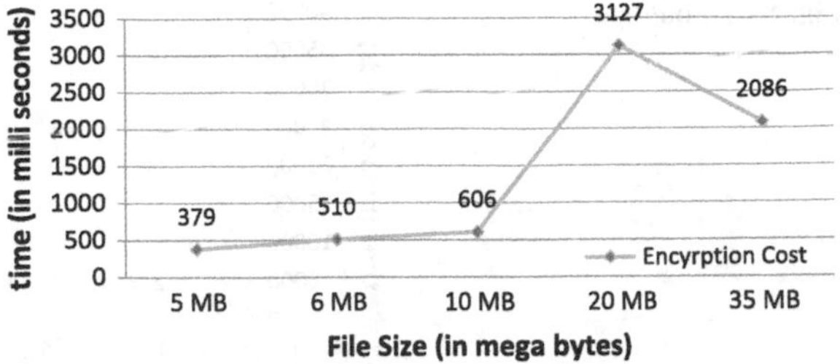

Fig. 5 Encryption cost

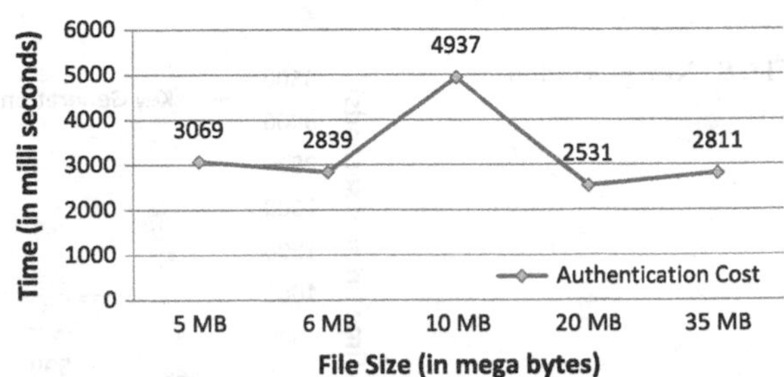

Fig. 6 Authentication cost

Authentication is an action that ensure other party that they were communicating to the legitimate user, this can be accomplished by adopting different authentication techniques. Authentication cost as its name indicates it is the cost or time required to authenticate a particular user in the network. In our case any client can able to move from one network to another. If this happens then server perform above discussed technique to authenticate the user.

A lot of research work is going on this field. Figure 6 shows that the different Authentication cost of different file size.

Network delay is one of the important parameters of performance of any type of network. We can define the network delay as the time needed to send bits of data to be traveled in the network from egress to ingress node. Delay is dependent on the location of nodes from the source to the destination node. A lot of research work is going on this field. Figure 7 shows that the different encryption costs of different file sizes.

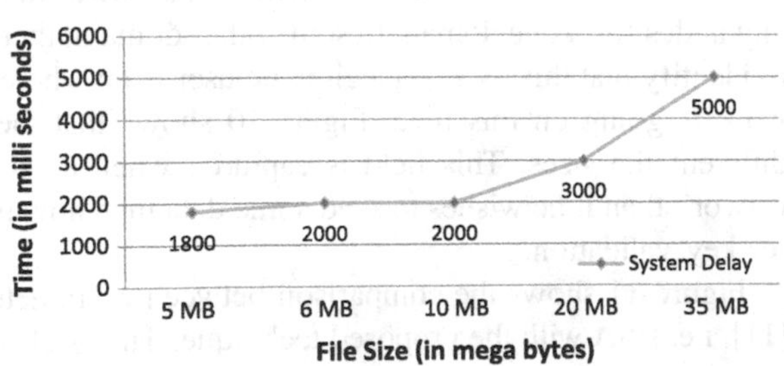

Fig. 7 System delay

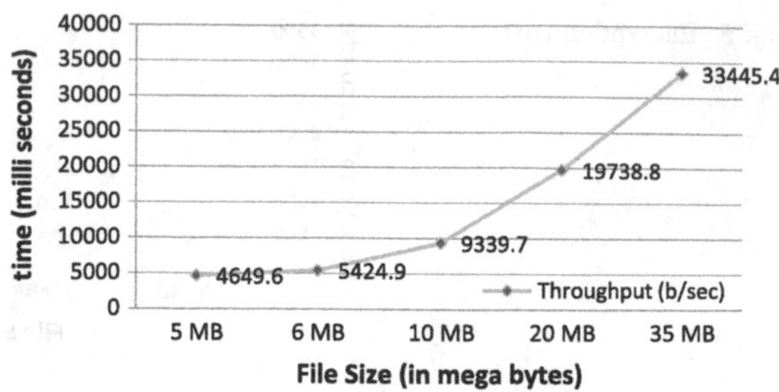

Fig. 8 Throughput

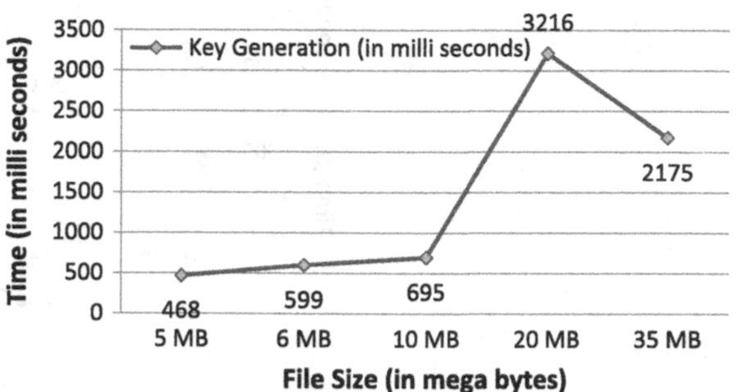

Fig. 9 Key generation

Throughput is referred as a number of bits or units of data is transferred or system is able to process in a given unit of time span. It can also be defined as a rate of successful transfer of information with the help of some channel. A lot of research work is going on this field. Figure 8 shows that the different throughput of different file sizes.

Key generation as its name indicates that any user is generating keys for further processing in the network. As we all know to communicate on the network, the public private key pair is more important. So in this key generation, we are more focused to calculate the time required to generate this public private key pair. A lot of research work is.

Going on this field. Figure 9 shows that the different Key Generation time needed for different file sizes, because here for each datum or file we compute different public private key pairs.

Key Validation as its name indicates that the truth or correctness of the keys. If any nodes from one domain to some other domain, then we perform some operation to identify that this user is a legitimate user from where it belongs. A lot of research work is going on this field. Figure 10 shows that the different key validation of different file sizes. This field is captured when a node is moving to some other network than if he wishes to send some data then how much time required is needed for key validation.

Figure 11 shows the comparison between asymmetric cryptographic techniques [11], i.e. RSA with the proposed technique. This is clearly shown that the SAWMN

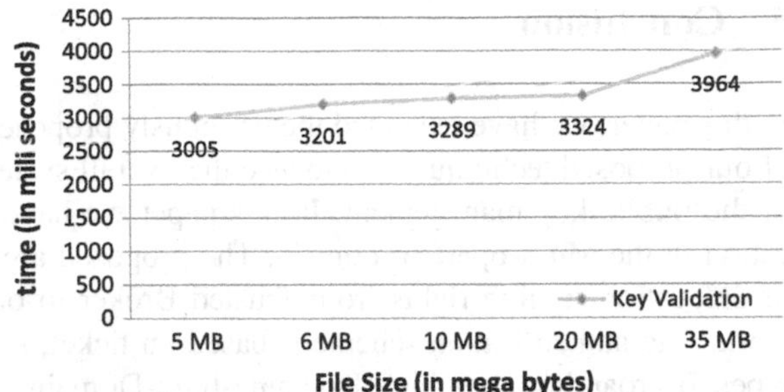

Fig. 10 Key validation

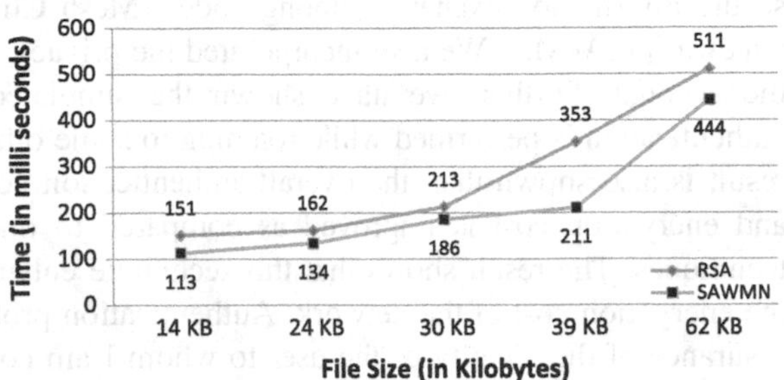

Fig. 11 Encryption technique comparisons between RSA

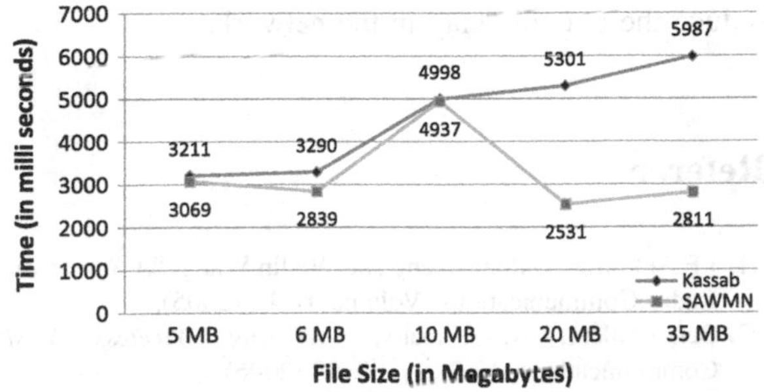

Fig. 12 Authentication cost comparison between Kassab and SAWMN

needed less time for encryption as compared to the RSA. First, we have taken the files of different sizes, then perform both the encryption on that file and record the time needed for encryption in milliseconds. Results show that our technique is much faster than the RSA.

Figure 12 shows the comparison between Authentication Cost techniques, i.e. Kassab with the proposed technique. This is clearly shown that the SAWMN needed less time for authentication as compared to the Kassab. First, we have taken the files of different sizes, then perform both the authentication on that file and record the time needed for authentication in milliseconds. Results show that our technique is much faster than the Kassab.

5 Conclusion

In this paper, we have extended the previously proposed technique in [6]. The goal of our proposed technique is to reduce the overall system complexity and overhead of the public key management. In this paper we have shown the secure authentication in the Multi-operator domain. The proposed architecture inherits the feature of delegation signing rights from Trusted Broker to other trusted node in the network. The authentication scheme is based on ticket, so it is best suited for various types of roaming, i.e. Inter-Domain. Intra-Domain and Inter-Operator domain. Furthermore, we have incorporated the identity-based encryption technique for secure information exchange among nodes (Mesh Client, Mesh Router and Mesh Gateway) in WMN. We also incorporated the privacy by utilizing fast HMAC into the account. Further, we have shown the simulated result which shows how authentication is performed while roaming to some other network. Our comparison result is also shown that, the overall authentication cost, system delay throughput and encryption cost is improved as compared to one of the previous proposed techniques. The result shows that this technique enhanced the authentication cost, the encryption cost of the network. Authentication protocols generally used for the assurance of the identity of the user to whom I am communicating. We have also considered the other parameters like securely generation, so that an attacker not able to do any type of attack in the network, apart from that we have also considered to reduce the overall delay in the network.

References

1. I.F Akyildiz, Xudong Wang and Weilin Wang, "*A Survey on Wireless Mesh Networks.*", IEEE Radio Communications, Volume 47(4), (2005).
2. Ben Salem, N. Hubaux, "*Securing Wireless Mesh Networks*", IEEE Wireless Communications, 13(2), pp 50–55 (2006).
3. L. Santhanam, B. Xie, D.P Agrawal, "*Secure and Efficient Authentication in Wireless Mesh Networks using Merkle Trees*", 33rd IEEE Conference on Local Computer Networks, LCN (2008).
4. Y. Fu, J. He, R. Wang, R. Li, "*Mutual Authentication in Wireless Mesh Networks.*" In: Proceedings of ICC (2008).
5. Y. Zhang, Y. Fang, "*ARSA: an Attack-Resilient Security Architecture for Multi-hop Wireless Mesh Networks*", IEEE Journal on Selected Areas in Communications, 24(10), (2006).
6. Ninni Singh and Hemraj Saini, "*Formal Verification on Secure Authentication in Wireless Mesh Network*", Second International Conference on Computer and Communication Technologies, in Springer AISC series, (2015).
7. Ford Long Wong, Hoon Wei Lim, "*Identity-based and inter-domain password authenticated key exchange for lightweight clients*", in: AINAW'07: Proceedings of the 21st International Conference on Advanced Information Networking and Applications Workshops, IEEE Computer Society, Washington, DC, USA, 2007, pp 544–550.
8. Her-Tyan Yeh, Hung-Min Sun, Password authenticated key exchange protocols among diverse network domains, Computers & Electrical Engineering, 31(3) (2005) 175–189.

9. Ren-Junn Hwang, Feng-Fu Su, "*A new efficient authentication protocol for mobile networks*", Computer Standards & Interfaces 28 (2) (2005) 241–252.
10. Summit R. Tuladhar, Carlos E. Caicedo, James B.D. Joshi, "*Inter-domain authentication for seamless roaming in heterogeneous wireless networks*", in: SUTC'08: Proceedings of the 2008 IEEE International Conference on Sensor Networks, Ubiquitous, and Trustworthy Computing, SUTC 2008, IEEE Computer Society, Washington, DC, USA, 2008, pp. 249–255.
11. Catherine Meadows, "*Formal methods for cryptographic protocol analysis: Emerging issues and trends*", IEEE Journal on Selected Areas in Communications 21 (2) (2003) 44–45.

A Wireless Sound Alert and Person Identity Notification System for People with Hearing Disabilities

K. Padma Vasavi

Abstract Over 10 % of the world's population, 800 million people, are suffering from hearing loss, and if they were to be taken into account it would be the third largest nation in the world. Current production of hearing aids meets less than 10 % of global need. About one out of forty people have hearing aids in developing countries. Moving in traffic without anybody's help and calling others who are not in distance reachable by hand is a major problem faced by hearing and speech impaired people. The aim of the paper is to enable the hearing and speech impaired people recognize the horn sound of vehicles and make their interaction easy with other people having similar disabilities. So, to address this problem there is a need for a device which can assist the hearing challenged people responding to important sounds. This paper proposes a sound alert system which can detect horn sound and gives an alert, and also it can help them in calling one another, when they are in indoor environment. The sound alert system works with the help of microphone, an analog circuit and a microcontroller unit in traffic mode, and on the other hand with the help of a wireless communication module in the interaction mode.

Keywords Hearing and speech impairment · Horn sound · Interaction · Vibration alert

1 Introduction

The major problem faced by hearing impaired people is going into traffic without anybody's help. This is because they cannot hear horn sounds. The person who whistles horn expects the other to move aside. As the hearing challenged person is not able to hear that sound, there is a chance for accidents. Also, the hearing impaired people cannot listen if someone is calling him from behind.

K.P. Vasavi (✉)
Department of ECE, SVECW, Bhimavaram, India
e-mail: padmavasaviece@svecw.edu.in

© Springer Science+Business Media Singapore 2017

S.C. Satapathy et al. (eds.), *Proceedings of the International Conference on Data Engineering and Communication Technology*, Advances in Intelligent Systems and Computing 468, DOI 10.1007/978-981-10-1675-2_36

So, the hearing challenged people need a device which can help them to respond to different sounds as well as a call from others. This purpose can be served using a device with a microphone and XBee modules to help respond to different sounds especially in traffic and to interact easily with other hearing impaired people, respectively.

The existing assistive devices for deaf are very less in number and they can help only those with less than 20 % hearing loss. In addition to this, most of them are very expensive. They are not available and affordable for the common man. As a result, most of the hearing challenged people from the developing countries are unable to get benefit out of any of these devices. The wristband that interprets sound for the deaf is a device which gives the pictorial depiction of wide variety of sounds like ringing of telephone, music, door bell, horn, etc. It also alerts deaf people in danger by vibrating and helps them view a pictorial depiction. The wrist band does not work alone; it is complemented by a micro device which can be worn on the belt. The micro device records the sound first, interprets it, and then sends a pictorial message, which can be viewed on the screen of the wrist watch. The watch even vibrates in case of danger [1].

The wristband discussed above cannot respond to more than one sound at a time and so ignores the other sounds heard at that time. So, there are maximum chances that the device ignores important sounds like horn. Also, the device vibrates for many kinds of sounds. So, it will be vibrating almost continuously causing great inconvenience to the user. This paper proposes a sound alert system for them which can detect horn sound and gives an alert, and also it can help them call one another, when they are in indoor environment. The sound alert system works with the help of microphone, an analog circuit, and a microcontroller unit in traffic mode, and with the help of a wireless communication module in the interaction mode. The rest of the paper is organized as follows: Sect. 2 gives the details of the proposed solution; Sect. 3 discusses the hardware and software implementation details of the proposed system; Sect. 4 describes the experiments and results obtained; and finally Sect. 5 concludes the paper.

2 Proposed Solution

The system level diagram of the sound alert system (SASY) is shown in Fig. 1. It has a display and some push button switches. The display shows the time as well as the number corresponding to the person who is trying to call this device.

The switches consist of 0–9 numbers for the purpose of calling any person in the interaction mode and also two additional switches ‘P’ and ‘M.’ The switch ‘P’ is used as a power button, i.e., to switch the device ON and OFF. The switch ‘M’ is used as mode control button, i.e., to switch the device from traffic mode to normal mode and viceversa. The device, when in the traffic mode, vibrates if there is any horn sound around it and, when in the normal mode, transmits a signal to the corresponding person when any switch is pressed and vibrates if it receives a signal.

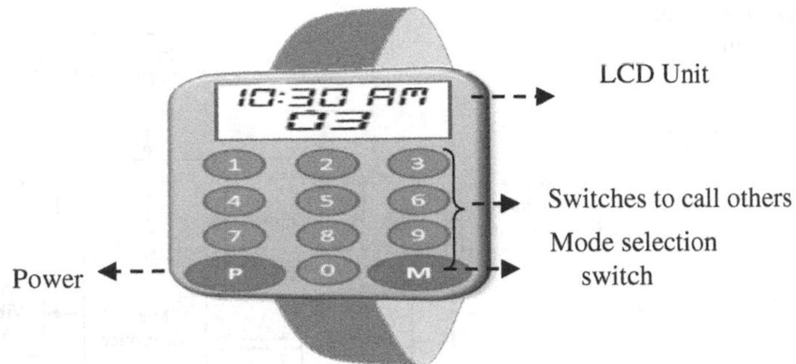

Fig. 1 Outer case design of proposed solution

The proposed solution has two modes of operation namely traffic mode and interaction mode. Mode selection switch is used to switch between the two modes.

The description of the subsystems in traffic mode is given below.

2.1 *Traffic Mode*

As the input need for the system is sound, there is a need to convert it into electrical signal, in order to perform any manipulations on the input and use it to control the output. So, the microphone is to be used as input device to convert the input sound into electrical signal. In case of using a microphone, noise comes into picture for sure. The easiest process to reduce noise is to eliminate the unwanted frequencies. A band-pass filter can efficiently perform this action. It can attenuate the lower and higher unwanted frequencies and allow only the range of frequencies that are selected. The amplitude of the signal obtained from the microphone is very less and is not enough to drive the microcontroller. So, the amplifier section should bc used to amplify the signal and bring it to the operating voltage range of microcontroller. The output is to be energized only when the input level is above the threshold level. Microcontroller is used to perform this control operation. As the user of the device is a hearing impaired person, vibration is the best way to give alert. So, micro vibration motor is required at the output.

2.2 *Interaction Mode*

In traffic mode, the microphone takes the sound as input and it converts it into voltage. A filter is used to eliminate the unwanted frequencies. The preamplifier amplifies this voltage signal and gives it as input to microcontroller. The micro-controller compares the input voltage level with the fixed threshold level, which is

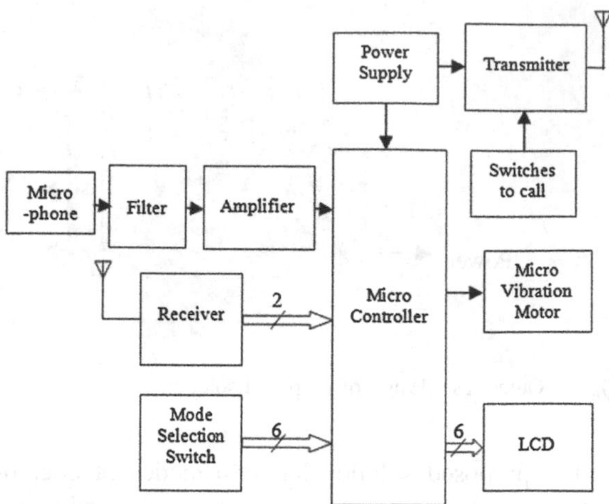

Fig. 2 Block diagram of sound alert system

fixed such that the microcontroller gives the output to the vibrator only when the input sound is too high, approximately above 85 db.

In interaction mode, the input from microphone will not affect any output of the microcontroller. In this mode, when a particular switch is pressed, the transmitter sends the signal to all the receivers but only the controller corresponding to that switch will respond by activating the vibrator and display, which shows the number corresponding to the person whose system transmitted the signal.

The block diagram of proposed system in interaction and traffic mode is shown in Fig. 2.

3 Implementation

The implementation of proposed system is divided into two parts:

Hardware implementation
Software implementation

3.1 Hardware Implementation

In this circuit, microphone is connected as input device. Its output is connected to the band-pass filter that is implemented using LM324. The output of the filter is obtained at pin 7 of LM324 [2]. The output of the filter is given to the amplifier which is implemented using the third opamp in the IC LM324. So, the input to the amplifier is given from the seventh pin of the same IC. Eighth pin of LM324 is the

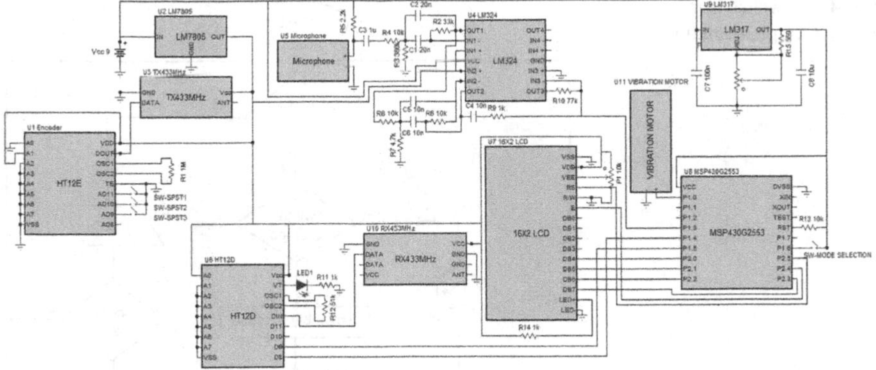

Fig. 3 Schematic of sound alert system

output of the amplifier. The output of the amplifier is given to P1.3 of the microcontroller MSP430G2553 [3].

The encoder HT12E [4] is connected to the RF transmitter [5]. All the data pins of the encoder are provided with switches connecting to ground except one data pin. This is done because the data pins are low when the receiver is in off state and also when the corresponding pins are connected to ground. If the receiver is ON and no signal is received they are in high state by default. So, avoid the confusion, one pin (pin 10) is left free as reference. Whenever the receiver is ON, this pin goes high and at the same time all the other data pins go high. This gives 11 condition at the microcontroller input. But when the switch corresponding to the device is pressed the data pin gets connected to ground and transmits a 0. So, when required switch is pressed the microcontroller gets 10 condition from 10th and 11th pins of encoder. The microcontroller is programmed to activate the vibrator and shows the senders the number on receiving this 10 condition. The vibrator is connected to P1.0 of MSP430G2553 and P2 is completely used to control the LCD (Figs. 3, 4, 5, and 6).

3.2 Software Implementation

The device works in two modes (traffic mode and interaction mode). The system first checks whether the traffic mode is ON or not. If the traffic mode is ON, the input sound is compared with the threshold and the output terminal is activated if input is greater than the threshold value. If the traffic mode is OFF, the device again checks whether any number is pressed. If the number is pressed, it transmits the corresponding data. If no number is pressed, it checks whether any signal is received. If any signal is received, the device performs the action implied by the data. If the device does not receive any signal, the device again goes to the initial state and checks for the mode again to continue the action.

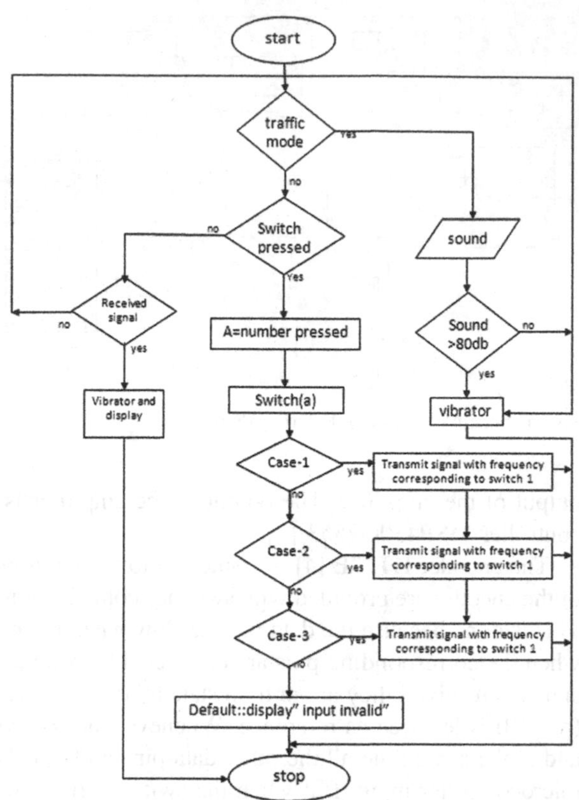

Fig. 4 Flowchart for software implementation

4 Results

Analog discovery kit (ADK) has been used to test the hardware at different stages of experiment. ADK is a portable device with oscilloscope, function generator, DC supplies, and all other basic devices that are required for an electronics lab. We have tested the microphone, filter, amplifier, and transmitter receiver pair using the oscilloscope available in ADK. Figures shown below are the waveforms obtained at different stages.

The output of the amplifier is shown in Fig. 7. The orange color waveform shows the signal from microphone and the blue color waveform shows the amplified output. Finally, the PCB of sound alert system is shown in Fig. 8 and the product with the end user is shown in Fig. 9.

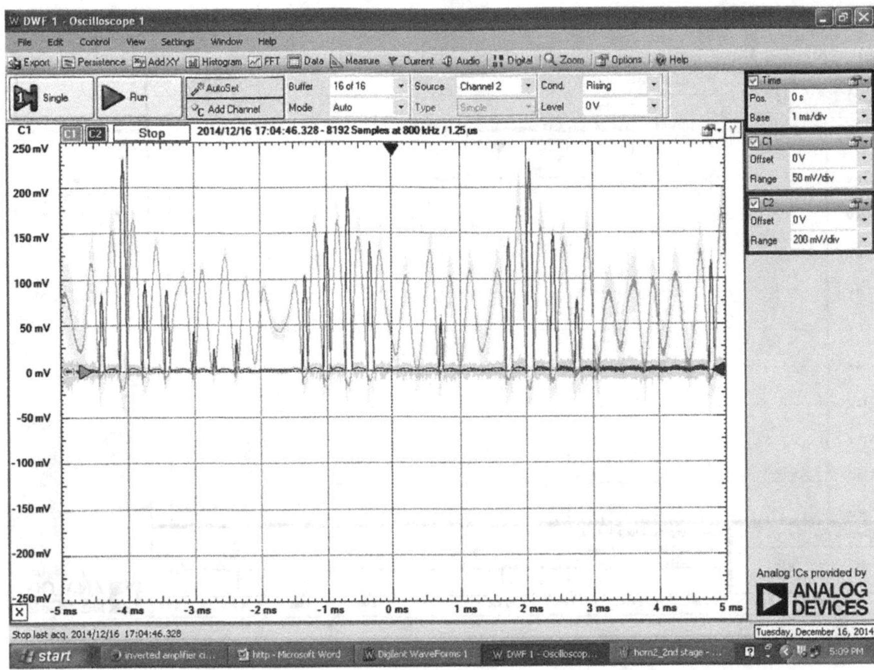

Fig. 5 Output of the amplifier

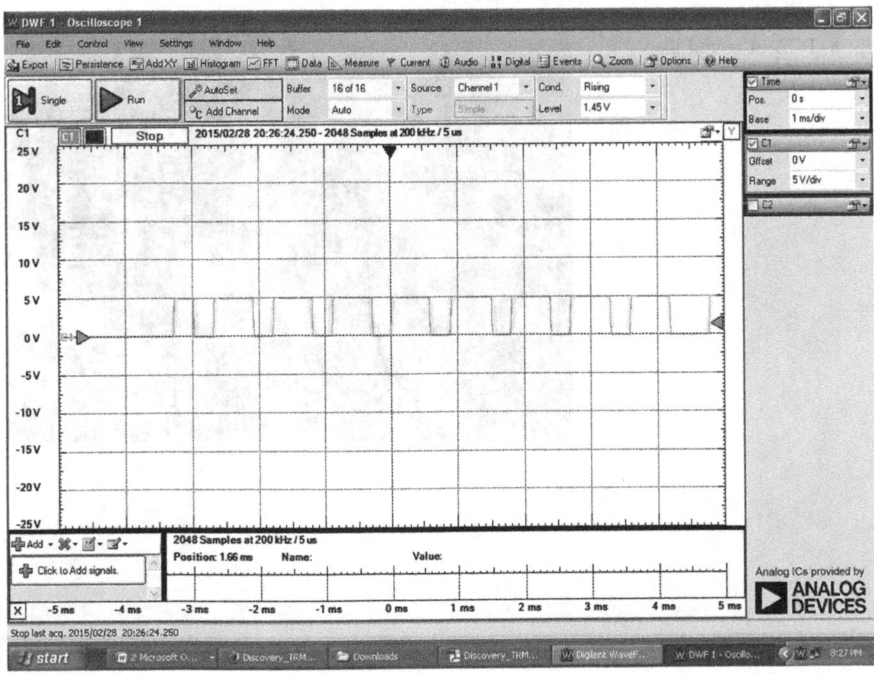

Fig. 6 Transmitter output

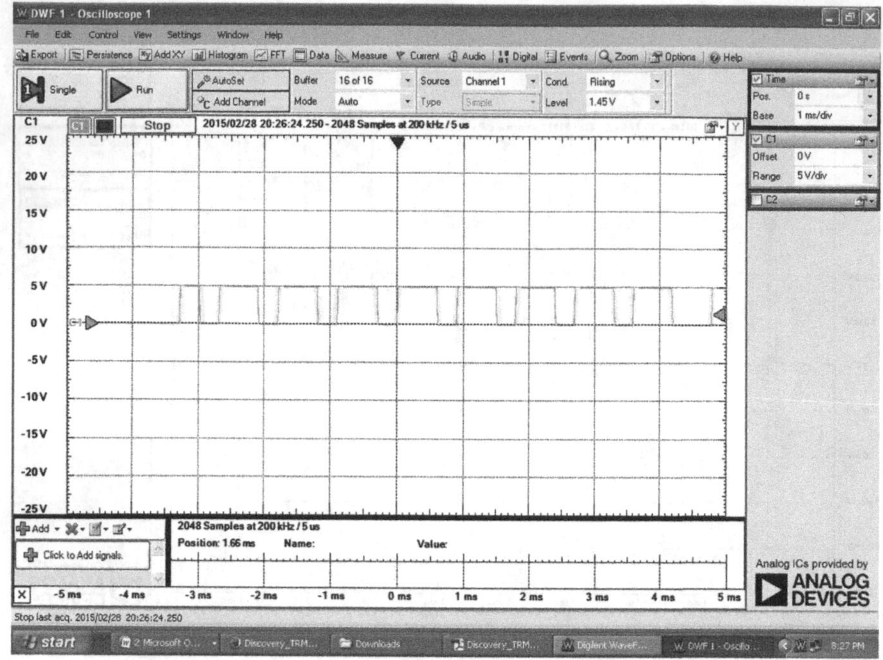

Fig. 7 Receiver output

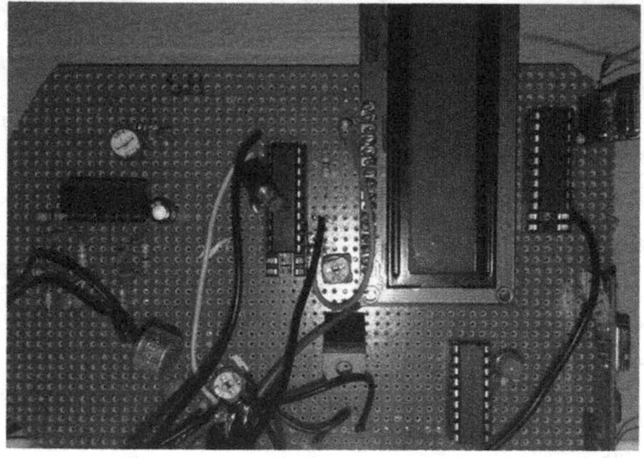

Fig. 8 PCB of sound alert system

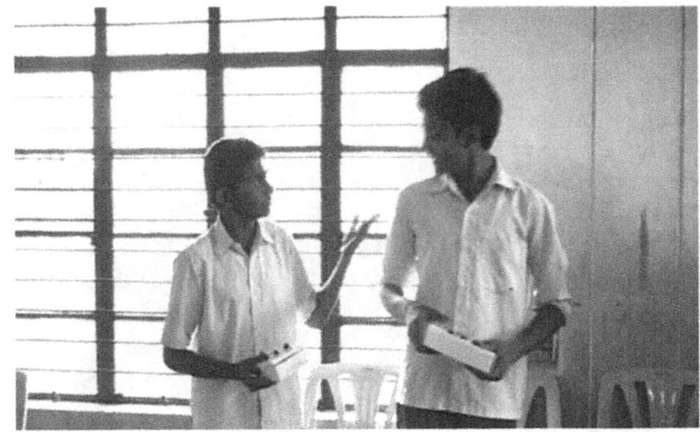

Fig. 9 Sound alert system with user

5 Conclusion

A wireless sound alert and person identity notification system for hearing challenged people is developed. It is tested for its working both under laboratory conditions and under real-world conditions and is found to be working satisfactorily. It is distributed to our clients at Sri Venkateswara Deaf School and they have been using it with full efficiency for the past 6 months.

Acknowledgments The author wishes to express her profound sense of gratitude to Sri. K. V. Vishnu Raju garu Chairman, SVES, for his encouragement toward innovation and research. She also would like to express her sincere thanks to Sri Ravichandran Rajagopal, Vice-Chairman, SVES, for supporting at all times of need. She also would like to thank Dr. G. Srinivasa Rao garu, Principal, SVECW for facilitating with a good research environment.

References

1. Konstantin Datz, "The wrist Band that Interprets sound for the Deaf" Deaf Times, 2010.
2. www.ti.com/lit/ds/symlink/lm124-n.pdf.
3. www.ti.com/lit/ds/symlink/msp430g2553.pdf.
4. www.farnell.com/datasheets/57850.pdf.
5. www.farnell.com/datasheets/56723.pdf.

Automatic Language Identification and Content Separation from Indian Multilingual Documents Using Unicode Transformation Format

Rajnish M. Rakholia and Jatinderkumar R. Saini

Abstract In Natural Language Processing (NLP), language identification is the problem of determining which natural language(s) are used in written script. This paper presents a methodology for Language Identification from multilingual document written in Indian language(s). The main objective of this research is to automatically, quickly, and accurately recognize the language from the multilingual document written in Indian language(s) and then separate the content according to types of language, using Unicode Transformation Format (UTF). The proposed methodology is applicable for preprocessing step in document classification and a number of applications such as POS-Tagging, Information Retrieval, Search Engine Optimization, and Machine Translation for Indian languages. Sixteen different Indian languages have been used for empirical purpose. The corpus texts were collected randomly from web and 822 documents were prepared, comprising of 300 Portable Document Format (PDF) files and 522 text files. Each of 822 documents contained more than 800 words written in different and multiple Indian languages at the sentence level. The proposed methodology has been implemented using UTF-8 through free and open-source programming language Java Server Pages (JSP). The obtained results with an execution of 522 Text file documents yielded an accuracy of 99.98 %, whereas 300 PDF documents yielded an accuracy of 99.28 %. The accuracy of text files is more than PDF files by 0.70 %, due to corrupted texts appearing in PDF files.

Keywords Gujarati • Indian language • Language Identification (LI) • Natural Language Processing (NLP) • Unicode Transformation Format (UTF)

R.M. Rakholia (✉)
R K University, Rajkot, Gujarat, India
e-mail: rajnish.rakholia@gmail.com

J.R. Saini
Narmada College of Computer Application, Bharuch, Gujarat, India
e-mail: saini_expert@yahoo.com

© Springer Science+Business Media Singapore 2017

S.C. Satapathy et al. (eds.), *Proceedings of the International Conference on Data Engineering and Communication Technology*, Advances in Intelligent Systems and Computing 468, DOI 10.1007/978-981-10-1675-2_37

1 Introduction

Language detection and language identification plays an important role in the field of Natural Language Processing (NLP). On the internet, written text is available in number of languages other than English and many document and web pages contain mix language text (more than one language in same document or same web page) such as Gujarati, Hindi, and English [1, 2].

1.1 Language Identification

Language identification is the task of automatically detecting the language present in a document based on the written text of the document and character encoding used in web page. Detecting multilingual documents and texts is also important for retrieve linguistic data from the internet. Language identification is the problem of classifying words and characters based on its language, it can be used for preprocessing stage in many applications (viz. parsing raw data, tokenizing text) to improve the quality of input data based on language specific model.

1.2 State of the Art (Language Identification)

Many methods and techniques with very high precision are available to identify popular languages in the world like English, German, Chinese, etc., from multilingual documents, but it cannot be applicable directly on resource poor languages due to its morphological variance and complex structure of framework such as Gujarati, Punjabi, and other Indian language.

1.3 Unicode Transformation Format

Unicode Transformation Format (UTF) is a standard character set which is used to display the character in proper format, which is written using different languages like: Gujarati, Hindi, Tamil, etc. These all are Indian languages which is not possible to display each character using American Standard Code for Information Interchange (ASCII), but it is possible to use in English. There are three different Unicode representations: 8-bit, 16-bit, and 32-bit encodings. UTF is supporting more diverse set of characters and symbols for different languages. We have used UTF for Indian language only, and it is mostly use in web technology and mobile application [3].

1.4 Essential of Language Identification

Many number of multilingual documents available on the internet in digital form in multilingual country like India. Different language has different framework and grammatical structure. Therefore, its need to automation tools which can identify the language(s) from written document and apply appropriate tool for further processing based on language(s) detect in document. A number of applications such as POS-Tagging, information retrieval, search engine, machine translation, accessibility of webpage, and other language processing activities require Language Identification as preprocessing step in multilingual environment.

1.5 Tools for Language Identification

Table 1 lists, number of tools (freely and commercially) available for automatic language identification.

2 Related Work on Language Identification

According to Verma and Khanna (2013) audio speech contains various information like gender, language spoken, emotion recognition, and phonetic information. They presented automatic language identification system using k-means clustering on MFCCs for features extraction and Support Vector Machine for classification. They

Table 1 Language identification tools

Sr. No.	List of tools available	Number of supported languages
1	Languid	72
2	Textcat	69
3	C# package for language identification of Microsoft	52
4	Xerox MLTT language identifier	47
5	Rosette language identifier by Basis Technology	30
6	SILC/Alis	28
7	Lid	23
8	Collexion	15
9	Stochastic language identifier	13
10	Langwitch by morphologic	7
11	Lextek language identifier	Many
12	Language identification program by ted dunning	2

tested proposed system on custom speech database for three Indian languages English, Hindi, and Tibetian. They achieved average classification accuracy of 81 % using small duration speech signals [4].

Anto et al. (2014) developed speech language identification system for five Indian languages, English (Indian), Hindi, Malayalam, Tamil, and Kannada. They had not used publicly available speech databases for these languages, but they created manually dataset by downloading YouTube audio file and remove the non-speech signals manually. They tested this system using created dataset consisting of 40 utterances with duration of 30, 10, and 3 s, in each of five target languages. They used 3, 4, and n g language models to implement this system. After experiment of this system, result shown that the use of 4 g language models can help enhance the performance of LID systems for Indian languages [5].

Yadav and Kaur S (2013) presented work related to identify different 11 regional Indian languages along with English from OCR corrupted text. They used distance measure-based metric to correct the text, naive Bayesian classification to identify the language of corrupted text and different n-gram model to represent the language. They tested this technique on different length of text, different n-gram (3, 4, and 5 g) language models and different percentage of corrupted texts [6].

Padma M et al. (2009) used profile features for language identify from multi-lingual document written in Indian languages. They have proposed to work on the prioritized requirements of a particular region, for instance in Karnataka state of Indian, English language used for general purpose, Hindi language for National importance, and Kannada language for State/Regional importance. They proposed very common concept in which they used bottom and top profile of character to identify languages from Indian multilingual document. In experimental setup they used 600 text lines for testing and 800 text lines for learning. They achieved average 95.4 % of accuracy [7].

Chanda S et al. (2009) proposed a scheme, to identify Thai and roman languages written in single document. They used SVMs-based method in proposed system to identify printed character at word level. They obtained accuracy of 99.62 %, based on the experiment of 10000 words [8].

According to Saha S et al. (2012), they studied and compared various feature reduction approaches for Hindi and Bengali Indian languages. They also studied different dimensionality reduction techniques which were applied on Named Entity Recognition task. Based on their analysis, they conclude that, Named Entity Recognition accuracy was poor for these languages. Performance of the classifier can be improved by dimensionality reduction [9].

Pati P et al. (2008) proposed algorithm for multi-script identification at the word level, they had started with a bi-script scenario which was extended to eleven-script scenarios. They used Support Vector Machine (SVM), Nearest Neighbor, and Linear Discriminate to evaluate Gabor and discrete cosine transforms features. They obtained accuracy of 98 % for up to tri-script cases, afterward they got 89 % accuracy [10].

Gupta V (2013) He had applied hybrid algorithm for Hindi and Punjabi language to summarize multilingual document. In proposed algorithm he had covered all most important features required for summarizing multilingual documents written in Hindi and Panjabi language and these features are: common part-of-speech tags like verb and noun, sentiment words like negative key word, position and key phrases, and named entities extraction. To identify weight of theses futures, he applied mathematical regression after calculating score of each features for each sentence. He got F-Score value of 92.56 % after doing experiment on 30 documents written in Hindi-Punjabi [11].

Hangarge M and Dhandra B (2008) they proposed a technique to identify Indian languages written in scanned version document based on morphological transformation features and its shape. They applied this technique on major Indian languages: Indian national language Hindi, old language Sanskrit and other two languages, and state languages Marathi, Bengali, and Assamese. They have created 500 blocks which contain more than two lines for each selected language. To decompose this blocks morphological transformation was used, after that they used KNN classifier and binary decision tree to classify these blocs. According to authors, this technique is quite different from other available technique for non-Indian language and they reported results were encouraging [12].

Padma M and Vijaya P (2010) they have proposed a method for language identification at the word level from trilingual document prepared using Hindi, English, and Kannada languages. The proposed method was trained by learning distinguish features of each language. After that they applied binary tree classifier to classify multilingual content. They obtained accuracy of 98.50 % for manually created database and average accuracy was found by 98.80 % [13].

3 Proposed Methodology

Based on the literature review and analysis of the tools available for Language Identification, we found that all researchers had used n-gram and other algorithm to identify particular language from multilingual document. We also analyzed that, these tools and methods cannot work for content separation. Existing work could not give proper and right output in case of mixed texts (for instance, "AअAઅAഅA") appears in single sentence of multilingual document.

But none of the researcher has used Unicode Transformation Format for Language Identification purpose. In our proposed methodology, we have used UTF-8 for language identification. Each character of each language written in multilingual document or in a webpage could be identifying by its unique Unicode value. In order to design a methodology for Indian languages, we created a list of few Indian languages with their range of Unicode value. This list is presented in Table 2, Unicode range is also covered vowel, consonant, reserved language specific characters, digit,s and various sign used in particular language [3].

Table 2 Unicode range for Indian Languages

Sr. No.	Indian languages	Unicode range
1	Gujarati	0A80–0AFF
2	Panjabi	0A00–0A7F
3	Tamil	0B80–0BFF
4	Oriya	0B00–0B7F
5	Telugu	0C00–0C7F
6	Kannada	0C80–0CFF
7	Malayalam	0D00–0D7F
8	Bengali and Assamese	0980–09FF
9	Kaithi	11080–110CF
10	Devanagari, Hindi, Marathi, Sindhi, Nepali and Sanskrit	0900–097F

Figure 1 shows diagrammatic representation of methodology and how to implement the proposed methodology for Indian languages.

3.1 Advantages

- This method can be applied for mixed texts that appear in single world or sentence (for instance, “AअAઅAഅA”).
- The proposed methodology is independent of font family of multilingual documents.
- It is also possible to implement this methodology in all most web technology other than JSP.
- It is free from the training phase.
- It can be extended for other language(s) by adding Unicode value in database.

3.2 Disadvantages

- It will lose the accuracy when multilingual document contain languages which has similar Unicode value, for instance languages Hindi, Marathi, and Devanagari (Table 2, Sr. No. 10).
- This methodology cannot be applied on scanned version of document.
- Loss of the accuracy in occurrence of mathematical sign, symbol, and special character appears in document.
- The proposed methodology losses the accuracy when corrupted text present in document.

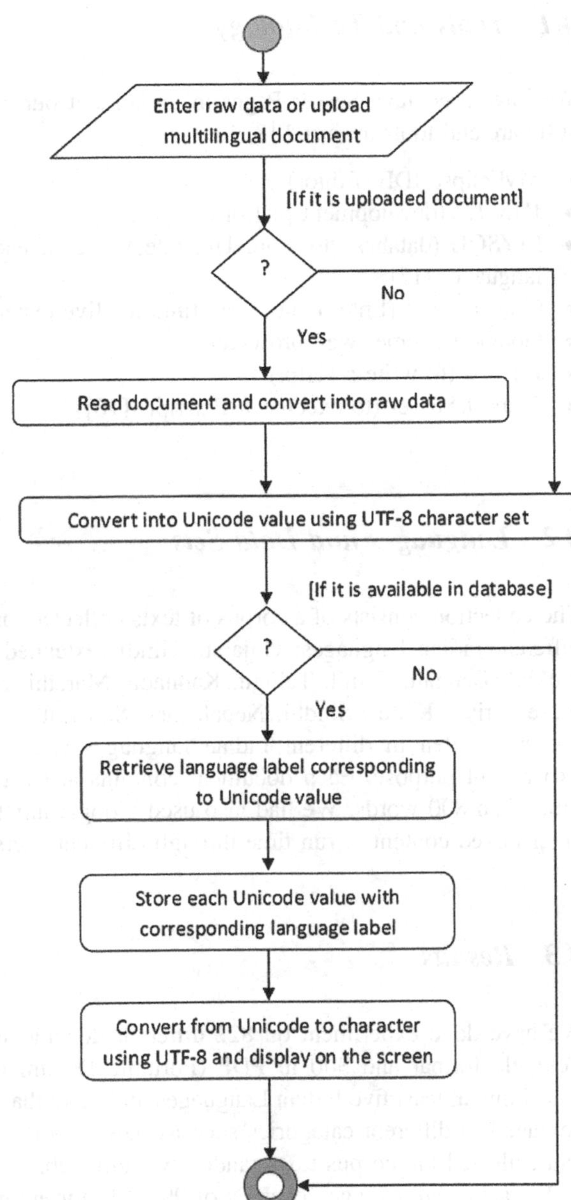
Enter raw data or upload multilingual document
[If it is uploaded document]
?
No
Yes
Read document and convert into raw data
Convert into Unicode value using UTF-8 character set
[If it is available in database]
?
No
Yes
Retrieve language label corresponding to Unicode value
Store each Unicode value with corresponding language label
Convert from Unicode to character using UTF-8 and display on the screen

Fig. 1 Flow of methodology

4 Experimental Results and Evaluation

We have described our methodology in this section; we constructed a matrix that contains all possible Indian language with their Unicode values of each character of each Indian language.

4.1 Tools and Technology

We have used Java Server Pages to implement our proposed methodology; other software and tools are used [8, 14]:

- MyEclipse IDE (Editor)
- JDK 1.7 (development platform)
- MYSQL (database to store Unicode value of each character of each Indian language) [15]
- Google input (Enter data at run time for live experiment)
- Google Chrome (web browser)
- JSP 2.0 (to write a script)
- Tomcat Server (to execute JSP script) [16]

4.2 Languages and Data Sets

The collection consists of a corpus of texts collected randomly from the web for 16 different Indian languages: Gujarati, Hindi (extended devanagari), Punjabi (Gurmukhi), Bengali, Tamil, Telugu, Kannada, Marathi, Malayalam, Kashmiri, Assamese, Oriya, Kaithi, Sindhi, Nepali, and Sanskrit. After that, we had mixed the content written in different Indian languages and prepared 822 documents for experiment purpose. Each document contains at least five Indian languages with more than 800 words. We had also used Google input tool for live experiment to enter mixed content at run time through different users.

4.3 Results

We have done experiment on 822 different documents in which 522 prepared in Text file format and 300 in PDF (Portable Document Format). Each document containing at least five Indian Languages and more than 800 words. The documents belonged to different categories such as news, sports, education, politics, etc. We had collected the corpus texts randomly from web.

We achieved average accuracy of 99.63 % for entire system in which accuracy obtain 99.98 % from text file format and 99.28 % from PDF format. Text file format losing average accuracy by 0.02 %, because of conjunctions appear in documents written in some Indian languages like Gujarat and Hindi. Sometime overwritten conjunctions cannot read by stream classes and such character get skipped by the system.

We have randomly selected four records from obtained result of entire system which is presented in Table 3. After analyzing the result for entire system, we found

Table 3 Results

Sr. No.	Indian languages	Total number of words		Word based accuracy (%)	
		Pdf file	*Text file*	*Pdf file*	*Text file*
1	Gujarati, Oriya, Tamil, Panjabi and Bengali	849	920	99.88	100
2	Panjabi, Hindi, Malayalam, Kannada and Gujarat	822	811	99.76	100
3	Marathi, Assamesem Bengali, Tamil and Gujarat	802	825	99.63	99.88
4	Hindi, Marathi, Devanagari, Gujarati and Panjabi	840	902	63.76	64.88

that text file accuracy was more than that of PDF file by 0.70 %. The reason of getting loss of accuracy in PDF file was corrupted text (character get overwritten at the time of PDF creation) appeared in portable documents which is not interpret by system and it will skip it.

5 Conclusion and Future Work

We have used 8-bit Unicode value for automatic Indian language identification and content separation from multilingual documents. The obtained results with an execution of 522 Text file document, we achieved accuracy of 99.98 % and for the PDF accuracy found 99.28 % with an execution of 300 documents. The accuracy of text files is more than PDF files by 0.70 %. The result showing that, proposed methodology can be applied for document classification and a number of applications such as POS-Tagging, information retrieval, search engine, and machine translation for Indian languages. In future, we will apply this proposed methodology in document classification for Indian language.

References

1. "Department of Electronics & Information Technology, India", *Indian Language Technology Proliferation and Deployment Center* [Online]. Available: http://www.tdil-dc.in/index.php?option=com_up-download&view=publications&lang=en [May 10, 2015].
2. "Ministry of Communication & Information Technology, India", *Technology Development for Indian Languages* [Online]. Available: http://ildc.in/Gujarati/Gindex.aspx [May 10, 2015].
3. "The Unicode Consortium, USA", *The Unicode Standard* [Online]. Available: http://www.unicode.org/standard/standard.html [May 10, 2015].
4. Verma, V. K., & Khanna, N. (2013, April). Indian language identification using k-means clustering and support vector machine (SVM). In Engineering and Systems (SCES), 2013 Students Conference on (pp. 1–5). IEEE.

5. Anto, A., Sreekumar, K. T., Kumar, C. S., & Raj, P. C. (2014, December). Towards improving the performance of language identification system for Indian languages. In Computational Systems and Communications (ICCSC), 2014 First International Conference on (pp. 42–46). IEEE.
6. Yadav, P., & Kaur, S. (2013, November). Language identification and correction in corrupted texts of regional Indian languages. In Oriental COCOSDA held jointly with 2013 Conference on Asian Spoken Language Research and Evaluation (O COCOSDA/ CASLRE), 2013 International Conference (pp. 1–5).IEEE.
7. Padma, M. C., Vijaya, P. A., & Nagabhushan, P. (2009, March). Language Identification from an Indian Multilingual Document Using Profile Features. In Computer and Automation Engineering, 2009. ICCAE'09. International Conference on (pp. 332–335). IEEE.
8. Chanda, S., Pal, U., & Terrades, O. R. (2009). Word-wise Thai and Roman script identification. ACM Transactions on Asian Language Information Processing (TALIP), 8(3), 11.
9. Saha, S. K., Mitra, P., & Sarkar, S. (2012). A comparative study on feature reduction approaches in Hindi and Bengali named entity recognition. Knowledge-Based Systems, 27, 322–332.
10. Pati, P. B., & Ramakrishnan, A. G. (2008). Word level multi-script identification. Pattern Recognition Letters, 29(9), 1218–1229. Chicago.
11. Gupta, V. (2013). Hybrid Algorithm for Multilingual Summarization of Hindi and Punjabi Documents. In Mining Intelligence and Knowledge Exploration (pp. 717–727). Springer International Publishing.
12. Hangarge, M., & Dhandra, B. V. (2008, July). Shape and Morphological Transformation Based Features for Language Identification in Indian Document Images. In Emerging Trends in Engineering and Technology, 2008. ICETET'08. First International Conference on (pp. 1175–1180). IEEE.
13. Padma, M. C., & Vijaya, P. A. (2010). Word level identification of Kannada, Hindi and English scripts from a tri-lingual document. International Journal of Computational Vision and Robotics, 1(2), 218–235.
14. H. Marti and B. Larry, "Accessing Database with JDBC," in *Core Servlets and Java Server Pages volume 1*, 2nd ed. Pearson Education, 2008, ch.17, pp. 499–599.
15. "MySQL", *Unicode Support [Online]. Available:* http://dev.mysql.com/doc/refman/5.5/*en/charset-unicode.html [September 6, 2014].*
16. "The Apache Software Foundation", *Apache Tomcat* 7 [Online]. Available: http://tomcat.apache.org/tomcat-7.0-doc/index.html [May 10, 2015].

Constrained Power Loss Minimization of DC Microgrid Using Particle Swarm Optimization

S. Angalaeswari and K. Jamuna

Abstract The optimal power flow has been considered as the important issue in the power system network. There are many ways to optimize the power flow out of which power loss minimization and voltage profile improvement are considered as the efficient ways. The power loss minimization for distributed system has been performed with particle swarm optimization (PSO) algorithm and the results are compared with forward/backward load flow method. The analysis is also performed with and without addition of renewable sources. The results show that the inclusion of DG at various buses reduces the power loss and improves the voltage magnitude profile

Keywords Microgrid · Distributed generation · Particle swarm optimization (PSO)

1 Introduction

The power generation from any country plays a vital role in improving the economy of that country. The maximum amount of power has been produced from conventional sources over long period of time. To satisfy the power demand and to make the world pollution free, the power sectors turn toward the nonconventional sources for the past one decade. When the renewable sources are placed nearer to the consumer, it greatly reduces the transmission loss and reduces the power electronic interfacing devices.

The sources that are located near to the load distribute the power to the load with the minimum distance and minimum effort; hence they are called as distributed

S. Angalaeswari (✉) · K. Jamuna
School of Electrical and Electronics Engineering, VIT University,
Chennai Campus, Chennai, Tamil Nadu, India
e-mail: angalaeswari.s@vit.ac.in

K. Jamuna
e-mail: jamuna.k@vit.ac.in

© Springer Science+Business Media Singapore 2017

S.C. Satapathy et al. (eds.), *Proceedings of the International Conference on Data Engineering and Communication Technology*, Advances in Intelligent Systems and Computing 468, DOI 10.1007/978-981-10-1675-2_38

energy sources. The cluster of distributed generators, energy storage devices, and loads are called as microgrids [1]. The microgrid is the best alternate solution for the conventional grid. The microgrid can be operated with public distribution grid in grid connected mode, or islanded mode, or transition mode. The microgrid can deliver power to the local loads more efficiently and reliably. If the power from the distributed generators in the microgrid is more than the load connected, the excess power can be given to the grid. Otherwise, the loads in the microgrid usually take power from the grid [2].

Even though the microgrid gives the reliable power to the consumers, the maximum efficiency from the sources is less due to the intermittent source availability of the nonconventional energy sources. From the available sources, the real power taken from the source to the load can be maximized using any optimization techniques. The best method to maximize the power flow is to minimize the real and reactive power losses in the system, thereby improving the voltage profile in the system. This paper is presented as follows: Sect. 2 gives the modeling of wind turbine, solar cell, and the fuel cell. Section 3 discusses the particle optimization algorithm which is used for minimization of the real power loss by considering the bus voltage limits as inequality constraint. Section 4 provides the load flow algorithm used in radial distribution system. Section 5 discusses the results of bus voltage and real power loss obtained from IEEE 33 bus system before and after introducing PSO. Section 6 gives the conclusion of the paper.

2 Modeling of Distributed Generators

The main advantages of distributed generators (DG) are transmission capacity relief, distribution capacity relief, grid improvement, improved grid asset utilization and increased reliability, reactive power (VAR) support, energy and load management, and voltage support.

3 Modeling of Wind Turbine

Wind energy is one of the most available and exploitable forms of nonconventional energy. The wind velocity will not be constant all the time. The maximum power extracted from the wind turbine has been estimated at around 59 % and it has been taken as Betz's limit [3].

(i) Area of the rotor (A) is given as

$$A = \frac{\pi d^2}{4}. \tag{1}$$

where d = diameter of the rotor.

(ii) Tip speed ratio λ is given as

$$\lambda = \frac{2\pi RN}{v_\infty}. \tag{2}$$

where R = radius of the swept area in m, N = rotational speed in revolutions per second and v_∞ = wind speed without rotor interruption in m/s

(iii) Torque coefficient is calculated from

$$C_T = \frac{Cp}{\lambda}. \tag{3}$$

C_p is assumed as 35 %.

(iv) The aerodynamic torque is calculated as

$$T_m = \frac{1}{2}\rho C_T \pi R^3 V\infty^2. \tag{4}$$

where ρ = air density $\approx$1.225 kg/m^3 at 15 °C and at normal pressure.

(v) The power available in the wind is given by

$$P_m = \frac{1}{2}\rho CpAV\infty^3. \tag{5}$$

4 Modeling of Solar Cell

A mathematical model is developed for the solar cell. The equivalent model of a solar cell is considered from Ref. [4]. There are many ways to represent the equivalent circuit of solar cell. Every circuit differs from others by the number of circuit components. In this paper, the single diode equivalent circuit is considered which includes current source in parallel with a diode, resistance R_s, and a shunt resistance R_{sh}. In the ideal condition, the R_s is very small, almost equal to zero and the R_{sh} is very large, almost equal to infinity. Then the current is given by the following equation: [using Kirchhoff's current Law] [5]

$$I = N_p I_{ph} - Np_{Id} \tag{6}$$

$$Iph = \{I_{sc} + k_1(T_c - T_{ref})\}\frac{\lambda}{1000}. \tag{7}$$

$$Id = Is\left[e^{\frac{qU}{NskTcA}} - 1\right]. \tag{8}$$

Table 1 Specifications of solar cell

I, Id	Output current from the solar cell, diode current in A
Iph, Isc	Current generated due to light illumination, short-circuit current (3.27)A
k_1	Cell's short-circuit current temperature coefficient, 0.0017A/°C
Tc, Tref	Cell temperature(350 K), reference temperature (301.18)K
λ, Is	Solar radiation, 1500 W/m^2, module saturation current, A
q, U	Electric charge of electron, 1.6 * 10^{-19} C, V_{oc}, voltage of PV cell, 400 V
k, A	Boltzmann constant, 1.38 * 10^{-23} J/K, p–n junction's ideality factor, 1.5
I_{rs}	Module reverse saturation current, 2.0793 * 10^{-6} A
E_g	Band gap energy of the semiconductor, 1.1 eV
N_s, N_p	Number of solar cells in series (800), in parallel (40)

$$Is = Irs\,[Tc/Tref]^3\, e^{\left[\frac{qEg[(\frac{1}{Tref}) - (\frac{1}{Tc})]}{kA}\right]}. \tag{9}$$

The power produced by the solar cell is (Table 1)

$$P = UI. \tag{10}$$

5 Modeling of Fuel Cell

Fuel cells are classified as power generators that can operate continuously when the fuel and oxidant are supplied. The reversible PEM (Poly ethylene membrane) fuel cell model is used in this paper. The theoretical cell voltage of the cell is given as [6]

$$U_{th} = -\frac{\Delta G}{nF}. \tag{11}$$

The theoretical (reversible) standard voltage is

$$U_0 = -\frac{\Delta G_0}{nF}. \tag{12}$$

$$\Delta G = \Delta H - T\Delta S. \tag{13}$$

The theoretical thermal voltage is given as

$$thermal = -\frac{\Delta H}{nF}. \tag{14}$$

Table 2 Specifications of fuel cell

ΔG, ΔH	Change in Gibbs energy, J/mol, change in enthalpy, −286 J/mol,
ΔS	Change in entropy, −0.1634 J/K/mol
n, F	Number of electrons per mole (2), Faraday's constant(96485.309 C/mol)
R, T	Universal gas constant(8.31451 J/K/mol), cell operating temperature, K
aH_2O,	Activity of species $H_2O \approx 1$
aH_2	Activity of species H_2 according to pressure, (1-thermal)0.8
$aO_2{}^{1/2}$	Activity of species O_2 according to pressure, (1-U_{th})0.2
I, Ns	Current from fuel cell (100)A, number of fuel cell stack(150)

The free energy can be expressed as

$$\Delta G = \Delta G_0 + RT \ln\left[\frac{aH_2O}{aH_2 \times aO_2{}^{1/2}}\right]. \tag{15}$$

The theoretical cell voltage is obtained as follows:

$$Uth = U_0 - \frac{RT}{nF}\ln\left[\frac{aH_2O}{aH_2 \times aO_2{}^{1/2}}\right]. \tag{16}$$

$$\text{Equivalent cell voltage} = 1.2297 + (T - 298.15)\frac{\Delta S_0}{nF} + \frac{RT}{nF}\ln\left[\frac{aH_2 \times aO_2{}^{1/2}}{aO_2^{3/2}}\right]. \tag{17}$$

The power generated = Ns × Equivalent cell voltage × I (Table 2).

6 Power Loss Minimization Using FW/BW Load Flow Method

The power loss is calculated on IEEE 33 bus system [7]. Since the nature of the system is radial, the normal load methods cannot be applied. The following assumptions are made in the considered system: (i) balanced radial distribution network for three phases; (ii) the system can be represented by the equivalent single-line diagram; and (iii) line charging capacitances are neglected at the distribution voltage levels [8].

In Fig. 1, V_i is the voltage at the sending end at an angle of δ_i; V_j is the voltage at the receiving end at an angle of δ_j; Z_{ij} is the series impedance of the transmission line; R_{ij} is the resistance of the line; and X_{ij} is the reactance of the line.

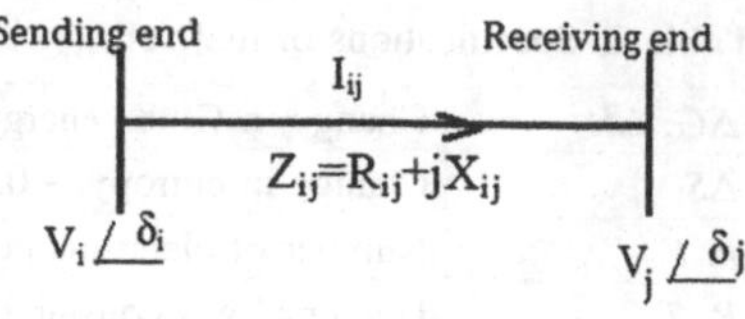

Fig. 1 Equivalent circuit of transmission line

The current flow in the transmission line can be calculated [7, 9] as

$$I_{ij} = \frac{V_i\angle\delta_i - V_j\angle\delta_j}{Z_{ij}}. \tag{18}$$

The real power and reactive power losses can be determined from the following equations:

$$P_{Lij} = r_{ij}\frac{Pij^2 + Qij^2}{Vj^2}. \tag{19}$$

$$Q_{Lij} = x_{ij}\frac{Pij^2 + Qij^2}{Vj^2}. \tag{20}$$

P_{ij}, Q_{ij} is the real, reactive power flowing in the line.

Algorithm for forward/backward load flow method:

1. Assume the flat initial voltages at all nodes for the first iteration.
2. Start with end node, and the node current can be computed using the following equation:

$$Iij = \left(\frac{Si}{Vi}\right)^*. \tag{21}$$

   * in Eq. (21) indicates the conjugate of the $\left(\frac{Si}{Vi}\right)$.
3. By applying KCL, the branch current from node i to node j can be calculated using the following equation in the backward sweep:

$$I_{i,i+1} = I_{i+1} + \sum \text{Currents in branches flowing away from node } i+1. \tag{22}$$

4. The voltage at ith bus(node) is computed by the following equation in the forward sweep mode:

$$V_i = V_{i+1} + (I_{i,i+1} \times Z_{i,i+1}). \tag{23}$$

5. The load current is updated with the new voltages and the real and reactive power loss is calculated. Repeat the above steps till the voltage difference between the successive iteration is less than the tolerance level.

7 Implementation of Particle Swarm Optimization (PSO)

PSO was first introduced by James Kennedy and Russell Eberhart in the year 1995. Particle swarm optimization is an optimization tool which is used to determine the optimum value of the objective function based on the particles position and velocity. Each individual in this algorithm is taken as particles and this particles move from the current position to new one by searching the optimum path. In every iteration, this particle identifies the best path to move from its current position from its own experience and its nearby particles movement. The direction of a particle is given as a set of nearby particles and its own experience [10].

Algorithm for PSO:

Step 1 Set the iteration count k as 0. Randomly generate NP particles, their initial position, and initial velocities.

Step 2 Evaluate the loss for each and every particle. If the voltage limits are satisfied, set the current position as particle best PB. Among all the particles, which one is having the minimum loss is considered as global best (GB). Otherwise, initialization should be repeated.

Step 3 Increment the Iteration count k = k + 1.

Step 4 Update the speed and update the current position.

Step 5 The particle best can be updated
if $fi(Xi^k) < fi(PBi^{k-1})$ then $PBi^k = Xi^k$ else $PBi^k = PBi^{k-1}$.

Step 6 The global best can be calculated as $f(GB^k) = \min\{fi(PBi^k)\}$
If $f(GB^k) < f(GB^{k-1})$ then $GB^k = GB^k$ else $GB^k = GB^{k-1}$.

Step 7 If the iteration reaches the maximum defined, then stop, or else go to step 3.

Flowchart: (Fig. 2).

8 Simulation Results

In an optimal power flow, the values of some or all of the control variables have to be determined so as to optimize the objective function. In this paper, the objective function is to minimize the real power loss in a radial distribution network (RDN) using PSO method. The algorithm is implemented on the IEEE 33 bus system. The bus voltages and the real power loss are calculated for the following three cases: (i) without PSO and without DGs; (ii) with PSO without DGs; and (iii) with PSO and DGs. For the analysis, the voltage limits considered are 0.95–1.05.

Case (i) Power Loss without PSO and without DGs (Load flow Analysis)

The real power loss has been calculated for IEEE 33 bus system by conducting forward/backward load flow analysis. Initially, the voltages at all buses are assumed with flat value of $1\angle 0°$. Then the load flow has been conducted by considering the

Fig. 2 Flowchart of particle swarm optimization

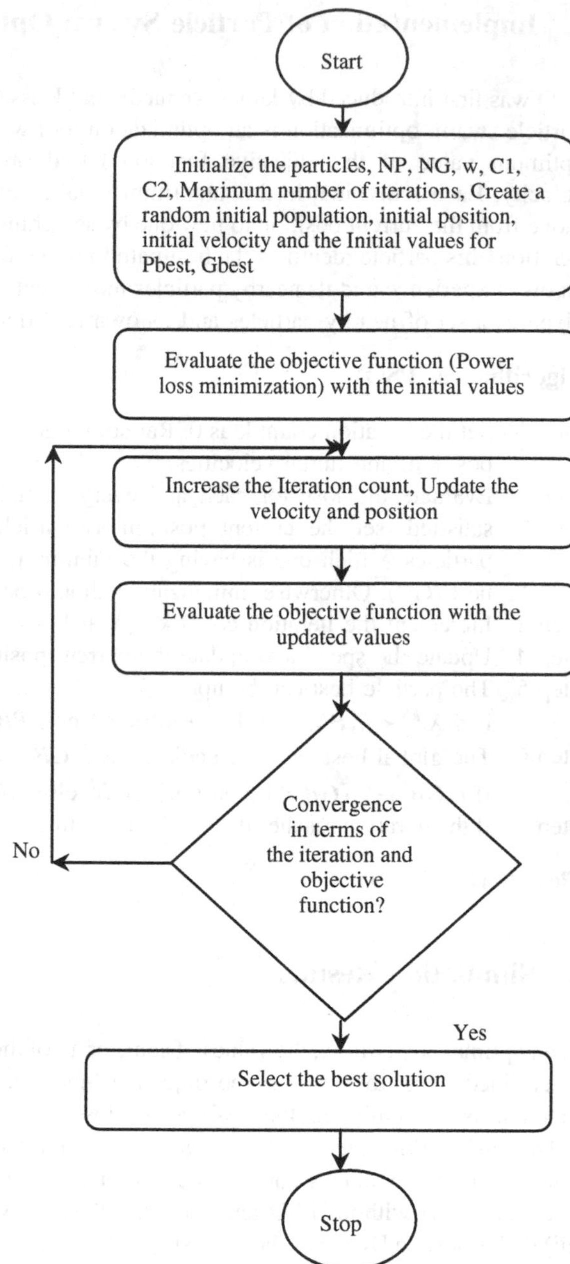

Table 3 Result of voltage magnitude profile for load flow (FBS method) and PSO without DG

Bus no	Voltage in p.u[a] without PSO and DGs (load flow)	Voltage in p.u[a] with PSO and without DGs	Bus no	Voltage in p.u[a] without PSO and DGs (load flow)	Voltage in p.u[a] with PSO and without DGs	Bus no	Voltage in p.u[a] without PSO and DGs (load flow)	Voltage in p.u[a] with PSO and without DGs
1	1.0000	1.0000	12	0.9177	0.9911	23	0.9793	0.9832
2	0.9970	0.9965	13	0.9115	0.9632	24	0.9726	0.9500
3	0.9829	0.9770	14	0.9092	0.9800	25	0.9693	0.9500
4	0.9754	0.9860	15	0.9078	0.9721	26	0.9475	0.9934
5	0.9679	0.9888	16	0.9064	0.9703	27	0.9450	0.9900
6	0.9495	0.9695	17	0.9044	0.9510	28	0.9335	0.9603
7	0.9459	0.9826	18	0.9038	0.9639	29	0.9253	0.9620
8	0.9323	0.9500	19	0.9965	0.9924	30	0.9218	1.0500
9	0.9260	0.9595	20	0.9929	0.9500	31	0.9176	0.9500
10	0.9201	0.9590	21	0.9922	0.9805	32	0.9167	0.9657
11	0.9192	0.9971	22	0.9916	0.9670	33	0.9164	0.9931

[a]p.u-per unit

constraints. The bus voltages are given in Table 3. The real power loss calculated from the load flow is 210.9 kW.

Case (ii) Power Loss with PSO and without DGs

The objective function of minimizing the real power loss has been determined by particle swarm optimization method in the radial network with the voltage constraints. The voltages are being improved by this algorithm. Even though this

Table 4 Result of voltage magnitude profile for PSO without and with DG

Bus no	Voltage in p.u[a] PSO without DGs	Voltage in p.u[a] PSO with DGs	Bus no	Voltage in p.u[a] PSO without DGs	Voltage in p.u[a] PSO with DGs	Bus no	Voltage in p.u[a] PSO without DGs	Voltage in p.u[a] PSO with DGs
1	1.0000	1.0000	12	0.9911	0.9911	23	0.9832	0.9833
2	0.9965	0.9965	13	0.9632	0.9637	24	0.9500	0.9500
3	0.9770	0.9770	14	0.9800	0.9804	25	0.9500	0.9500
4	0.9860	0.9860	15	0.9721	0.9725	26	0.9934	0.9970
5	0.9888	0.9889	16	0.9703	0.9705	27	0.9900	0.9900
6	0.9695	0.9699	17	0.9510	0.9532	28	0.9603	0.9609
7	0.9826	0.9843	18	0.9639	0.9643	29	0.9620	0.9625
8	0.9500	0.9500	19	0.9924	0.9925	30	1.0500	1.0500
9	0.9595	0.9600	20	0.9500	0.9500	31	0.9500	0.9500
10	0.9590	0.9594	21	0.9805	0.9807	32	0.9657	0.9665
11	0.9971	0.9971	22	0.9670	1.0204	33	0.9931	0.9963

[a]p.u-per unit

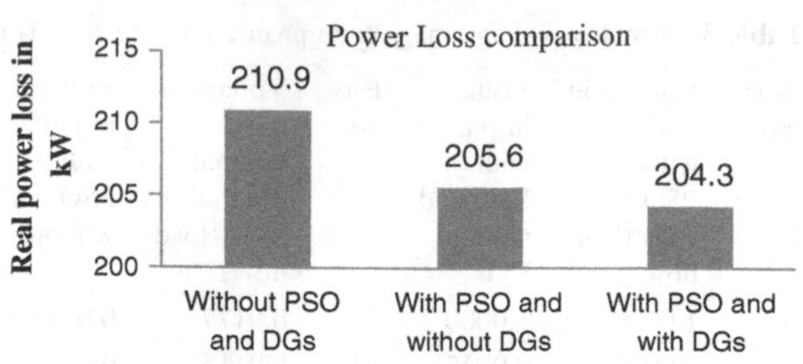

Fig. 3 Comparison of real power loss

algorithm gives unique result for every run due to the selection of random numbers in the velocity, the real power loss is always less than the value obtained from normal load flow. The average real power loss determined by this method is 205.6 kW. The voltage profile is given in Table 4.

Case (iii) Power loss with PSO and with DGs

In the third case, the distributed generators of three sources are considered. The DGs are placed randomly at three buses in the network. A wind turbine is connected at bus 32. Solar panel of series and parallel combinations is connected at bus 21. Fuel cell stack is connected at bus 25. The ratings of all DGs are calculated from the given specifications. After introducing DGs in the network, the real power loss has been reduced to 204.3 kW. The voltage profile is also improved a lot (Fig. 3).

9 Conclusion

In this paper, the real power loss has been minimized using PSO which has maximized the power flow of the network. Due to the introduction of the DG at specified buses, the power losses have been further reduced. The implementation of PSO algorithm and the incorporation of DG in the system reduce the real power loss from 2.51 % (Reduction in real power loss of PSO compared to load flow) to 3.12 % (Reduction in real power loss of PSO and DGs compared to load flow). Hence, the power flow is also maximized.

References

1. Pranay S. Shete., NirajkurnarS. Maurya., Dr. R. M. Moharil., Abhijit A. Dutta.,: Analysis of Micro-grid Under Different Loading Conditions. International Conference on Industrial Instrumentation and Control (ICIC) College of Engineering Pune, India. May 28–30, pp. 1120–1124 (2015).
2. Parimita Mohanty., G Bhuvaneswari., R Balasubramanian: Optimal planning and design of Distributed Generation based micro-grids. Industrial and Information Systems (ICIIS), 2012 7th IEEE International Conference. Chennai, India. 6–9 Aug, pp. 1–6 (2012).

3. S.N. Bhadra., D. Kastha., S. Banerjee: Wind Electrical Systems, Oxford University Press, 2005.
4. Martin Gahid, Pavol Spanik: Design of Photo voltaic Solar Cell Model for Standalone Rcnewable System, pp. no. 285–288, (2014).
5. N. Pandiarajan and Ranganath Muthu: Mathematical Modeling of Photovoltaic Module with Simulink, International Conference on Electrical Energy Systems (ICEES 2011), 3–5 Jan, pp. 314–319(2011).
6. Zehra Ural., Muhsin Tunay Gencoglu: Mathematical Models of PEM Fuel Cells, 5th International Ege Energy Symposium and Exhibition (IEESE-5), Pamukkale University, Denizli, Turkey 27–30 June (2010).
7. A. Michline Rupa., S. Ganesh: Power Flow Analysis for Radial Distribution System Using Backward/Forward Sweep Method, International Journal of Electrical, Computer, Electronics and Communication Engineering Vol: 8, No: 10, pp. 1540–1544(2014).
8. D. Das, H. S. Nagi, D. P. Kothari.: Novel Method for solving Radial distribution networks. IEE, Prot, Gener, Transm, Distrib, Vol. 141, No. 4 (1994).
9. A. D. Rana., J. B. Darji., Mosam Pandya: Backward/ Forward Sweep Load Flow Algorithm for Radial Distribution System, IJSRD Vol. 2, Issue 01, pp: 398–400(2014).
10. Qinghai Bai: Analysis of Particle Swarm Optimization Algorithm, Vol. 3, No. 1. Computer and Information Science. pp. 180–184 (2010).

Selective Cropper for Geometrical Objects in OpenFlipper

B. Maonica, Priyanka Das, Pravin B. Ramteke and Shashidhar G. Koolagudi

Abstract Computer graphics remains one of the most exciting and rapidly growing computer fields. It includes geometry processing as a major part of it. Every element in Graphics can be processed using different algorithms for acquisition, reconstruction, analysis, manipulation, simulation, and transition of simple, primitive, and complex structures. One such commonly used function is cropping/clipping of geometrical objects. In this paper, an approach has been proposed for cropping a 3D object. This algorithm allows users to crop out selective portions of geometrical objects based on certain constraints like the axis, position, and amount to be cropped. The proposed algorithm has been provided as a plugin to the open-source software OpenFlipper and the results of the crop algorithm have been presented.

Keywords Cropping · Clipping · Selective cropper · OpenFlipper

1 Introduction

Computer graphics has become a common element in user interfaces, data visualization, motion pictures, and many other applications. Hardware devices and algorithms have been developed for improving the effectiveness, realism, and speed of picture generation. One of the widely researched areas of graphics includes geometry processing or mesh processing. A mesh can be defined as a representation of a geometric object as a set of finite elements [1]. Processing of such elements includes

B. Maonica (✉) · Priyanka Das · P.B. Ramteke · S.G. Koolagudi
National Institute of Technology Karnataka, Surathkal, India
e-mail: maonicarao@gmail.com

Priyanka Das
e-mail: priyanka.d9359@gmail.com

P.B. Ramteke
e-mail: ramteke0001@gmail.com

S.G. Koolagudi
e-mail: koolagudi@nitk.edu.in

© Springer Science+Business Media Singapore 2017

S.C. Satapathy et al. (eds.), *Proceedings of the International Conference on Data Engineering and Communication Technology*, Advances in Intelligent Systems and Computing 468, DOI 10.1007/978-981-10-1675-2_39

knowledge from interdisciplinary fields involving concepts from computer science, applied mathematics, mechanics, physics, and engineering to develop efficient algorithms for the acquiring, reconstructing, analyzing, manipulating, simulating, and transmitting simple, primitive, and complex geometrical models which can be 2D or 3D. The results of such processing are applied in classical CAD, interactive editing of shapes, and simulating movie productions or medical applications.

Many functionalities are dealt within computer graphics, ranging from transformations, to circle drawing to cropping. A cropping algorithm is very useful in cropping the object precisely and as per user's requirement. Cropping can be used in editing physical photograph, artwork or film footage, or in image editing software. The selection of a proper dimension and axis is a crucial in efficient cropping of the object.

In this paper, an approach for cropping a 3D object is proposed which selectively crops out a section of an object or all of objects in the scene based on the constraints such as the choice of axis, position, and amount. The algorithm is tested on the open-source software OpenFlipper developed by Prof. Leif Kobbelt as it provides a platform for processing a variety of geometry processing algorithms [2].

The rest of the paper is organized as follows. A detailed review of the research studies in this area is explained in Sect. 2. Methodology of the proposed algorithm is discussed in Sect. 3. Results are analyzed in Sect. 4. Section 5 concludes the paper with some future directions.

2 Literature Survey

Over the years, as computer graphics algorithms progressed, there have been a plethora of operations that can be applied to geometrical objects.

In case of 3D computer graphics, there are two kinds of projections for viewing 3D objects (refer Figs. 1 and 2). The clipping plane is the parallel projection, while the perspective projection is the way 3D graphics is viewed. One important function is the 'Cropping/Clipping' function which selectively removes a part of the object from the view. In case of cropping, the region of the object lying outside the boundaries of the 3D projection is cropped or removed.

Most widely used algorithm for clipping lines in 2-dimension or 3-dimension space is "Cohen-Sutherland" line-clipping algorithm. The algorithm was developed in 1967 during flight simulator work by Cohen and Sutherland [3]. It removes partially visible objects against another bounding object. The conditions for the visibility of lines depend on the coordinates of the line when compared to the maximum and minimum constraints of the bounding window coordinates. The 2D space is divided into nine regions as shown in Fig. 3 and outcodes are generated to efficiently determine the lines and portions of line that are in the center region of interest and need to be retained. The outcodes are computed as per the position of the coordinates when compared to the above 7-section design. The outcode is defined by four bits for two-dimensional clipping. The bits in the 2D outcode represent top, bottom, right,

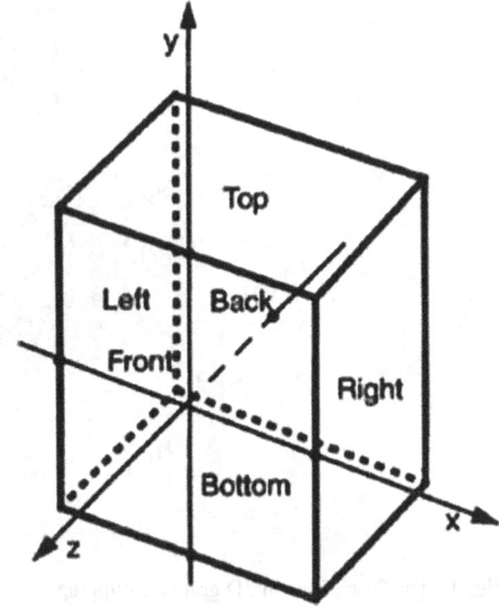

Fig. 1 The parallel projection [3]

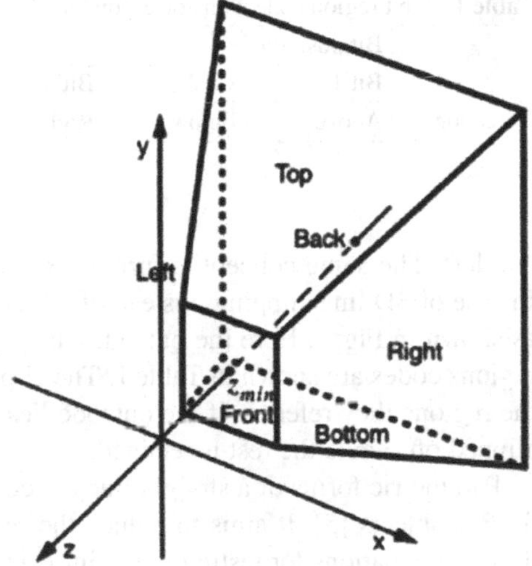

Fig. 2 The perspective projection [3]

Fig. 3 The nine regions in 2D graphics clipping

1001	1000	1010
0001	0000	0010
0101	0100	0110

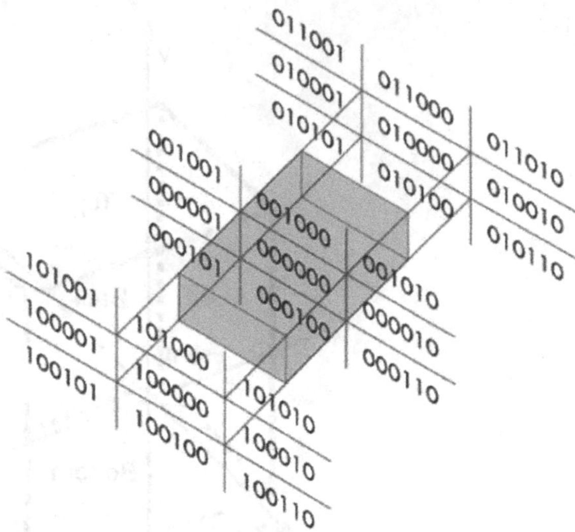

Fig. 4 The 27 regions in 3D graphics clipping

Table 1 6-bit region code with corresponding directions

	Bit position					
	Bit 1	Bit 2	Bit 3	Bit 4	Bit 5	Bit 6
Direction	Above	Below	Right	Left	Behind	Front

and left. The same concept is further extended to implement 3D line clipping [4]. In case of 3D line clipping, instead of 7 blocks, the region is divided into 27 blocks as shown in Fig. 4. Here the outcodes are represented using 6 bit. The bits in these regions codes are shown in Table 1. The clipping is done based on the outcodes and the regions they refer to. If the outcode lies beyond the clipping window, then it is clipped off, while the rest is retained.

Parametric forms of a straight line are considered to design a clipping algorithm for 3D objects [5]. It aims to reduce the redundant calculations. It is followed by deriving equations for testing if a point is inside the window or not. After this step, the new parameter values for visible portion of line segments are computed. Finally, the visible portion of the line segment is displayed. It is claimed to be more efficient than Cohen–Sutherland algorithm. Another classical approach for polygon clipping is the Sutherland–Hodgeman polygon-clipping algorithm [6]. It takes a list of all vertices in the subject polygon as the input. As the next step, one side of the clip polygon is extended infinitely in both directions, and the path of the subject polygon is traversed. Vertices from the input list are inserted into an output list if they lie on the visible side of the extended clip polygon line, and new vertices are added to the output list where the subject polygon path crosses the extended clip polygon line. This process is repeated iteratively for each clip polygon side, using the output list from one stage as the input list for the next. Once all sides of the clip polygon have been processed, the final generated list of vertices defines a new single polygon that is entirely visible. If the subject polygon is concave at vertices outside the clipping polygon, the new polygon may have coincident (i.e., overlapping) edges—this will be acceptable for rendering, but not for other applications such as computing shadows using this algorithm.

The Weiler–Atherton clipping algorithm is used in computer graphics. It can be extended to 3D graphics through Z-ordering which improves efficiency through visible surface determination [7]. The objects must be present in a clockwise manner and should be non self-intersecting. Coordinates of the clip region as well as the polygon to be clipped are saved in a two lists which also show the position of the coordinates, either inside or outside. The lists are used to match all the polygon intersections and linking the lists at the intersections. If there are no intersections, then clipping or merging is required. Some polygon combinations may be difficult to resolve, especially when holes are allowed.

From the literature it is observed that the previous approaches are not easily applicable for the 3D object or polygon processing. Also, all constraints needed for the cropping are not covered. We have proposed the 'Selective Cropper' algorithm which operates in three-dimensional space and on many types of primitive objects. It also includes three specific constraints based on which the cropping is done, namely, the choice of axis, position, and amount to be cropped.

3 Methodology

The proposed algorithm selectively crops out portions of an object, group of objects, or all the objects in the scene. The process of cropping has been depicted in Fig. 5.

The various constraints imposed on cropping are choice of axis, position, and amount of cropping.

1 **Choice of axis**: Cropping can be performed by choosing any one of the axes—"X axis," "Y axis," and "Z-axis." Cropping by choosing the X-axis involves the cropping of objects along YZ plane (i.e., the plane perpendicular to the X-axis). Similarly, cropping by choosing the Y-axis involves the cropping of objects along

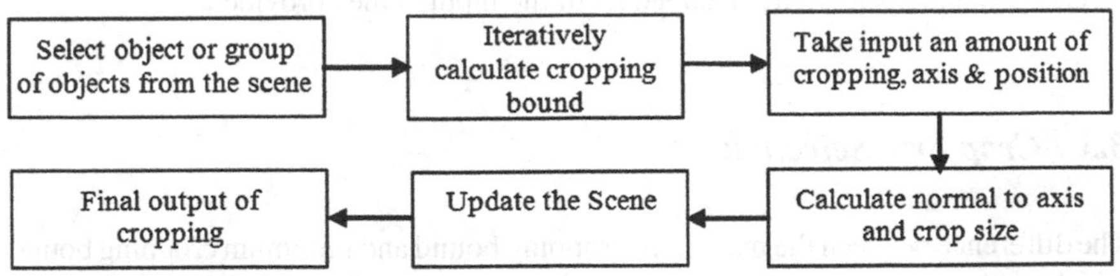

Fig. 5 Flow chart depicting the process of cropping

XZ plane and cropping by choosing the Z-axis involves the cropping of objects along XY plane. X-axis is considered for cropping if axis is not provided.
2 **Position**: This constraint allows to choose whether to crop from both sides, just from the right side, or just from the left side. If the position slider is in the middle, cropping is done from both sides. On moving the slider to the right, cropping is done from the left side, the amount increasing as the slider moves further right. Similarly, on moving the slider to the left, cropping is done from the right side, the amount increasing as the slider moves further left. By default, the value of the position constraint is moderate.
3 **Amount of cropping**: This constraint allows to define the amount of cropping to be performed. By default, the amount slider is at the right, no cropping is done, and entire object is retained. As the slider is moved left, the amount of cropping done gradually increases. If the slider is at the left extreme, maximum amount of cropping is done as most of the object is cropped.

3.1 Cropping Bound Selection

Cropping bound and cropping window are calculated by iterating over all objects. The minimum cropping bound and maximum cropping bound are calculated for various axes in the following way by considering the current scene size, both minimum and maximum values of each.

The minimum cropping bound along the X-axis is calculated by finding out the minimum of the current value of minimum cropping bound and the current minimum scene size along the X-axis. The maximum cropping bound along the X-axis is calculated by finding out the maximum of the current value of maximum cropping bound and the current maximum scene size along X-axis. Similar process is followed for cropping along Y-axis and Z-axis respectively. The scene variables are updated on multiple passes on the object which is specified in a 'for' loop.

3.2 Calculating Normal

The normal to the selected axis of cropping is calculated. The values of amount to be cropped and position are obtained from the input values provided.

3.3 Crop Size Selection

The difference between the maximum cropping bound and minimum cropping bound, which represents the size of the object in the direction of the cropping bound, is calculated. This difference is multiplied with the size obtained from the size slider.

Position is calculated using normal position value from the values of the position constraint and offset.

Crop size along X-axis is calculated by multiplying the current crop size along the X-axis and the difference between the maximum cropping and the minimum cropping bound along the X-axis. Similarly, the crop size along the Y-axis is calculated by multiplying the current crop size along the Y-axis and the difference between the maximum cropping bound and the minimum cropping bound along the Y-axis.

Similarly, the crop size along Z-axis is calculated by multiplying the current crop size along the Z-axis and the difference between the maximum cropping bound and the minimum cropping bound along the Z-axis.

3.4 Updation

The cropped section is calculated and the view is updated so that it gets reflected on the scene.

4 Results

The functionality of the selective cropper is applied to a variety of geometrical objects. Test cases include existing primitives such as cylinder, pyramid, sphere, cube, cuboid, octahedron, tetrahedron, etc. The cropping has been done along all the three axes, and also amount and position constraints have been applied to get different sets of results. The open-source software OpenFlipper has been used to test the proposed algorithm. The algorithm is implemented as an additional plugin. In order to load and unload plugins at runtime, they are required to be in Qt plugin implementation, which involves the usage of QtScript [8]. Further, to build the system, place the plugins in the right directories cMake [9]. cMake is cross-platform, free, and open-source software that is used to manage build process of a software [10]. The implementation of this plugin is based on the documentation available online for another plugin, the mesh smoother [11].

Figure 6a shows the original cylinder under consideration, with default amount and position constraints. On applying selective cropping along the X-axis, it is observed that cropping has been performed on the YZ plane as shown in Fig. 6b. The constraints of position and amount are kept as default. Then selective cropping along the Y-axis is performed, which leads to cropping in the XZ plane as shown in Fig. 6c. The cropping axis is kept as the Z-axis, while the position constraint is slid to the left while the amount slider is in the middle. The cropped image is depicted in Fig. 6d. The cropping axis is kept as the Z-axis while the position constraint is slid to the left while the amount slider is in the middle. The cropped image is depicted in Fig. 7a. And in the next case, the position slider is slid to the right while the amount slider is left in the middle as shown in Fig. 7b. Now, the position constraint is left as

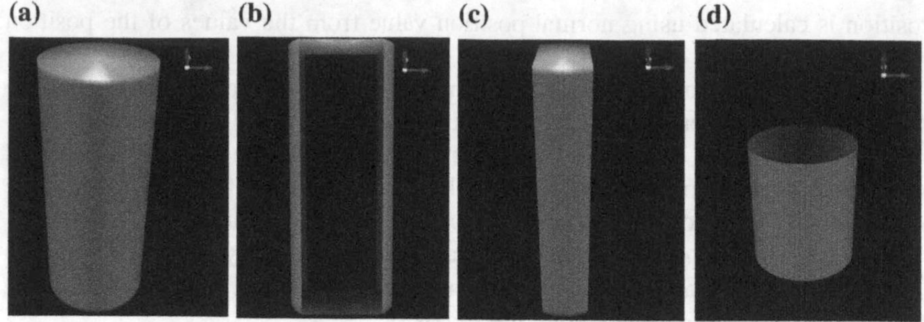

Fig. 6 **a** Original cylinder (with default position and amount constraints). **b** Cylinder after selective cropping along X-axis (with default position and amount constraints). **c** Cylinder after selective cropping along Y-axis (with default position and amount constraints). **d** Cylinder after selective cropping along Z-axis (with default position and amount constraints)

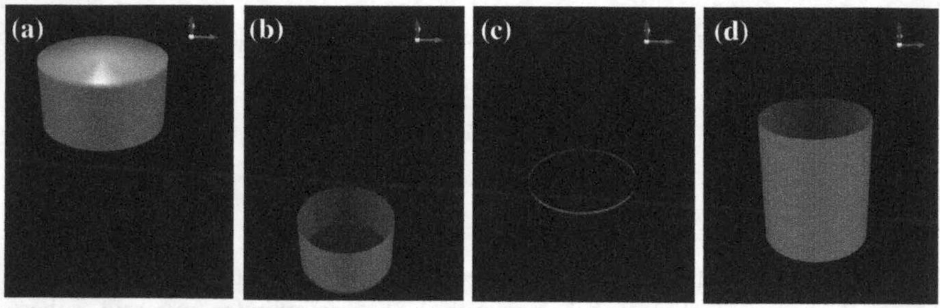

Fig. 7 **a** Cylinder after selective cropping along Z-axis with position slider at the *left* and amount slider in the *middle*. **b** Cylinder after selective cropping along Z-axis with position slider at the *right* and amount slider in the *middle*. **c** Cylinder after selective cropping along Z-axis with amount slider at the *left* and default position. **d** Cylinder after selective cropping along Z-axis with amount slider to the *right* and default position

default and the amount slider is slid to the extreme left as shown in Fig. 7c. Further, the position constraint is left as default and the amount slider is slid to the extreme right. The cropped image is shown in Fig. 7d.

5 Conclusion and Scope for Future Work

Our proposition includes the implementation of a new functionality called the selective cropper for geometrical objects which performs the task of cropping out selective regions of an object, a group of objects or all the objects in the scene based on certain constraints such as choice of axis, position, and amount. We have used the plugin programming feature of OpenFlipper and tested the functionality in the software in the form of a plugin. Results of the same have also been shown in the above sections.

In this work, the algorithm was designed to work as a plugin for the software OpenFlipper. Further, the algorithm can be tested on other graphics softwares. The functionality provided by the plugin can be improved by specifying more constraints of cropping in the future. Cropping based on the user-specified coordinates for a bounding region has not been implemented and is a direction for future work.

References

1. Bielefeld University (2015, Jan 12) Graphics and Geometry Group [Online]. Available: http://graphics.uni-bielefeld.de/research/
2. Jan Mobius and Leif Kobbelt, "OpenFlipper: An Open Source Geometry Processing and Rendering Framework", Curves and Surfaces: 7th International Conference, Avignon, France, June 24-30, 2010, pp. 488–500.
3. Cohen, Y.D., and Sutherland, B., "A New Concept and Method for Line Clipping", ACM Transactions on Graphics, 3(1):1–22, January 1984
4. R. Kodituwakku, K. R. Wijeweera, M. A. P. Chamikara, "An efficient line clipping algorithm for 3D Space", International Journal of Advanced Research in Computer Science and Software Engineering, May 2012
5. Liang, Y.D., B.A., Barsky, and M. Slater, Some Improvements to a Parametric Line Clipping Algorithm, CSD92-688, Computer Science Division, University of California, Berkeley, 1992
6. James, Andries van Dam, Steven Feiner, and John Hughes. "Computer Graphics: Principle and Practice". Addison-Wesley Publishing Company. Reading, Massachusetts: 1987. pages 689–693.
7. Weiler and Atherton. "Computer Graphics: Principle and Practice". Addison-Wesley Publishing Company. Reading, Massachusetts: 1987. pages 595–605
8. Qt Documentation, Qt Script [Online] Available: http://doc.qt.io/qt-5/qtscript-index.html
9. CMake, [online] Available: http://www.cmake.org/overview
10. Ogre Forums, Introduction to Cmake-The Ogre Build System [online] Available: http://www.ogre3d.org/tikiwiki/tikiindex.php?page=getting+started+wi-th+cmake
11. Developer Documentation, Implementing a Mesh Smoother Plugin, [Online], Available: http://www.openflipper.org/media/Documentation/OpenFlipper-1.2/ex2.html

Opinion Mining Feature Extraction Using Domain Relevance

Jawahar Gawade and Latha Parthiban

Abstract The feature extraction of opinions from online user reviews is a task to identify on which features user is going to write a review. There are number of existing approaches for opinion feature identification but, they are extracting features from a single review corpus. These techniques ignore the nontrivial disparities in distribution of words of opinion features across two or more corpora. This proposed work discusses a novel method for opinion feature identification from online reviews by evaluation of frequencies in two corpora, one is domain-specific and other is domain-independent corpus. This disparity is measured using domain relevance. The first task of this proposed work is to extract candidate features in user reviews by applying a set of syntactic dependence rules. The second task is to measure intrinsic domain relevance and extrinsic domain relevance scores on the domain-independent and domain-dependent corpora, respectively. The third task is to extract candidate features that are less generic and more domain-specific, are then conformed as opinion features.

Keywords Intrinsic domain relevance · Domain-dependent · Domain-independent · And candidate features · Opinion feature · Extrinsic domain relevance

1 Introduction

The opinion mining refers to collect different types of opinions of people over the same issue from web. Mostly, the reviews are the best way to express users opinion on web. The sale of product is depending on the opinion features extracted from

Jawahar Gawade (✉)
Computer Science Engineering Bharath University, Chennai, India
e-mail: jawahar009@gmail.com

Latha Parthiban
Department of Computer Science, Pondicherry University CC, Pondicherry, India

© Springer Science+Business Media Singapore 2017
S.C. Satapathy et al. (eds.), *Proceedings of the International Conference on Data Engineering and Communication Technology*, Advances in Intelligent Systems and Computing 468, DOI 10.1007/978-981-10-1675-2_40

user reviews. This technique provides accurate results which are helpful in customer decision-making process.

Nowadays, opinion mining is a challenging task with increased use of social media. The sale of any product and its marketing is based on the customer reviews about that product. This opinion mining is done in different levels. For example, consider a review for a "cell phone" given by some customer as "the external is very beautiful, also not costly, though the battery is poor; I still resolutely recommend this cell-phone." For this review document-level opinion mining [1] detects the overall subjectivity expressed on product in a comment or review document and it does not associate review and opinion with product-specific component. Although above review expresses a general positive opinion on the cell phone, it also contains contradictory opinions associated with different components of the cell phone. The opinion orientations for the "cell phone" itself and its "external" are positive, but the opinion polarity for the component of "battery" is negative. Such opinions may very well tip the balance in purchase decisions. Consumers are unsatisfied with just the overall opinion of product rating. They want to know why it receives the rating [2]. In opinion mining, an opinion feature indicates a product or a component of a product on which customer express their opinions. In current paper, this work recommends a novel method to the detection of such features from unstructured textual reviews. A novel technique is proposed to detect opinion features by exploiting their distribution discrepancies across different corpora. This work proposes how to evaluate the domain relevance of an opinion feature across two domains [3].

2 Literature Review

There are number of existing techniques for mining opinion features. These techniques involves: allowing Latent Dirichlet allocation (LDA), Association rule mining (ARM), Mutual reinforcement clustering (MRC), Dependency parsing (DP). This work will discuss each of these techniques bellow:

i. Latent Dirichlet Allocation (LDA) [4]:
This technique is generative probabilistic model for gathering of discrete data such as text domains. It is three-level hierarchical Bayes model in which each item of gathered items is modeled as a finite combination over an underlying topics set. Each topic is modeled as an infinite combination over an underlying topic probabilities set. The probabilities of topic provide an explicit presentation of a document in the context of text modeling. It presents efficient approximate inference techniques based on variation methods and for empirical Bayes parameter estimation an EM algorithm is used. This work report results in text classification, document modeling, comparing to a mixture of unigrams model, collaborative filtering, and LSI probabilistic model.

ii. Association Rule Mining (ARM) [5]:
Company selling products on the web site often ask their users to review the products that they have bought and the services associated. Our task is performed in

various steps: (1) Product features mining that have been commented on by users; (2) Opinion sentences in every review are identifying and deciding whether every opinion sentence is negative or positive; (3) summarization of outputs. There are various novel methods to perform these work; proposed system experimental outputs using opinions of a number of products sold online determine the effectiveness of the methods.

iii. Mutual Reinforcement Clustering (MRC) [6]:
The activities on feature-level sentiments trust on determining the explicit relation between opinion words in reviews and product feature words. The sentiment relations are difficult between the two objects. In this paper, this work proposes a novel mutual reinforcement approach to operate with the feature-level opinion mining. Specially,

(1) The approach clusters product features and opinion words iteratively and concurrently by fusing both their content details and information of sentiment links.
(2) Based on the product feature classifications and opinion word classifications, it make the sentiment association set between the two categories of data objects by strongest sentiment links determination.

Based on the preconstructed association set proposed system approach can predict opinions relating to different product features. It generates opinion evaluation more accurately.

iv. Dependency Parsing (DP) [7]:
User-generated content is a kind of novel media content created by end customers. In this paper, there are two main subtasks of opinion mining which are topic extraction and sentiment categorization. This task inputs the blog data from web which is a typical application of UGC. The information evaluate in our experiments and the outcomes show that our methods to the two tasks are challenging. A many approaches have been proposed to extract features in opinion mining. Supervised learning model may work better in a given corpus and if it is applied to other corpus then the model must be retrained [8, 9]. Unsupervised natural language processing (NLP) methods [7, 10, 4] identify opinion features by defining syntactic rules for domain-independent that capture the local context of the feature terms and dependence roles.

However, rules do not work well on real-life reviews, which lack formal structure. Existing extract opinion features by mining statistical patterns of feature terms only in the single review corpus, without considering their distributional characteristics in another different corpus [11, 12]. This technique is to find the distributional structure of an opinion feature in a given domain-dependent review corpus, for example, cell phone reviews, is different from that in a domain-independent corpus. For instance, the opinion feature battery occurs frequently in the domain of cell phone reviews, but not as frequently in the domain-irrelevant culture article collection.

Ho and Jin define a Novel Lexicalized HMM-Based Learning Framework for Web Opinion Mining [8]. Company selling products on the web regularly request their customers to share their reviews and experiences on products they have purchased. In this research, sentiment expressions and sentences are also recognized and orientations of opinion for each product entity are categorized as negative or positive [8]. This system suggests a novel machine learning framework using HMMs (lexicalized). This system approach naturally integrates linguistic features such as part of speech.

In Ho and Jin [8], this system Novel Lexicalized HMM-Based Learning Framework is used for Web Sentiment Analysis Mining, this model works problem as an information extraction task, which will be based on Conditional Random Fields [9].This work employ the supervised algorithm, which represents the state of the art on the engaged information. Finally, this work investigates the performance of our Conditional Random Fields-based technique. Conditional Random Fields-based system has performance improvement by 0.077, 0.126, 0.071, and 0.178 regarding F-Measure in the single-domain extraction in the four corpuses. In the cross-domain setting, our system has performance improvement by 0.409, 0.242, 0.294, and 0.343 regarding F-Measure over the baseline.

In Bu, Liu, Wang, Qiu, and Chen, this system has two subtasks of opinion mining: extraction of topic and sentiment categorization [7], which incorporates the Syntactic Information in Opinion Mining in User-Generated data. This system proposes techniques based on the consideration of syntactic knowledge; also this work takes the blog data on web. It is a typical application of UGC which has evaluating data in our experiments and the outcomes describe that our system techniques to the two tasks are challenging.

Liu, Bu, Qiu, and Chen review Word Expansion and Target Feature Extraction through Double Propagation [10], Opinion targets (targets, for short) are products and their components opinions are expressed. To perform the job, this work discovers that there are some syntactic relationships that link opinion words and targets. These relationships can be recognized using a dependency parser and then utilized to expansion of initial opinion lexicon and to target extraction [4]. This proposed system is depending on bootstrapping process. This work is nothing but double propagation which propagates knowledge between opinion words and final targets. The benefit of the proposed system is that it only requires first opinion lexicon to initiate the bootstrapping. This method is semi-supervised because of opinion word.

Chen, Qiu, Bu, and Liu expand sentiment Word and Extraction of Target through Double Propagation process [10]. Our work is performed in various: (a) Product features mining that have been commented on by users; (b) Considering sentiment sentences in each review and decide whether each sentiment sentence is negative or positive; (c) Summarizing the outcomes. For the above issue nowadays customers are updating their knowledge and try to select proper product at minimum price. To perform this task they utilize network services to check features of a particular product, their ratting. Savvy users nowadays are not much satisfied with the overall opinion of product rating. They want to know why it receives the rating, that is, which positive or negative attributes contribute to the final rating of the product. A customer first goes to check online reviews of other customer for the same product, which provides quality of product to the purchasing customer. Most important is that these techniques work with a single given corpus [13]. The overhead comes across when they read the reviews, as reading these reviews manually and making the decision is a critical and time consuming task. The need is to have an application which will extract the most dominating opinion features and their corresponding public sentiments.

There are number of existing approaches as discussed above for opinion feature identification but they are extracting features from a single review corpus. These techniques ignore the nontrivial disparities in distribution of words of opinion features across two or more corpora. This proposed work discusses a novel method for opinion feature identification from online reviews by evaluation of frequencies in two corpora, one is domain-specific and other is domain-independent corpus. Proposed method is summarized as follows: Initially, several syntactic dependence rules are applied to produce a group of candidate features from the given domain opinion corpus, for example, hotel, cell phone, movies. Next, for each accepted feature candidate, its domain relevance score regarding the domain-unspecific and domain-specific corpora is majored, which is called as the extrinsic domain relevance (EDR) and intrinsic domain relevance (IDR) scores, respectively. In the final step, candidate features with small IDR and high EDR scores are pruned. Thus, call this interval thresholding the intrinsic and extrinsic domain relevance (IEDR) technique.

3 Proposed System

The proposed system allows to identify opinion features from online text reviews by exploiting the difference in opinion feature statistics across two corpora as shown in Fig. 1. It is done using a measure called domain relevance. In this a list of candidate opinion features is extracted from the domain review corpus by defining a set of syntactic dependence rules. For each extracted candidate feature, its intrinsic domain relevance (IDR) and extrinsic domain relevance (EDR) scores are estimated on the domain-dependent and domain-independent corpora, respectively. Candidate features that are less generic (EDR score less than a threshold) and more domain-specific (IDR score greater than another threshold) are then confirmed as opinion features. This will be done by different modules as discussed below:

1. **Review Corpuses Loading**: First of all, application have to load the Review corpuses including Domain-dependent corpora and Domain-independent corpora. The statement is usually provided with XML or TEXT version. Users have to design a XML or a TEXT parser to read the statement serially. These statements should be loaded into arrays of string data type.
2. **POS Tagging**: A Part-Of-Speech Tagging is apart of software that reads the text in some language and assigns parts of speech to each word, like noun, verb, adjective, noun phrase etc. This task use open source Stanford Natural language programming parser for POST. This parser is instantiated with English Model.

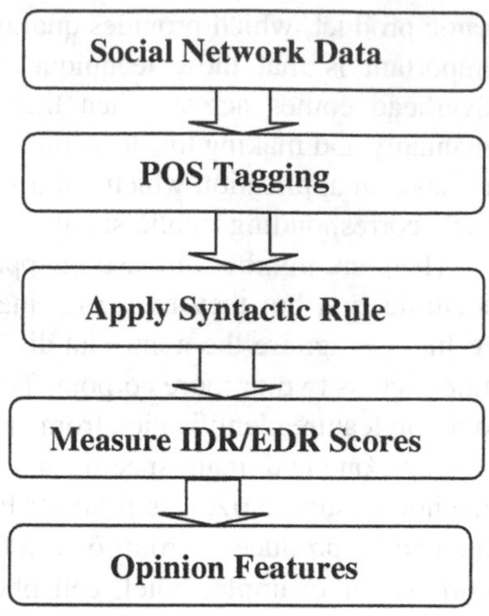

Fig. 1 Proposed system

3. **Identification of Candidate Feature**: For the opinions reviews, this work will identify the candidate features by extracting the frequent noun, noun phrase, and adjectives to design syntactic rules.
4. **Domain Relevance (DR)**: The difference in opinion feature called disparity is evaluated using domain relevance, which is characterized by the term relevance text collection.
5. **Intrinsic Domain Relevance(IDR)**: The Intrinsic Domain Relevance score is majored for each extracted candidate feature on domain-specific corpora. It represents how much the candidate feature is domain-specific with given domain.
6. **Extrinsic Domain Relevance(EDR)**: The Extrinsic Domain Relevance score is majored for each extracted candidate feature on domain-independent corpora. It represents how much the candidate feature is generic with given other external domains.
7. **Opinion Features Mining using Intrinsic Extrinsic Domain Relevance (IEDR)**: The Intrinsic Extrinsic Domain Relevance is a technique to identify and extract opinion features. A candidate features with EDR scores less than a threshold and IDR scores greater than another threshold are conformed as opinion features.

4 Mathematical Model of Proposed System

$D = \{d_1, d_2, d_3, \ldots\ldots, d_r\}$ is a set of review document.

Each document '*di*' in the review corpus may be either domain-dependent or domain-independent reviews.

So, let D_d = set of domain-dependent review

And D_{ind} = set of domain-independent review.

Each document 'd_i' where $d_i \in D_d$ or $d_i \in D_{ind}$, contains no of words. Let W is the set of words, represented as $W_i = \{w_1, w_2, w_3, \ldots\ldots, w_N\}$, where $'i'$ means ith document and Wij means from ith document jth word.

Let POS be the set of some parts of speech, i.e., $\boldsymbol{POS} = \{\boldsymbol{N}, \boldsymbol{V}, \boldsymbol{S}, \boldsymbol{Adj}, \boldsymbol{Prep}\}$, where $N = Noun$, $V = Verb$, $S = Subject$, $Adj = Adjective$, $Prep = Preposition$.

Let S be the statement of POS mentioned above.

For finding out opinion features, we have to separate candidates features from each statement.

So, CF is the set of candidate features

Any phrase $'P'$ is candidate feature by following mathematical equation,

$$P = \begin{cases} CF & if \quad N + SV \\ CF & if \quad N + VO \\ CF & if \quad N + PO \\ \text{Null} & \quad \text{Otherwise} \end{cases} \tag{1}$$

From the above-selected candidate features, we have to find final opinion features in both domain-dependent and domain-independent reviews.

For this we have to find each term $'T'$, Which is the word in document, belongs to which $'aspect'$ or $'opinion'$ with the help of term frequency and inverse document frequency.

TF = term frequency, i.e., how many times term t_i occurs in given document.

IDF = Inverse Document Frequency,

i.e., how significantly a term t_i is mentioned across all documents.

TF-IDF means the term weight T_W,

i.e., [14] represented as Tw_{ij} and can be found as

$$Tw_{ij} = \begin{cases} (1 + \log TF_{ij}) * \log \frac{N}{DF_i} & if \;\; TF_{ij} > 0 \\ 0 & \quad \text{otherwise} \end{cases} \tag{2}$$

From this Tw_{ij} we can find similarity between terms using following formula:

$$S_i = \sqrt{\frac{\sum_{j=1}^{N} (W_{ij} - W_i)^2}{N}} \tag{3}$$

According to the value of S_i we can group as domain-dependent and domain-independent.

5 Dataset and Expected Results

i. We are using dataset for movie reviews from below-mentioned URL: https://snap.stanford.edu/data/web-Movies.html
for movie reviews expected feature extraction shown in Table 1.

Table 1 Movie review features

Scene	Director	Plot	Actor	Story

Table 2 Cell phone features

Screen size	Processor	Camera	Battery	Wireless carriers
Screen technologies	Operating system	Price		

ii. We are using dataset for various products like cell phones from below-mentioned URL:
https://snap.stanford.edu/data/web-Amazon.html
http://www.gsmarena.com/
for cell phone reviews expected feature extraction shown in Table 2.

6 Conclusion

This system propose a novel inter-corpus statistics technique to feature extraction of opinion based on the IEDR feature filtering method, which utilizes the disparities in distributed characteristics of features across two domains, which are domain-specific and domain-unspecific or independent. IEDR determines features of candidate that are specific to the given review corpus and yet not overly generic, i.e., domain-unspecific. Finally, a good quality domain-unspecific or independent domain is less important for the proposed system. This task is nothing but evaluating the influence of domain size and selection of topic on performance of feature extraction.

References

1. L. Qu, G. Ifrim, and G.Weikum, "The Bag-of-Opinions Method for Review Rating Prediction from Sparse Text Patterns", Proc. 23rd Intl Conf. Computational Linguistics, pp. 913–921, 2010.
2. D. Bollegala, D.Weir, and J. Carroll, "Cross-Domain Sentiment Classification Using a Sentiment Sensitive Thesaurus", IEEE Trans. Knowledge and Data Eng., vol. 25, no. 8, pp. 1719–1731, Aug. 2013.
3. P.D. Turney, "Thumbs Up or Thumbs Down?: Semantic Orientation Applied to Unsupervised Classification of Reviews", Proc. 40th Ann. Meeting on Assoc. for Computational Linguistics, pp. 417–424, 2002.
4. D.M. Blei, A.Y. Ng, and M.I. Jordan,"Latent Dirichlet Allocation", J. Machine Learning Research, vol. 3, pp. 993–1022, Mar. 2003.
5. W.X. Zhao, J. Jiang, H. Yan, and X. Li, Jointly Modeling Aspects and Opinions with a Maxent-Lda Hybrid", Proc. Conf. Empirical Methods in Natural Language Processing, pp. 56–65, 2010.
6. F. Fukumoto and Y. Suzuki, Event Tracking Based on Domain Dependency", Proc. 23rd Ann. Intl ACM SIGIR Conf. Research and Development in Information Retrieval, pp. 57–64, 2000.

7. G. Qiu, C.Wang, J. Bu, K. Liu, and C. Chen,"Incorporate the Syntactic Knowledge in Opinion Mining in User-Generated Content", Proc. WWW 2008 Workshop NLP Challenges in the Information Explosion Era, 2008.
8. W. Jin and H.H. Ho,"A Novel Lexicalized HMM-Based Learning Framework for Web Opinion Mining", Proc. 26th Ann. Intl Conf. Machine Learning, pp. 465–472, 2009.
9. N. Jakob and I. Gurevych,"Extracting Opinion Targets in a Singleand Cross-Domain Setting with Conditional Random Fields", Proc. Conf. Empirical Methods in Natural Language Processing, pp. 1035–1045, 2010.
10. G. Qiu, B. Liu, J. Bu, and C. Chen,"OpinionWord Expansion and Target Extraction through Double Propagation", Computational Linguistics, vol. 37, pp. 9–27, 2011.
11. V. Hatzivassiloglou and J.M. Wiebe, "Effects of Adjective Orientation and Gradability on Sentence Subjectivity", Proc. 18th Conf. Computational Linguistics, pp. 299–305, 2000.
12. B. Pang, L. Lee, and S. Vaithyanathan,"Thumbs up?: Sentiment Classification Using Machine Learning Techniques," Proc. Conf. Empirical Methods in Natural Language Processing, pp. 79–86, 2002.
13. R. Mcdonald, K. Hannan, T. Neylon, M. Wells, and J. Reynar,"Structured Models for Fine-to-Coarse Sentiment Analysis", Proc. 45th Ann. Meeting of the Assoc. of Computational Linguistics, pp. 432–439, 2007.
14. Zhen Hai, Kuiyu Chang, Jung-Jae Kim, and Christopher C. Yang,"Identifying Features in Opinion Mining via Intrinsic and Extrinsic Domain Relevance", IEEE TRANSACTIONS ON KNOWLEDGE AND DATA ENGINEERING, VOL. 26, NO. 3, MARCH 2014.

Imaging-Based Method for Precursors of Impending Disease from Blood Traces

Basant Singh Sikarwar, Mukesh Kumar Roy, Priya Ranjan and Ayush Goyal

Abstract This study calculates the rheological disease factor from stain patterns of blood microfluidic drop samples of patients. The work explores programmed recognition of infection from the specimen of dried miniaturized scale drop blood stains from a patient's pathological examination. In this novel pathological examination proposed in this study less than 10 micro liters of blood is required. This strategy has the benefit of being significantly affordable and low cost, effortless and less intrusive, entirely robust for disease screening in infants and the old. Infection affects the physical and mechanical properties of biological fluid (blood), as reported by many medical and fluid mechanics fraternity previously, which thus influences the specimen of dehydrated blood miniaturized scale beads. For instance, low platelet index causes a drop in viscosity (one of the mechanical properties of the biological fluid—blood) due to blood thinning. Thus, the blood miniaturized scale drop stain specimen can be utilized for screening infections. This study proposes a programmed investigation of the dehydrated small-scale droplet blood stain specimen utilizing machine vision and pattern and feature extraction and recognition. The specimens of small-scale drop blood stains of ordinary healthy people are discernible from the specimen of miniaturized scale drop blood stains of unhealthy people. As a contextual investigation, the miniaturized scale drop blood stains of TB infected have been contrasted with the small-scale drop blood stains of healthy noninfected people. This study dives into the fundamental fluid mechanics behind how the specimen of the dehydrated small-scale drop blood stain is shaped. A thick circular boundary in the dehydrated small-scale drop blood stains of healthy

B.S. Sikarwar (✉) · M.K. Roy
Department of Mechanical and Automation Engineering, Amity University Uttar Pradesh, Noida, Uttar Pradesh, India
e-mail: bssikarwar@amity.edu

Priya Ranjan
Department of Electrical and Electronics Engineering, Amity University Uttar Pradesh, Noida, Uttar Pradesh, India

Ayush Goyal
Department of Electronics and Communication Engineering, Amity University Uttar Pradesh, Noida, Uttar Pradesh, India

© Springer Science+Business Media Singapore 2017

S.C. Satapathy et al. (eds.), *Proceedings of the International Conference on Data Engineering and Communication Technology*, Advances in Intelligent Systems and Computing 468, DOI 10.1007/978-981-10-1675-2_41

people and thin crack or ridge lines in the dehydrated small-scale drop blood stains of those infected with tuberculosis have been observed. The circular boundary is because of microfluidic channel stream, an outward current conveying blood cells suspended in plasma to the periphery. Concentric formed circles (brought about by internal Marangoni stream) and inner depositions are a few of the other stain patterns that were seen in the dehydrated small-scale drop blood stain specimen of typical noninfected people.

Keywords Blood stain • Blood drop • Tuberculosis • Disease • Screening • Microfluidics

1 Introduction

The method of drying microfluidic drops of blood on blotting surface paper for the measurement of molecular compounds has developed as a novel negligibly intrusive alternative for the screening of illness from blood tests, which can be utilized for reasonable inference of patients' infection. Investigation of dehydrated blood microdrop stains has been known in prior art [1, 2, 3], and therapeutic scientists are utilizing this procedure for detection of different ailments circumspectly after medical experimental trials on this strategy [4, 5, 6, 7]. When approval is obtained, demonstrated with medical test trials, this sickness detection system utilizing dehydrated microdrop dried blood stain patterns can give a low-cost diagnostic device for fast screening of malady from a patient's blood specimen. Experimentally, any chemical in either the plasma or serum segment of blood will likewise be detectable and measurable in and can be distinguished and measured from a dried out miniaturized scale bead of that same blood specimen, given the assumption that the molecular compounds to be measured in the dried out blood stain will have robustness to drying out. Examination of whether a molecular compound is steady over drying is likewise important if this technique is to be utilized for broad study in epidemiology of an infection in a populace. Once the steadiness investigation and exploratory acceptance and medical tests validate this system, it can encourage development of a novel illness screening or disease recognition instrument comprising of a microfluidic gadget that gathers a microdrop from a patient's blood and measures the physical properties of blood and distinguishes patterns in the dried out blood stain to perceive which malady the patient has. Earlier publication showing that the gathering of a blood smaller scale bead with blotter filter paper is presently as exact as microcapillary pipettes and microfluidic tubes [8] which makes this examination plausible to distinguish unhealthy from the physical properties of blood miniaturized scale beads of extremely exact volume acquired from a patient blood test. The channeling of an exact amount of blood volume as a micro level bead, in a dried blood spot on blotter filter paper and its handling is presently a controlled method [9]. Amid the drying procedure of blood specimen on blotter filter paper, cytoplasmic organelles of blood cells rupture. To solve this issue extra separation strategy may be required for certain molecular compounds. There exist two different problems which may accompany the investigation of dehydrated blood micro drops—elution efficiency of molecular compounds being detected and exactness of small-scale bead volume of the gathered blood specimen. Despite these downsides, dried out blood spot technique has numerous advantages. The first advantage is minimal physical discomfort due to the negligibly obtrusive drawing of blood with only a little prick in a finger capillary vein when contrasted with the profoundly intrusive and agonizing gathering of blood through a vein punctured by needle. What makes the dehydrated blood spot test uniquely cost-effective is that it avoids the requirement of a centrifugal machine, which is an important part for standard blood tests in pathological laboratories. Thus the dried blood spot test can be performed outside a pathological laboratory without centrifugation and separation of the blood. At room temperature, most chemicals remain robust to drying up to 1 week. The dried blood stain technique likewise has low infection risk in light of the fact that each one of those infections which may be dynamic in the fluid condition of blood and its constituents get to be noninfective because of breakage of their capsid upon being dried. Albeit naturally one may consider dried out blood test performed with blood microdrop sample from finger pricking has distinctive properties than blood test suctioned from a needle puncturing a vein by means of a syringe. Dried out blood stain sickness identification is an exceptionally practical and quick technique for nationwide epidemiological study in remote areas with minimal human intervention. In this study, an economical cost-effective method for disease detection from blood stain is proposed. This method will eliminate the requirement for blood tests which requires sample collection and transport of blood serum. Blood is a non-Newtonian, natural liquid, half of which is plasma containing red blood cells and white blood cells [10, 11, 12]. It fulfills shear-thinning condition, i.e., thickness diminishes upon expanded shear strain. Blood concentration of particles can be estimated by hematocrit, calculated as volume of RBC plasma over total blood volume [13, 14]. Measurement of the physical properties of blood has been the topic of exploration globally for over 20 years. This exploration has an immediate motivation in medical research in light of the fact that it has been observed that physical properties of blood assume a critical part in numerous human and animal infections [15, 16, 17, 18]. Clinical and medical publications [19, 20, 21] have shown that the specimen patterns of dehydrated micro beads of blood and other such liquids were specific to typical versus sick patients. This sort of examination [22] concentrates on the dehydration procedure of a microdrop of blood and the specimen blood drop pattern of stain resulting after the dehydration for recognizing patients experiencing a specific infection from healthy people. This research has discovered that the dehydrated blood microdrop stain pattern is totally distinct for infected individuals as compared to the typical healthy people (see Fig. 1). The aforementioned scientific research papers published results that showed how blood's physical properties vary for normal and unhealthy patients. In consideration of this foundational setup, this work investigates the blood microfluidic drop stains of patients with different diseases and relates the illness with the specimen pattern of the dried microfluidic drop of blood.

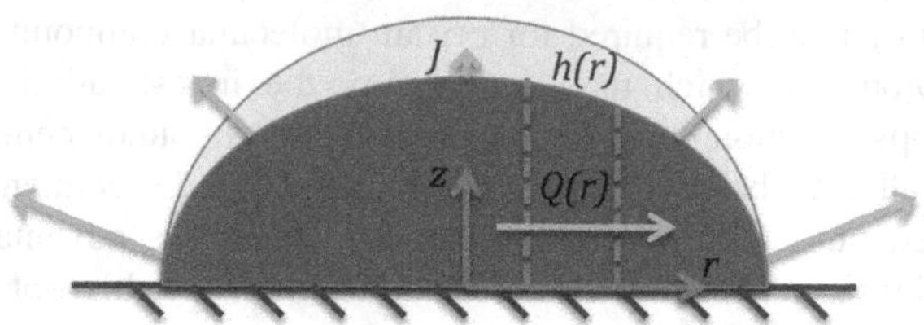

Fig. 1 Axisymmetric drying drop. Polar coordinates-shaped direction framework (*r*; *z*). Flow stream *Q*(*r*) in the bead caused by the flux (*J*) drying from the surface of the bead with height h(r). The lines with dashes demonstrate the control width volume (∂r)

2 Methodology

Medical research papers have shown that diverse specimen patterns of dried microfluidic blood spots were seen in healthy and ailing people. Researchers in fluid mechanics have found that the dried blood spot pattern is correlated with blood's physical properties. Subsequently, the dehydration procedure of blood is to a great extent impacted by blood's physical properties. Physical properties rely upon the ratio of RBC volume of blood with respect to the net volume of blood. Our exploration consolidated these established researchers' examination that allows relating illnesses such as tuberculosis (TB) with the dried blood microdrop spot pattern. For such a novel method, straightforward, rigorous, and affordable, a microfluidic device was developed and manufactured to obtain dried blood microdrop spots of different specimen of blood. Hypothesis and exploratory points of interest of the dehydrating procedure are detailed as follows.

2.1 *Microfluidic Blood Drop Spot Drying*

In dehydrating blood drops, the evaporative flux of the drop surface results in an isotropic stream flowing outwards radially in all directions in the drop, which moves the WBCs and RBCs to the periphery. At the point when a pinned line of contact of the drop is shaped, the blood that vanishes from the line of contact locale must be restored by fluid from the inside of the microdrop. A capillary phenomenon causes the stream to keep up the drop's semispherical cap-like shape created by surface tension, a flow of compensating properties is needed to replenish the fluid that vanishes from the pinned line of contact. The pinning of the line of contact happens due to the paper's surface roughness. Furthermore, the red and white blood cells suspended in the blood flow toward the contact line because the blood microfluidic drop spot boundary is fortified by the pinning, caused by a self-pinning phenomenon.

The measure of stream in a drying blood microdrop can be computed from mass preservation with the flux with evaporative properties as given in Eq. (1) below:

$$J = -D\nabla C \tag{1}$$

The vapor's coefficient of diffusion in the encompassing gaseous material is termed as (*D*). This diffusion coefficient is distinct for distinctive blood specimen (normal and sick people groups). To locate the evaporative flux from the vapor field of concentration, the term on the left in Eq. (2) is solved given the following boundary conditions:

$$\frac{\partial c}{\partial t} = D\nabla^2 C \tag{2}$$

To represent dehydrating procedure of blood bead drop, we take a drop that is axisymmetric with a pinned line of contact, which can be characterized utilizing the polar coordinate system, as illustrated in Fig. 1. In this microdrop, we characterize an imperceptibly little control width volume (∂r), situated at a separation distance (r) from the bead center. Law of mass preservation requires that the derivative in the measure of fluid in this control volume rises to the net inflow of fluid into the control volume, minus the measure of fluid that has vanished from the bead surface. This can be written as

$$\frac{\partial h}{\partial t} = -\frac{1}{r}\frac{\partial Q}{\partial r} - \frac{1}{\rho_b}J \tag{3}$$

where the drop height of the bead is termed as h(r, t), time is termed as (t), volume stream is termed as (Q), and density of blood is termed as (ρ_b). Consequently, the lessening in drop tallness of a volume component levels with the evaporative flux through the drop free surface in addition to the net outflow to the adjacent components.

The author's research solves the above equation in [23, 24, 25, 26]. These papers reported that amid evaporation, the liquid flows radially out from the center of the bead to the subsiding contact line. The most extreme grouping of liquid was seen near the contact line.

2.2 Drying Microfluidic Blood Drop Experiment

In this current work, a micro drop was placed on filter paper in a controlled conditional environmental (see Fig. 2) for the evaporation. Samples of blood were gathered from patients with TB and without TB. They were gathered in 10 ml sterilized vials having 1 % anticoagulant (sodium heparin). The vials were placed in a cold environment at 5 °C. The blood specimens utilized as a part of this research were gathered from nondiseased, TB patients, and those suffering from anemia.

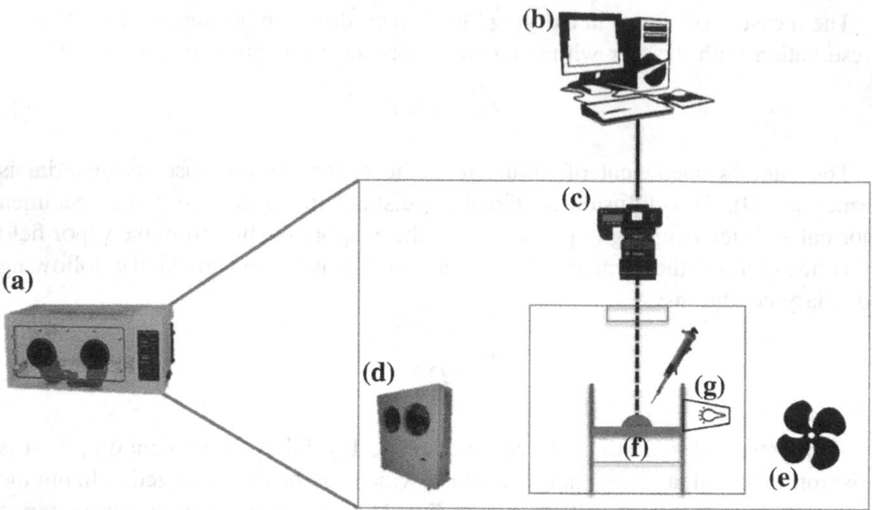

Fig. 2 Flow chart of hardware utilized for gathering the specimen of blood spots of a blood microfluidic drop on a material comprised of filter paper—**a** chamber with controlled conditions **b** computer system, **c** EOS 70 HD camera, (d) cooling refrigerator for maintaining constant environmental conditions during the experiment, **e** exhaust system to recycle air, **f** filter paper substrate, and (g) cooled visualization sources

A flow diagram of the experimental test ring utilized as a part of this exploratory research is illustrated in Fig. 2. The microdrops for each blood specimen were delicately set on the filter paper substrate in a controlled environment in the chamber at 50\% relative humidity and 32 °C chamber temperature. The filter paper substrate was cleaned using compressed air. The roughness of the filter paper is measured using a 3D microscope and it was 0.1 micrometers. The physical properties such as viscosity, hematocrit, wettability, and surface tension of the blood were estimated.

A balance having a resolution of 0.001 milligrams was used to measure the stain after complete drying. Rate of evaporation J (t) was estimated using Eq. (1). Simultaneously, a Canon camera (EOS 70D) with macro lens was utilized to image the drying bloodstain microdrop. The captured image dimensions were 22.3 mm (width) by 14.9 mm (length), which were caught at a rate of 2 frames per second with a standard resolution of 5000 pixels by 3584 pixels, sufficient for capturing the dehydration, gelation, and crack formation of the microdrop of sessile mode.

Toward the beginning of the drying of the microdrop, the liquid is complex, shear thinning, and heterogeneous. Subsequently, a slim periphery forms after dehydration on the filter paper substrate. All the while, Marangoni convection makes the particles such as the red blood cells to gather at the subsiding triple-phase contact line of the footprint of the microdrop. These results in gathering of solids alluded to as the center or corona. The remainder is serum that does not have any microparticles and makes small-scale designs and that adhere to the filter paper substrate consequent to dehydration.

Pictures were taken using a computer that controlled the camera and then were stored in the computer memory. They were then examined as displayed in the results with the programmed automated algorithm for recognition of patterns in the blood microdrop stains. The acquisition system of the images was high enough in resolution for capturing the patterns of the blood drop drying. A cooled cathode backdrop illumination source was utilized on the blood microdrop to avoid reflection. Density of light likewise influenced the dissipation and drying of the blood microdrop.

2.2.1 Experimental Protocols

(1) Cut the filter paper (filter paper COLORBOK®, neutral white, original size 8.5in X 11in) as 60 mm x 60 mm squares.
(2) Attach the filter paper squares to aluminum tile and place on the work table with a cover on top to minimize foreign particles on their surfaces.
(3) Approximately 2 hours before the experiment, wash the aluminum tile with isopropanol, then deionized water, and dry them with compressed air.
(4) Check with the level that the target platform is horizontal.
(5) Calibrate the video image system by recording several images with an appropriate graduate scale, for the conversion of pixels to mm. Save the images mentioning the objective type, camera, and magnification in the file name.
(6) Set the volume and the flow rate and on the syringe pump.
(7) Connect rigid 1/16" plastic tubing, with the appropriate fittings, to the needle.
(8) Check that the needle is perpendicular to the impact surface. The alignment should be checked by first ensuring that the needle is perpendicular to the holder bar by viewing the needle from multiple angles, then with a spirit level, check that the bar itself is horizontal.
(9) Flush the tube attached to the needle with distilled water to remove any obstruction in the tube and needle. Use a separate syringe and small beaker for this step.
(10) Purge the water from the tube by forcing air in using a 10 mL syringe.
(11) Clean the needle with Kimwipe® to remove any possible residue or obstruction.
(12) The plastic test tubes (15 ml) containing blood should be stored in the fridge when not used. The temperature inside the fridge should be kept around 3.5 °C (between 3 and 4 on the knob inside the lab refrigerator) in order for the blood to be maintained within 2–8 °C recommendation. (Note: The blood should not be used more than a week after the bleeding date).
(13) Put the plastic tubes containing the blood on the rocker approximately 60 min before starting experiment. Before starting experiment, measure temperature with a traceable thermometer to verify blood temperature is the same as room temperature (±0.5C).

2.2.2 Experimental Procedure for Preparation of Blood

The experiment in this study is done in sync with measuring the properties of blood. This experiment requires a refrigerated temperature setting of 3.5 °C for cooling the 10-mm vials of plastic with the blood specimen. This is to keep the temperature of the blood samples between 2–8 °C. The blood is utilized within 7 days of taking the sample. The vials of plastic with the blood specimen are placed atop the rocker an hour prior to beginning the experiment, before which the blood sample's temperature is measured to confirm that it is within ±0.5° error of the ambient temperature.

2.2.3 Measurement of Physical Properties of Blood

On beginning the examination, the ambient temperature and relative humidity are recorded along with the time and date of the measurements. Also, the blood sample's temperature is measured with a traceable thermostat and the blood sample's hematocrit is measured using the HemataSTAT® II separation technology. A microcapillary syringe-pump viscometer is utilized to measure the blood sample's viscosity at various quantities of shear rate (10, 100, 500, 1000, 5000, and 10000). Finally, the surface tension of the blood is measured from a pendant drop of the blood sample.

The schematic diagram of the screening of disease using the physical properties of blood and the features of the bloodstain patterns is shown in Fig. 3.

As discussed above, the methodology of disease screening cum diagnostic is based on the physical properties of blood and stain pattern recognition.

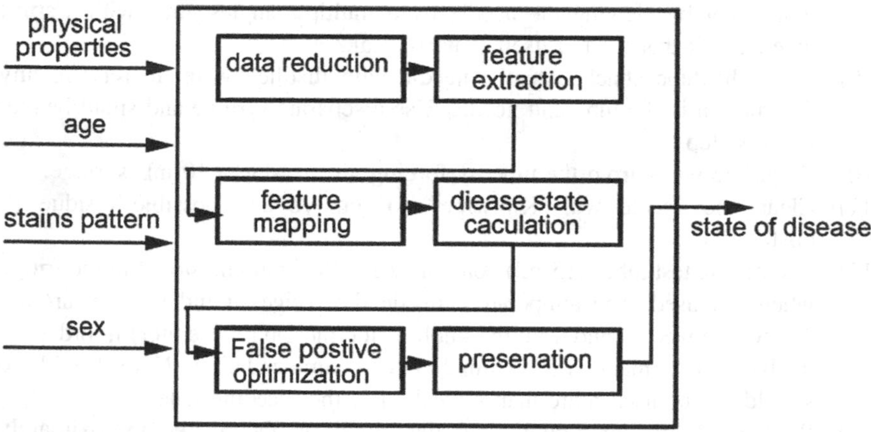

Fig. 3 Schematic diagram of blood disease detection device using feature extraction of bloodstain and physical properties of the blood

2.3 Dehydrated Blood Microfluidic Drop Pattern Crack Formation

The crack formation was researched from dehydration of different specimens of blood on filter paper substrate. In the blood microfluidic drop specimen of the normal healthy patients, strong peripheral flow was seen conveying live RBCs to the three-phased contact line, where evaporative flux was at its highest created by a wedge geometry. A circle of particles (RBCs, WBCs, and platelets) is built up over the deposition (adhering of the peripheral line of contact) stage, followed by the wetting (subsiding of the line of contact) stage. It was shown that chemical and physical nonuniformity of substrate affected the features of the blood microfluidic drop stain. Also, Marangoni convection inside the drop brought on by inclinations of surface tension impact the flux, staining, and deposition of the blood microfluidic drop. This causes gathering at the drop's periphery instead of the corona. The mechanism of physical deposition causes a thick peripheral boundary to be formed in the dehydrated blood microfluidic drop of nondiseased healthy individuals.

2.4 Dried Blood Spot Specimen

Blood microfluidic drop stain samples were collected of normal healthy individuals, TB patients, and those suffering from anemia. Figure 4 shows two sample blood stains of each group.

The exploratory method for gathering the blood microfluidic drop stain pattern information from blood samples of patients in this work was given legitimate assent by the ethical review board of the authors' institute.

3 Results

Figure 5 outlines the programmed image analysis of the blood microfluidic drop stains of healthy, anemic, and TB patients. The stain patterns are different for the tuberculosis infected, anemic, and normal blood samples. Specifically, the main feature of the blood microfluidic drop stain pattern of normal healthy individuals is a thick contact line periphery around the boundary as compared to the thin stain circle in the center. For anemic patients, the blood stains have the main feature of thick cracks. The blood stains of TB patients have thin cracks and a uniform crack formation from center to boundary as compared to the nonuniform patterns of the other groups.

Figure 5 illustrates feature extraction of the patterns from the blood microfluidic drop stains for each of the three groups of patients. An aggregate of 28 blood

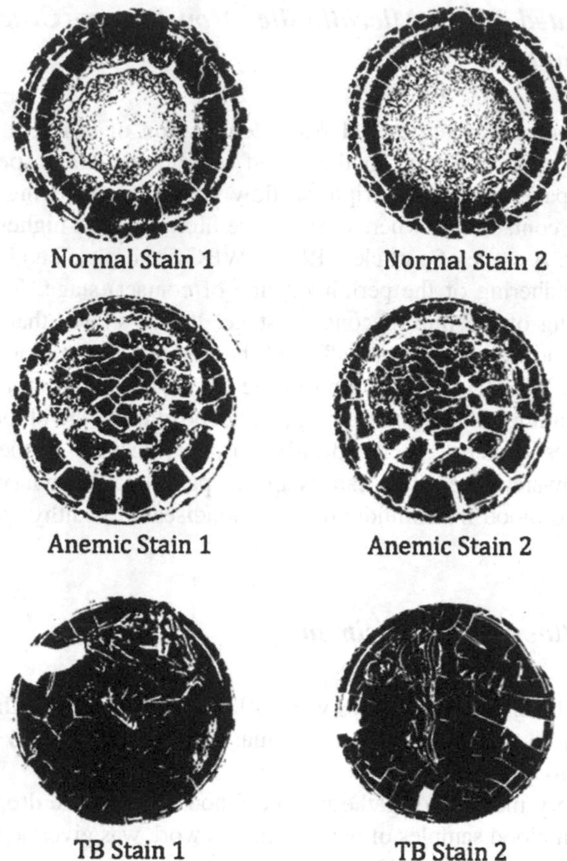

Fig. 4 Blood stain patterns of TB patients, anemic patients, and normal healthy individuals

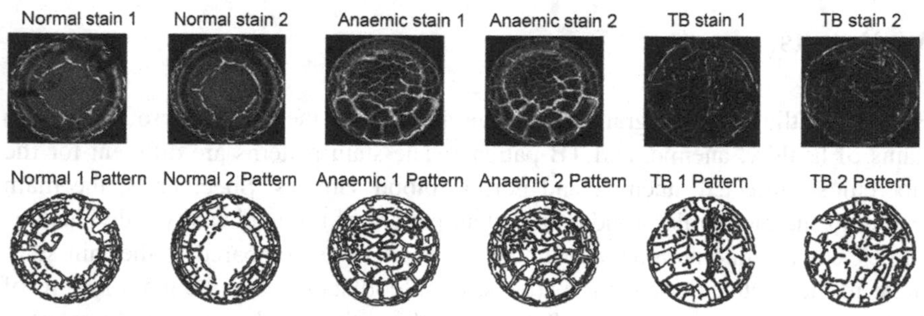

Fig. 5 Patterns extracted from the blood microdrop stains

microfluidic drop stain specimen were gathered and captured by the camera. They were handled in two trials or experiments. One sample of each of these experiments is shown for each blood disease group.

Table 1 Image distance (average ± std) in between the dehydrated blood microfluidic drop stain patterns of diseased and nondiseased blood samples

Patient	RMSE	MSE	χ2
Normal	6.8 ± 1.05	46.55 ± 13.95	201.35 ± 5.35
Anemic	9.5 ± 0.15	90.6 ± 3.55	217.65 ± 5.85
TB	12.2 ± 0.15	147.8 ± 3.55	340.75 ± 9.45

Image distance calculated as root-mean-square error (RMSE), mean-square error (MSE), and chi-squared distance (χ^2), calculated as mean distance of all pixels

Table 1 gives the image differences between the average image of the normal blood microfluidic drop stain and the images of the diseased blood microfluidic drop stains. This allows calculation of a blood microfluidic drop stain-based measure of infection or disease in blood. Stain patterns of the dehydrated blood microfluidic drop were collected, visualized, and processed over two sets (Set 1 and Set 2) of examinations for normal healthy, anemic pallid, and the TB infected.

The outcomes in Table 1 show the image distance/separation calculated in between the stain patterns of dehydrated blood microfluidic drops of the nondiseased (average) and diseased. This image distance can be translated into a rheological disease factor (RDF), which is plotted in Fig. 6 and can be utilized as a blood disease level measure to differentiate between diseased and nondiseased blood samples.

In Fig. 6, rheological disease factor, a measure of disease in blood, is calculated as the normalized mean square error between the blood microdrop stain pattern of a patient and an average of normal blood microdrop stain patterns. Figure 6 shows the mean ± standard deviation of the rheological disease factor for the three disease groups of patients—normal (control) group, TB group, and anemic group (there were a total of 28 blood microdrop stain data samples in this study).

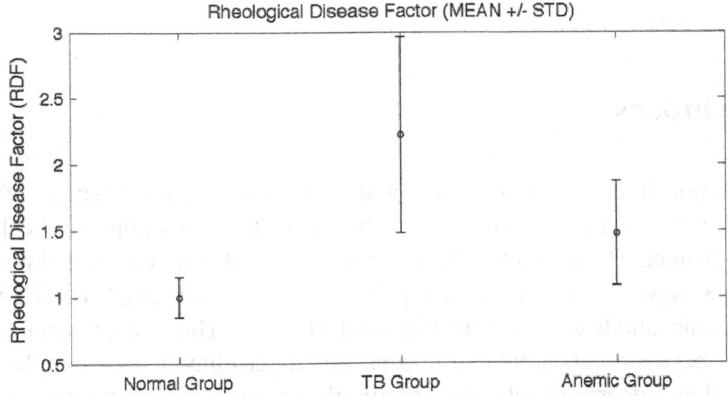

Fig. 6 Rheological disease factor (RDF), a measure of disease in blood, plotted for the three disease groups—the normal group, the TB group, and the anemic group. RDF was calculated as the normalized mean square error between the blood microdrop stain pattern of a patient and an average of normal blood microdrop stain patterns

4 Discussion

The analysis during this study confirms the clinical examination that diverse patterns of blood stains have been discovered for the nondiseased versus the pathologically infected. Fluid mechanics analysis published research shows that microfluidic drop stain patterns are largely affected by the physical and mechanical properties of the fluid. Our work in this study merged the analyses of these two fields to predict illnesses like anemic condition and infectious disease to the physical and mechanical properties and stain patterns of blood. With the distinctive stain patterns of microfluidic drops of blood samples of TB infected, nondiseased individuals, and anemia patients, the stain pattern of the microfluidic drop of a blood sample may be utilized to determine if a patient is unhealthy and which illness the person has. Using this novel method of disease detection from the physical properties and stain patterns of a microfluidic blood drop, rapid, minimally invasive, and affordable micropipette devices like microfluidic viscosity meters and a tensiometer based on the pendant drop technique may be developed and manufactured to estimate physical properties such as viscousness, wettability, and surface tension of blood. In this study, we examined the dehydrated stain patterns of blood microfluidic drops as an alternate technique to screen disease from unhealthy and healthy blood samples. The untreated specimen of blood were gathered with an extremely precise micropipette instrument, and the microfluidic droplets were deposited onto a filter substrate and the stain patterns formed were imaged and captured. Physical properties and mechanical properties of blood like surface tension, wettability, viscosity, and viscoelasticity, measured in the controlled environment chamber, affect the staining process of the microfluidic blood droplets upon collection, dehydration, and formation of crack patterns. Stain deposition patterns of the dehydrated microfluidic blood droplets formed by the deposition, dehydration, and fracturation were discovered to vary for diseased versus nondiseased blood samples.

5 Conclusions

The distinction in stain patterns among the infected and noninfected dehydrated microfluidic blood drops can actually be used to differentiate a healthy and unhealthy patient. The distinction between the infected and noninfected blood stain patterns was most evident for the stain patterns of the blood samples of TB infected, less for anemia, and least for the healthy control group. Therefore, the pattern image distance in between unhealthy and normal (averaged) blood stain samples may be investigated as a degree of infection (rheological disease factor) measured from the stain patterns of dehydrated microfluidic blood drop samples. This provides for a minimally invasive approach that needs to be investigated with a greater set of infected and noninfected blood stain sample data. Moreover, the distinction within the stain patterns of dehydrated microfluidic blood drops of blood infected by malaria, chikungunya, dengue, etc., needs to be researched in order to determine if the stain patterns of dehydrated microfluidic blood drops may be utilized to screen infection from the blood samples of patients.

This paper proposes a low-cost medical screening device for infectious diseases, specifically TB, which is the number one infection worldwide and a global epidemic threat. This screening device can help various agencies who are working day and night to eradicate such diseases. In developing countries such as India, it can aid DOTS and tuberculosis treatment centers by facilitating fast screening and level indicator of TB infection for effective diagnosis, right treatment, and on-time medication. With such innovative screening tools that detect disease from blood, sputum, and/or urine [27], the detection of infections can be made more rapid, affordable, and accessible to remote areas where there is no pathological laboratory or power.

Acknowledgments We would like to thank Amity University, Uttar Pradesh for providing us the resources and basic facilities in the campus and Baramdev Medical Centre, Noida, India for the pathological blood samples.

References

1. Ross, N. E., Pritchard, C. J., Rubin, D. M., Duse, A. G.: Automated image processing method for the diagnosis and classification of malaria on thin blood smears. Medical and Biological Engineering and Computing, 44(5), pp. 427–436 (2006)
2. Roy, M., Sikarwar, B., Goyal, A.: Design and Fabrication of Microfluidic Device for Measuring Surface Tension of Biological Fluids, India International Science Festival, IIT Delhi, Department of Science & Technology, Govt. of India, (2015)
3. Sikarwar, B., Roy, M., Ranjan, P., Goyal, A.: Automatic Pattern Recognition for Detection of Disease from Blood Drop Stain obtained with Microfluidic Device, Advance Intelligent& Computing System, Springer, vol-425, pp. 655–667 (2015)
4. Guthrie, R., Susi, A.: A simple phenylalanine method for detecting phenylketonuria in large populations of newborn infants, Pediatrics, 32(3), pp. 338–343 (1963)
5. Lakshmy, R.: Analysis of the use of dried blood spot measurements in disease screening. Journal of diabetes science and technology, 2(2), pp. 242–243 (2008)
6. Mei, J. V., Alexander, J. R., Adam, B. W., Hannon, W. H.: Use of filter paper for the collection and analysis of human whole blood specimens. The Journal of nutrition, 131(5), pp. 1631S-1636S (2001)
7. Kapur, S., Kapur, S., Zava, D.: Cardio metabolic risk factors assessed by a finger stick dried blood spot method. Journal of diabetes science and technology, 2(2), pp. 236–241 (2008)
8. Chamoles, N. A., Niizawa, G., Blanco, M., Gaggioli, D., Casentini, C.: Glycogen storage disease type II: enzymatic screening in dried blood spots on filter paper. Clinica chimica acta, 347(1), pp. 97–102 (2004)
9. Zytkovicz, T. H., Fitzgerald, E. F., Marsden, D., Larson, C. A., Shih, V. E., Johnson, D. M., Grady, G. F.: Tandem Mass Spectrometric Analysis for Amino, Organic, and Fatty Acid Disorders in Newborn Dried Blood Spots A Two-Year Summary from the New England Newborn Screening Program. Clinical Chemistry, 47(11), pp. 1945–1955 (2001)

10. Jinks, D. C., Minter, M., Tarver, D. A., Vanderford, M., Hejtmancik, J. F. McCabe, E. R.: Molecular genetic diagnosis of sickle cell disease using dried blood specimens on blotters used for newborn screening. Human genetics, 81(4), pp. 363–366 (1989)
11. Crossle, J., Elliot, R. B., Smith, P.: Dried-blood spot screening for cystic fibrosis in the newborn. The Lancet, 313(8114), pp. 472–474 (1979)
12. Chace, D. H., Kalas, T. A., Naylor, E. W.: Use of tandem mass spectrometry for multianalyte screening of dried blood specimens from newborns. Clinical Chemistry, 49(11), pp. 1797–1817 (2003)
13. Alvarez-Muñoz, M. T., Zaragoza-Rodríguez, S., Rojas-Montes, O., Palacios-Saucedo, G., Vázquez-Rosales, G., Gómez-Delgado, A., Muñoz, O.: High correlation of human immunodeficiency virus type-1 viral load measured in dried-blood spot samples and in plasma under different storage conditions. Archives of medical research, 36(4), pp. 382–386 (2005)
14. Gelb, M. H., Turecek, F., Scott, C. R., Chamoles, N. A.: Direct multiplex assay of enzymes in dried blood spots by tandem mass spectrometry for the newborn screening of lysosomal storage disorders. Journal of inherited metabolic disease, 29(2–3), pp. 397–404 (2006)
15. Li, Y., Scott, C. R., Chamoles, N. A., Ghavami, A., Pinto, B. M., Turecek, F., Gelb, M. H.: Direct multiplex assay of lysosomal enzymes in dried blood spots for newborn screening. Clinical chemistry, 50(10), pp. 1785–1796 (2004)
16. Zhang, X. K., Elbin, C. S., Chuang, W. L., Cooper, S. K., Marashio, C. A., Beauregard, C., Keutzer, J. M.: Multiplex enzyme assay screening of dried blood spots for lysosomal storage disorders by using tandem mass spectrometry. Clinical chemistry, 54(10), pp. 1725–1728 (2008)
17. Brutin D., Sobac. B. and Nicloux C.: Influence of Substrate Nature on the Evaporation of a Sessile Drop of Blood, AME, Journal of Heat Transfer, vol. 134, pp. 1–8 (2012)
18. MacDonell, H. L.: Bloodstain Patterns, Laboratory of Forensic Sciences, Corning, NY USA (2005)
19. Yakhno, T. A.: Drying Drop Technology as a Possible Tool for Detection Leukemia and Tuberculosis in Cattle," Journal Biomedical Science and Engineering, vol. 8, pp. 1–23 (2015)
20. Savina L.: Crystalline Structures of Serum of Healthy and Ill Patients, Soviet Kuban, Krasnodar, p. 96 (1999)
21. Sikarwar, B.S., Sharma, S.K., Shukla, R.K., and Ranjan P.: Parametric study of sessile drop evaporation at atmospheric condition, Proceedings of the 17th ISME Conference, ISME-17, IIT Delhi, India (2015)
22. Shabalin, V. N. and Shatokhina, S.N.: Morphology of Biological Fluids, Khrisostom, Moscow, p. 304 (2001)
23. Deegan, R.D.: Pattern Formation in Drying Drops," Physical Review E, vol. 61, pp. 475–485, (2000)
24. Rapis, E.: Changing the Physical Non equilibrium Phase of Complex Plasma Proteins Film in Patients with Carcinoma, Technical Physics, vol. 47, pp. 510–512. (2002)
25. Ghosh, A., Sikarwar, B.S., and Attinger, D.: Microfluidic Measurements of Physical Properties of Blood for Forensic Studies, ASME 12th International Conference on Nanochannels, Microchannels, and Minichannels, Chicago Illinois, USA, (2014)
26. Sikarwar, B.S., Ghosh, A., and Attinger, D.: Simple, Low Cost Microfluidic Measurements of Physical Properties of Blood for Forensic Studies, ASME 12th International Conference on Nanochannels, Microchannels, and Minichannels, Chicago Illinois, USA (2014)
27. Goyal, A., Roy, M., Gupta, P., Dutta, M. K., Singh, S., Garg, V.: Automatic Detection of Mycobacterium tuberculosis in Stained Sputum and Urine Smear Images, Archives of Clinical Microbiology, 6 (3) (2015)

A Hybrid of DEA-MLP Model for Stock Market Forecasting

S.S. Panigrahi and J.K. Mantri

Abstract Being a burning area of interest among the researchers the stock market forecasting draws attention worldwide for new innovations of models, ways of interpretation, wider aspects, and multidimensional analysis. The present paper proposed a hybrid of Data Envelopment Analysis and Multilayer Perceptron model for stock market forecasting. The proposed model is experimented with real-world data from BSE Sensex. The DEA model has assigned the responsibility to filtered predicators for MLP model to enhance its prediction capabilities. Separate empirical investigations are conducted for existing MLP and then for the proposed model to justify the superiority of the proposed model. It was observed that the proposed one has outperformed the existing one in most of the front.

Keywords Data envelop analysis · Artificial neural network · Multilayer perceptron · Stock market forecasting

1 Introduction

Risk and Returns are the two faces of stock market. The responsibility of forecasting stock market is to "minimize risk" and "optimize the returns". To be precise, the motto of the stock market forecasting is to optimize the level of returns by minimizing the risk associated with the investment in the stock market. The model having better forecasting efficiency has wider acceptability in comparison with other available models for similar task. Further, model with better accuracy and well before prediction can prevent unexpected loss to an investor in particular and sudden downfall in the market trend in general. Hence, there always scope exists for

S.S. Panigrahi (✉) · J.K. Mantri
Department of Computer Science & Applications,
North Orissa University, Baripada, India
e-mail: panigrahisasankasekhar@gmail.com

J.K. Mantri
e-mail: jkmantri@gmail.com

© Springer Science+Business Media Singapore 2017

S.C. Satapathy et al. (eds.), *Proceedings of the International Conference on Data Engineering and Communication Technology*, Advances in Intelligent Systems and Computing 468, DOI 10.1007/978-981-10-1675-2_42

further research to innovate new models or enhance the capabilities of existing models to face the challenges put forth by the real-world volatility in the financial market.

Soft computing approaches paved the way ahead for forecasting financial market and bring this area into the forefront with certain accepted level of accuracy. Artificial Neural Network (ANN) model is one of the major contributors in this field of research. An artificial neural network is composed of a harmonic blend of computing elements literally known as artificial neurons designed in simulation with architecture of the cerebral cortex of the brain. Multilayer perceptron is a type of neural network which consists of nodes separated into multiple layers. The MLP network counters the problem in a forward direction via its layers.

Proposed by Charnes et al. [2] in 1978 the Data Envelopment Analysis (DEA) emerged as a unique comparative efficiency measurement technique to measure the relative efficiencies of similar kind of units known as Decision Making Unit (DMU) where multiple inputs and output(s) are involved. Schefczyk [9] praised the model by writing that the DEA is the "only viable methodology that links all factors of efficiency by evaluating the relationships between each input and output to arrive at a scalar measure of performance."

In the present work MLP and DEA are used to made a hybrid for better forecasting accuracy relating to the stock market. The design of the model has explained and experimented with available data and relative efficiencies are compared. The conclusion has drawn at par with result obtained out of the experiment.

2 Review of Related Literatures

Neural network, specifically the MLP has widely and versatile used in the field of financial market forecasting independently and also combined with other existing soft computing approaches. The literature is enriched with new findings by numerous contributors working in the area. Few contributions in near past are briefly reviewed in this context.

A novel approach to input selection through neural network for forecasting stock market behavior was proposed in 2006 by Huang et al. [5]. The experiments with data from S&P 500 and NIKKEI 225 proved the efficacy of the proposed system. Tsai and Wang [11] combined ANN and decision trees to construct a stock price forecasting model in 2009. The experiment was done over the electronic industry and the output reveals the supremacy of the proposed model over the existing models. Maciel and Ballini [6] in 2010 contributed by taking the pain to compare ANN with traditional forecasting method, e.g., GARCH and conclude in favor of ANN with remarks that properly trained neurons can improve robustness depending on the network structure. A tree-based data mining algorithm and model of stream data time series pattern in a dynamic stock market was proposed by treating market behavior and interest as input in the year 2011 by Tiwari and Gulati [10]. In the same year, an experiment was conducted over the collected data from the German Stock Exchange by Mandziuk and Jaruszewicz [7] for the development of neuro-genetic system for short-term stock index prediction. In this system, the Genetic Algorithm is applied to find an optimal set of input variables out of mutually related input variables for a 1-day prediction.

A hybridized approach with variables from both technical and fundamental indicators of stock market was made to predict stock market through neural network and succeed to improve prediction accuracy of future price of stock by Ayodele et al. [1] in 2012. Liang-Ying and Ching-Hsue [12] proposed another hybrid model in the same year by utilizing synthesis feature selection to optimize the recurrent network (RNN) for predicting stock price trends. For the experiment, the Taiwan stock exchange capitalization weighted stock index is employed and the result proved the superiority of the proposed model. Panigrahi and Mantri [8] made an effort to put a comparative analysis of MLP and SVM in the field of financial time series analysis in 2014. The empirical analysis indicates SVM as a handy tool over MLP in various fronts for financial forecasting of similar nature. Another contribution comes up from Chih-Ming [4] who integrated Data Envelopment Analysis, Ant Colony Optimization, and Gene Expression Programming Procedure for Resolving Portfolio Optimization Problems. In the same year, Hegazy et al. [3] present a paper to predict the gold price in the Forex market. The major contribution was made by introducing the use of Quantum Differential Evolution Algorithm in a Neuro-fuzzy system composed of an Adaptive Neuro-Fuzzy Inference System controller to predict stock market. The algorithm was evaluated with actual financial data and proved to be better than the existing one.

3 Objectives and Methodologies

Multilayer perceptron is one of the landmark models to forecast stock market. Perhaps, there is no further need to prove it again. The only thing rests is to enhance its efficacy by supplementing certain add-ins in the form of extension of the model to suit the real-world situation or to merge with other established techniques, etc. The basic objective of this paper is to experiment with DEA and MLP to make a hybrid for prediction with comparatively less number of variables without compromising with the level of accuracy of prediction. The real usefulness of the model rests on the logic that the less number of variables under consideration will lessen the calculations to be made which leads to both time and memory saving approaches.

The collected historical data for the stated period in Section-4 forms the base for calculation of technical indicators under the study. All the variables considered are continuous in nature. Except “close” all are used as predicators where as “Close” is used as the target variable of the model for prediction in both existing and proposed models.

Table 1 Attributes in use

Attribute	Remarks	Attribute	Remarks
Open	Current day open	EMA	Calculated for the last 100 days
High	Current day maximum	PPO	The Price Oscillator in percentage
Low	Current day minimum	PAIN	The Price Action Indicator calculated for the current day
Close	Current day close	MACD	Shows the difference between 9 days and 26 days EMA
Volume	Volume of transaction for current day	RSI	Calculated from 14 days avg. max and avg. min
Highest high	Calculated for the last 21 days	Momentum	It is the difference between previous and current close
Lowest low	Calculated for the last 21 days	%K	Represents the percentage alert line
SMA	Calculated for the last 21 days	%D	Represents the percentage definite line
WMA	Calculated for the last 65 days	–	–

4 Preparation of Data

The selection of data set have very crucial role to play for experimental basis. It is the launch paid from where the model proposed has to be tested for their desired credibility. The study considered the historical BSE Sensex data from January 2008 to July 2015 to represent the real-world stock market data for empirical experiment of the model. The historical data of Open, High, Low, Close, and Volume are collected from Yahoo finance on daily basis and 12 number of technical indicators, viz., SMA, WMA, EMA, etc., are taken into consideration for experimental work. Table 1 gives a quick look of the attributes used.

5 Models in Use

Data Envelopment Analysis (DEA) and Multilayer Perceptron (MLP) are the two models which are converged to make a hybrid which gives the viability of the present study. One has already proved its efficiency to handle multiple inputs and outputs with easy to calculate comparative edge of the DMUs and the other one made a revolution in the field of forecasting by its remarkable footsteps. The following subsections are contributed to put forward a bird's eye view on these models.

5.1 *Data Envelopment Analysis*

Data Envelopment Analysis gets quick response as a fruitful and easy to handle technique for peers' performance evaluation form researcher across the globe. It is used to develop efficiency scores for Decision making units having similar kind of nature. The score ranges from zero to one with the logic that the DMU having scored closer to 1 is more efficient relating to the DMUs closer to zero. It has been extensively applied for performance evaluation and benchmarking of both non-profitable and profit-bearing units having qualitative or quantitative data. This paper implemented CCR model of the DEA for experiment. The efficiency score of DMUs in this model are calculated by the formula,

$$\text{Efficiency} = \frac{\text{Weighted sum of outputs}}{\text{Weighted sum of inputs}}$$

5.2 *Multilayer Perceptron*

Multilayer perceptron (MLP) is a form of Artificial Neural Network (ANN) which processes information in the way as the biological nervous systems does. MLP has remarkable ability to derive meaning from complicated or imprecise data to extract patterns and detect trends that are too complex to be noticed by either humans or other computer techniques. The common feature of a multilayer perceptron may be summarized as

i. A nonlinear active function is associated with each neuron of the network.
ii. Apart from input and output layers it is constituted by one or more number of hidden layers.
iii. The network of MLP shows very high degrees of connectivity within the network.

6 The Proposed Model

The model proposed is a DEA-based MLP model which filtered the input variables of MLP through DEA. It differs from the usual MLP in the sense that the input variables are not arbitrarily drawn rather they have to pass through DEA to prove their comparative efficiencies to be included in the model. A schematic representation of the model is reflected in Fig. 1. At the beginning the inputs of the model have to be selected via personal judgment. Then the researcher has to decide which

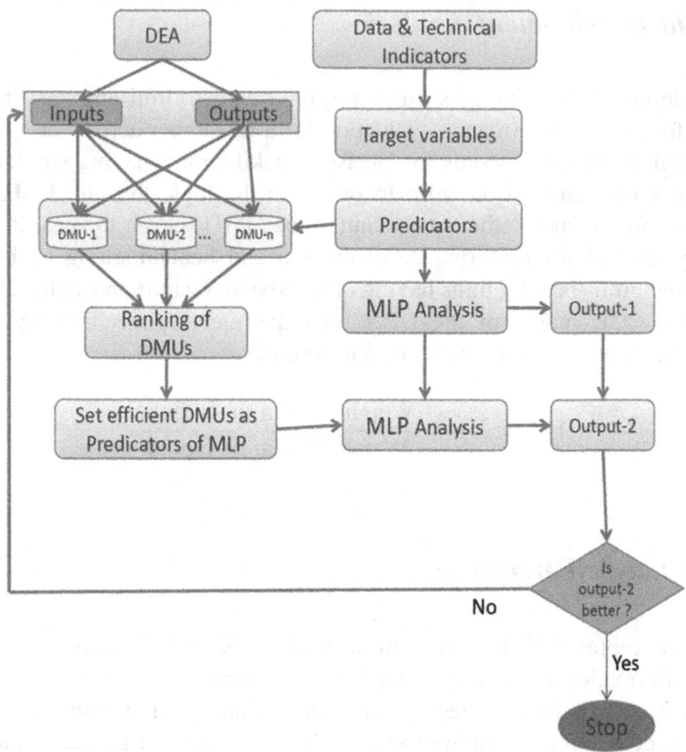

Fig. 1 Schematic view of the proposed model

one will act as the target variable. All the predicators have to be set as per model requirement. When all things are set one can run the MLP model to get the outputs and record it to facilitate comparative study with the outputs of the proposed model latter on. Now all the predicators used in previous model have to be taken as DMUs for test of comparative efficiencies. To test the DMUs again the inputs and output(s) of the DEA model have to be selected. Weighted inputs and output(s) have to be used for calculation of efficiencies of DMUs. On the basis of efficiency score top "n" number of DMUs has to be selected to be used as predicator for the proposed model. Taking the selected predicators again MLP is run over the data set. The output of the proposed model has to be noted and compared with the previous one to know the improvement if any. If there is improvement and the researcher is satisfied with it then stop further analysis and decide in favor of the proposed model else repeat the process of comparative efficiencies calculation through the DEA.

For DEA model the inputs considered include Mean, Median, Mode, Range, Variance, and Standard Deviation. The z-score is taken as the single output. The inputs are assigned random weight where the sum of weights assigned to all inputs of a DMU is equal to unity. The CCR model of the DEA is used for calculation of comparative efficiencies of DMUs. All total 16 numbers of predicators are tested for relative efficiencies and the top 10 are considered for the proposed model. Finally, outputs of both the model are compared and conclusion is drawn.

7 Output Analyses

The outputs of both of the model are analyzed with help of statistical parameters for training and validation and graphs. Table 2 gives a clear picture of comparative predictive capabilities of both the models.

So far as the statistical parameters are concerned, we take the assistance of parameters, viz., R^2, CV, NMSE, RMSE, MAE, etc. R^2 is a statistical parameter which gives us ideas about the proportion of variance explained by the model. Higher the R^2 value better the model explained the proportion of variance. Here we can notice that the proposed model bits the existing model. In case of the measurement of variation, i.e., the coefficient of variation the proposed model outperforms the existing model in both training and validation. The remaining parameters in Table 2 measure the correctness of the prediction in terms of levels and the deviation between the actual and predicted values. The smaller the values are the better is the prediction. Here we can observe that except for the MAE of the training data set the proposed model performs better than the existing model which proves its efficacy even with less number of predicators.

If we put a close look into Figs. 2 and 3 it can be observed that the predicted values of close are much closer and well distributed in Fig. 3 then Fig. 2 reflects the credibility of the proposed model against the existing MLP.

Table 2 Statistical parameters compared

Parameters	MLP		DEA based MLP	
	Training	Validation	Training	Validation
R^2	88.706 %	–	91.871 %	–
Coefficient of variation	0.082237	0.095766	0.06977	0.079426
NMSE	0.11294	67.38544	0.081294	46.3519
RMSE	1540.2373	2692.6514	1306.745	2233.217
MSE	2.3723e + 006	7.2504e + 006	1.7076e + 006	4.9873e + 006
MAE	1302.3053	2680.8787	1070.0865	2219.0081
MAPE	7.202036	9.5463774	5.8764275	7.9034725

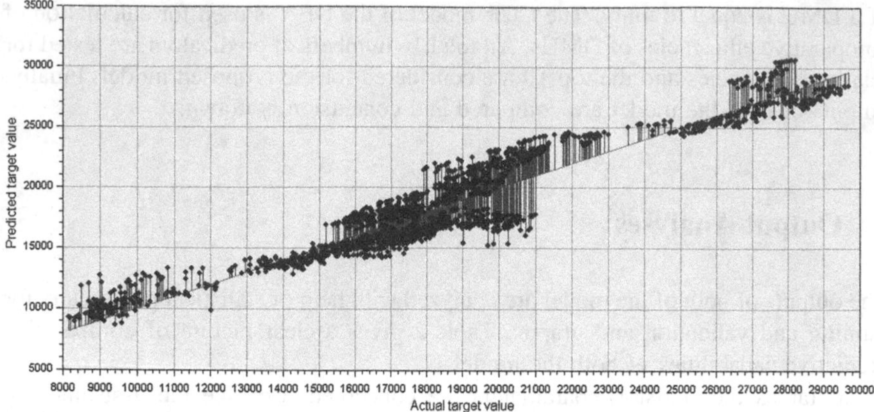

Fig. 2 Actual versus predicted values of close for MLP

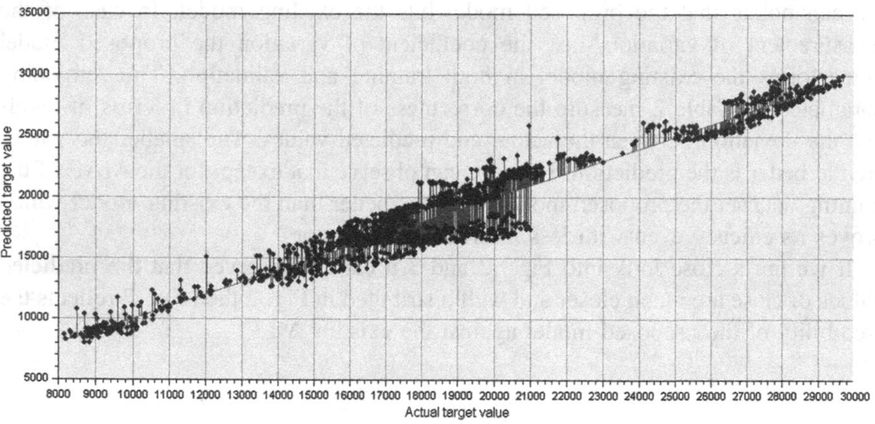

Fig. 3 Actual versus predicted values of close for proposed Model

8 Findings and Concluding Remarks

The trend of financial time series data is very hard to be explained by a single method. Thanks to soft computing approaches for allowing hybrids to be made of existing methods to enhance prediction capabilities. The study ends with the conclusion that the implementation of DEA for filtering predicators into the MLP certainly enhances the prediction capabilities of the MLP. Further due to comparatively less number of predictors are in use the proposed model can run in less time by saving both analysis time and memory. The DEA model if used judiciously may improve the efficiencies of other existing forecasting model.

References

1. Ayodele, A. A., Charles, A. K. et al: Stock Price Prediction using Neural Network with Hybridized Market Indicators. Journal of Emerging Trends in Computing and Information Sciences, VOL. 3, NO. 1(2012).
2. Charnes, A., Cooper, W.W., Rhodes, E.: Measuring the Efficiency of Decision Making Units, European Journal of Operational Research, 26th November 429–444 (1978).
3. Hegazy, O., Soliman, O. S. et al: Neuro-Fuzzy System Optimized Based Quantum Differential Evolutionary for Stock Market Forecasting. International Journal of Emerging Trends & Technology in Computer Science (IJETTCS), Volume 3, Issue 4, July-August 2014.
4. Hsue, Chih-Ming: An Integrated Procedure for Resolving Portfolio Optimization Problems using Data Envelopment Analysis, Ant Colony Optimization and Gene Expression Programming. International Journal of Computer Science and Business Informatics, Vol. 9, No. 1(2014).
5. Huang, W., Wang, S. et al: A New Computational Method of Input Selection for Stock Market Forecasting with Neural Networks. V.N. Alexandrov et al. (Eds.): ICCS 2006, Part IV, LNCS 3994, pp. 308 – 315, (2006).
6. Maciel, L. S. l, Ballini, R.: Neural Networks Applied To Stock Market Forecasting: An Empirical Analysis. Journal of the Brazilian Neural Network Society, Vol. 8, ISS. 1, pp. 3–22 (2010).
7. Mandziuk, J., Jaruszewicz, M.: Neuro-genetic system for stock index prediction. Journal of Intelligent & Fuzzy Systems, 22(2), 93–123, IOS Press (2011).
8. Panigrahi, S. S., Mantri, J. K.: Accuracy Driven- Financial Forecasting using MLP and SVM. International Journal of Advanced Trends in Computer Science and Engineering, Vol. 3, No. 4, Pages: 33–37 (2014).
9. Schefczyk, M.: Operational performance of airlines: An extension of traditional measurement paradigms. Strategic management Journal, pp. 301–317 (1993).
10. Tiwari S., Gulati, A.: Prediction of Stock Market from Stream Data Time Series Pattern using Neural Network and Decision Tree. IJECT Vol. 2, Issue 4, Oct. - Dec (2011).
11. Tsai, C.F., Wang, S.P.: Stock Price Forecasting by Hybrid Machine Learning Techniques. Proceedings of the International MultiConference of Engineers and Computer Scientists 2009 Vol I, IMECS 2009, March 18 – 20, Hong Kong (2009).
12. Wei, Liang-Ying, Cheng, Ching-Hsue: A Hybrid Recurrent Neural Networks Model Based On Synthesis Features to Forecast the Taiwan Stock Market. International Journal of Innovative Computing, Information and Control, Volume 8, Number 8 (2012).

Low-Noise Tunable Band-Pass Filter for ISM 2.4 GHz Bluetooth Transceiver in ±0.7V 32 nm CNFET Technology

S.K. Tripathi, Md. Samar Ansari and Amit M. Joshi

Abstract This paper presents a CNFET-based low-noise band-pass filter for application in 2.4 GHz Bluetooth transceivers. The proposed circuit exploits the ultrawide voltage and current bandwidths of CNFET-based analog building blocks as compared to their CMOS counterparts. Tuning of the center frequency is demonstrated via variation of a grounded resistor. SPICE simulations show that the CNFET-based building block consumes only 15 % power as compared to its CMOS equivalent. Noise analysis and Monte Carlo analysis also demonstrate the high performance characteristics of the proposed circuit.

Keywords Carbon Nanotube Field Effect Transistor (CNFET) · CMOS Scaling · Current mode (CM) · Dual-output inverting current conveyor (DOICC-II) · Voltage mode (VM)

1 Introduction

CMOS has been the dominating technology since last 30 years due to its high noise margin, small chip area, and extremely low power requirements. But continued downscaling of device size, up to sub-10 nm regime, dictated various limitations in the performance of MOSFET [1]. Source-to-Drain tunneling and short-channel effects are the key factors that degrade the performance of a MOS transistor. This is expected to result in preventing further scaling of the standard MOS device and

S.K. Tripathi (✉) · A.M. Joshi
Department of Electronics & Commmunication Engineering, Malaviya National Institute of Technology, Jaipur, India
e-mail: shailendra.amu@gmail.com

A.M. Joshi
e-mail: amjoshi.ece@mnit.ac.in

Md. S. Ansari
Department of Electronics Engineering, Aligarh Muslim University, Aligarh, India
e-mail: mdsamar@gmail.com

© Springer Science+Business Media Singapore 2017

S.C. Satapathy et al. (eds.), *Proceedings of the International Conference on Data Engineering and Communication Technology*, Advances in Intelligent Systems and Computing 468, DOI 10.1007/978-981-10-1675-2_43

expected to reach its limit by end of current decade [1, 2]. Therefore, the semiconductor industry is looking for alternate devices to incorporate with current CMOS technology.

The Carbon Nanotube Field Effect Transistor (CNFET) is an emerging technology which can be a viable alternative to cope with the limitations of CMOS technology. A CNFET is very similar to conventional MOSFET, except it has carbon nanotubes (CNTs) as a channel. These CNTs have near ballistic transport of charge carriers which make them excellent channel materials for high-speed and ultralow power electronic design [3]. The CNFET has outstanding properties like higher drive current, higher carrier mobility, higher transconductance, and good control over channel, as compared to its CMOS counterparts [4, 5].

Current conveyors (CCs) have gained a lot of consideration for their key advantages such as simpler circuit realizations, wide bandwidth, and large dynamic range [6, 7]. It is a very versatile building block and can be used to realize both voltage-mode (VM) and current-mode (CM) circuits. The design of active filters using current conveyor, has been of great interest from last two decades. Additionally, CM circuits received widespread attention due to their benefits weighed against VM circuits in terms of higher linearity, larger dynamic range, and lower power requirements. Further, circuits based on CM approach are also appropriate for implementation with MOS technology [8, 9]. The second generation inverting current conveyor (ICC-II) has become attractive for implementing a wide range of electronic functions either in voltage mode or current mode because of their high performance and versatility [10]. The design of active filter requires following main properties in the circuits: use of minimum active and passive components (preferably grounded ones), availability of most/all standard filter functions, and tuning mechanisms.

Several VM and CM circuits have been discussed in literature [9–15]. In 2012, Horng et al. proposed a cascadable current-mode universal filter using only two active elements and grounded capacitors [9]. Also, a current-mode universal filter with low THD and tuning through external currents is reported [11]. A voltage-mode fully differential biquadratic filter is proposed with the advantage of digital control of filter parameters [12]. Also, a high-order current-mode filter is proposed in [13] using CC-II. The circuit obtained all filter functions without any component matching conditions. A differential voltage current conveyor (DVCC) based VM filter has been proposed having only grounded components with orthogonal control of resonant angular frequency and condition of oscillation [14]. The proposed filter circuit has attempted to match the requirements of active filter design, as mentioned above.

The paper is organized as follows. Section 2 provides a brief overview of the CNFET. A comparative study of the performance of the chosen current conveyor for different technology nodes is presented in Section 3. Section 4 illustrates the proposed voltage-mode filter using CNFETs. The performance and results are depicted in Sect. 5. Finally, Sect. 6 concludes the paper.

2 Overview and Design Issues of Nanotube Transistor

A Carbon Nanotube (CNT) is a cylindrical nanostructure of Graphene lattice with excellent electrical, thermal, and mechanical properties [13]. The carbon nanotubes have commonly two forms, (a) Single-walled carbon nanotubes (SWCNTs) and (b) Multiwalled carbon nanotubes (MWCNTs) as shown in Fig. 1a. The SWCNTs are tubes of Grephene layer rolled in cylindrical form. The diameter is 1–2 nm and length a few microns (0.2–5 μm). The MWCNTs emerge as a coaxial tube with layers of SWCNTs. Their diameter ranges from 2 to 25 nm and interlayer spacing is about 0.36 nm. The performance of SWCNTs is better as compared to MWCNTs (although being costlier) but MWCNTs are easy to produce in large quantities [13].

CNTs are of three types, based on chiral vector (C) and chiral angel (θ). A CNT is known as armchair, if $m = n$ and $\theta = 0$. If chiral vector $m \neq n$ and chiral angel lies between 0^0 and 30^0, CNT is said to be chiral type and known as zigzag, if $m = 0$ and $\theta = 30^0$ as depicted in Fig. 1b, c. The major design parameters of carbon nanotube FET, are number of CNTs (N) in channel area, space between CNTs called pitch (S) and nanotube's diameter (D_{CNT}) [13]. The gate length and width are illustrated as L_{gate} and W_{gate}, respectively [14]. The CNT's diameter (D_{CNT}) and threshold voltage (V_{th}), the width of CNFET-based transistor (W), number of CNTs in channel (N), inter-CNT spacing (S), and energy gap (P_g) are related by Eqs. (1), (2), (3), and (4).

$$D_{CNT} = \frac{a\sqrt{n^2+m^2+nm}}{\Pi} \tag{1}$$

$$V_{th} = \frac{aV_{\Pi}}{qD_{CNT}\sqrt{3}} \tag{2}$$

$$W = (N-1)*S + D_{CNT} \tag{3}$$

$$\sum g = \frac{0.84eV}{D_{CNT}} \tag{4}$$

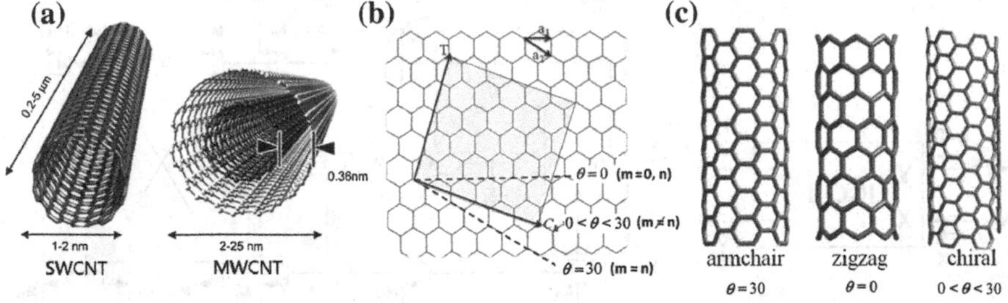

Fig. 1 **a** Single & multiwalled structure of CNTs, **b** Representation of chiral vector & chiral angel, **c** Types of CNTs

Here, m and n are the indices of chiral vector of the lattice and $a = 2.49$°A (constant of lattice structure); and $V_{\pi} = 3.033$ eV is the Carbon π–π bond energy.

3 CMOS and CNFET-Based Designs: Performance and Discussion

The block diagram of inverting current conveyor is shown in Fig. 2a which has one low-impedance, one high-impedance input terminal, and two high-impedance output terminals [15]. It makes the building block appropriate for VM and CM applications [6]. The circuit realization of a Dual-Output (DO) ICC-II of Fig. 2b is based on (5). The transistors M_5 and M_6, work as a current mirror which is set to drive two differential amplifiers consisting of transistors M_1 & M_2 and M_3 & M_4. Additionally, transistors M_7 and M_{11} provide the necessary feedback action to make the voltage V_X independent of current drawn from the terminal X. The current in terminal X is conveyed to the Z+ terminal with the help of transistors M_7, M_8, M_{11} and M_{12}. Using extra current mirror stage (M_{13}–M_{18}), the current is conveyed in an inverted manner to the Z−terminal. Moreover, the sum of drain currents of M_1 and M_4 is equal to drain currents of M_2 and M_3. Further, since transistors M_9 and M_{10} are biased with equal gate voltages (and since their source voltages are also equal), they would have equal drain currents. For matched M_9 and M_{10}, this would result in $V_{P1} = V_{P3}$. Routine analysis of the two differential pairs yields (6) and (7). In addition to this, some important CNFET design parameters for DOICC-II are shown in Table 1.

$$\begin{bmatrix} I_Y \\ V_X \\ I_{Z-} \\ I_{Z+} \end{bmatrix} = \begin{bmatrix} 0 & 0 & 0 \\ -1 & 0 & 0 \\ 0 & -1 & 0 \\ 0 & 1 & 0 \end{bmatrix} \begin{bmatrix} V_Y \\ I_X \\ V_Z \end{bmatrix} \tag{5}$$

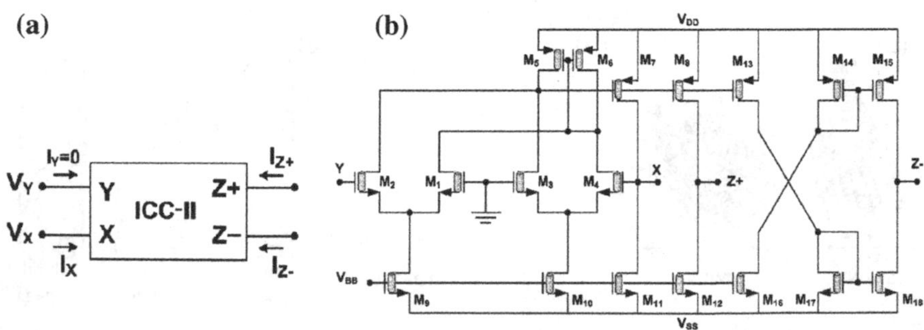

Fig. 2 **a** Circuit symbol of DOICC-II, **b** Transistor-level realization of DOICC-II

Table 1 Design parameters of a CNFET

Parameter	Value	Parameter	Value
Oxide thickness (T_{OX})	4 nm	Transistor channel length (L_{ch})	32 nm
Dielectric constant (K_{OX})	16	No. of CNTs (N)	6
Power Supply	0.9 V	Inter-CNT space (S)	20 nm
CNT's chirality (n1; n2)	19, 0	Diameter of CNT (D_{CNT})	1.5 nm

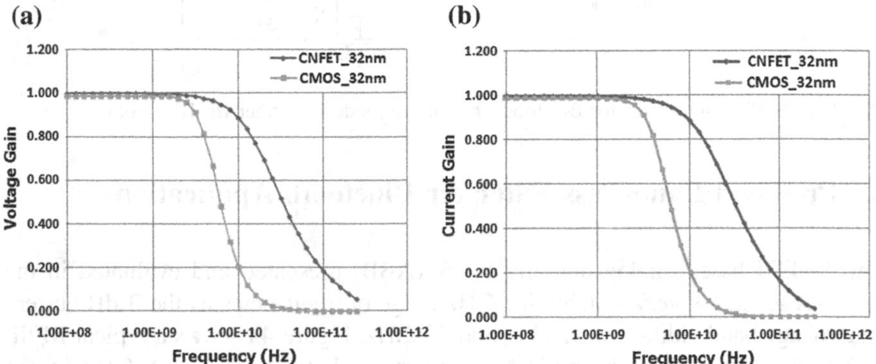

Fig. 3 **a** Voltage gains, **b** Current gains for CMOS & CNFET DOICC-II

$$V_Y - V_{P1} + V_{P1} - 0 = 0 - V_{P3} + V_{P3} - V_X \tag{6}$$

$$V_Y = -V_X \tag{7}$$

A comparative analysis of the performance of the DOICC-II at different technologies and feature sizes of CMOS & CNFET is presented next. Figure 3a highlights that 3 dB voltage bandwidths obtained from HSPICE simulation are 4.75 GHz and 19.29 GHz, for 32 nm CMOS and 32 nm CNFET, respectively. Additionally, Fig. 3b shows the current bandwidths for the different designs. The frequency response depicts that the current bandwidth of the CNFET DOICC-II is also extended to 20 GHz, which exceeds that of its CMOS counterparts. Moreover, HSPICE simulations match the theoretical agreements of the design by confirming (7). Comparison for other relevant parameters for the two technologies is presented in Table 2.

Table 2 Performance comparison of CMOS and CNFET-based DOICC-IIs

Designs/Factors	CMOS 32 nm	CNFET 32 nm
Model	PTM	Stanford
Transistor count	18	18
Current bandwidth	4.88 GHz	20.09 GHz
Voltage bandwidth	4.75 GHz	19.29 GHz
Supply voltage	±0.75 V	±0.70 V
Power dissipation	3 mW	0.43 mW

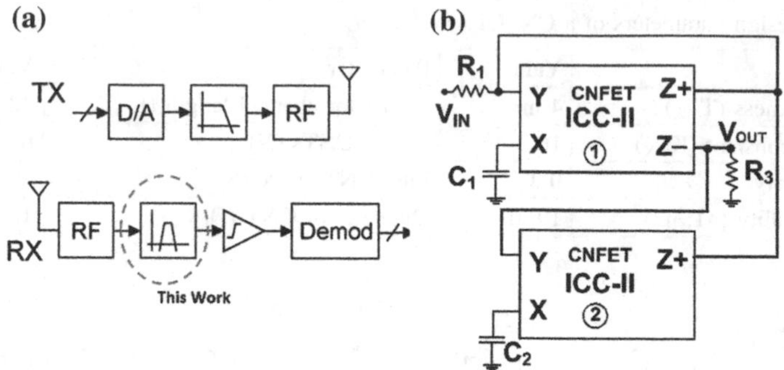

Fig. 4 **a** RF/IF architectures for Bluetooth, **b** Voltage-mode BPF filter for Bluetooth

4 Proposed Band-Pass Filter for Bluetooth Applications

The CNFET-based analog building block (ABB) presented and evaluated in the previous section is very suitable for GHz range of frequencies as the 3 dB current and voltage bandwidths are well beyond 20 GHz. Figure 4a shows a typical RF/IF architecture for Bluetooth applications operating between 2.4–2.48 GHz in the unlicensed industrial, scientific, and medical (ISM) radio bands. In the present work, a voltage-mode band-pass filter (BPF) is proposed which is an essential part of the Bluetooth receiver. The circuit, presented in Fig. 4b, uses two resistors and two grounded capacitors along with two ABBs. Frequency tuning is performed by making the resistor R_2 variable. Routine analysis of the filter circuit results in the band-pass filter function represented by Eq. (8). The pole frequency ω_0 and the quality factor Q can be derived as given in (9) and (10).

$$\frac{V_{OUT}}{V_{IN}} = \frac{s/C_2R_1}{s^2 + s/C_2R_2 + 1/C_1C_2R_1R_2} \tag{8}$$

$$\omega_0 = \sqrt{\frac{1}{C_1C_2R_1R_2}} \tag{9}$$

$$Q = \sqrt{\frac{C_2R_2}{C_1R_1}} \tag{10}$$

5 Simulation Results

In this section, the performance of the proposed filter is explored. SPICE simulation results are shown in Fig. 5a, b. The complete range of Bluetooth frequencies can be realized by the proposed voltage-mode filter. To achieve this, the passive elements

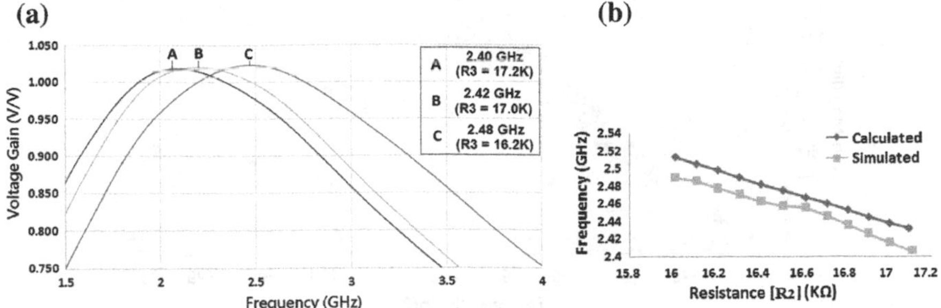

Fig. 5 **a** Tuning of entire Bluetooth range of frequencies, **b** Calculated and simulated center frequencies of band-pass filter

are taken as $C_1 = C_2 = 5$ f F, $R_1 = 10$ K and R_2 is varied from 17.2 to 16.2 K, to cover the full range of Bluetooth band from 2.4 to 2.48 GHz as illustrated in Fig. 5a. Furthermore, Fig. 5b reflects that calculated and simulated frequencies are in close conformity for concerned range and validates that filter frequency decreases with increase in resistance (R_2) as predicted by (9).

In addition to this, noise analysis of proposed filter has been also performed as illustrated in Fig. 6a. It is depicted from the results that input and output noises are below 90 nV. Moreover, Fig. 6b illustrates the Monte Carlo analysis for proposed band-pass filter. The analysis has been done for 30 trials with AC sweep environment. For the purpose of HSPICE simulations, CNFET parameters developed by Stanford University were used (Fig. 7).

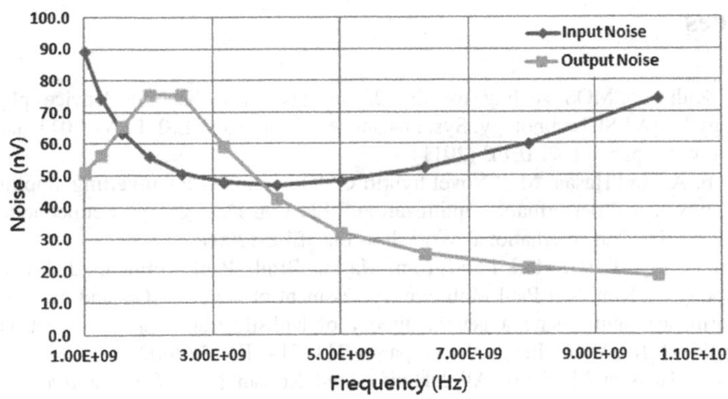

Fig. 6 Noise analysis for the proposed filter

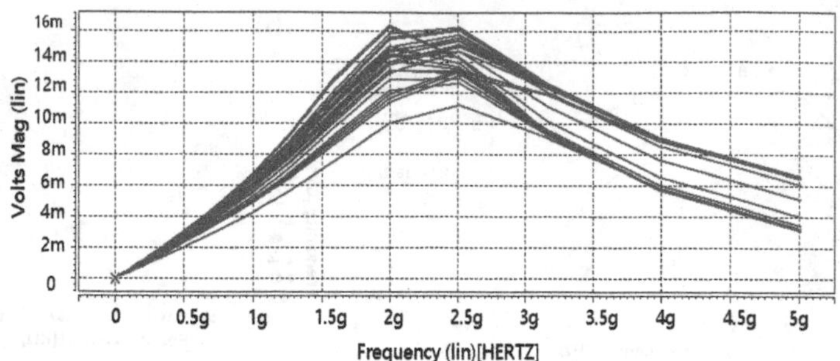

Fig. 7 Monte Carlo analysis for the proposed band-pass filter

6 Conclusion

In this work, it has been attempted to test the performance of an analog active block (Current Conveyor) based on CMOS and CNFET technological nodes. The building block based on CNFET has 3 dB current and voltage bandwidths exceeding 19 GHz while bandwidth approached to around 4.75 GHz for CMOS design. The CNFET-based circuit consumes seven times less power as compared to CMOS. Moreover, a tunable voltage-mode filter is presented, which can operate in the entire Bluetooth frequencies (2.40–2.48 GHz). Extensive SPICE simulations have been carried out to test the integrity of design. Such CNFET-based high frequency applications are expected to be worthy replacements of the contemporary CMOS designs once Moore's law saturates.

References

1. Kelin J Kuhn. "CMOS scaling for the 22 nm node and beyond: Device physics and technology". In VLSI Technology, Systems and Applications (VLSI-TSA), 2011 International Symposium on, pages 1–2. IEEE (2011).
2. Usmani, F. A., and Hasan, M., "Novel hybrid CMOS and CNFET inverting amplifier design for area, power and performance optimization." Electron Devices and Semiconductor Tech, 2009. IEDST'09. 2nd International Workshop on. IEEE (2009).
3. Jing Guo, Supriyo Datta, Mark Lundstrom, Markus Brink, Paul McEuen, Ali Javey, Hongjie Dai, Hyoungsub Kim, and Paul McIntyre. Assessment of silicon MOS and carbon nanotube fet performance limits using a general theory of ballistic transistors. In Electron Devices Meeting, 2002. IEDM'02. International, pages 711–714. IEEE (2002).
4. Mohammad Hossein Moaiyeri, Ali Jahanian, and Keivan Navi. Comparative performance evaluation of large FPGAs with CNFET-and CMOS-based switches in nanoscale. Nano-Micro Letters, 3(3):178–177 (2011).
5. Sharma, Jyoti, Mohd Ansari, and Jankiballabh Sharma. "Electronically Tunable Resistor-less Universal Filter in ±0.5 V 32 nm CNFET." Electronic System Design (ISED), 2014 Fifth International Symposium on. IEEE (2014).

6. Sedra, A. S., Gordon W. Roberts, and F. Gohh. "The current conveyor: history, progress and new results." IEEE Proceedings G (Circuits, Devices and Systems) 137.2, 78–87, (1990).
7. Beg, Parveen, M. A. Siddiqi, and Mohd Samar Ansari. "Multi output filter and four phase sinusoidal oscillator using CMOS DX-MOCCII." International Journal of Electronics 98.9 (2011): 1185–1198.
8. Bhopendra Singh, Abdhesh Kumar Singh, and Raj Senani. New universal current-mode biquad using only three ZC-CFTAs. Radioengineering, 21(1) (2012).
9. Jiun-Wei Horng, Chun-Li Hou, Ching-Yao Tseng, Ryan Chang, and Dun-Yih Yang. Cascadable current-mode first-order and second-order multifunction filters employing grounded capacitors. Active and Passive Electronic Components (2012).
10. Ehab Ahmed Sobhy and Ahmed M Soliman. Novel CMOS realizations of the inverting second-generation current conveyor and applications. Analog Integrated Circuits and Signal Processing, 52(1–2):57–64, (2007).
11. Sudhanshu Maheshwari and Iqbal A Khan. Novel cascadable current mode translinear-c universal filter. Active and passive electronic components, 27(4):215–218 (2004).
12. Parveen Beg, Sudhanshu Maheshwari, and Muzaffer A Siddiqi. Digitally controlled fully differential voltage-and transadmittance-mode biquadratic filter. Circuits, Devices & Systems, IET, 7(4) (2013).
13. Jiun-Wei Horng et al. High-order current-mode and transimpedance mode universal filters with multiple-inputs and two-outputs using MOCCIIs. Radioengineering, 18(4):537–543 (2009).
14. Shahram Minaei and Erkan Yuce. All-grounded passive elements voltage-mode DVCC-based universal filters. Circuits, Systems and Signal Processing, 29(2):295–309 (2010).
15. Inas A Awad. Inverting second generation current conveyors: the missing building blocks, CMOS realizations and applications. Int Journal of Electronics, 86(4):413–432 (1999).

Classification of Punjabi Folk Musical Instruments Based on Acoustic Features

Inderjeet Singh and Shashidhar G. Koolagudi

Abstract Automatic musical instrument classification can be achieved using various features extracted such as pitch, skewness, energy, etc., from extensive number of musical database. Various feature extraction methods have already been employed to represent data set. The crucial step in the feature extraction process is to find the best features that represent the appropriate characteristics of data set suitable for classification. This paper focuses on classification of Punjabi folk musical instruments from their audio segments. Five Punjabi folk musical instruments are considered for study. Twelve acoustic features such as entropy, kurtosis, brightness, event density, etc., including pitch are used to characterize each musical instrument from 150 songs. J48 classifier is used for the classification. Using the acoustic features, recognition accuracy of 91 % is achieved.

Keywords Algoza · Dhol · Dilruba · Punjabi folk musical instruments · Sarangi · Tumbi · J48 · Pitch

1 Introduction

Punjabi musical instruments are getting much more popular nowadays. In earlier times, only few musical instruments were used for some religious purposes or for cultural events. However, people have started showing interest toward music and a lot of musical instruments came into existence. Classification of musical instruments is discipline of its own right, and it has been used over the years. Punjabi musical instruments can be classified based on certain features like on their size, composition, range, etc. These are all the external features. We use to classify musical instrument based on internal features, i.e., the music it generates. Practical applications involve

Inderjeet Singh (✉) · S.G. Koolagudi
National Institute of Technology Karnataka, Surathkal, India
e-mail: inderjeet231990@gmail.com

S.G. Koolagudi
e-mail: koolagudi@nitk.ac.in

© Springer Science+Business Media Singapore 2017

S.C. Satapathy et al. (eds.), *Proceedings of the International Conference on Data Engineering and Communication Technology*, Advances in Intelligent Systems and Computing 468, DOI 10.1007/978-981-10-1675-2_44

multimedia for organizing and examining the data, retrieving music information, and music transcription. There are varieties available in musical instruments, we have mainly focused on music containing the instruments that are frequently occurring in popular folk music. These traditional musical instruments particularly belong to Punjab region and widely used there. These are algoza, dhol, dilruba, tumbi, and sarangi. We classify these instruments based on certain features.

The rest of paper is organized as below. An assessment of existing task is given in Sect. 2. The proposed method is explained in Sect. 3. Experimentation and results along with explanation of database are given in Sect. 4. Section 5 concludes the work with some further research directions.

2 Literature Review

Many researches have focused on the analysis of patterns observed in musical instrument analysis, various feature extraction schemes have been proposed and adopted. Different computational models or classification algorithms have been employed for the purposes of instrument detection and classification. Mel-frequency cepstral coefficients (MFCC) features are commonly employed in music genre and instrument classifications. Research findings report [1] that Gaussian mixture models and support vector machine classifiers can be able to distinguish eight musical instruments with 70 % accuracy. Eronen [2] assessed the performance of MFCC, spectral, and temporal features such as amplitude envelope and spectral centroids for instrument classification. Karhunen [3] has conducted experiments to decorrelate the features. k-nearest neighbor (k-NN) classifiers is used with fivefold cross-validation. The results favored the MFCC features. Monophonic musical instrument classification in [4] also uses k-NNC using different feature sets with improved accuracy. Instrument classification on solo music is shown in [5] where high accuracy is achieved using cepstral features. The standardization of the extraction and description of different audio features [6] is targeted by MPEG-7 audio framework. Livshin and Rodet [7] consider music samples from different performers and uses GDE feature selection algorithm for selecting a small feature set. In [8], the sound description of MPEG-7 audio features based on their perceived features similarity is accessed. Xiong [9] compared the MFCC and MPEG-7 audio features for the purpose of music classification, adopting the hidden Markov models (HMMs)and k-NN. A study conducted by Brown et al. [10] on identifying four instruments based on various musical features such as cepstral coefficients, spectral centroid, and autocorrelation coefficients. The classification of various instruments is also studied by Kostek [11].

3 Proposed Method

The proposed method is based on classifying five musical instruments so that its accuracy will get increased. The proposed method is shown in Fig. 1. The various steps involved in this approach are shown below.

A. Database Collection

In the proposed method, as shown in Fig. 1 we firstly collected different samples of above mentioned instruments. There are total 150 samples out of which 66 different samples of algoza, 32 samples of dhol, 22 samples of dilruba, 14 samples of sarangi, 18 samples of tumbi.

B. Feature Extraction

In order to extract the required features, the audio signal is first decomposed using bank of filters, which is used for selecting frequencies ranging from low frequency values to high frequency values so that each of the channels can be examined separately. These channels are further decomposed into frames with default frame length of 50 ms and half overlapping. Now the decomposition is performed using a discrete Fourier transform, which, for an audio signal x can be computed as:

$$X_k = \sum_{n=0}^{N-1} x_n e^{\frac{-2\pi i}{N} k_n} \quad (1)$$

The output of this step is the graph which highlights the repartition of the amplitude of the frequencies. The result obtained is the spectrum of the audio signal. Now we have to evaluate periodicities in signals, we have to local correlation between the samples. If we take a signal x, the autocorrelation function is computed as follows:

$$R_{xx}(j) = \sum_{n} x_n \bar{x}_{n-j} \quad (2)$$

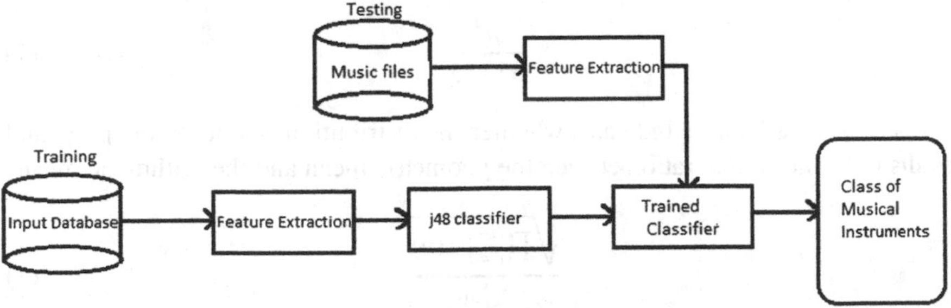

Fig. 1 Proposed system

For a given lag j, the autocorrelation R_{xx}(j) is computed by multiplying point par point the signal with a shifted version of it of j samples.

Root Mean Square Energy (rms): The global energy of the signal x can be computed simply by taking the root average of the square of the amplitude, also called root mean square (RMS):

$$x_{rms} = \sqrt{\frac{1}{n}\sum_{i=1}^{n} x_i^2} = \sqrt{\frac{x_1^2 + x_2^2 + \cdots + x_n^2}{n}} \tag{3}$$

Irregularity: The irregularity of a spectrum is the degree of variation of the successive peaks of the spectrum. It is calculated as the sum of the square of the difference in amplitude between adjoining partials.

$$x_{irreg} = \frac{(\sum_{k=1}^{N}(a_k - a_{k+1})^2)}{\sum_{k=1}^{N} {a_k}^2} \tag{4}$$

Zero Crossing Rate: zero cross counts the number of times the signal crosses the X-axis. It is an indicator for noisiness of the signal. Zero crossing rate can be calculated by using the below equation:

$$ZCR = \frac{\sum_{n=1}^{N} |sign(F_n) - sign(F_{n-1})|}{2N} \tag{5}$$

Entropy: Given a prediscretized feature set, the noisiness of the feature X can be measured as the entropy, which is defined as

$$H(X) = -\sum_i P(x_i)lgP(x_i) \tag{6}$$

Kurtosis: It is defined as the fourth cumulant divided by the square of the variance of the probability distribution, equivalent to:

$$\frac{\mu^4}{\sigma^4 - 3} \tag{7}$$

Flatness: The flatness indicates whether the distribution is smooth or spiky, and results from the simple ratio between the geometric mean and the arithmetic mean:

$$\frac{\sqrt{\prod_{n=0}^{N-1} x(n)}}{(\frac{\sum_{n=0}^{N-1} x(n)}{N})} \tag{8}$$

Some of the sample features of different instruments are shown in Table 1. Comparison of five musical instruments is shown in Fig. 2 where rows represent histograms for various features and columns represent the name of the instruments. The comparisons show that instruments belong to string category produces higher range values for different features as compared to dhol and algoza.

C. Instrument Classification model

In the proposed work, J48 classifier is used for modeling and classifying different musical instruments. This classifier generates a decision tree for the given data set by recursive partitioning of data. The decision is grown using depth-first strategy. The main functionality of this algorithm is to split the data sets by considering all the possible tests and selecting a test that gives the best information gain. In order to calculate the entropy gain of all these binary tests efficiently, the training data set belonging to the node is sorted for the values of the continuous attributes. The designed algorithm takes the different music samples and after feature extraction process, these samples are inputted to classifier to give an accuracy of 71 %. On the second time after improving the input data samples, again the same procedure follows which gives an accuracy of about 91 % as shown in Fig. 3.

Algorithm 1: Algorithm for musical instrument classification
Data: Music signal in .wav format
Result: confusion matrix

```
while .wav file getting processed do
    read input file;
    p = extract pitch;
    rms = extract rms;
    reg = extract irregularity;
    zcr = extract zero cross(zcr);
    eneg = extract energy;
    skew = extract skewness;
    ent = extract entropy;
    kurt = extract kurtosis;
    bri = extract brightness;
    ed = extract eventdensity;
    m = extract mode;
    flat = extract flatness;
    store total 14 features into a file ;
end
give input to J48 in arff file format;
train classifier with training data;
test with testing data to determine classified musical instrument;
calculate accuracy based output;
```

Table 1 Features of different musical instruments

	Pitch	rms	Irregularity	Zcr	Energy	Skewness	Entropy	Kurtosis	Brightness	Eventdensity	Mode	Flatness
Algoza	637.385	0.07823	0.39419	1385.1	5.22E−01	1.34E + 07	0.68487	2.12E + 10	0.54783	1.6667	−0.04748	0.016
Dhol	104.93	0.36123	1.1763	498.09	0.53731	1.55E + 09	0.82233	4.90E + 12	0.37249	2.9433	−0.1292	0.049
Dilruba	347.11	0.19862	0.95941	956.65	0.54455	2.99E + 08	0.77821	6.14E + 11	0.40306	0.38873	0.048191	0.048
Sarangi	179.967	0.52231	0.87463	1842.3	0.51695	1.74E + 10	0.88704	1.59E + 14	0.60209	0.33459	0.15388	0.195
Tumbi	556.28	0.15705	0.64985	817.71	0.56731	8.32E + 08	0.75795	5.71E + 12	0.21633	3.4286	−0.07203	0.006

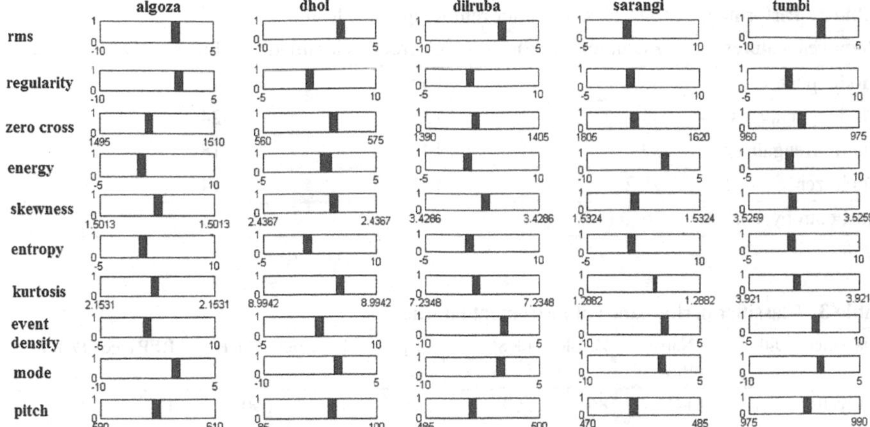

Fig. 2 Comparison of different features for musical instruments. Columns represent various musical instruments and rows represent histogram for different features

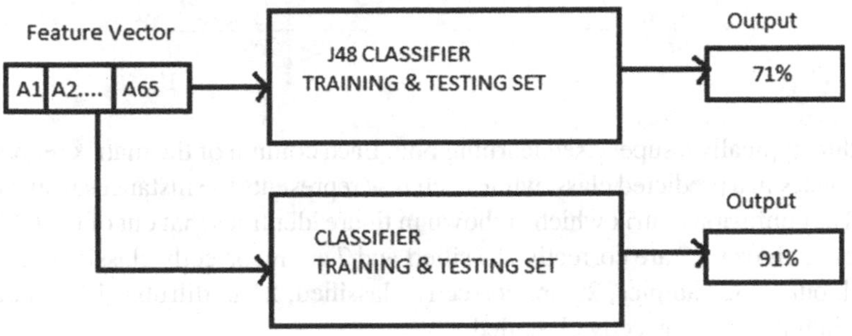

Fig. 3 Illustration of output calculation using J48 classifier

4 Experiments and Results

Various classifiers give different results with different accuracies. The J48 classifier has tree algorithm which gives result with more accuracy as compared to different classifiers. It treats input as nodes of the tree and depending on certain conditions, it creates decision tree of the inputs. This is efficient and easier to calculate the output which leads to more accurate results. Table 1 shows the five instruments which we used and their corresponding feature values. From the analysis we can easily see that pitch value of dhol is very low as it belongs to brass family, where as the pitch value of tumbi is high which belongs to string family. After then algoza comes which has almost similar pitch values as compared to tumbi but its pitch value gets lower in some cases. As both sarangi and dilruba also come under string family and sarangi is a variant to dilruba, they both gives almost similar results in every case. Table 4 gives the confusion matrix of the result we obtained from J48 classifier. Confusion matrix is a specific table layout that allows visualization of the performance of an

Table 2 Performance of J48 classifier after removing some features

Removed features	Accuracy (in %)	Correctly classified	Incorrectly classified
Mode, pitch	90	135	15
Mode, pitch, rms	85	136	24
Pitch, irreugalrity	89	142	18
Pitch, zcr	91.2	146	14
rms, entropy, zcr	76	127	33

Table 3 Classifier performance of instrument families

Classifier	J48	Naive Bayes	ZeroR	KStar	JRip	Logistics	OneR	REPTree	PART
Correctly classified	137	141	86	90	136	143	90	129	137
Incorrectly classified	13	9	64	60	14	7	60	21	13
Accuracy (in %)	91.3	94	57.3	60	90.7	95.3	60	86	91.3

algorithm, typically a supervised learning one. Each column of the matrix represents the instances in a predicted class, while each row represents the instances in an actual class. The confusion matrix which is shown in figure identifies that out of the 64 input samples of algoza, 57 are correctly classified and 7 are incorrectly classified. In case of dhol, out of 32 samples, 29 are correctly classified, 21 in dilruba, 13 in sarangi, and 17 in tumbi are correctly classified.

Table 2 shows the performance of J48 classifier after removing certain features. The performance of various classifiers are shown in Table 3. When we supplied input to Naive Bayes classifier, it uses 10 folds for cross-validation, and gives an accuracy of 94 % and 141 are correctly specified. For ZeroR classifier which comes under rules family, 57.3 % accuracy and 86 are correctly specified. For Kstar classifier which comes under lazy family, 60 % accuracy and 90 are correctly specified. For JRip classifier which comes under rules family, 90.7 % accuracy and 136 are correctly specified. Considering One-R classifier, it gives almost to K-star classifier. This One-R comes under rules family. Logistics have the highest accuracy of 95 % in which correctly classified are 143. This comes under function family. REP tree comes under tree family having 86 % accuracy and PART is a class for generating a PART decision list. Uses separate-and-conquer. Builds a partial C4.5 decision tree in each iteration and makes the "best" leaf into a rule. It gives accuracy of 91.3 % (Table 4).

Table 4 Confusion matrix for instrument classification

	Algoza	Dhol	Dilruba	Sarangi	Tumbi
Algoza	**57**	1	6	0	0
Dhol	1	**29**	2	0	0
Dilruba	1	0	**21**	0	0
Sarangi	0	1	0	**13**	0
Tumbi	0	1	0	0	**17**

5 Summary and Conclusion

In this paper, we have considered those musical instruments that belong to different categories which make it easier to classify them based on their acoustic features. First we collected the music files and then after extracting various features from it, we examined the use of J48 classifier for better classification. Also we have considered some music files for training and testing the classifier including 10-fold cross-validation which leads to the output much more accurate. The scope of some current studies and performance achieved are given in Table 3, where the classification accuracy and classifier performance of various instrument families are listed. It can be seen that our results are better than or comparable with those obtained by other classificrs.

As a future work, the system shall be tested with a large database and including mixed audio clips of various instruments from which we have to classify each and every instrument using different classifiers and compare result and determine which classifier give more accurate result.

References

1. J. Marques and P. Moreno, A study of musical instrument classification using Gaussian mixture models and support vector machines, Compaq Comput. Corp., Tech. Rep. CRL 99/4, 1999.
2. A. Eronen, Comparison of features for musical instrument recognition, in Proc. IEEE Workshop Appl. Signal Process. Audio Acoust., 2001, pp. 1922.
3. G. Agostini, M. Longari, and E. Poolastri, Musical instrument timbres classification with spectral features, EURASIP J. Appl. Signal Process., vol. 2003, no. 1, pp. 514, 2003.
4. I. Kaminskyj and T. Czaszejko, Automatic recognition of isolated monophonic musical instrument sounds using kNNC, J. Intell. Inf. Syst., vol. 24, no. 2/3, pp. 199221, Mar. 2005.
5. S. Essid, G. Richard, and B. David, Efficient musical instrument recognition on solo performance music using basic features, presented at the Audio Engineering Society 25th Int. Conf., London, U.K., 2004, Paper 25. accessed 22.11.2005. [Online]. Available: http://www.tsi.enst.fr/%7Eessid/pub/aes25.pdf
6. ISO/IEC Working Group, MPEG-7 Overview, 2004. accessed 8.2.2007. [Online]. Available: http://www.chiariglione.org/mpeg/standards/mpeg-7/mpeg-7.htm

7. A. A. Livshin and X. Rodet, Musical instrument identification in continuous recordings, in Proc. 7th Int. Conf. Digital Audio Effects, 2004, 222226. [Online]. Available: http://dafx04.na.infn.it/
8. G. Peeters, S. McAdams, and P. Herrera, Instrument sound description in the context of MPEG-7, in Proc. Int. Comput. Music Conf., 2000, pp. 166169.
9. Z. Xiong, R. Radhakrishnan, A. Divakaran, and T. Huang, Comparing MFCC and MPEG-7 audio features for feature extraction, maximum likelihood HMM and entropic prior HMM for sports audio.
10. J. C. Brown, O. Houix, and S. McAdams, Feature dependence in the automatic identification of musical woodwind instruments, J. Acoust. Soc. Amer., vol. 109, no. 3, pp. 10641072, Mar. 2001.
11. B. Kostek, Musical instrument classification and duet analysis employing music information retrieval techniques, Proc. IEEE, vol. 92, no. 4, pp. 712729, Apr. 2004.

Combined DWT–DCT-Based Video Watermarking Algorithm Using Arnold Transform Technique

Amit M. Joshi, Shubham Gupta, Mohit Girdhar, Pranshu Agarwal and Ranabir Sarker

Abstract The communication has become faster and easier through the Internet. The creation and delivery of various data (video, images, speech and audio) have grown by many fold. However, the processing of these data in the digital domain raises the issues like the content protection and providing the ownership of original content. Digital watermarking is a very useful technique which can overcome the shortcomings of current copyright laws of the digital data. Video is one of the most popular multimedia objects. Video watermarking is used to embed the ownership logo into the digital video content for the protection. The proposed algorithm uses energy compression property of DCT and multiresolution of DWT. The Arnold transformation technique is applied, which enhances the security level and also increases the robustness. Experiment results show that the proposed algorithm has the larger embedding capacity and the excellent robustness against all the attacks.

Keywords Arnold transform · Blind detection · Copyright protection · Human visual system · Normalized correlation

1 Introduction

The growth of Internet technology has enabled the faster sharing of image/video with great ease [1]. With the help of various handheld devices, peoples are exchanging information rapidly from anywhere in the world. The easy access of editing tool has explored the possibilities of manipulation of original content [2]. The digital watermarking is a useful data hiding technique where an ownership information is covertly inserted in original video/image content [3]. The paper

A.M. Joshi (✉) · S. Gupta · M. Girdhar · P. Agarwal · R. Sarker
Department of Electronics & Communication Engineering,
Malaviya National Institute of Technology, Jaipur, India
e-mail: amjoshi.ece@mnit.ac.in

© Springer Science+Business Media Singapore 2017

S.C. Satapathy et al. (eds.), *Proceedings of the International Conference on Data Engineering and Communication Technology*, Advances in Intelligent Systems and Computing 468, DOI 10.1007/978-981-10-1675-2_45

covers an efficient watermarking scheme which has robustness against all types of attacks and hides the content as imperceptible manner. The retrieval process is blind and extracted watermark is checked by correlation measurement with original watermark. The proposed algorithm is useful for copyright protection and ownership verification [4]. Sinha et al. [5] developed a robust and imperceptible video watermarking method using Discrete Wavelet Transform (DWT) and Principal Component Analysis (PCA). After the decomposition of DWT, the wavelet coefficients are highly correlated and they are decorrelated with the help of PCA technique. The binary watermark is inserted in principal components of low-frequency wavelet coefficients. The algorithm has robustness against the spatial attacks, but the performance humiliates against temporal attacks such as frame dropping, frame removal, frame swapping, etc. Deshpande et al. [6] designed various approaches based on Discrete Cosine Transform (DCT) with the concept of inserting an assorted watermark in the video. The algorithms were used to embed the invisible watermark as audio, video, and image in original video sequences. The algorithm showed limited performance against various noise-based attacks. Agarwal et al. [7] explained DWT–DCT-based video watermarking scheme where the watermark is embedded in some randomly selected frames. The middle frequency bands (LH and HL) of the wavelet transform were chosen and DCT is applied to these bands. The mid-frequency components of DCT are altered in order to insert the watermark. The algorithm posses excellent robustness against the attacks, but it is nonblind scheme where the original watermark is required at the time of extraction. This limits the application domains of the algorithm. In this paper, an efficient and novel watermark embedding technique is developed which provides the robustness against different attacks and also inserts the watermark as invisible manner. The blind watermark detection technique is able to extract the watermark effectively from watermarked video. Then extracted watermark is checked with correlation measurement with the original watermark and is useful to prove ownership and copyright of the originator.

2 Background Theory

Generally, the video watermarking algorithms are categorized into two main groups known as spatial domain and transform domain watermarking [8]. The spatial domain watermarking methods are used to embed the watermark data directly in the pixel values of the image/frame. On the other hand, transform domain techniques such as Discrete Fourier Transform (DFT), DCT, DWT are able to insert the watermark into transforms coefficient. Compared to spatial domain methods, transform domain techniques are effective in terms of greater robustness and higher imperceptibility.

2.1 Discrete Wavelet Transform (DWT)

DWT is a multiresolution and orthogonal transform technique for an image/frame. This is a very useful transformation technique for converting the spatial domain signal to frequency domain coefficients [9]. The wavelet transform is helpful to decompose any input image/frame into four subbands known as LL, LH, HL, and HH which provide the information of horizontal, vertical, and diagonal directions. LL is the low frequency band which represents the approximate part of an image/frame while the remaining three bands LH, HL, and HH provide the detailed parts. The watermark can be inserted into low frequencies in order to achieve higher robustness against geometric manipulation, lossy compression, and low pass filtering. However, the modification in higher frequency bands may result in perceptual distortion. The watermark can be inserted in the high frequency band by less visible effects, but at the same time it suffers in terms of the robustness against the compression and filtering attacks.

2.2 Discrete Cosine Transform

A Discrete Cosine Transform (DCT) is useful to generate a finite sequence in the forms of the cosine function at various frequencies of oscillation. It is a method which converts any signal in the basic frequency components [10]. DCT is basically a block-based transformation technique and divides the image/frame in size of N × N nonoverlapping blocks. For an input image X and the output image Y (of size N × M), the DCT coefficients can be calculated as per Eq. (1):

$$Y(u, v) = \sqrt{\frac{2}{m}}\sqrt{\frac{2}{n}}\alpha \sum_{m=0}^{M-1}\sum_{n=0}^{N-1} X(m, n) \cos\frac{(2m+1)u\pi}{2m} \cos\frac{(2n+1)v\pi}{2n}, \quad (1)$$

where

$$\alpha = \frac{1}{\sqrt{2}}\ for, u\,\&\,v = 0$$
$$\alpha = 1\ for, u\,\&\,V = 1, 2, \ldots, M-1.$$

The DCT transformation breaks the image into different bands of frequency as high frequency, low frequency, and middle frequency. The low frequency components are sensitive for HVS (Human Visual System). At the same time, the high frequency components are susceptible against several attacks. Therefore, the middle frequency components are an obvious choice to have a tradeoff between robustness and imperceptibility. DCT has better performance than DFT in terms of continuities. DCT is useful to represent the signal with minimum number of coefficients while DFT is a periodic representation of signal with truncation of coefficients.

DCT is one of the most popular standards for image and video compression standard. It can represent a particular signal in more compact form. DFT has another drawback that it truncated the coefficient of the signal and therefore there is a loss of information. DCT is continuous periodic structure and has a close approximation of the original signal.

2.3 Arnold Transform

Arnold transform is periodic in nature and is used to scramble the values of the original signal [11]. Therefore, it has potential for a wide range of application in the information hiding schemes. It is defined as Eq. (2):

$$\begin{pmatrix} x' \\ y' \end{pmatrix} = \begin{pmatrix} 1 & 1 \\ 1 & 2 \end{pmatrix} \begin{pmatrix} x \\ y \end{pmatrix} \bmod N \quad \mathrm{x, y} \in \{0, 1, \ldots, \mathrm{N} - 1\} \tag{2}$$

Here, (x, y) is the value of pixel of the original image and (x', x') is the pixel value of the transformed image. N indicates the order of the image matrix and depends on image size. In general, the image is in the form of $N \times N$, i.e., square image otherwise you can always resize into the square size image. It tries to rearrange the pixels of the original frame/image. As the image is treated as a matrix of discrete point, there the repetition of the Arnold transforms to a certain step or repeating the inverse transformation at the same instance will result in the rebuilding of the image. This reveals that the transformation is a cyclical nature. The circle of the transform will be increased by equal increments of image size N. Arnold transform is easier for the implementation and is useful to increase the video security.

3 Proposed Video Watermarking Algorithm

The proposed video watermarking method combines the advantages of energy compression and multiresolution properties of DCT and DWT, respectively [12]. The addition of Arnold transform helps to enhance the robustness of the algorithm.

The watermark embedding steps for proposed algorithm are as follows:

Step 1 The original video is divided into frames. Then, each frame is transformed using DWT at the second level of decomposition to have sub-band LH_2 (or HL_2). This middle frequency components are chosen for embedding as it has an optimized performance between robustness and invisibility

Step 2 The obtained sub-band LH_2 (or HL_2) is further divided into 4×4 blocks and DCT is applied to each individual block

Step 3 The original watermark is scrambled with Arnold transform using secret key K. Here, K is defined as the number of times it scrambles. This has resulted in a series of vector which comprises values of zeros and ones

Step 4 Two uncorrelated sequences of pseudorandom values are generated. One sequence is used for embedding the watermark bit '0' (PN-0) whereas second sequence helps to embed a watermark bit '1' (PN-1). The size of pseudorandom sequence is defined by number elements of middle frequency band of 4 × 4

Step 5 In a middle frequency band of 4 × 4 block, two random sequences are inserted with constant gain factor. The embedding is as per Eqs. (3) and (4):

If the watermark bit is 0 then,

$$X^{'} = X + \alpha \times PN - 0 \quad (3)$$

Otherwise, If the watermark bit is 1 then,

$$X^{'} = X + \alpha \times PN - 1 \quad (4)$$

Where, X refers to coefficient matrix of middle frequency of transformed block. X' is a coefficient matrix of middle frequency of DCT after watermark embedding

Step 6 After the completion of embedding process, Inverse DCT (IDCT) is applied to get watermarked 4 × 4 size blocks

Step 7 Finally, second level inverse DWT (IDWT) is applied to get LH_2 (or HL_2), then all obtained subsequent watermarked frames are combined to get secure watermarked video

The steps of the watermark extraction algorithm are as follows:

Step-1 Apply two-level DWT on watermark frame to have sub-band LH_2 (or HL_2)

Step-2 The sub-band of LH_2 (or HL_2) is divided into 4 × 4 blocks and DCT is applied to each block

Step-3 Two pseudorandom sequence PN-0 and PN-1 are generated with same technique which was used at the time of watermark embedding

Step-4 The watermark is calculated by correlation between each block middle frequency DCT coefficient and two random sequences (PN0 and PN1). If a value of correlation with PN-0 is higher than PN-1, the watermark bit is retrieved as '0' otherwise the watermark bit is of value '1'

Step-5 The watermark image is reconstructed with extracted watermark bit and apply N times (where N is a period of Arnold transform)

4 Results

The algorithm is tested using MATLAB for the performance evaluation. We have used akiyo.yuv (frame size of 720 × 486) as an original test video and the first frame of same is as shown in Fig. 1a. The binary logo is used as a watermark and is shown in Fig. 1b. Figure 1c shows the first watermarked frame of the original video.

Normalized cross-correlation (NC) is the most commonly used metric to calculate the degree of similarity (or dissimilarity) between two images/frames [13]. It measures the correlation between the original watermark and the extracted watermark. The robustness of the algorithm is verified against different attacks and NC is calculated after the consideration of noise, compression and filtering attack. NC may take any value between 0 and 1 [14]. NC should be ideally 1 and higher the value of NC defines more similarity between the original watermark and extracted watermark. The definition of NC is as follows in Eq. (5):

$$NC = \frac{\sum_{p=0}^{N-1} \sum_{q=0}^{M-1} w(m,n) \times w'(m,n)}{\sqrt{\sum_{p=0}^{N-1} w(m,n) \times w(m,n) \sum_{q=0}^{M-1} w'(m,n) \times w'(m,n)}} \tag{5}$$

w(m, n) is the original frame of video, w′(m, n) is the watermarked frame of video.

The normalized cross-correlation has an advantage over the normal cross-correlation. It has a lesser sensitivity toward linear changes in amplitude of illumination for two images. Mean Squared Error (MSE) is quantified parameter which measures the error value after the averaging of the squares of two images. MSE is given by the following Eq. (6):

$$MSE = \frac{1}{M * N} \sum_{k=0}^{M-1} \sum_{l=0}^{N-1} [I_o(k,l) - I_w(k,l)]^2 \tag{6}$$

I_o = Original image and I_w = watermarked image.

Fig. 1 **a** First frame of original video, **b** original watermark and **c** first frame of watermarked video

Table 1 Proposed video watermarking against frame based attacks

Type of attack	NC	MSE	PSNR
Without attack	0.9973	0.8725	48.7260
Gaussian noise (0 mean and 0.01 variance)	0.9845	73.8074	29.9987
Salt and pepper noise (density 0.05)	0.9947	49.3210	31.2343
Gamma correction	0.9233	205.62	25.0020
Median filtering (3 × 3)	0.9943	1.907	46.4927
Sharpening	0.9342	10.07	38.1005
Histogram equalization	0.9762	46.82	22.2345

Table 2 Proposed video watermarking against temporal attacks

Type of attack (10 %)	NC	MSE	PSNR
Frame averaging	0.9211	23.3120	32.4332
Frame swapping	0.9355	1.7021	46.8774
Frame removal	0.9331	0.8725	48.7260

MSE is inversely related to Peak Signal-to-Noise Ratio (PSNR) [15]. PSNR and MSE are determined between original frame and watermarked frame to verify perceptual distortion after the watermark embedding process. PSNR is the ratio of maximum value of the pixel of the frame/image and MSE. It signifies the power of corrupting noise which influences the fidelity representation. PSNR is generally useful for the measurement of quality after the reconstruction of the original signal. The signal is the original frame, whereas the noise is the error commenced by various attacks. The proposed algorithm is evaluated against different attacks which commonly occur in the particular frame of the video and values are noted in Table 1. Similarly, the performance is measured against various temporal attacks and reported in Table 2.

Tables 1 and 2 show that the proposed algorithm has an excellent robustness against almost all the attacks. However, the performance is degraded in the presence of gamma correction and histogram equalization. However, the overall performance of the algorithm is admirable. The algorithm is also compared with previous similar work for spatial and temporal domain attacks and is shown in Tables 3 and 4, respectively. The proposed algorithm outperforms all the related schemes and has better NC against all types of attacks.

Table 3 Comparison of NC for proposed video watermarking with previous work for spatial domain attacks

Spatial attacks	Proposed work	Deshpande et al. [5]	Agarwal et al. [6]	Sinha et al. [7]
Without attack	0.9973	0.9658	0.9785	0.9680
Gaussian noise (0 mean and 0.01 variance)	0.9845	0.8993	0.9154	0.9250
Salt and pepper noise (density 0.05)	0.9947	0.9156	0.9357	0.9250
Median filtering (3 × 3)	0.9943	0.9345	0.9526	0.9490
Histogram equalization	0.9762	0.9456	0.9563	0.9497

Table 4 Comparison of NC for proposed video watermarking with previous work for temporal domain attacks

Temporal attacks	Proposed work	Deshpande et al. [5]	Agarwal et al. [6]	Sinha et al. [7]
Frame swapping	0.9973	0.9634	0.9796	0.9556
Frame removal	0.9754	0.8756	0.8863	0.7958

5 Conclusion

A novel video watermarking algorithm is proposed based on Discrete Wavelet Transform (DWT) and Discrete Cosine transform (DCT). The proposed algorithm has the combined properties of both transformation techniques. The imperceptibility of the algorithm increases with higher decomposition level of DWT, which can improve the performance in terms of robustness. DCT is useful to have energy compaction and helps to reduce the invisibility feature during watermark embedding. The watermark is also scrambled and then inserted as a spread spectrum pattern using Arnold transform technique which largely improve the security and robustness. The performance of the proposed algorithm is validated on MATLAB and also compares with previous work. The simulation results show that the proposed algorithm has better performance against spatial and temporal attacks. The proposed video watermarking algorithm is useful in copyright protection and ownership identification applications. The algorithm embeds the watermark in each 4 × 4 size of block, thus embedding capacity is also higher compare to all exiting schemes.

References

1. Wei, Z., Wu, Y., Deng, R. H., & Ding, X.: A hybrid scheme for authenticating scalable video codestreams., vol. 9, No.4, pp. 543–553, IEEE Transactions on Information Forensics and Security (2014).
2. Joshi, A. M., Darji, A., & Mishra, V.: Design and implementation of real-time image watermarking, IEEE pp. 1–5, IEEE International Conference on Signal Processing, Communications and Computing (ICSPCC) (2011).
3. He, Xuansen, Tao Zhu, and Gaobo Yang.: A geometrical attack resistant image watermarking algorithm based on histogram modification, 26.1, pp. 291–306, Multidimensional Systems and Signal Processing (2015).
4. Joshi, A. M., Mishra, V., Patrikar, R. M..: FPGA prototyping of video watermarking for ownership verification based on H. 264/AVC, pp. 1–24, Multimedia Tools and Applications, (2015).
5. Deshpande, N., Rajurkar, A., Manthalkar, R.: Robust DCT based video watermarking algorithms for assorted watermarks., vol. 1, pp. V1-320, IEEE, 2010 2nd International Conference on Signal Processing Systems (2010).
6. Agarwal, A., Bhadana, R., Chavan, S.: A robust video watermarking scheme using DWT and DCT, vol. 2, no. 4, pp. 1711–1716. International Journal of Computer Science and Information Technologies (2011).
7. Sinha, S., Bardhan, P., Pramanick, S., Jagatramka, A., Kole, D. K., Chakraborty, A.: Digital video watermarking using discrete wavelet transform and principal component analysis, vol. 1, no.2, pp. 7–12 International Journal of Wisdom Based Computing (2011).
8. Tareef, Afaf, and Ahmed Al-Ani: A highly secure oblivious sparse coding-based watermarking system for ownership verification. 42.4, pp. 2224–2233. Expert Systems with Applications (2015).
9. Dragoi, Ioan-Catalin, and Dinu Coltuc: On Local Prediction Based Reversible Watermarking, 24.4, pp. 1244–1246, IEEE Transactions on Image Processing (2015).
10. Joshi, A. M., Mishra, V., & Patrikar, R. M.: Design of real-time video watermarking based on Integer DCT for H. 264 encoder, 102(1), pp. 141–155, International Journal of Electronics (2015).
11. Arnold, M., Chen, X. M., Baum, P., Gries, U., & Doerr, G.: A phase-based audio watermarking system robust to acoustic path propagation, 9(3), pp. 411–425, IEEE Transactions on Information Forensics and Security (2014).
12. Huai-bin, Wang, Yang Hong-liang, Wang Chun-dong, and Wang Shao-ming.: A new watermarking algorithm based on DCT and DWT fusion, IEEE, pp. 2614–2617, 2010 International Conference on Electrical and Control Engineering (2010).
13. LI, L., Dong, Z., Lu, J., Dai, J., Huang, Q., Chang, C. C., & Wu, T.: An H. 264/AVC HDTV watermarking algorithm robust to camcorder recording, 26, pp. 1–8, Journal of Visual Communication and Image Representation (2015).
14. Zhu, X., Ding, J., Dong, H., Hu, K., & Zhang, X.: Normalized Correlation-Based Quantization Modulation for Robust Watermarking, 16(7), pp. 1888–1904. IEEE Transactions on Multimedia (2014).
15. Joshi, A., Patrikar, R. M., & Mishra, V.: Real Time Implementation of Digital Watermarking Algorithm for Image and Video Application. INTECH Open Access Publisher (2012).

A Basic Simulation of ACO Algorithm Under Cloud Computing for Fault Tolerant

Virendra Singh Kushwah and Sandip Kumar Goyal

Abstract Fault is not a new thing; it is a kind of a problem which is dealt at every level of resource. Tolerance of the fault is required mechanism else smooth functioning of the defined system cannot be stand up. There are many algorithms for fault tolerant during balancing loads in the cloud computing. In this article, selected fault-tolerant algorithms have been discussed and cloudsim tool is a better tool for implementation purpose. But ACO is an approach by which we could achieve our desired objectives. The main motive of this article is to understand involvement of ACO algorithm into fault tolerance techniques. At last, needs of load balancing is required for managing faults while equalization loads of resources.

Keywords Cloud computing · Fault tolerant · ACO · Cloudsim

1 Foundation

1.1 Cloud Computing

Cloud computing is extremely intriguing issue in IT field. Numerous looks into Cloud Computing are going on. It implies at whatever point we require it, for a few applications or some product, we interest for it and we instantly get it. We need to pay just that we utilize. This is the fundamental maxim of distributed computing. Distributed computing has fundamentally two sections, the first part is of Client Side and the second part is of Server Side. The Client Side solicitation to the Servers and the Server reacts to the Clients. The solicitation from the customer first goes to the Main Processor of the Server Side. The Main Processor is joined to

V.S. Kushwah (✉) · S.K. Goyal
Department of Computer Science & Engineering, Maharishi Markandeshwar University, Mullana, Ambala, India
e-mail: kushwah.virendra248@gmail.com

S.K. Goyal
e-mail: skgmmec@gmail.com

© Springer Science+Business Media Singapore 2017

S.C. Satapathy et al. (eds.), *Proceedings of the International Conference on Data Engineering and Communication Technology*, Advances in Intelligent Systems and Computing 468, DOI 10.1007/978-981-10-1675-2_46

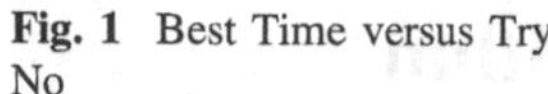

Fig. 1 Best Time versus Try No

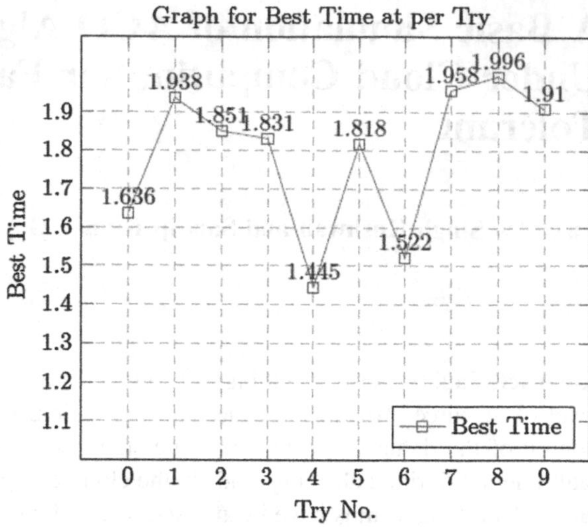

Table 1 Parameter settings

Sr. no.	Parameters	Value
1	No. of Ants	20
2	No. of Tours	02
3	No. of Tries	10
4	Ant Alpha	1.0
5	Ant Beta	2.0
6	Ant Rho	0.1
7	Ant q0	0.98
8	ACO Variant	Ant Colony System
9	Instance	d1291.tsp

numerous different Processors, the Main Processor sends that demand to any of the Other Processors which have free space. All Processors are occupied in their allocated occupation and none of the Processors gets Idle. The procedure of doling out occupation from Main Processor to the Other Processor and after finish the employment, then coming back from the Other Processor to the Main Processor is much the same as Ant takes their sustenance and come back to their home. Presently, the counterfeit ants are utilized as a part of distributed computing. The distributed computing is made out of three administration models, five fundamental attributes, and four arrangement models [1] (Fig. 1), (Table 1).

1.2 Importance of Fault Tolerant

Fault tolerance refers to a correct and continuous operation even in the presence of faulty components. In the greater part of the actual time-based cloud applications, processing of computing nodes are done remotely. So, there are more chances of errors. So there is an improved and expanded prerequisite for adaptation to non-critical failure to accomplish higher dependability for the ongoing figuring on cloud-based base [2].

Fault tolerance is forwarded by error computing which has two constituent phases. The phases are *Effective Error Computing* which targeted at bringing the effective error back to a dormant state, i.e., before the occurrence of error and *Latent Error Computing* aimed at ensuring that the error does not become effective again [3]. There are so many fault-tolerant techniques and are used by cloud computing:

- **Self-Healing**: In this method, by making use of divide and conquer strategy a huge task is distributed into several components. This division is done for better improvement and performance. In this, different instances of an application are running on distinguishable virtual machines and failure of all these individual instances are handled automatically.
- **Job Migration**: Sometimes it happens that due to some reason a particular machine fails and cannot execute jobs. On such a failure, a task is migrated to working machine using HA-Proxy. Also, there are algorithms that automatically determine the fault and migrates batch applications within a cloud of multiple datacentres.
- **Check Pointing**: This is a proficient task level fault tolerance technique for large applications. In this method, check pointing is done in the system. When a task or more than one task is failed down, instead of initiating from beginning, the tasks are restarted from the recently checked pointed state.
- **Replication**: Replication means copying. Several replicas of tasks are created and they are running on various kinds of resources, for effective execution and for getting the desired result. Hadoop, HA-Proxy, Amazon EC2 like tools are there in which replication can be implemented.
- **Task Resubmission**: Due to high network traffic or due to heavy work load, a task may fail whenever such failed task is detected at runtime the job is reassigned either to the same system or dissimilar working network facility for execution. For these, certain algorithms are designed, which assigns tasks to resources on the basis of certain properties.

1.3 Ant Colony Optimization

Dorigo [4] presented the insect calculation in light of the conduct of genuine ants in 1996; it is another heuristic calculation for the arrangement of combinatorial enhancement issues. Ant has the capacity of discovering an ideal way from home to food. On the method for ants moving, they lay some aura on the ground; while a disconnected ant experience a previous main laid trail, this ant can distinguish it and choose with high likelihood to tail it. Subsequently, the trail is strengthened with its own aura. The likelihood of ant picks a way is extent to the convergence of a way aura. To a way, the more ants pick, the way has denser aura, and the denser aura draws in more ants. Through this positive input component, subterranean insect can locate an ideal way at long last.

Each and every ant is uncomfortable insect by behaviorally. They have an extremely constrained storage and show singular conduct that seems to keep a huge arbitrary part. Going about as an aggregate then again, ants figure out how to perform an assortment of confounded undertakings with extraordinary unwavering quality and consistency. In spite of the fact that this is basically self-association instead of learning, ants need to adapt to a wonder that looks all that much like over preparing in fortification learning methods. The intricate social practices of all ants have been quite concentrated by science based subject, and PC researchers are currently searching that these conduct examples can give models to taking care of troublesome combinatorics enhancement issues. The endeavor to create calculations propelled by one part of subterranean insect conduct, the capacity to discover what PC researchers would call briefest ways, has turned into the area of Ant Colony Optimization (ACO), the best and broadly perceived algorithmic procedure in view of insect conduct [5]. Here we have following a simple and sweet Ant Colony Optimization algorithm:

Algorithm 1: Pseudo Code for Ant Colony Optimization (ACO)

```
Start;
Introduce the aura;
while ceasing model not fulfilled do
    Position every ant in a beginning virtual machine;
    while ceasing when each ant has construct an answer do
        for each ant do
            Choose virtual machine for next undertaking by aura trail power;
        end
    end
    Overhaul the aura;
end
Stop;
```

1.4 CloudSim Simulation Tool

We are using CloudSim simulation tool [6]. The modularity of tool created it the right selection. Every element is enforced as a Java classes and may be further enhanced terribly simple. CloudSim could give a proceeding recreation structure summed up by the most assets of the cloud concept. The substances in CloudSim impart through messages [7]. Since host and VM are static elements, every adjustment in their state ought to be acknowledged by the datacenter. The agent, taking into account the reenactment arrangement (number of cloudlets and their particular) will ask for the VM creation, cloudlets booking and it will hold up to be educated by the datacenter when the cloudlets consummation is figured it out.

As we have noticed and know that there are two types of algorithms: scheduling algorithms and planning algorithms. In theory they are the same, however, in terms implementation they are quite different. Simulation can be done on the WorkflowSim, which is an extended version of CloudSim [8]. Remember that WorkflowSim have three layers: Workflow Planner, Engine, and Scheduler. Inside *Workflow Planner*, we have a global view of the whole workflow (all the tasks and their dependencies) and in each iteration *Workflow Engine* release tasks those are free (which means their parents have completed successfully) to *Workflow Scheduler*. Workflow Scheduler matches these free tasks to resources (Condor VMs in WorkflowSim) and submits them for execution [9].

2 Literature Review

Mishra et al. [10] added to a successful burden adjusting calculation utilizing insect settlement streamlining method to boost or minimize distinctive execution parameters like CPU burden, memory limit, defer, or system load for the billows of diverse sizes. They utilized a heuristic calculation taking into account subterranean insect state streamlining and have been proposed to start the administration load appropriation under distributed computing structural planning. The aura upgrade system has been demonstrated as a productive and compelling instrument to adjust the heap. This adjustment backings to minimize the make compass of the distributed computing based administrations and transportability of overhauling the solicitation likewise has been focalized utilizing the subterranean insect settlement streamlining strategy. This strategy does not consider the adaptation to internal failure issues.

Li et al. [11] have been suggested a Load Balancing Ant Colony Optimization (LBACO) to locate the ideal asset distribution for every errand in element cloud framework. Not just does it reduce the schedule of a given undertakings set however it likewise adjusts to the dynamic distributed computing framework and parity the whole framework load. The new booking system was reenacted utilizing the CloudSim toolbox bundle. Examinations results demonstrated the suggested LBACO calculation outflanked FCFS and the ACO.

Song et al. [12] have been proposed another supporting middle information adaptation to internal failure distributed computing system, named IDF Support structure so as to adequately handle the testing cloud halfway information adaptation to noncritical failure issue. The errands are partitioned into the key undertaking and regular assignment as indicated by their attributes, and are doled out three diverse adaptations to internal failure levels. As indicated by the adaptation to internal failure level of diverse undertakings, they likewise proposed inward assignment middle information flaw-tolerant calculation (Inner errand IDF) and external errand transitional information issue tolerant calculation (Outer assignment IDF).

Gan et al. [13] proposed a viable and secure information honesty check plan with blunder tolerant in light of TP (Third Party) in the distributed computing. So, their plan accomplishes the TP manages per-preparing and information recuperation. The mystery equality era network created repetition open to TP and customer, so that the TP can trade the client for information recouping.

Lu et al. [14] used queuing Petri nets for modeling hybrid cloud platform. At the same time, they analyzed the system behavior and performance of the hybrid cloud platform according to the established model. They additionally investigated the flaw sorts in half and half distributed computing, utilizing the model to figure the ongoing execution of the framework. At last, they utilized this procedure as a part of a flame crisis cloud to break down the impact of the technique to the execution of the framework and confirm the system. The consequences of the analyses demonstrate that the methodology can utilize both under the low load state and high load condition of half breed distributed computing stage.

Prakash et al. [15] proposed a plan that is sufficiently successful to ideally utilize the assets. The outcomes examination demonstrated that the execution time of the proposed scheduler is not exactly the current methodologies. However, there are still numerous limitations that should have been overcome for more successful and fearful results. Proposed methodology can be incorporated with other asset use ways to deal with acquire a general asset usage model/approach.

Haidri et al. [16] presented a heuristic-based burden adjusted booking model for effective execution of assignments. The suggested model adjusts the heaps originating from a few clients among datacenters and thus it offers better asset use and high accessibility as enhanced reaction time and turnaround time. The proposed calculation is executed utilizing CloudSim test system and the outcome demonstrates that the proposed calculation outflanks to existing calculations on comparable targets. The proposed load balancer shrewdly ties the cloudlets to virtual machines to minimize the turnaround time and reaction time so that the wanted goal is satisfied.

The model proposed depends on incorporated burden adjusting procedure. Before tying cloudlets to VM, load balancer first figures the remaining limits of all VMs and after that dispatch the cloudlet to all the more capable VM. Also, as burden is initially corrected among VMs, the heap redistribution procedure is reasonable. With a specific end goal to gauge execution of the proposed model recreation study has been put through different test conditions. It has been found that the model functions admirably in guaranteeing an even appropriation of the workload. The work should likewise be possible by utilizing transformative strategies, for example, GA, ACO and PSO, and so forth.

3 Analysis of Results

We are using ACO algorithm to simulate Ant Colony Optimization System under cloudsim for understanding basic concept. The following parameters are used to run algorithm:

Initialization took 0.442 s

try 0, Best 51456, found at iteration 5, found at time 1.636
try 1, Best 51536, found at iteration 8, found at time 1.938
try 2, Best 51380, found at iteration 7, found at time 1.851
try 3, Best 51358, found at iteration 7, found at time 1.831
try 4, Best 51613, found at iteration 6, found at time 1.445
try 5, Best 51339, found at iteration 7, found at time 1.818
try 6, Best 51124, found at iteration 6, found at time 1.522
try 7, Best 51156, found at iteration 8, found at time 1.958
try 8, Best 51016, found at iteration 9, found at time 1.996
try 9, Best 51152, found at iteration 9, found at time 1.910

From the above calculations, it is easy to calculate the total average best try time and total average time by n-ants, as given below:

t_avgbest = 1.7905000000000002 (Total Average Best Try Time by Ants)
t_avgtotal = 2.0080999999999998 (Total Average Time by Ants)

For the simplicity, we have taken ACO variant as ACS (Ant Colony System) to understand the basic concept of ACO algorithm. We have chosen only 20 ants and two times of try the tours as we can see the above.

Here we have a graph for showing the relation between best time and number of try. The graph is generated using ACO algorithms followed by ACS instance.

The above figure is the summary of results which are generated by ACO under ACS instance as given in the various steps above. Best time is calculated by various times, those are generated by after specified iterations.

4 Conclusion

This article is most useful for understanding the basic concepts on fault tolerant and its selected techniques of fault tolerant. We have focused on conceptualized study on fault tolerance with ACO algorithms. Various algorithms are proposed by many authors but this article is integration of all.

Using this average best time, we can enhance or design a new algorithm to improve the durability and reliability of the cloudlets. In the next task, we will implement such thing into fault tolerant for cloud computing.

Acknowledgments Authors are thankful to Department of Computer Science & Engineering at Maharishi Markandeshwar University, Ambala for giving highly motivational supports.

References

1. Dikaiakos, M.D., Katsaros, D., Mehra, P., Pallis, G. and Vakali, A.: Cloud computing: Distributed internet computing for IT and scientific research. Internet Computing, IEEE, 13(5), pp. 10–13. (2009).
2. Deng, J., Huang, S.C.H., Han, Y.S. and Deng, J.H.: Fault-tolerant and reliable computation in cloud computing. In GLOBECOM Workshops (GC Wkshps), 2010 IEEE (pp. 1601–1605). IEEE. (2010).
3. Bala, A. and Chana, I.: Fault tolerance-challenges, techniques and implementation in cloud computing. IJCSI International Journal of Computer Science Issues, 9(1), pp. 1694–0814. (2012).
4. Dorigo, M., Birattari, M., Blum, C., Clerc, M., Stützle, T. and Winfield, A. eds.: Ant Colony Optimization and Swarm Intelligence: 6th International Conference, ANTS 2008, Brussels, Belgium, Proceedings (Vol. 5217). Springer. (2008).
5. Xue, S., Li, M., Xu, X., Chen, J. and Xue, S.: An ACO-LB Algorithm for Task Scheduling in the Cloud Environment. Journal of Software, 9(2), pp. 466–473. (2014).
6. Calheiros, R.N., Ranjan, R., De Rose, C.A. and Buyya, R.: Cloudsim: A novel framework for modeling and simulation of cloud computing infrastructures and services. arXiv preprint arXiv:0903.2525. (2009).
7. Nita, M.C., Pop, F., Mocanu, M. and Cristea, V.: FIM-SIM: Fault Injection Module for CloudSim Based on Statistical Distributions. Journal of Telecommunications and Information Technology, (4), pp. 14–23. (2014).
8. Calheiros, R.N. et al,.: Cloudsim a toolkit of cloud computing environments and evaluation of resource provisioning algorithms. SOFTWARE PRACTICE AND EXPERIENCE, 41 (1):23–50 (2011).
9. Chen, W. and Deelman, E.: Workflowsim: A toolkit for simulating scientific workflows in distributed environments. In E-Science (e-Science), 2012 IEEE 8th International Conference on (pp. 1–8). IEEE. (2012).
10. Mishra, R. and Jaiswal, A.: Ant colony optimization: A solution of load balancing in cloud. International Journal of Web & Semantic Technology (IJWesT), 3(2), pp. 33–50. (2012).
11. Li, K., Xu, G., Zhao, G., Dong, Y. and Wang, D.: Cloud task scheduling based on load balancing ant colony optimization. In Chinagrid Conference (ChinaGrid), 2011 Sixth Annual (pp. 3–9). IEEE. (2011).
12. Song, B., Ren, C., Li, X. and Ding, L.: An Efficient Intermediate Data Fault-Tolerance Approach in the Cloud. In Web Information System and Application Conference (WISA), 2014 11th (pp. 203–206). IEEE. (2014).
13. Gan, H. and Chen, L.: An Efficient Data Integrity Verification and Fault-Tolerant Scheme. In Communication Systems and Network Technologies (CSNT), 2014 Fourth International Conference on (pp. 1157–1160). IEEE. (2014).
14. Lu, M. and Yu, H.: A Fault Tolerant Strategy in Hybrid Cloud Based on QPN Performance Model. In Information Science and Applications (ICISA), 2013 International Conference on (pp. 1–7). IEEE. (2013).
15. Prakash, V. and Bala, A.: A novel scheduling approach for workflow management in cloud computing. In Signal Propagation and Computer Technology (ICSPCT), 2014 International Conference on (pp. 610–615). IEEE. (2014).
16. Haidri, R.A., Katti, C.P. and Saxena, P.C.: A load balancing strategy for Cloud Computing environment. In Signal Propagation and Computer Technology (ICSPCT), 2014 International Conference on (pp. 636–641). IEEE. (2014).

A Neuromorphic Majority Function Circuit with $O(n)$ Area Complexity in 180 nm CMOS

Juhi Faridi, Mohd. Samar Ansari and Syed Atiqur Rahman

Abstract An artificial neuron with a step activation function is first designed and verified. Thereafter, a synaptic weight generation circuitry is designed to provide a suitable sum current to the activation function neuron to achieve the task of majority function generation for digital logic inputs. HSPICE simulations are performed to verify the proposed theoretical framework, with the proposed network correctly yielding the appropriate low or high digital logic state corresponding to the input combinations applied. Superiority of the proposed circuit in terms of transistor area required is also demonstrated. The transistor count increase *linearly* with the number of variables in the case of the proposed circuit; whereas for conventional static CMOS implementations an *exponential* increase in transistor count is exhibited.

Keywords Artificial neural networks (ANN) · Neuron · Majority function generator · Synapse · Synaptic weights

1 Introduction

Work on artificial neural networks has been motivated right from their inception by the recognition that the human brain computes in an entirely different way from the conventional digital computer. The brain is a highly complex, nonlinear, and parallel information processing system. It has the capability to organize its structural constituents, known as neurons, so as to perform certain computations many times

Juhi Faridi
Z.H. College of Engineering & Technology, Aligarh Muslim University, Aligarh, India
e-mail: faridij.262@gmail.com

Mohd. Samar Ansari (✉) · S.A. Rahman
Department of Electronics Engineering, Aligarh Muslim University, Aligarh, India
e-mail: mdsamar@gmail.com

S.A. Rahman
e-mail: atiqamu@gmail.com

© Springer Science+Business Media Singapore 2017

S.C. Satapathy et al. (eds.), *Proceedings of the International Conference on Data Engineering and Communication Technology*, Advances in Intelligent Systems and Computing 468, DOI 10.1007/978-981-10-1675-2_47

faster than the fastest digital computer in existence today. The brain routinely accomplishes perceptual recognition tasks, for instance recognizing a familiar face embedded in an unfamiliar scene, in approximately 100–200 ms, whereas tasks of much lesser complexity may take minutes on a conventional computer.

Formally, a neural network is a machine that is designed to model the way in which the brain performs a particular task [1–8]. The network is implemented by using electronic components or is simulated in software on a digital computer. A biological neuron receives input from other neurons (generally thousands in number) from its synapses and the inputs are summed up. When the input exceeds a threshold, the neuron sends an electrical spike that travels from the body, down the axon, to the next neurons. Similarly, in artificial neural networks, each neuron receives a number of inputs and an activation function is applied to these inputs which results in activation level of neuron (output value of the neuron). A similar technique is used in this paper to design a neuron suitable for the chosen problem.

In Boolean logic, the majority function is a function with n inputs and a single output. The value of the operation is *true* (*1*) when more than half the number of arguments are true, and *false* (*0*) otherwise. For circuit level implementation of the n–input majority function, an activation function is required for comparing the input current coming to the comparator with certain threshold and giving the output as a step function, i.e., *0* or *1,* while the synaptic weights control the input going to the comparator according to the requirement. On combining them both we get the desired neural circuit for a majority function generator.

In this paper, a 5-input majority function neuron is proposed, for which first an activation function with threshold logic was implemented and then the design of the synaptic weights and switching transistors was performed. On comparing the transistors needed for implementing 3-input, 5-input, and 7-input majority functions using the neuron designed and the hugely popular static CMOS design, it is found that the neuron designed was far more area efficient at higher number of inputs.

The main aim of this paper is to find an efficient implementation of the neuron for which implementation of the activation function was followed by the design of the synaptic weights. The application of the neuron as a 5-input majority function circuit is discussed. Comparison of the neuron with the CMOS implementation proves the neuron designed to be beneficial for higher number of inputs. Lesser number of transistors will be needed as compared to the CMOS implementation of the majority function. Significant advantage can be seen from 7-input majority function in comparison with CMOS. It is faster and linearly separable function so there is constant increase of four transistors for increasing number of inputs by two, rather than exponential increase as can be seen in case of static CMOS implementations.

This paper is organized as follows. Section 2 contains details of the proposed circuits, i.e., the neuronal activation circuit and the synaptic weight generation circuit. Section 3 presents verification of the proposed technique via HSPICE simulations. Some conclusive remarks appear in Sect. 4.

2 Proposed Neuron and Synaptic Weight Generation Circuit

Activation Function Circuit The circuit to implement the activation function with a step output is shown in Fig. 1, from where it can be seen that the current I_{in} is the signal coming from the synapse [1]. As is explained in the next paragraph, it is the resultant current which is obtained according to the input conditions and the designed weights of the circuit. The resulting current I_{in} is provided to a transresistance activation circuit build up of the transistors M_{14} through M_{29}. More specifically, current mirrors formed by M_{14} & M_{19} and M_{15} & M_{18} form the biasing stages for transistors M_{16}, M_{17}, M_{20}–M_{23}. The three sets of inverters M_{24}–M_{25}, M_{26}–M_{27} and M_{28}–M_{29} form a positive feedback.

Implementation of Synaptic Weights The circuit designed for implementing the synaptic weight generation for a 5-input majority function circuit is presented in Fig. 2, where it is evident that M_1–M_2 form a PMOS current mirror and M_{12}–M_{13} form an NMOS current mirror. Due to current mirroring, transistors M_2 through M_6 act as current sources, and M_{13} forms a current sink. I_{SS} is the reference current flowing in M_1. Transistors M_7 through M_{11} act as switches with the logical inputs applied as gate voltages, V_{in1} through V_{in5}, which control the input current I_{in} entering the activation circuit. Next, the aspect ratios for the various pertinent transistors are discussed.

Switch transistors (M_7 through M_{11}): Input voltages V_{in1} through V_{in5} control the current I_{in} with the help of the input combinations applied to the transistors *M_7 through M_{11}*. Therefore, the current entering the activation function block can be controlled by the values of the various inputs.

Sink transistors ($M_{12}-M_{13}$): The W/L ratio of M_{13} is kept 2.5 times of M_{12} because for the case when all the inputs are *high*, the current going into the sink transistor M_{13} will be $I_p = 5 \times I_{SS}$ and $I_n = 2.5 \times I_{SS}$ hence the current entering into the activation function (I_{in}) will be $2.5 \times I_{SS}$. Similarly, for the case when inputs are

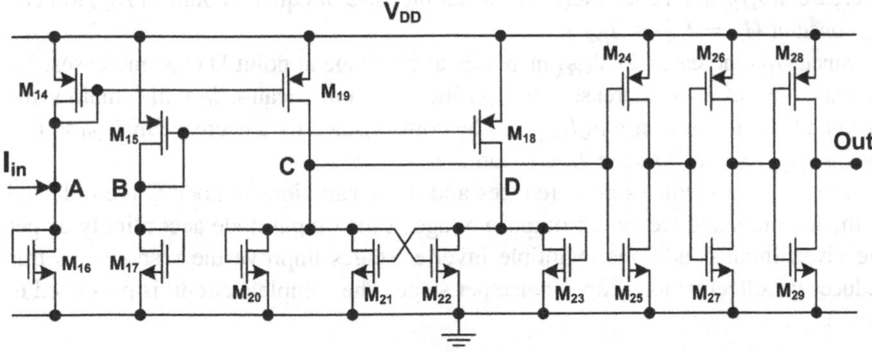

Fig. 1 Implementation of the activation function circuit

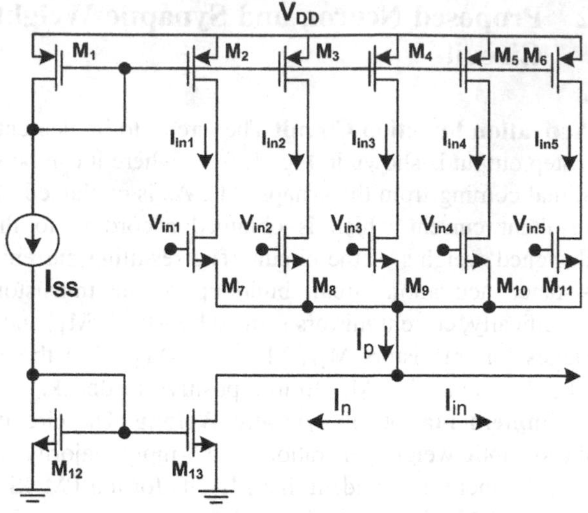

Fig. 2 Implementation of the synaptic weight generation circuitry

low, the all–zero state current will be $-2.5 \times I_{SS}$ since no current will be coming from the source transistor matrix, but still a current will be sinking into M_{13}. Therefore, a uniform range of current from $2.5 \times I_{SS}$to$-2.5 \times I_{SS}$ is achieved.

Since the reference current is kept constant, only the weights and the input conditions will decide the current entering in the activation function (I_{in}). Next, the effect of this I_{in} on the activation circuit is explained.

The drain voltage of M_{15} (V_{D15}) is constant and due to this gate voltage of M_{15} (V_{G15}) and the gate voltage of M_{18} (V_{G18}) will be constant and a constant current I_D flows in the drain of M_{18}. As I_{in} varies so V_A varies and this alters V_{G14} which in turn alters V_{G19} and current I_{D19} changes due to which current I_C changes.

As I_{in} increases, V_A increases and therefore V_{G19} increases, and since M_{19} is PMOS the drain current decreases and this reduces the voltage at node C (V_C) due to a decrease in I_C (or I_{D20}). This decrease changes the gate voltage of M_{22} due to cross coupling effect and V_{G22} decreases. This change in V_{G22} will be reflected as a decrease in I_{D20} and hence there will be an increase of equal amount in I_{D23} to keep I_D constant ($I_D = I_{D22} + I_{D23}$).

Since I_{D23} increases so V_{G23} increases and voltage at point D (V_D) increases. V_D increases but as V_{G23} increases so V_{G21} increases but because I_C will remain same as initially so this increase in I_{D21} will be compensated by a decrease in I_{D20} so this means I_{D20} reduces to keep I_C constant.

Therefore, V_C reduces as I_C reduces and these variations at node C are reflected at the inverters and we get an output voltage at the output node accordingly as per the given input conditions. Multiple inverter stages improve the response as this reduces the effective load capacitance per stage. The complete circuit is presented in Fig. 3.

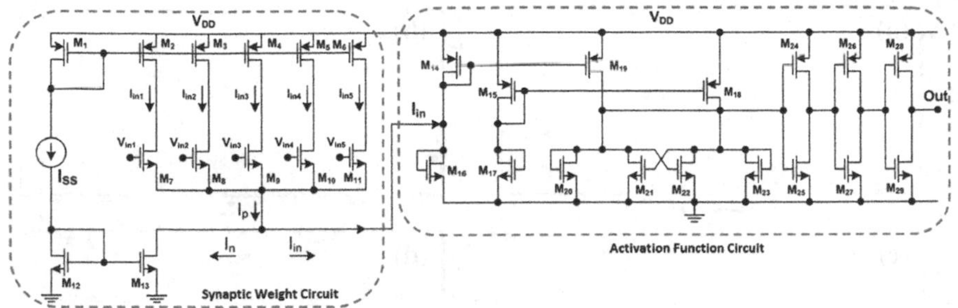

Fig. 3 Proposed neural circuit for 5–input majority function generation

3 Simulation Results

The operation of the proposed circuit was tested using HSPICE simulations. The aspect ratios for the various transistors used in the circuit are listed in Table 1. PTM 180 nm CMOS process model parameters were used for the purpose of simulations. The biasing supply was kept at V_{DD} = 3 V. The value of the bias current, I_{SS}, was fixed at 10 μA, as is explained below.

First, the activation circuit was tested separately without the synaptic weight generation circuit and the results of HSPICE simulations are presented in Fig. 4a from where it is evident that the activation function is faithfully produced in the current range of −60– +60 μA. Therefore, for a 5–input majority function generation circuit, it is prudent to set the biasing current, I_{SS} equal to 10 μA so that the maximum current entering into the activation function circuit (when transistors M_7 through M_{11} are all conducting) does not exceed 60 μA.

Results of HPSICE simulations are presented in Fig. 4b–f, and pertinent inferences drawn from the plots obtained are discussed below.

Figure 4b*: Majority zero (four inputs low and 5th input is swept)* From the obtained graph, it is seen that initially when the 5th transistor (M_{11}) is OFF the majority is zero and the obtained output is also zero (2.7 μV) and as the 5th transistor gets ON the majority still remains zero and the obtained output is also zero (2.82 μV).

Table 1 Aspect ratios of the transistor

Transistors	W (μm)	L (μm)	Transistors .	W (μm)	L (μm)
M1, M2, M3, M4, M5, M6	0.72	1.26	M19, M18	1.19	0.18
M7, M8, M9, M10, M11	0.36	1.26	M20, M23	0.21	0.18
M12	0.36	1.26	M21, M22	0.34	0.18
M13	0.9	1.26	M24, M26, M28	0.54	0.18
M14, M15, M16, M17	0.18	0.72	M25, M27, M29	0.18	0.18

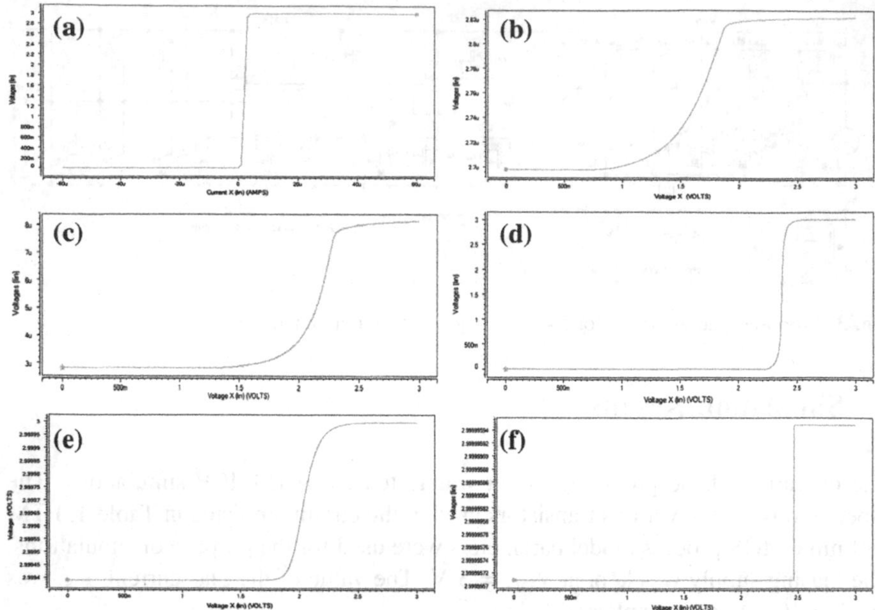

Fig. 4 Simulated outputs for **a** Activation function circuit **b** Majority zero (four inputs low and 5th input is swept) **c** Majority zero (three inputs low, one high and 5th input is swept) **d** Majority switching from 0 to 1 as fifth input is swept from low to high (two inputs are low, two inputs are high and 5th input is swept from low to high) **e** Majority one (three inputs are high and one input is low and 5th input is swept) **f** Majority one (four inputs high and 5th input is swept)

Figure 4c*: Majority zero (three inputs low, one high and fifth input is swept)* From the obtained graph, it is observed that initially when M_{11} is OFF, the majority is zero and the obtained output is also zero (3 μV) and as the fifth transistor becomes ON the majority still remains zero and the obtained output is also zero (8 μV).

Figure 4d: *Majority switching from zero to one as fifth input is swept from low to high (two inputs are low, two inputs are high and fifth input is swept from low to high)* From the plot obtained, it is clear that when the fifth transistor is OFF, the majority is zero and the obtained output is also zero (0 V) and as the fifth transistor gets ON, the majority becomes one and the obtained output is also one (3 V).

Figure 4e: *Majority one (three inputs are high and one input is low and fifth input is swept)* Initially when M_{11} is OFF, the majority is one and the obtained output is also high (~2.99 V) and as the fifth transistor gets ON the majority still remains one and the obtained output is also one (~2.99 V).

Figure 4f: *Majority one (four inputs high and fifth input is swept)* When the fifth transistor is OFF, even then the majority is one and the obtained output is also high (~2.99 V) and as the fifth transistor gets ON the majority still remains one and the obtained output retains its value (~2.99 V).

Table 2 Results of HSPICE simulations for the proposed circuit of Fig. 5 for different input combinations

Inputs					Output	
Vin1	Vin2	Vin3	Vin4	Vin5	Desired	Obtained
0	0	0	0	0	0	2.7 μV
0	0	0	0	1	0	2.8 μV
1	0	0	0	0	0	3 μV
1	0	0	0	1	0	8 μV
1	1	0	0	0	0	0 V
1	1	0	0	1	1	3 V
1	1	1	0	0	1	2.99 V
1	1	1	0	1	1	3 V
1	1	1	1	0	1	2.99 V
1	1	1	1	1	1	2.99 V

Several other input combinations were tested using HSPICE simulations. Table 2 presents the desired and obtained values of the voltage outputs for different input combinations. It can be seen that the circuit has two broad categories of outputs: (i) *high*, when the output voltage is ~2.99 V (for a circuit operating on a +3 V supply), (ii) *low*, when the output voltage is less than 10 μV. It is readily verified from the table that the circuit performs correctly for all input combinations tested.

Next, a comparison is drawn between the industry standard static CMOS design style, and the proposed neural implementation. Table 3 presents an estimate of the number of transistors required for designing the majority function circuit for various input counts. As can be seen from the tabular comparison, the proposed circuit is more area efficient when the number of inputs is equal to or more than 7.

More importantly, the transistor count increases *linearly* with the number of variables in the case of the proposed circuit; whereas for conventional static CMOS implementations an *exponential* increase in transistor count is expected. Therefore, the proposed technique would result in more and more area efficient implementations of the majority function generator as the number of variables scale up.

Table 3 Comparison of transistor counts for the proposed neural circuit and conventional static CMOS implementation

Number of inputs	Conventional CMOS	Proposed neural implementation
3	12	25
5	32	29
7	78	33

4 Conclusion

A neuron with a step activation function was first designed and verified using computer simulations. Thereafter, a synaptic weight generation circuitry was designed to provide a suitable sum current to the activation function neuron to achieve the task of majority function generation. HSPICE simulations verified the proposed theoretical framework, with the proposed network correctly yielding the appropriate *low* or *high* state corresponding to the input combinations applied. Superiority of the proposed circuit in terms of transistor area required over static CMOS implementations was also demonstrated.

References

1. Tisan, Alin, and Jeannette Chin. "An end user platform for implementing Artificial Neutron Networks on FPGA." In Industrial Informatics (INDIN), 2015 IEEE 13th International Conference on, pp. 856–859. IEEE, (2015).
2. Chasta, Neeraj K. "High Speed, Low Power Current Comparators with Hysteresis." arXiv preprint arXiv:1203.2999 (2012).
3. Suykens, Johan AK, Joos PL Vandewalle, and Bart L. de Moor. Artificial neural networks for modelling and control of non-linear systems. Springer Science & Business Media, (2012).
4. Nedjah, Nadia, Rodrigo Martins da Silva, and Luiza de Macedo Mourelle. "Compact yet efficient hardware implementation of artificial neural networks with customized topology." Expert Systems with Applications 39.10 (2012): 9191–9206.
5. Basu, Jayanta Kumar, Debnath Bhattacharyya, and Tai-hoon Kim. "Use of artificial neural network in pattern recognition." International journal of software engineering and its applications 4.2 (2010).
6. Wojtyna, R., and T. Talaska. "Transresistance CMOS neuron for adaptive neural networks implemented in hardware." Technical Sciences 54.4 (2006).
7. Meireles, Magali RG, Paulo EM Almeida, and Marcelo Godoy Simões. "A comprehensive review for industrial applicability of artificial neural networks." Industrial Electronics, IEEE Transactions on 50.3 (2003): 585–601.
8. Goser, Karl F. "Implementation of artificial neural networks into hardware: Concepts and limitations." Mathematics and computers in simulation 41.1 (1996): 161–171.

Human Action Recognition: An Overview

Pravin Dhulekar, S.T. Gandhe, Harshada Chitte and Komal Pardeshi

Abstract In this paper, an approach for detection of human actions from video is given. It has many applications, such as surveillance, human healthcare systems, border security, virtual reality, HMI, etc. This paper includes the detection of human actions and labeling them on an LCD. The frames are made from video and then median filter is applied. The feature extraction is done by applying PCA. The segmentation process is done using Gabor filter. Finally, KNN classifier is applied for classification. Different datasets are also provided.

Keywords Action recognition · Surveillance · Median filter · KNN classifier · Gabor filter · PCA

1 Introduction

Human action recognition is one of the most important topics in computer vision. Action recognition aims to recognize the actions and goals of one or more agents from a series of observations on the agent's action and environmental conditions. The goal of action recognition is to automatically detect the activities from a video and label them on an LCD display. Human action recognition has a wide range of applications, such as security, surveillance, entertainment, video annotation &

Pravin Dhulekar (✉) · S.T. Gandhe · Harshada Chitte · Komal Pardeshi
Department of Electronics and Telecommunication Engineering, Sandip Institute of Technology and Research Center, Nashik, Savitribai Phule Pune University, Pune, India
e-mail: pravin.dhulekar@sitrc.org

S.T. Gandhe
e-mail: stgandhe@gmail.com

Harshada Chitte
e-mail: chitte.harshada12@gmail.com

Komal Pardeshi
e-mail: komalpardeshi30@gmail.com

© Springer Science+Business Media Singapore 2017

S.C. Satapathy et al. (eds.), *Proceedings of the International Conference on Data Engineering and Communication Technology*, Advances in Intelligent Systems and Computing 468, DOI 10.1007/978-981-10-1675-2_48

retrieval and human computer interaction, border security. A security system and surveillance were operated by human years ago. It will be beneficial to use human action recognition. Video surveillance is an effective tool for today's applications. As this topic is growing very fast, human action detection has naturally become a step to be followed. The applications for human action detection are as follows.

1. Security systems: Many years ago, the security and surveillance systems were monitored by a person and was very time consuming. Our method will be very useful in this field.
2. Healthcare systems: In a healthcare system the rehabilitation process can be monitored by the action recognition.

Our proposed method involves the recognition of human action which involves the preprocessing, segmentation, feature extraction, and classification. Our method will recognize the action through these steps and will display it on LCD display. For the process of preprocessing median filter is applied. Median filter removes the noise. In the process of segmentation, we will take out the silhouettes and will apply the Gabor filter. After the process of segmentation, features are extracted by using PCA. PCA reduces the large dimensions of data space. The classification process is applied for which KNN classifiers are used.

In Sect. 2, we have given some related work. We present our action representation technique and algorithm in Sect. 3, and then we talk about the result in Sect. 4. Finally, we bring a close to this paper in Sect. 5.

2 Related Work

The paper [1] focuses on challenges and gives the overview of current advances in the field. It discusses the features that are extracted by using the global and local representations. Some surveys provide an overview of existing methods to handle the challenges. The author in [1] focuses on the research challenges that can be faced in action recognition, such as variation in viewpoint, occlusions, execution rate, and camera motion. The human Markov models (HMM) is described here. In [2] a method for human action recognition from multi-view image sequences is presented. The method uses the combined motion and shape flow information with variable considerations. Author used a combined local–global optic flow for extracting motion. Flow feature and invariant moments with flow deviations which extracts the global shape flow feature from the image sequences. Ronald Poppe [3] provided a detailed overview of current advances in the field. The limitations for the state of the art and the outline promising directions of research are also discussed. Masoud el. [4] described the article that deals with the problem of classification of human actions from video and uses the motion features. These motion features are computed very easily and they are projected into a lower dimensional space and here matching is performed. The suggested method for classification is very accurate. The article also gives the recovery of two-dimensional and three-dimensional properties of the person. The various levels of action recognition are described. The levels are split into three categories as low level, mid-level, and high level. Low level consists of core technology. Mid-level consists of human action recognition. High level consists of applications. In paper [5] author concentrated on the approaches that aim on classification of full body motions, such as kicking, punching, and waving. In paper [6], author presents a method for human action recognition from multi-view image sequences that uses the combined motion and shape flow information with variability consideration. A combined local–global (CLG) optic flow is used to extract motion flow feature and invariant moments with flow deviations are used to extract the global shape flow feature from the image sequences. In paper [7] the sampling strategies for detecting the actions are described. It gave the recent trend for sampling for the better performance. The sampling with the high density on action recognition is also explored. They have also investigated the impact of random sampling over dense grid for computational efficiency. Real time action recognition is also given. Vemulapalli et al. [8] represented the 3d skeleton detection. In this paper, a new body part-based skeletal representation for action recognition is given. Inspired by the observation that for human actions, the relative geometry between various body parts provides a more meaningful description than their absolute locations we explicitly model the relative 3D geometry between different body parts in our skeletal representation.

3 Action Representation

The action recognition has following stages:

1. Input Video 2. Image Conversion 3. Preprocessing 4. Segmentation 5. Feature Extraction 6. Classification 7. Action Detection.

1. Input Video:

In this step, the video is given as an input (Fig. 1).

2. Image conversion:

The video which is given as an input is then converted into number of images. Theseimages are then considered as test images. These images are then processed through the further stages.

3. Preprocessing:

Preprocessing enhances the visual appearance of images. It improves the manipulation of datasets. If the enhancement techniques are not used correctly, it can emphasize image artifacts or it can even lead to loss of information. In this paper, we have converted the colored image into gray image. After that we have resized the image. We will get a threshold image by adjusting the threshold value. Preprocessing involves image resampling, grayscale contrast enhancement, noise removal, mathematical operations, manual correction.

Where, image resampling reduces or increases the number of pixels of the dataset. Grayscale contrast enhancement improves the visualization by brightening

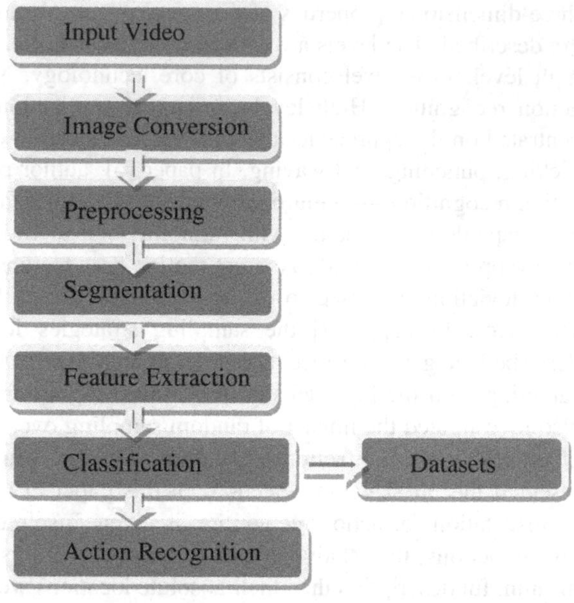

Fig. 1 Work flow of action recognition process

the dataset. Noise removal is typical preprocessing step to improve the result of later preprocessing. It involves some techniques like low-pass filtering, high-pass filtering, band-pass filtering, mean filtering, median filtering.

Median filtering is very widely used in digital image processing because under certain condition, it preserves edges while removing noise. In this paper, we are using median filtering. In median filtering, the 3 × 3 subregion is scanned over the entire image. At each position, the center pixel is replaced by the median value. Low-pass filtering, high-pass filtering, and band-pass filtering are efficient only in some cases. Most of the times, they blur the image. Median filtering is slower to compute than mean filtering. It prevents edges. It can remove noise.

Mathematical operators such as dilation and erosion are also applied. Dilation is used to connect feature in an image whereas erosion is used to disconnect features in an image and remove small ones. Manual correction tunes an image by editing it.

4. Segmentation:

Image segmentation is an important part in the action recognition system. Segmentation removes unwanted regions as well as finds the boundaries between the regions. Segmentation involves partitioning of an image into the nonoverlapping regions in a meaningful way. It identifies separate objects within an image. Silhouettes are taken out from the images. The Gabor filter can be used for this process of segmentation. Gabor filter can serve as excellent band-pass filter for unidirectional signals. Gabor filters [9] is linear and local. Its convolution kernel is a product of a Gaussian and a cosine function. The filter is characterized by a preferred orientation and a preferred spatial frequency. A 2-D Gabor filter acts as a local band-pass filter with certain optimal joint localization properties in the spatial domain and in the spatial frequency domain.

5. Feature Extraction:

Feature extraction involves reducing the amount of resources required to describe a large set of data. In this paper, we have used the Principle Component Analysis (PCA) for extracting the features. PCA is sensitive to the relative scaling of original variables. The PCA are orthogonal because they are the Eigen vectors of the co variant matrix which is symmetrical. The purpose of PCA is to reduce the large dimensionality of the data space. The advantages of PCA are that it is a powerful method in image formation, data patterns, and dimension will be reduced by avoiding redundant information without any loss. PCA in signal processing can be described as a transform of a given set of 'n' input vector with the same length 'k' formed in the 'n' dimensional vector.

$$x = [\,x1, x2, \ldots\ldots, xn]^T \tag{1}$$

into a vector y according to

$$y = A(x - mx) \tag{2}$$

The vector mx in above equation is the vector of mean values of all input variables defined by relation,

$$mx = E\{x\} = 1/k \sum xk \tag{3}$$

The rows in matrix A are formed from the Eigen vectors of C_x matrix is possible according to relation

$$C_x = E\left\{(x - mx)(x - mx)^T\right\} \tag{4}$$

The element $C_x(i, i)$ lying in its main diagonal are the variances of x.

$$C_x(i, i) = E\{(x_i - m_i)^2\} \tag{5}$$

The other values $C_x(i, i)$ determine the covariance between input variable x_i and x_j.

$$C_x(i, i) = E\{(x_i - m_i)(\,x_j - m_j)\} \tag{6}$$

PCA is possible according to

$$x = A^T y + mx \tag{7}$$

6. Classification:

For the purpose of classification we are using here KNN classifier. KNN classifier is a minimum distance classifier. KNN algorithm is a nonparametric method used for classification. K-Nearest Neighbors (KNN) algorithm uses a database in which the data points are separated into several separate classes for predicting the classification of new sample point. In KNN classification, the output is a class membership. Here an object is classified by a majority voter of its neighbors. If K = 1, then the object is simply assigned to the class of that single nearest neighbor. The algorithm is that a positive integer k is specified, along with a new sample then we select the k entries in our database which are closest to the new sample. We find the most common classification of these entries and this is the classification we give to the new sample.

Classification involves the identification of images. The classification algorithm assumes that the image to be classified has some features and those features belong to one or more classes. The design cycle of the classification involves following steps.

- Collect the data and the labels.
- Choose the features out of them.
- Select a classifier for classification.
- Train the classifier and evaluate it.

4 Experimental Analysis

Experimental analysis gives the comparison between various techniques used for detecting the human actions. After doing the experiment, our method gives us about 90–95 % result (Table 1).

We have also included the graph for ROC (Fig. 2).

Table 1 Comparison of different techniques with proposed techniques

Sr no.	Name of the paper	Result (approx.) (%)	Techniques used
1.	Behavior histograms for action recognition and human detection	86.66	Only classification is done. HOG technique is used
2.	Detecting human action in active video	40–50	Matching of frames is done
3.	Sampling strategies for real time action recognition	83	Dense sampling is used for better performance
4.	Our proposed work	90–95	Median filter, Gabor filter, PCA, KNN classifier

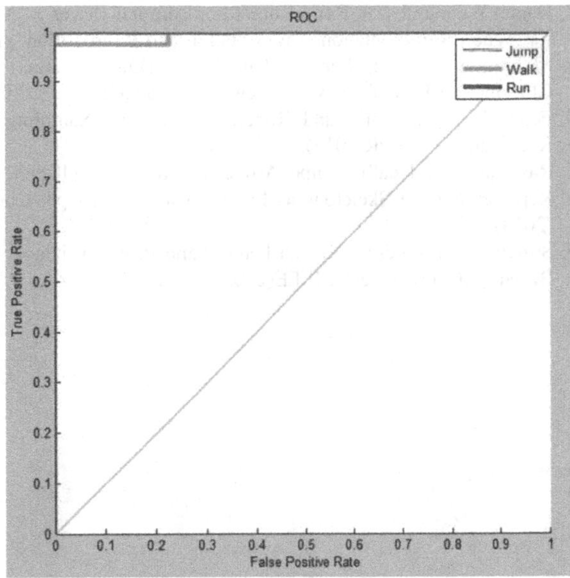

Fig. 2 ROC for some actions like jump, walk and run

5 Conclusion

In this paper, we have explained the methods for feature extraction, segmentation, and classification which are involved in action recognition process. We have provided the methods to recognize human actions from a video. Our method provides the action recognition accurately. The median filter used for preprocessing gives the good result. Other methods also provide good results.

Many more methods can be derived from future work. More methods will be derived to solve the problems involved during detecting actions. The variety of datasets is also discussed.

References

1. Manoj Raman thanz, Student Member, IEEE, Wei-Yun Yau, Senior Member, IEEE and Eam Khwang Teoh, Member IEEE,"Hman Action Recognition with Video Data: Research and Evaluation Challenges", (2014).
2. Hao Jiang, Ze-Nian Liand Mark S. Drew, "Detetcting Human Action on Active Video", IEEE (2006).
3. Ronald Poppe, "A Survey on Vision Based Human Action Recogntion", Elsevier, (2010).
4. Christian Thurau, Czech Technical University, Faculty of Electrical Engineering Department for Cybernetics, Center for Machine Perception 121 35 Prague2, Karlovo n'am'est'$_1$, Czech Republic, "Behavior Histograms for Action Recognition and Human Detection", Springer-Verlag Berlin, (2007).

5. Daniel Weinland a, ft, Remi Ronfard b, Edmond Boyer c "A Survey of Vision Based Methods for Action Representation, Segmentation and Recognition", Elsevier, (2011).
6. Mohiuddin Ahmad, Seong-Whan Lee, "Human Action Recognition Using Shape and CLG-motion Flow from Multi-view Image Sequences, (2008).
7. Feng shi, Emil Petriu and Robert Lagani'ere, "Sampling Strategies for Real time action Recognition", CVPR(2013).
8. Raviteja Vemulapalli, Felipe Arrate and Rama Chellappa, "Human Action Recognition by Representing 3D Skeletons as Points in a Lie Group", Computer Vision Foundation, IEEE (2014).
9. Simona E. Grigorescu, Nicolai Petkov, and Peter Kruizinga, "Comparison of Texture Features Based on Gabor Filters", IEEE(2002).

Performance Analysis of CREIDO Enhanced Chord Overlay Protocol for Wireless Sensor Networks

N. Bhalaji, S. Jothi Prasanna and N. Parthiban

Abstract Wireless sensor networks (WSN) are alluring to the researchers because of its creditable and commendable wide variety of applications in the new era. The main and major impediment for sensor networks is its overall architecture and its coexistence with the other established networks. This paper discusses about the application of chord overlay protocol in mobile wireless sensor networks enhancing the robustness of overlay architecture for the betterment of packet delivery ratio by amending CREIDO packet that changes the traditional operation of stabilize function in the chord protocol. This has been extensively simulated by the help of OMNeT++ simulator and its additional frameworks for mobile wireless sensor networks and overlay networks. The result obtained from the simulation expounds that the enhanced chord accomplishes the traditional chord protocol's packet delivery ratio by 13.7 %.

Keywords Mobility · Overlay · Stabilize · Fingers · Chord

N. Bhalaji (✉)
Department of IT, SSN College of Engineering,
Kalavakkam, Chennai 603 110, Tamil Nadu, India
e-mail: drnbhalaji@gmail.com

S.J. Prasanna
Department of CSE, SRM University,
Kattankulathur, Chennai 603 203, Tamil Nadu, India
e-mail: jp_drums@ymail.com

N. Parthiban
School of Computing Science and Engineering,
VIT University, Chennai 600 127, India
e-mail: parthiban24589@gmail.com

© Springer Science+Business Media Singapore 2017

S.C. Satapathy et al. (eds.), *Proceedings of the International Conference on Data Engineering and Communication Technology*, Advances in Intelligent Systems and Computing 468, DOI 10.1007/978-981-10-1675-2_49

Fig. 1 CREIDO packet specification

VERSION 4 bits	SOURCE ID 6 bits	DEST. ID 6 bits	LAYER 1 bit	RESERVED 2 bits

1 Introduction

Wireless sensor networks (WSN) is gaining new dimensions in the recent time owing to its potential societal application [1]. All the compelling and effective recent solutions are utilized to deploy a complete application over larger scale [2].

Even though it has a wide range of applications and advantages, the WSN falls back in the most important characteristic of integration with other networks. The researchers tend to develop colossal solutions in a vertical manner that may work independently but not with other network architectures [3]. This independent non-acclimatizing architecture of sensor networks impede unswervingly limits the research progress [4]. To surmount the above mentioned problem of integration, many authors propose impending solution as implementation of overlay network protocols in sensor networks (Fig. 1).

Due to the growing nature of consumer-based applications, sensor networks primarily focuses on data collection. They were assumed only to extract data from an isolated network and push data into the main stream network like Internet or LAN. But in the recent applications like body area networks, urban sensing, etc. sensor networks were not deployed in an isolated manner as data collection nodes, also deployed as data generating, people centric node.

In this article, Chord overlay protocol basically designed for P2P networks is implemented over sensor networks. As an amendment to the traditional chord protocol, the CREIDO packet is introduced in order to find the node joins and exits from the topology. The periodical stabilize function and fix_fingers operates only when notified on node joins and exits by the CREIDO packet.

2 Literature Review

Ali et al. primarily establishes the fact that P2P protocols holds a greater potential in sensor networks for making the architecture integral with other networks [4]. Also, they elucidate the advantages of overlapping and propose a DHT-based clustering protocol that places only the more powered master node in the chord ring. Even though this makes the protocol vary from application-specific development, this is not a complete solution satisfying the future focus of sensor networks. There are many pitfalls yet to be rectified as suggested by the authors in the open questions section.

H. Dai et al. focuses on developing an application level overlay networking into the gateway that encapsulates the sensor packets into the IP packets making the sensor network transparent for every device in the network through the database API. This article [5] throws a thought provoking traditional integration method on the readers only in a theoretical perspective. The techniques in [6] are majorly devised to make the autonomous sensor networks cooperate with the de facto TCP/IP standards [6]. They propose spatial IP addressing scheme that provides semi-unique IP addresses to the nodes to overcome the unfeasible dynamic configuration of IP address over a large scale. This article clearly defines the solution for some problems that impedes the overlapping of IP networks and sensor networks. But, there are many other dynamic dimensions like power consumption, etc. yet to be clearly debunked. In [7], authors emphasize that making sensor networks communicable with the internet is inevitable due to its imperative need. They arrive at the requirements needed to develop an optimal solution and finally proposes a system that is more energy efficient for communication with IP networks. But, this system is not for crucial packet delivery scenarios.

3 Chord Protocol

Chord is a resource routing kind of protocol founded on the top of distributed hashing table (DHT) that arranges the nodes contiguously in a one-dimensional ring based on identifiers. Node identifier is obtained by hashing nodes unique address. A node 'B' is said to be a successor of another node 'A' if its node identifier is the next greatest in the ring and vice versa denotes A is the predecessor of 'B' [8].

All nodes maintain a routing table with m entries called finger table [8]. The information about the other nodes are stored in this system. The kth entry in the finger table of the node K is the smallest node r that is greater than $r + 2(k - 1)$. The existing structure of chord protocol solves the issues like Flexible naming, load balancing, decentralization, availability. and flexibility. A search for the node '*f*' at node '*r*' begins by determining whether '*f*' is the immediate successor of '*r*'. If So, The search is terminated. Otherwise, '*r*' forwards the search request to the largest node in the finger table that precedes '*f*'.

To improve the lookups, predecessor, and successor pointers of the node should be up-to-date. The following three behaviors will be unveiled if a node join or node exit affects the chord ring before stabilization occurs [8].

1. The nodes in the affected region may have inaccurate successor and predecessor pointers or inaccurate keys impelling to failure of lookups.
2. In other case, successor and predecessor pointers may be correct but the keys may be inaccurate.
3. The final case may be the nodes may have accurate successors, predecessors and keys yielding to a successful lookup.

When a new node joins the finger table, it must initiate its finger table and the other existing nodes must update their finger table to reflect the existence of '*r*' [9]. A new node into the network accomplishes the following three tasks.

1. Instigates its predecessors and fingers.
2. Renovates the fingers of existing nodes to manifest the inclusion of new node in the network.
3. Acquaint higher layer software to renew the values associated with the new node.

4 Operation of CREIDO Packets

In the basic chord overlay protocol, the node entry and exit is exposed with the help of a stabilize function [8]. The archetype discussed in [8] elucidates the following scenario. Node n becomes the member of a chord system with Node ID resting between n_p and n_s. n would procure n_s as its successor. When n notifies n_s, the later takes n as its predecessor. After the run of next stabilize function, n_p queries ns for its successor. n_s expounds its successor as n to n_p. Now, n_p gets hold of n as its successor. Finally n_p will notify n and n acquires n_p as its predecessor. At this juncture, all the successor and predecessor pointers are up-to-date. By this method, chord makes its architecture robust. Even though this makes the chord ring stable, the drawback of this kind is perceiving node enters and departs in the periodical run of stabilize function. The message being sent in between the failure and join of one node and the run of stabilize function will not reach the destination from source in case of inaccurate fingers or keys.

Node entry and exit are notified periodically by the stabilize function in the existing protocol. This is the main detriment. If a node joins or departs from the overlay ring, it is notified to the neighbors only after the next run of stabilize function. Till then the successor, predecessor pointer, and fingers are erroneous priming to lookup insolvency. For example, ruminate a scenario of a node joining or departing the overlay abruptly after the run of stabilize function. Until the next run, the pointers and fingers are inaccurate leading to lookup insolvency. This can be mended with the help of periodic CREIDO packets being passed between the neighbors alluded in this article. Whenever the CREIDO finds a change in the topology, stabilize function is run and the finger table is fixed so as to cope up the dynamic topological changes. The CREIDO packet structure is as given below.

VERSION—Indicates the version of Protocol.
SOURCE ID—Holds the 6 bit Source ID.
DEST. ID—Holds the 6 bit Destination ID.
RESERVED—2 bits are reserved for future use.

5 Simulation Setup

To simulate the proposed scheme of this article, OverSim [10] a framework of OMNeT++ [11], is used for extensive simulation. The most important reason for choosing OMNeT++ is it provides extensible features enabling developers to create custom underlay network for simulating in OverSim framework. In this simulation, a custom underlay network with sensors nodes is configured using the INETMANET modules. Second, in the OverSim framework, chord protocol is appended with the CREIDO packet specification. Finally, with the simulation parameters as specified in Table 1 the custom underlay configured is imported into OverSim for simulation and the results has been obtained.

5.1 Network Model

In this simulation model simulated area is a free space with nodes moving according to the mobility model. In the beginning of simulation, all nodes are placed in a manner that all nodes are connected in the network. Nodes are connected through wireless links and said to be in contact, if they fall within the specified sensing range of each other.

5.2 Performance Metrics

For the comparison and performance analysis of the existing and proposed system, the packet delivery ratio, throughput, and control packet overhead were taken into consideration.

Table 1 Simulation parameters

Parameter	Value
Examined protocol	Chord and CREIDO chord
Routing type	Recursive routing
Transmission range	150 m
Message packet size	512 bytes
CREIDO periodicity (s)	10, 15, 20
No. of nodes	25 nodes
Area	3000 m * 3000 m
Simulation time	1500 s
Propagation model	Free space
Movement model	Random waypoint model
Pause time	5 s
Maximum speed	10 m/s

5.3 Mobility Model

Mobility plays a vital role in ad hoc networks. It increases the capacity and packet delivery ratio [12]. Mobility model incorporated for this experiment is the random waypoint model.

Random Waypoint Mobility model: This mobility model mimics erratic movement. In this model, a node moves from its extant location to a new location by randomly choosing a direction and speed. Each node moves in the random walk mobility model occurs in either a constant time interval t or a constant distance traveled d.

5.4 Simulation Parameters

This section describes the parameters smeared for the simulation analysis for the purported protocol. The performance analysis is studied by the simulation of the conventional protocol and the titivated protocol by insinuating the parameters in the below table.

6 Performance Analysis and Results

The proposed technique is simulated with a scattered network consisting of 25 mobile sensor nodes parameters. When two nodes encounter each other in their transmission range, communication is ensued. The influence of packet delivery ratio, throughput, and control packet overhead of the protocol put forward in this article is swotted by varying the CREIDO periodicity facilitated by the parameters as given in the Sect. 5.4 and its performance is analyzed with the metrics throughput, packet delivery ratio, and control packet overhead as defined in the Sect. 5.2.

Figure 2 depicts packet delivery ratio while the CREIDO periodicity is set as 10 s. While random waypoint (RWP) mobility model shows an average 12 % of increase in delivery ratio. Packet delivery ratio while the CREIDO periodicity is set as 15 s is illustrated in Fig. 3. RWP mobility model delivers a 19 % increase in delivery ratio. Figure 4. portrays the Packet delivery ratio while the CREIDO periodicity is set as 20 s. RWP mobility model shows an average 17 % of increase in delivery ratio.

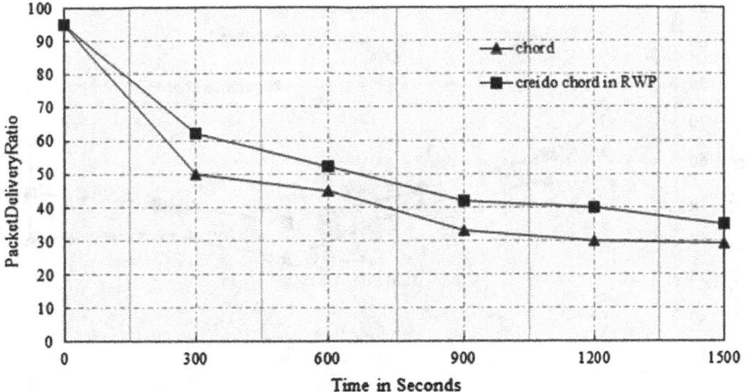

Fig. 2 PDR for 10 s interval

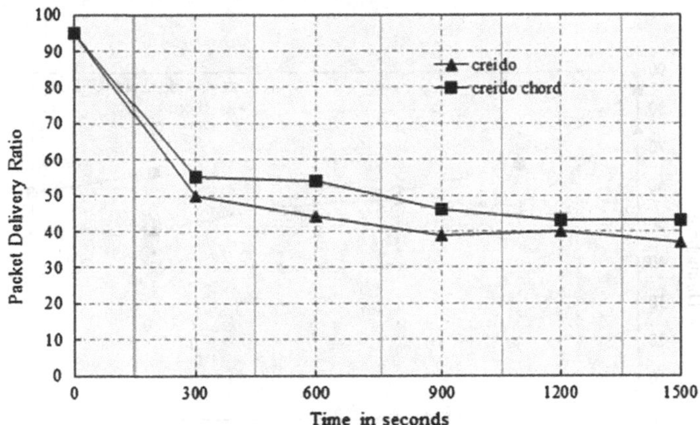

Fig. 3 PDR for 15 s interval

Figure 5 portrays throughput while the CREIDO periodicity is set as 10 s. In RWP model throughput is increased to 10 % in average. Throughput obtained for the CREIDO periodicity set at 15 s is illustrated in Fig. 6. RWP model has 14 % increased average. The throughput decreases slightly when the transitions of CREIDO packet falls off due to comparative increase in CREIDO periodicity. Figure 7 describes throughput while the CREIDO periodicity is set as 20 s. In RWP model throughput is increased to 13 % in average. The throughput slightly decreases as the number of transitions of CREIDO packet is decreased due to comparative increase in CREIDO periodicity.

Figure 8 describes Control packet overhead when CREIDO packets are transmitted at a period of 10 s. Control overhead rate is averagely increased to 9 % in RWP. Even though increase in Control Packet overhead is havoc, it is tolerable because average Control Packet overhead is only 9 %. Figure 9 illustrates control

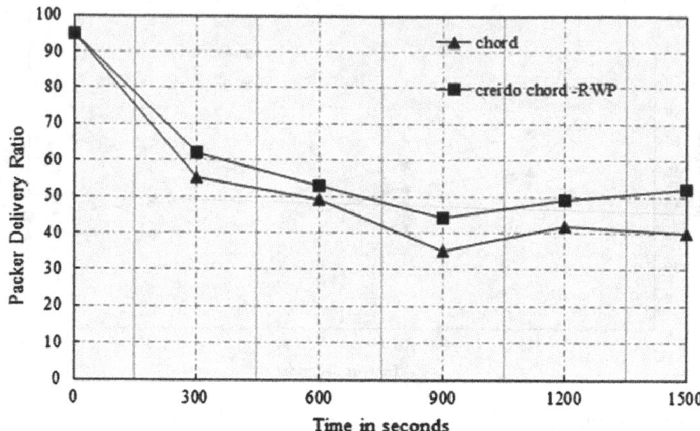

Fig. 4 PDR for 20 s interval

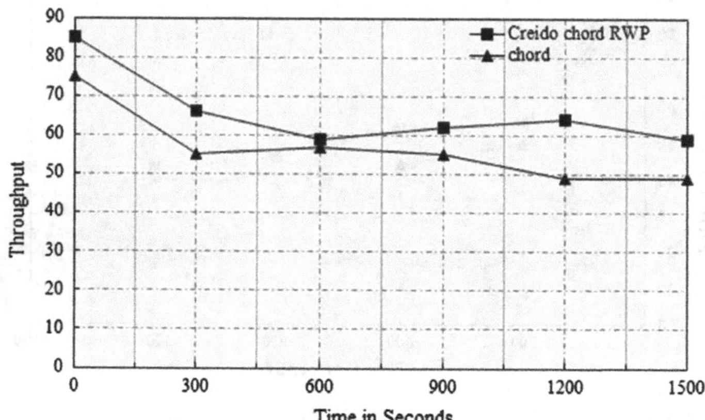

Fig. 5 Throughput for 10 s interval

packet overhead when CREIDO packets are transmitted at a period of 15 s. Control overhead rate is averagely elevated to 12 in RWP model. Control Packet overhead decreases significantly as the CREIDO periodicity increases. Figure 10 gives a picture of control packet overhead when CREIDO packets are transited at a period of 20 s. Control overhead rate is averagely increased to 9 % in RWP. Control packet overhead decreases comparatively because of the increase in CREIDO periodicity.

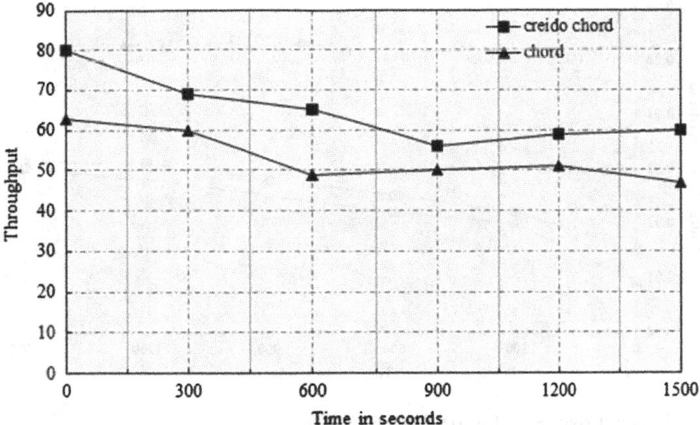

Fig. 6 Throughput for 15 s interval

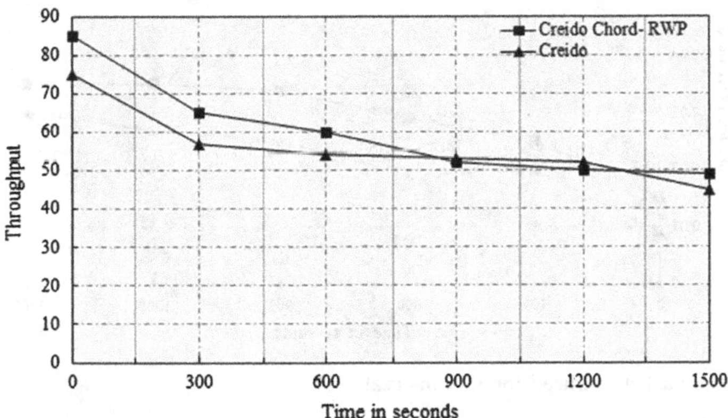

Fig. 7 Throughput for 20 s interval

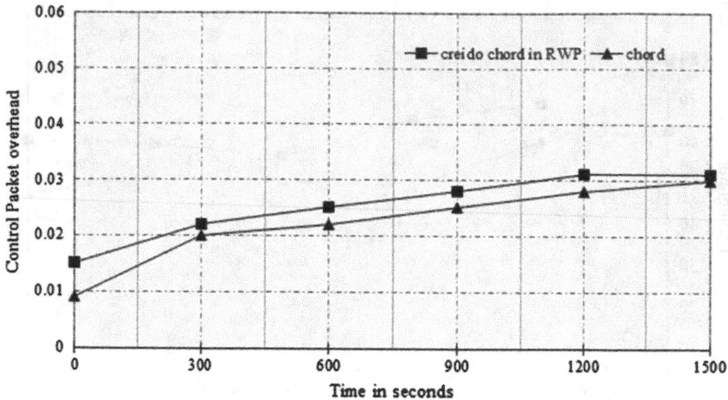

Fig. 8 Control packet overhead for 10 s interval

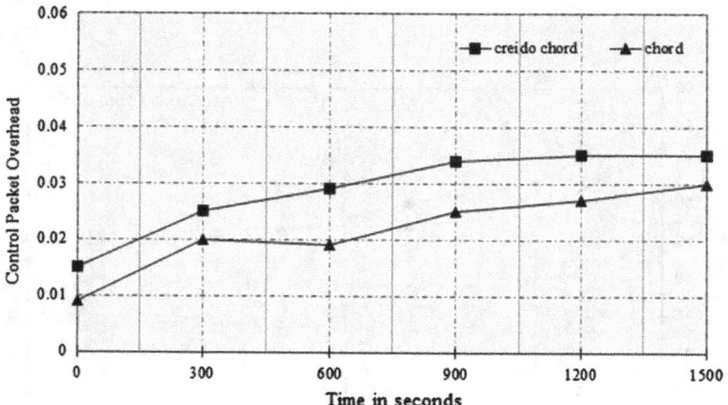

Fig. 9 Control packet overhead for 15 s interval

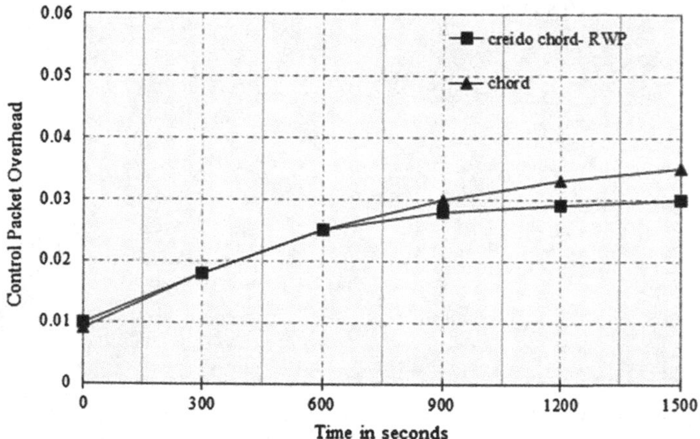

Fig. 10 Control packet overhead for 20 s interval

7 Conclusion

This article fabricates periodic CREIDO packet over chord protocol for Mobile sensor networks. Extensive simulations were carried out with the frameworks of OMNeT++. This improvised the aforementioned performance metrics and noteworthy recuperated results were obtained. An average of 13.7 % step-up in packet delivery ratio, 14.6 % proliferation in throughput and 14 % escalation in control packet overhead is realized. The imminent prospect will be extensively studied for performance of the devised protocol for various movement models and mitigating the security issues that may occur due to the amendment made by the CREIDO packet.

References

1. Dutta, P., Grimmer, M., Arora, A., Bibyk, S., Culler, D.: Design of a wireless sensor network platform for detecting rare, random, and ephemeral events. In: Proceedings of the 4th international symposium on Information processing in sensor networks, p. 70 (2005).
2. Mainwaring, A., Culler, D., Polastre, J., Szewczyk, R., Anderson, J.: Wireless sensor networks for habitat monitoring. In: Proceedings of the 1st ACM international workshop on Wireless sensor networks and applications, pp. 88–97 (2002).
3. Culler, D., Dutta, P., Ee, C., Fonseca, R., Hui, J., Levis, P., Polastre, J., Shenker, S., Stoica, I., Tolle, G., others: Towards a Sensor Network Architecture: Lowering the Waistline. In: HotOS (2005).
4. Ali, M., Langendoen, K.: A case for peer-to-peer network overlays in sensor networks. In: International Workshop on Wireless Sensor Network Architecture (WWSNA'07), pp. 56–61 (2007).
5. Dai, H., Han, R.: Unifying micro sensor networks with the internet via overlay networking [wireless networks]. In Local Computer Networks, 2004. 29th Annual IEEE International Conference on, pp. 571–572 (2004).
6. Dunkels, A., Alonso, J., Voigt, T.: Making TCP/IP viable for wireless sensor networks. Tech. rep. (2003).
7. Kosanovi, Stoj: Connecting wireless sensor networks to Internet. FACTA UNIVERSITATIS, Mechanical Engineering 9(2), 169–182 (2011).
8. Stoica, I., Morris, R., Karger, D., Kaashoek, M., Balakrishnan, H.: Chord: A scalable peer-to-peer lookup service for internet applications. ACM SIGCOMM Computer Communication Review 31(4), 149–160 (2001).
9. Balasubramanian, A., Levine, B., Venkataramani, A.: Replication routing in DTNs: a resource allocation approach. IEEE/ACM Transactions on Networking (TON) 18(2), 596–609 (2010).
10. Baumgart, I., Heep, B., Krause, S.: OverSim: A flexible overlay network simulation framework. In: IEEE Global Internet Symposium, 2007, pp. 79–84 (2007).
11. Varga, A., Hornig, R.: An overview of the OMNeT++ simulation environment. In: Proceedings of the 1st international conference on Simulation tools and techniques for communications, networks and systems & workshops, p. 60 (2008).
12. Davies, V., others: Evaluating mobility models within an ad hoc network. Master's thesis (2000).

Implementation of Hashing in Virtual Tour

Kuhu Gupta, Aishwarya Shastry, Devesh Krishnani, Mridu Sahu and Govind P. Gupta

Abstract Virtual environments are being used in a wide range of academic and commercial applications. It can give users a natural feeling of the environment by creating realistic virtual worlds. Virtual tours are used to familiarize a user with his surroundings. But most of them are just images rotating and do not provide end user the functionality of navigating around the tour. This leads to the user terminating the virtual tour. Many virtual tours are large in size because of repetition of images in several places, this leads the virtual tour to lag, thus, affecting end user experience. In this work, virtual tour application is made using graphics rendering by Three.js, hashing and cube mapping. The application allows to take a virtual tour of NIT Raipur. The article finds a relationship between the graphics rendering of the scene and creation of navigation paths to ease virtual touring.

Keywords Virtual tour · Three.js (JavaScript Library) · Skybox · Hashing · Navigation

Kuhu Gupta (✉) · Aishwarya Shastry · Devesh Krishnani · Mridu Sahu · G.P. Gupta
Department of Information Technology, National Institute of Technology Raipur, Raipur, India
e-mail: kuhu.gupta08@gmail.com

Aishwarya Shastry
e-mail: aishwarya.shastry231@gmail.com

Devesh Krishnani
e-mail: dev.krish23@gmail.com

Mridu Sahu
e-mail: mrisahu.it@nitrr.ac.in

G.P. Gupta
e-mail: gpgupta.it@nitrr.ac.in

© Springer Science+Business Media Singapore 2017

S.C. Satapathy et al. (eds.), *Proceedings of the International Conference on Data Engineering and Communication Technology*, Advances in Intelligent Systems and Computing 468, DOI 10.1007/978-981-10-1675-2_50

1 Introduction

Nowadays, many virtual tours are available on the Web and this gives us an evidence that virtual tours are becoming accessible to a larger and more diverse audience [1]. The important aspect is that students from all over the country are interested to take admission in colleges. Therefore, there is a lot of interest in the public for visiting educational institutes. For those who want to experience the aura of the college sitting at home, virtual tour is an asset. This virtual tour application would also be helpful for the college website as it can be taken online by anyone in any part of the world who wants to know about the campus and wants to explore all the facilities that exist in the campus. The web application navigates through all the amenities that are present in the campus. It takes the viewer on a tour of each department starting from the entrance of the institute. The significance of the virtual tour is that we get a panoramic view of the campus which will enhance our journey on the Internet. The users can navigate and interact with the system, yet they are not allowed to walk through the walls. Different camera views are set at different locations and at different angles to provide natural views of a scene.

The paper describes an approach that allows a layman to take a virtual tour of a building. In addition, the tour is the combination of images rendered on the sides of a 3D cube and navigation paths which can easily adapt to the needs of a user [2]. Three.js and cube mapping generates 3D cube map by first rendering the scene six times from a viewpoint, with the views defined by an 90° view frustum representing each cube face [3]. The optimization of the entire virtual tour is achieved by the concept of hashing. A lookup table is created to store images in a local database and a hash function to append some random digits to increase the security. The paper illustrates the results by means of a real case, which is a reconstruction of National Institute of Technology, Raipur. In the next section an overview of the related work is presented. Then, some of the components of the system are discussed more in detail: Three.js (Sect. 3) Skybox and Cube mapping (Sect. 4), Thumbnail navigation using OrbitControls.js (Sect. 5) and Hashing (Sect. 6) Selective Navigation (Sect. 7), Methodology (Sect. 8), Future developments (Sect. 9). The last two sections contain some result analysis for research and the conclusions.

2 Related Work

In recent years many virtual campuses have been built [4]. A virtual model for Avcilar campus of Istanbul University has been built [5]. A virtual tour model of IIT Gandhinagar has been built which consists of a panoramic image rotating at a 360° where clickable images take us to another panorama [6]. Similarly, IIT Kanpur and IIM Bangalore Virtual tour consists of a menu of the landmarks present in the campus giving a spherical view of each landmark [7]. Amherst college and Harvard college virtual tours provides a tourist guide who describes the sliding images which show us the important landmarks of the college [8]. The virtual tour of Harvard college has a menu which contains 360° panoramas, videos, and photos [9]. The University of Cambridge provides virtual tour of its every college by stitching together panoramas to form a cube enabling real-time viewing [10]. The renowned Oxford University gives the satellite view of its landmarks in its virtual tour by stitching panoramic images to form a moving cube [11]. Nanyang Technological University also has its own virtual tour [12]. In the present scenario, virtual reality and web technologies in the field of distance education for web-based medical simulations play a significant role [13].

3 Three.Js (JavaScript Library)

WebGL (Web Graphics Library) is a JavaScript API used to render 2D and 3D graphics to the screen in a compatible browser [14]. Programming directly in the WebGL API can be very complicated at times but there are libraries that simplify this and one such library is Three.js [15]. Three.js is a lightweight JavaScript library which is used to create 3D computer graphics on a web browser [16]. It is a cross browser JavaScript API Cross Browser is the ability of a web application to work well in browsers that it is compatible with and to stop in a dignified way when its features do not match the browser [17]. Three.js supports renderers, scenes, lights, loaders, cameras, and animation [18]. The sequence of steps which occur while rendering the skybox to make a virtual tour using the different features of Three.js is shown in Fig. 1 [19]. First of all, a scene is setup with a camera and renderer [20].

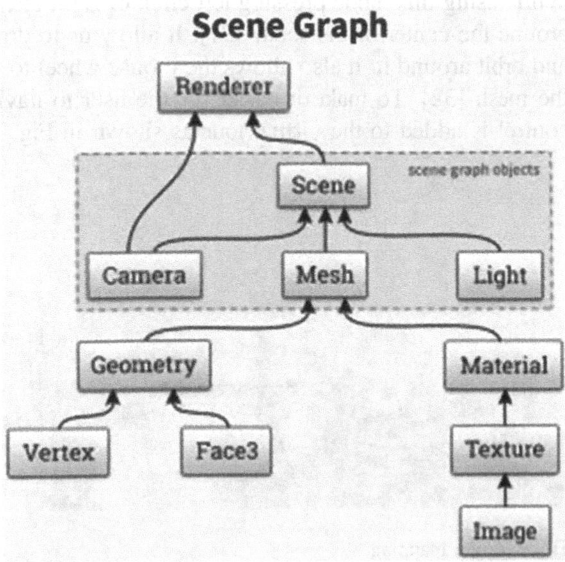

Fig. 1 Rendering of a scene

The Renderer generates the final detailed images, based on current camera, 3D geometries, materials, and lighting setup [21]. The Camera determines what is actually rendered on the screen while lights have an effect on how the materials are shown and used when creating shadow effects [22]. Every object that is rendered in Three.js is a Mesh consisting of geometry and material [23]. A geometry holds all data necessary to describe a 3D model consisting of vertices and faces [24]. Materials describe the appearance of objects by supporting textures for adding detail to a 3D object by applying an image to one or more of the faces of that object [25].

4 Skybox and Cube Mapping

Skybox is a procedure for creating backgrounds in computer and video games, The Sky, distant mountains, buildings, and other objects are put across at the cubes faces using cube mapping technique [26]. The cube map is created by first rendering the scene six times from a viewpoint, with the views defined by a 90° view camera placed inside the cube [27]. The movement of the cube along with the camera creates a vision of three-dimensional surroundings [28]. The concept of cube mapping is clearly represented in Fig. 2 [29].

5 Thumbnail Navigation Using OrbitControl.Js

To view all sides of the skybox it becomes crucial to add Orbit controls mode [30]. While using this mode pressing left click or right click in mouse rotates the camera around the center of the scene [31]. It allow us to drag the mouse across the mesh and orbit around it. It also allows the mouse wheel to be used to zoom in and out of the mesh [32]. To make it easier for the user to navigate inside the skybox, orbit control is added to the virtual tour as shown in Fig. 3.

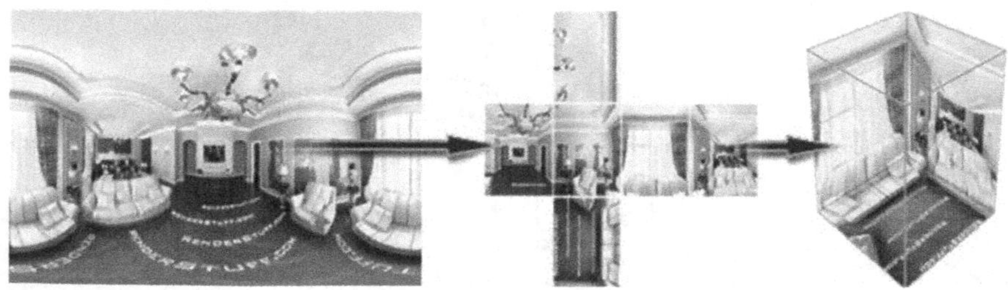

Fig. 2 Cube mapping

Fig. 3 Arrow keys are used to navigate inside the skybox

6 Hashing

Hashing is used to index and retrieve items in a database [32]. Using hashed value the items can be retrieved faster [33]. Here a $n * n$ matrix is created, the first column contains the hashed value of initial skybox of every place and their corresponding columns contains the hashed value of the places that can be visited through them [34]. MD5 hashing functions is used to reduce the size of array, it provides cryptic address which leads to more security [35]. The flow of movement in the campus in an optimal way is shown in Fig. 4.

7 Selective Navigation

JQuery is a cross-platform JavaScript library designed facilitate to the client-side scripting of HTML [36]. Tipue Search is a search engine jQuery plugin, It's free, open source, responsive, and fast [37]. The live mode of Tipue Search indexes the list of pages, but no other content and the entry in the search box lists out all the related pages on the site to ease selective navigation [38].

8 Methodology

The article used methodology for virtual touring application and the flow diagram for this is shown in Fig. 5 and step by step walkthrough is mentioned in Table 1. The figure clearly shows the exact features available in our virtual tour. It consists

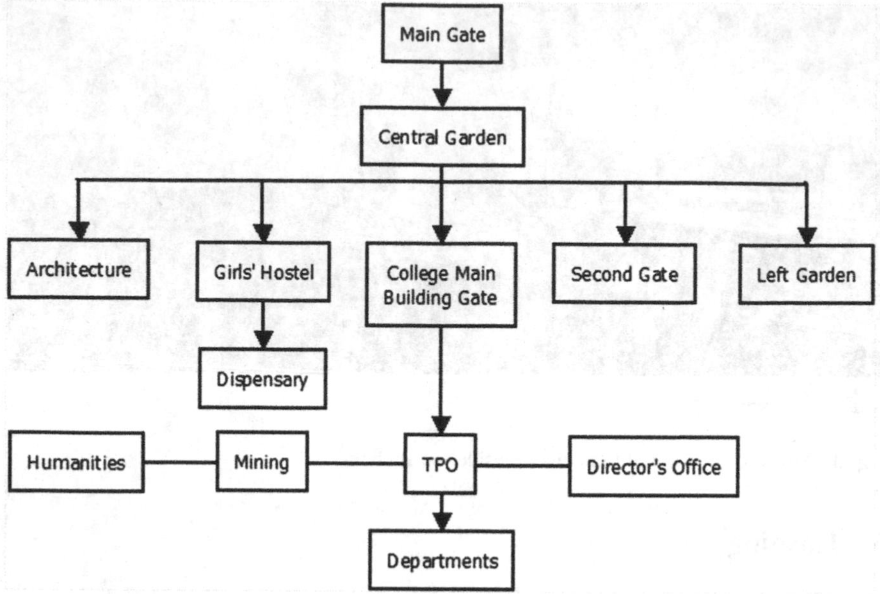

Fig. 4 Flow diagram of navigation

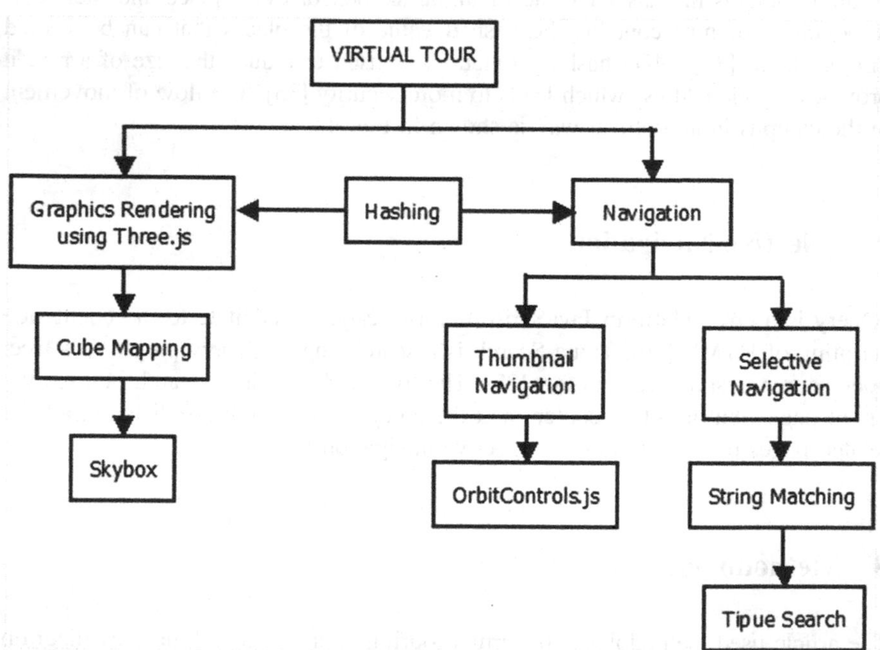

Fig. 5 Virtual touring application flow diagram

Table 1 Font sizes of headings

Steps	Description
Step 1	Use graphics rendering available in three.js
Step 2	Cube mapping is used for formation of a cube where the camera is put
Step 3	Camera is kept at inside of the cube thus making it a skybox
Step 4	Navigation is used to help user navigate through the campus
Step 5	Selective navigation is used for navigating through landmarks and department
Step 6	Orbit controls.js control the arrow keys which are used to navigate inside the skybox
Step 7	Thumbnail navigation is used to see the sides of the cube
Step 8	The string entered in search box is first matched with the database
Step 9	Tipue search helps in taking the user to the specific location

of two parts, graphics rendering and navigation. Hashing is used to connect both. Navigation available is of two types, selective Navigation and thumbnail navigation.

The outcomes of methodology are discussed in the result section.

9 Future Developments

Future work consists of using a real person for the virtual guide. This virtual guide rotates 360° instead of the camera. The user can see through the eyes of the Virtual guide. The goal is to include text to speech and speech to text for easy navigation for physically disabled. The application can be extended to aim to include live event viewing. Application can include videos instead of images and let the user experience a real time experience. Application can be modified to use digital image processing for modeling the surrounding and thus create a virtual tour with more effective real life effect, the object in the tour would have freedom to go anywhere in the tour without changing the instance of application.

10 Result Analysis

Virtual Tour provides a simulation of a surrounding thus allowing them to view it while not being physically present. It allows the user to observe and experience the place. Virtual tour is built using three.js which is a library used for many other computer graphics application over the world. It is simple and efficient to use. The virtual tour uses images of environment it is based on thus providing a unique experience. The users are allowed to freely navigate the virtual tour using an object

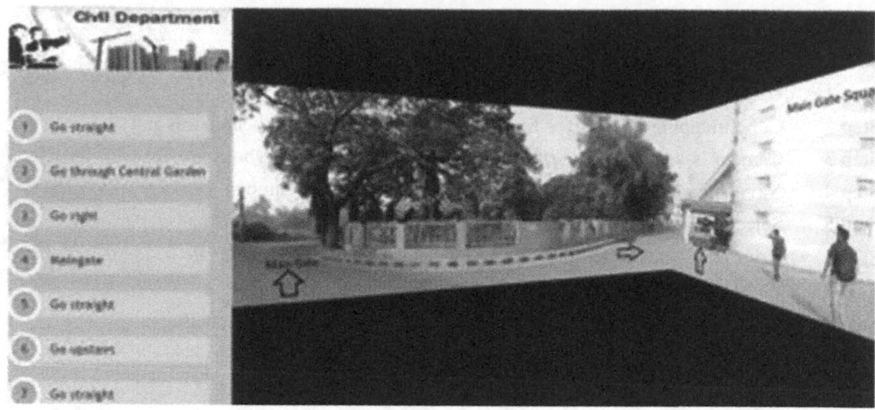

Fig. 6 The *left* side of the image shows the navigation options

Fig. 7 The *right* side of the image shows the skybox of the entrance of civil department

but cannot go through walls. Virtual tour also has a search facility for selective navigation plus speech to text and text to speech navigation for physically disabled.

The virtual tour has its limitations. It requires high speed internet to function properly. The browser should support JavaScript and WebGL. Images of high quality and low size are required to make the tour efficient. Since three.js is a constantly evolving online library therefore proper maintenance is required. Figures 6 and 7 are the screenshots of the virtual tour of a department navigating the user from the entrance to the department.

11 Conclusion

Proposed article shows, a successfully created virtual tour implemented along with hashing. NIT Raipur is used as a data set for the project. Every Landmark and Department is covered. If a user wants to know about NIT Raipur. He/she can tour the campus using our Virtual Tour. The Technique "Skybox" made using Three.js was the core part of our project. Hashing algorithm was used for selective navigation. Our analysis gave a space complexity of O (n * n) and time complexity of O (n). Using a cube proved to be more efficient and beneficial both in term of space and time as time taken to render square images is quite less when compared to rendering a sphere as only two panorama images are required to create the Cube, whereas more than ten images were required for a sphere.

References

1. Frederic Kleinermann, Olga De Troyer, Christophe Creelle, Bram Pellens: Adding Semantic Notaions, Navigation Paths and Tour Guides to Existing Virtual Environments 13th International Conference VSMM (2007).
2. Marc Pollefeys, Luc Van Gool, Ive Akkermans, Dirk De Becker, Kris Demuynck: A Guided Tour to Virtual Sagalassos. Proceedings of the conference on Virtual reality, archeology, and cultural heritage (2001).
3. Fernando, R., Kilgard, M.: The Cg tutorial. Addison-Wesley, Boston, Mass. (2003).
4. Khan, F. et al.: Using VRML to Build a Virtual Reality Campus Environment. World Congress on Engineering. (2008).
5. M. Nusret SARISAKAL, K. Gökhan CEYLAN: "*A Virtual Reality For Avcilar Campus Of Istanbul University Using VRML*," Istanbul University- Journal of Electrical and Electronics Engineering, Vol 3, no 2 (2003).
6. dea.com,: 4Deal Experience beyond 3D Toursl 360 Interactive Virtual Tours, http://www.4dea.com/index.html#.
7. Fotobubbles.com,: IIT Kanpur Virtual Tour - see India's premier institute!, http://www.fotobubbles.com/portfolio/iit-kanpur-virtual-tour/.
8. Amherst.edu,: Virtual Tourl Amherst College, https://www.amherst.edu/aboutamherst/visiting/virtualtour.
9. College.harvard.edu,: Virtual Tourl Harvard College, https://college.harvard.edu/admissions/visit/virtual-tour.
10. Kings.cam.ac.uk,: Virtual tour of the college, http://www.kings.cam.ac.uk/visit/virtual-tour/.
11. Chem.ox.ac.uk,: Virtual Tour of Oxford, http://www.chem.ox.ac.uk/oxfordtour/.
12. Sourin, A.: Nanyang Technological University virtual campus [virtual reality project]. IEEE Comput. Grap. Appl. 24, 6, 6–8 (2004).
13. Dimitropoulos, K. et al.: Building Virtual Reality Environment for Distance Education on the Web: A case study in Medical Evaluation. INTERNATIONAL JOURNAL OF SOCIAL SCIENCES. 2, 1, (2008).
14. Tavares, G.: WebGL Fundamentals. (2012).
15. Steir, G.: Getting started with WebGL and Three.js, http://solutiondesign.com/blog/-/blogs/getting-started-with-webgl-and-three-1, (2012).
16. GitHub,: mrdoob (Mr.doob), http://github.com/mrdoob.
17. Jibbering.com,: Browser Detection (and What to Do Instead), http://jibbering.com/faq/notes/detect-browser.

18. GitHub,: mrdoob/three.js, https://github.com/mrdoob/three.js/wiki/Features.
19. Davidscottlyons.com,: Intro to WebGL with Three.js, http://davidscottlyons.com/threejs/presentations/frontporch14/#slide-16.
20. Packtpub.com,: Working with the Basic Components That Make Up aI PACKT Books, https://www.packtpub.com/books/content/working-basic-components-make-threejs-scene.
21. Clara.io,: Renderering Basic Learn Clara.io, https://clara.io/learn/user-guide/rendering/rendering_basics.
22. Packtpub.com,: Working with the Basic Components That Make Up aI PACKT Books, https://www.packtpub.com/books/content/working-basic-components-make-threejs-scene.
23. Steir, G.: WebGL and Three.js: Lighting, http://solutiondesign.com/blog/-/blogs/webgl-and-three-js-lighting/, (2014).
24. Geometry of 3-D model, http://threejs.org/docs/Reference/Core/Geometry.
25. WebGL, L. et al.: Three.js - Materials and TextureI PACKT Books, https://www.packtpub.com/books/content/threejs-materials-and-texture.
26. Greene, N.: Environment Mapping and Other Applications of World Projections. IEEE Comput. Grap. Appl. 6, 11, 21–29 (1986).
27. Fernando, R., Kilgard, M.: The CG Tutorial-The Definitive Guide to Programmable Real-Time Graphics. Addison-Wesley Longman Publishing Co, Boston (2003).
28. Etienne, A., Etienne,J.: Lets Do a Sky, http://learningthreejs.com/blog/2011/08/15/lets-do-a-sky/, (2011).
29. Renderstuff.com,: 360 panorama in 3ds Max tutorial, http://renderstuff.com/creating-virtual-360-panorama-cg-tutorial/.
30. Dirksen, J.: Three.js essentials. Packt Pub., Birmingham, UK (2014).
31. Dun, R., Fan, Z., Yao, H., Zhang, S.: A Three-dimensional Scene Simulation System for Evaluating Textures of Fabrics. International Conference on Logistics Engineering, Management and Computer Science (LEMCS 2015). pp. 138–143. Atlantis Press (2015).
32. Pettit, N.: The Beginner's Guide to three.js, http://blog.teamtreehouse.com/the-beginners-guide-to-three-js, (2015).
33. H. Cormen, T., E. Leiserson, C., L. Rivest, R., Stein, C.: Introduction to Algorithms. The MIT Press, Massachusetts Institute of Technology, Cambridge (2009).
34. Konheim, A.: Hashing in computer science. Wiley, Hoboken, N.J. (2010).
35. Peter Kankowski.: "Hash functions: An empirical comparison".
36. Schneier, B.: Cryptanalysis of MD5 and SHA: Time for a New Standard, https://www.schneier.com/essays/archives/2004/08/cryptanalysis_of_md5.html, (2004).
37. jquery.org, j.: jQuery, https://jquery.com/.
38. Tipue.com,: Tipue, http://www.tipue.com/?d=1.

RENT: Regular Expression and NLP-Based Term Extraction Scheme for Agricultural Domain

Niladri Chatterjee and Neha Kaushik

Abstract This paper addresses the task of automatic term extraction in agricultural domain. There is a paramount call for applying effective data processing on a huge amount of agricultural data lying unprocessed. The method is based on basic techniques in Named-entity recognition, and involves a resequencing of the conventional procedure of automatic term extraction. Several domain-specific patterns identified by the domain experts have been used for this purpose in the baseline algorithm. After evaluating the performance of baseline, several improvements have been proposed by observing the obtained results on a given agricultural text. These improvements have been incorporated into the RENT algorithm. Both the algorithms have been applied on more than 1400 pages of agricultural text. It is concluded that the RENT algorithm significantly outperforms the baseline algorithm with a precision of more than 80 %, recall more than 60 % and f-measure more than 68 % on random samples. A comparison with the Termine, a well-known software for term extraction, is also presented which shows that RENT has a better precision.

Keywords Automatic term extraction • Pattern analysis • Regular expression • Agricultural term extraction • Natural language processing • Hypergeometric distribution

---

Niladri Chatterjee (✉) · Neha Kaushik
Indian Institute of Technology Delhi, Hauz Khas, New Delhi 110016, India
e-mail: niladri.iitd@gmail.com

Neha Kaushik
e-mail: swami.neha@gmail.com

Neha Kaushik
Kasturba Institute of Technology, Directorate of Training
and Technical Education, Government of N.C.T. of Delhi, New Delhi 110088, India

© Springer Science+Business Media Singapore 2017

S.C. Satapathy et al. (eds.), *Proceedings of the International Conference on Data Engineering and Communication Technology*, Advances in Intelligent Systems and Computing 468, DOI 10.1007/978-981-10-1675-2_51

1 Introduction

Automatic term extraction is a key step toward many recent developments and applications in text processing, such as query expansion [1], keyword tagging in research documents, sentiment analysis [2], and ontology construction [3, 4]. Understanding automatic term extraction requires the knowledge about 'termhood,' and various term extraction methods, where termhood is defined as the degree of relationship of a linguistic unit with domain-specific context [5]. Term extraction methods are typically classified into four categories: statistical methods, distributional methods, contextual methods, and linguistic methods. In this work, our focus is on extracting agricultural terms from texts. In particular, we have taken the documents available over different repositories of Government of India.

The reason for choosing agricultural domain is that in recent time there is a paramount call of effective data processing methods to be applied on the agricultural data. Several initiatives have already been taken by the central (national) and the state (provincial) governments to meet the various challenges facing the agricultural sector in the country. Mission mode projects (MMPs) to be operationalized by Department of Agriculture and Cooperation[1] aims to provide services such as:

- information for farmers on fertilizers, seeds, pesticides
- information for farmers on Government schemes
- information for farmers on soil recommendations
- information on crop management
- information on weather and marketing of agriculture produce

Terms extracted from the available agricultural data can be used to populate agricultural thesauri, and agricultural ontologies which in turn can be used for effective information dissemination.

Due to the vastness of agricultural data and lack of processed data, there is also a need of employing efficient knowledge management techniques, and drawing useful inference from this data. The proposed approach emphasizes on extracting the domain-specific terms with minimum human intervention. The method espouses the idea from named-entity recognition. However, in the proposed approach, we reorder some of the steps involved in existing term extraction methods for effective application of the algorithm.

Automation of term extraction in agricultural domain is quite challenging because of the following reasons:

(a) *Vastness of data*—Agriculture as a domain is often categorized into a large number of subdomains. For illustration, several categories are crops, fertilizers and seeds, livestock, soil, irrigation, farming systems and urban farming,

[1]http://deity.gov.in/content/agriculture.

weather, agrobiodiversity, soils, water resources, and utilization, water management and crop production, irrigation and drainage management, soil fertility, fertilizers, cropping patterns, farm machinery, diseases of crops, insect pests and their management, seed production and technology, agroforestry, forage crops and grasses, sericulture, crop biotechnology. Large agricultural data associated with each of these subdomains is quite vast in themselves. The focus of the present work has been on the following subdomains: crops, fertilizers and seeds, livestock, soil, irrigation, farming systems and urban farming only.

(b) *Composite terms* are very difficult to be extracted from the agricultural data. For example, the term "tuber crops" refers to certain type of crops. Princeton WordNet[2] provides the meaning for "tuber" and "crops," but it does not mention about the composite term "tuber crops". Indian WordNet[3] also does not provide for the meanings of composite terms, often relevant to Indian agriculture. Most of these terms are also not present in AGROVOC,[4] a multilingual agricultural thesaurus, NAL[5] (National Agricultural Library) Thesaurus.

(c) *Difficult categorization*—Categorizing terms is a challenging task in agricultural domain. For example, "coconut" may be considered as a "fruit," "vegetable" as well as a "seed"; "pumpkin" is a "fruit" which is used as a "vegetable." The term "vegetable" refers to "any plant whose fruit, seeds, roots, tubers, bulbs, stems, leaves, or flower parts are used as food." Moreover, being polysemous it also means "a dull, spiritless, and uninteresting person." The terms "Rabi," "Zaid," and "Kharif" can be categorized as three types of crops as well as three crop seasons in Indian context.

(d) *Format of data*. Most of the data available in the agriculture domain is in the tabular form, with no relations or constraints specified on the fields, it becomes very difficult to find out the relations between the various concepts when only the tabular data is available. The available data do not reflect the conceptual relationships that can be used by a system to suggest concepts for expanding the query or making it more specific.

The paper is organized as follows. Section 2 highlights the need of a novel approach for automatic term extraction, and Sect. 3 discusses the proposed approach. Section 4 consists of results and analysis of the proposed approach. Section 5 concludes the paper.

[2]http://wordnetweb.princeton.edu/perl/webwn.

[3]http://www.cfilt.iitb.ac.in/wordnet/webhwn/wn.php.

[4]http://aims.fao.org/standards/agrovoc/functionalities/Search.

[5]http://agclass.nal.usda.gov/agt.shtml.

2 Need of a Novel Approach for Automatic Term Extraction

The existing resources although being supported by ICT, are all developed and maintained by human experts based upon the expert knowledge. In the context of India, we have a huge amount of agricultural data in the form of textual documents, tables, and spread sheets. However, the data is often underutilized because of lack in data processing applied to agricultural data.

The existing methods [5–7] for automatic term extraction are based on linguistics and statistics for extracting the terms. These methods make use of POS tagging and stop words removal as a preprocessing step for automatic term extraction. For example, usually nouns are considered as good candidate for being a valid term. But it is not always the case, like "irrigate/cultivate/sow" are verb but are quite important agricultural term. Also it is not possible to include all the unwanted words of the domain in the stop list. Stop list more often contains general words [8] and articles and helping verbs like 'a,' 'an,' 'the,' 'of,' 'is,' 'and are', etc. which in any domain do not qualify to be a term. Moreover, some words in the stop list may actually help in extracting some important terms in the domain, one such word is 'of'. 'Of' not only helps in identifying the important terms but is very much useful in relationship extraction as well. For example, consider the phrases: production of *wheat*, production of *rice*, production of *maize*, production of *milk,* if we had already omitted the word 'of' from the text, we would not be able to extract these phrases, and what we observed here is that the words following 'of' in each of these phrases are important terms.

Typical sequence of events in the existing methods for automatic term extraction is POS Tagging, stop words removal, linguistic filters, Application of Statistical/Distributional/Contextual/Hybrid Measures.

So what is needed to resolve the problems identified above is a resequencing of these steps and some additional information along with the statistical/distributional/contextual/hybrid measures to extract the important terms of the domain. Section 3 presents the proposed scheme.

3 The Proposed Scheme

The proposed approach borrows the concept from the field of Named Entity Recognition. It also takes into account part-of-speech and statistical information associated with the words contained in the text. This is a domain-specific approach as it involves knowledge related to the underlying terms, their roles, and categories and subcategories, and semantic roles in a context. The method extracts single word, two words, and three words terms from the agricultural domain.

We use some domain-specific patterns in the form of regular expressions, listed in Table 1, identified by domain experts to extract the important terms from the

Table 1 Regular expressions used in the algorithm along with the corresponding patterns

S. No.	Regular expression	Corresponding pattern
1	(production), (of), (\w+), (), (\w+), 1, phrase	Production of________ ________
2	(\w+), (production), 2, word, $, $, $	____________Production
3	(\w+), (season), 2, word, $, $, $	____________season
4	(\w+), (crop), 2, word, $, $, $	____________crop
5	(\w+), (cultivation), 2, word, $, $, $	____________cultivation
6	(\w+), (revolution), 1, word, $, $	____________revolution
7	(\w+), (hybrid), 2, word, $, $, $	____________ hybrid
8	(use), (of), (\w+), 1, word, $, $	Use of __________
9	(\w+), (sector), 2, word, $, $, $	________sector
10	(\w+), (systems), 2, word, $, $, $	________systems
11	(\w+), (consumption), 2, word, $, $, $	__________consumption
12	(consumption), (of), (\w+), 1, word, $, $	Consumption of ___________
13	(\w+), (productivity), 2, word, $, $, $	__________productivity
14	(include), (\w+), 1, word, $, $	Include ___________
15	(growth), (of), (\w+), 1, word, $, $	Growth of ___________
16	(such), (as), (\w+), 1, word, $, $	Such as _________
17	(growth), (in), (\w+), 1, word, $, $	Growth in __________
18	(cultivation), (of), (\w+), 1, word, $, $	Cultivation of________
19	(including), (\w+), 1, word, $, $	Including __________
20	(millions), (of), (\w+), 1, word, $, $	Millions of_________

input documents. The regular expressions have been written using the following rules:

(a) '\w+'—one or more occurrence of alphanumeric character
(b) '$'—matches end of string, hence to identify '*k*' words, we need to put '$k + 1$' '$' signs at the end
(c) 'phrase'—will look for phrase
(d) 'word'—will look for word
(e) Number preceding 'word'—looks for this many number of words preceding or proceeding the particular pattern

As in named-entity recognition, entities can be identified by following certain rules, for example, names always start with capital letters, and are mostly proper noun. Similarly, this method is also based on certain rules. The baseline algorithm is given in Fig. 1.

This algorithm uses some assumptions to weight the terms. A word is preferred over the others if it is a noun. A word (excluding the stop words) with high raw frequency of occurrence is more likely to be an important term. A word which occurs with different patterns is given more weightage as compared to a word which occurs with a single pattern only.

S.1 Identify the single words and bigrams occurring with these patterns and put them in a list, called key_*list*.

- Remove any spaces before or after the extracted words in case of single words.
- Exclude the names of countries, states, city, and numbers whether in numerical or textual form.

S.2 Remove the stopwords from this list. The resultant list is named as candidate_list.

S.3 In the candidate_list, lets represent each word as W_i, i=1 to n.

S.3.1 For each W_i, assign a weight in the following way:

Weight(W_i)=1, initially, Weight(W_i)= Weight(W_i)+1, if W_i is a noun

S.3.2 For each pattern P_j in which W_i occurs, Weight (W_i) = Weight (W_i) +1

Weight (W_i) =a*weight (W_i) +b*frequency (W_i) //value of a and b to be decided experimentally

S.4 Sort candidate_list as per weight (W_i), i=1 to n, in descending order.

Fig. 1 The baseline algorithm

agriculture is an important sector of Indian economy as it contributes about 17% to the total GDP and provides employment to over 60% of the population. Indian Agriculture has registered impressive growth over last few decades. The *foodgrain* production has increased from 51 million tonnes (MT) in 1950-51 to 250 MT during 2011-12 highest ever since independence. the production of *oilseeds* (nine-major oilseeds) has also increased from 5 MT to 28 MT during the same period. The rapid growth has helped Indian agriculture mark its presence at global level. India stands among top three in terms of production of various agricultural commodities like *paddy, wheat, pulses, groundnut, rapeseeds, fruits, vegetables, sugarcane, tea, jute, cotton, tobacco leaves,* etc (GOI, 200809).

Fig. 2 Sample text

After observing the results and further analyzing the input text, it has been realized that the baseline method can be improved with the incorporation of following points.

Words occurring consecutively separated by commas are likely to be important terms. For example, from the text given in Fig. 2, we infer that there are two such consecutive occurrence of comma separated terms, as given below, from the Agriculture domain.

A. "paddy, wheat, pulses, groundnut, rapeseeds, fruits, vegetables, sugarcane, tea, jute, cotton, tobacco leaves"
B. "harvesting, thrashing, winnowing, bagging, transportation, storage, processing, and exchange"

Some more terms are identified by using part-of-speech information associated with the text. This is based on the observation that a noun followed by a noun, and an adjective followed by a noun constitute important keywords of the domain. Out

agriculture/NN is/VBZ an/DT important/JJ sector/NN of/IN Indian/JJ economy/NN as/IN it/PRP contributes/VBZ about/RB 17/CD %/NN to/TO the/DT total/JJ GDP/NN and/CC provides/VBZ employment/NN to/TO over/IN 60/CD %/NN of/IN the/DT population/NN ./.Indian/NNP Agriculture/NNP has/VBZ registered/VBN impressive/JJ growth/NN over/IN last/JJ few/JJ decades/NNS ./.The/DT *foodgrain/NNproduction/NN* has/VBZ increased/VBN from/IN 51/CD million/CD tonnes/NNS -LRB-/-LRB- MT/NNP -RRB-/-RRB- in/IN 1950-51/CD to/TO 250/CD MT/VBP during/IN 2011-12/CD highest/JJS ever/RB since/IN independence/NN ./.the/DT production/NN of/IN oilseeds/NNS -LRB-/-LRB- nine-major/JJ oilseeds/NNS -RRB-/-RRB- has/VBZ also/RB increased/VBN from/IN 5/CD MT/NN to/TO 28/CD MT/NN during/IN the/DT same/JJ period/NN ./.

The/DT rapid/JJ growth/NN has/VBZ helped/VBN *Indian/JJ agriculture/NN* mark/VB its/PRP$ presence/NN at/IN global/JJ level/NN ./.India/NNP stands/VBZ among/IN top/JJ three/CD in/IN terms/NNS of/IN production/NN of/IN various/JJ *agricultural/JJ commodities/NNS* like/IN paddy/NN ,/, wheat/NN ,/, pulses/NNS ,/, groundnut/NN ,/, rapeseeds/NNS ,/, fruits/NN ,/, vegetables/NNS ,/, sugarcane/NN ,/, tea/NN ,/, jute/NN ,/, cotton/NN ,/, tobacco/NN leaves/NN ,/, etc/FW -LRB-/-LRB- GOI/NNP ,/, 200809/CD -RRB-/-RRB- ./.

Fig. 3 Sample text after POS tagging

of all such pairs of words extracted as terms, we keep only those in which at least one of the words constituting the candidate terms have already been extracted by our algorithm in previous steps. For illustration, Fig. 3 shows the text of Fig. 2 after having tagged the parts of speech of the text. Terms fulfilling these criteria have been italicized. There are three such terms as shown in Fig. 3.

Keeping in mind the observations noted above, we present an improved algorithm named RENT (Regular Expression & Natural language processing-based Term extraction scheme). This is given in Fig. 4.

4 Results and Analysis

We executed the proposed baseline algorithm on approximately 1400 pages of input text pertaining to agricultural domain, collected from Department of Agriculture's website including the links *farmer.gov.in*, *agricoop.nic.in*, handbooks available from FAO, *nios.ac.in* etc. In Fig. 1, step S.3.2 specifies two constants 'a' and 'b' whose values are to be decided experimentally. After experiments with different values of 'a' and 'b,' we found that the algorithm works well when 'a' is taken to be five times 'b'. The program outputs the extracted terms, and their scores sorted in descending order.

We have calculated the precision, recall and F-measure using random samples of input data as it is not feasible to calculate recall on such a large amount of data used in the experiment. Here we proceeded as follows. We randomly selected pages from different input documents. First we calculated recall on 5 pages, 10 pages, and continued the experiment till randomly selected 50 pages of text. Figure 5 shows the precision, recall and f-measure for the Baseline Algorithm.

S.1 Identify the single words and bigrams occurring with these patterns and put them in a list, called Key_list.

- Remove any spaces before or after the extracted words in case of single words.
  - Exclude the names of countries, states, city, and numbers whether in numerical or textual form.

S.2 Remove the stopwords from this list. The resultant list is named as Candidate_list.

S.3 In the Candidate_list, lets represent each word as W_i, i=1 to n.

S.3.1 For each W_i, assign a weight in the following way:

Weight(W_i)=1, initially, Weight(W_i)= Weight(W_i)+1, if W_i is a noun

S.3.2 For each pattern $P_j$in which W_i occurs, Weight (W_i) = Weight (W_i) +1

Weight (W_i) =a*weight (W_i) +b*frequency (W_i) //value of a and b to be decided experimentally

S.4 Sort Candidate_list as per weight (W_i), i=1 to n, in descending order.

S.5 Extract the words in the text that occur separated with commas. Let this list be C_list.

Remove stopwords from C_list. Append C_list to Candidate_list using the regular expression '(?<=,)[^,\b\.]+'.

S.6 Use following linguistic filters to extract more terms:

- (NNP,NNP)
- (NNP,NNS)
- (NNP,NN)
- (JJ, NNP)
- (NNS,NNS)
- (NN,NN)
- (JJ,NN)
- (NN,NN,NNS)
- NN,NN,NN)
- (JJ,NN,NN)

S.7 for each string CS extracted in S.6

if either of the word is in Candidate_list and length of CS<=15

append CS to Candidate_list

Fig. 4 The RENT algorithm

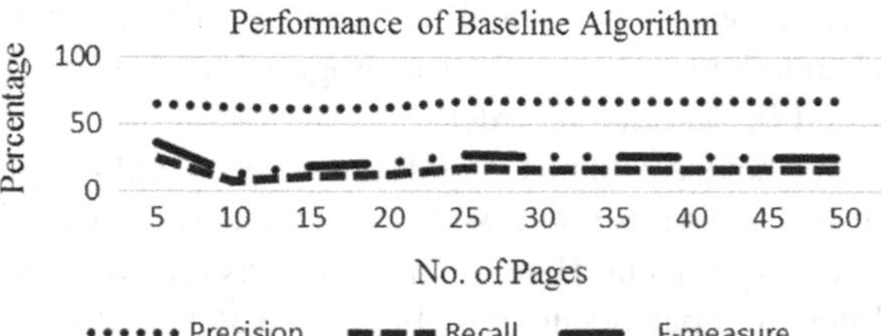

Fig. 5 Performance of baseline algorithm

It can be seen from Fig. 5 that the baseline algorithm has a significant precision, but lacks in recall. It has further been observed that if we would have calculated recall of the method on complete data set, it would have been low, in spite of having a fairly good precision. We did not analyze the baseline algorithm further, as it did not show favorable recall values on random sample data.

We have done the analysis of the proposed RENT algorithm extensively in three ways:

(a) Human evaluation for precision and recall on random sample data. Here, we have manually calculated the cumulative frequency of agricultural terms as the pages have been selected in sets of 5. The RENT algorithm has clearly outperformed the baseline algorithm. It can be seen in the chart given in Fig. 6 that the approximate precision of the RENT algorithm on sample documents is 80 %, average recall is 60 % and average *f*-measure is 70 %.

(b) Estimation of the overall precision of RENT using Hypergeometric Distribution

We use hypergeometric distribution [9] to estimate the number of terms per set of five pages, as explained earlier, and draw inference from there to analyze the performance of the RENT algorithm. The underlying assumptions are the following:

- t is the number of valid terms out of the total terms extracted by the algorithm
- t_s is the number of valid terms in the random sample taken to calculate precision
- n is the total number of terms extracted by the algorithm
- n_s is the total number of terms in the random sample taken to calculate precision.

Our aim is to estimate the value of t, when the values of n and n_s are known; and the value of t_s is manually determined using human judgement. Then t can be estimated as follows:

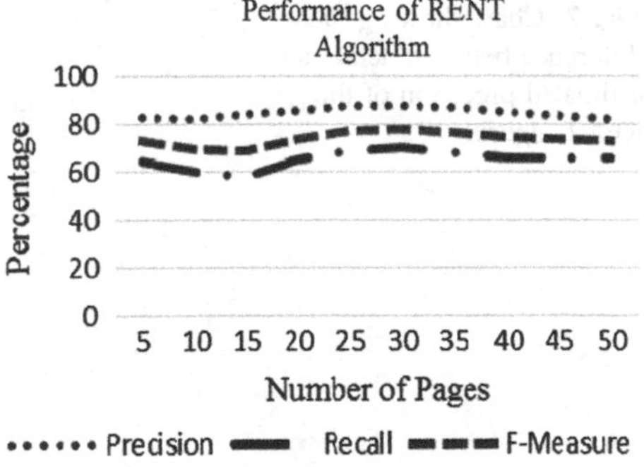

Fig. 6 Performance of RENT algorithm

$$t \approx \frac{n}{n_s} t_s \tag{1}$$

We have calculated the precision in two ways:

(1) first, determined the number of input terms manually in varying number of input pages, and then
(2) by using the hypergeometric distribution based estimation of the number of terms present in the document.

For illustration, with the sample of 10 pages we have the following statistics:

- t_s is the number of valid terms extracted with 5 pages as the random sample is 118.
- n is the total number of terms extracted by the algorithm on 10 pages is 357
- n_s is the total number of terms in the random sample of 5 pages is 142.

Hence substituting the values in Eq. (1), we get

$$t \approx \frac{357}{142} 118 \approx 296$$

The chart given in Fig. 7 shows that one can use hypergeometric distribution quite effectively to estimate the overall precision of the algorithm as there are only marginal differences in the actual and estimated precision for almost all the random samples.

Thus, we have estimated the overall precision of the RENT algorithm using Eq. (1) on different samples. We have applied Eq. (1) on nine different samples in order to have a good approximation of the overall precision. The overall precision of the proposed algorithm is 83.94 %.

(c) Manual Comparison of RENT with Termine in terms of precision, recall, and F-measure

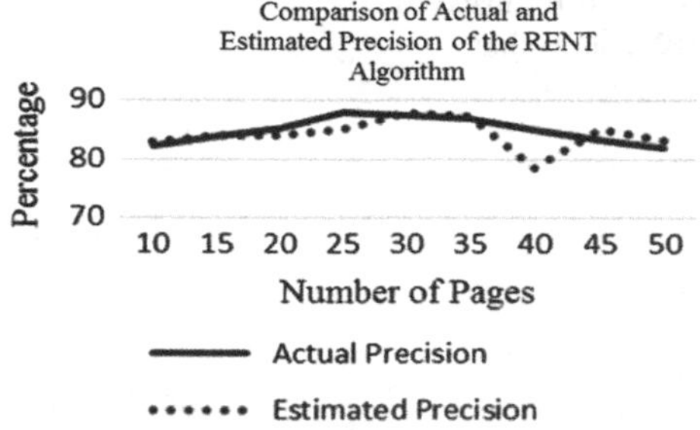

Fig. 7 Chart showing the difference between actual and estimated precision of the RENT algorithm

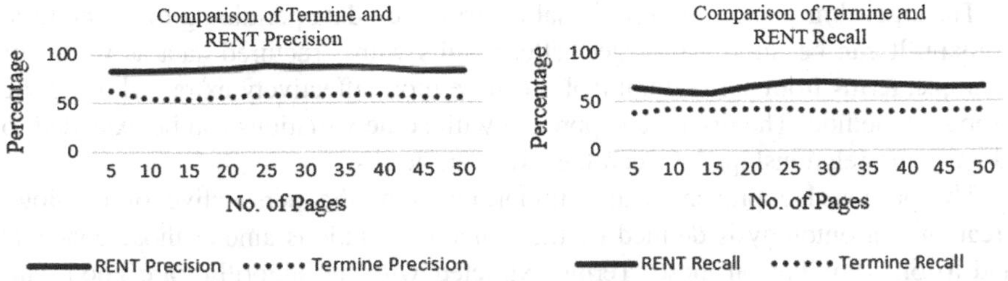

Fig. 8 Comparison of Termine and RENT performance

Termine[6] is a well-known software for automatic term extraction [6]. Figure 8 presents a comparison of RENT and Termine. The RENT algorithm has a much better precision and recall as compared to Termine.

5 Conclusion and Future Work

This paper proposes a Baseline algorithm and an improved version of the baseline, named RENT for automatic erm extraction in Agricultural domain. The idea for the baseline algorithm comes from the field of named-entity recognition where typically experts discover rules for identifying important entities. In a similar vein, we have designed regular expressions for extracting terms from a document. The baseline algorithm is improved by incorporating NLP techniques. Hence, the name RENT has been given to the new algorithm—Regular Expression &Natural Language Processing-based Term extraction.

The RENT algorithm significantly outweighs the baseline method. An extensive study has been done to carry out the performance analysis of the RENT algorithm. It is shown experimentally that the hypergeometric distribution can be used to estimate the precision of the algorithm. A comparison of RENT algorithm with the baseline algorithm and with Termine [6] has been presented. Termine uses C-value measure to extract terms. RENT is shown to have outperform both the baseline as well as Termine. The strength of RENT algorithm lies in the initial list of terms identified using patterns selected by domain experts, as it uses this list to validate the keywords identified in subsequent steps.

The RENT algorithm is designed with a focus on Indian Agriculture, however, its application need not be limited to Indian Agriculture only. It can be applied to Agriculture Domain in general with minor changes in the patterns used to identify the initial list of terms. It can, in principle, be applied to any other domain as well with the help of domain experts to identify the domain relevant patterns.

---

[6]http://www.nactem.ac.uk/software/termine/#form.

The algorithm works well on textual unstructured data pertaining to agricultural domain. It can be improved to get better results with structured data as well. For example, terms from agricultural tables can be more effectively extracted using the proposed method. The discussed approach with some variations can be extended to identify the relationships between the extracted terms.

The proposed approach is also important from the perspective of ontology creation. An ontology is defined by the concepts, relations among those concepts and axioms holding on these. Terms extracted with the algorithm are good candidates for seed concepts of the underlying ontology. These along with some more regular expressions, identified by domain analysis, can be used for extracting the relations and axioms.

More terms can be identified using the seed terms, as identified by this method, from the lexical resources in the agricultural domain. Such resources include WordNet, AGROVOC, and NAL thesaurus.

We also propose to extend this work for Hindi language as well. Doing this is quite challenging. First and foremost challenge in this regard is identification of domain-specific patterns in Hindi. Other challenges include lack of agricultural knowledge resources and data in Hindi.

References

1. Shaila S.G., Vadivel A.,: TAG Term Weight-based N gram Thesaurus Generation for Query Expansion in Information Retrieval Application. J. Inf. Sc. Vol. 41 no. 4, pp. 467–485 (2015).
2. Nasukawa T., Yi J.,: Sentiment Analysis: Capturing Favorability using Natural Language Processing. In: Proceedings of the 2nd International Conference on Knowledge Capture, ACM, pp. 70—77 (2003).
3. Maynard D., Li Y., and Peters W.: NLP Techniques for Term Extraction and Ontology Population. In: Proceedings of the 2008 Conference on Ontology Learning and Population: Bridging the Gap between Text and Knowledge, pp. 107–127 (2008).
4. Velardi P., Fabriani P. & Missikoff M.,: Using Text Processing Techniques to Automatically Enrich a Domain Ontology. In: Proceedings of the International Conference on Formal Ontology in Information Systems- Volume 2001, ACM, pp. 270—284 (2001).
5. Kageura K., and Bin U.: Methods of Automatic Term Recognition: A Review. Terminology: International Journal of Theoretical and Applied Issues in Specialized Communication 3.2, 259—282 (1996).
6. Frantzi, K., Ananiadou, S., Mima, H.: Automatic Recognition of Multi-Word Terms: the C-Value/NC-Value Method. International Journal on Digital Libraries 3.2, 115–130 (2000).
7. Zhang Z., Iria J., Brewster C., and Ciravegna F.: A Comparative Evaluation of Term Recognition Algorithms. In: 6th Language Resources and Evaluation Conference, pp. 2108–2113, European Language Resource Association, Paris (2008).
8. Fox C.: A Stop List for General Text. ACM SIGIR Forum, Volume 24. No. 1–2, ACM, pp. 19–21 (1984).
9. Hayter J., Anthony,: Probability and Statistics for Engineers. Cengage Learning India Private Limited, 2009.

Review of Clustering Techniques

G. Sreenivasulu, S. Viswanadha Raju and N. Sambasiva Rao

Abstract Clustering is the procedure of consortium a set of entities in such a manner those similar entities should in the same group. Cluster analysis is not one specific approach, but the general process to be observed. Clustering can be viewed by different algorithms that differ independently, in their view what is meant by a cluster and how to find them perfectly. Popular notions of clusters include groups with minimum distances among the cluster members. The clustering problem has been discussed by researchers in different things with respective domain. It reveals broad scope of clustering and it is very important in the process of data analysis as one step. However, it is very difficult because of the researchers may assume in different contexts. Clustering is one of best approach of data mining and a common methodology for statistical data analysis. It is used in all major domains like Banking, Health care, Robotics, and other disciplines. This paper mainly aims to discuss about limitations, scope, and purpose of different clustering algorithms in a great detail.

Keywords Data mining · Clustering (un-supervised) algorithms · Categorical data

G. Sreenivasulu (✉)
Department of CSE, ACEEC, Hyderabad, India
e-mail: gvsreenu@gmail.com

S.V. Raju
JNTUH, Hyderabad, India
e-mail: svraju.jntu@gmail.com

N.S. Rao
SITW, Warangal, India
e-mail: snandam@gmail.com

© Springer Science+Business Media Singapore 2017

S.C. Satapathy et al. (eds.), *Proceedings of the International Conference on Data Engineering and Communication Technology*, Advances in Intelligent Systems and Computing 468, DOI 10.1007/978-981-10-1675-2_52

1 Introduction

In data mining, clustering [1] is considered to be an active research area for exploratory data analysis in recent years. Clustering is an important in several multidisciplinary areas, such as data mining, machine learning, telecommunications, pattern recognition, biomedical, Information theory, etc. The use of Data mining is to mine informative knowledge from legacy data and it going to produce knowledge about the data by using one of the prominent approaches called knowledge discovery in databases (KDD). In Data mining Knowledge discovery refers to "Developing new knowledge from the historical data." Data mining mainly focuses on four components; such as preprocessing, association, classification, and clustering analysis.

Preprocessing is aims to remove the redundancies, filling the missing values and making the data into bins based on their boundaries. Association mining is used to find out most frequently occurring data item or item set from the given data sets. It is one kind of greedy approach to implement business logic over the data.

Classification is fairly known as a supervised learning. In this approach by gaining enough knowledge from training data we are going to test the test data in the form of rules. Those rules are called as decision rules. If the rules are properly working on test data so, we can claim for inducing of decision tree. Otherwise the tree will be over fit. This method is mainly used to inference rules from existing data.

The best component in data mining is cluster analysis [2]. It is a technique where in a collection of similar objects from a given dataset are gathered such that the objects in different collection are dissimilar. These collections are known as clusters. Clustering is a major role in data mining. This is going to formed into various group based on their similar behavior. Where the similarity distance [3] must be low in intra cluster similarity and it must be high in inter cluster, In such way we have to keep the data into proper groups. The way of approach is going to vary for various kinds of data. In general, data might be mixed with numerical and categorical. Majority of the clustering techniques that are developed for numerical data could be simple to use in normal circumstances but not to categorical data (Fig. 1).

The figure shows the typical process of how the data mining execution process for the generation of insights from the data sets.

Fig. 1 Data mining typical process

1.1 Importance of Clustering

In data mining clustering is considered one of the important unsupervised learning approach; so, In other problems of this type, it handles with choosing a structure in a group of unlabeled data. One of the known of clustering will be "The procedure of keeping objects into groups whose entities are similar in some group." So, a cluster contains collection of objects who are "similar" and are "dissimilar" to the entities belong to some other cluster. The main importance of clustering is while working with different types of data, it is always desired situation for us to group the data based on similar behavior.

1.2 The Goals of Clustering

The major objective of clustering is grouping the set of unlabeled data with its intrinsic values. However, it is always a major problem for users to decide what constitute an truth clustering? It can be addressed that, there is no perfect "best" part to which would be the final objective of the clustering independently. Subsequently, it is the user who must supply this situation, so that, clustering will fulfill this situation. For instance, we are interested in finding label values for similar groups, and in finding "natural clusters" and explain their properties (different *types*), in guessing useful and useful groupings (*"useful" data classes*) or in finding uncommon data entities.

1.3 Clustering Applications

- *Marketing*: From the large dataset finding various groups of customers with similar buying pattern and making them as a group.
- *Biology*: finding classification of seeds and breeds given their features;
- *Libraries*: book issues, book returns, and registering for new titles;
- *Insurance*: It is finding similar groups of various insurance holders with a high allege cost; identify customers who are turns out as faultier;
- *City-planning*: Based on demographic survey identifying groups of people according to their residential type, value, and location;
- *Earthquake studies*: It is helpful to study earthquake symptoms and informs to epicenters to identify dangerous zones;
- *WWW*: document categorization; grouping weblog data to find out similar access patterns.
- *Image Mining*: grouping of images based on similar features.
- *Medical diagnosis*: grouping of similar symptoms for identification of disease.
- *Aadhar verification:* To eliminate duplicates if any data found redundant.

Clustering is supporting various kinds of algorithms, likely hierarchical, density based, grid based, and centroid etc., all the these of algorithms are having their limitations in terms of size, approach, and time. We will discuss in detail about all parameters in cluster analysis section.

2 Literature Survey

Data mining supports various types of data. These are based on the domain size and the measurement scale. Data is classified into many types based on its features. They are numerical, categorical, and mixed data types.

2.1 Numerical Data

The data which is identified on a numerical scale is known as numerical data. Statistical methods are used to measure the numerical data, and their results can be displayed using tables, charts, and histograms.

2.2 Categorical Data

This data is identified on nominal scale. The analysis here is statistical and these statistical techniques are used when one or more variables are involved in nominal scale. For example: names such as 'male' and 'female' rather than measurements are used as categorical attributes.

2.3 Mixed Data

This data which is identified on structure and union scale. This analysis here is used to store different types of data types like numeric, char, and other types. We can able to store and process them in relevance to combination of different types. To process mixed data various generic approaches made available.

2.4 Cluster Analysis

The goal of clustering is to group data points based on the available information, which describes the data points and their associations. "The main aim is to group similar objects together, the greater the similarity within a group the better and more the difference between them or diverse the clustering." Clustering is an unsupervised method because the dataset does not contain labeled data points to train the current data.

There are various types of clustering [1, 4] algorithms, such as partitioning-Based, hierarchical, and density-Based methods. In most of the cases, partitioning-based clustering techniques discover all the clusters at a time. The examples for partitioning-based clustering algorithms are *K*-means clustering, *K*-medoids clustering, EM (Expectation Maximization) clustering, etc. The hierarchical clustering techniques use past clusters in order to discover successive clusters. The approaches for hierarchical clustering are divisive clustering which works based on top-down approach and agglomerative clustering which works based on bottom-up approach. The density of data objects leads to the arbitrary-shaped clusters that are formed using density-based algorithms. Here, a cluster is explained as a collection of data objects in a region where the density of the data objects satisfies some criteria. The DBSCAN and OPTICS are the examples for density-based clustering.

2.5 *Clustering Algorithms*

Clustering methods can be classified using cluster model, as mentioned above. The most important examples of clustering algorithms, particularly time-evolving data clustering algorithms, out of many existing clustering algorithms are reviewed below.

The time-evolving data [5, 6] clustering task has been examined to a great extent in the numerical domain [7]; the dataset which consistently changes with time is introduced. Based on the requirement of application, a different approach for data stream clustering is presented. The idea behind this method is to periodically store detailed summary statistics as an online component and later used it an offline component. The analyst utilizes the offline component types of inputs, such as time stamp, number of clusters to understand the cluster details in the data stream. For a fast data stream, it is quite difficult to make use of statistical data as an efficient choice for processing and storing. For this purpose, the ideas of a pyramidal time stamp frame are used with a micro-clustering approach.

2.6 *Clustering Algorithms Discussions*

S. No.	Clustering algorithm	Type of data supports	Base algorithm	Data size	Feature learning	Time complexity	Applications	Software implemented
1	*K*-Means	Numeric, vector quantization	Lioyd's algorithm	Small	Semi-supervised	$O(n^{dk+1} \log_n)$	Market segmentation, astronomy	Apache, Spark etc.
2	*K*-Mediods	Numeric, list	*K*-Means and medoids	Limited	Unsupervised	$O(K^n)$	Disk storage, heuristic search	Java, Scala
3	BIRCH	Real value vectors	Tree data structure	Large	Unsupervised	2-Phase process	Genetics, market analysis	Julia
4	CLARA	Vectors, array list	*K*-mediods	Large	Unsupervised	$O(k^h.d)$	Intelligence, law enforcement	R, IAV automotive
5	CLARANS	Spatial values	SLINK, CLINK	Large	Unsupervised	$O(n^2)$, $O(n^3)$	Web log analysis	IAV automative
6	COOLCAT	Categorical	Jaccord's co-efficient	Limited and large	Unsupervised	$K(h, D) = k(h) + K(D \text{ using } h)$	Synthesis data generator	Java, SAS
7	LIMBO	Categorical	Information bottleneck	Limited and large	Unsupervised	$O(n^2 d^2 \log n)$	Market-basket	Source Forge
8	BIRCH	Categorical, real valued vector	DBSCAN	Large	Unsupervised	2-Phase process (CF-Tree)	Auto encoder, deep learning	Java, R
9	PAM	Categorical, numeric	*K*-medoids	Large	Unsupervised	$O(n^2)$	CRM, marketing, medical diagnostics	Scala
10	MARDL	Categorical	Centroid based	Large	Unsupervised	$O(n^2)$	Web search analysis	Java
11	CURE	Categorical	Partitional	Large	Unsupervised	$O(n^2 \log n)$	Pattern recognition	Scala

(continued)

(continued)

S. No.	Clustering algorithm	Type of data supports	Base algorithm	Data size	Feature learning	Time complexity	Applications	Software implemented
12	OPTICS	Spatial data, categorical	DBSCAN	Large	Unsupervised	$O(n * \log n)$	Log analyzer, analytics	Java, R
13	DBSCAN	Categorical	Euclidian coefficient	Large	unsupervised	$O(n * \log n)$	Genetic comparison	Malthus
14	ROCK	Categorical	Jaccord's Coefficient,	Limited and large	Unsupervised	NA	Image processing	Java, SAS
15	CNIR	Time evolving, categorical	DCD	Large	Unsupervised	$O(\log n)$ $O(C^{1}, C^{t-1})$	Network streaming	Source Forge

2.6.1 *K*-Means Algorithm

The *K*-Means algorithm [8, 9] uses a recursive technique. So, its functionality it is known like *k*-means algorithm; it is defined from the core algorithm as Lloyd's algorithm, especially in the Data mining community.

Given an initial set of *k* means $m_1\ldots, m_k$, the algorithm proceeds by two steps:

Assignment step (step-I): It allows each Data point to the cluster based on whose mean the least in the same cluster calculated as sum of squares. So, the Euclidean distance is calculated as squared of sum squares., this is also represents the "nearest" mean.

$$S_i^{(t)} = \{x_p: \left\|x - m_i^{(t)}\right\|^2 \le \left\|x_p - m_j^{(t)}\right\|^2 \forall j, 1 \le j \le k\},$$

where each x_p is assigned to exactly $S^{(t)}$, even if it and sometimes it assigns two or more of them.

Update step (step-II): It calculates the new means to be the centroids of the data points in the new clusters. $mi^{(t+1)} = \frac{1}{\left|S_i^{(t)}\right|} \sum_{xj \in S_i^{(t)}} xj$. The algorithm has combines when there is no change in assignments, Since all the steps choose the objective of algorithm, and there will be only a finite number of such divisions, the approach converge to a local optimum.

The algorithm [10, 11] is presented as assigning objects to the nearest cluster by calculating their distance. The algorithm aims at minimizing the objective, and thus it assigns by "least sum of squares," which is exactly equivalent to assigning by the least Euclidean distance. *k*-medoids [12] have been proposed to allow using other distance measures [13].

Categorical data clustering problem is first addressed in the customer transaction data of a market database is used. In the task of clustering data was assigned to nonlinear dynamic scheme called STIRR in the categorical domain. Most widely used *K*-means clustering algorithm is extended to *K*-modes in the categorical domain where all the clusters are represented by the value called Node, which is the most frequent value in that cluster. In clustering procedure, a regularity-based technique is used to revise modes based on the minimum clustering cost function. In K-prototypes algorithm, the dissimilarity measure is combined by its definition and it is further combines with *K*-means algorithm to allow objects described by both type of attributes (Mixed and Categorical). By using a general matching dissimilarity approach for categorical and other modes instead of calculating their means, a new procedure is developed, it will allows the use of the *K*-means [8] algorithm to efficiently cluster categorical datasets.

The CACTUS [14] is a very fast summarization-based method and it is used to find a novel approach of a cluster for categorical data. It is not only fast but also scalable [15] as it requires only two data scans. This algorithm called CLICK is introduced which it finds clusters in categorical values based on a method name of k-partite. It scales better with respect to data dimensionality rather than the other existing techniques. It is also capable to sense subspace clusters and outperforms past methods by a ratio of two by three.

The COOLCAT [16] and LIMBO [5], ROCK [17] are statistical-based clustering algorithms where the grouping of data points is done through statistical measures In COOLCAT algorithm, the predictable entropy of the complete arrangement of data points is minimized based on the division of the data points. The connection between categorical attribute and entropy of clustering is used to design an heuristic algorithm based on incremental approach. COOLCAT is to cluster large datasets of records with categorical attributes and data streams efficiently. Hereby, the results are very stable for various sample sizes and constraint settings are compared to all other existing clustering methods for categorical data. It is an incremental algorithm because it has the ability to cluster new data points without having referring to every point that has been clustered so far and this is usually called as data stream clustering.

The LIMBO, a hierarchical clustering algorithm belongs to categorical type, that constructs the framework called Information Bottleneck (IB) for quantifying the related information defended when clustering is introduced. LIMBO generates clusters with different shapes in a single execution. The IB framework is used to create a new distance quantity for categorical attribute values and this is applied in clustering. The comparison in the quality of clusters is presented with other clustering algorithms. It supports a trade-off between quality of the clusters obtained and efficiency (in terms of space and time) of the algorithm. But most of the authors observed that, It allows for considerable developments in effectively with a minor decrease in cluster quality.

In recent years, researchers mainly focused on improving the time complexity of the existing algorithms. As a result, CLARANS and BIRCH were developed which perform clustering efficiently. The recent need is to generate clusters based on the semantic meaning by processing massive datasets (known as big data) efficiently. From the development of pre-clustering techniques which can process large datasets efficiently. To obtain the rough clusters by which the performance of these methods are greater than other methods such as *K*-means clustering.

BIRCH is popularly known as Balanced Iterative Reducing and Clustering using Hierarchies, is a hierarchical clustering technique developed for large datasets where a set of multi-dimensional data points are given. The clustering feature (CF) of the defined set as the triple CF = (*N*, LS, SS), where Linear sum is shown as LS and square sum is shown as SS is the square sum of data points. In the first step of the algorithm, an initial CF tree is constructed by scanning the entire dataset and in the second step a small CF tree is reconstructed by the pendent nodes of the initial CF tree after removing the outliers. In the third step, any clustering algorithm is used to cluster the pendent nodes. The clusters are represented by their cluster feature on which agglomerative hierarchical clustering algorithms are applied to directly find the clusters with capture maximum distribution in the data. BIRCH is also flexible because the user can confine the number of clusters to form and desired diameter thresholds for clusters. The forth level discords outliers and seeds are produced as the centroids of clusters. Those seeds are used to obtain new clusters by reallocating the data points to its nearest seeds.

The partitioning around mediods (PAM) algorithm developed by L. Kaufman etc., it is comparable to the *K*-means approach. This algorithm finds a sequence of objects called "Medoids" that are located in the middle of cluster. The goal of this technique is to minimize the sum of dissimilar objects to their closed medoids. This has two phases, the one is Build phase which consist of the following steps.

(1) k entities are selected as medoids.
(2) The dissimilarity matrix is constructed if it not informed.
(3) Allocate every entity to its nearest medoid.

The second phase is Swap, which contains below steps.

I. The search for each cluster is average dissimilarity based on the coefficient of all the entities in the cluster and selection of the entity that lower most this coefficient as the medoid for this cluster is performed.
II. If there is change in any one cluster medoid then go back to step 3, otherwise stop.

To make *K*-medoids to be more scalable and efficient for large database, a sample-based method, called Clustering LArge Applications, it is also known as CLARA, it can be used. The basic idea behind CLARA is as represented like this: a small portion of data is chosen for representation rathen than taking the complete set of data into consideration. The medoids from the samples are chosen using partition around medioids. The original dataset is approximately represented by a sample when it is chosen from a reasonably random method. In the whole dataset, the representative objects are selected by checking similarities. CLARA outputs the best clustering result only by using dataset samples and applying PAM on each sample.

CLARANS was developed by Ng and Han in 1994 which is an improved version of CLARA method. It stands for Clustering LArge Applications based on a RAndomized Search. The process of CLARANS for finding K–medoids is similar to that of searching in a graph, nodes are the sets of medoids and the edges are connected if the two nodes differ exactly one medoid. CLARANS [18] starts with an arbitrary node and randomly check with maximum neighbors to generate neighbors where as very few neighbors are compared related to desired small sample in CLARA. If that partition is considered to be a better one then it could continue by selecting same node otherwise it will process by local node with minimum. The best node is returned for the formation of partitioning result and the complexity is $O(N^2)$ is considered in the form of number of points. Finally, their application is restricted to numerical data of lower dimensionality with inherited and well-separated clusters of high density.

In NIR and NNIR [19, 20] are introduced whose basic idea is to characterize the clusters using attribute values with their importance values as representatives. As a result, the importance of attribute values summarize and characterize the given data and this information is utilized to perform data labeling for clustering data in a better manner than clustering using these representatives. The MARDL (MAximal Resemblance Data Labeling) [21] is presented which examines the importance of all attribute values with respect to NIR or NNIR. In NIR a framework is created to carry the clustering on time-evolving data in the categorical data. This framework approach [22] notices the drifting of concepts over time period and generates a collection of subcluster as an outcomes based on current concepts using MARDL.

Furthermore, the technique known as data labeling is applied to judge proper cluster label to the unlabeled data point and which is discussed in Clustering Using REpresentatives (CURE [23]). However, it generates nonspherical clusters in the numerical domain. In this, an explicit data labeling method is used which works based on the delay or cost in between the representative objects of the cluster and an data point which is unlabled. The CURE is an algorithm, not appropriate to large databases but the improvements in it are introduced. The OPTICS [24] finds an enhanced cluster ordering for regular and interactive cluster study. The runtime complexity of OPTICS is same as DBSCAN because both are structurally equivalent. If n is the number of objects and when a spatial kind of index is used, then OPTICS time complexity becomes $O(n * \log n)$. DENsity-based CLUstEring (DENCLUE) [25, 26] is a clustering technique which works on a collected works of density distribution functions. The method is defined on the following steps:

(1) The impact of each data point can be properly model by using a mathematical function, titled an influence function, which explains the effect of a data point within its region.
(2) The overall compactness of the data space can be model systematically as the total of the influence function functional to all data points;
(3) Clusters can then be strong-minded mathematically by identify thickness attractors, where thickness attractors are local maxima of the on the whole density function.

The major compensation of DENCLUE are:

(1) It has a strong mathematical foundation and generalize various clustering methods, including hierarchical, density-based, and partitioned-based methods.
(2) It has good clustering property for datasets with huge amount of clamor.
(3) It allows a dense of mathematical model of subjectively shaped clusters in high dimensional datasets. It uses lattice cells, and keeps information about grid cells that actually contain data points.

3 Conclusion

Clustering is considered one of the important features in knowledge discovery process. In this paper more than 15 clustering algorithms studied and identified differences for numerical and categorical data. This study helps the readers to better understand the concept of clustering and helps them to use better as per their requirement. This detailed explanation will give more detailed information on variety of algorithms.

4 Future Work

In future, we are going to give more detailed study in terms of application-and domain-specific results of above mentioned 15 clustering algorithms for better understanding of both numerical and categorical clustering algorithms in a great detail.

References

1. A Tutorial on Clustering Algorithms http://www.elet.polimi.it/upload/matteucc/Clustering/tutorial_html/.
2. Zhexue Huang, "Clustering large data sets with mixed numeric and Categorical values", Mathematical and Information Sciences, 2005.
3. Yang, Y. T. Pierce, and Carbonell, J., "A study on retrospective and on-line event detection", In Proceedings of the 21st ACM International Conference on Research and Development in Information Retrieval, pages 28–36, 1998.
4. Aggarwal, C., Wolf, J.L., Yu, P.S., Procopiuc, C. and Park, J.S., "Fast Algorithms for Projected Clustering", ACM SIGMOD'99, pp. 61–72, 1999.
5. Andritsos, P., Tsaparas, P., Miller, R.J. and Sevcik, K.C., "Limbo: Scalable Clustering of Categorical Data", Extending Database Technology (EDBT), 2004.
6. Han, J. and Kamber, M., "Data Mining Concepts and Techniques", Morgan Kaufmann, 2001.
7. Anil, K. Jain, and Richard, C. Dubes., "Algorithms for Clustering Data", Prentice-Hall International, 1988.
8. Anjan Goswami, Ruoming Jin, and Gagan Agrawal, "Fast And Exact K-Means Clustering", International Conference on Data Mining (ICDM), 2004.
9. Huang, Z., "Extensions to the k-Means Algorithm for Clustering Large Data Sets with Categorical Values", Data Mining and Knowledge Discovery, 1998.
10. Hartigan, J. A. and Wong, M.A., "A k-means clustering algorithm", Applied Statistics, pp. 100–108, 1979.
11. Subramanian, D.K., Narasimha Murthy, M., Vijaya, P.A., "Leaders-Subleaders: An Efficient hierarchical clustering algorithm for large data sets", ELSEVIER, PP. 505–513, 2004.
12. Danyang Cao, Bingru Yang, "An improved k-medoids clustering algorithm", Computer and Automation Engineering (ICCAE), 2010.
13. Chen, H.L., Chuang, K.T and Chen, M.S., "Labeling Un clustered Categorical Data into Clusters Based on the Important Attribute Values", IEEE International Conference on Data Mining (ICDM), 2005.

14. Ganti, V., Gehrke, J. and Ramakrishnan, R., "CACTUS—Clustering Categorical Data Using Summaries", ACM SIGKDD, 1999.
15. Joydeep Ghosh, "Scalable clustering methods for data mining", In Nong Ye, editor, "Handbook of Data Mining", chapter 10, pp. 247–277, Lawrence Ealbaum Assoc, 2003.
16. Barbara, D., Li, Y. and Couto, J., "Coolcat: An Entropy-Based Algorithm for Categorical Clustering", ACM International Conf. Information and Knowledge Management (CIKM), 2002.
17. Guha, S., Rastogi, R. and Shim K., "ROCK: A Robust Clustering Algorithm for Categorical Attributes", International Conference on Data Engineering, (ICDE), 1999.
18. Ng, R.T., Han, J., "CLARANS: a method for clustering objects for spatial data mining", IEEE Transactions on Knowledge and Data Engineering, 2002.
19. Venkateswara Reddy, H., Viswanadha Raju, S. et al., "Our-NIR: Node Importance Representative for Clustering of Categorical Data", International Journal of Computer Science and Technology, 2(2), pp. 80–82, 2011.
20. Venkateswara Reddy, H., Viswanadha Raju, S. et al., "POur-NIR:Modified Node Importance Representative for Clustering of Categorical Data", International Journal of Computer Science and Information Security, 9(4), pp. 146–150, 2011.
21. Venkateswara Reddy, H., Viswanadha Raju, S. et al., "A Threshold for clustering Concept – Drifting Categorical Data", IEEE 3 rd International Conference on Machine Learning and Computing, 3, pp. 383–387, 2011.
22. Venkateswara Reddy, H., Viswanadha Raju, S. et al., "Comparing N-Node Set Importance Representative results with Node Importance Representative Results for Categorical Clustering: An Exploratory Study", International Journal of Computer Engineering Science, 2(8), pp. 1–15, 2012.
23. Guha, S., Rastogi, R., Shim, K., "CURE: An Efficient Clustering Algorithm for Large Databases", ACM SIGMOD International Conference on Management of Data, pp. 73–84, 1998.
24. Mihael Ankerst, Markus, M., Breunig, Hans-Peter Kriegel, Jörg Sander. "OPTICS: Ordering Points To Identify the Clustering Structure". ACM SIGMOD international conference on Management of data. ACM Press. 1999, pp. 49–60.
25. Sudipto Guha, Adam Meyerson, Nina Mishra, Rajeev Motwani, and Liadan O'Callaghan, "Clustering data streams: Theory and practice", IEEE Transactions on Knowledge and Data Engineering, pp. 515–528, 2003.
26. Usama, M., Fayyad, Gregory Piatetsky-Shapiro, and Padhraic Smyth, "From data mining to knowledge discovery: An overview" In Advances in Knowledge Discovery and Data Mining", American Association for Artificial Intelligence, pp. –34.1996.

Novel Zero Voltage Transition-Based Interleaved Boost Converter

V. Ramesh and Y. Kusuma Latha

Abstract A soft switching-based boost converter is presented in this paper. This converter comprises of elementary boost conversion proposed modules along with auxiliary inductor. This converter facilitates switching of MOSFET's at zero voltage. This process of operation of switches reduces the switching losses and improves the conversion efficiency. The converter topology with open loop and closed loop configuration is investigated.

Keywords Interleaved boost converter · Zero voltage switching · Zero current switching · Zero voltage transition

1 Introduction

Non linear loads are increasing in recent past. They are abundant source of harmonics and causes voltage and current distortions [1]. Improvement in power factor and reduction in harmonics is essential for delivering quality power. Boost converter is widely used for various high power applications [2]. Especially requiring higher voltage than line input. These are generally regulators and rectifiers with integrated circuits [3]. There are abundant renewable energy sources that produce low voltage output and boost converter is most preferred to obtain the required voltage specifications.

The proposed converter operates in continuous conduction mode [4]. At higher order frequencies, for improved efficiency and performance characteristics many soft-switching techniques such as zero voltage switching (ZVS) or zero current switching (ZCS) techniques are gaining importance [5]. The power factor correction is efficiently achieved by interleaved boost converters [6, 7]. To avoid damping,

V. Ramesh (✉) · Y.K. Latha
Department of Electrical and Electronics Engineering, K L University, Guntur, AP, India
e-mail: sitamsramesh@gmail.com

Y.K. Latha
e-mail: kusumalathayp@gmail.com

© Springer Science+Business Media Singapore 2017

S.C. Satapathy et al. (eds.), *Proceedings of the International Conference on Data Engineering and Communication Technology*, Advances in Intelligent Systems and Computing 468, DOI 10.1007/978-981-10-1675-2_53

interleaved boost converter is provided with coupled winding soft switching which is enabled using extra switches [8]. The proposed converter archive improved performance characteristics. The proposed closed loop interleaved boost converter with ZVT has improved switching technique with PI controllers which reduce the ripple current and increases the stability of system for closed-loop configuration.

In this paper, the modeling and circuit configuration of proposed interleaved boost converter is presented Sect. 2. The operation and analysis of the interleaved boost converter are presented in the Sect. 2.1. The open-loop and closed-loop model of interleaved boost converter and result are presented in Sect. 3 and finally the paper is conclude in Sect. 4.

2 Proposed Interleaved Boost Converter

The conventional circuit diagram of the interleaved boost converter is shown in Fig. 1 shows the proposed closed-loop interleaved boost converter in shown in Fig. 2 shows the simplified version of the main circuit diagram is shown in Fig. 3 shows the two MOSFETs are shunted with inductor Ls to achieve the ZVS characteristics. For the purpose of analysis, the inductors and capacitors are replaced by their equivalent current and voltage sources.

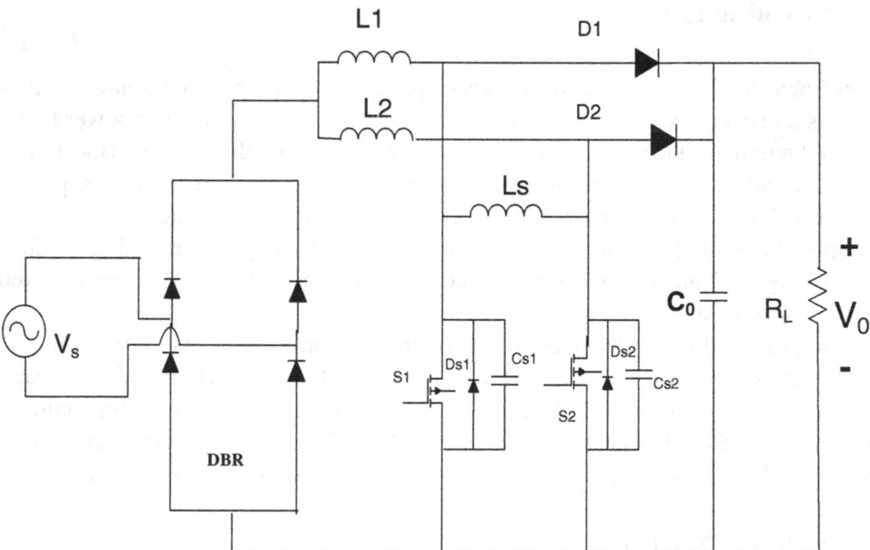

Fig. 1 Conventional interleaved boost converter

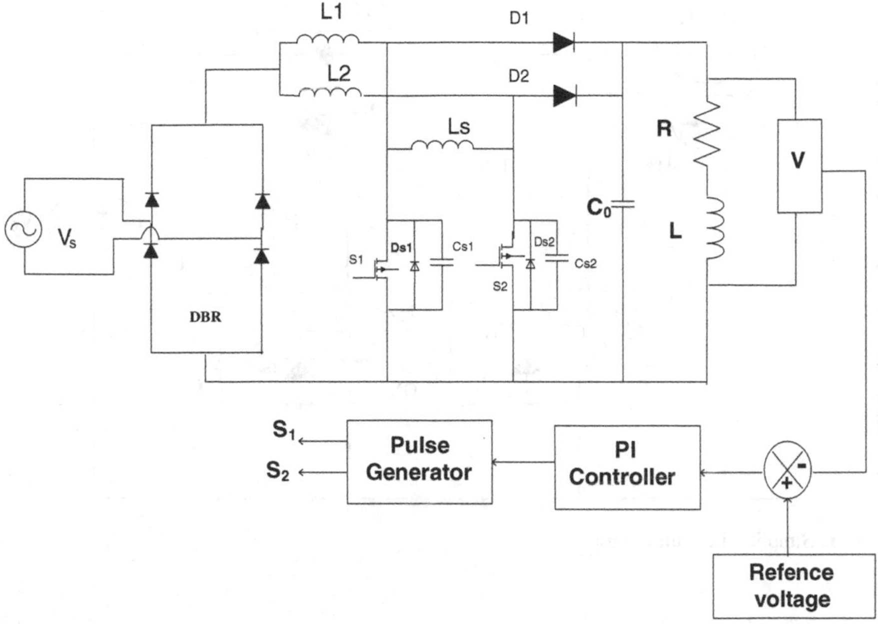

Fig. 2 Proposed closed-loop interleaved boost converter

2.1 Operation and Analysis of Interleaved Boost Converter

The following are the preliminary assumptions made during analysis of proposed converter:

(1) The voltage ripple across the output capacitor C_0 is negligible.
(2) The forward voltage across the switches is neglected.
(3) The inductances are identical and current through the inductors is similar.
(4) The capacitance is identical.

The two MOSFETs are operated in the pulse width modulation (PWM) control signals. The signals are having similar duty cycles and frequencies. The converter operation is split into different modes. The equivalent circuit of the interleaved boost converter and corresponding wave forms are illustrated in Figs. 4 and 5.

Mode-1: In mode-1, S_1 is conducting and S_2 non conducting. C_{s1} and C_{s2} not charging. The diode D_{s1} is conducting where as D_{s2} non conducting.

$$i_{Ls}(t) = I_L - \frac{V_0}{Ls\,t} \tag{1}$$

$$i_{s1}(t) = \frac{V_0}{Lst} \tag{2}$$

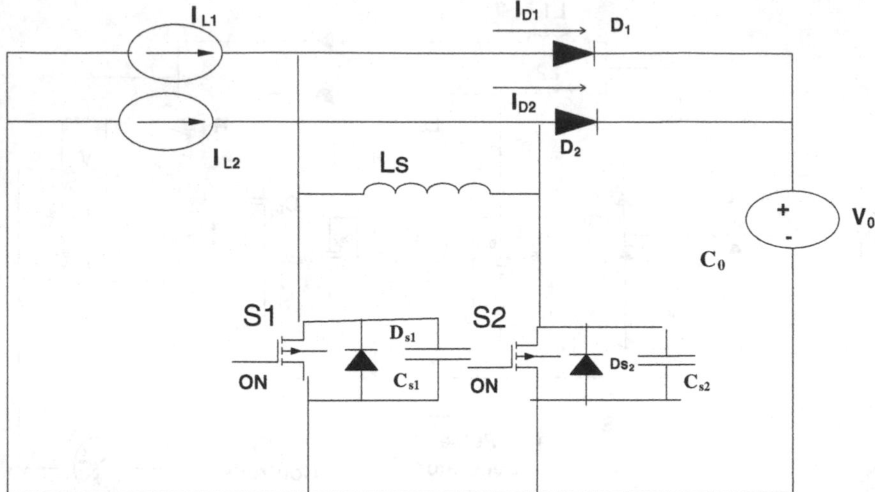

Fig. 3 Simplified circuit diagram

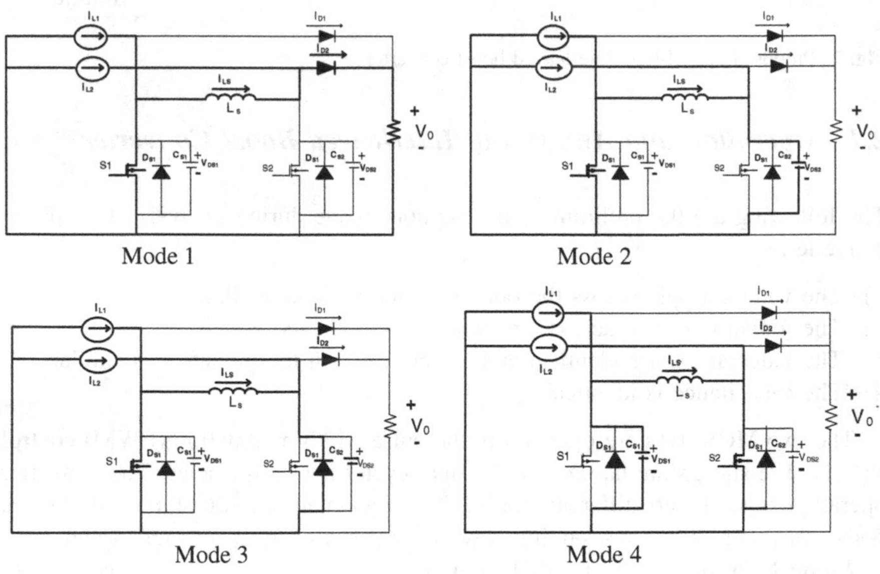

Fig. 4 Various modes of operation of proposed interleaved boost converter

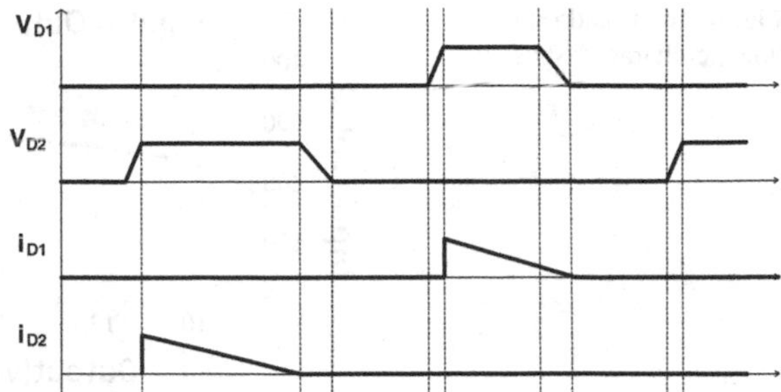

Fig. 5 Theoretical wave form

$$i_{D2} = 2I_L - \frac{V_0}{Lst} \tag{3}$$

$$t_{01} = (\frac{3}{4} - D_{\text{eff}})T_S - \sin - 1\frac{\frac{V_0}{(V_0 + \frac{2I_L}{wCs})}}{w} \tag{4}$$

Mode-2: In mode-2, S_1 is conducting S_2 nonconducting. C_{s2} corresponding to S_1 charging and Cs_1 not charging. The diode D_{s1} is conducting, where as D_{s2} nonconducting.

$$v_{Ds2}(t) = V_0 \cos wt \tag{5}$$

$$i_{LS}(t) = -V_0 wCs\sin(wt) - I_L \tag{6}$$

$$t_{12} = \frac{\pi}{2w} \tag{7}$$

Mode-3: In mode-3, S_1 is conducting and S_2 nonconducting C_{s1} and C_{s2} capacitor not charging D_{s2} is conducting where as D_{s1} notconducting.

$$t_{23} = (D_{\text{eff}} - \frac{1}{2})Ts \tag{8}$$

Mode-4: In mode-4, S_1 and S_2 conducting. C_{s1} capacitor is charging and C_{s2} not charging. D_{s2} notconducting and D_{s1} is conducting (Fig. 6).

$$v_{Ds1}(t) = (V_0 + \frac{2I_L}{wCs})\sin(wt) \tag{9}$$

$$t_{34} = (\sin - 1(V_0/(V_0 + \frac{2I_L}{wCs})))/w \tag{10}$$

Fig. 6 Input and output voltage characteristics

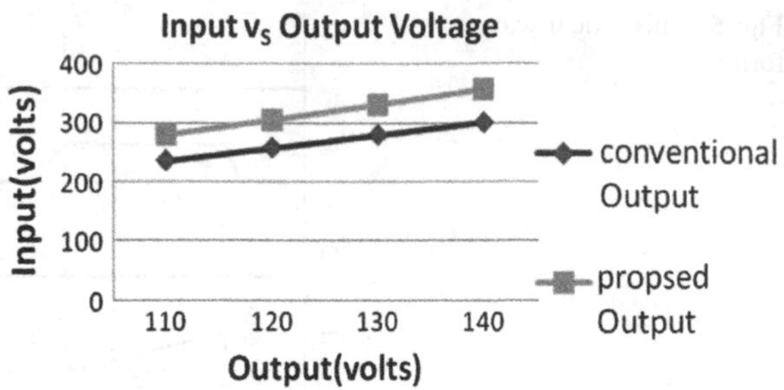

3 Simulation Results

3.1 Open Loop Interleaved Boost Converter with Zero Voltage Transitions

Figure 7 shows the circuit diagram of interleaved boost converter with zero voltage transition. Figure 8 shows the input voltage and current in open loop configuration. The input voltage is 120 V and input current is 65 Amps. For Open loop configuration of interleaved boost converter with ZVT. Figure 9 shows that the output voltage raises to 335 V and at t = 0.4 s, while the input voltage is varied in order to observe the robustness of the converter topology. Figure 10 shows the output current for open-loop Interleaved Boost Converter with Zero Voltage Transitions is 12 Amps. And at t = 0.4 s raises to 13.5 Amps and does not steady state condition. illustrate the Table 1 for simulation circuit parameters and Table 2 conventional and proposed circuit input voltage and output voltage characteristics.

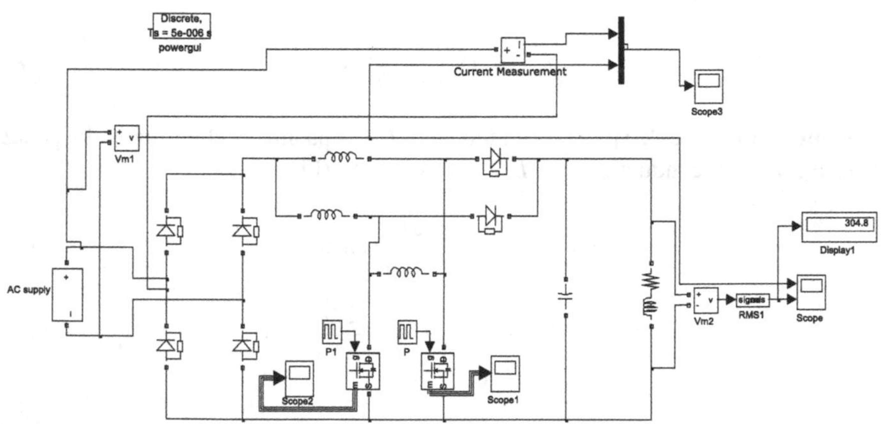

Fig. 7 Circuit diagram for open loop interleaved boost converter with ZVT

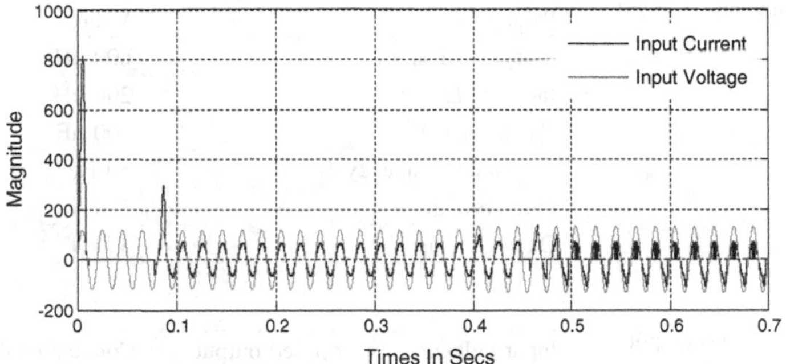

Fig. 8 Input voltage and current for open loop

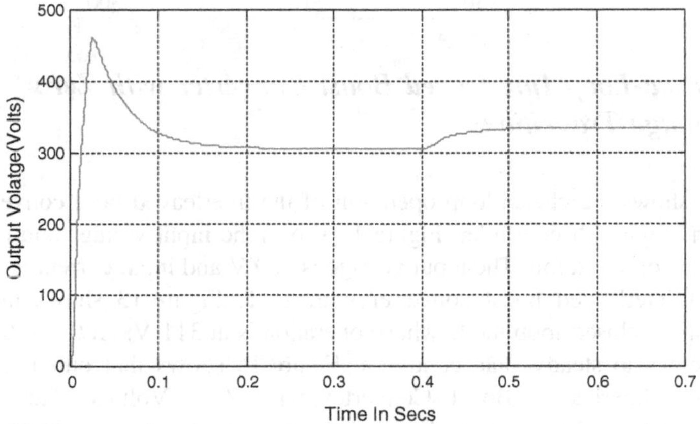

Fig. 9 Output voltage for open loop

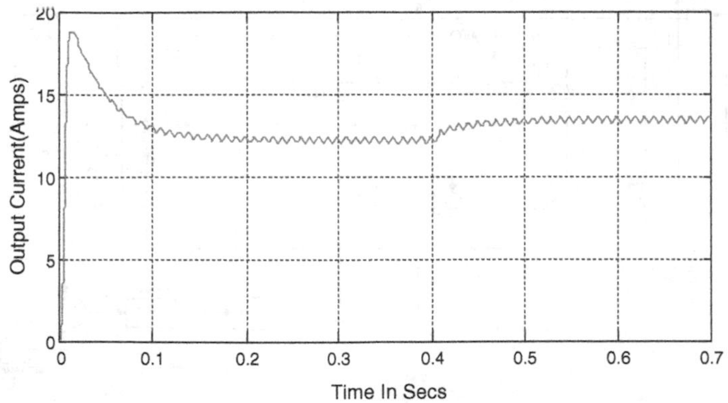

Fig. 10 Output current for open loop

Table 1 Simulation circuit parameters

Parameter	Value
Inductor *L*1 and *L*2	600 μH
Inductor *Ls*	200 μH
Capacitor *C*0	300 μF
Switching frequency *fs*	50 Hz
Input voltage	120 V
Output voltage	304 V

Table 2 Input and output voltage characteristics

Input voltage	Proposed output	Conventional output
110	247	235
120	270	257
130	293	279
140	316	300

3.2 Closed-Loop Interleaved Boost Converter with Zero Voltage Transitions

Figure 11 shows the closed-loop operation of the interleaved boot converter with ZVT is employing PI controller. Figure 12 shows the input voltage and current in closed-loop configuration. The input voltage is 120 V and input current is 65 Amps in case of interleaved boost converter with ZVT. Figure 13 shows the output voltage for the closed-loop mode whose operation is at 311 V. At 0.4 s, the output voltage comes to steady state condition. Figure 14 shows that output current is closed-Loop Interleaved Boost Converter with Zero Voltage Transitions at 12 Amps. At $t = 0.4$ s, it raises after that come to steady state condition.

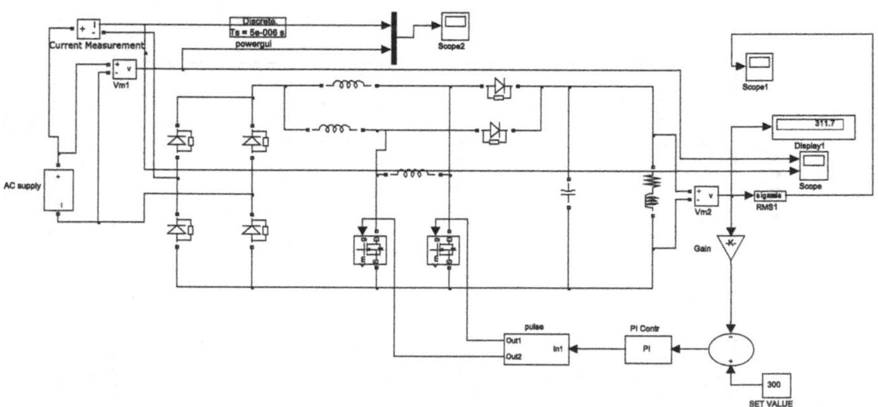

Fig. 11 Circuit diagram for closed-loop interleaved boost converter with ZVT

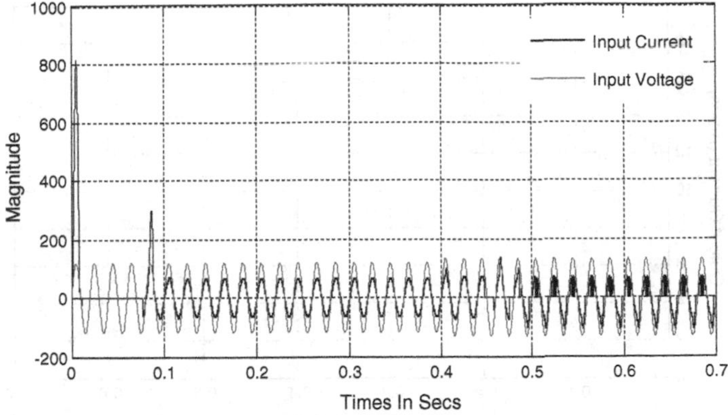

Fig. 12 Input voltage and current for closed loop

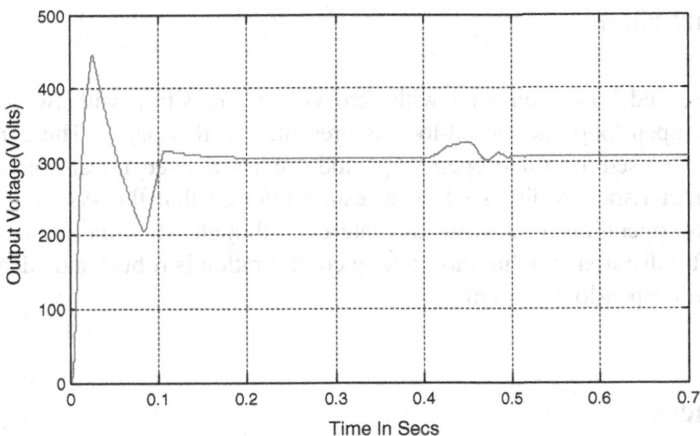

Fig. 13 Output voltage for closed-loop

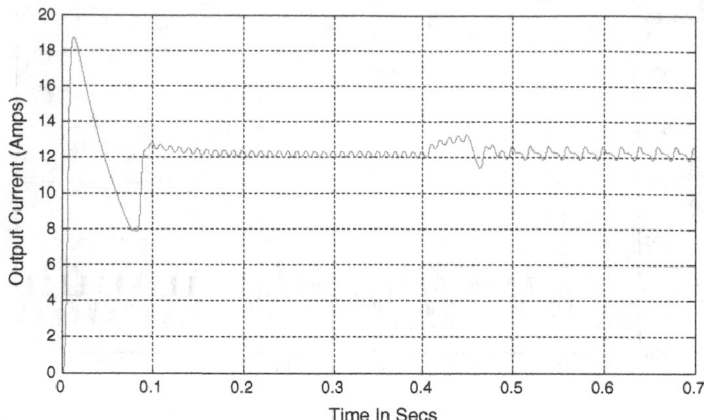

Fig. 14 Output current for closed-loop

4 Conclusion

The interleaved boost converter with zero voltage transition with two configurations, i.e., open-loop and closed-loop is presented in this paper. The comparative analysis of closed-loop and open-loop mode of interleaved boost converter with ZVT characteristics is discussed. The results inferred that the system employing closed-loop operation reaches steady state even though when the input voltage is varied (after disturbance) thus closed loop configuration is robust and advantageous compared to open-loop system.

References

1. G. Yao, A. Chen, and X. He, "Soft switching circuit for interleaved boost converters," IEEE Trans. Power Electron., vol. 22, no. 1, pp. 80–86, Jan. 2007.
2. Jang and M. M. Jovanović, "Interleaved boost converter with intrinsic voltage-doubler characteristic for universal-line PFC front end," IEEE Trans. Power Electron., vol. 22, no. 4, pp. 1394–1401, Jul. 2007.
3. H. Bodur and A. F. Bakan, "A new ZVT-PWM DC-DC converter," IEEE Trans. Power Electron., vol. 17, no. 1, pp. 40–47, Jan. 2002.
4. V.Ramesh,U. Haribabu "Simulate The Implementation Of Interleaved Boost Converter With Zero Voltage Transition" International Journal of Engineering Research and Applications (IJERA) ISSN: 2248–9622, Vol. 2, Issue 6, November-December 2012, pp. 1129–1135.
5. C. M.Wang, "A new single-phase ZCS-PWM boost rectifier with high power factor and low conduction losses," IEEE Trans. Ind. Electron., vol. 53, no. 2, pp. 500–510, Apr. 2006.
6. Q. Ting and B. Lehman, "Dual interleaved active-clamp forward with automatic charge balance regulation for high input voltage application," IEEE Trans. Power Electron., vol. 23, no. 1, pp. 38–44, Jan. 2008.

7. H. Mao, O. A. Rahman, and I. Batarseh, "Zero-voltage-switching DC-DC converters with synchronous rectifiers," IEEE Trans. Power Electron., vol. 23, no. 1, pp. 369–378, Jan. 2008.
8. J.-H. Kim, D.-Y. Jung, S.-H. Park, C.-Y. Won, Y.-C. Jung, and S.-W. Lee, "High efficiency soft-switching boost converter using a single switch," J. Power Electron., vol. 9, no. 6, pp. 929–939, Nov. 2009.

A Survey of Cloud Based Healthcare Monitoring System for Hospital Management

G. Shanmugasundaram, P. Thiyagarajan and A. Janaki

Abstract Cloud computing is an emerging new technology that can be integrated with healthcare monitoring system. The current survey of all healthcare organizations shows that it needs the assist of cloud computing to store the patients PHI and to get help in emergency condition using cloud-based virtual server. For an effective monitoring system cloud uses BSN to monitor the patient health conditions. Here the cloud acts as a virtual server and stores the patient information in third party server which causes serious threats to security and privacy. The healthcare center contains various approaches which have been used to monitor the healthcare information based on cloud environment. The objective of this study is to discuss about various techniques and taxonomy about the current methods of cloud computing used in hospital healthcare monitoring system. Moreover the strengths and weaknesses of the healthcare monitoring system approaches are also discussed.

Keywords Healthcare system · Hospital monitoring · Cloud-assisted healthcare · System · Mobile cloud solutions

1 Introduction

Mobile healthcare system is an important application for pervasive computing. Mobile cloud computing system improves the service provided in health care [1] by deploying wearable body sensor node and by providing remote health care moni-

G. Shanmugasundaram (✉) · A. Janaki
Department of Information Technology, Sri Manakula Vinayagar Engineering College, Pondicherry, India
e-mail: anitguy2006@gmail.com

A. Janaki
e-mail: janakimtech@gmail.com

P. Thiyagarajan
Department of Computer Science and Engineering, Mahindra Ecole Centrale, Hyderabad 43, India
e-mail: thiyagu.phd@gmail.com

© Springer Science+Business Media Singapore 2017

S.C. Satapathy et al. (eds.), *Proceedings of the International Conference on Data Engineering and Communication Technology*, Advances in Intelligent Systems and Computing 468, DOI 10.1007/978-981-10-1675-2_54

toring to help people who are in need of an emergency condition in far off places. Each patient's health details are manually stored and maintained by authorized person and they monitoring them. The personal health records (PHRs) and electronic health monitoring (EHM) are the electronic versions of patient health information [2].

Today several healthcare organizations started to shift the patient health information to the cloud environment. Healthcare monitoring device is used for monitoring the patient's health conditions. This health monitoring information can be stored and communicated with the medical user or healthcare providers (doctor or nurse). In emergency situation health monitoring device will provide the service to client who registered in healthcare center using mobile cloud computing. In health care center each and every patient have their own Personal Health Information(PHI) such as Heart Beat (HB), Blood Sugar Level (BSL), Blood Pressure (BP) and Body Temperature (BT) are collected by wearable sensor and those information are send to medical user such as doctor or nurse by using cloud computing technologies [3].

1.1 Mobile Cloud Computing

Mobile cloud computing is the combination of three main abstractions of cloud computing and mobile computing and wireless network. These three are combined to facilitate rich computational experience for mobile users. Mobile computing: is a technology that allows transmission of data from one place to another place, here the data can be of any type such as voice, video, and document or files. Cloud computing: Is a network of remote servers which is hosted on the internet to store, manage, and process data, it is an internet-based computing, where by sharing the resource software and information are provided to the computer and it delivers hosted service's over the internet. Mobile cloud computing is just used to share or store the information from one place to another without using any wired media. This computing technique serves as a platform for the effective sharing of health information's. The rest of the paper is organized as follows, Sect. 2 describes about the taxonomy-based study on existing cloud healthcare systems. Section 3 presents the various opportunities to work cloud-based health solutions.

2 Review on Existing Works

The objective of the study is to explore various cloud-based healthcare solutions. Here the concentration focuses on the security aspects addressed in existing healthcare systems. The taxonomy of the cloudbased healthcare monitoring system is illustrated in Fig. 1.

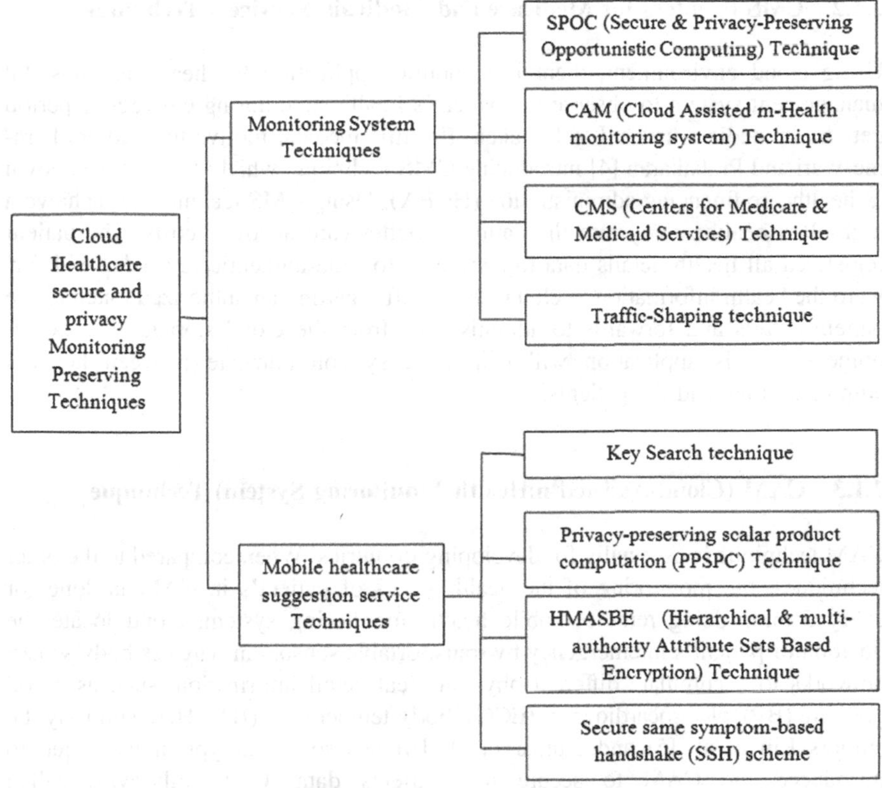

Fig. 1 Taxonomy of the cloud-based healthcare monitoring system

2.1 Secure and Privacy Healthcare Monitoring Preserving Techniques in the Cloud Computing

There are various techniques in literature to monitor the patient's health conditions. However, there is no clear classification of the healthcare monitoring and providing security techniques. Therefore, here we classify the healthcare monitoring system into two main taxonomy (A) Monitoring System Techniques and (B) Mobile healthcare suggestion service Techniques at the top level.

2.1.1 Monitoring System Techniques

The above technique is commonly used in the hospital healthcare cloud-based system to monitor the patient's health condition with the help of wearable body sensor network (BSN) and some sensor gateway. This technique is classified into several approaches, here we briefly define and present the approaches based on monitoring system techniques.

2.1.2 CMS (Centers for Medicare and Medicaid Services) Technique

Using cloud environment, there is a mobile application for healthcare hospital management system to observe the patient's health state during emergency period gets an immediate first aid to be taken. For this medical motive the authors Parameswari and Prabakaran [4] introducing CMS technique which it otherwise known as healthcare financing administration (HCFA). Using CMS technique to achieve a high-class healthcare system they aim for better care at lower costs. The patient registered all health details data to administrator, the authenticate sender will forward the health information to cloud storage after getting an authorized token of the patient details and forwards to administrator from the cloud storage. The development of this application will help to easy communicate between hospital administrations and the patients.

2.1.3 CAM (Cloud-Assisted mHealth Monitoring System) Technique

CAM technique is especially for developing countries. When compared to the other techniques, the monitoring of the health status of patient's in CAM is done for every 5 min. Using remote mobile health monitoring system, cloud locate the position of patient's in emergency by transportable sensors in wireless body sensor networks to accumulate different physiological detail information, such as blood pressure (BP), electrocardiogram (ECG), body temperature (BT). Here similarly the authors Lin et al. [5] and Kumar et al. [6] use some encryption technique to introduced the CAM to secure the patients data, CAM otherwise called anonymization technique which contains issue of addressing security and privacy problem to cloud servers from resource constrained without compromising the privacy policy.

2.1.4 Traffic-Shaping Technique

In cloud computing system, the mobile health monitoring system should operate with least service delay and it should provide privacy preservation by utilizing geographically distributed clouds. Here the authors Shen et al. [7] introducing the traffic-shaping algorithm to convert the health data traffic into the non-health data traffic through reduced traffic analysis (TA) attacks and to minimize the service delay. Another name of Traffic Shaping (i.e.) Packet Shaping is to manage and maintain the computer network traffic.

2.1.5 SPOC (Secure and Privacy-Preserving Opportunistic Computing) Technique

SPOC technique can monitor the high-intensive process and senses PHI by wireless body sensor network (BSN) and store all the health information in cloud storage environment. SPOC is based on attribute-based access control (ABAC). Authors here present a reference technique for hospital health monitoring system with some internal attackers. They used the miniaturized wearable body sensor nodes and hospital systems for monitoring the patient's health information (PHI) and the sensor used to sense the PHI such as heart beat, body temperature, blood pressure, and blood sugar level. By providing high-quality pervasive healthcare monitoring system for emergency medical condition, the authors Ambika et al. [8] use the cloud computing and wireless sensor network, to get the emergency help using sensor by monitoring patient's condition for each and every second.

2.2 *Mobile Healthcare Suggestion Service Techniques*

This technique is used in the hospital healthcare center cloud-based system to monitor and view the patient health using wearable body sensor network (BSN) and update the suggestion by required doctor. This technique is classified into several approaches, here briefly define and present the approaches based on online health suggestion service techniques.

2.2.1 Secure Same Symptom-Based Handshake (SSH) Scheme

Using mobile application to know who have the same diseases or same symptom to share their mutual experiences by communicate each other, when two patients meet, only if they have the same diseases or same symptom and they can use their private keys to make mutual authentication communication to share their own experiences. There exist many challenging security issues in MHSN (Mobile Healthcare Social Network) to overcome these issues the authors Lu et al. [9] launch SSH technique for exchanging their private key securely to the authorized person.

2.2.2 HMASBE (Hierarchical and Multi-authority Attribute Sets-Based Encryption) Technique

The patients will have full control of their medical details and they can efficiently share the health report to doctors, including other users, like healthcare providers, family members or friends. Here the authors Chen et al. [10] initiate HMASBE technique to ensure flexible and scalable access control to PHR data in the cloud environment. The patients share their communication connection between Health

Care Provider (HCP), Cloud Service Provider (CSP), Trusted Central Authority (TCA), Domain Authority (DA), Lower Level Attribute Authority (LLAA), after connection the patient exchange and collect their health information and stored in cloud electronic medical records.

2.2.3 Privacy-Preserving Scalar Product Computation (PPSPC) Technique

The smart phone is an important part in mobile healthcare monitoring system. Similarly the authors Sushrutha [11] established a privacy-preserving scalar product computation technique (PPSPC) and applied this technique in smart phones to allow an authorized user to decide who can join in the opportunistic computing whenever an authorized user is in an emergency condition. The smart phone accumulate and stores the data for every 5 min by the sensors and the collected information are encrypted using the session key Mownika and Pradeep [12], those encrypted data are transmitted to healthcare center using Wi-Fi with 3G transmissions.

2.2.4 Key Search Technique

The health industry is the growing environment due to huge population and now a day's healthcare system combine with cloud computing to shift the patient data into EMR (Electronic Medical Records) is being in use. Here the authors Rana and Bajpayee [13] introducing the key search algorithm for tracking the patient's health and during crisis state the key search technique will generate an alert message and send to certain doctor and medical professionals along with the location (address) of the patient and those patients location address are sense through the GPS sensor system. The entire summary of the technique discussed here is tabulated in Table 1.

3 Discussion

Cloud computing technology used in hospital environment it creates interesting possibilities for sharing information with patients about their conditions for treatment process. The healthcare providers can use various new technologies to update the health conditions for their patient. Using cloud service the health organization reduces the storage place and maintaining cost. The major advantage of using cloud computing in healthcare is easy to storing and accessing information. The healthcare center can help the physicians to stay in touch with their patients and examine their health condition effectively. In above tabular column we discuss about the challenges, advantage, and disadvantage. Here we discuss about the major issues security preservation in cloud computing.

Table 1 Summary of existing techniques

S. no.	Author and year	Proposed techniques	Advantage	Disadvantage
1.	Rana and Bajpayee [13]	Key search technique	Emergency alert SMS send only to the authorized person	No accurate measurements of environmental parameters No efficient use of energy for remote site, mobile devices and vehicle based data
2.	Ambika et al. [8]	SPOC (Secure and Privacy-Preserving Opportunistic Computing)	Monitoring patients health in each 5 min	Just monitor while emergency no any indications and internal attacker
3.	Sushrutha [11]	Privacy-preserving scalar product computation (PPSPC)	Calculate the scalability problem	Faces many problems in security and privacy preservation
4.	Shen et al. [7]	Traffic-shaping technique	Reduced communication costs Minimized service delay Detected by TA (traffic analysis) attackers	Complicated case where users have random medical request and diverse privacy preservation requirements
5.	Lin et al. [5]	CAM (Cloud Assisted mHealth monitoring system)	The intellectual property of mHealth service providers	Problem in addressing security and privacy is the computational to cloud servers
6.	Chen et al. [10]	HMASBE (Hierarchical and Multi-Authority Attribute Sets Based Encryption)	To reduce the potential computational load	No prevention from unauthorized access control
7.	Parameswari and Prabakaran [4]	CMS (Centers for Medicare and Medicaid Services)	Registered customers have their accounts with unique ID and password	More integration and security issues
8.	Lu et al. [9]	Secure same symptom-based handshake (SSH) scheme	Develop a SSH scheme for MHSH Improve the processing of medical devices by connection them to cloud	There is no patients 'selfish incentive mechanism for MHSN

From the above review, we found various issues related to security and privacy aspects in cloud-based health systems. The first issue in cloud service is traffic analysis attacker [5] and second issue there is no prevention from unauthorized access control [4, 10, 13]. Another most important issue in security and privacy is internal attacker [8, 9, 11]. These problems still carry in healthcare cloud-based service, so the cloud computing is still a developing technology which implies that in the future year the services it provide will be greater than our expectations or beyond on imagination. Those issues are briefly explained below.

Internal Attacker: An internal attack occurs when an individual or a group within an organization seeks to disrupt operations or exploit organizational assets. In many cases, the attacker employs a significant amount of resources, tools, and skill to launch a sophisticated computer attack and potentially remove any evidence of that attack as well. Highly skilled and disgruntled employees (such as system administrators and programmers) or technical users who could benefit from disrupting operations may choose to initiate an internal attack against a company through its computer systems.

Traffic analysis attacker: Traffic analysis attacks are aimed at deriving critical information by analyzing the statistics of traffic flows. For example, in a military communication network, by intercepting traffic using sniffing tools and monitoring pattern change of link load, Traffic analysis attacks challenge the design of traditional systems where encryption is typically used as the main method for protecting security and privacy. However, it is obvious that encryption cannot protect many other important characteristics of traffic which may be mission critical and require protection.

Unauthorized access control: Unauthorized access is the act of gaining access to a network, system, application, or other resource without permission. Unauthorized access could occur if a user attempts to access an area of a system they should not be accessing. Unauthorized access could be result of unmodified default access policies or lack of clearly defined access policy documentation.

4 Conclusion

A survey is presented about cloud-based healthcare monitoring system for hospital management. The privacy of the electronic health data in mobile cloud computing environment is a serious issue that requires special considerations. We presented a review on the technologies and approaches that are currently being used to deal with the important issue of healthcare monitoring system and mobile healthcare suggestion service. We have categorized the cloud healthcare secure and privacy monitoring preserving techniques into monitoring system techniques and mobile healthcare suggestion techniques. We have presented the various security and privacy issues which can be used as new research avenue in cloud-based healthcare monitoring system.

References

1. http://docs.media.bitpipe.com/io_10x/io_108673/item_650544/cloud%20computing%20wp.pdf.
2. http://www.aspenadvisors.net/results/whitepaper/cloud-computing-healthcare-there-silver-lining.
3. http://www.csc.com/health_services/insights/60186-cloud_computing_in_the_healthcare_industry.
4. R. Parameswari, N. Prabakaran, "An Enhanced Mobile Healthcare Monitoring System in Mobile Cloud Computing", International Journal of Advanced Research in Computer and Communication Engineering, Vol:1, Issue:10, (2012).
5. Huang Lin, Jun Shao, Chi Zhang, Yuguang Fang, "CAM: Cloud-Assisted Privacy Preserving Mobile Health Monitoring, Information Forensics and Security, IEEE Transactions, Vol:8, (2013).
6. Amit Kumar, Siba Prasad Ojha, Anuradha.C, "CAM:Cloud Assisted Privacy Preserving Mobile Health Monitoring", International Journal For Technological Research In Engineering, Vol:2, Issue:8, (2015).
7. Qinghua Shen, Xiaohui Liang, Xuemin (Sherman), Xiaodong Lin, Henry Y.Luo, "Exploiting Geo-Distributed Clouds for a E-Health Monitoring System with Minimum Service Delay and Privacy Preservation", IEEE Journal of Biomedical and Health Informatics, Vol:18, (2014).
8. Ambika. S, HamsaPriyaa. M, Lalitha. R, Rajakuamari. K, "Secure and Privacy Approach in Mobile Healthcare SPOC Secure & Privacy-Preserving Opportunistic Computing)", IJIRST-International Journal for Innovative Research in Science & Technology, Vol:1, Issue:11, (2015).
9. Rongxing Lu, Xiaodong Lin, Xiahui Liang, Xuemin, " Secure Handshake with symptoms-matching: the essential to the success of mhealthcare social network", Proceedings of the Fifth International Conference on Body Area Networks, (2010).
10. Chen Danwei, Chen Linling, FAN Xiaowei, HE Liwen, PAN Su, Hu Ruoxiang, "Securing Patient-Centric Personal Health Records Sharing System in Cloud Computing", China Communication-Supplement, (2013).
11. M.B. Sushrutha, "Cloud-Assisted Privacy Preserving Mobile Health Monitoring Using PPSPC technique", International Journal of Advanced Research in Computer and Communication Engineering, Vol:3, Issue:3, (2014).
12. Mownika. K, K.C. Pradeep, "Secure and Privacy approach in mobile healthcare emergency using PPSPC technique, International Journal of Research in IJRCCT Computer and Communication Technology, Vol: 2, Issue:10, (2013).
13. Jubi Rana, Abihijeet Bajpayee, "HealthCare Monitoring and Alerting System Using Cloud Computing", "International Journal on Recent and innovation Trends in Computing and Communication, Vol:3, Issue:2, (2015).

Design of Common Subexpression Elimination Algorithm for Cyclotomic Fast Fourier Transform Over GF (2^3)

Tejaswini Deshmukh, Prashant Deshmukh and Pravin Dakhole

Abstract The frequency transformation methods like fast fourier transform algorithms can be competently used in realization of discrete fourier transforms over Galois field, which have broad applications in network security and digital communication in error correcting codes. The cyclotomic fast fourier transform (CFFT) is a type of fast fourier transform algorithm over finite fields This method utilizes the benefit of cyclic decomposition. The cyclotomic breakdown of input data is used to reduce the number of operations which can be equally exploited to get a set by set treatment of the input sequence. Common subexpression elimination (CSE) is an useful optimization process to solve the multiple constant multiplication problems. In this paper, common subexpression elimination algorithm for cyclotomic fast fourier transform over fixed field 2^3 is designed. Using CSE algorithm, we reduce the additive complexities of cyclotomic fast fourier transform. The design of CSE is to spot regular patterns that are present in expressions more than once and replace them with a single variable. Using above method every regular pattern calculates only once, thus minimizing the area of CFFT architecture required in VLSI implementation.

Keywords Common subexpression elimination (CSE) · Cyclotomic fast fourier transforms (CFFTs) · Multiple constant multiplication (MCM)

Tejaswini Deshmukh · Prashant Deshmukh (✉) · Pravin Dakhole (✉)
Department of Electronics, Yeshwantrao Chavan College of Engineering, Nagpur, India
e-mail: pr_deshmukh@yahoo.com

Pravin Dakhole
e-mail: pravin_dakhole@yahoo.com

Tejaswini Deshmukh
e-mail: tejaswini.deshmukh@gmail.com

© Springer Science+Business Media Singapore 2017

S.C. Satapathy et al. (eds.), *Proceedings of the International Conference on Data Engineering and Communication Technology*, Advances in Intelligent Systems and Computing 468, DOI 10.1007/978-981-10-1675-2_55

1 Introduction

A fast fourier (FFT) is an algorithm that computes the discrete fourier transform (DFT) of a sequence, or its inverse fourier transform. The fast fourier transform decomposes a function of time into frequencies that make it up. In the composite field, the applications of the fourier transform occur throughout the subject of signal processing. Discrete fourier transform over finite field has wide applications in digital communication system. The fast fourier transform over finite fields is extensively used in various channel coding. Classical FFT algorithms developed for the case of complex field are not well suited for the case of finite fields. The cyclotomic fast fourier transform [1] is used over finite field to generate fast fourier transform. A finite field or Galois field is a field that contains a finite number of elements. This algorithm first decomposes a DFT into several circular convolutions, and then derives the DFT results from the circular convolution results. The advantage of cyclotomic fast fourier transform over conventional fast fourier transform is that this algorithm has very low multiplicative complexity, but they have high additive complexities. Common subexpression elimination technique is used to reduced the additive complexity of cyclotomic fast fourier transform. To solve the multiple constant multiplication (MCM) problems in matrix vector multiplication common subexpression elimination (CSE) is an valuable method to optimized the design. The idea of common subexpression elimination is to recognize common variables that are present in expressions more than once and replace them with a single variable. However, how to select common elements to remove for obtaining minimal results is an NP complete problems [2, 2]. Common subexpression elimination (CSE) is a serious process in many multipliers less implementation of digital signal processing algorithms. The aim of CSE is to decrease the number of logic operators used and to minimize the logic depth of the DSP algorithm implemented in VLSI. CSE algorithm is used to reduce the additive complexities of cyclotomic fast fourier transforms (CFFTs) because of which CFFTs achieve smaller total operational complexities than other FFTs, requiring fewer additions. Steps of cyclotomic fast ourier transform algorithm is explained in Sect. 2. Common subexpression elimination algorithm and illustration of CSE is shown in Sect. 3.

2 Algorithm for CFFT

The Fourier transform of a polynomial is given in [1], defined by equation

$$f(x) = \sum_{i=0}^{n-1} f_i x^i, \tag{1}$$

where degree of Eq. 1 is $f(x) = n - 1$ and $n = (2^m - 1)$, in the field GF (2^m) is the collection of elements

$$F_j = f(\alpha^j) = \sum_{i=0}^{n-1} f_i \alpha^{ij}, j \in [0, n-1], \tag{2}$$

where α is an element of order n in the field GF(2^m). A linearized polynomial over finite field [4] is defined as $L(x) = \sum_i l_i x^{2^i}, l_i \in GF(2^m)$.

Let $x \in GF(2^m)$ and $\gamma_0 \gamma_1 \cdots\cdots \gamma_{m-1}$ forms a basis of the field.

If $x = \sum_{i=0}^{m-1} x_i \gamma_i, x_i \in GF(2), then L(x) = \sum_{i=0}^{m-1} x_i L(\gamma_i)$.

Consider the cyclotomic cosets over GF (2)

$$\{0\}, \{k_1, k_1 2, k_1 2^2, \ldots..k_1 2^{m1-1}\}, \ldots \{k_l, k_l 2, k_l 2^2, \ldots k_l 2^{m_l - 1}\}, \tag{3}$$

where $ki \equiv ki 2^{m_i} \bmod n$ and l is count of cosets.

Method of cyclic breakdown of inputs to fast fourier transform algorithm is divided into four steps [5] stated as follows,

Step 1: Find out the cyclotomic cosets of the given input elements. Cosets of the same size will be processed by same processing block.
Step 2: Formulation expansion of $f(\alpha^j)$, where j is number of input elements, using the formula

$$f(\alpha^j) = \sum_{i=0}^{l} L_i(\alpha^{jk_i}), \tag{4}$$

where

l = length of cosets
k_i = first element of coset
j = 0 to $n - 1$
$n = (2^m - 1)$
2^m = Galois field (GF2^m)

Hence figure out the coefficients 'αijs' using equation defines in step 2.
Step 3: Carry out cyclic convolution Li(y) between normal basis and input elements [6].
Step 4: The same analysis is applied to other coset elements and realize the equation F = A L f.

Fourier transform equation F = A L f is converted into three different stages using cyclic convolution [1, 5].

$$\mathrm{F} = (\mathrm{AQ})(\mathrm{C}(\mathrm{Pf})) \tag{5}$$

$$F = \begin{pmatrix} F_0 \\ F_1 \\ F_2 \\ F_3 \\ F_4 \\ F_5 \\ F_6 \end{pmatrix} = \left(\begin{pmatrix} 1&1&1&1&1&1&1 \\ 1&0&1&1&1&0&0 \\ 1&1&0&1&0&1&0 \\ 1&1&0&0&1&0&1 \\ 1&1&1&0&0&0&1 \\ 1&0&0&1&0&1&1 \\ 1&0&1&1&1&1&0 \end{pmatrix} \begin{pmatrix} 1&0&0&0&0&0&0&0&0 \\ 0&1&0&1&1&0&0&0&0 \\ 0&1&1&1&0&0&0&0&0 \\ 0&1&1&0&1&0&0&0&0 \\ 0&0&0&0&0&1&0&1&1 \\ 0&0&0&0&0&1&1&1&0 \\ 0&0&0&0&0&1&1&0&1 \end{pmatrix} \right) \times \left[\begin{pmatrix} 1 \\ 1 \\ \gamma^2+\gamma^4 \\ \gamma+\gamma^4 \\ \gamma+\gamma^2 \\ 1 \\ \gamma^2+\gamma^4 \\ \gamma+\gamma^4 \\ \gamma+\gamma^2 \end{pmatrix} \left[\begin{pmatrix} 1&0&0&0&0&0&0 \\ 0&1&1&1&0&0&0 \\ 0&0&1&1&0&0&0 \\ 0&1&1&0&0&0&0 \\ 0&1&0&1&0&0&0 \\ 0&0&0&0&1&1&1 \\ 0&0&0&0&0&1&1 \\ 0&0&0&0&1&1&0 \\ 0&0&0&0&1&0&1 \end{pmatrix} \begin{pmatrix} f_0 \\ f_1 \\ f_2 \\ f_4 \\ f_3 \\ f_6 \\ f_5 \end{pmatrix} \right] \right] \tag{6}$$

$$\begin{pmatrix} 1&0&0&0&0&0&0 \\ 0&1&1&1&0&0&0 \\ 0&0&1&1&0&0&0 \\ 0&1&1&0&0&0&0 \\ 0&1&0&1&0&0&0 \\ 0&0&0&0&1&1&1 \\ 0&0&0&0&0&1&1 \\ 0&0&0&0&1&1&0 \\ 0&0&0&0&1&0&1 \end{pmatrix} \begin{pmatrix} f_0 \\ f_1 \\ f_2 \\ f_3 \\ f_4 \\ f_5 \\ f_6 \end{pmatrix} = \begin{matrix} f_0 \to f_0 \\ f_1+f_2+f_4 \to v_1+f_1 \\ f_2+f_4 \to v_1 \\ f_1+f_2 \to v_2 \\ f_1+f_4 \to v_3 \\ f_3+f_6+f_5 \to v_4+f_3 \\ f_6+f_5 \to v_4 \\ f_3+f_6 \to v_5 \\ f_3+f_5 \to v_6 \end{matrix} \tag{7}$$

From Eq. 6 matrix multiplication of stage 1(p.f) requires eight adders.

$$\begin{pmatrix} 1 \\ 1 \\ \gamma^2+\gamma^4 \\ \gamma+\gamma^4 \\ \gamma+\gamma^2 \\ 1 \\ \gamma^2+\gamma^4 \\ \gamma+\gamma^4 \\ \gamma+\gamma^2 \end{pmatrix} \begin{pmatrix} f_0 \\ v_1+f_1 \\ v_1 \\ v_2 \\ v_3 \\ v_4+f_3 \\ v_4 \\ v_5 \\ v_6 \end{pmatrix} \to \begin{pmatrix} f_0 \\ v_1+f_1 \\ v_1(\gamma^2+\gamma^4) \\ v_2(\gamma+\gamma^4) \\ v_3(\gamma+\gamma^2) \\ v_4+f_3 \\ v_4(\gamma^2+\gamma^4) \\ v_5(\gamma+\gamma^4) \\ v_6(\gamma+\gamma^2) \end{pmatrix} \to \begin{pmatrix} f_0 \\ w_1 \\ w_2 \\ w_3 \\ w_4 \\ w_5 \\ w_6 \\ w_7 \\ w_8 \end{pmatrix} \tag{8}$$

Stage 2 Multiplication by constant (C.(P.f)) requires six multipliers.

$$F = \begin{pmatrix} 1&1&0&0&0&1&0&0&0 \\ 1&0&0&1&1&1&0&1&1 \\ 1&0&1&1&0&1&1&1&0 \\ 1&1&0&1&1&0&1&1&0 \\ 1&0&1&0&1&1&1&0&1 \\ 1&1&1&0&1&0&0&1&1 \\ 1&1&1&1&0&0&1&0&1 \end{pmatrix} \begin{pmatrix} f0 \\ w1 \\ w2 \\ w3 \\ w4 \\ w5 \\ w6 \\ w7 \\ w8 \end{pmatrix} \tag{9}$$

Stage 3 is matrix vector multiplication (A.Q).(C(Pf))

$$\begin{pmatrix} f0+w1+w5 \\ f0+w3+w4+w5+w6+w7 \\ f0+w2+w3+w5+w6+w7 \\ f0+w1+w3+w4+w6+w7 \\ f0+w2+w4+w5+w6+w8 \\ f0+w1+w2+w4+w7+w8 \\ f0+w1+w2+w3+w6+w8 \end{pmatrix} = \begin{bmatrix} F0 \\ F1 \\ F2 \\ F3 \\ F4 \\ F5 \\ F6 \end{bmatrix} \quad (10)$$

Above matrix vector product gives cyclotomic fast fourier transform for GF (2^3). CFFT achieves very high additive complexity. To reduce the number of adders common subexpression elimination algorithm is used in CFFT. Complete illustration is explained in the next section.

3 Common Subexpression Elimination Algorithm

Common subexpression elimination (CSE) is an optimization method that search for happening of similar variables and replacing them with a single variable that holds the computed value [7]. Algorithm involves following steps:

(1) Recognize common present in the CFFT matrix
(2) Choose a regular pattern for exclusion
(3) Eliminate all possible combinations of the chosen pattern
(4) Removed pattern is computed only one time
(5) Do again the process until not any of several patterns are present

Consider the cyclotomic fast fourier transform equation for computation of reduced additive complexity. The CSE algorithm identifies "*f*0 + *w*3 + *w*5 + *w*7" as the pattern in initial iterative, and generates new variable "w9" to restore it. The identified pattern is substituted to the base of constant matrix as an additional row to be further optimized by CSE algorithm. In the subsequent iterative, the patterns "*f*0 + *w*1 + *w*4 + *w*8" and "*f*0 + *w*2 + *w*6 + *w*8" are recognized and replaced by "*w*10" and "*w*11," respectively. A simple recognition of calculation in Eq. 10 requires 32 XOR gates. However, after CSE optimization, CFFT requires 21 XOR gates. The reduction of XOR gate up to 34 %. Complete illustration of the CSE algorithm applied to Eq. 9 is shown below. Consider the computations as follows (Figs. 1, 2, 3, 4, and 5):

Fig. 1 Pattern 1 elimination

$$\begin{bmatrix} 110001000 \\ 100111011 \\ 101101110 \\ 110110110 \\ 101011101 \\ 111010011 \\ 111100101 \end{bmatrix} \quad \begin{aligned} F0 &= f0 + w1 + w5 \\ F1 &= w4 + w8 \\ F2 &= w2 + w6 \\ F3 &= f0 + w1 + w3 + w4 + w6 + w7 \\ F4 &= f0 + w2 + w4 + w5 + w6 + w8 \\ F5 &= f0 + w1 + w2 + w4 + w7 + w8 \\ F6 &= f0 + w1 + w2 + w3 + w6 + w8 \end{aligned}$$

Fig. 2 Pattern 2 elimination

$$\begin{bmatrix} 1100010000 \\ 0000100010 \\ 0010001000 \\ 1101101100 \\ 1010111010 \\ 1110100110 \\ 1111001010 \\ 1001010100 \end{bmatrix} \quad \begin{aligned} F0 &= f0 + w1 + w5 \\ F1 &= w4 + w8 \\ F2 &= w2 + w6 \\ F3 &= f0 + w1 + w3 + w4 + w6 + w7 \\ F4 &= f0 + w2 + w4 + w5 + w6 + w8 \\ F5 &= w4 + w7 \\ F6 &= w3 + w6 \\ w9 &= f0 + w3 + w5 + w7 \end{aligned}$$

Fig. 3 Pattern 3 elimination

$$\begin{bmatrix} 11000100000 \\ 00001000110 \\ 00100010010 \\ 00010010001 \\ 10101110100 \\ 00100000101 \\ 11110010100 \\ 10010101000 \\ 11001000100 \end{bmatrix} \quad \begin{aligned} F0 &= f0 + w1 + w5 \\ F1 &= w4 + w8 + w9 \\ F2 &= w2 + w6 + w9 \\ F3 &= w3 + w6 + w10 \\ F4 &= f0 + w2 + w4 + w5 + w6 + w8 \\ F5 &= w2 + w8 + w10 \\ F6 &= f0 + w1 + w2 + w3 + w6 + w8 \\ w9 &= f0 + w3 + w5 + w7 \\ w10 &= f0 + w1 + w4 + w8 \end{aligned}$$

Fig. 4 Pattern 4 elimination

$$\begin{bmatrix} 110001000000 \\ 000010001100 \\ 001000100100 \\ 000100100010 \\ 000011000001 \\ 001000010010 \\ 010100000001 \\ 100101010000 \\ 110010010000 \\ 101000101000 \end{bmatrix} \quad \begin{aligned} F0 &= f0 + w1 + w5 \\ F1 &= w4 + w8 + w9 \\ F2 &= w2 + w6 + w9 \\ F3 &= w3 + w6 + w10 \\ F4 &= w4 + w5 + w11 \\ F5 &= w2 + w7 + w10 \\ F6 &= w1 + w3 + w11 \\ w9 &= f0 + w3 + w5 + w7 \\ w10 &= f0 + w1 + w4 + w7 \\ w11 &= f0 + w2 + w6 + w8 \end{aligned}$$

Fig. 5 Pattern 5 elimination

$$\begin{bmatrix} 01000000000010 \\ 00001000110000 \\ 00000000010001 \\ 00010010001000 \\ 00001100000100 \\ 00100001001000 \\ 01010000000100 \\ 00010001000010 \\ 11001001000000 \\ 10000000100001 \\ 10000010000000 \\ 00100010000000 \end{bmatrix} \quad \begin{aligned} F0 &= w1 + w12 \\ F1 &= w4 + w8 + w9 \\ F2 &= w9 + w13 \\ F3 &= w3 + w6 + w10 \\ F4 &= w4 + w5 + w11 \\ F5 &= w2 + w7 + w10 \\ F6 &= w3 + w1 + w11 \\ w9 &= w3 + w7 + w12 \\ w10 &= f0 + w1 + w4 + w7 \\ w11 &= f0 + w8 + w13 \\ w12 &= f0 + w5 \\ w13 &= w2 + w6 \end{aligned}$$

$$F = \begin{pmatrix} 1 & 1 & 0 & 0 & 0 & 1 & 0 & 0 & 0 \\ 1 & 0 & 0 & 1 & 1 & 1 & 0 & 1 & 1 \\ 1 & 0 & 1 & 1 & 0 & 1 & 1 & 1 & 0 \\ 1 & 1 & 0 & 1 & 1 & 0 & 1 & 1 & 0 \\ 1 & 0 & 1 & 0 & 1 & 1 & 1 & 0 & 1 \\ 1 & 1 & 1 & 0 & 1 & 0 & 0 & 1 & 1 \\ 1 & 1 & 1 & 1 & 0 & 0 & 1 & 0 & 1 \end{pmatrix} \begin{bmatrix} f0 \\ w1 \\ w2 \\ w3 \\ w4 \\ w5 \\ w6 \\ w7 \\ w8 \end{bmatrix} = \begin{pmatrix} f0 + w1 + w5 \\ f0 + w3 + w4 + w5 + w6 + w7 \\ f0 + w2 + w3 + w5 + w6 + w7 \\ f0 + w1 + w3 + w4 + w6 + w7 \\ f0 + w2 + w4 + w5 + w6 + w8 \\ f0 + w1 + w2 + w4 + w7 + w8 \\ f0 + w1 + w2 + w3 + w6 + w8 \end{pmatrix} \tag{11}$$

Table 1 Additive complexity of CFFT for GF (2^3)

n = 7	CFFT without CSE	CFFT with CSE
Additive complexity of stage 1	8	8
Additive complexity of stage 3	32	21
Total additive complexity of CFFT	40	29

4 Result and Conclusion

The algorithm offered for the calculation of cyclotomic fast fourier transform over Galois field with reduced additive complexity. The operational complexity of cyclotomic fast fourier transform using Common subexpression elimination algorithm is stated in Table 1. The straight forward realization of computation in cyclotomic fast fourier transform requires 32 adders. After applying CSE algorithm additive complexity of cyclotomic fast fourier transform is reduced. CFFT with CSE requires 21 adders. The reduction of adders up to 34 %.

Acknowledgments The authors would like to show gratitude Prof. P. Trifonov and Ali AL GHOUWAYEL for providing details of CFFTs. They are thankful to Ning Wu, Xiaoqiang Zhang, Yunfei Ye and Lidong Lan for providing details of common subexpression elimination algorithm.

References

1. S. V. Fedorenko and P. V. Trifonov: A method for Fast Computation of the Fast Fourier Transform over a Finite Field. Problems of Information Transmission, 39(3):231–238, July –September 2003.
2. Ali AL GHOUWAYEL, Yves LOUET, Amor NAFKHA and Jacques PALICOT: On the FPGA Implementation of the Fourier Transform over Finite Fields GF (2^m). SUPELEC-IETR Avenue de la Boulaie CS 4760135576 CESSON-SEVIGNE Cedex, FRANCE-2007.
3. Truong, T.-K., Jeng, J.-H., and Reed, I.S.: Fast Algorithm for Computing the Roots of Error Locator Polynomials up to Degree 11 in Reed–Solomon Decoders. IEEE Trans. Commun., vol. 49, no. 5, pp. 779–783, 2001.
4. N. Chen, and Z. Y. Yan: Cyclotomic FFTs With Reduced Additive Complexities Based on a Novel Common Subexpression Elimination Algorithm. IEEE Tranc. Signal Processing, Vol. 57, no.3, pp. 1010–1020, Mar.2009.
5. M. M. Wong, and M. L. D. Wong: A new common subexpression elimination algorithm with application in composite field AES S-box. Tenth Int. Conf. Information Sciences Signal Processing and their Applications (ISSPA 2010), pp. 452–455, May 2010.

6. R. Blahut: Theory and Practice of Error Control Codes. Reading, Massachusetts: Addison-Wesley, 1983.
7. Ning Wu, Xiaoqiang Zhang, Yunfei Ye, and Lidong Lan: Improving Common subexpression Elimination Algorithm with A New Gate-Level Delay Computing Method. Proceedings of the World Congress on Engineering and Computer Science, Vol II 23–25 October, San Francisco, USA, 2013.

On Analysis of Wheat Leaf Infection by Using Image Processing

Rittika Raichaudhuri and Rashmi Sharma

Abstract In present scenario, agriculture forms a vital part in India's economy. More than 50 % of India's population is dependent (directly or indirectly) on agriculture for their livelihood. In India many crops are cultivated, out of which wheat being one of the most important food grain that this country cultivates and exports. Thus it can be seen that wheat forms a major part of the Indian agricultural system and India's economy. Hence, maintenance of the steady production of above stated crop is very important. The main idea of this project is to provide a system for detecting wheat leaf diseases. The given system will study the leaf image of a wheat plant through image processing and pattern recognition algorithms. Former algorithms are used for extracting vital information from the leaf and the latter is used for detecting the disease that it is infected with. For image processing and segmentation usage of k-means algorithm and canny filter has been suggested. Pattern recognition is achieved through PCA or GLCM and classification through SVM or ANN.

Keywords Wheat leaf · Image processing · Probability theory · Feature extraction and classification

1 Introduction

India being an agricultural-based country, more than half of its population earns their living through agriculture. Also agriculture has an important role to play in India's economy in the form of export of agricultural products. Hence it has become mandatory to maintain a consistency level when it comes to production of crops.

Rittika Raichaudhuri (✉) · Rashmi Sharma
Centre of Information Technology, College of Engineering Studies,
University of Petroleum and Energy Studies, Dehradun, India
e-mail: rittika.rc@gmail.com

Rashmi Sharma
e-mail: rashmi.sharma@ddn.upes.ac.in

© Springer Science+Business Media Singapore 2017

S.C. Satapathy et al. (eds.), *Proceedings of the International Conference on Data Engineering and Communication Technology*, Advances in Intelligent Systems and Computing 468, DOI 10.1007/978-981-10-1675-2_56

Numerous varieties of crops are produced in India, with rice and wheat constituting almost half of the entire crop production. Therefore it has become necessary to maintain a steady production rate for both these crops. However it is very hard to maintain this consistency in production as many a times the entire harvest gets destroyed due to various reasons.

Most of the time the cause of crop destruction is due to some kind of disease that has affected the plant harvest. A farmer is very much prone to human error for example one may become late in detecting the disease or might miss the fact that certain portion of the harvest has already been infected, etc. Hence in order to avoid this kind of human error, an intelligent automated system is needed that can alert the farmers if any portion of the harvest is affected by diseases. Detection of disease can be made by studying the leaf since they clearly show the symptoms.

This paper proposes an intelligent system that can detect the diseases for a wheat crop by processing the image of a wheat leaf. The proposed system will perform mainly two operations: (a) Image processing for extracting necessary features and (b) pattern recognition for studying the extracted features for detecting the type of disease.

2 Literature Survey

A good volume of work has already been performed pertaining to this field.

Kiran R. Gavhale et al. [1], in this study an overview about detecting plant leaf diseases based on texture analysis of a leaf has been discussed. In this paper, a summary of different image processing techniques has been discussed for detecting leaf diseases for several plants. Detection methods such as BPNN, SVM, k-means clustering, and SGDM are discussed in details.

Jyotismita Chaki et al. [2], a comparative study between three different techniques viz. moment-invariant model, centroid-radii model, and binary-superposition model has been provided in this paper. The above stated image processing techniques are applied for identifying the plant species by processing the plant leaf image. The experimental space consisted of three different types of plants species and around 180 leaf images that were divided among three classes each class consisting of 60 images.

Ji- Xiang Du et al. [3], this paper proposed a new image classification technique known as move median center (MMC) hypersphere classifier for identifying the plant species by processing a plant leaf image. The core idea of the above-stated method is to treat each pattern class as a sequence of hypersphere which is in turn treated as a set of points.

Khushal Khairnar et al. [4], through this paper the author provides an overview regarding the various feature extraction and classification methods used in image processing for detecting the type of plant disease. The above-stated methods are applied on leaf images for detecting diseases and classifying them accordingly. A comparative study of different feature extraction methods such as CCM, SGDM,

Wavelet transform, etc., and the two classification methodss SVM and ANN have been provided.

JyotismitaChaki et al. [5], in this paper an automated image processing system has been designed for detecting the plant type from the given leaf image of the respective plant. According to this study, Gabor filter has been used for performing feature extraction and Manhattan distance classification technique is used for performing the function of classification.

3 Background

In order to detect plant diseases through leaf images four major steps must be followed. First step being image acquisition followed by image pre-processing, next being feature extraction, and last classification.

3.1 Image Acquisition

In this step, digital images of leaf that are of high resolution are acquired. An image database needs to be developed as per the requirements of the application.

3.2 Image Preprocessing

This step can be further broken down into three parts: first being image enhancement, second being color-space conversion, and third being image segmentation.

Image Enhancement: in this process, the quality of the digital image is improved for providing a better perception of the image. The main aim is to modify the essential attributes of the image so that the image becomes suitable for a specific task. Image enhancement can be done at both spatial as well as at frequency level of the image [6]. Techniques such as median filter, histogram equalization, image smoothening, image sharpening, etc., can be used for performing image enhancement.

Color-Space Conversion: it is the means through which a color management module (CMM) transforms color from one device's space into another. The conversion process may have the need of approximating the color values so that the image's most important color qualities are preserved and not lost. For example in certain cases for some softwares, there might be a need of converting images from RGB color space to HSV (Hue Saturation Value) color-space representation.

Image Segmentation: is the practice of partitioning a digital image into multiple fragments. The main goal of this step is to identify objects or extract other related information from digital images. There are numerous image segmentation techniques such as:

- Region based
- Edge based
- Threshold based
- Feature-based Clustering
- Color based

3.3 Feature Extraction

It is the procedure of outlining a set of necessary features, or image characteristics that form the core element which when represented in an efficient or meaningful manner give the required information that is important for analysis and classification purpose. Feature extraction technique can be based on color, shape, or texture features. The most commonly used feature extraction technique is the texture extraction technique. Before texture extraction can be performed texture analysis must be performed. Texture analysis can be further broken down into statistical, structural, fractal, and signal processing techniques. Some of the algorithm techniques that can be used for texture feature extraction are gabor filter, color co-occurrence methods, wavelet transform, etc.

3.4 Classification

It is the process of understanding the meaning of the feature extracted from the image and matching the extracted information with the predefined set of rules and thus coming to a conclusion. This stage provides the result, i.e., in this stage the plant disease is identified and made known to the user. The different techniques that may be used for classification purpose are radial basis function, artificial neural network, support vector machine, etc.

3.5 Wheat Crop

Wheat constitutes one of the oldest forms of crops to be cultivated in India. 'Gramineae' is the scientific name given to wheat plant and it belongs to the genus, 'Triticum'. India cultivates mainly four types of wheat plant viz. hard red winter wheat, soft wheat, durum, white wheat. Wheat too like any other crop is prone to diseases. Detection of these diseases becomes a mandatory task so as to protect and preserve the wheat harvest. This paper suggests an idea for developing an automated system than can detect the disease that has infected the plant by processing the image of the plant leaf. Some diseases that can affect the crop are [7]:

- **Powdery Mildew**: the symptoms are formation of white powdery substance over the surface of the leaf [7] (Fig. 1).
- **Brown Rust**: in this brown pustules are formed over the leaf that are either circular or elliptical in nature [7] (Fig. 2).
- **Stripe Rust/Yellow Rust**: the leaf may show bright yellow or dull yellow to orange yellow stripe pustules (Uredia) [7] (Fig. 3).
- **Black Rust**: Pustules of dark reddish brown nature will occur on both sides of the leaf [7] (Fig. 4).

Fig. 1 Wheat leaf infected with powdery mildew

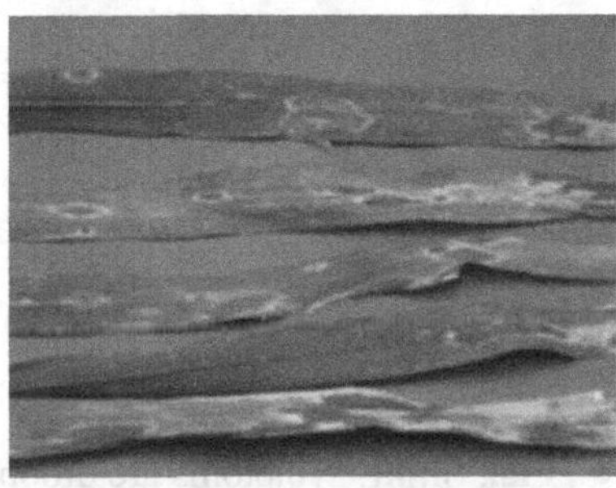

Fig. 2 Wheat leaf infected with brown rust

Fig. 3 Wheat leaf infected with stripe ruts/yellow rust

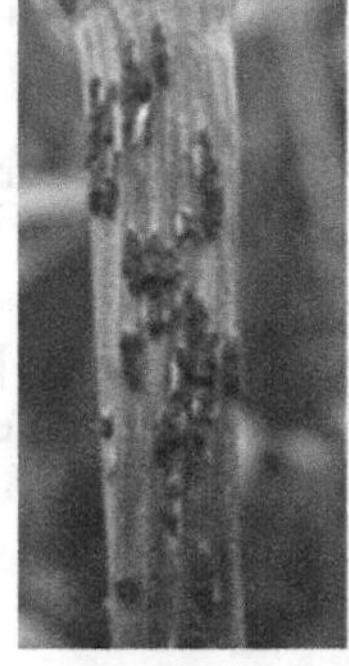

Fig. 4 Wheat leaf infected with black rust

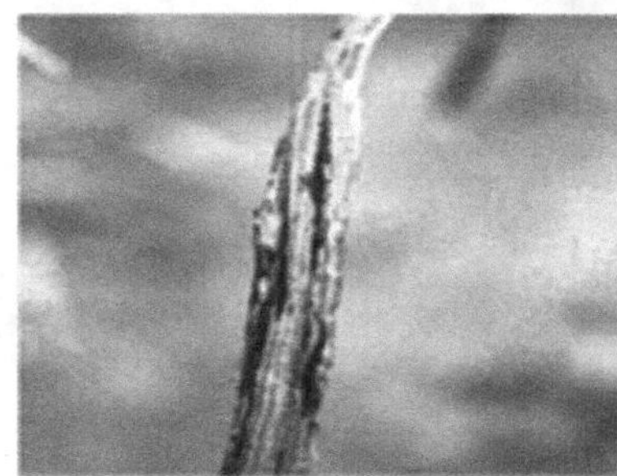

Fig. 5 Wheat leaf infected with flag smut

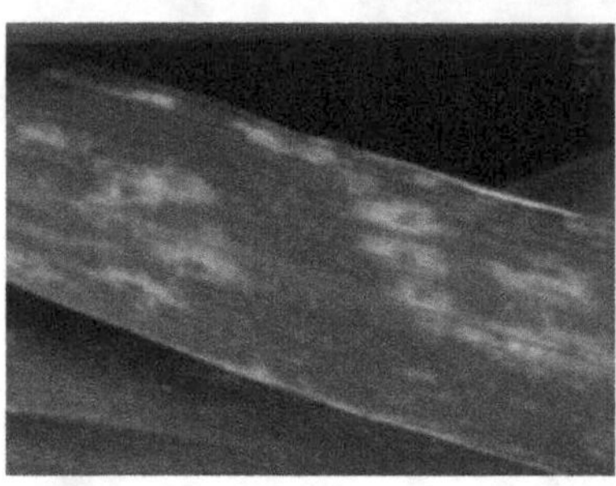

Fig. 6 Wheat leaf infected with leaf blight

- **Flag Smut**: symptoms are drooping leaves with formation of grayish black sori (black powdery mass of spores) followed by withering of leaves [7] (Fig. 5).
- **Leaf Blight**: Reddish brown oval spots with yellow margin may appear on the leaf [7] (Fig. 6).

4 Proposed Work

The methodology of the proposed system can be broken down into two segments: (a) Image Processing segment: where the properties of the leaf image will be enhanced segmented from the background and (b) Pattern recognition segment:

where the required features will be extracted and this information will be matched with the predefined knowledge about the plant diseases for detecting which disease has actually affected the plant.

4.1 Image Processing Techniques

Image Acquisition: a database with the required wheat leaf image will be created for performing the necessary experiment.

Image Enhancement: if needed clipping algorithm may be applied as it might become necessary to separate the required region from its background [6].

Color-Space Conversion: in this step if the image is in RGB color space it would be converted into either HSI (Hue/Saturation/Intensity) or HSV (Hue/Saturation/Value) color space. In RGB color space of a digital image, the color components of an object's color are all correlated with the amount of light hitting the object. This makes object discrimination difficult in RGB color space. However in HIS or HSV the chroma (color), components of an object is separate from the intensity components of the object thus making it easier to differentiate between the objects [1, 8].

Image Segmentation: Many methods such as K- means clustering, threshold method, and edge detection can be used for performing image segmentation. For this automated system, edge detection technique and feature-based segmentation will be applied. Canny filter can be used for edge detection purpose and K-means Clustering can be used for feature-based segmentation. Compared to the rest of the edge detection techniques, Canny filter is immune to noise and can produce perfect results as per the requirement [9, 10]. This method may be used with Markov Random field-based segmentation which is basically used for color segmentation. This combination of algorithm will help in identifying the edges accurately.

4.2 Pattern Recognition Technique

Feature Extraction: in this step color feature, and texture feature will be extracted from the leaf. After extracting the necessary feature, it will be used by the classification step for identifying the disease. For this purpose either GLCM (Gray-level co-occurrences matrix) or PCA (Principal Component analysis) method may be used for texture extraction [1, 11]. PCA technique is mainly used for statistical pattern recognition for dimensionality and feature extraction. GLCM is also a statistical approach, which gives information about the relative position between two pixels in an image. Also for color feature extraction CIE $L^*a^*b^*$ color space can be used, where L^* is the measure for brightness, a^* is the chromatic layer which indicates the color lying along the red-green axis and b^* is the chromatic layer which indicates the color lying along the blue-yellow axis [12] (Fig. 7).

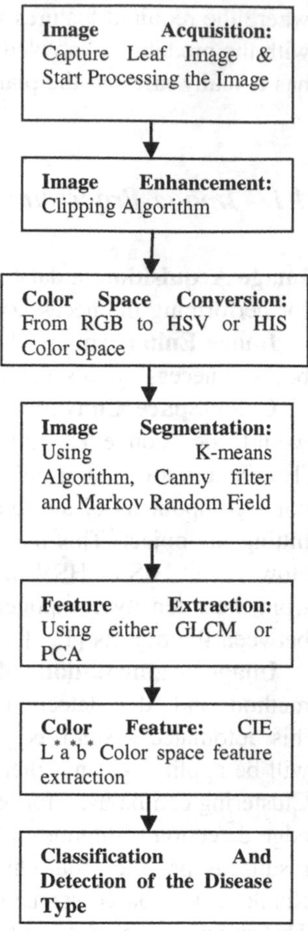

Fig. 7 Step-wise image processing and pattern recognition algorithm

Classification: For this segment SVM (Support Vector Machine) classifier long with probability theory will be used instead of neural networks. SVM is a supervised learning algorithm that helps in recognizing the pattern according to the given data. The problem with using ANN is that the number of layer and nodes is uncertain, also the fact that training certain neural networks takes a lot of time [1, 4, 9].

5 Conclusion

The proposed automated system will be able to detect the wheat leaf disease through image processing. This system when fed with a leaf image will be capable of detecting the earlier mentioned diseases of the wheat plant. In near future, this suggested system will be capable of showing physical results. The future scope of this project is the real-time application of this system.

References

1. Kiran R. Gavhale, Ujwalla Gawande, "An Overview of the Research on Plant Leaves Disease detection using Image Processing Techniques", IOSR Journal of Computer Engineering (IOSR-JCE), vol.16, Issue.1, pp: 10–16, 2014.
2. Jyotismita Chaki, Ranjan Parekh, "Designing An Automated System For Plant Leaf Recognition", International Journal of Advances in Engineering & Technology, vol.2, Issue.1, pp: 149–158, 2012.
3. Ji-Xiang Du, Xiao-Feng Wang, Guo-Jun Zhang, "Leaf shape based plant species recognition", Applied Mathematics and Computation, ScienceDirect, pp: 883–893, 2007.
4. Khushal Khairnar, Rahul Dagade, "Disease Detection and Diagnosis on Plant using Image Processing—A Review", International Journal of Computer Applications, vol.108, Issue.13, 2014.
5. Jyotismita Chaki, Ranjan Parekh, "Plant Leaf Recognition using Gabor Filter", International Journal of Computer Applications, vol. 56, Issue.10, 2012.
6. S.S.Bedi, Rati Khandelwal, "Various Image Enhancement Techniques—A Critical Review", International Journal of Advanced Research in Computer and Communication Engineering, vol. 2, Issue.3, 2013.
7. Indian agricultural research institute, http://iari.res.in.
8. G. Anthonys, N. Wickramarachchi, "An Image Recognition System for Crop Disease Identification of Paddy fields in Sri Lanka", International Conference on Industrial and Information Systems, IEEE, pp: 28–31, 2009.
9. Jagdeesh D. Pujari, Rajesh Yakkundimath, Abdulmunaf S Byadgi, "Image Processing Based Detection Of Fungal Diseases In Plants", International Conference on Information and Communication Technology, Science Direct, 2014.
10. G.T. Shrivakshan, Dr. C. Chandrasekar, "A Comparison of various Edge Detection Techniques used in Image Processing", International Journal of Computer Science, Vol. 9, Issue 5, 2012.
11. Alaa Eleyan, Hasan Demirel, "Co-occurrence matrix and its statistical features as a new approach for face recognition", Turk J Elec & Comp Sci, Vol. 19, Issue 1, 2011.
12. G. Anthonys, N. Wickramarachchi, "An Image Recognition System for Crop Disease Identification of Paddy Fields in Sri Lanka", International Conference on Industrial and Information Systems, 2009.

Divisive Hierarchical Bisecting Min–Max Clustering Algorithm

Terence Johnson and Santosh Kumar Singh

Abstract The inspiration for the Divisive Hierarchical Bisecting Min–Max Clustering Algorithm came from the Bisecting K-Means clustering Algorithm. To obtain K clusters, bifurcate the set of input values into two clusters, select one of these clusters to split further (each time bisect the selected cluster using the Min–Max Clustering Algorithm), and so on, until K clusters have been produced. The Min–Max Clustering Algorithm initially computes the minimum of the input set and then finds a point which is at the greatest distance from the minimum. The remaining values from the set of data items are then accumulated into twoclusters formed by the maximally disjoint min and max values.

Keywords Data mining • Hierarchical clustering • Bisecting K-Means clustering • Min–Max Clustering • Bisecting min max clustering

1 Introduction

Clusters are set of input items in which the maximum separation between each pair of items is as little as can be [1]. Clustering deals with the grouping together of data items which are similar amongst itself and differ to a greater extent in terms of proximity to items of other groups [2]. The clustering process categorizes input data

Terence Johnson (✉)
AMET University, Chennai, India
e-mail: ykterence@rediffmail.com

Terence Johnson
Agnel Institute of Technology and Design, Assagao, Goa, India

S.K. Singh
Thakur College of Science and Commerce Kandivali (E), Mumbai, India
e-mail: sksingh14@gmail.com

© Springer Science+Business Media Singapore 2017

S.C. Satapathy et al. (eds.), *Proceedings of the International Conference on Data Engineering and Communication Technology*, Advances in Intelligent Systems and Computing 468, DOI 10.1007/978-981-10-1675-2_57

into groups whose values are similar according to some criteria [3]. Clustering is extensively used in Market Segmentation, grouping of genetic code for medical research, web applications and in almost all organizations relying on automated decision support and analysis [4].

1.1 Literature Review

The conventional clustering problem in databases, its allied fields, and inter disciplinary domains of statistics, machine learning, pattern recognition, high performance computing, visualization, etc. is to find a grouping of points from a given input set of values into smaller sets called clusters so that the items within each cluster are alike to each other and unfamiliar to values in other clusters [5]. This unfamiliarity of clusters is based on the separation computed using standard distance measures [6]. The fundamental methods in clustering are based on partitioning algorithms in which a set of seed values are used in order to partition the data [7]. Many different techniques of clustering exist like the K-Means and K-Medoids algorithms [8]. The K-Means algorithm uses a sample mean as a representation for each of the K clusters, a square error score function that measures the proximity of each data item to the mean of the closest cluster and a greedy search approach to iteratively move the centers so that the square error is gradually reduced to a minimum. In the traditional instantiation of the algorithm, at each iteration input values are assigned in a greedy strategy to the nearest mean generally in a Euclidean distance way and the resulting new means are then computed. This process continues until no values change assignment, or equivalently, the means do not move [9], i.e., until either the cluster means remain unchanged in the next iteration or until the clusters remain the same in the next iteration. Many times, the K-Means method takes an exponentially large time to converge and so the clustering process is forced to halt after a specified number of iterations [10]. In medoid-based methods, the points from the repository are used as representative seed values as the algorithm searches for the best set of k representative data items which yield exemplary clustering results. A pragmatic technique in this class called CLARANS [11], improves the efficiency of the algorithm by restricting the search space. Another class of clustering algorithms is the Hierarchical clustering algorithms like the Bisecting K-Means Clustering algorithm. In the Bisecting K-Means Clustering Algorithm, to obtain K clusters, the set of all values is split (bisected) into two clusters and then one of these clusters is selected for further bisection, each time the selected cluster is bisected using the K-Means Clustering Algorithm, and so on, until K clusters have been output [12].

2 Issues with Bisecting K-Means Clustering

Since the Bisecting K-Means Clustering Algorithm uses the K-Means algorithm to split the input set into two clusters, the issues related to the K-Means algorithm also affect the Bisecting K-Means Clustering Algorithm. The complexity of the K-Means method is O(nKtd) [13] where n represents the number of values in the input set, K is the number of required clusters, and t denotes the number of iterations the process should undertake if the means are not the same in the next iteration or if the clusters themselves are not the same in the next iteration and d designates the number of dimensions. The K-Means algorithm culminates when either the same cluster means show up in the next iteration or if the same clusters are formed in the next iteration. If neither of this happens then the K-Means algorithm halts the clustering process abruptly after a number of prespecified iterations t. In such situations, it is a very highly probable that the clusters that are formed will be inaccurate [14]. This is so, as an abrupt termination cuts off the possibility that the means may have converged at a later stage providing a fair chance for accurate formation of clusters.

3 Proposed Alternative

The proposed alternative presented in this paper, is the Divisive Hierarchical Bisecting Min Max Clustering Algorithm. The divisive hierarchical clustering algorithm [15] belongs to the category of hierarchical clustering algorithms in which we start at the top with the entire input set in one cluster and then the cluster is split using a clustering algorithm after which the procedure is applied recursively until each point in the dataset is its own singleton cluster or until the number of user-defined clusters are obtained. The idea used in this paper is to obtain K clusters by splitting the set of all input values into two clusters initially and then selecting one of these clusters for further bisection, each time bisecting the selected cluster using the Min–Max Clustering Algorithm, and so on, until K clusters have been output. The Min–Max Clustering Algorithm first computes the minimum of the dataset and then finds a data item which is at maximum distance from the minimum. The values in the input set other than the min and max are pushed into two clusters formed by the min and max values. The complexity of the Min–Max Clustering algorithm is O(n + m) where O(n) is the time required to find the minimum of n data points and O(m) is the time taken to find the point at maximum distance from the minimum. As n is equal to m, the complexity reduces to O(n + n) = O(2n) O(n). Thus, there is a considerable improvement in the running time of the Bisecting Min Max Clustering Algorithm when compared with the Bisecting K-Means clustering Algorithm. The Bisecting Min–Max Algorithm begins by

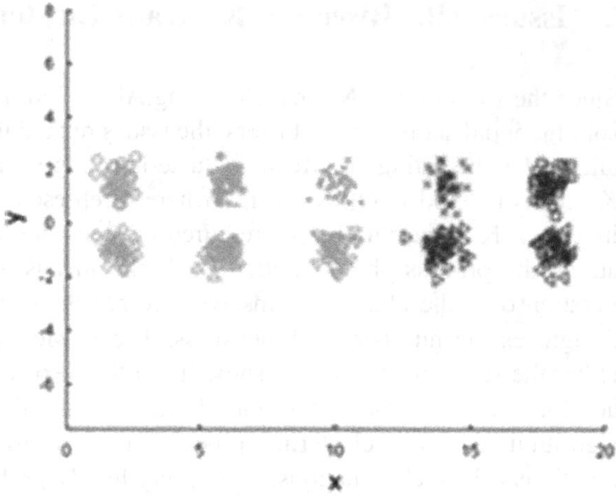

Fig. 1 Dataset

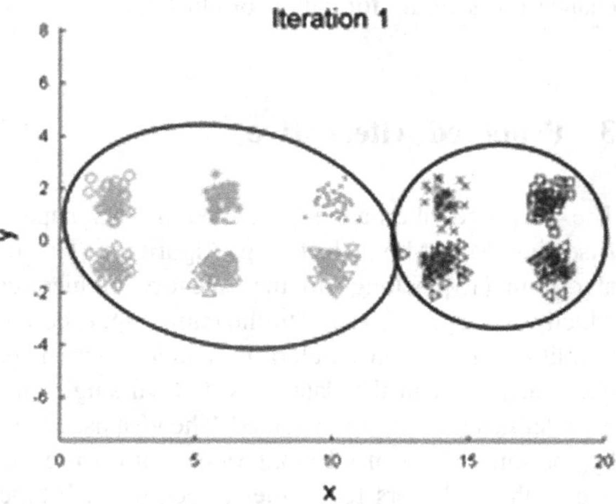

Fig. 2 Formation of two clusters with Min–Max Clustering

splitting the dataset shown in Fig. 1 initially into two clusters. The two clusters formed by Min–Max Clustering can be depicted as shown in Fig. 2. If the number of clusters prespecified by the user is K = 3, then the cluster having the larger Sum of Squared Error (SSE) among the two clusters is selected for further bisection and the selected cluster is again bisected into two clusters using Min–Max Clustering. Thus, the three required clusters are obtained as shown in Fig. 3. If the cluster count given by the user apriori is of K = T clusters then continue the procedure of finding the cluster with the larger value of SSE and split that cluster using Min Max Clustering until the user specific K = T clusters are obtained.

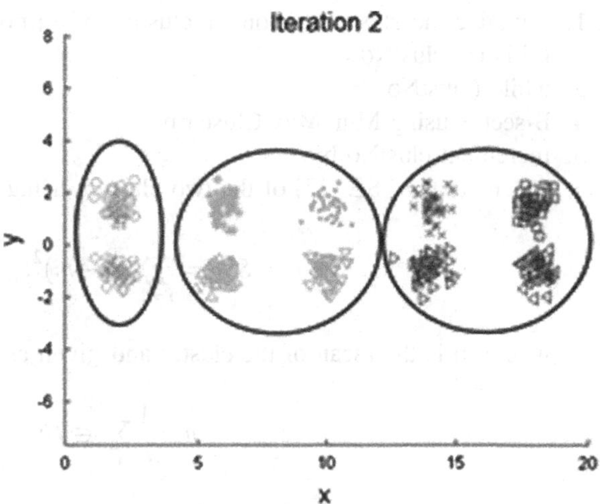

Fig. 3 K = 3 clusters after bisecting cluster1 which has larger SSE

4 The Min–Max Clustering Algorithm

Input: The number of user specified count of clusters K = T and a data repository having n data items or objects.

S = {S_1, S_2, S_3, S_4, …, S_n}

Output: A set of 2 clusters.

Step 1: First the minimum of the input set, min, is computed.
Step 2: Next, locate a point, max, which is further most from the min by using the Euclidean distance measure. The Euclidean distance [16] between the points a(p_1, q_1, r_1, s_1) and b(p_2, q_2, r_2, s_2) is given by:

$$d(a,b) = \sqrt{(p_1 - p_2)^2 + (q_1 - q_2)^2 + (r_1 - r_2)^2 + (s_1 - s_2)^2}$$

Step 3: Put the values which remain into clusters formed by these min and max points using the Euclidean distance measure.
Step 4: Output the result showing two clusters.

5 The Divisive Hierarchical Bisecting Min–Max Clustering Algorithm

Input: The number of user-specified count of clusters K = T and a data repository having n data items or objects.

S = {S_1, S_2, S_3, S_4, …, S_n}

Output: A set of K clusters.

1. Initialize the clusters to contain clusters of all points.
2. Initialize clustNo = 1;
3. while (clustNo != K)
4. Bisect S using Min–Max Clustering
5. Increment clustNo by 1
6. Calculate the SSE [17] of the two clusters using

$$SSE = \sum_{i=1}^{n} (xi - m)^2,]] >$$

where m is the mean of the cluster and given as

$$m = \frac{1}{n} \sum_{x} \in Ctx$$

7. if (SSE1 > SSE2)
8. assign Cluster1 as data S
9. else
10. assign Cluster2 as data S
11. end while.
12. Output K clusters.

5.1 *Implementation of the Proposed Algorithm*

Consider a user-specific cluster requirement of three clusters for any arbitrary input set.

Step I: Computing the minimum of the input set
The minimum of the input set is computed by pseudocode 1 given below.

```
for (loop identifier <= no. of data values){
euclidDist = distance (origin, dataItem);
    if(euclidDist <minim)
         minim = euclidDist;
}

Min=minim;
```

If the data size is considered as n, then the rate at which the above loop is growing is of the order of O(n).
Step II: Locating maximum value from the minimum

All values in the input set are checked with the minimum to locate a point which is further most from the minimum using the pseudocode 2 shown below.

```
for(loop identifier <= no. of data values){
     dist = distance(min, dataItem);
     if(dist>maxim)
           maxim=dist;
}
Max=maxim;
```

Considering the data size as n, the rate at which the above loop is growing is of the order of O(n).
Step III: Computing the mean of the bisected clusters
The pre-requisite to finding the larger of the SSEs from the bisected clusters is to find the mean of the clusters. This is done using the pseudocode 3 given below.

```
for(loop identifier <= size of cluster){
       sum + = clust[identifier];
}
Mean = sum / size of cluster;
```

The rate at which the above loop is growing is of the order of O(n) for the cluster of size n.
Step IV: Calculating the SSE of the bisected clusters
The most important step of identifying the cluster to bisect in the next iteration is determined by calculating the sum of squared error (SSE). The cluster having a larger value for the SSE is selected as the cluster for further bisection. Finally, this is implemented as shown below in pseudocode 4 and 5.
pseudocode 4:

```
for (loop identifier <= no. of points in cluster1){
         a =clust1[i];
         d1+=sse(m1, a);
}
for (loop identifier <= no. of points in cluster2){
         a =clust2[i];
         d2+=sse(m2, a);
}
sse(double x1,double y1){
  SSE=0;
  SSE=SSE+((x1-y1)*(x1-y1));
  return SSE;
}
```

pseudo-code 5:

```
if (SSE1 > SSE2){
        data = cluster1;
        size of data = size of cluster1;
}
else {
        data = cluster2;
        size of data = size of cluster2;
}
```

Step V: Output K clusters

```
if((dist(minim,p)<=dist(maxim,p))
       Assign p to Cluster 1
else
       Assign p to Cluster 2
```

where p = input value.

5.2 Experimental Results

The algorithm is tested with a 1D and 4D array input set of 1,000,000 points and also a 13D array input set of 500,000 points arbitrarily generated by a pseudocode such as the one shown below for finding three clusters. The pseudocode below specifically shows the generation of a random set of 100,000 one dimensional values.

```
int arrayRow = 100000;
int data[]= new int[arrayRow;
for(int i=0; i<arrayRow; i++){
  data[i]= (int)(Math.random()*10 +10);
}
```

On running the algorithm, the outputs obtained for Divisive Hierarchical Bisecting Min Max Clustering for a randomly generated 1D, 4D, and 13D array dataset for three clusters are shown in Figs. 4, 5 and 6, respectively.

The output clearly illustrates that the values of the input set, initially considered as one cluster is bisected using the Divisive Hierarchical clustering strategy with Min–Max clustering and then the sum of squared error (SSE) of the two clusters is calculated to determine which of the two clusters is chosen for further bisection. Then the chosen cluster is again bisected using Min–Max clustering to get the three

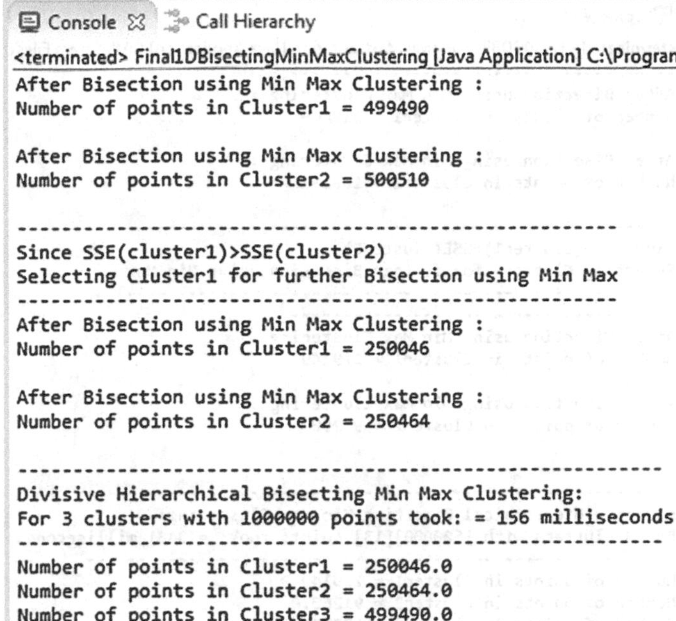

Fig. 4 Output of Divisive Hierarchical Bisecting Min–Max Clustering for 1D data

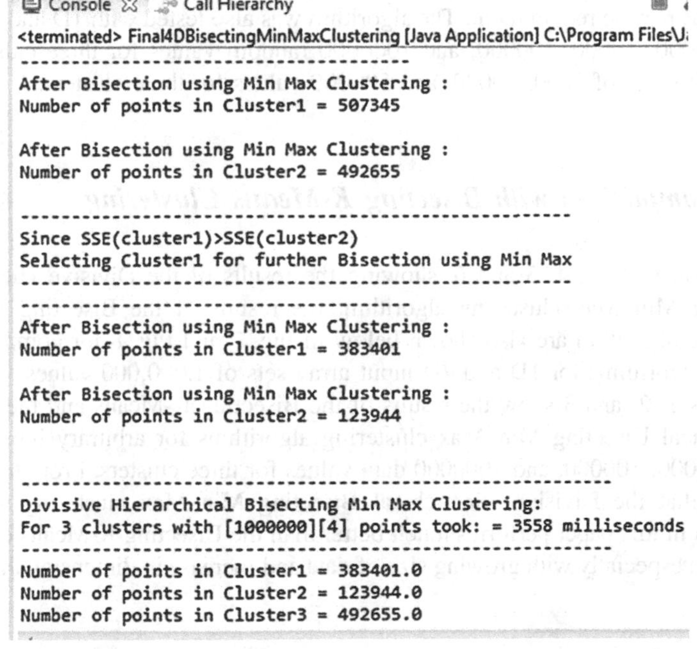

Fig. 5 Output of Divisive Hierarchical Bisecting Min–Max Clustering for 4D data

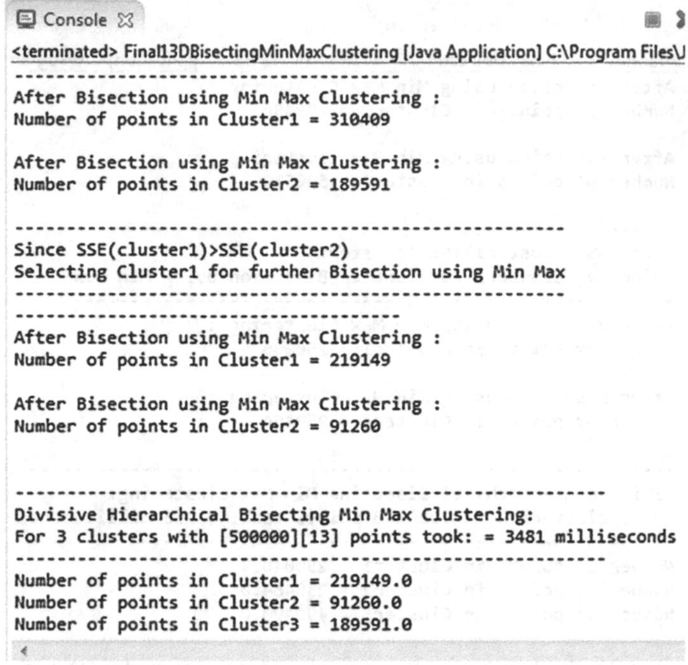

Fig. 6 Output of Divisive Hierarchical Bisecting Min–Max Clustering for 13D data

clusters as per the requirement. The algorithm was also tested with 1D and 4D input sets of 1000, 10000, 100000, and 1000000 random values for three clusters and with 13D input of 1000, 10000, and 500000 values for three clusters.

5.3 Comparison with Bisecting K-Means Clustering

In addition to Figs. 4, 5 and 6, showing the results of the Divisive Hierarchical Bisecting Min–Max clustering algorithm, the results of the Bisecting K-Means clustering algorithm are also shown below in Fig. 7 and Fig. 8 for comparison of the two algorithms for 1D and 4D input array sets of 1,000,000 values.

Tables 1, 2, and 3 show the results of the Bisecting K-Means and the Divisive Hierarchical Bisecting Min–Max clustering algorithms for arbitrary input sets of 1000, 10000, 100000, and 1000000 data values for three clusters. From the table it is clear that the Divisive Hierarchical Bisecting Min–Max clustering algorithm proposed in this paper performs much better than the Bisecting K-Means clustering algorithm especially with growing size of data and increase in dimensionality. It can

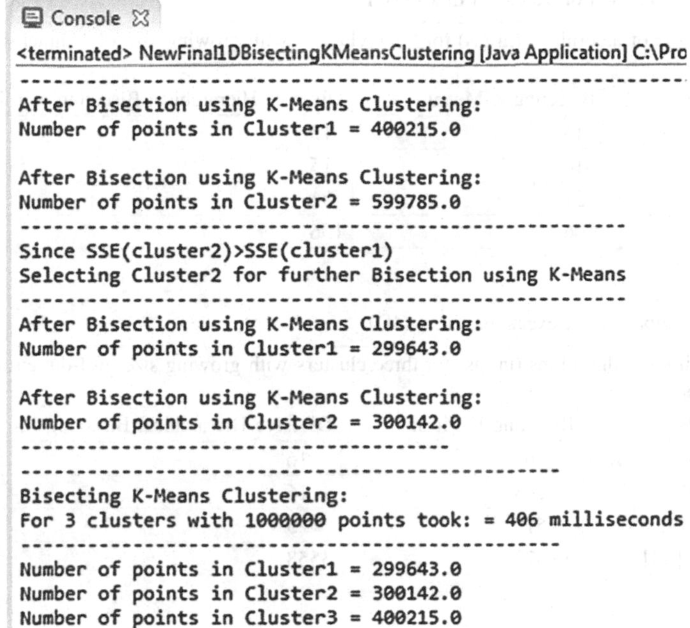

Fig. 7 Output of Bisecting K-Means Clustering for 1D data

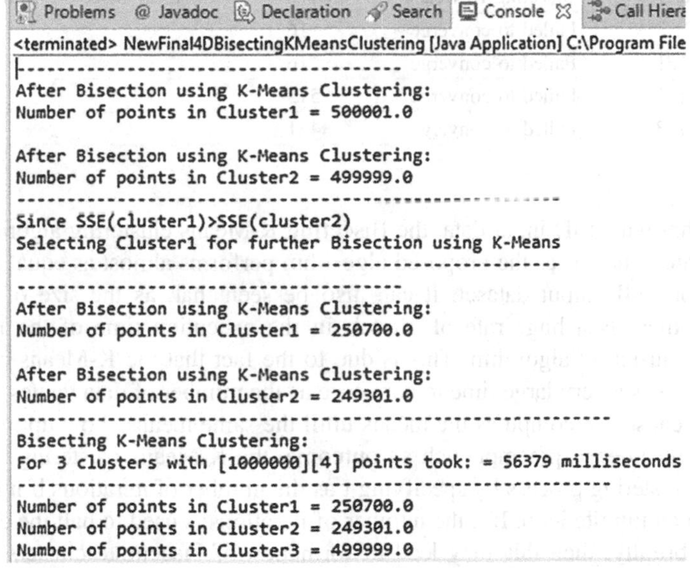

Fig. 8 Output of Bisecting K-Means Clustering for 4D data

Table 1 Comparison of execution times (1D)

Running time of algorithms (in ms) for three clusters with growing size of 1-dimensional data points		
Data points	Bisecting K-Means	Divisive Hierarchical Bisecting Min Max
1000	15	15
10,000	16	15
100,000	31	31
1,000,000	406	156

Table 2 Comparison of execution times (4D)

Running time of algorithms (in ms) for three clusters with growing size of 4-dimensional data points		
Data points	Bisecting K-Means	Divisive Hierarchical Bisecting Min Max
[1000] [4]	63	16
[10,000] [4]	203	16
[100,000] [4]	1578	125
[1,000,000] [4]	56,379	3558

Table 3 Comparison of execution times (13D)

Running time of algorithms (in ms) for three clusters with growing size of 13-dimensional data points		
Data points	Bisecting K-Means	Divisive Hierarchical Bisecting Min Max
[1000] [13]	Failed to converge	16
[10,000] [13]	Failed to converge	16
[100,000] [13]	Failed to converge	515
[500,000] [13]	Failed to converge	3481

be seen that with 13D input data, the Bisecting K-Means clustering algorithm fails to converge, whereas as the proposed algorithm performs almost as equally well as it does for a 4D input dataset. It can also be seen that, as the size of the data increases there is a huge rate of growth in the execution time of the Bisecting K-Means clustering algorithm. This is due to the fact that the K-Means clustering approach takes a very large time to converge as the number of data values increases mainly because it recomputes the means until the same means show up in the next iteration. In order to preempt such an outcome, the K-Means clustering algorithm halts the clustering process by specifying t as the number of iterations if it threatens to go into an infinite loop. If t, the number of iterations is used to halt the clustering process abruptly, then this may lead to formation of inaccurate clusters with the Bisecting K-Means clustering algorithm.

6 Conclusion

This paper provides an alternative to the Bisecting K-Means clustering algorithm by replacing the K-Means method for bisecting the selected cluster to split with the Min–Max Clustering algorithm. This algorithm convincingly outperforms the Bisecting K-means method as is evident from the comparison of the two algorithms for growing data size and increasing dimensionality. It is well known that the Bisecting K-Means is recognized as a superior performing algorithm than the K-Means and agglomerative hierarchical clustering algorithms. As such, it can be inferred that the Bisecting Min–Max Clustering algorithm which is a divisive hierarchical clustering algorithm will also perform much better than the K-Means and agglomerative clustering algorithms as it is already seen to be performing better than the Bisecting K-Means clustering algorithm.

Acknowledgments We are thankful to Ms. Valerie Menezes, Asst. Prof. in the Dept. of Computer Engineering of Agnel Institute of Technology and Design, Assagao, Goa, India, for her cooperation in testing and debugging the software which led to various improvements with respect to the simulation and comparison of the Bisecting K-Means Clustering algorithm with the Divisive Hierarchical Bisecting Min–Max Clustering algorithm.

References

1. Terence Johnson and Santosh Kumar Singh, "Improved Collinear Clustering in Lower Dimensions", Proceedings of 'Second International Conference on Emerging Research in Computing, Information, Communication and Applications', ERCICA 2014, ISBN 9789351072638, Vol III, pp 343–348, © Elsevier Publications 2014.
2. Sudipto Guha, Adam Meyerson, Nina Mishra, Rajiv Motwani and Liadon O'Callaghan, "Clustering Data Streams: Theory and Practice", IEEE transactions on Knowledge and Data engineering, Vol 15, No 3, pp 515–528, May/June 2003.
3. Terence Johnson, "Bisecting collinear clustering algorithm," International Journal of Computer Science Engineering and Information Technology Research, © TJPRC Pvt. Ltd, ISSN: 2249–6831, vol. 3, Issue 5, Dec. 2013, pp. 43–46.
4. Sunita Jahirabadkar and Parag Kulkarni, "SCAF-An efficient approach to classify subspace clustering," International Journal of Data Mining and Knowledge Management Process, vol. 3, no. 2, March 2013.
5. M. Ester, H.P. Kriegel, J. Sander, M. Wimmer and X. Xu Incremental Clustering for Mining in a Data Warehousing Environment, Proc. Very Large Databases Conf., 1998.
6. Charu C. Agarwal and Philip S. Yu, Redefining clustering for high dimensional applications, IEEE transactions on Knowledge and Data Engineering, Vol. 14, No. 2, March/April 2002.
7. A. Jain and R. Dubes, Algorithms for Clustering Data, New Jersey, Prentice Hall, 1998.
8. L. Kaufman and P. Rousseuw, Finding Groups in Data-An Introduction to Cluster Analysis. Wiley, 1990.
9. Willi Klosgen and Jan M. Zytkow, Hand Book of Data Mining and Knowledge Discovery, Oxford University Press, 2002.
10. Terence Johnson and Santosh Kumar Singh, "K-Strange Points Clustering Algorithm", Proceedings of 'Computational Intelligence in Data Mining' – Vol I, Print ISBN

978-81-322-2204-0, Online ISBN 978-81-322-2205-7, and 'Smart Innovation, Systems and Technologies' - Vol 31 2015, pp 415–425, © Springer Series 2014.

11. Martin Ester, Hans-Peter Kriegel, Xiaowei Xu. "A database interface for clustering in large spatial databases", Proceedings of 1st International Conference on Knowledge Discovery and Data Mining (KDD-95).
12. J.A.S. Almeida, L.M.S. Barbosa, A.A.C.C. Pais and S.J. Formosinho (2007), "Improving Hierarchical Cluster Analysis: A new method with outlier detection and automatic clustering", Chemo metrics and Intelligent Laboratory Systems, 87, pp. 208–217.
13. D. J. Hand, Heikki Mannila and Padhraic Smyth, Principles of Data Mining, MIT Press, 2001.
14. Terence Johnson and Santosh Kumar Singh, "Enhanced K Strange Points Clustering Algorithm", Proceedings of the '2nd International Research Conference on Emerging Information Technology and Engineering Solutions' EITES 2015, 978-1-4799-1838-6/15, pp 32–37, IEEE Computer Society Conference Publishing Services, © 2015 IEEE, DOI 10. 1109/EITES.2015.14.
15. Jiawei Han and Micheline Kamber, Data mining-concepts and techniques (San Francisco CA, USA, Morgan Kaufmann Publishers, 2001.
16. A. Alfakih, A. Khandani, and H. Wolkowicz, "Solving Euclidean distance matrix completion problems". Comput. Optim. Appl., 12(1999), pp. 13–30.
17. Pang-Ning Tan, Michael Steinbachand Vipin Kumar, Introduction to data mining (Addison Wesley, 2006).

Open Source Big Data Analytics Technique

Ishan Sharma, Rajeev Tiwari and Abhineet Anand

Abstract In this mobile computing and business era, a huge amount of data is generated, which is Big Data. Such a large data becomes unmanageable and cannot be used for analytics using traditional methods. A large number of fields and sectors, ranging from economic and business activities, involve with Big Data problems. Big Data analytics is extremely valuable to make decisions for increasing productivity in businesses, which gives us a lot of opportunities to make great progresses in many fields. So, this paper discusses approaches and environments for carrying out analytics for Big Data applications. It revolves around important areas of analytics, Big Data, tools, and data base used. A comparative study is done and tabulated on parameters like Data Base used, real-time analytics, size, etc. Then on open source technology Kibana, Elastic search and JASON Query, a big data analytics experimental setup is done. Analytics is done in many dimensions like domain counts, percentile gross margins, sector-wise count, etc. Their drawn results are recorded and reported in form of graphs.

Keywords Big data · Open source · Kibana · JASON · Analytics · Stocks · Profit margins · Counts

1 Introduction

In this era of mobile computing, a huge amount of data is generated by day to day mobile applications and by activities of our smart devices. So many platforms available for social media, social interaction, and social computing, the amount of

---

Ishan Sharma · Rajeev Tiwari (✉) · Abhineet Anand
Department of Computer Science, UPES, Dehradun, India
e-mail: rajeev.tiwari@ddn.upes.ac.in

Ishan Sharma
e-mail: i2s7h5@gmail.com

Abhineet Anand
e-mail: aanand@ddn.upes.ac.in

© Springer Science+Business Media Singapore 2017

S.C. Satapathy et al. (eds.), *Proceedings of the International Conference on Data Engineering and Communication Technology*, Advances in Intelligent Systems and Computing 468, DOI 10.1007/978-981-10-1675-2_58

data generated is so large in scales of PentaBytes, ExaBytes and even more. Such sizes of generated data cannot be processed by our traditional computing devices. The commercial and business data is very precious. It can sail a large giant to its destination or can sink it to wipe its existence. Such precious, sensitive and commercial information hidden in data can lead an organization to its success, if found and used correctly. Generating and storing this large amount of data is not a big problem with current technology development, but the processing and analysis of huge data to infer useful information is one major challenge. Organizations gain important insights from this data which help them to sustain in competitive market. Such huge collected data is Big Data.

Big Data [1] can be termed as the collection of large and complex datasets such that traditional data processing techniques are not able to process it. With its growing scope in three dimensions, i.e., volume, velocity, and variety, there is a necessity emerging as big data analytics which can be defined as the process of collecting, organizing, and analyzing large sets of datasets so as to discover and generate a useful pattern and other information. Whether it is a big enterprise, scientific laboratory, or any organization, all needs big data analytics to increase their work efficiency and improve operations. Chen [2] proposed a new scientific paradigm for data intensive scientific discovery (DISD), also known as one of big data problems. Retail giants need big data to enhance their business and to improve customer retention help with product development and gain a competitive advantage. Even the cloud computing is extensively utilized with Big Data problem. Agarwal [3] found that Cloud computing brings its own set of novel challenges like scalable data management that must be addressed to ensure the success of data management solutions in the cloud environment. This can be done by applying various analytic (open source or non-open source) software on (real time and offline) data stream to analyze and visualize it accordingly to meet the requirement. Singh [4] suggested that building the analytics-based solution on a particular platform is inherently dependent on the ability of the platform to adapt to increased data processing demands. Many applications like Jaspersoft, Tableau Desktop, Pentaho Big Data, Business Analytics and BI, FICO Big data analyzer and Skytree, etc., are some of the paid applications while ELK stack, Apache Hadoop, etc., are some of the application freely available in the market which can be used in big data analytics. These applications can be used to analyze, predict, and generate solutions to various complex business problems.

2 Literature Review

A number of research proposals are proposed in Big Data Analytics by researchers. A big issue explored was sending bid data on networks to reach to server for analytics. It was very time consuming and inefficient activity. So, a viable solution was developed in the form of Google File System (GFS) [5] and the MapReduce [6] paradigm in the early 2000s. Google File System and MapReduce distributes data across among the commodity servers so that the computational activities can be done where the data is stored. This method removes the data transferring overhead. Furthermore, methods for ensuring the resilience of the cluster and load balancing of processing were specified. GFS and MapReduce are the main components of Apache Hadoop project, which consist of two main architectural components: the Hadoop Distributed File System (HDFS) and Hadoop MapReduce (Apache Hadoop).

HDFS [7] is the main storage component of Hadoop with distributed architecture which consist of different nodes participating in a master–slave architecture. Files stored in HDFS are known as data nodes which are split into blocks and are replicated and distributed across different slave nodes on the cluster. Master node which is called as name node, maintains metadata in blocks in a file. In a cluster this metadata is stored, which is used to distribute workload to servers.

Duggal and Paul [8] suggested various approaches for the Big Data problems in hand through Map Reduce framework over Hadoop Distributed File System (HDFS) which is supported by Apache. Map Reduce technique minimizes, file indexing with mapping, sorting, shuffling and finally reducing. While HDFS is the data storage application to store the large amount of data.

Job Tracker [9] controls the jobs of MapReduce which is a software daemon. JobTracker service distributes MapReduce tasks to specific nodes in the cluster. A job is a full MapReduce program, including a complete execution of Map and Reduces tasks over a dataset. MapReduce is based on a master–slave architecture. It is a programming model for processing datasets in computer clusters on a large scale. It comprises of two basic functions, map() and reduce(). Customized processing logic can be implemented by specifying map() and reduce() function specifically. The map() function takes key-value pair as input pair and produces a list of intermediate key-value pairs. The MapReduce runtime system groups together all intermediate pairs based on the intermediate keys and passes them to reduce() function for producing the final results. The standard templates of MapReduce are given below:

```
Map (in key, in value) —>list(out key,intermediate value)
Reduce (out key,list(intermediate value)) – ->list(out value)
The signatures of map() and reduce() are as follows : map
(k1,v1) ! list(k2,v2)and reduce (k2,list(v2)) ! list(v2)
```

Bai [10] proposed a technique to query on (online and offline) log files. Size of the log files was up to 7 GB scale and 148928992c log events were generated. Tools and techniques used were Hbase, ElasticSearch and Flume to search, store and fetch the log files, respectively. Hbase is based on Hadoop and is a NOSQL Database. Flume is used to collect log events from end user and can be termed as a collector agent. Search time varies from 6 to 63 s while wanted log events varies from 25 to 4000.

In Big Data Analytics, Advanced Analytics in Oracle Database [11], there has been a use of oracle database for storage of data, For a data discovery, Oracle Endeca Information discovery has been used, there has been use of specialized

Business Intelligence (BI) tool as Oracle Exalytics and finally for decision management purposes Oracle Real time Decisions has used.

Cohen et al. [12] has proposed a new practice of data analysis, known as MAD skills. These skills are magnetic, agile, deep (MAD) where Magnetic means including all type of data sources and not restricting up to traditional data sources like tables, etc., agile means to be able to easily ingest, digest, produce, and adapt data at a rapid pace and last deep means to be able to query sophisticatedly and deeply in the data repository. The magnetic property is implemented using various storing mechanisms like Greenplum parallel database system. However using sophisticated statistical techniques, with a focus on density methods Deep property can be achieved and finally data parallel algorithms enabled agile feature of the technique.

In Starfish: A self-tuning system for Big Data Analytics by Herodotou et al. [13], a new concept of self-tuning model for data analytics has been used where tools like extensible MapReduce Execution Engine and HDFS have been used as storage systems. Flume and Scribe has been used for inputting the stream and SQL Client and other DB systems have been used as the Data outputs. For analytics a starfish system has been created which takes help of various applications like pig, hive, oozie, elastic MR, etc. It also proposed MADDER technique which is one step ahead of MAD technique discussed earlier. This technique includes all the concept of MAD, i.e., magnetic, agile, deep, and adds data lifecyle awareness, elasticity and robustness. So, these were few related works which does analytics based on various parameters and matrices. Their context comparison is given in Table 1.

Most of the reviewed technologies and techniques are licensed and can incur a huge cost to organization. However technology and technique used in our work is open source and freely available. Simultaneously it is equally effective as compared to other peer techniques. The technique used in this paper is quite easy to setup and is very efficient for quick analytics of data. While other techniques can require a professional knowledge of the domain. Our technique can be easily used by everyone. It can be applicable to real time as well as to off line data also.

Table 1 Comparison table of various existing schemes

Authors (reference) attributes	Jun Bai et al. [10]	Oracle Whitepaper [14]	Herodotou [11]	Our work
DB used in technique	Hbase	Oracle data	Ex-HDFS	Document based
Analytic tool	ElasticSearch	Oracle Exalytics	Starfish	Kibana
Data source type	Log Files	–	–	JSON
Size of data	7 GB	–	–	6756 records
Real time analytics of data	No	Yes	Yes	Yes
Offline analytics of data	Yes	Yes	Yes	Yes
Technology used	–	Oracle advanced analytics	MADDER	ElasticSearch
Ownership	Open source	Proprietary	Self-tuning	Open source

3 Proposed Work and Its Implementation

In this research we present a collection of applications, which are completely open source, for the analysis of big data. We present design, techniques and experience and we will apply it on the stock data in JSON format and analyze and visualize it for prediction and extract relevant information which is beneficial for the end user. This research work covers a new scope of analyzing the various trends in industry using Elastic search and Kibana utilizing the concepts of data warehousing and data mining and visualization. It analyzes the data using Elasticsearch for making the random table and template and compare the results along with visualization which is given in coming sections.

3.1 Tools Used

The tools used in the techniques are: 1. ElasticSearch [14] which is a search server. It is based on Lucene which is a open source information retrieval software library. 2. Kibana [15, 16] which is an open source data visualization plugin for ElasticSearch. It uses various types of visual structure like bar chart, pie chart, line and scatter plot on top of large volume of data and 3. Fiddler [17] which is a HTTP debugger proxy server application which captures HTTP and HTTPS traffic and logs it to review by the user.

4 Implementation Outline

Our technique consist of four major steps:

4.1 Flow Chart

See Fig. 1

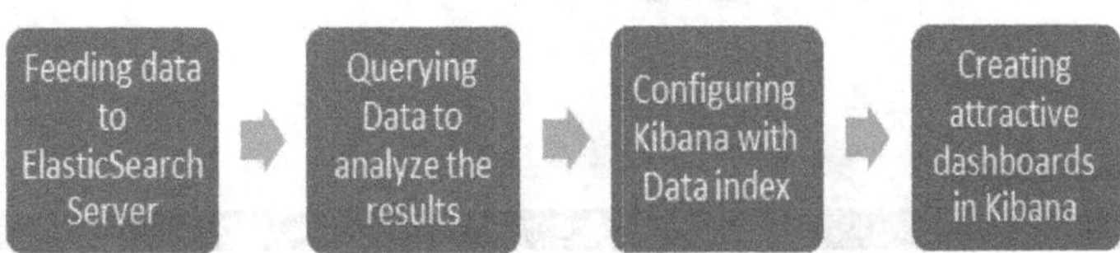

Fig. 1 Our work flow

5 Detailed Description

To begin with one should have ElasticSearch and Kibana on his computer. If one is using windows environment he can install ELK Stack along with Fiddler to feed data to ElasticSearch. Our technique starts with feeding data to ElasticSearch. Installing process will go beyond the scope of this paper. To start open ElasticSearch.bat file in bin folder of Elastic search and start feeding the data.

1. First go to the composer tab of fiddler, then fill the address of server and method as "POST". Then put the data in "Request Body" section and finally execute using execute button as shown in Fig. 2.
   Here stockteste is nodes name and infoe is index name.
2. Now after feeding data to ElasticSearch one can query the data also using JSON queries and feeding it to the ElasticSearch Server.
   First put the address in the address bar along with index name and add "\ search" after it. Then place the query in the "Request Body" section and finally execute it using execute button.
3. Now to visualize the data fed to the ElasticSearch Server start Kibana using Kibana.bat file in bin folder of Kibana and it will start on port 5601 (which is default). After logging into Kibana, it will ask to configure an index pattern. Now this index is same as the one which is provided at the time of feeding data to ElasticSearch. Kibana syncs with ElasticSearch to visualize the data provided.
4. After configuring an index pattern move to Visualize tab to select the kind of graphical interface you want to see and then select the source (to begin start from a new search) and then configure the metrics to achieve the result.

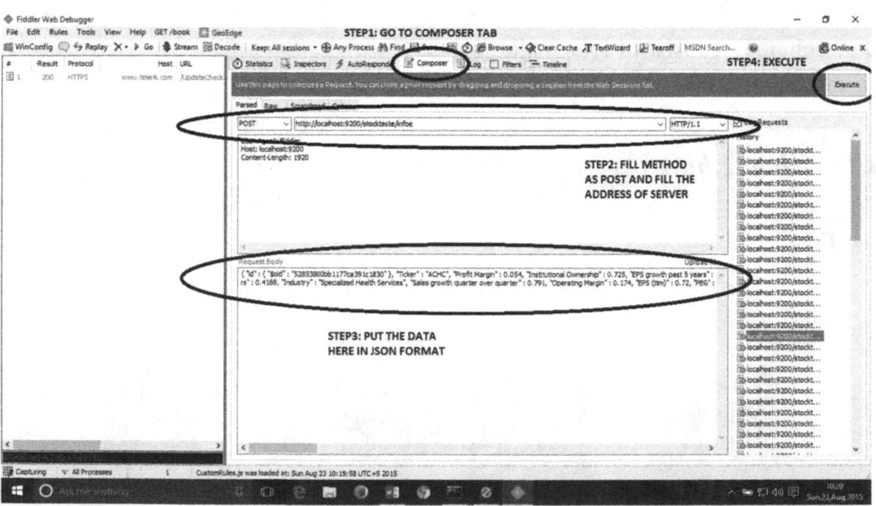

Fig. 2 Feeding data to ElasticSearch server using Fiddler

6 Analytics Results

From the stocks data we fed to elastic search, the visualizations on our result parameters are:

6.1 Domain Name Based Count

Figure 3 represents number of companies by their domain names (like inc., corp., ltd., etc.). Graph shown has stocks organizations on X axis and their count on Y axis. Which shows the top fifty domain names on basis of their counts.

6.2 Percentile Gross Margin of Stock Companies

Figure 4 represents gross margin of top 50 stock companies in tabular form.

6.3 Sector-Wise Gross Margin of Companies

Figure 5 depicts line graph of top 50 sector-wise companies along with their gross margins.

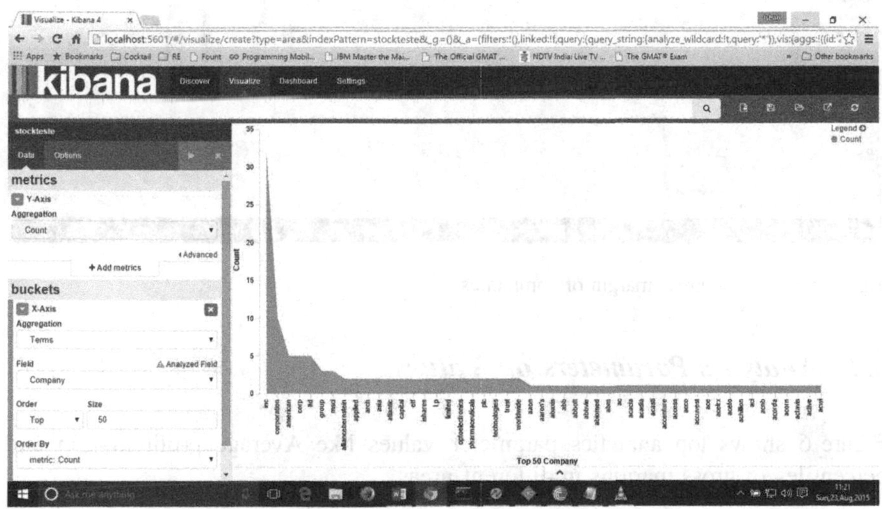

Fig. 3 Domain name based count

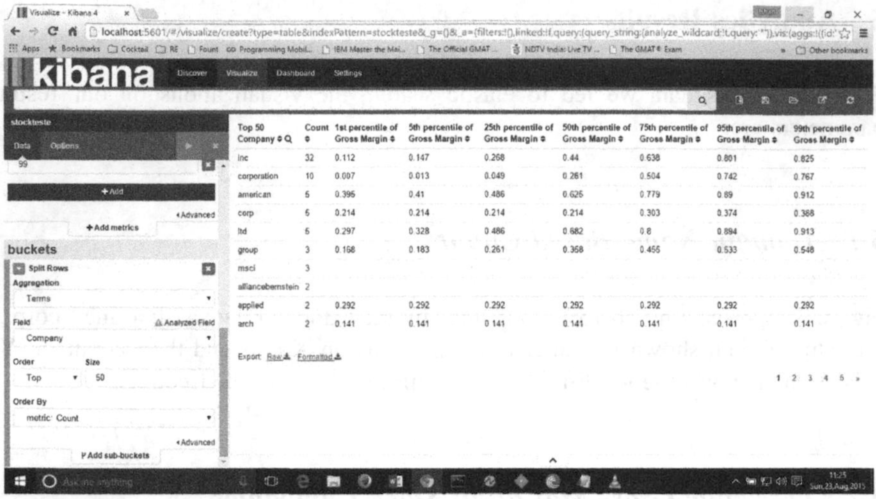

Fig. 4 Percentile gross margin of stock companies

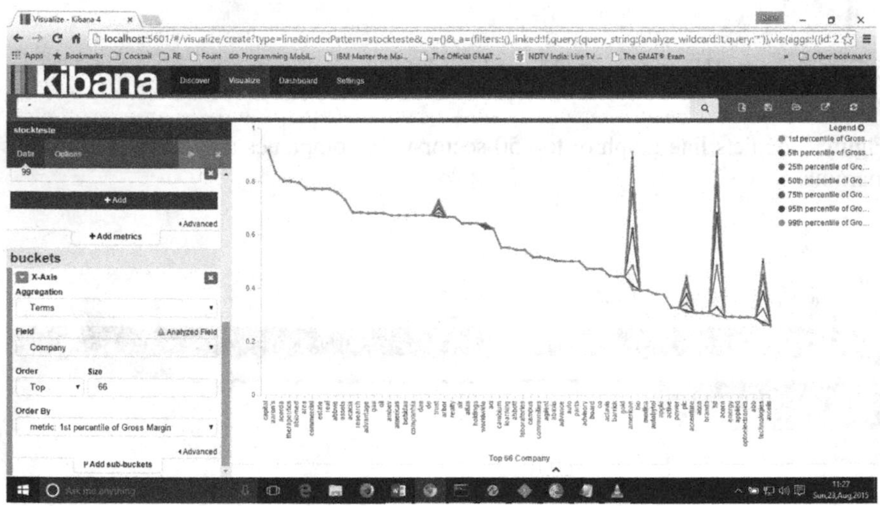

Fig. 5 Sector-wise gross margin of companies

6.4 Analytics Parameters on Values

Figure 6 shows top analytics parameter values like Average profit margin and percentiles of gross margins in different areas.

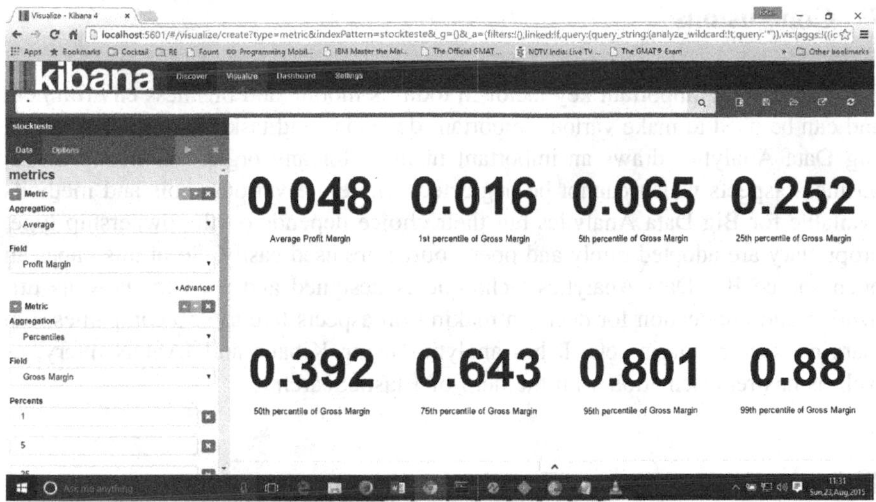

Fig. 6 Analytics parameters on values

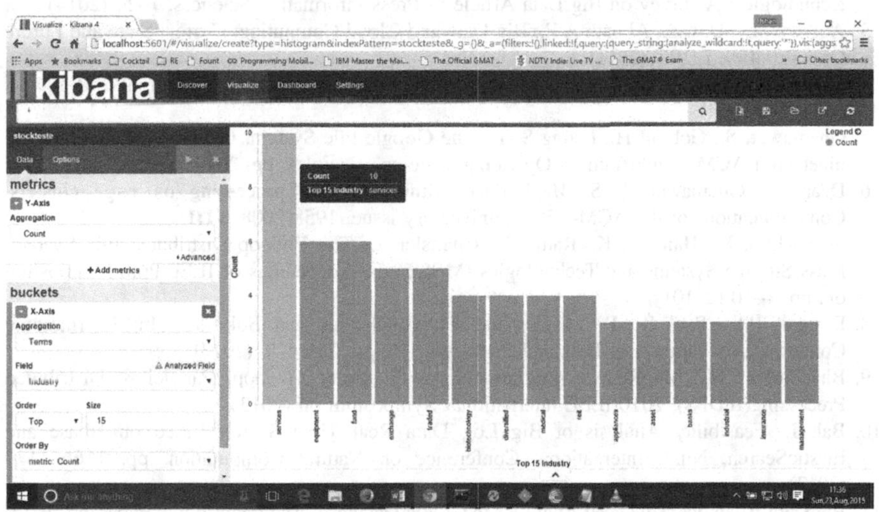

Fig. 7 Sector-wise companies count

6.5 Sector-Wise Companies Count

Figure 7 shows a bar chart of number of companies in different sectors.

7 Conclusions

Big Data is very important key factor in today's mobile and business environment and can be used to make various important decisions and task for major industries. Big Data Analytics draws an important insights for any organization on various business aspects which cannot be neglected. There are various tools and methods available for Big Data Analytics but their choice depends on the ownership type, proprietary are adopted rarely and open source are used easily. So in this paper an open source Big Data Analytics technique is designed and used to show its utilization and application for decision making on aspects like top 50 companies, top margins, sector margins etc. It has analytics using Kibana and JASON query, as well as has representation with the help of ElasticSearch.

References

1. Wikipedia the free encyclopedia, https://en.wikipedia.org/wiki/Bigdata.
2. Chen, C. L. P., Zhang, C.-Y.: Data-intensive applications, challenges, techniques and technologies: A survey on Big Data Article in Press Information Sciences, 1–24 (2014).
3. Agarwal, D., Das, S., Abbadi A.E.: Big Data and Cloud Computing: Current State and Future Opportunities. EDBT (2011).
4. Singh, D., Reddy, C. K.: A survey on platforms for big data analytics. In: Journal of Big Data, pp. 1–20 (2014).
5. Ghemawat, S., Gobioff H., Leung S.-T.: The Google File System. In: 03 Proceedings of the nineteenth ACM symposium on Operating systems principles. pp. 29–43 (2014).
6. Dean, J., Ghemawat, J. S. *M*apReduce: simplified data processing on large clusters, Communications of the ACM—50th anniversary issue: 1958 2008, **51:1**.
7. Shavachko, K., Hairong, K., Radia, S., Chansler R.: The Hadoop Distributed File System, Mass Storage Systems and Technologies (MSST). In Proceedings of IEEE 26th Symposium on. pp. 1–10 (2010).
8. Duggal, P.S., Paul S.: Big Data Analysis: Challenges and Solutions. In: International Conference on Cloud, Big Data and Trust, vol. 15, pp. 269–276 (2013).
9. Bhandarkar, M.: MapReduce programming with apache Hadoop. Parallel & Distributed Processing(IPDPS), 2010 IEEE International Symposium on (2013).
10. Bai, J.: Feasibility Analysis of Big Log Data Real Time Search Based on Hbase and ElasticSearch, Ninth International Conference on Natural Computation. pp. 1166–1170 (2013).
11. Big Data Analytics Advanced Analytics in Oracle Database, An Oracle White Paper. pp. 1–13. 2013.
12. Cohen, J., Dolan, B., Dunlap, M., Hellerstrin, J., M., Welton, C.: MAD Skills: New Analysis Practices for Big Data. In: Proceeding of the VLDB Endowment, vol. 2, no. 2, pp. 1481–1492 (2009).
13. Herodotou, H., Lim, H., Luo, G., Borisov, N., Dong, L., Cetin, F., B., Babu, S.: Starfish: A Self-tuning System for Big Data Analytics. In: 5th Biennial Conference on Innovative Data Systems Research (CDIR 11). pp. 261–273 (2011).
14. Wikipedia the free encyclopedia, https://en.wikipedia.org/wiki/Elasticsearch.
15. Wikipedia the free encyclopedia, https://en.wikipedia.org/wiki/Kibana.
16. Kibana: Explore, Visualize, Discover Data and Elastic, https://www.elastic.co/products/kibana.
17. Wikipedia the free encyclopedia, https://en.wikipedia.org/wiki/Fiddler (software).

Security Analysis of OnlineCabBooking Android Application

Nishant Grover, Jyotsna Saxena and Vikas Sihag

Abstract Android devices are not just phones, they are necessity for its users. Android's open-source nature leads to massive development of applications for end users to use which also create lots of vulnerable applications that are ready to be exploited. In this paper we discuss and analyse the OnlineCabBookings Android application which is used for booking cabs online through smartphones and present vulnerabilities in its implementation which make it a playground for attackers. With insight to the vulnerabilities of application under analysis, we proposed a security framework that can be used by other applications to authenticate users and conduct transactions in a secure manner.

Keywords Android OS · Mobile application analysis · WinPcap · OnlineCab-Booking · Wireshark · Security analysis

1 Introduction

In last few years, the rise of smartphones has proved that the mobile phones are not merely a medium of communication, the applications of smartphones are limitless [1]. With the growth of smartphone market, the world has contracted down to a hand-held device which can be easily accessed with a single touch, Smartphone. Android being an open-source OS, there is no scarcity of development tools, hence plenty of applications are created everyday all around the world for its users [1]. This greatly inspired people to use Android not only for usage, rather to tweak it as well. But the technology in smartphones does not safeguard itself as data transfered over network

Nishant Grover · Jyotsna Saxena · Vikas Sihag (✉)
Sardar Patel University of Police Security and Criminal Justice, Jodhpur, India
e-mail: vikas.sihag@policeuniversity.ac.in

Nishant Grover
e-mail: spu1411216@policeuniversity.ac.in

Jyotsna Saxena
e-mail: spu1411231@policeuniversity.ac.in

© Springer Science+Business Media Singapore 2017

S.C. Satapathy et al. (eds.), *Proceedings of the International Conference on Data Engineering and Communication Technology*, Advances in Intelligent Systems and Computing 468, DOI 10.1007/978-981-10-1675-2_59

has to be secured by application. This paper analyzes the requests and responses that are generated in OnlineCabBooking application and how they can be exploited for attacks. The paper is divided under five sections, first section provide a brief introduction to Android OS architecture, second section gives description of activity structure of OnlineCabBooking application and the tools used to analyze it [3]. Third section explains vulnerabilities found and the potential consequences on the end users and company. Fourth section presents the improvements that can be made to current application to make it secure and the final section presents conclusion.

2 Related Work

In analysis of Webview vulnerabilities of Android Application [5], two critical vulnerabilities were found. The first was Excess Authorization vulnerability which allowed injected javascript to execute Java code of Android applications. Through this vulnerability, more data than the intended Webview can be loaded which actual application does not want to display to user. The second vulnerability was File Based Cross Zone vulnerability which allowed malicious javascript through “file://” URL to read any file from the system. The manual analysis was done on Alive, AIM, Ad Libraries, and an automated analysis was conducted on 65 applications. It was found that 54 % applications were vulnerable to at least one of the vulnerabilities mentioned above. Multiple security vulnerabilities were found in FortiClient of Android and iOS [6]. FortiClient is an endpoint security suite, intended to provide all-in-one security solution. On decompiling the application, Hardcoded encryption keys were found that can be used to decrypt all the network traffic from the application.

Discuss [7] the Android security framework which describes how different application components communicate with each other. Android uses a plain permission assignment model label for applications to restrict access to resources and other applications of system, but for necessary and convince reasons, the designers of android have added several security refinements during its evolution. Application components are allowed to interact with components of other applications, known as Inter Component Communication (ICC). As security focal point, the Android middleware controls all ICC establishment for labels assigned to applications and their components. In simple terms, access is restricted by permission access label in form of string, which do not need to be unique itself. These bundles of permission labels are assigned by developers. When ICC is initiated, a reference monitor looks at permission labels of application, if component of target application with access permission label is in the collection, ICC allows establishment to proceed. If not found, the ICC establishment is denied even when the components are of same application. The permission labels are written in XML manifest file that is available with every application package. This defines the security policy of application which is set at the time of installation.

3 Android OS Architecture and Application Security

Android OS can be further grouped into four main layers, i.e., the kernel at bottom, Android libraries over the kernel, applications framework over the libraries, and applications on the top [2]. As previously mentioned the kernel, i.e., heart of the system is Linux. It was chosen since it has a proven track record in desktop systems and in many cases does not require drivers to be rewritten. The libraries that come with Android provide much of the graphics, data storage, and media capabilities. The main core of libraries is the Dalvik Virtual Machine (DVM), which is a part of this layer. The DVM is a bytecode interpreter which is highly optimized for executing on the mobile platform [4]. Embedded inside, the library layer is the Android runtime, which powers the applications. The applications framework provides all the major APIs that will include things like sharing of data, accessing the telecommunication system, receiving notifications, and more. The framework is the API that all applications will use to access the lowest level of the architecture. At the application layer, all of Androids software is written in Java, which is interpreted by the DVM. Even the most basic features such as the phone and the contacts application reside in this layer. This layer contains software written by the Android team as well as any third-party software that is installed on the device.

All applications over Android execute in an Application Sandbox. By default, only a few of system resources are allocated to the application installed. When executing, Android applications run on their own Linux process with a unique assigned ID. While running inside a sandbox, Android prevents applications to access system or its resources directly. Android applications ask for authorization, at the time of installation through its manifest file. Android's sandbox uphold inter-application isolation and authorizations to permit or deny an application access to the device resources such as file system, network status and access, sensors information and APIs in general. For this, Android utilizes Linux features such as process-level security, user and group IDs that are attached with the application and permissions of operations that an application have to access the system (Fig. 1).

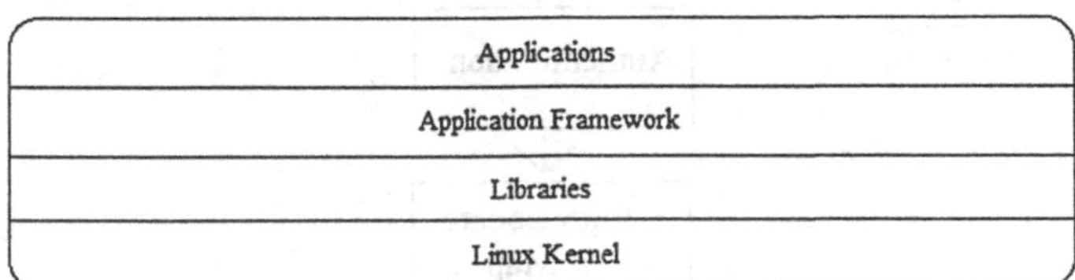

Fig. 1 Android OS architecture

4 Activity Structure of Application

OnlineCabBooking application allows you to book cabs with a few taps on your smartphone. It serves in more than 50 cities to its customers on regular basis. The activity structure of application is mentioned in Fig. 2 OnlineCabBooking first authenticates the user by mobile number and password. After authentication, Google Maps display user's location with the help of GPS. If GPS is not available or working, booking a cab is not possible. Now the available options for booking are to book by directly selecting the car through map or by the "Book Later" option. Other options such as View Past Bookings, Wallet and Recharge, Farechart, and Check Transactions are also provided in the left menu. All of the requests generated by the application are HTTP GET request means they can be easily replayed. The response generated by server is in form of JSON (JavaScript Object Notation). With the response and request in plain text, one could easily analyze the whole application by simply going through every available option and capturing all the requests and responses generated by application and server. For testing the application, Android Operating System Version 4.0.4 with Linux Kernel 3.0.8 is used as platform. Two tools namely Wireshark and Maryfi are used for complete traffic analysis of the application.

4.1 Wireshark

Wireshark is a network traffic analyzer that can capture live packets or from a pcap file on the disk. Wireshark supports more than 100 protocols and provide filters so only packets with desired rules and protocols can be viewed.

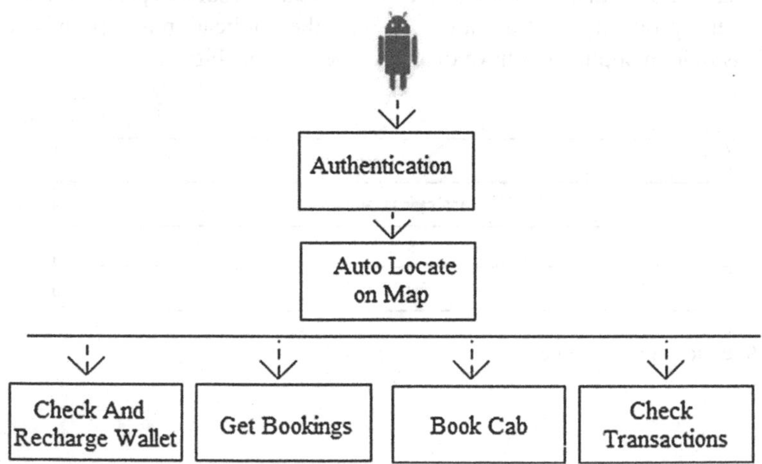

Fig. 2 Activity diagram

4.2 Maryfi

Maryfi creates a Wi-Fi hotspot on your laptop device with the help of Microsoft Virtual Adapter. The hotspot can be used as repeater which allows sharing of Internet connection for devices that are not in range of home Wi-Fi or it can act as a proxy for network traffic analysis of connected devices.

Connect mobile device to Maryfi hotspot that shared the Internet connection and Wireshark was used to capture and monitor all traffic on the virtual adapter.

5 Vulnerabilities Found

The vulnerabilities found in the application are mentioned below in Fig. 3

5.1 User Profile

User profile display various details of customer which is stored at OnlineCabBooking server.

```
:~$ http://iospush.OnlineCabBooking.com/../?$customer_number=9829
######
```

Fig. 3 Activity diagram

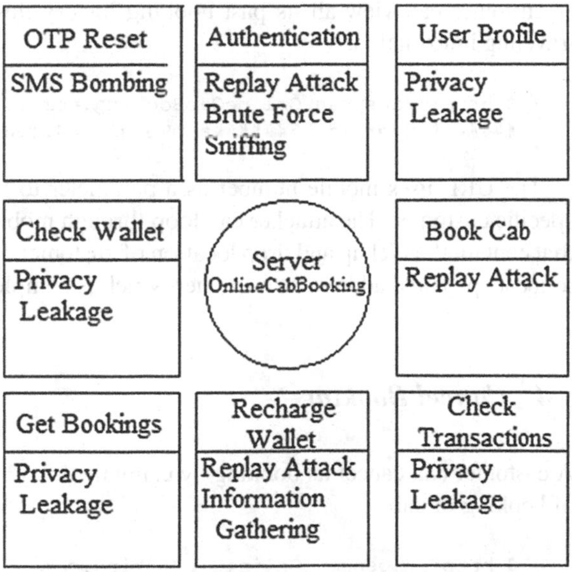

OTP Reset
SMS Bombing
Authentication
Replay Attack
Brute Force
Sniffing
User Profile
Privacy
Leakage
Check Wallet
Privacy
Leakage
Server
OnlineCabBooking
Book Cab
Replay Attack
Get Bookings
Privacy
Leakage
Recharge
Wallet
Replay Attack
Information
Gathering
Check
Transactions
Privacy
Leakage

The URL contains parameters such as mobile numbers, latitude and longitude for pickup and drop point, and time of pickup. The URL can be used directly to book cab at any specific time without opening up the application. The attacker can exploit the URL by creating fake requests for booking cabs on random mobile numbers. This will create chaos and it will not be possible to differentiate genuine requests from fake ones.

5.2 Cab Search

The URL shown below is used to search for nearby cabs available in the area.

```
:~$ http://iospush.OnlineCabBooking.com/.../?$longitude=72.9####
&$latitude=26.2#####&$customer_number=9829######&$user_id=15####
&$city=J*R&density=160&source=android&appVersion=4.1.8&appVersio
nCode=37
```

The URL contains parameters such as latitude and longitude of customer around which cabs would be located. The attacker can exploit the URL by creating fake requests and can cause Denial of Service attack that can block the service for customers.

5.3 Customer Bookings

A customer can view all its past booking history through which he can track his traveling information.

```
:~$ http://iospush.OnlineCabBooking.com/.../?$customer_number=98
29######&$user_id=15#####&appVersion=4.1.8&appVersionCode=37
```

The URL uses mobile number as a parameter to fetch the booking details of a specific customer. The attacker can loop through mobile numbers to catch response that contain the pickup and drop location of customers. This information can be used to locate physical address of customer which is complete privacy risk.

5.4 Cancel Booking

A customer can cancel its booking by using the job id that was allocated at the time of booking of cab.

```
:~$ http://iospush.OnlineCabBooking.com/.../?user_id=&$booking_
id=OCB-PP-C15#####&appVersion=4.1.8&cancellation_reason=Chose+in
```

```
correct+time&appVersion=4.1.8&appVersionCode=3
```

The URL passes the job id which is incremental in nature and is of 8 digits. The attacker can create a loop that can cancel any booking of a random customer. This will cause disruption of service among customers and the company.

5.5 View Transactions

A customer can view all the transactions that it has conducted to recharge OnlineCab-Booking wallet.

```
$~$ http://iospush.OnlineCabBooking.com/.../?$user_id=15#####
```

The URL passes the user id which display back all the transactions that are conducted by customer. The attacker can loop through ids to list of potential victims that do frequent high-valued transactions. This list can be utilized to conduct further attacks.

5.6 View Wallet

A customer can view the amount of credit he has in wallet to pay the cost of trip directly from his wallet.

```
$~$ http://iospush.OnlineCabBooking.com/.../?$user_id=15#####
```

The URL passes the user id which display the current amount of credit in the wallet of customer. The attacker can loop through id parameter to create a list of potential victims that haVE high credit in their wallet. This list can be used to attack authentication details of victims.

5.7 Recharge Wallet

A customer can recharge his OnlineCabBooking wallet through either Credt/Debit Card or by Netbanking options.

```
:~$ http://iospush.OnlineCabBooking.com/.../?user_id=&$booking_
id=OCB-http://iospush.OnlineCabBooking.com/.../?auth_token_id=g
hjkls***sdfg&user_id=15#####&name=fname%20lname&email=s**@gmail.
com&phone=9829######&amount=100.0&$txn_id=18####&txn_source_id=
1&txn_payment_mode_id=3&appVersion=4.1.8&appVersionCode=37
```

The URL passes the transaction id, mobile number, and user id which credits the wallet of a customer. On analysis, the request can be replayed with same transaction id to recharge the wallet again without spending the money from card again. Further, the attacker can change txn_id paramater to view transaction details of any random customer to that would reveal sensitive information of customer.

6 Security Framework

On observing the found vulnerabilities, it can be easily seen that the application lacks the basic session authentication and maintenance mechanism. After authentication of user, a limited time token can be allocated to user for its further requests in the application and this token would be sent along with each request. Furthermore, this token has to be verified for validity of user or whether it has expired or not before generating any response corresponding to request. The generation of token can be done from various details of user such as device id, Android version, user id and mobile number, etc. To further secure the token itself, instead of sending the varying length token with each request, a hash of token could be used instead, which server could map in active session database of users. This would prevent reverse engineering of token generation mechanism for attacks.

The above method will prevent any unauthenticated requests but in case the token (or hash) is sniffed out of the network, the attacker could replay the token to impersonate the user. To prevent it, lightweight encryption can be used to encrypt the requests between application and server that would bring confidentiality to user requests or publicly available HTTPS can be used to encrypt the application traffic.

7 Conclusion

New technologies always create new areas of concern for information security experts. Early adoption of such new technology is slow and it takes time for the development of effective security controls for it. Android is a victim of its own success. Over half of Android devices are vulnerable to known security flaws that can be exploited and can affect privacy of user. More and more applications are being developed on the daily basis which are none to less concerned about security vulnerabilities and are easily distributed in the market through Google Play Store. Our analysis of OnlineCabBooking application listS various vulnerabilities present in the application that can be used to perform attacks such as Session Hijacking, Replay Attacks, SMS Bombing, Privacy leakage leading Information Gathering for Social Engineering attacks and Network Sniffing. All these attacks are possible because of absence of session token. The proposed security framework for such applications can remove several vulnerabilities and will protect users from many attacks.

Google should set some baseline of security that applications should provide to their customers before publishing the application in store. OnlineCabBooking has over a million downloads in Google Play Store but still has critical vulnerabilities that are not paid any attention. Slow patching of vulnerabilities on mobile devices is a serious issue which needs to be in public notice. Just because the application is being used from a mobile phone, we cannot consider it secure. The huge smartphone market is already the playground for cyber criminals which affect individual users, it is up to the developers how they protect their users and stand against cyber attacks.

Note: For security of company, application name has been changed.

References

1. Li Ma, Lei Gu and Jin Wang: Research and Development of Mobile Application for Android Platform: International Journal of Multimedia and Ubiquitous Engineering, Vol. 9, No. 4 (2014), pp. 187–198 ed
2. Paul Michael Kilgo: Android OS: A robust, free, open-source operating system for mobile devices, pp. 2–3 (2010)
3. You Joung Ham et. al.: Android Mobile Application System Call Event Pattern Analysis for Determination of Malicious Attack. International Journal of Security and Its Applications, Vol. 8, No. 1 (2014), pp. 231–246
4. Lu Cheng: Analysis and Comparison with Android and iPhone Operating System, pp. 14–15 (2010)
5. Erika Chin and David Wagner: Bifocals: Analyzing WebView Vulnerabilities in Android Applications
6. Denis Andzakovic: FortiClient Mulitple Vulnerabilities
7. William Enck, Machigar Ongtang, Patrick McDaniel: Understanding the Android Security. Pennsylvania State University

Analyzing Security Schemes in Delay Tolerant Networks

Lalit Kulkarni, Debajyoti Mukhopadhyay and Jagdish Bakal

Abstract As the feature of delay tolerant networks (DTNs) is open, and intermittent, any node can participate in this type of network. Trusting a DTN node becomes a critical task for the routing algorithms. For the past few years, security techniques have evolved and a lot of research has been carried out in this area. Among these security techniques, trust-based, reputation-based systems, and incentive-based security schemes are discussed in this paper. These techniques are useful to detect selfish DTN nodes and noncooperative DTN nodes. These schemes for DTN nodes are the most common methods to detect malicious, selfish DTN nodes.

Keywords Delay tolerant networks · Selfish nodes · Opportunistic network

1 Introduction

Delay tolerant networks (DTNs) work in intermittent connectivity environment tolerating large delays and high loss rates [1, 2]. These features forces DTN to follow "store-and-forward" routing strategy. In this strategy if a DTN node, 'A' is connected with any other DTN nodc 'B' then, 'A' has to forward the data (generally referred to as a *bundle* in DTN) to 'B'. As shown in Fig. 1, the data will not be secured, if 'B' is a compromised node and 'A' will send the bundle to 'B' without checking B's trust value.

Lalit Kulkarni (✉)
GH Raisoni College of Engineering, Nagpur, India
e-mail: lvkulkarni@gmail.com

Debajyoti Mukhopadhyay
Maharashtra Institute of Technology, Pune, India
e-mail: debajyoti.mukhopadhyay@gmail.com

Jagdish Bakal
SSJ College of Engineering, Dombivali, India
e-mail: bakaljw@gmail.com

© Springer Science+Business Media Singapore 2017
S.C. Satapathy et al. (eds.), *Proceedings of the International Conference on Data Engineering and Communication Technology*, Advances in Intelligent Systems and Computing 468, DOI 10.1007/978-981-10-1675-2_60

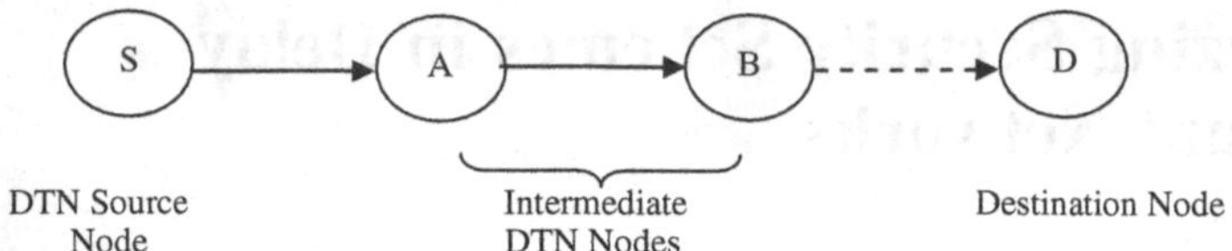

Fig. 1 Bundle forwarding in DTN

In such transactions, trusting DTN node 'B' is very important. If the node 'B' is a selfish DTN node, the packet delivery ratio will be affected significantly. A trusted authority [3] is one of the solutions for populating the trust information to all DTN nodes. Incentive-based schemes are also used to force selfish nodes to participate in data forwarding [4]. These approaches can be applied towards building relationships and credit history between DTN nodes through a central entity. There are many other schemes to monitor DTN nodes that will be discussed in detail in the following sections.

2 Reputation-Based and Trust-Based Schemes

The main purpose of Opportunistic Networks is to get help from various resources and forward the bundle. To achieve this goal the nodes need to contact the stranger nodes. The system that provides security architecture which believes that these stranger nodes are not trusted, with the help of trust and trust management in the opportunistic networks. The COmposite Trust and Trust Management in Opportunistic Networks (COTTON) model solution [5], software agents make the security decisions on the basis of trust values that are calculated by the authorities in the system, based on the Symantec Web Framework. The trust management system collects the node history and calculates the trust value using some mathematical model. Thus this system proposed a distributed trust model in which each mobile device specifies a group of users on trust values, just like a friend circle. This model calculates trust values depending on the peer review and trust which are based on reliable factors with the contact history. The system also suggests trust-based cryptography. The COTTON model which is suggested here, introduces with two concepts that are, the Helper Registry and the Helper Advertisement. The first one maintains the database of all identified services that formulates the degree of functional and nonfunctional characteristics that match with the user's requirements. The Helper Advertisement gives the detail about the capabilities and services provided by an entity. The most important advantage of the methods and approaches suggested in this scheme is to build up trust management in Opportunistic networks [5].

The common language which is used in the computerized world for communication, the Semantic Web, helps trust management systems by integrating it with the existing communication technologies. In this scheme, a novel architecture of Semantic Web solution, COTTON is presented and evaluated.

Some of the security issues which are faced in establishing the opportunistic network are basically due to heterogeneity, high mobility, and delay tolerance [5]. The system discusses about the selfishness issues which are observed in cooperation enforcement. The communicating nodes need more incentive to cooperate with each other for the better bundle delivery ratio. The system introduces the *'hot potato'* approach where nodes have to take a decision of accepting to receive a packet and whether to blindly pay for it or not. Second, according to the system, the context and content-based forwarding approaches present in the conventional communication are the challenging data privacy issues. Hence the author suggests the method called as Privacy preserving context-based forwarding. In context context-based forwarding the destination's context is private information that should be protected and not to be sent to the network cloud. In context-based forwarding, the intermediate nodes need to compare their context with the destination's context. Thus the system suggests identity-based encryption in which the intermediate nodes that are taking part in communication try to discover the various attributes that are specifically matching with the destination's attribute's value. In this process, the source node does not learn any additional information except these attributes to maintain destination's privacy. The solution is given by suggesting the method of secure content-based routing. In content-based communication, there is complete decoupling between the sender and the receiver. The intermediate nodes update their routing tables based on the interests advertised by the receivers. To overcome this limitation multiple layer commutative encryption is used. The idea is for the receiver to encrypt its receiver advertisement with 'r' layers corresponding to the 'r' next hops using 'r' different keys, and for the publishers to do the same with their published event. By rotating the encryption layers this solution enables content-based routing that preserves privacy of the receivers very efficiently [6].

Dynamic Trust Management for Delay Tolerant Networks [7] discusses various trust management protocols. Once the DTN is actually deployed, then the initial trust is calculated for the initial phase. For the second phase the trust is calculated for further communication by maintaining the history of all the nodes; if a new node is encountered in the network then the trust for that node will be calculated explicitly. All these aspects are covered in this scheme. This approach is verified on the simulator and the performance is measured against non-trust-based protocols and proved that the dynamic trust management helps to improve the message delivery in case of DTN. The heart of the scheme is Trust Management Protocol which calculates the trust factor of every node dynamically. There are two parts of the protocol i.e. how to update the trust of existing nodes and how to calculate and update the trust factor of newly encountered nodes. While trust maintenance is the main aspect of the algorithm, it also takes the parameters like location, selfishness, maliciousness and energy into consideration while calculating trust.

Iterative Trust and Reputation Management Mechanism (ITRM) explains trust management and adversary detection for DTNs [3, 8]. This scheme proposes an iterative scheme for trust management and malicious node detection. In the presence of Byzantine attackers, the proposed scheme provides high data availability with low latency. The algorithm also computes the reputation of the network nodes accurately in a short amount of time in the presence of attackers without any central authority. The algorithm's computational complexity is linear with the number of nodes in the network. The scheme is scalable and suitable for large scale implementations. The system also emphasizes on ITRM. As in every trust and reputation management mechanism, the system has two main goals: (A) Computing the service quality (reputation) of the peers who provide a service (SP) by using the feedbacks from the peers who used the service (raters) [3, 8], and (B) Determining the trustworthiness of the raters by analyzing their feedback about SPs. The system assumes the following two sets: (i) The set of SPs, and (ii) The set of service consumers.

In this mechanism, a malicious node may be misjudged as a rater node that can cause the malicious node detection algorithm to fail. This can be overcome by designing a robust and fool-proof algorithm to identify potential rater nodes that can help the judge-nodes to analyze the SP feedbacks.

3 Incentive-Based Schemes

The concept of giving incentives is introduced for forcing selfish nodes to participate in bundle forwarding [2, 9]. But it is not necessarily done only for the DTN networks. The cellular networks face traffic issues due to heavy overload with the data. The scheme of incentive framework makes use of the concept of DTN for the data traffic. The issue of large delay in DTN hampers the performance of cellular networks. Hence the concept of incentive comes in the picture. If the end user is satisfied with the services of the intermediate user then it indicates that the intermediate user has done a good job of forwarding the data in cellular network. In such cases, the incentive is given to the intermediate nodes that have forwarded the data in negligible amount of time and to the end users satisfaction.

Auction algorithms are used in this scheme to decide, which intermediate node will be selected for forwarding the packet to the end user [9]. The auction algorithm may introduce the delay for selecting the forwarding nodes from the available nodes, but the scheme claims that this delay is considerably small and can be neglected. There are two main steps of auction algorithm: First *Allocation Step* where the participating nodes are decided and pricing is also fixed. Then the actual delay is forecasted and the incentive which has to be given to the intermediate nodes is also decided. This scheme finds the balance between the traffic (that is offloaded) and the users' satisfaction value. The incentive framework is developed such that it forces users to control the delay tolerance in traffic offloading of cellular networks. The experimentation was done in the university campus. Initially a temporary network was created for a small duration and then later the same experiment was done for a longer duration. The threshold for incentive units was set i.e. the node was made to pay the threshold value for offloading one unit of traffic. If more traffic has to be offloaded then this threshold has to be increased. This threshold is called as reserve price. If the reserve price is high as discussed, the node is willing to pay more to intermediate forwarding nodes and more participation is encouraged with the help of this scheme.

The System in [10] suggests the different techniques such as Oblivious forwarding protocols, Context-based forwarding protocols which can be used so that the security about content of the message can be maintained. The system also suggests that there should be decentralized approach of trust in distributed systems which is very important and necessary aspect for the nodes which are taking part in communication. The layers of a networking architecture are beneficial as there is no dependency between the intermediate layers in respect of security.

Practical Incentive (Pi) protocol [2] works on layered coin architecture to enhance incentive scheme. Every node adds a layer after forwarding the packet to the next DTN node and gets the incentive from TA. The scheme is based on SMART and ITRM discussed in previous section and it is the extension of the trust development work.

4 Game-Theory-Based Schemes

MobiGame [11] is a reputation-based system which deals with user-centric approach that calculates the dynamic reputation. It also uses Bayesian Equilibrium of game theory that makes this approach secure. In addition to reduce the overheads on the DTN nodes it uses short signature & batch verification techniques. Every DTN node has to register to Offline System Manager (OSM) to own the public key encryption. This public key will be used to encrypt the bundles. The system rewards the bundle forwarding node using Reward Depreciation Function (RDF). RDF takes the current time as input and calculates remaining rewards for bundle forwarding. Threat model of a system works on selfish and malicious nodes.

This algorithm works with *spray and wait* routing protocol [11]. The approach is based on trust probability value of a node in DTN. It works in two steps. In the first iteration (called as *period*), source node or forwarding node sprays the bundle to some selected nodes with specific trust probability. And the scheme only creates few copies of the bundles for these selected nodes, i.e., the minimum number copies (*Lmin*). To calculate this value following parameters are considered.

dr—desired delivery rate	n—number of attackers
td—time constant	Pn trust probability value of a node
λ—rate of exponentially distributed time between DTN nodes	

If the bundle is not delivered in the *desired delivery time*, the second iteration starts. In this period the partially trusted nodes are contacted and aggressive *spray and wait* routing is done to increase the delivery rate. This technique is secured in the first period. But due to insufficient resources or intermittent problem of DTN, the second less secured period is used to deliver the bundles.

Monitoring misbehavior detection becomes very difficult because of lack of evidence and witness in DTN [12]. The probabilistic misbehavior detection system uses inspection game theory-base approach which deals with the inspector node called as trusted authority (TA). TA verifies the trust value of the inspectee. TA decides some rules for the inspectee and do the partial verification or partial inspection. If the inspectee violates the rules, the TA can punish this misbehavior with the help of partial verification scheme to complete verification scheme. TA checks and keeps the track of forwarding history from all the nodes and assigns a reputation to each node. As the reputation of the node increases, it gets benefit from the network, like verification by TA with lower probability value. The bad reputation nodes will be checked with high probability value. The system uses packet loss ratio (PLR) to identify the misbehaving node. The system also indicates that the node can induce blackhole attack if the PLR value is 1, and it induces a gray hole attack if the PLR value is between 0 and 1. It considers a node as a normal DTN node only if PLR is 0 (which is a high possibility in DTN). The problem with this approach is that it works only with First Contact Routing protocol. It also works only with homogeneous transmission range of nodes as it assumes the finite transmission range of every node and that is static. It may work with MaxProp and ProPHET [12] routing protocols if there exists any two nodes are in same time slot at any given time, which is another difficult assumption in case of DTN.

Table 1 Security schemes with detection and performance parameters

Sr no	Security scheme	Selfish node detection	Misbehavior detection	Contact history	Incentive based scheme	Game theory based	Improvement in PDR
1	iTrust (2014)	√		√		√	√
2	Wormhole Detection (2010)		√				
3	Dynamic Trust Mgmt (2014)	√		√		√	√
4	Pi (2010)	√	√		√		
5	SMART (2009)	√			√		
6	MobiGame (2011)			√		√	
7	ITRM (2010)		√				
8	Probabilistic Detection (2014)		√				√
9	Adversary Detection (2010)		√				
10	Mitigating Misbehavior (2012)			√			√

5 Summary

As discussed in previous sections there are various schemes for security in DTN. All these schemes are analyzed in following table with the factors depending on techniques they use, and their performance outcomes. Out of these schemes, the game- theory-based approaches have good results for misbehavior detection, but the overhead on the network is high. These schemes also introduce more delay in the detection process. The schemes which use the contact history for detection of misbehavior nature of a node perform well with few assumptions (Table 1).

6 Future Work

The distributed approach is very necessary for the security schemes as in the central approach one node is overloaded and, if the central node is not connected with the intermittent connectivity then the trust establishment process gets delayed further resulting in more delay. If this trust calculation is distributed among various trusted authorities then the trust establishment process can be finished faster and secure packet delivery ratio will be increased with improved delay. This kind of approach may have few issues like, if the trusted authorities gets compromised then whole process of trust establishment will be compromised. To overcome this issue a new incentive-based scheme for trusted authorities has to be developed.

References

1. H. Zhu, X. Lin, R. Lu, Y. Fan, and X. Shen, "SMART: A Secure Multilayer Credit-Based Incentive Scheme for Delay-Tolerant Networks", IEEE TRANSACTIONS ON VEHICULAR TECHNOLOGY, VOL. 58, NO. 8, OCTOBER 2009.
2. R. Lu, H. Zhu, X. Lin, Y. Fan, and X. Shen, "Pi: A Practical Incentive Protocol for Delay Tolerant Networks" IEEE Transactions on Wireless Communications, vol. 9, no. 4, April 2010.
3. E. Ayday and F. Fekri, "Iterative Trust and Reputation Management Using Belief Propagation," IEEE Transactions On Dependable and Secure Computing, vol. 9, no. 3, pp. 375–386, May. 2012.
4. H. Zhu, X. Lin, R. Lu, Y. Fan, and X. Shen, "SMART: A Secure Multilayer Credit-Based Incentive Scheme for Delay-Tolerant Networks", IEEE Transactions On Vehicular Technology, vol. 58, no. 8, October 2009.
5. E. Tamez, I. Woungang and L. Lilien, "Trust Management in Opportunistic Networks: A Semantic Web Approach", IEEE conference on, Security and Trust and the Management of e-Business, vol. 13 no. 5 pp 235–238. 2009.
6. A. Shikfa, M. Onen and R. Molva, "Privacy in Content-Based Opportunistic Networks", IEEE conference on Advanced Information Networking and Applications Workshops, vol. 12 no. 3, pp 832–837.
7. I. Chen, F. Bao, M Chang and J. Cho, "Dynamic Trust Management for Delay Tolerant Networks and Its Application to Secure Routing", IEEE TRANSACTIONS ON PARALLEL AND DISTRIBUTED SYSTEMS, VOL. 25, NO. 5, MAY 2014 pp 1200–1210.

8. E. Ayday, H. Lee, and F. Fekri, "Trust Management and Adversary Detection for Delay Tolerant Networks," Military Communication Conference, pp. 1788–1793, Oct. 2010.
9. A. Shikfa, M. Onen and R. Molva, "Privacy in Context-Based and Epidemic Forwarding", IEEE International Symposium on World of Wireless, Mobile and Multimedia Networks & Workshops, 2009, WoWMoM 2009.
10. X. Zhuo, W. Gao, G. Cao, and S. Hua "An Incentive Framework for Cellular Traffic Offloading", IEEE TRANSACTIONS ON MOBILE COMPUTING, VOL. 13, NO. 3, MARCH 2014 pp. 541–555.
11. L. Wei, Z. Cao and H. Zhu "MobiGame: A User-Centric Reputation based Incentive Protocol for Delay/Disruption Tolerant Networks", IEEE Globecom 2011 Proceedings.
12. H. Zhu, S. Du, Z. Gao, M. Dong and Z. Cao "A Probabilistic Misbehavior Detection Scheme toward Efficient Trust Establishment in Delay-Tolerant Networks", IEEE TRANSACTIONS ON PARALLEL AND DISTRIBUTED SYSTEMS, VOL. 25, NO. 1, JANUARY 2014.

An Approach to Design an IoT Service for Business—Domain Specific Web Search

Debajyoti Mukhopadhyay and Sachin Kulkarni

Abstract The efforts are made to extract the most relevant results while searching for the specific query in huge web storage available in the WWW. Majority of the Web pages are written in such a way that the crawler finds it difficult to extract any specific domain. The concept of ontology is used to find domain specific results. The purpose of this work is to design a crawler which crawls through the domain specific web pages in the web according to the stated ontologies. To minimize the bias on accessing the highly relevant web links in a deep web, we are proposing an intelligent crawling mechanism. The intelligent crawler makes use of effective page ranking algorithm which leads the user to the most relevant results in a specific domain. The proposed Internet of Things (IoT) service will incorporate this mechanism to an effective extent.

Keywords WWW · Crawler · Deep web · Ontology · Ranking · IoT

1 Introduction

In this paper, we have put forth the concept of intelligent crawler which extracts the most relevant results for user query. This crawler crawls through the deep web for a specific domain of the query and displays the results to the user. The new ranking algorithm is applied on to the links extracted by crawler and are then given to the user. The deep web [1] contains the content which is much larger than surface web. In the approach we are following, we first will process the user query. The query processing module identifies the specific business domain. The next module will

Debajyoti Mukhopadhyay · Sachin Kulkarni (✉)
Department of Information Technology, Maharashtra Institute of Technology, Pune, India
e-mail: sachin8030@gmail.com
URL: http://www.mitpune.com

Debajyoti Mukhopadhyay
e-mail: debajyoti.mukhopadhyay@gmail.com

© Springer Science+Business Media Singapore 2017

S.C. Satapathy et al. (eds.), *Proceedings of the International Conference on Data Engineering and Communication Technology*, Advances in Intelligent Systems and Computing 468, DOI 10.1007/978-981-10-1675-2_61

harvest the surface web as well as deep web for finding out the highly related web links. On these links, page ranking algorithm is applied and the leads will be interfaced with the user. This work has the basic intention to provide help for the business drivers to analyze the current market requirements and to predict their clients precisely so as to increase the growth.

2 Domain Specific Web Search

Traditional crawler parses through all the web pages in breadth first search (BFS) strategy [2] which is not an efficient way. A focused crawler [3] crawls through domain specific pages only. The pages which are not related to the particular domain are not considered. In our approach we crawl through the web and add web pages to the database, which are related to a specific domain (i.e., a specific ontology) and discard web pages which are not related to the domain.

Identification of a domain of a particular web page [4, 5] which fits into the desired domain is done in a way mentioned in Fig. 1.

Calculation of Relevance Score

First of all activities, we need to assign weights to each term in the document [6, 7]. This weight assignment should be done as per the importance of that term in that particular domain. We have to initialize the weight table with the values of term frequency. The table will be having the details for the domain, specific term of that domain, and its weight. Once the table gets initialized, the relevance of the web

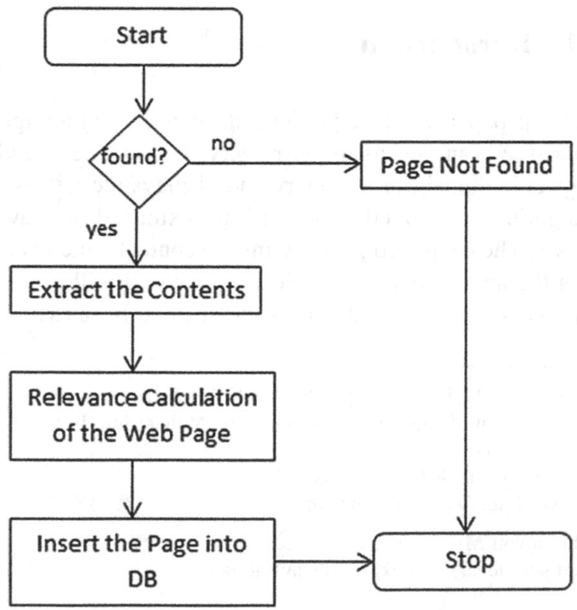

Fig. 1 Web page domain checking

Table 1 Domain specific weight table

Domain	Sr. no.	Term	Weight
Pharmacy	1	Rantac D	5
	2	Antacid	3
	3	Tablet	2
	4	Medicine	1

page is calculated. Count of the term frequency gets multiplied with the weight of that term and the result is stored in one particular variable. In this way, all the terms of the document and their respective weights are then clubbed together and this cumulative addition will be treated as the relevance score of the web page. The example of the domain specific weight table is as follows:

Consider

W = web page, T = set of tokens, F = set of Term Frequency, wt = weight, R_Score = relevance score

When web page W is given as input to tokenize, the web page gets parsed and tokens are identified. Let, T is the set of generated tokens. Term frequency of each token will be determined and stored into set F (Table 1).

$$T = \{t1, t2, t3, t4 \ldots, tn\}$$
$$F = \{f1, f2, f3, f4 \ldots .fn\}$$

Now we have to calculate relevance score as follows:

$$R_{Score} = \sum_{i=1}^{n} (wt_i{}^{*}t_i)f_i$$

Crawling for Multiple Domains

The crawler crawls through multiple domains simultaneously for different ontologies [8]. The main challenge in this traversing is to identify the irrelevant web pages and keep them aside. The general scenario of multi-domain crawling can be understood from Fig. 2.

The major challenge in the structure given in Fig. 2 is to crawl to the relevant but unreachable web pages. This problem will be resolved by the technique to harvest the deep web pages present in the WWW. It is a big challenge to detect the deep web pages because they are distributed sparsely. Also their position keeps changing. To solve this problem, focused crawlers emerged. Form-Focused Crawlers (FFC) [3] is designed with parameters like link, page & form classifiers. But the link classifiers are used to only guess the distance to the relevant page which is hard to estimate, which decreases the efficiency of focused crawler.

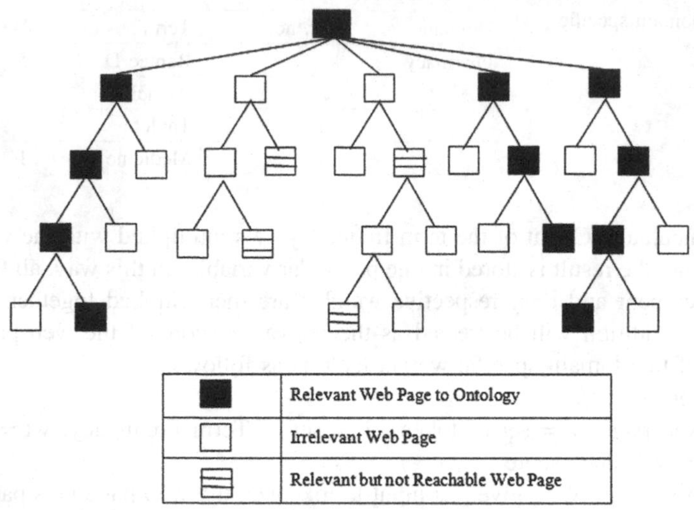

Fig. 2 Multi-domain crawling

3 Proposed Approach

3.1 *Target Business*

The system will be designed in a robust way meaning it covers all the business segments related to the stakeholders. The target of our system has the business segments like Web, IT & Software, Design & Multimedia, Data Entry & Admin, Engineering & Architecture, Sales & Marketing, HR, Legal & Finance, Translations, Customer Services, etc.

3.2 *Intelligence*

Custom Crawling is carried out by intelligent crawler which traverses through social media, forums, wiki, press releases, classifieds, corporate blogs, industry publications, job sites and market places.

Personal Insights will be collected as a part of raw data in user acquisition process. This will help us understand "who is our next client".

Verification Technology is needed to make the system authenticated.

3.3 User Acquisition & Business Model

User Acquisition goes through phases like Launch, Growth & Maturity. From FB pages, forums, LinkedIn Profiles we collect the user information.

Business Model basically has three stages. Traffic generation phase will generate & collect data about users from the emails, social media, etc. Trial Registration helps us Tele-calling the users for readvertisement. The final stage is Paid Registrations. In this stage, sales training, support, subscription, long term discount, etc. things are covered.

In short, our system answers the very important question of business i.e. who is my next client? For this, we are proposing this approach which uses domain specific intelligent crawler.

4 Ranking

4.1 Hyperlink Structure of the Web

A set of pages in the web may be modeled as nodes in a directed graph [8]. The edges between nodes represent links between pages. A graph of a simple 8-page web is depicted in Fig. 3 below. The directed edge from node two to node three signifies that page two links to page three. However, page three does not link to page two, so there is no edge from node three to node two.

Hubs and Authorities
The concept of 'authorities' and 'hubs' uses the link structure of the web to identify pages that can be considered to be the most 'authoritative'. In this algorithm a page is considered to be an *authority* if it is referenced by many pages relevant to that subject. Pages that are linked to such related authorities are called as *hubs*. The concept is very similar to PageRank [9] algorithm but here the emphasis is not just to count the references from all pages in WWW but only from pages relevant to the

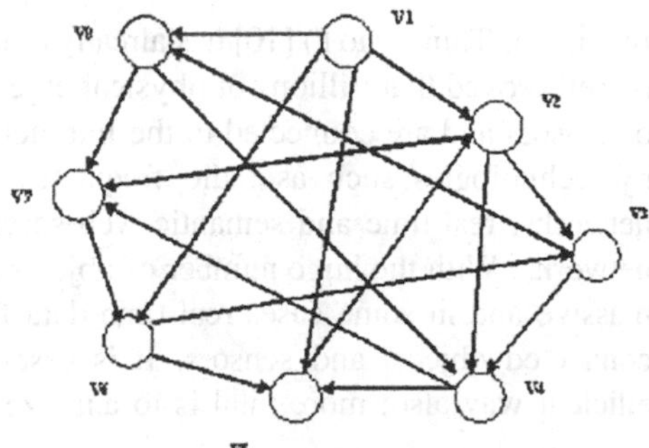

Fig. 3 Hyperlink graph

topic. To sum up, ***Authority*** is the page that provides important, trustworthy information on a given subject and ***Hub*** is the page that contains links to authorities.

4.2 Algorithm

```
1. Initialize authority weights to 1.
2. Repeat until the weights converge.
3. For every hub  i ε H
```

$$Hb = \left(\sum_{j \in O(i)} \frac{1}{I(j)} At_j{}^{3} \right)^{\frac{1}{3}}$$

```
4. For every authority  i ε A
```

$$At_i = \left(\sum_{j \in I(i)} \frac{1}{O(j)} Hb_j{}^{3} \right)^{\frac{1}{3}}$$

```
5. Normalize the weights.
6. According to the normalize weights rank the web pages
```

Where, **At** = Authority weight, **Hb** = Hub weight, **O** = Number of outgoing links, **I** = Number of incoming links

5 Internet of Things Corner

Internet of Things (IoT) [10] has already emerged as the next big thing in Internet. It is well proved that trillions of physical objects are outfitted with numerous amounts of sensors and are connected to the Internet via access networks which are enabled by technologies such as radio frequency identification (RFID), wireless sensor networks, real-time and semantic web services, etc. IoT is basically a network of networks. With the huge number of objects and sensors connected to the Internet, a massive and in some cases real-time data flow will be automatically produced by connected objects and sensors. It is essential to collect correct raw data in an efficient way also; more vital is to analyze that raw data to extract more valuable

Table 2 Features

Sr. no.	Features
1	Query input panel
2	Business domain identification
3	Deep web harvesting interface
4	Relevant document collection
5	Application of ranking algorithm
6	Displaying the leads to user

information such as correlations among objects and services which is nothing but the Internet of services. In our proposed approach, an Android application will be developed which inculcates all the system features discussed in Table 2. This mobile application facilitates the user access via access points/hotspots. Business domain specific search engine will be deployed on the device connected to the hotspot and then it will be operated.

6 Conclusions

The approaches of searching in WWW are evolving constantly but the growth rate of these improvements is not that fast. The search engine proposed in this work effectively handles the challenges of relevant not-reachable web pages. In this approach, we have proposed a prototype that uses multiple ontologies to perform multiple domains specific crawling for businesses to identify their clients in the market. The intelligent crawler parses the relevant not-reachable web pages specified in given ontologies. The ranking algorithm proposed here is a novel approach based on the scaling factor. Our design is scalable and can be easily adopted by other enterprises as their tool to identify the clients.

References

1. Feng Zhao, Jingyu Zhou, Chang Nie, Heqing Huang, Hai Jin, "SmartCrawler: A Two-stage Crawler for Efficiently Harvesting Deep-Web Interfaces", DOI 10.1109/TSC.2015.2414931, IEEE Transactions on Services Computing.
2. Leiserson, Charles E.; Schardl, Tao B., "A Work-Efficient Parallel Breadth-First Search Algorithm (or How to Cope with the Nondeterminism of Reducers)", ACM Symp. on Parallelism in Algorithms and Architectures, 2010.
3. Luciano Barbosa and Juliana Freire, "An adaptive crawler for locating hidden-web entry points", Proceedings of the 16th international conference on World Wide Web, pages 441–450, ACM, 2007.
4. Debajyoti Mukhopadhyay, Anirban Kundu, Sukanta Sinha, "Introducing Dynamic Ranking on Web-Pages Based on Multiple Ontology Supported Domains", T. Janowski, H. Mohanty, and E. Estevez (Eds.): ICDCIT 2010, LNCS 5966, pp. 104–109, 2010. © Springer-Verlag Berlin Heidelberg 2010.

5. Ruma Dutta, Anirban Kundu, Rana Dattagupta, Debajyoti Mukhopadhyay, “Mining the web with hierarchical crawlers – a resource sharing based crawling approach”, International Journal of Intelligent Information and Database Systems, Vol. 3, No. 1, 2009.
6. Debajyoti Mukhopadhyay, Arup Biswas, Sukanta Sinha, “A New Approach to Design Domain Specific Ontology Based Web Crawler”, 10th International Conference on Information Technology, DOI:10.1109/ICIT.2007.20.
7. Debajyoti Mukhopadhyay, Sukanta Sinha, “A New Approach to Design Graph Based Search Engine for Multiple Domains Using Different Ontologies”, International Conference on Information Technology, DOI 10.1109/ICIT.2008.46.
8. M. Najork, H. Zaragoza and M. Taylor, “HITS on the Web: How does it compare?” in the *proceedings of ACM,* pp. 471–478, 2007.
9. L. Page, S. Brin, R. Motwani, and T. Winograd. “The PageRank citation ranking: Bringing order to the web”, Technical report, Stanford Digital Library Technologies Project.
10. Guest Editorial Special Issue on Internet of Things (IoT): Architecture, Protocols and Services, IEEE Sensors Journal, Vol. 13, No. 10, October 2013.

Weather Forecasting Using ANN with Error Backpropagation Algorithm

Meera Narvekar, Priyanca Fargose and Debajyoti Mukhopadhyay

Abstract Weather forecasting is very necessary for making strategic plans and executing daily tasks in multiple application areas like airport, agriculture, electricity, water reservoir, tourism, and daily life. Accurate prediction of weather conditions is the challenge in front of meteorological departments. Many techniques such as data mining, k-means, decision tree, etc. have been used to predict the weather. The technique presented in the paper uses artificial neural network with error back propagation. Here we generate weather forecast for next day. The results are obtained in acceptable range of accuracy. ANN suits best classification technique that works on complex systems like Weather forecasting, in which datasets are nonlinear. The forecasting model used is limited to only certain geographical area which comes under Mumbai, India.

Keywords Artificial neural network · Error backpropagation algorithm · Weather forecasting · ANN

1 Introduction

Weather forecasting is a probabilistic system, where the future outcome is difficult to state in advance. The weather depends on multiple climatic conditions like temperature, air pressure, humidity, wind flow speed and direction, cloud height and density, and rainfall. Change in any of this parameter can have more or less impact on the future weather.

Meera Narvekar (✉) · Priyanca Fargose
D. J. Sanghvi College of Engineering, Mumbai, India
e-mail: narvekarmeera@gmail.com

Priyanca Fargose
e-mail: priyancafargose@gmail.com

Debajyoti Mukhopadhyay
Maharashtra Institute of Technology, Pune, India
e-mail: debajyoti.mukhopadhyay@mitpune.edu.in

© Springer Science+Business Media Singapore 2017

S.C. Satapathy et al. (eds.), *Proceedings of the International Conference on Data Engineering and Communication Technology*, Advances in Intelligent Systems and Computing 468, DOI 10.1007/978-981-10-1675-2_62

The weather forecasting is a real time system used in several applications such as airport, agriculture, electricity, water reservoir, and tourism. Thus, accuracy of the weather forecasts is of prime importance for decision-making.

Many classification techniques are such as support vector machine, K-nearest neighbor, naive Bayes classifier, decision tree, learning vector quantization, artificial neural network (ANN), etc., are used to classify whether in classes such as sunny, partly cloudy, thunderstorm etc. Out of these ANN uses artificial intelligence to understand the complex and nonlinear input data and generates the frequent patterns of similar samples together which associate with one class of output parameter.

Artificial neural network with error back propagation technique is used for classification of weather conditions. The success of the model depends on multiple criteria like set of input data, structure of the model, activation function of each layer, number of input and output parameters, number of hidden layers and neurons in each layer. A proper arrangement of all these parameters will give more accurate results. The meteorological centers record the observations in terms of different Input parameters. Each of the parameter is different than the other one and has different set of values. Handling such nonlinear data is a very challenging task. ANN is a special classification technique that deals efficiently with the nonlinear data.

This paper uses huge dataset of seven years with an interval of 15 min for training and testing the ANN Model. The tansig and purelin activation functions are used in hidden layers. Use of Back propagation technique will help the model to learn and adapt the necessary adjustments in weight and bias value depending on the error generated in each epoch. Thus, after finishing with the training process, the model will generate more accurate forecasts. The model is created considering only a specific geographical area. Scope of the model is limited to weather forecasting in Mumbai city only.

Following section covers previous work done on Weather Forecasting using multiple classification models.

2 Related Work

Many researchers have used different classification techniques for multiple purposes. Lai et al. [1] describes a methodology to short-term temperature and rainfall forecasting over the east coast of China based on some necessary data preprocessing technique and the dynamic weighted time delay neural networks (DWTDNN), in which each neuron in the input layer is scaled by a weighting function that captures the temporal dynamics of the biological task. Results have shown that neural networks with as few as a single hidden layer and an arbitrary bounded and nonconstant activation function can approximate weather forecast prediction for rainfall and temperature. The next step aims to show that it can also predict wind direction.

Shuxia [2] forecasted power consumption using a nonlinear network model between power consumption and the affected factors were obtained through training the relative data of power consumption from 1980 to 2005 in China. The result shows that the BP neural network with immune algorithm is more accurate than optimized neural network by genetic algorithm. Marzi et al. [3] studies solutions for forecasting option prices in a volatile financial market. It reviews a mathematical model based on traditional Black-Scholes parametric solution. The neural network was trained using the Levenberg–Marquardt backpropagation algorithm. When the NNHybrid model was compared against both the Black-Scholes model and the NNSimp model, a number of observations were made. The first outcome was that the NNHybrid's Mean Absolute percentage error for the entire data set was better than both the Black-Scholes and NNSimp models. Gao et al. [4] performs sales forecasting for accurate and speedy results to help e-commerce companies solve all the uncertainty associated with demand and supply and reduce inventory cost. The model is proposed with extreme learning machine. By determining the coefficient of ELM, that is the number of hidden nodes, and conducting the experiments with different variables, the proposed method does have a certain effect for reducing the forecasting RMSE and improving the accuracy and speed. Harshani et al. in [5] has used Ensemble neural network where, finite numbers of ANN are trained for the same task and their results are combined. The performance is compared with Back propagation neural network (BPN), radial basis function network (RBFN), and general regression neural network GRNN). The results show that, ENN model predicts rainfall more accurately than individual BPN, RBFN and GRNN. The weakness of the technique proposed in paper is that it predicts only the rainfall. Hayati et al. in [6] has used ANN for "one day ahead" (the next day) temperature prediction. The training was done on 65 % of dataset. Testing performed on 35 % of dataset. Network has one hidden layer and sigmoid transfer function is used in that layer. Number of neurons and epochs were decided by trial and error method. The strengths of this technique are, the structure used has minimum prediction error, good performance and reasonable prediction accuracy. The model is limited to forecast only temperature. Hall et al. [7] developed a neural network and initially two years data was used consisting nineteen variables. In this application, two networks are created, a QPF network for predicting amount of precipitation and a PoP network for probability or confidence in the forecast. This technique improves precipitation forecast dramatically, particularly for applications requiring accurate results. The main drawback of this model is, it focused only on precipitation forecasting.

Above survey shows that ANN is most suitable to predict the weather conditions.

3 Techniques Used

This section covers the detailed information about the techniques used in this model to forecast weather. First section covers ANN and second section describes error backpropagation algorithm.

3.1 Artificial Neural Network

An artificial neuron is a computational model inspired in the natural neurons. For each input received, the neuron is activated and emits a signal. This signal might activate other neurons. ANN basically consists of inputs, which are multiplied by weight and then computed by a mathematical function which determines the activation of the neuron. Depending on the weights, the computation of the neuron will be different [8]. By adjusting the weights of an artificial neuron we can obtain the output we want for specific inputs.

3.2 Error Backpropagation Algorithm

The backpropagation algorithm uses supervised learning, which means that we provide the algorithm with examples of the inputs and outputs we want the network to compute, and then the error is calculated as difference between actual and expected results. The idea of the backpropagation algorithm is to reduce this error, until the ANN learns the training data. The training begins with random weights, and the goal is to adjust them so that the error will be minimal [8].

4 System Design

The system is designed with proper arrangement of sub modules working together to achieve the final output, that is the weather forecast. Following Fig. 1 shows the block diagram for system; followed by the detailed explanation on working of system.

4.1 Data Preprocessing and Cleaning

The acquired data needs to be preprocessed. The input data is revised in various ways for each input parameter as mentioned further. The values for each parameter

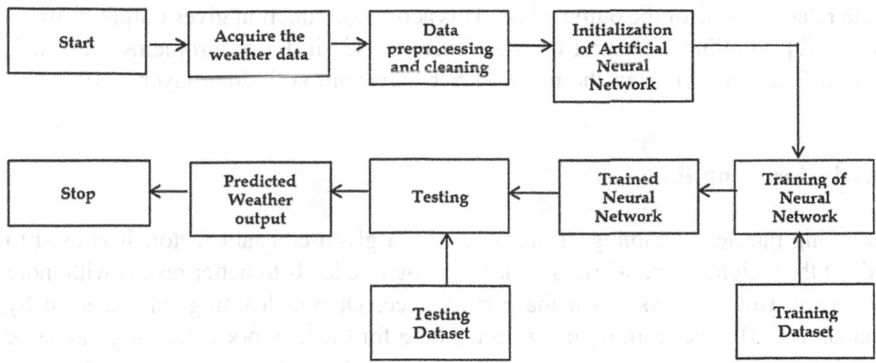

Fig. 1 Block diagram of the system

are checked for allowed range of values in that parameter and the outliers which do not fit in the range are removed. Records with missing data are removed. Also the nonnumeric data is converted to numerical acceptable data.

4.2 *Initialization of Artificial Neural Network*

4.2.1 Architecture

The architecture of neural network is of 28-10-10-5 type. That means, the network has 28 input parameters, two hidden layers each with 10 neurons and 5 output neurons.

4.2.2 Weights and Bias

The values of bias and weights connecting individual neurons in all layers are assigned randomly using a random function with a seed value. Change in seed value for random function also has a great impact on the efficiency of system. Following Table 3.2 shows the summary of results achieved using different seed values for the random function used for assigning values to weights.

4.2.3 Activation Function

The transfer functions normally used in multilayer neural network are tansig, logsig, and purelin. Tansig is a neural transfer function used to calculate a layer's output from its net input. It is used because meteorological observed data is nonlinear in nature, from which it is difficult to generate the frequent patterns of similar type of data related to one of the output class. This activation function gives output from −1 to +1 [9]. Purelin is a neural transfer function used in fitting problems. This activation function is used in the middle layer between two hidden layers [10].

4.2.4 Learning Rate

Learning rate leads training of network with a given constant factor. It is used to adjust the weights in network. Smaller learning rate leads to better results with more accuracy. But is slows down the performance. Default learning rate selected by matlab is 0.01. The learning rate selected here for the network is 0.5. It gives same accuracy as given by value 0.1, but the number of epoch increases when learning rate is kept low.

4.2.5 Number of Epochs

In neural networks numerous iterations(epochs) of the input and output processes are carried out by adjusting internal weights and bias till desired target values and output is acceptable. Maximum numbers of epochs set for the network are 1000. Less number of epochs such as 500 may limit the training process in case of high amount of dataset.

4.3 Training Neural Network

The model learns by itself using backpropagation algorithm where, the achieved output is compared with desired output and error is calculated from which the weights of individual links to neurons are updated. Thus, for the system, 70 % of the input dataset is given as input to the network. Also, the network is provided with target output for the given input dataset for applying backpropagation algorithm with supervised learning.

4.4 Testing Neural Network

Once training of neural network with training data is done, the model needs to be tested and validated. Thus, remaining 30 % of the input dataset is used for testing.

Here, the target output is not provided to the network. The network produces the output for the given dataset of testing, and then the error is calculated using the mean squared error (MSE) formula. The MSE measures the average of the squares of errors that is, the difference between the actual output and the predicted output of the model.

The estimated MSE of the predictor is given as follows:

$$MSE = \frac{1}{N}\sum_{I=1}^{N}(\hat{Y}_i - Y_i)^2$$

where n is number of epochs, $\hat{Y}$ is the vector of n predictions (output), and Y is vector of the true values. Lesser the MSE value of the model, more accurate the results are.

The network is now ready to forecast the weather. Thus a random unseen sample with all 28 input parameters from same area is given to the network, and its output is displayed as the weather forecast for the next day.

5 Performance and Results

The performance of the system is calculated using mean squared error, accuracy of output and time taken for execution. We divided the data collected from various sources and divided them into six datasets. Each set varied in number of parameters, size of data. This was done to identify the most influential parameters to determine accuracy of the system.

The above Table 1 shows that, the dataset number six is giving better performance (mean squared error as 0.0786 and accuracy as 90.6351 %) as compared to other datasets. It has 8 years data with 90 daily readings which makes more than 2.5 lakhs number of records. Thus, it is observed that more number of records and more number of input parameters used for training increases the performance of the system.

Above Table 2 shows the testing performed on various combinations of ANN structure that are further compared with respect to their performance. Here, the fourth case with 2 hidden layers with 10 neurons each is more feasible; thus it is selected to be used in the system.

The values of bias and weights connecting individual neurons in all layers are assigned randomly using a random function with a seed value. Change in seed value for random function also has a great impact on the efficiency of system.

Table 1 Performance comparison with respect to various dataset

Number of attributes	Number of years	Daily observations	Total records	Mean squared error	Accuracy
28	1.5	08	4320	0.6552	54.5362
05	1.5	08	4320	2.8047	38.5942
28	0.8 summer	08	1920	0.8323	21.2029
18	05	01	1825	1.6999	47.6349
18	20	01	7300	0.2461	67.2139
07	08	90	262800	0.0786	90.6351

Table 2 Performance comparison for different number of hidden layer and neurons

Number of hidden layers	Number of neurons	Mean squared error	Accuracy
1	5	0.0850	90.3640
1	10	0.0802	91.9435
2	5–5	0.0827	90.0585
2	10–10	0.0757	92.0249
2	15–15	0.0759	91.3160

Following Table 3 shows the summary of results achieved using different seed values for the random function used for assigning values to weights. The most accurate seed value is chosen after comparing the efficiency with other seed values. In following Table 4, the seed value 491218382 has more accuracy and less MSE.

Table 4 shows the comparison among all combinations of activation functions. The best combination which gives lowest MSE and highest accuracy is selected. This system uses tansig function at hidden layer and purelin function at output layer between the layers because of their special functionality.

Learning rate leads training of network with a given constant factor. It is used to adjust the weights in network. Smaller learning rate leads to better results with more accuracy. But is slows down the performance.

Table 3 Summary of results achieved using various seed values used for random function

Seed value for random function	Number of epochs	Running time (hh:mm:ss)	Mean squared error	Accuracy
101	10	00:00:57	12.756	31.7144
4912	290	01:11:11	0.0773	90.6023
456457	215	00:19:13	0.0779	91.4178
491218382	127	00:11:54	0.0774	90.6351
738745463523	11	00:01:09	6.3146	37.7523

Table 4 Performance comparison for various combinations of activation functions

Activation function at hidden layer	Activation function at output layer	Running time (hh:mm:ss)	Number of epochs	Mean squared error	Accuracy
Purelin	Purelin	00:00:30	5	0.1632	88.7553
Purelin	Logsig	00:05:17	55	0.0773	91.4280
Purelin	Tansig	00:05:23	59	0.0773	92.0695
Logsig	Purelin	00:22:00	225	0.0757	92.0249
Logsig	Logsig	00:21:43	225	0.0757	92.0249
Logsig	Tansig	00:12:56	123	2.9830	62.7199
Tansig	Purelin	00:19:29	210	0.0767	91.9002
Tansig	Logsig	00:22:34	262	4.6776	35.5925
Tansig	Tansig	00:15:47	169	0.0787	91.2129

Table 5 Performance comparison for various learning rates

Learning rate	Training time
0.01	00:11:30
0.1	00:11:39
0.3	00:12:50
0.5	00:11:22
0.8	00:13:09
1	00:14:53

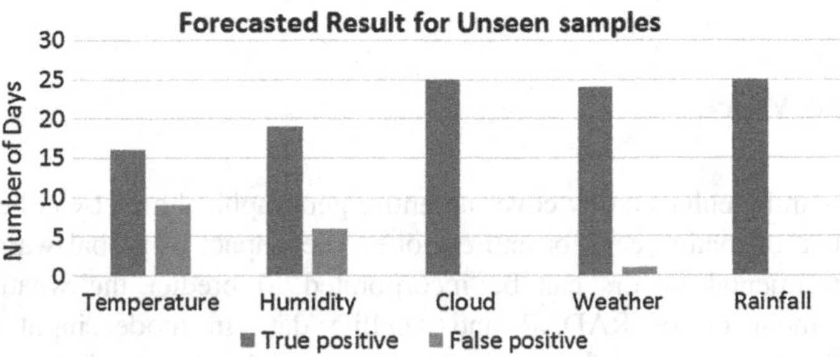

Fig. 2 Forecasted result for unseen dataset

Following Table 5 shows the comparison among various learning rate within the range of 0 to 1. Default learning rate is 0.01. The learning rate selected here for the network is 0.5. It gives same accuracy, mean squared error and number of epochs as given by other learning rates.

Finally the model is selected with proper values for each part of it. The 70 % of the input dataset is selected as training dataset and remaining 30 % is used as testing and validation dataset. The model is later used for forecasting the weather conditions of next day when the current day's weather is given as input. Test case of 25 unseen samples each is used to forecast the weather for day next to those days. Results denoting accuracy of the algorithm are tabulated below in Fig. 2. The parameters of false positives and true positives are used as references for comparison as well as for benchmarking accuracy levels. Accuracy levels are checked for forecasting next day weather conditions.

6 Conclusion

The weather forecasting system using artificial neural network with error back-propagation algorithm has been implemented. The system forecasts next day weather by accepting input parameters of previous day. The output accuracy obtained is high. The seed value used for random function should be chosen properly to get the more accurate results, because the seed value defined for random function is used to assign the weights to the neurons of each layer of the ANN. The initial weights used in the model have a great impact on the performance of the system. The activation function used in the middle layers also has an impact on the performance. The tansig and purelin are best suited for the forecasting problems. The learning rate directly doesn't affect the results of the system, but it is responsible for delaying or speeding up the training time. The implemented system gives the error rate at 0.0773 and the accuracy as 90 % which is very near to accurate results. Thus, this system can be used to forecast the next day weather for Mumbai city.

7 Future Work

The model can be enhanced by covering entire geographical area by collecting the historical data of many years for entire globe. The impact of global warming and other environmental factors can be incorporated to predict the weather more accurately. Inclusion of RADAR and satellite data in model input can also improvise the performance of weather forecast especially the clouds report in such images can help in rainfall prediction.

Acknowledgment We express our sincere thanks to the Indian meteorological center, Mumbai, India. The center has collected the data under harsh climatic conditions and makes it available for us.

References

1. L.L. Lai, H. Braun, Q.P. Zhang, Q. Wu, Y.N. Ma, W.C. Sun, L. Yang, "Intelligent Weather Forecast", Proceedings of the Third International Conference on Machine Learning and Cybernetics, Shanghai, 26–29 August 2004.
2. Yang Shuxia, "Power Consumption Forecast by Combining Neural Networks with Immune Algorithm", Eighth ACIS International Conference on Software Engineering, Artificial Intelligence, Networking, and Parallel/Distributed Computing.
3. Hosein Marzi, Senior Member, IEEE, Mark Turnbull, and Elham Marzi, "Use of Neural Networks in Forecasting Financial Market", 2008 IEEE Conference on Soft Computing in Industrial Applications (SMCia/08), June 25–27, 2008, Muroran, JAPAN.
4. Ming Gao, Wei Xu, Hongjiao Fu, Mingming Wang, Xun Liang, "A Novel Forecasting Method for Large-scale Sales Prediction Using Extreme Learning Machine", 2014 Seventh International Joint Conference on Computational Sciences and Optimization.
5. Ch. Jyosthna Devi, B. Syam Prasad Reddy, K. Vagdhan Kumar, B. Musala Reddy, N. Raja Nayak, "ANN Approach for Weather Prediction using Back Propagation," International Journal of Engineering Trends and Technology- Volume 3Issue1- 2012.
6. Mohsen hayati and Zahra mohebi, "Temperature Forecasting based on Neural Network Approach", World Applied Sciences Journal 2(6): 613–620, 2007, ISSN 1818-4952, IDOSI Publications, 2007.

7. Tony Hall, Harold E. Brooks, Charles A. Doswell," Precipitation Forecasting Using a Neural Network", Weather and Forecasting, Volume 14, June 1999, pp 338–345.
8. Carlos Gershenson, "ArtificialNeural Networks for Beginners", Gershenson@sussex.ac.uk.
9. http://in.mathworks.com/help/nnet/ref/tansig.html.
10. http://in.mathworks.com/help/nnet/ref/purelin.html.

Coherent Rule Mining Using AIR

Meera Narvekar, Apurva Kulkarni and Debajyoti Mukhopadhyay

Abstract In the field of data mining, there are multiple data mining algorithms. This paper focuses on the proposed algorithm which works on logical coherence between data items and mines positive as well as negative rules. The proposed system overcomes limitations of a priori algorithm like reducing need of support and minimizing need of domain expert to decide that support and confidence. Indexing mechanism and set theory is used to enhance the performance. After generating coherent rules, recall method is applied for pruning. Comparative study of a priori algorithm and proposed algorithm is also discussed.

Keywords Coherent rules · Indexing · Set theory

1 Introduction

The association rule mining algorithms uses two values, i.e., support and confidence. Support is the value which acts as key element to start algorithm and Confidence value actually discovers the association rules. In depth domain knowledge is required to decide minimum support and confidence value. Thus, domain experts are required to determine these values. These values of support and confidence are dependent on level of knowledge domain experts are having which may vary many times. The value of confidence threshold will decide amount of rules which will be generated [1]. If threshold value is too high then algorithm may

Meera Narvekar (✉) · Apurva Kulkarni
D J Sanghvi College of Engineering, Vile Parle (w), Mumbai 400056, India
e-mail: narvekar.meera@gmail.com; meera.narvekar@gmail.com

Apurva Kulkarni
e-mail: apurva475@gmail.com

Debajyoti Mukhopadhyay
WiDiCoRel Research Lab, Department of Information Technology, Maharashtra Institute of Technology, Kothrud, Pune 411038, India
e-mail: debajyoti.mukhopadhyay@gmail.com

© Springer Science+Business Media Singapore 2017
S.C. Satapathy et al. (eds.), *Proceedings of the International Conference on Data Engineering and Communication Technology*, Advances in Intelligent Systems and Computing 468, DOI 10.1007/978-981-10-1675-2_63

filter important rules and mine very few rules where as if threshold value is too low then multiple rules will get generated which will dilute the strong rules of importance.

This raises the need of system which can eliminate the need of varying confidence and support. Also it should work independent of databases which reduces dependency on domain experts.

2 Literature Review

Association rule learning is very popular and interesting method which works on logical relation between data items and finds association rule [2]. With increase in the amount of data being collected and stored in databases, there has been increase in the demand of discovering correlation among the data with the help of association rules. Association rules find their key applications in the process of making decisions about marketing activities such as planning promotional schemes [2]. Traditional data mining process have problems such as less repeatability, no particular interest to business and lack of end user understand ability [3]. Thus, it lacks soft power in solving real-world complex problems [4]. Proposed System provides Attribute selection feature which overcomes these weaknesses. Association rule mining generates very large set of rules which may even distract user to focus on important rules. Thus a novel notion of combined patterns is proposed to extract useful and actionable knowledge from a large amount of learned rules. It gives an account of two kinds of redundancy in combined patterns: (1) the repetitive combined rules within a rule cluster, and (2) Generating repeated rules for same data items [5].

When system is dealing with very large amount of data performance and efficiency is major factor that should be considered [6]. The efficiency of association rule algorithm can be increased by focusing on reducing number of passes, sampling, parallel execution and constraint based association rule mining [6]. But all these methods do not eliminate need of scanning dataset completely. To increase performance proposed system suggests use of indexing method and set theory.

3 Proposed System

3.1 Architecture of Proposed System

3.1.1 Dataset and Attribute Selection

Nowadays the databases are dense [7]. They have multiple attributes and different class labels. Generating rules therefore consumes more time and memory. But many times user is interested in only few attributes and class labels and therefore processing irrelevant data becomes unnecessary. To overcome this disadvantage proposed system introduces attribute selection phase where user can select one or many attributes according to their frequency of occurrence or by choice.

3.1.2 Index File Generation

For each combination of attributes the whole database is scanned and the record containing that set of attributes is linked to index of that record in dataset which forms an index file. Index file reduces repeated scanning of dataset which reduces time complexity of the system.

3.1.3 Applying Set Theory

Using Set Theory [8] to determine common elements. To generate intersection C, $C = A \cap B$ and to calculate only A the formula would be A–C and similarly to calculate elements in only B the formula is B–C.

3.1.4 Coherent Rule Mining Algorithm

As name implies the algorithm [7] finds logical coherence between data items and then maps them to coherent rule. Here system will be calculating support matrix which will help algorithm to map equivalences to coherent rule. For example, we can express the equivalence [7] of p which is *Customer buys bread* and q which is *Customer buys butter* as $(p \rightarrow q)$ i.e. *Customer buys bread if and only if he buys butter* and vice versa (Fig. 1).

To decide this implication to coherent rule we will form support matrix for above example. Here 1 indicates presence of item and 0 indicates absence of item.

	p1	p0
q1	S(q1, p1)	S(p0, q1)
q0	S(p1, q0)	S(p0, q0)

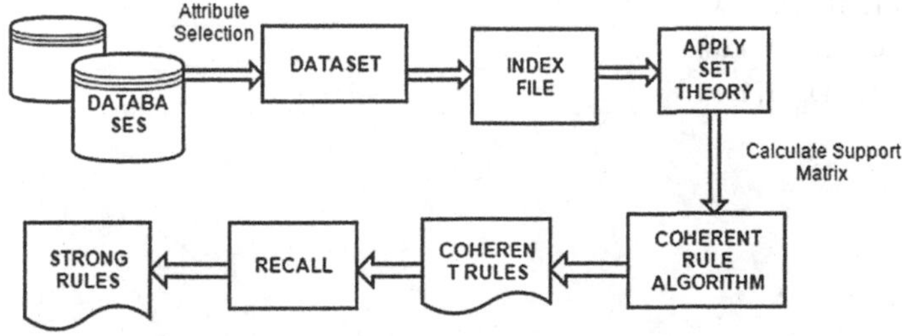

Fig. 1 Architecture of coherent rule mining using AIR system

S(q1, p1), Support for q1 and p1 indicates number of records containing q1 and p1.
S(q1, p0), Support for q1 and p0 indicates number of records containing q1 and not containing p.
S(q0, p1), Support for q10 and p1 indicates number of records containing p1 and not containing q.
S(q0, p0), Support for q0 and p0 indicates number of records not containing q and p.
Now, the conditions necessary for forming coherent rules are checked as follows [1]:

1. (S (p, q)) > (S (¬p, q))
2. (S (¬p, ¬q)) > (S (¬p, q))
3. (S (p, q)) > (S (p, ¬q))
4. (S (¬p, ¬q)) > (S (p, ¬q))

If above conditions are true then we can form coherent rule as:
p → q and ¬p → ¬q

4 Illustration with Example

4.1 Dataset

Dataset is having total 12 records with 5 attributes {a, b, c, d, e} where a1 represents occurrence of data item a and a0 indicates absence of data item a and two class labels High and Low.

Dataset is as follows:

z0 a1, b1, c0, d1, e0, High
z1 a0, b1, c0, d1, e0, High
z2 a1, b1, c1, d1, e1, High
z3 a1, b0, c1, d1, e0, High
z4 a1, b1, c0, d1, e0, High
z5 a0, b1, c1, d0, e1, High
z6 a1, b0, c1, d1, e0, Low
z7 a1, b1, c0, d0, e0, Low
z8 a1, b0, c1, d0, e1, Low
z9 a1, b1, c1, d1, e0, Low
z10 a0, b1, c0, d1, e1, Low
z11 a0, b1, c1, d0, e1, Low

4.2 Attribute Selection

Here user can select any number of attributes. Let us assume that we have selected 2 attributes a1, d1.

And then system will generate combinations of selected attribute. As we have selected 2 attributes there are only four combinations are possible {a1, d1, a1d1}.

```
a1 →, a1, $
a2 →, d1, $
a3 →, a1, d1, $
```

Note the last combination a1, d1.

4.3 Index File

The index file will contain all three combinations—{a1, d1, a1, d1}. There will be two arrays maintained one array will have total count of records containing that combination of attribute and other array will store index positions of records containing that attribute combination. Index file will look as follows:

```
1→ cnt 8→ 0 2 3 4 6 7 8 9
2→ cnt 8→ 0 1 2 3 4 6 9 10
3→ cnt 6→ 0 2 3 4 6 9
```

Now dataset is scanned for data item a1, d1. There are total 6 records indexed at {0, 2, 3, 4, 6, 9} and total occurrences of Class High is 6 and Low is also 6.

S(a1, d1) = 6
S(High) = 6
S(Low) = 6

4.4 Set Theory

Now support matrix is computed. Let A = a1, d1 and B = High. And A1 indicates records containing a1d1 i.e. 6, Hence A1 = 6, Similarly B1 = 6. Now we will calculate A ∩ B. That means number of records containing both A and B i.e. a1d1

Table 1 Support matrix

	A1	A0
B1	S(A1, B1) = 4	S(A0, B1) = 2
B0	S(A1, B0) = 2	S(A0, B0) = 4

and High. For that we will access index file, access a1d1 index and for all records containing a1d1 scan them for High. Here rather than scanning whole dataset, i.e., 12 records, only index file is accessed with 6 records. After scanning index positions {0, 2, 3, 4, 6, 9} we can see High is present at only four positions {0, 2, 3, 4} Hence a1d1 present and High present this count is 4 i.e. A ∩ B = 4

A1B1 = 4
A1B0 = (A1-A1B1) = (6 – 4) = 2
A0B1 = (B1-A1B1) = (6 – 4) = 2
A0B0 = Total records–(A1 U B1 – A ∩ B) = 12 – (12 – 4) = 4

These values forms the support matrix shown in Table 1

4.5 *Coherent Rule Mining Algorithm*

From the support matrix following conditions are checked and as all conditions are true algorithm maps data items to coherent rules.

- (S (A1, B1)) > (S (A0, B1))
- (S (A1, B1)) > (S (A1, B0))
- (S (A0, B0)) > (S (A0, B1)
- (S (A0, B0)) > (S (A1, B0))

Hence Coherent rule is A → B i.e. a1d1 → High.

4.6 *Pruning Using Recall*

Pruning helps to generate Strong rules. It also uses support matrix to calculate recall, precision and accuracy. Precision or positive predicted rule is the ratio of retrieved instances that are relevant to the total retrieved instances, while recall or sensitivity is the ratio of relevant instances that are retrieved to the total relevant items in dataset. Proposed system mines the rules dependent on recall and precision and analyses their relevance.

5 Algorithm

5.1 Terminologies

String z[]//csv database
Int zt//total records in dataset
String ab[]//attribute array
String a[]//selected attribute array
Int at//total selected attributes
String c[]//class array
Int ct//total class count
String ps[]//power set array
Int pst//total number of elements in powerset; logically pst=pow(2,at)
int index[i][j]//ith element in powerset occurs at positions j where j=0 to zt and i=0 to pst index[i][j] indicates element i in powerset which is present at j positions in record set z[].
int app[i]//total number of records where ith element of powerset occurs in dataset, where i=0 to pst and app[i]=0 to zt
if app[i]=0 then (ith element doesn't occur at single record)
if app[i]=zt then (ith element present at all records)
int cp[i]//for each class i, cp[i] indicates total occurrence of that class label.
For each a $\rightarrow$ b rule we will find following four values
Int aa//number of records not containing 'a' as well as not containing 'b'
Int ap//number of records not containing 'a' but containing 'b'
Int pa//number of records containing 'a' but not containing 'b'
Int pp//number of records containing 'a' as well as containing 'b'
Where 'a' is one element of power set having combination of selected attributes and 'b' is class label

5.2 Algotithm

```
for(i=1;i<pst;i++)//for each element in powerset array ps[], pst is total number of elements in ps[]
{
for(int j=0;j<ct;j++)//for each class label c[], ct is total number of classes
{
pp=ap=aa=pa=0;//initializing count variables
if(app[i]!=0)//if ith element in powerset occurs at least at 1 record then do following
{
for(k=0;k<app[i];k++)//traversing each record where i is present
{
```

```
if(z[index[i][k]].contains(c[j]))//z[index[i][k]] indicates kth record where ith element is present
pp++;//if z[[index][k]] contains class label c[j] then both 'a' and 'b' are present
else
pa++;//'a' is present but 'b' is absent
}
ap=cp[j]-pp;//cp[j]total class count –pp will give ap count
aa=zt-(app[i]+cp[j]-pp);//
if(((pp>ap)&&(pp>pa))&&((aa>ap)&&(aa>pa))//if it is coherent rule
{
System.out.println("pp="+pp+"ap="+ap+"pa="+pa+"aa="+aa
+"acc="+acc+" "+ps[i]+" → "+c[j]);
}
}
}
}
```

6 Result Analyses

6.1 Result Analysis of Zoo Dataset

The proposed system has been tested on two datasets. The zoo dataset containing 16 attributes has value true or false and total 7 class labels. Other real time dataset containing 10 attributes and 3 class labels. The existing algorithm is implemented in weka 3.6.0 and proposed system is implemented in java. The result analysis is shown in Table 2 and Fig. 2.

Table 2 Zoo dataset result analyses

Attribute	Rules by proposed system	Total rules by existing system	Unique rules by existing system	Strong rules by existing system
1. (Hair)	1	8	4	1
2. (Venomous)	0	6	3	0
3. (Fins)	1	8	4	1
4. (Backbone)	1	8	4	1
5. (Feather)	1	8	4	1
6. (Breath)	1	6	3	1
7. (Leg)	1	6	3	1

(continued)

Table 2 (continued)

Attribute	Rules by proposed system	Total rules by existing system	Unique rules by existing system	Strong rules by existing system
8. (Airbone)	1	8	4	1
9. (Tail)	0	6	3	0
10. (Aquatic)	1	6	3	1
11. (Toothed)	1	8	4	1
12. (Milk)	1	8	4	1
13. (Egg)	1	6	3	1
14. (Preadator)	0	6	3	0
15. (Catsize)	1	4	2	1
16. (Domestic)	0	8	4	0

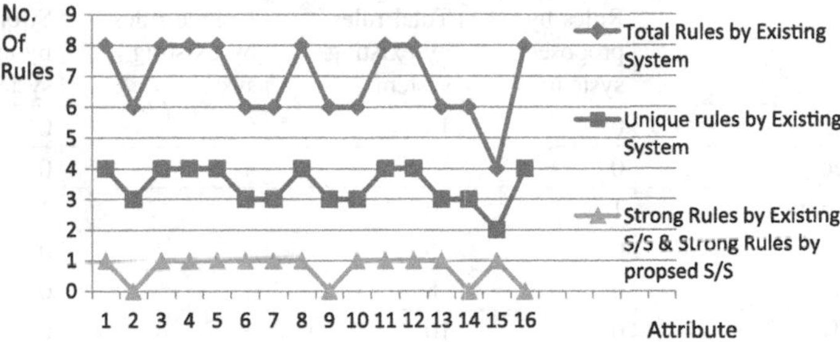

Fig. 2 Result analysis of zoo dataset

In Table 2 we can see that existing system generates duplicate rules which are highlighted and generates rules having confidence greater than zero but proposed system generates rules having confidence greater than 50 %. That clearly shows proposed system generates only more logically strong rules. Strong rules generated by existing system are exactly same as number of rules generated by proposed system which clearly shows proposed system does not miss out any single strong rule (Fig. 3).

6.2 *Result Analysis of Retailer Dataset*

Another example is real time transaction data taken from electronics goods shop having total 102 records and 10 attributes having values true or false denoted by 0 or 1, respectively. The result comparison is shown below (Table 3):

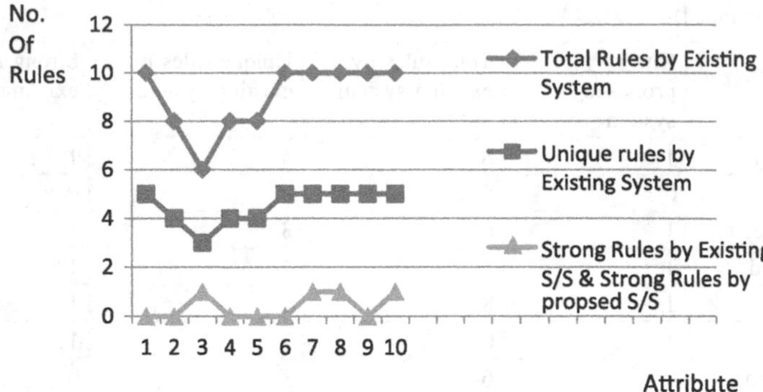

Fig. 3 Result analysis of retailer dataset

Table 3 Result analysis of retailer dataset

Attribute	Rules by proposed system	Total rules by existing system	Unique rules by existing system	Strong rules by existing system
1. (Male)	0	10	5	0
2. (Carvideo)	0	8	4	0
3. (Recordingpen)	1	6	3	1
4. (Dvd)	0	8	4	0
5. (Tv)	0	8	4	0
6. (Married)	0	10	5	0
7. (Bluetoothheadphones)	1	10	5	1
8. (Hometheatre)	1	10	5	1
9. (Female)	0	10	5	0
10. (mp3mp4)	1	10	5	1

6.3 Performance Analysis

Proposed system also prunes data based on recall, precision, accuracy, and confidence. To decide generated rule is strong or weak we need to decide threshold value.

The above Table 4 and Fig. 4 shows detailed working of existing system and we can see that for confidence value 55–80 total number of rules vary drastically but still number of rules for recall = 1, accuracy = 1 and precision = 1 varies for very small values that clearly shows that: Reduction in confidence value decreases total number of rules generated but does not affect Strong logically coherent rules. Hence threshold value can be selected between window 50–80.

Table 4 Threshold value calculation for zoo dataset

Confidence support	Rules	Recall = 1	Precision = 1	Accuracy = 1
0	5313	2556	649	192
5	5313	2556	649	192
10	5313	2556	649	192
15	5313	2556	649	192
20	5313	2556	649	192
25	5313	2556	649	192
30	5313	2556	649	192
35	5313	2556	649	192
40	5313	2556	649	192
45	5313	2556	649	192
50	5313	2556	649	192
55	5175	2556	613	192
60	4826	2556	446	192
65	4199	2556	372	192
70	3924	2556	359	192
75	3156	2556	321	192
80	2890	2556	252	192
85	2743	2556	225	192
90	2640	2556	192	192
95	2556	2556	192	192
100	2556	2556	192	192

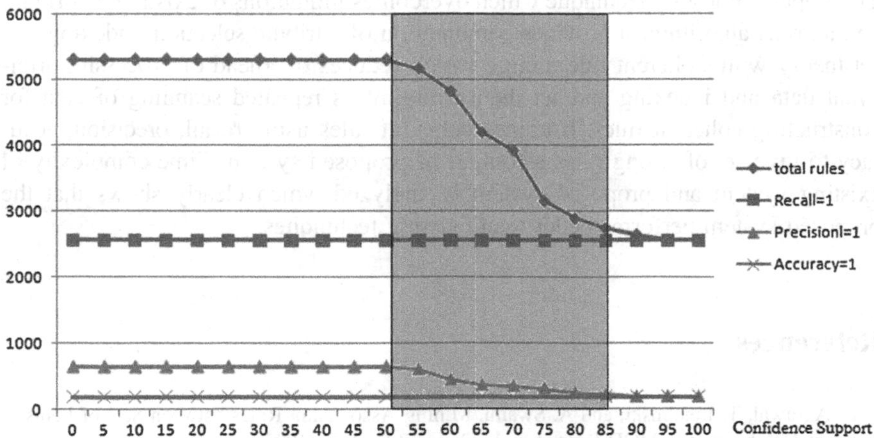

Fig. 4 Threshold calculation for zoo dataset

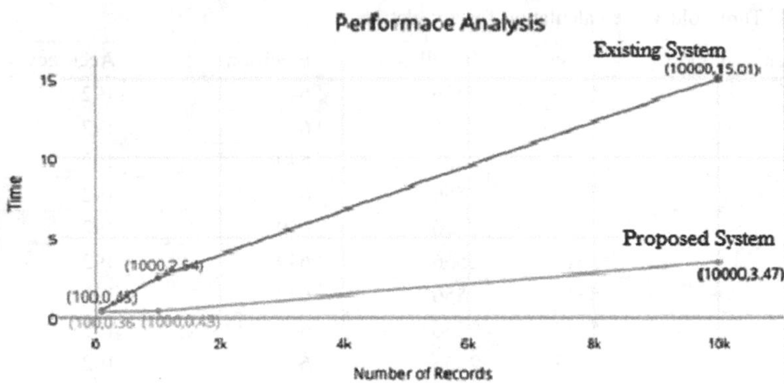

Fig. 5 Performance analysis

6.4 *Performance Analysis*

To compare time complexity system is run on 3 different datasets having 16 attributes and 100 records, 1000 records and 10000 records. For 10000 records time required by proposed system is one third of existing system. Analysis is shown below (Fig. 5).

7 Conclusion

This paper proposes a technique which overcomes limitations of existing algorithm, i.e., a priori algorithm. It provides combination of attribute selection, indexing and set theory with Coherent rule mining which reduces overhead of processing irrelevant data and indexing and set theory minimizes repeated scanning of data for constructing coherent rules. It prunes coherent rules using recall, precision, accuracy to give set of strong rules as output of proposed system. Time complexity of existing system and proposed system is analyzed which clearly shows that the proposed system performs better than existing techniques.

References

1. R. Agrawal, T. Imielinski, and A. Swami. Mining Association Rules between Sets of Items in Large Databases, SIGMOD Record, vol. 22, pp. 207–216, (1993).
2. Apurva Kulkarni, Aarti Deshpande.domain Driven Approach for Coherent Rule Mining. IEEE, 978-1-4799-2432-5, pp. 109–114, (2013).
3. H. Mannila, Database Methods for Data Mining, Proc. Fourth Int'l Conf. Knowledge Discovery and Data Mining Tutorial, (1998).

4. Longbing Cao, Domain Driven Data Mining: Challenges and Prospects, *IEEE Transaction on knowledge and data Engineering,* Vol. 22, No. 6, (June 2010).
5. Y. Zhao, H. Zhang, L. Cao, C. Zhang and H. Bohlscheid Combined pattern mining: From learned rules to actionable knowledge, *Proc. AI*, pp. 393–403 (2008).
6. Apurva Kulkarni, Meera Narvekar, S S Mantha: Comparative Study of Techniques to improve Efficiency of Association Rule Mining. Elsevier, volume 49 (2015).
7. C. Longbing, Introduction to Domain Driven Data Mining, Data Mining for Business Applications, L. Cao, P.S. Yu, C. Zhang, and H. Zhang, eds., pp. 3–10, Springer, (2008).
8. G.I. Webb, Association Rules, The Handbook of Data Mining, pp. 26–39. Mahwah, (2003).

Enhanced Data Dissemination in a Mobile Environment

Meera Narvekar, Heena Mukadam and Debajyoti Mukhopadhyay

Abstract In a mobile environment, data broadcast is a preferred and promising technique to disseminate multiple data items with high request rates to multiple mobile users. The sole objective of a good broadcast strategy is to reduce the waiting time of the mobile users and conserve battery power consumption of mobile devices. The existing system stands as a huge scope of improvement to overcome issues such as reducing access time, tuning time and power consumption. In this paper, the proposed system incorporates optimal design and a dynamic technique that aims to reduce the average access time, tuning time, and subsequently save battery power consumption of mobile users for the build of an optimal environment. The proposed system performs efficient ordering and effective allocation of data items over multiple channels in a data broadcast at the server side which provides better performance and optimal results compared to the existing system.

Keywords Data broadcast · Multiple channels · Mobile users · Mobile devices · Access time · Battery power consumption

1 Introduction

In mobile computing environment, we study two fundamental information delivery methods. The two methods are point-to-point access and broadcast. In comparison to point-to-point access, data broadcast is a more attractive method. Though the

Meera Narvekar (✉) · Heena Mukadam
Computer Department, DJ Sanghvi College of Engineering, Mumbai 400056, India
e-mail: meera.narvekar@gmail.com

Heena Mukadam
e-mail: hmukadam91@gmail.com

Debajyoti Mukhopadhyay
WiDiCoRel Research Lab, Department of Information Technology, Maharashtra Institute of Technology, Kothrud, Pune 411038, India
e-mail: debajyoti.mukhopadhyay@gmail.com

© Springer Science+Business Media Singapore 2017

S.C. Satapathy et al. (eds.), *Proceedings of the International Conference on Data Engineering and Communication Technology*, Advances in Intelligent Systems and Computing 468, DOI 10.1007/978-981-10-1675-2_64

broadcast concept is not new and has been applied in our surroundings for decades including the radio programs, TV programs, and newspaper to name a few. Along with reducing access time, what also makes it more efficient and effective is that any data broadcast program can scale up to an arbitrary number of users. Hence, data broadcasting has become a promising technique to disseminate data items to mobile clients due to its high scalability factor.

With broadcast as the medium of data delivery various information systems can be served to reciprocate services. Few of the many industrial fields that can benefit and utilize advantages of data broadcast are stock markets, weather forecast services, security alerts, and telecommunication service providers as discussed in [1].

2 Literature Review

In this section, the prior research work and the analysis of the existing systems in the field of mobile computing environment is covered. In [2], the authors studied the working of three basic information delivery models, i.e., push data broadcast, on-demand data broadcast and hybrid data broadcast. In case of push data broadcast, the information is disseminated without the intervention of mobile users. While in pull-based data broadcast the information is disseminated on the request the mobile users. Since, a hybrid model is a combination of pull and push, the advantages and disadvantages of each are equally complemented and balanced to provide optimal solution.

In [2], the research provides a comparative analysis of various broadcast scheduling techniques in a mobile environment. The authors design approach in [1] is closely relevant to the idea of the proposed system presented in this paper. However, there is a huge scope of improvement in terms of allocation of data items and bandwidth utilization to reduce the access time of mobile users. Various channel allocation methods have been proposed in [3] and have been improved and covered during the design and implementation of the proposed system. While in [4], the authors provide architecture for wireless information dissemination of broadcast and the on-demand services. However, the existing system model and approach are different from proposed, i.e., the prior work considers only limited parameters like limited number of channel, limited workload, limited bandwidth.

In [5], different schemes for the allocation of data on broadcast and on-demand channels are explored. However, it fails to demonstrate scenarios of replication of data items with higher request rates, which is overcome by incorporating the proposed mechanism. While [6], provides working of channel allocation for heterogeneous data broadcasting but with limited number of channels and bandwidth.

Since, prior research topics in the field of data broadcasting are mainly based on the assumption that the disseminated data items are of the same size and sequential ordering of data items [7]. Also noticed patterns of existing research have been restricted to limited resources of channel [8]. Motivated by the fact that various kinds of data items may be disseminated in advanced mobile information systems, we explore the scope of generating dynamic broadcast programs in a heterogeneous mobile environment. While heterogeneous refers to, data items of different sizes, different frequencies that can be disseminated over multiple channels with varying bandwidth. This approach serves to be the base idea for the proposed system discussed in this paper.

3 Proposed Work

To differentiate proposed research from traditional and conventional methods, proposed system considers data items with varying sizes and frequencies to be termed as heterogeneous data items. The flow of proposed system works in the following sequence: First, identify hot and cold items amongst mobile user preferences. Then generate two arrays that hold index of only hot data items to be broadcast. This is done considering the access frequency of data items at the end of every hour and similarly at the end of every day. This will give an understanding of data items with high access frequency amongst users. The arrangement is based on the descending order of the access frequency. Second, is to allocate these arrays over multiple channels in a broadcast cycle followed by calculated replication broadcast.

The ability to maintain multiple information queries in spite of the high frequency of requests and large number of mobile clients, data broadcast mechanism proves to be a desirable technique. Such efficient data dissemination will reduce the average waiting time of mobile user to access the desired data item. The average access time is made of up two components: the probing time and downloading time. The probing time is defined as the time that a mobile user is expected to wait until the item of interest appears on the broadcast channel. While the downloading time, is when a mobile user spends on downloading the desired data item from the broadcast channel.

To access the desired data item, the mobile users are expected to wait and capture the data of interest once it arrives on the broadcast channel. As our proposed functionality, the index of the data items to be broadcast will give a prior idea of arrival time. This will enable mobile users to switch to between active and doze mode. Hence, it is fulfilling the purpose of saving the battery power of mobile devices.

3.1 Proposed System Architecture

The system architecture in Fig. 1 is divided into three modules A, B, and C; which includes one controller, multiple local servers, and zero or more clients connected to each local server at any given time. A local server is also known as BSS i.e. Module ‘A’. For instance, A controller will handle only one region say Mumbai. A region

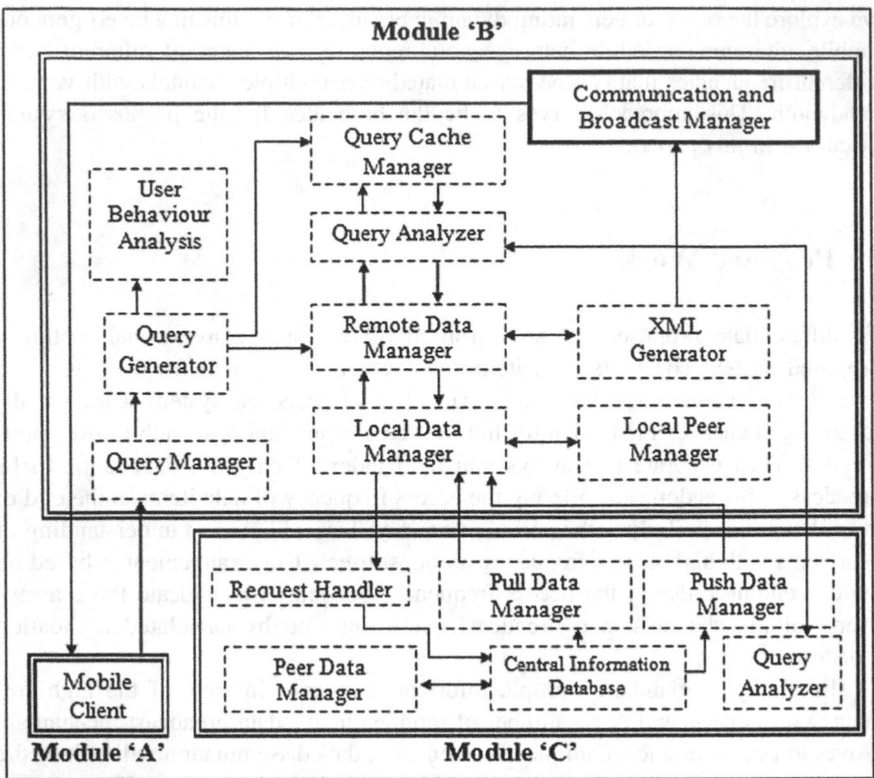

Fig. 1 Proposed broadcast mechanism

will contain multiple cells say Dadar, Bandra, Andheri, and so on. While another region, could be Vasai-Virar. This region will have its own controller. This region may also have multiple cells say Vasai, Virar, Palghar, and so on. A controller may be connected to multiple neighboring controllers. Consider a single region to understand the working of the proposed system.

Module 'A' functions via a request from a mobile user is initiated. If the requested item is not found in the cache the user request is forwarded to the local server. Module 'B' that is the Local server, which is the most important entity of the architecture. Each local server is connected to the controller. A local server is connected to zero or more neighboring local servers. For the purpose of broadcasting data, the system deals with services and not individual data items in those services. For the purpose of storing and indexing data, the system deals with individual data items of the services. Wherever the term 'data items' is used, it is assumed that these data items are properly categorized into different services. Services may be replicated across multiple Local Servers. The user request is processed by the local server. Next the query generator converts the data item request into a query format to be forwarded to the query manager. The query cache responds on the existence in cache. While query analyzer, will look for history of a requested data item and its affinity towards the threshold of high request rate. Thc local data manager will maintain the local data of the services and also stores the data from a remote server for a response for some particular time interval. This module is looked for, after the response for a particular query is not present in query cache manager. If found it forwards to the xml generator. The xml generator converts the response to an xml format to make it independent of any platform and feeds input to communication and broadcast manager. The communication and broadcast manager is the heart of the proposed system. This module helps in dissemination of data in an efficient manner to the user. The working is explained in detail in Sect. 3.2. While the Module 'C' that is the central server acts as head of all the local servers to provide data and to manage all the local servers using request handler, pull and push data manager and central information database. This sums up at the controller where all the data items of all the services will always be present.

3.2 Proposed Broadcast Mechanism

The broadcast mechanism in Fig. 2, is the heart of the system. It is the core functionality proposed to disseminate data over multiple channels to mobile clients. The major modules of the proposed broadcast scheme based on the algorithm are:

INPUTS: Data items with high frequency will be termed as 'Hot' data items, while with low frequency is termed as 'Cold' data items. Consider the following example, If data items d1, d2, d3, d4, and d5 were accessed very frequently at BSS1 during 9 a.m. to 10 a.m. on 1st January; then the same data items are expected to be accesses on January 2 in the same time duration. Hence, these items will be considered 'Hot' items in that time slot and will be pushed to clients during that period on 2nd January. Also an algorithm will execute at the end of every hour. This algorithm will remove 'cold' items from the local servers. An item will be tagged as cold item if its requests (of the whole day, i.e., 24 h) drop below a certain threshold value. The two inputs are

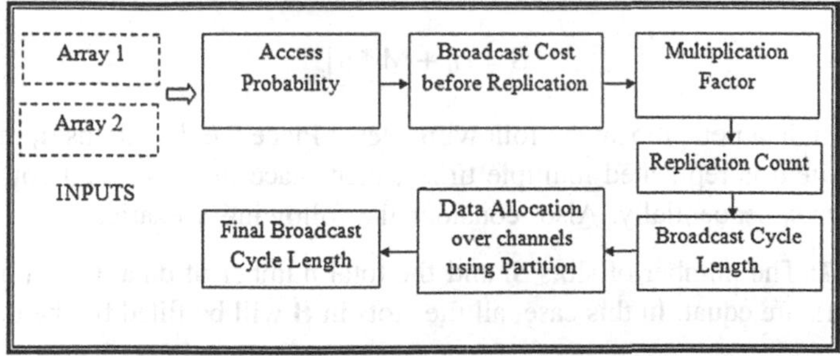

Fig. 2 Proposed broadcast algorithm

- **A1** holds an array of data items sorted in descending order as per their access frequency and is generated in form of one full list in a day.
- **A2** holds an array of data items sorted in descending order as per their access frequency of that hour and is generated after every hour.

The arrays are converted into XML file by XML generator to make it independent of any platform for further compilation, processing and execution.

Step 1: Calculating Access Probability of Data Items: Calculate the access probability $\mathbf{p_i}$ for each data item $\mathbf{d_i}$ with their corresponding frequency $\mathbf{f_i}$ till all data items **n** in array **A2**, have not be covered. The value of $\mathbf{p_i}$ will be in the range [0, 1].

$$p_i = \frac{f_i}{\sum_{j=1}^{n} f_j} \tag{1}$$

Step 2: Cost of Broadcast Cycle: The initial total $\mathbf{cost_i}$ of broadcasting all the data items can be calculated based on the access frequency and size prior to any partitioning or replication. It is defined by the formula

$$cost = \left(\sum_{j=1}^{n} f_j\right) * \left(\sum_{j=1}^{n} z_j\right) \tag{2}$$

Step 3: Calculating Multiplication Factor M: The Multiplication Factor is denoted by **M** and lies in the range of [0, 2] which is selected arbitrarily. The value of **M** is adjusted in such that the length of Broadcast cycle **B** is shorter, when **cost** is higher, and longer when **cost** is lower.

Step 4: Calculating Replication Factor: Since, the access probability $\mathbf{p_i}$ for each data item $\mathbf{d_i}$ and the Multiplication Factor **M** is determined. Calculate the replication count $\mathbf{r_i}$ for each data item $\mathbf{d_i}$. It is computed using the following formula

$$r_i = \lceil M * n * p_i \rceil \tag{3}$$

Step 5: Create broadcast cycle B: Calculate the length **S** (no. of slots) of **B** with the given formula

$$S = \lceil n + M * n \rceil \tag{4}$$

For each data item, repeat the following steps: Place the data items $\mathbf{d_i}$ in the slots $\mathbf{s_i}$. If data item is replicated multiple times, then place the replicated copies in the empty slots $\mathbf{s_i}$ sequentially. Also, consider the following scenarios:

Scenario 1: The number of slots **S**, and the total number of data items (along with replication), are equal. In this case, all the slots in **B** will be filled by the data items.

Scenario 2: The number of slots **S**, are more than the total number of data items (along with replication). In this case, we have empty slots in **B** (after placing all data items from **A2** in B). These empty slots will be filled by data items from Array **A1**.

Step 6: **Allocation of data items over multiple channels:** Consider Access Probability and the Size of the Data Item as two inputs for the further processing of allocation of data items over multiple channels using Partitioning Procedure [6].

Step 7: Calculate final cost: Consider **k** to be the total no. of channels, $\mathbf{cost_i}$ to be the cost of $\mathbf{Channel_i}$ and $\mathbf{N_i}$ to be the data Items in $\mathbf{Channel_i}$

$$\mathrm{Cost} = \sum_{i=1}^{K} \mathrm{cost_i} \tag{5}$$

$$\mathrm{cost_i} = \left(\sum_{j=1}^{c} f_j^{(i)} \right) * \left(\sum_{j=1}^{N_i} z_j^{(i)} \right) \tag{6}$$

4 Performance and Result Analysis

During performance and result analysis, study shows that the proposed system overcomes issues of the existing systems adapting the conventional method. Performance testing on the systems is done to obtain stability in results, via multiple test runs to prove optimal stability. Load testing on proposed algorithm has been tested by including, data items of varying sizes and access frequencies over multiple channels and replication. In existing system, considers static sizes of data items, static access frequencies, single channel and no replication. Figures 3, 4, and 5 demonstrate the comparison of proposed method versus conventional method. Regression Testing is performed where input of data items varies to achieve optimal results. Consider input of data items as 20 in Fig. 3, input of data items as 30 in Fig. 4, input of data items as 45 in Fig. 5 along the x axis. While the y axis demonstrates average access time obtained.

Based on results of proposed system it is noticed the average access time of the mobile users is reduced. Hence, the purpose of enhanced data dissemination is

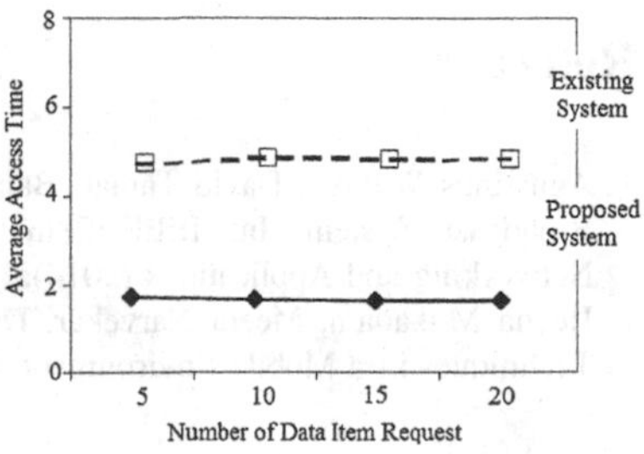

Fig. 3 Proposed system versus existing system

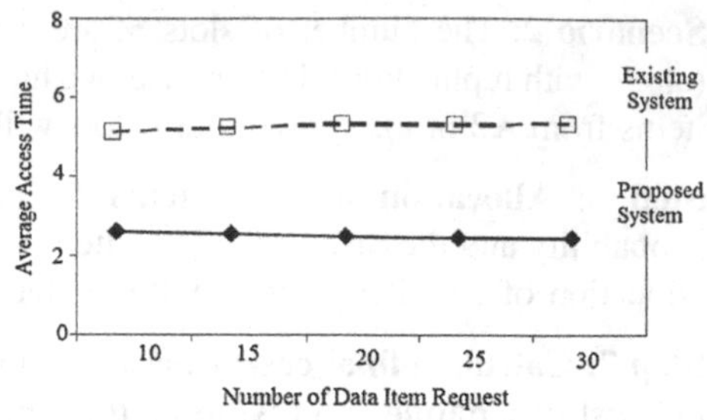

Fig. 4 Proposed system versus existing system

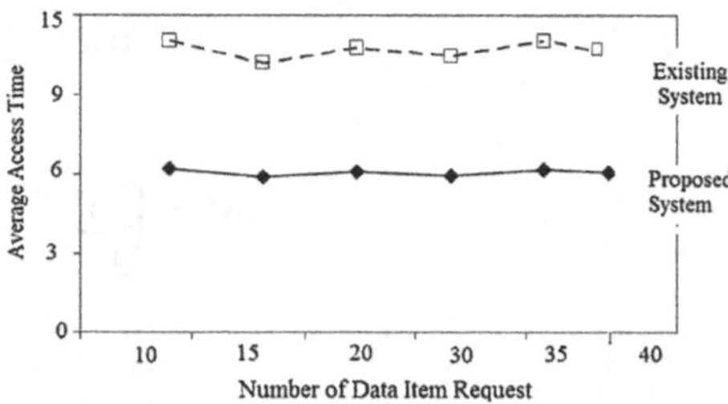

Fig. 5 Proposed system versus existing system

fulfilled. The results show the proposed system obtains optimal and stable results over multiple runs of the system.

5 Conclusion and Future Work

In this paper, the proposed system aims to achieve efficient data dissemination in a mobile environment. The performance and results analysis show our proposed system is an optimal solution to reduce the waiting time of the mobile user, achieve high scalability and save batter power consumption. The work majorly focuses on having developed a dynamic broadcast algorithm over multiple channels corresponding to dynamically changing access patterns in real time. Some of the future work will include an optimal index scheme along with the proposed broadcast scheme to achieve better, balanced and optimal results in a wireless environment.

References

1. Agustinus Waluyo, David Taniar, Bala Srinivasan.: Design and Implementation of a Mobile Broadcast System. In: IEEE 28th International Conference on Advanced Information Networking and Applications (2014).
2. Heena Mukadam, Meera Narvekar, Dr. S.S. Mantha.: Comparison of Broadcast Scheduling Techniques in a Mobile Environment. Elsevier, ICACTA (2015).

3. Agustinus Waluyo, Bala Srinivasan, David Taniar.: Allocation of Data Items for Multi Channel Data Broadcasting in a Mobile Computing Environment, pp. 409–418 Springer, Heidelberg (2004).
4. Hsiao-Ping Tsai, Hao-Ping Hung.: On Channel Allocation for Heterogeneous Data Broadcasting. IEEE Transactions on Mobile Computing, VOL. 8, NO. 5 (2009).
5. Wang-Lee, Qinglong Hu.: A study on channel allocation for data dissemination in mobile computing environments. Mobile Networks and Applications, 117–129 (1999).
6. Qinglong Hu, Wang-Chien Lee.: Optimal Channel Allocation for Data Dissemination in Mobile Computing Environments. IEEE (1998).
7. Agustinus Borgy Waluyo, Bala Srinivasan.: Optimizing Query Access Time over Broadcast Channel in a Mobile Computing Environment, Springer Heidelberg (2004).
8. J. Beaver, P.K. Chrysanthis, K. Pruhs, V. Liberatore.: To Broadcast Push or Not and What? In 7th International Conference on Mobile Data Management, IEEE (2006).

Adaptive Ontology Construction Method for Crop Pest Management

Archana Chougule, Vijay Kumar Jha and Debajyoti Mukhopadhyay

Abstract Knowledge represented as ontologies can be accessed easily by automated systems in Semantic Web. If ontologies are used to represent agricultural knowledge, e.g., crop pest information, it can be shared by many existing expert systems in agricultural field. Different languages are used by farmers all over India. As knowledge in the form of ontologies can be converted easily into different languages, farmers in various states in India can benefit from expert's knowledge. Developing crop pest ontology from scratch will be a difficult task for agricultural experts and it will consume lot of time. We provide user-friendly interface in which agricultural expert can upload text descriptions of crop pests. The system will extract keywords from text files by applying keyphrase extraction steps and comparing it with AGROVOC thesaurus. For this purpose, we propose a Pest Keywords Extraction Algorithm described in detail in the paper. Agricultural expert can add new pest types, pest examples, and details of each pest such as reason, symptom, and remedy for pest. All the details will be automatically saved as pest ontology in OWL format. The system is adaptive as the expert can see pest type hierarchy, add or remove a pest type and pest details at any point in crop pest ontology. Once complete, ontology for crop pests is ready to be used by expert systems as part of inference engine.

Keywords Ontology · Agriculture · Expert system · Key phrase extraction

Archana Chougule (✉) · Debajyoti Mukhopadhyay
Maharashtra Institute of Technology, Pune, India
e-mail: chouguleab@gmail.com

Debajyoti Mukhopadhyay
e-mail: debajyoti.mukhopadhyay@gmail.com

V.K. Jha
Birla Institute of Technology, Mesra, Ranchi, India
e-mail: vkjha@bitmesra.ac.in

© Springer Science+Business Media Singapore 2017
S.C. Satapathy et al. (eds.), *Proceedings of the International Conference on Data Engineering and Communication Technology*, Advances in Intelligent Systems and Computing 468, DOI 10.1007/978-981-10-1675-2_65

1 Introduction

Agriculture is the largest economic sector in India and has a significant role in the economy. Use of expert knowledge and advanced technology for increasing agricultural production will be a great help to farmers. Many times farmers lose their crops because of lack of knowledge of using correct pest management technique. The knowledge of crop pest management held by agricultural experts if it is represented in formal terms can be easily provided to farmers through mobile or Web-based software applications. The expert knowledge should be represented in ontologies as ontologies can be shared among diverse applications including Semantic Web applications [1, 2]. Once knowledge base is ready in terms of ontologies, it will be easier to convert it into the languages understood by farmers. More interoperability between various agricultural systems can be achieved by expressing expert knowledge in terms of ontologies.

With domain ontologies, expert knowledge can be shared efficiently between researchers in various agricultural research centers/communities [3]. Both declarative and procedural representation of crop pest knowledge is possible using ontologies. Procedural representation of expert knowledge will help to build rich knowledge base and find new facts of crop pests.

Support of inference rules and semantic reasoners like Blossam, Cyc, KAON2, Cwm, Drools, Flora-2, Jena [4] and Prova2 are available for ontologies. Better indexing of crop pests and treatment knowledge is possible if it is mentioned in terms of ontologies. Artificial intelligence applications for crop pest management can be developed by analyzing reasons for crop pests and their effects mentioned in ontologies. We can define categories for pest management like reasons for pests, symptoms and treatments for pests. Compared to other resources of agricultural knowledge like Internet, thesaurus, PDF documents, it becomes easier to provide desired specific information with ontologies [5]. Formal and specific representation of pest management knowledge helps farmers in easier understanding of expert knowledge. Farmers can have a look at treatment options available for particular crop by exploring ontology and can minimize yield loss.

Pests develop resistance against controlling measures over time, so new pest control measures are required and it is very important to update farmers on it. Remedies for pests which are not based on recent knowledge will not be relevant. So dynamic updating of pest knowledge base is required. It is possible with our system named as CropPestOntoGenerator. CropPestOntoGenerator assists to generate ontologies from text descriptions of particular crop pests [6]. It helps agricultural expert to add new pest types, pest examples, symptoms, reasons and remedy for pests. Agricultural expert can build new ontology or update existing one using CropPestOntoGenerator.

2 Design and Implementation

The core idea of CropPestOntoGenerator is to construct ontology dynamically for crop pest management. The agricultural experts have many description documents related to crop pest management with them. These documents are generally in text format. These text descriptions cannot be used directly by automated systems. For directly using this kind of text descriptions, it must be represented in well-structured format as ontology. With ontologies, we can represent hierarchy of crop pests [7, 8]. Once pest ontology is prepared, it can be easily converted into any language, which farmer can understand [9]. The overall workflow of CropPestOntoGenerator is shown in Fig. 1.

The result of workflow is integrated ontology denoted as $\mathbf{U_o = <F_o, K_c, M_o>}$, where

U_o—Integrated ontology
F_o—Foundation Ontology element
K_c—Retrieved knowledge from Corpus
M_o—Mapping rules for ontology

The system provides the foundation ontology to agricultural experts as shown in Fig. 2. K_c is extracted keywords from rules given by experts for adding keywords to

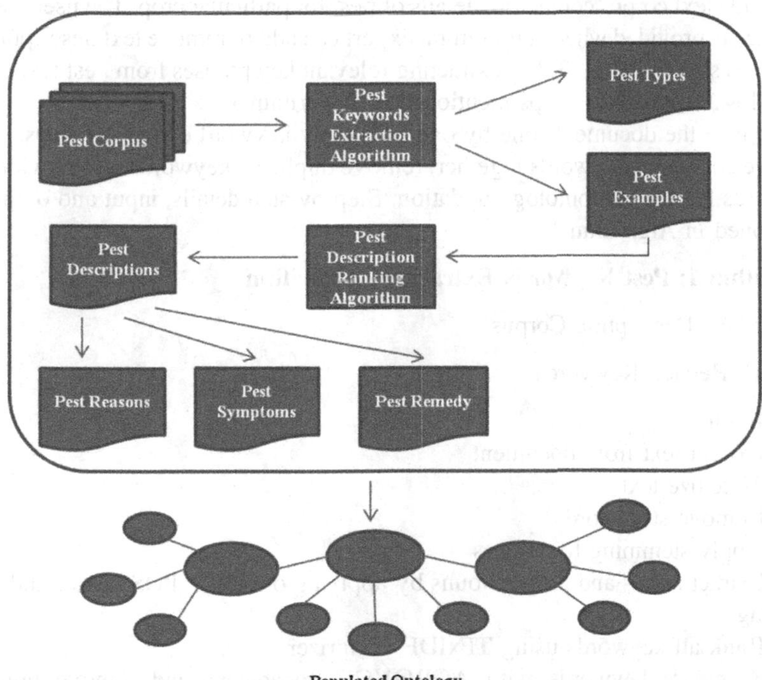

Fig. 1 Workflow of the CropPestOntoGenerator

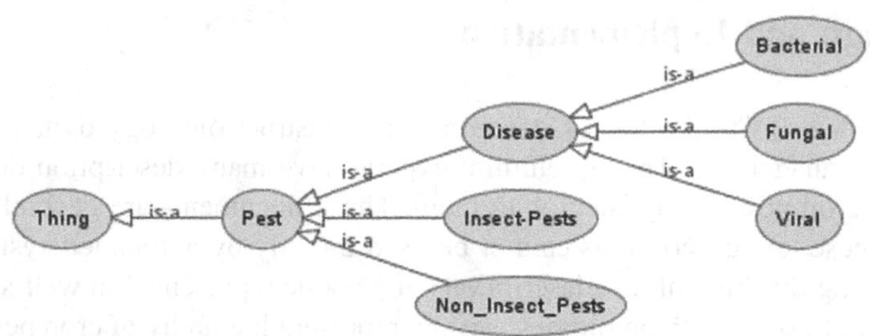

Fig. 2 Foundation ontology for crop pest management used by CropPestOntoGenerator

ontologies. We have chosen Web Ontology Language (OWL) as the language for constructing ontology.

As OWL is meant to represent information on Internet; once complete crop pest knowledge is stored as OWL document, it can be accessed using Internet, or can be used by any other agricultural expert systems with ease. The foundation ontology mentions basic categories of crop pests as insect pests, non-insect pests and diseases. Diseases are again divided into three subcategories as fungal diseases, bacterial diseases and viruses.

To add more categories of pests, pest examples and pest details agricultural expert is assisted with extracted keywords from text descriptions. Agricultural expert needs to provide text corpus containing details of pest for particular crop. The user-friendly interface is provided where agricultural expert can add or remove text descriptions in corpus as shown in Fig. 3. For extracting relevant key phrases from pest text corpus we follow various NLP steps mentioned in Algorithm 1.

We take the documents one by one and apply keyword extraction steps. At the end, we collect all keywords together, remove duplicate keywords and provide final keywords list for pest ontology updation. Step-by-step details, input and output are mentioned in Algorithm 1.

Algorithm 1: Pest Keywords Extraction Algorithm

Input: Pest Description Corpus

Output: Related Keywords

1: Begin
2: Extract text from document
3: Tokenize text
4: Remove stop words
5: Apply stemming to phrases
6: Extract nouns and proper nouns by applying openNLP POS tagger and retain POS tags
7: Rank all keywords using TFxIDF vectorizer
8: Compare keywords with AGROVOC vocabulary and remove unrelated keywords

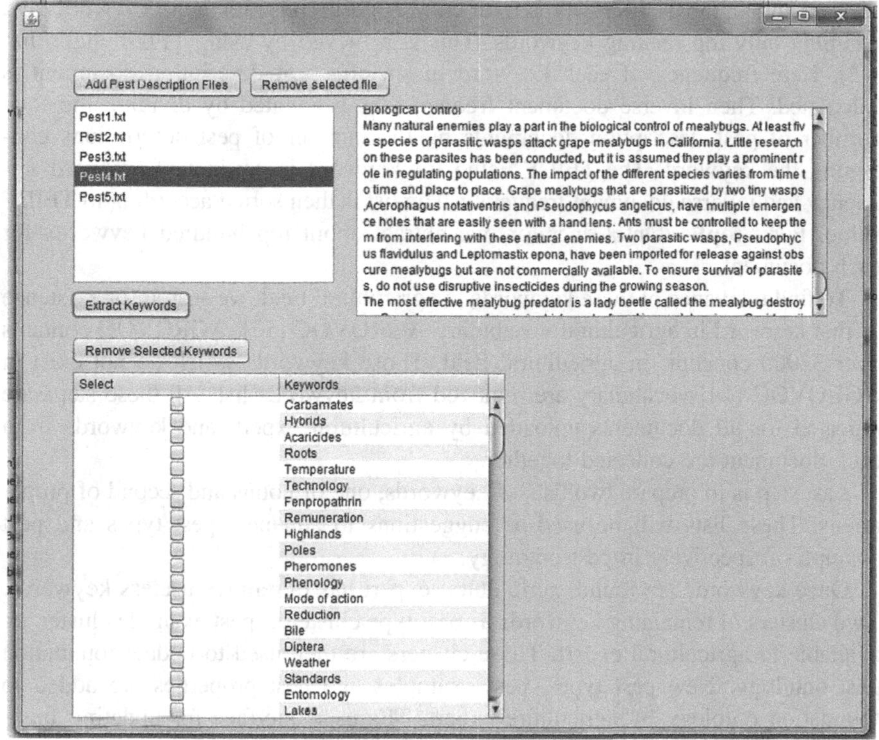

Fig. 3 Keywords extraction from crop pests' corpus

9: Repeat from step 2 until all documents are extracted
10: Remove duplicate keywords
11: Divide the list into two lists one containing nouns and another containing proper nouns
12: End

The text from each document related to crop pest is extracted and segmented in tokens along the word boundaries using StringTokenizer provided by Java programming language. Stopwords in document are not useful for ontology building. Therefore such words are identified and removed from tokens collection by comparing tokens with the list of stopwords in English. Then stopwords' frequency of each keyword is counted and those keywords whose frequency is less than three are removed from the list. After stopwords removal, the keywords in list are stemmed by using Porter Stemmer Algorithm [10], which transforms a word to its root form.

The pest type is stored as owl:class in pest ontology and pest example as owl: individual. The classes and individuals from ontology are generally nouns and proper nouns in English sentences. We take advantage of this fact and extract only nouns and proper nouns from stemmer output using openNLP POS tagger [11]. Keywords with tags NN, NNP and NNPS are passed to next step of algorithm.

Next step in the pest Keywords Extraction Algorithm is ranking of keywords and retaining only top ranking keywords. This is achieved by using TFIDF algorithm [12]. Here frequency of each keyword in specific pest description document is calculated. Then inverse document frequency is calculated by dividing the total number of pest description documents by the number of pest descriptions containing the keyword. The TFIDF value for keyword is product of keyword frequency and inverse document frequency. The list is then sorted according to TFIDF value. For CropPestOntoGenerator, we extract about top hundred keywords for each document.

To find relevance of these keywords to agricultural field, we search for existence of that keyword in agricultural vocabulary AGROVOC [13]. AGROVOC contains over 32000 concepts in agricultural field. Those keywords which do not exist in AGROVOC [14] vocabulary are removed from keywords list. All these steps are repeated for all documents uploaded by agricultural expert, and keywords from each document are collected together.

Last step is to prepare two lists of keywords, one of nouns and second of proper nouns. These lists will be used as suggestions to add new pest types and pest examples respectively in pest ontology.

Once keywords are found, agricultural expert has to remove useless keywords. Two clusters of remaining keywords as pest type cluster or pest example cluster are available to agricultural expert. These clusters are then used to update foundation pest ontology. New pest types, pest examples and pest properties are added to foundation ontology by agriculture expert. The user interface for updating foundation pest ontology is shown in Fig. 4.

Not only keywords extracted by the system can be added to crop pest ontology, but also new keywords from experts' own knowledge can be added. Such keywords are maintained in separate keywords ontology named as EXPERTVOC.

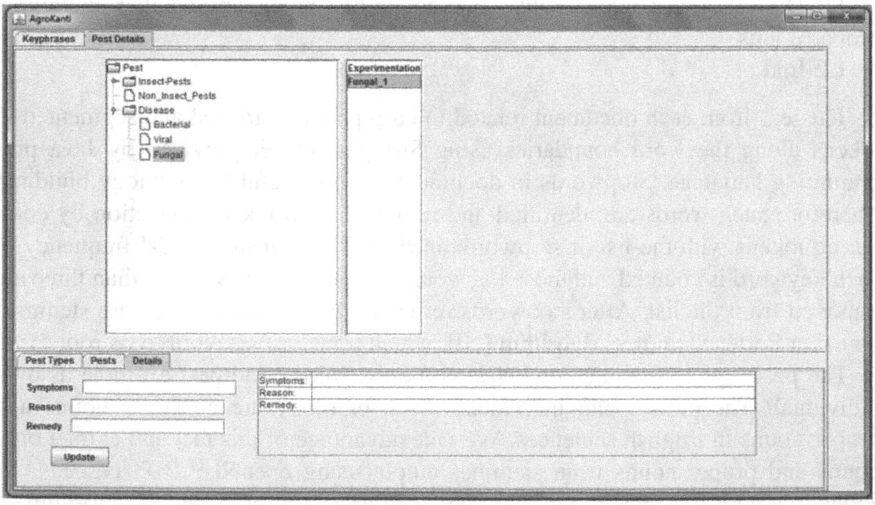

Fig. 4 Updating foundation pest ontology

EXPERTVOC will be dynamically updated whenever crop pest ontology is constructed or updated. EXPERTVOC will be used along with AGROVOC every time after first use of system. So our system is adaptive and becomes more and more robust periodically. It helps in improvement of precision values.

Agricultural expert can add or update symptoms, reasons and remedy for each pest example. We provide assistance to add these details by Pest Description Ranking Algorithm. Here user can provide more than one text description document for specific crop pest, and the system ranks these documents by counting number of keywords that exists in that document matching to words in currently constructed crop pest ontology. The document with maximum matching keywords is suggested as best document to be used for adding symptoms, reasons and remedy for each pest example.

For storing the information as OWL document we have used Protégé 3.48 APIs [15]. Protégé APIs provide reach collection of classes and methods for adding and removing OWL classes, individuals, data properties and object properties [16]. We store pest types as OWL classes and pest examples as OWL individuals. Symptoms, reasons and remedy for pest example are stored as data properties of individuals in string format.

3 Performance Evaluation

We tried pest keyword extraction algorithm on around 100 documents and calculated precision and recall for the algorithm.

Results of Pest Keywords Extraction Algorithm applied on six sample documents describing crop pests are given in Table 1.

Precision is calculated as

$$\text{Precision} = \frac{\text{Retried keywords} * \text{Useful keywords}}{\text{Uscful keywords}}$$

Table 1 Results of PKEA algorithm

Doc Id	Retrieved keywords	Useful keywords	Keywords classified as pests	Keywords classified as pest types
1	250	88	34	54
2	300	101	45	56
3	159	75	28	47
4	88	40	19	21
5	274	79	27	52
6	311	121	56	65

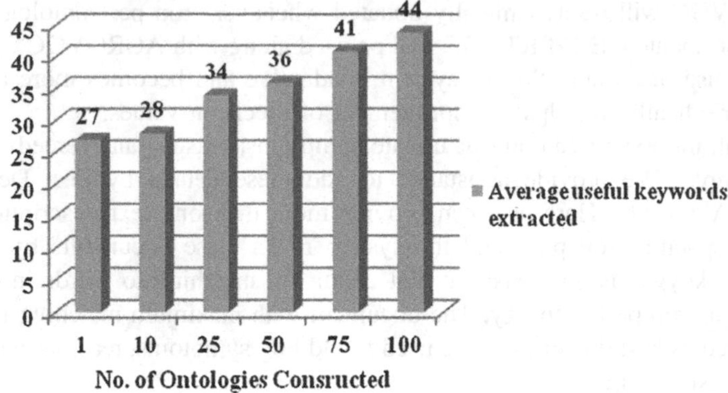

Fig. 5 Improvement in number of useful keywords extracted

Recall for the algorithm is calculated as

$$\text{Precision} = \frac{\text{Retried keywords} * \text{Useful keywords}}{\text{Useful keywords}}$$

Average precision of CropPestOntoGenerator is 1.4 and average recall is 3.7.

We used CropPestOntoGenerator to generate hundred ontologies. We retrieved top hundred keywords from each document and recorded number of useful keywords for each. We found improvement in number of useful keywords as a result of adaptive algorithm. Improvement in results is shown as graph in Fig. 5.

4 Conclusion

As ontologies play major way in Semantic Web and it is an easy way to represent knowledge in various languages, representing agricultural knowledge in terms of ontologies is also important. The system described in the paper called as CropPestOntoGenerator serves this purpose. Using natural language processing techniques, the system assists agricultural expert to generate hierarchy of pests for particular crop and also to identify pest examples of specific crop type. With CropPestOntoGenerator, agricultural expert does not need to know anything about ontologies. Using GUI provided by CropPestOntoGenerator, agriculture expert has to just fill in details about pest types, pest examples and reasons, symptoms and remedy for pest example. Assistance for filing this information is provided by extracting knowledge from text descriptions of pests provided by agricultural expert. With CropPestOntoGenerator, we demonstrate that using NLP techniques and specific representation capability of OWL, knowledge with experts can be converted into structured format and provided to farmers in various ways. We plan to extend this work to use generated pest ontologies to help farmers.

We plan to use these ontologies as part of an agricultural expert system, where knowledge from ontologies will be used to provide expert advice to farmers. We will analyze weather data using Big Data analytics and find out relation between weather changes and crop pests in specific location of country. We will use crop pest knowledge from generated ontologies and results from weather data analysis and develop an inference engine. This inference engine will be perfect combination of expert knowledge and information technology. It will be a core part of our agricultural expert system.

References

1. Sukanta Sinha, Rana Dattagupta, Debajyoti Mukhopadhyay; Designing an Ontology based Domain Specific Web Search Engine for Commonly used Products using RDF; CUBE 2012 International IT Conference, CUBE 2012 Proceedings, Pune, India; ACM Digital Library, USA; September 3–5, 2012; pp. 612–617; ISBN 978-1-4503-1185-4.
2. Debajyoti Mukhopadhyay, Aritra Banik, Sreemoyee Mukherjee, "A Technique for Automatic Construction of Ontology from Existing Database to Facilitate Semantic Web", 10th International Conference on Information Technology, ICIT 2007 Proceedings; Rourkela, India; IEEE Computer Society Press, California, USA; December 17–20, 2007; pp. 246–251, IEEE Xplore.
3. Debajyoti Mukhopadhyay, Archana Chougule; An Approach to Manage Ontology Dynamically based on Web Service Composition Requests; CUBE 2012 International IT Conference, CUBE 2012 Proceedings, Pune, India; ACM Digital Library, USA; September 3–5, 2012; pp. 653–658; ISBN 978-1-4503-1185-4.
4. Jena Ontology API, "http://jena.apache.org/documentation/ontology/".
5. Debajyoti Mukhopadhyay, Rituparna Kumar, Sourav R. Majumdar, Subhobroto Sinha, "A New Semantic Web Services to Translate HTML pages to RDF", 10th International Conference on Information Technology, ICIT 2007 Proceedings; Rourkela, India; IEEE Computer Society Press, California, USA; December 17–20, 2007; pp. 292–294, IEEE Xplore.
6. Chris Biemann, "Ontology Learning from Text: A Survey of Methods", LDV forum, 2005.
7. Hoifung Poon and Pedro Domingos, "Unsupervised Ontology Induction from Text", ACL'10 proceedings of the 48th Annual Meeting of the Association for Computational Linguistics, pp. 296–305, ACM Digital Library, 2010.
8. René Witte, Ninus Khamis, and Juergen Rilling, "Flexible Ontology Population from Text: The OwlExporter", International workshop on Ontology Dynamics-IWOD2007, 2007.
9. Toader Gherasim, Mounira Harzallah, Giuseppe Berio, and Pascale Kuntz, "Methods and Tools for Automatic Construction of Ontologies from Textual Resources: A Framework for Comparison and Its Application", Advances in Knowledge Discovery & Management, SCI 471, pp. 177–201, Springer-Verlag Berlin Heidelberg, 2013.
10. Porter, "An algorithm for suffix stripping", Program, Vol. 14, no. 3, pp 130–137, July 1980.
11. OpenNLP, "https://opennlp.apache.org/".
12. TFIDF Algorithm, http://en.wikipedia.org/wiki/Tf-idf.
13. Medelyan, O., Witten I. H., "Thesaurus-based index term extraction for agricultural documents." In: Proc. of the 6th Agricultural Ontology Service (AOS) workshop at EFITA/WCCA 2005, Vila Real, Portugal, 2005.

14. AGROVOC Thesaurus, http://aims.fao.org/agrovoc#.VF29AvmUc2U.
15. Protégé, "http://protege.stanford.edu/".
16. Paul Buitelaar, Daniel Olejnik, Michael Sintek, "A Protégé Plug-In for Ontology Extraction from Text Based on Linguistic Analysis", The Semantic Web:Research and Applications, LNCS Volume 3053, DOI 10.1007/978-3-540-25956-5_3, pp. 31–44, Springer Berlin Heidelberg, 2004.

Deployment of Data Aggregation Technique in Wireless Sensor Network

Rutusha Patel and Shailaja Kanawade

Abstract Data aggregation is implemented in wireless sensor network to recapitulate the requisite information and send it to sink by reducing redundancy. We have developed energy-balanced real-time data aggregation technique, which prolongs the era of the network by trivializing power consumption and enhancing data transmission. Data coming from sensor node is accumulated at Intermediate node in an energy-efficient way with minimum data latency and this data is available at the base station. Sensor nodes are battery operated and efficiency of a battery hinges on number of bits transmitted. In our algorithm, we have lessen the number of bits transmitted. In order to surge lifetime of a wireless sensor network, we have to develop an energy-efficient sensor node for this we have to develop an intermediate node which accumulates data processes that and make it available to the base station precisely. We have used VB.NET as a front end programming and SQL database. We have attained here 90 % of energy saving.

Keywords Data aggregation · Wireless sensor network · Sensor node · Energy efficiency

1 Introduction

When we think at data-centric network, all messages must be delivered to all sinks. The real power lies in potential to process on data. The facile example of processing of data is aggregation of data i.e., computation of data and broadcasting the aggregated value. More advanced example inculcates measuring environment and battery life. There are many aggregation techniques implemented such as

Rutusha Patel (✉) · Shailaja Kanawade
Electronics and Telecommunication Sandip Institute of Technology and Research Center (SITRC), Nashik, India
e-mail: rutushap@gmail.com

Shailaja Kanawade
e-mail: kanawade.shailaja@sitrc.org

© Springer Science+Business Media Singapore 2017
S.C. Satapathy et al. (eds.), *Proceedings of the International Conference on Data Engineering and Communication Technology*, Advances in Intelligent Systems and Computing 468, DOI 10.1007/978-981-10-1675-2_66

fuzzy-based logic for structure-free aggregation, concealed data aggregation, hybrid, etc. The actual perk lies on the location of sink that is weather sink is located in harsh and unpredictable environment or in capturable environment. Intuitively, if all data sink is located in craggy environment, then there is very little possibility that they are able to interact with intermediate node. The mechanism of data aggregation is quiet transparent which includes the formation of network in tree form where there is an intermediate node which applies some form of aggregation of data, which it has collected from its children and it broadcasts the data to sink. The potential of data aggregation is arbitrated on the basis of accuracy, latency, and message overhead (1). In this paper, we have shown real-time data collected at the intermediate node. The data which is collected is highly correlated with each other as per spatial correlation, the correlated data save more energy and provide maximum accuracy with reduced delay. Flow of our paper shows block diagram of our system, algorithm to implement data aggregation, result, calculation, and comparison with other system.

2 Related Work

In contemporary years, many explorations have been carried out in data aggregation and data gathering at intermediate node. A variety of protocols have been proposed that permits routing and data aggregation simultaneously. Data aggregation can be realized by collective knowledge of artificial intelligence and networking concepts (8). Real-time data aggregation protocol (RDAG) proposed by M. Yeganeh in structure-free network has controlled parameters aggregation gain, miss ratio, and energy efficiency but it has one drawback that is latency is not evaluated (6). Structure-free and energy-balance data aggregation protocol (SFEB) (7) have controlled delay and message overhead, but end-to-end delay and energy consumption is increased for longer message packets. By considering the above drawbacks, we have developed an energy-balanced real-time data aggregation technique which reduces delay, increases energy efficiency, reduces message overhead, and promises accuracy (7).

3 Block Diagram of a System

Figure 1 represents block diagram of our system. Here typically we are developing an intermediate node whose main work is to process data. At the transmitting side, we are using here MQ7 gas sensor, which senses data from real environment it basically provides analog output. Analog output is given to Arduino microcontroller that converts it into 10-bit data (digital data). Arduino microcontroller contains 10-bit ADC, so it converts data into 10 bit. Visual basic.net will perform the aggregation algorithm that we have developed. It uses SQL as a data base. Aggregated data is

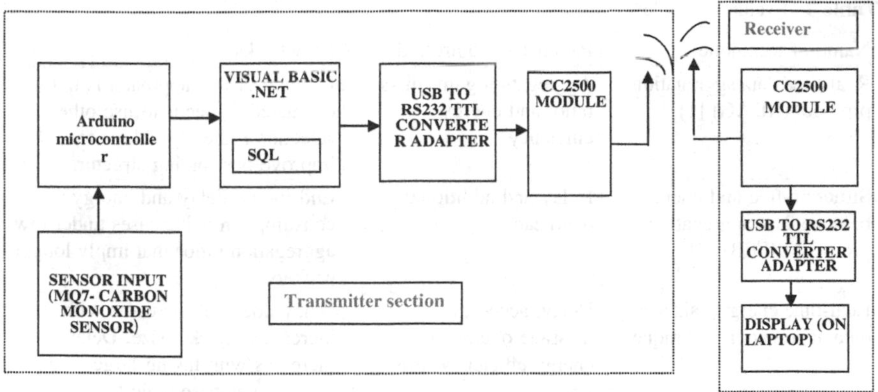

Fig. 1 Block diagram of a system

radiated through CC2500 transreceiver here cc2500 works as a transmitter. At the receiver section, cc2500 transreceiver is used to receive the data and the data is displayed on com port 17 in laptop. Here cc2500 transreceiver works as a receiver. USB to RS232 TTL converter is used to configure input and output pins and convert into TTL logic at transmitter side and vise-versa at receiver side (1, 2).

4 Algorithm of System

Steps for algorithm are as follows:

Step 1 Make hardware connections and check Arduino port is connected to com port 17
Step 2 Convert the analog data into digital by ADC
Step 3 Create database in SQL for storing real-time data from MQ7 carbon monoxide sensor
Step 4 Give name to the database table as AGRITBLE5
Step 5 Declare ID
Step 6 Read ID by descending order
Step 7 Check table has entry with ID index = 0
Step 8 If ID is equal to 0 then insert values with 0 index and row has entry then select ID from aggritable with decending order
Step 9 Set value of ID to 10 check the value of ID
Step 10 If ID < 10 then insert values and increment ID = ID + 1 till ID >= 10
Step 11 If ID >= 10 Then calculate the aggregated value of inserted data
Step 12 Print the aggregated data and radiate in wireless medium
Step 13 Delete aggritbl5 to release memory for new values

Table 1 Evaluation table

Name of technique	Parameters controlled	Drawbacks
Real-time data aggregation protocol (RDAG) [1]	Aggregation gain, miss ratio, and energy efficiency	Latency of this approach is not evaluated. It needs to use other selection route algorithms to improve our routing structure
Structure-free and energy balance data aggregation protocol (SFEB) [2]	Delay and additional overhead	End-to-end delay and energy consumption is increases under low aggregation ratios that imply longer packets
Real-time energy-balanced data aggregation technique	Delay, accuracy, message overhead, energy efficiency, miss ratio	Delay does not increases with increase in packet size. Delay increases with taking large data entry in database table for computing aggregation

First and foremost thing we are doing is to make hardware connections and connect arduino board to com port 17 in laptop. Then the algorithm creates the database to store real-time data from sensor and gives name AGRITBL5. Database table contains 10 entry it has ID pointer as named kp which is set to 10. That means we are using chunk of 10 data. Its selects order of ID in decending order; the moment data is stored in database event occurres and it checks kp pointers position, when kp pointer has reached to 10th position it will stop accepting values and fires aggregation queries to calculate average of collected data. After calculating average this data is sent to transmitter to transmitt in wireless medium. Once the aggregated data has been sent, it clears the AGRITBL5 (ame of database table) to store new values in it. If kp pointer has not reached to 10th position, then insert the data till it reaches to 10th position (Table 1).

5 Result and Discussion

There are number of sensors like light, temperature, gas, sound power, etc., we have used here MQ7, i.e., carbon monoxide sensor. Basic application of our hardware is in pollution monitoring. Sensor motes are equipped with atmega328 microcontroller which has clock speed 16 MHz and 32 KB flash memory, our scheme uses carrier sense multiple access media access protocol. Efficiency of a sensor nodes are determined by power supply. In our database, we are taking chunk of 10 data and average of 10 data is calculated and average is transmitted. Transmitter consumes 4000 nJ of energy, while transmitting a single bit uses (2).

A. *Calculations*:

Energy calculations:
1 bit uses 4000 nJ of energy
We are transmitting 10 bit, so

$$4000 * 10 = 40000 \text{ nJ} \tag{1}$$

Average of 10 data is taken and send only once instead of sending 10 times
Hence, transmitter is on only for sending average data instead of sending all the data in database.
40000 * 9 = 360000 nJ of energy is required to transmit data 9 times
But since we are using only one transmission, we are saving 360000 nJ of energy which results in 90 % of energy saving from non aggregated data (1, 2, 3, 4, 5).

Delay calculations

$$T_{\text{total}} = t_{\text{process}} + t_{\text{transmit}} \tag{2}$$

T_{total} is total time required
t_{process} time required to process data
t_{transmit} is time required to transmitt data

$$\begin{aligned} \text{total time required} &= 9000 \text{ ms} + 50 \text{ ms} \\ &= 9050 \text{ ms} \end{aligned} \tag{3}$$

Total delay is 9050 ms
Figure 2 shows the display of average data calculation. Chunk of 10 data is taken and their average is calculated and transmitted in wireless medium, for example, in above case aggregated data is 1013.
Evaluation table showing comparison with two techniques and drawbacks of two techniques is covered in real-time energy-balanced data aggregation protocol.

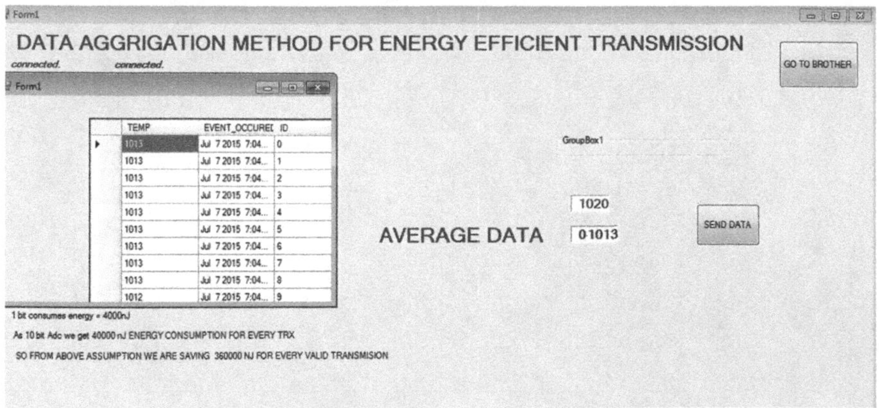

Fig. 2 Display of average data calculation

6 Conclusion

In this paper, we have developed a real-time energy-balanced data aggregation technique that promises the accuracy and reduces delay. For evaluation, we have considered data from carbon monoxide sensor as a real-time data. While comparing our technique with nonaggregated data, it shows that we are saving approximately 90 % of energy also message overhead and network data traffic is decreased as we are sending aggregation of 10 chunks of data instead of sending it one by one. The evaluation table shows that our technique covers drawbacks of real-time data aggregation as well as of structure-free energy-balanced data aggregation. We have developed an algorithm and hardware arrangements that is beneficial to the society as a pollution monitoring system. Even if we increase the packet size of each data in database table, delay is not increased and accuracy is also promised. If we increase the number of data entry in data base table then delay increases. Amount of delay increased is just equal to time required to fill the database table. Future work includes taking large number of data entry in database table by balancing accuracy, delay, and message overhead.

References

1. Mohammad Hossein Yeganeh, Hamed Yousefi, Naser Alinaghipour and Ali Movaghar, "RDAG: A Structure-free Real-time Data Aggregation Protocol for Wireless Sensor Networks", 17th IEEE International Conference on Embedded and Real Time Computing Systems and Applications, 15332306/11 $26.00 © 2011 IEEE, DOI 10.1109/RTCSA.2011.70.
2. Chih-Min Chaoy and Tzu-Ying Hsiaoz, "Design of Structure-Free and Energy-Balanced Data Aggregation in Wireless Sensor Networks", IEEE HPCC, 2009. High Performance Computing and Communications, 2009. HPCC'09. 11th IEEE International Conference, 25–27 June 2009, Page(s): 222–229 E-ISBN: 978-0-7695-3738-2, Print ISBN: 978-1-4244-4600-1, INSPEC Accession Number: 10789244.

Energy-Efficient Self-organization Wireless Sensor Network for Traffic Management in Smart Cities

Sumedha Sirsikar and Manoj Chandak

Abstract People are delimited by abundance use of wireless computing and communication devices that are connected via wireless technology in an overlay network. Wireless sensor networks (WSNs) have gained much attention in both public and research communities as they are expected to bring the communication between humans, environment, and machines to a new paradigm. Wireless networks gradually need to self-organize and become accustomed to new conditions for ease in management and operation. In this article, we have considered an application of traffic management for smart city. The implementation of sensors for such an application needs long life of sensors. Clustering of sensor nodes can be used for energy conservation in WSN. We have proposed a framework of self-organized wireless sensor network (SOWSN) implemented the algorithm for clustering of nodes based on residual energy of nodes and node distance to minimize the average energy consumption. We will also study the effect of node distance on energy consumption.

Keywords Wireless sensor networks · Traffic management · Smart city · Clustering · Energy efficiency · Network lifetime

1 Introduction

Recently, there is a huge increase in the traffic in all the metro cities due to large number of vehicles. Currently, time-based traffic signals are used to control the traffic. But still the existing system is inefficient to manage the traffic.

Sumedha Sirsikar
Maharashtra Institute of Technology, Pune, Maharashtra, India
e-mail: sirsikarsd@gmail.com

Manoj Chandak (✉)
Ramdeobaba College of Engineering and Management, Nagpur, Maharashtra, India
e-mail: chandakmb@gmail.com

© Springer Science+Business Media Singapore 2017

S.C. Satapathy et al. (eds.), *Proceedings of the International Conference on Data Engineering and Communication Technology*, Advances in Intelligent Systems and Computing 468, DOI 10.1007/978-981-10-1675-2_67

This era is the technology based. The concept of smart cities is now in development. This can include the traffic management as well, because managing the traffic smartly will contribute to development of smart city. This can be achieved through the use of wireless sensor network. As the sensors are cheaply available, they can be installed in the vehicles easily.

Our system consists of sensors deployed on the roadways as well as the vehicles. A central node at every junction receives data from sensors place on the road and vehicles. The sensors detect the presence of vehicle and the transmitter wirelessly transmits the traffic data to the central node.

2 Solution for Traffic Management

The sensor nodes are deployed at the roads or the junctions. Also sensor nodes are deployed on each vehicle as seen in Fig. 1. As the vehicles come near the junction, their presence is sensed by the sensors present there. This will make use of counting the vehicles. When it goes beyond some value, the traffic system will become active. The person sitting on the vehicle will be notified about the traffic and will be suggested an alternate route through GPS.

When the traffic increases, the sensors will also increase and this sensor network is battery powered. Hence, the energy of the sensor nodes is to be saved. This is done through the clustering approach. When the sensor network is long lasting, the traffic management system could be more effective.

3 Long Life of WSN

WSN is large-scale network of sensor devices, each possess sensing, computation, and communication capabilities. Sensor nodes are constrained in terms of processing power, communication bandwidth, and energy and memory. Thus, WSN

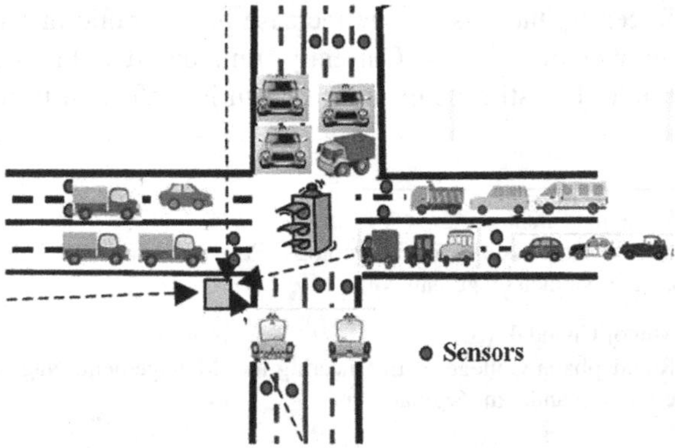

Fig. 1 Deployment of sensors at roadways

requires very efficient resource utilization. Energy usage is an important issue in the design of WSNs and self-organization can handle this issue prominently. Clustering of sensor nodes is also one form of self-organization.

4 Self-organization

'Self-organization (SO)' is a process in which systems become more organized over the period, with self-induced changes in organization, without applying any external control [1]. Molleman considered SO as: the local autonomy to make decisions [2]. According to Dressler, SO is a process in which structure and functionality (pattern) at the global level emerge solely from interactions among the local components without any external or centralized control [3]. Haken, the founder of Synergetics, denotes SO as: A system achieves spatial, temporal, or functional structure when there is no interference provided from outside [4]. Kevin L. Mills defined Self-organization as: mechanism that allows devices to build topologies that can monitor and adapt to environmental changes without human intervention [5].

4.1 Mapping of Self-organization in WSN

The working of sensor nodes can be compared with that of human brain, as shown in Table 1. The human brain starts thinking, i.e., it senses the activities in the environment and instructs our body to react accordingly. Same is the case with sensor nodes. Applying SO to wireless sensor networks is just like providing it with the ability to take its own decisions.

Table 1 Comparison between human brain and WSN

Sr. no.	Human brain	WSN
1	Physical mechanism of thought operation in brain	Sensing capacity of sensor nodes
2	Gaging human brain's storage requirement (i) Average rate of information acquisition (ii) Total amount of accumulated data	Sensor nodes filter the sensed data and remove redundancy after processing

5 Design Policies for Self-organization Functionality in WSN

Applying self-organization to sensor networks requires designing of the rules for interactions among the nodes.

Policy 1: Derive global properties from local behavior rules:

WSN have no centralized entities that could inform the nodes about environmental or topological changes; hence, each node has to continuously monitor its local environment and react accordingly. For example, individual sensor nodes are unaware of global environment, i.e., who are their neighboring nodes or which node is able to be a cluster head; but through local interactions, they get knowledge about entire network [6]. This is how the network takes a global decision.

Policy 2: Utilize implicit coordination:

Implicit coordination means that coordination information is not communicated explicitly by signaling messages, but is inferred from the local environment [7]. A node observes other nodes in its neighborhood; based on these observations, it determines the status of the network and reacts accordingly.

5.1 Model for Designing a Self-organized WSN

The model for designing a self-organized sensor network integrates the policies (that are mentioned as above) and it is applied iteratively to obtain a protocol that implements the desired self-organization function. Figure 2 shows the design process for the model of self-organized sensor network.

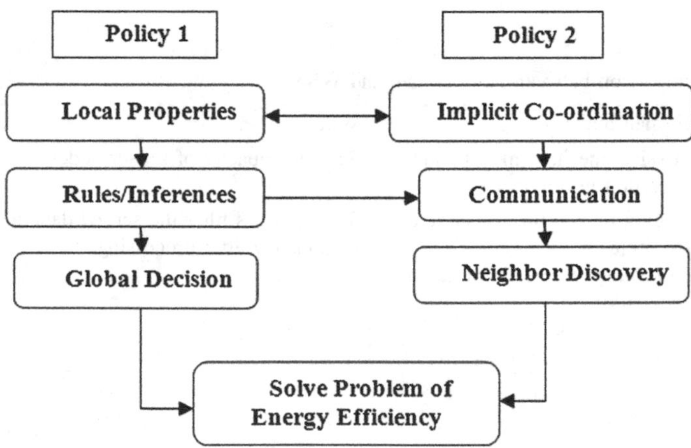

Fig. 2 A design process for a self-organized WSN

6 Proposed Work

We have applied the concept of self-organization through clustering algorithm to WSN, in order to make it energy efficient. Following sections describe the same.

6.1 Framework of Self-organized Wireless Sensor Network (SOWSN)

Several new design themes have emerged for sensor networks. Each sensor has limited battery power, thus the network must minimize total energy usage. A WSN with sensor nodes communicating with each other through local interaction, implicit coordination and neighbor discovery is able to take global decision. The WSN now becomes self-organized wireless sensor network (SOWSN). Still SOWSN is not energy efficient. Hence, clustering technique is applied to it so that it becomes energy-efficient SOWSN. This whole process is reflected in Fig. 3.

6.2 Proposed Algorithm

Energy is the major constraint of WSN. In this research work, this problem is kept in mind and a solution to overcome energy constraint has been proposed. This proposed solution is based on clustering and consists of five phases:

1. Division of Network into Fixed Regions
Divide complete network area into fixed number of rectangular regions.

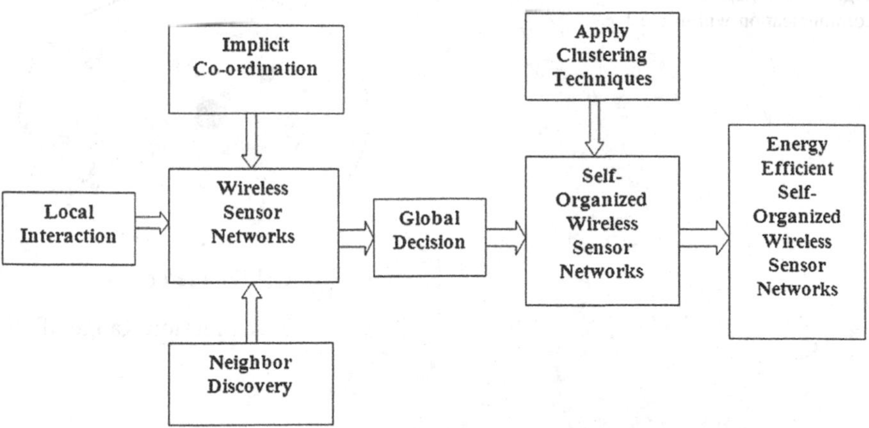

Fig. 3 Framework of SOWSN

2. Cluster Head Selection

The first step toward clustering is cluster head (CH) selection. Sensor nodes are selected as CH based on two parameters: highest residual energy and node degree. The node degree refers to number of 1-hop neighbors of a node.

3. Cluster Formation

The clusters are formed on the basis of node distance.

Steps for cluster formation are as follows:

1. The distance between nodes is calculated by Euclidian distance formula.
2. Nodes at single hop within the transmission range of the CH join that CH.
3. Next hop nodes which are within cluster diameter are allowed to join cluster.
4. The nodes which are in the cluster diameter they join cluster with single hop member nodes by two-hop communication as shown in Fig. 4.

4. Actual Data Transmission

In this phase, the CHs create time slots for data transmission. The node which is 2-hop away sends its data to the 1-hop neighbor, which adds its data to the received one and finally transmits it to the CH. This reduces the number of transmissions to the base station and also energy consumption. The CH nodes send their data to the sink node or base station through multi-hop communication.

5. Re-clustering

During the continuous operation of sensor network, the battery level of sensor nodes as well as cluster heads goes down. Due to this, sensor nodes and cluster heads become inactive. If the CH energy is below the threshold value, the sensor network is reorganized into new clusters. In this process, the node with highest residual energy and degree will become new cluster head. This process shows the self-organization behavior. Based on all cluster reformation process, a network

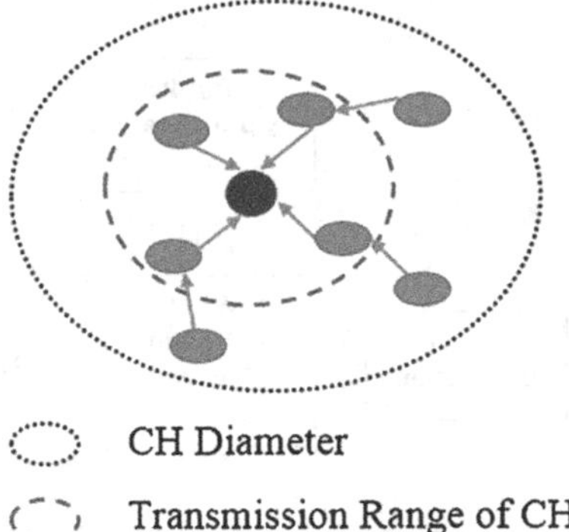

Fig. 4 Two-hop communication within cluster

model is defined. Entities of our model are defined in this section. Let M be the network model which can be described as follows:

$M = \{N, C, CH, BS\}$
where
N: Number of sensor nodes,
C: Cluster of sensor nodes,
CH: Cluster head BS: Base station

6.3 Mathematical Modeling

Mathematical model is described in this section.
N: Set of all nodes present in the network
$N = \{n1, n2, n3, \ldots.\}$
Ei = Initial energy of node,
Er = Residual energy of node,
Ec = Energy consumed by node = Initial energy – Residual energy

- TCE: Total Consumed Energy

  $$\mathrm{TCE} = \sum_{i=1} (Ei - Er)$$

- AEC—Average Energy Consumption:
  AEC = TCE/Number of Nodes

D: Distance between two nodes

The distance 'D' between any two nodes with x & y coordinates (x_1, y_1) & (x_2, y_2) is calculated using the Euclidean distance formula:

$$D = \sqrt{(x_2 - x_1)^2 + (y_2 - y_1)^2}$$

7 Simulation and Results

The simulation of proposed two top clustering algorithm is carried out with NS2. Table 2 shows the simulation parameters.

Figure 5 shows the cluster formation in the network based on residual energy and node degree. Different colors show different clusters.

The proposed algorithm is run for different number of nodes at different transmission ranges. The average energy consumption for every simulation is calculated. Table 3 shows the obtained values of Average Energy Consumed and Network

Table 2 Simulation parameters

Parameter	Values
Network area	3400 m × 1800 m
Channel type	Wireless
Number of nodes	40, 60, 80, 100
Initial energy	1000 J
Threshold energy	100 J
Transmission range	250, 350, 400 m
Transmitting and receiving power	1 W
Simulation period	100 s

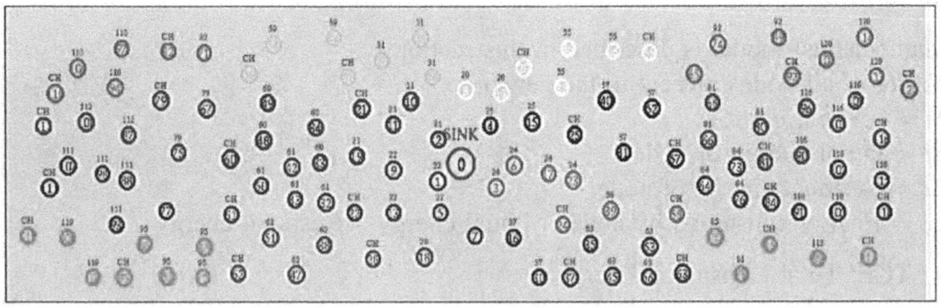

Fig. 5 Cluster Formation

Table 3 Average energy consumed and network lifetime

No. of sensor nodes	Average energy consumed (J)			Network lifetime (S)		
	250 m	350 m	400 m	250 m	350 m	400 m
40	7.51	11.73	13.79	3667.23	2254.67	2014.18
60	8.56	11.23	13.45	4431.19	2980.53	2755.35
80	8.54	10.72	12.87	5853.03	4429.31	3725.03
100	8.54	10.45	12.31	6367.97	5105.35	3935.22

Lifetime for different number of nodes like 40, 60, 80, and 100 at transmission ranges 250, 350, and 400 m. Figure 6 shows corresponding graphs.

As we can see in the graph and table, as we increase the number of nodes from 40 to 100, the average energy consumption is remaining constant at range 250 m. For range 350 and 400 m, there is reduction in energy consumption. Also the network lifetime is increasing at every simulation run. Again when the CH's energy goes down, nodes mutually decide which node will be the new CH. The ability to take this decision comes through Self-Organization and clustering.

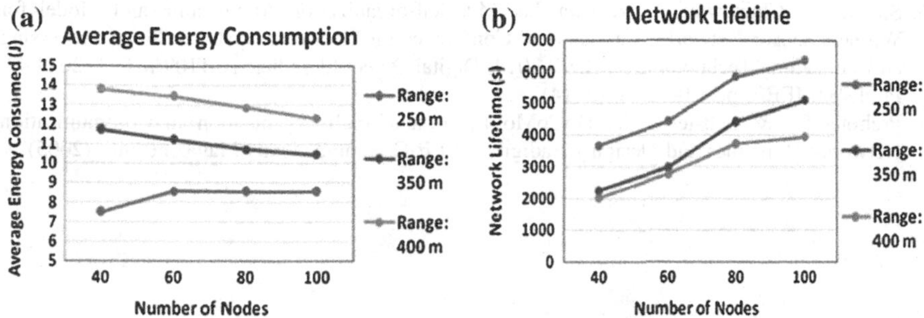

Fig. 6 **a** Average energy consumption versus number of nodes **b** Network lifetime versus number of nodes

8 Conclusion

Self-organization is necessary for the operation of sensor networks without human intervention. The clustering of sensor nodes proves to be a self-organizing behavior of WSN. In this paper, we have proposed a framework of self-organized wireless sensor network (SOWSN) and implemented the algorithm for two-hop clustering of nodes based on residual energy of nodes and node degree. We have compared performance of our algorithm for calculation of average energy consumption by taking different number of nodes and at different transmission ranges.

The simulation results show that the proposed algorithm is able to reduce average energy consumption. This reduction in average energy consumption increases the network lifetime. Also when the number of nodes is increased, the energy consumption is reduced at particular range, which shows the scalability of the network. The extension of this work is to compare our algorithm with various benchmark protocols.

References

1. Ashby W.R.: Principles of the self-organizing dynamic system, Journal of General Psychology, pp. 125–128, (1947).
2. YANG Yi, YAN Hong-yan, YANGZe-yun: Research on Mutual Guarantee System of SMEs Based on Self-organization Theory, 2010 International Conference on Management Science & Engineering, Melbourne, Australia November (2010).
3. Dressler: A Study of Self-Organization Mechanisms in Ad Hoc and Sensor Networks, Article published in Elsevier Computer Communications 31, pp. 3018–3029 (2008).
4. Wang-Yan: Study on Model and Architecture of Self-Organization Wireless Sensor Network.
5. Kevin L. Mills: A brief survey of self-organization in wireless sensor networks, Wirel. Commun. Mob. Comput. 2007; 7: 823–834 published online Wiley InterScience May (2007).

6. Sirsikar S., Chunawale A., & Chandak M.: Self-organization Architecture and Model for Wireless Sensor Networks, International Conference on Electronic Systems Signal Processing and Computing Technologies, ICESC 2014, Digital Object Identifier: 10.1109/ICESC.2014.42, Publisher: IEEE, pp. 204–208, (2014).
7. Prehofer C., & Bettstetter C., DoCoMo Euro-Labs.: Self-Organization in Communication Networks: Principles and Design Paradigms, IEEE Communications Magazine July (2005).

Inter Area Frequency Monitoring in Large Power Grid Network

Gresha S. Bhatia and J.W. Bakal

Abstract Power grids are considered to be the backbone of modern society. This grid, however, has remained unchanged over many decades and is collapsing. Blackouts, outages are becoming inevitable and the demand for electricity has been exceeding the supply implying a significant risk of system wide failures. Hence, monitoring the grid operations and determining mechanisms to prevent the collapsing scenario are the requirement of the hour. In this paper, we focus on monitoring the frequency component of the power grid. Fluctuations in the frequency parameter can cause catastrophic scenario. To prevent this, in this paper, we have proposed a mechanism of variance monitoring to determine the frequency fluctuations and later on apply the load distribution mechanism to redistribute the load of the faulted node through a combination of Intentional Removal (IR) and avcrage distance between the nodes.

Keywords Failures · Variance monitoring · Frequency component · Capacity of a node · Operating modes · Intentional removal section

1 Introduction

Power grids are regarded as one of the most critical infrastructures on earth. A successful attack on these critical power grids can cause the equipments to behave irrationally causing a rippling effect, in turn affecting a wide area, creating a long-term outage that can eventually pass on these failures in other crucial infrastructures that are highly dependent on these power grids. Therefore, any fluctuations in the operating parameters of the grid can lead to the entire grid

G.S. Bhatia (✉)
Thadomal Shahani COE, Mumbai, India
e-mail: itsgresha@yahoo.co.in

J.W. Bakal
Shivajirao S Jondhale COE, Mumbai, India
e-mail: bakaljw@gmail.com

© Springer Science+Business Media Singapore 2017

S.C. Satapathy et al. (eds.), *Proceedings of the International Conference on Data Engineering and Communication Technology*, Advances in Intelligent Systems and Computing 468, DOI 10.1007/978-981-10-1675-2_68

becoming unstable and tending toward moving the subgrid toward cascading failure and finally into a blackout state. To add to this the operators of the system, clueless of the scenario, create panic. Since these operators are unaware of the reasons behind, the catastrophic failures are unable to communicate it to the maintenance and field personnel. Therefore, there is a need to provide a monitoring service in the control center which would provide system-level information based on dependency graphs, event logs to increase the situational awareness between the control centers and the utility personnel [1, 2].

The remainder of this paper is organized as follows. Section 2 presents the operating modes of the grid in the existing energy management systems. Frequency as one of the components of operation is considered as the focus area and described in Sect. 3. Frequency monitoring through variance measures and its details are discussed in Sect. 4. This section further elaborates the mechanisms involved for monitoring the frequency component parameter of the grid and specifies ways to prevent the cascading failures. Further conclusion drawn is specified in Sect. 5.

2 Modes of Operation of the Grid

The existing energy management systems (EMS) have been designed to operate the grid in multiple modes or states during which specific operational and control actions need to be initiated. These modes can be classified into the normal mode in which all the equipment's are expected to be operating within the operating limits and no contingencies cause real-time operating limit violations [1, 2]. This mode majorly focuses on the load frequency control. Alert mode forms the normal insecure state where in changes in one or more contingencies create operating limit violations [3].

Failure mode is activated when some of the equipment's operate outside the real-time operating limits. In this mode, there is a need to identify the node that has failed and identify the appropriate action that needs to be undertaken quickly so as to restore the power in the shortest response time. Thus, restoration mode has to be activated as fast as possible to steer the system back to the normal state [1].

Furthermore, the operating modes are selected such that at any given instant of time, the power grids exists in any one of the states mentioned above [4].

3 Focus Area

Blackouts refer to the total loss of power to an area and occur due to power station tripping. These outages may last from minutes to even days, completely shutting down critical infrastructures such as the financial services, telecommunication networks, etc., to name a few [5].

Thus, since the power grids form the foundational critical infrastructure, our focus is on building a robust transmission grid system capable of detecting and recovering from cascading failures due to drop in frequency parameter in an effective manner [5]. This curbing can thus prevent the spreading of failures and enable faster system wide recovery in a timely and cost-effective manner [5].

Though a lot of work has been done in the domain of cascading failures, these failures are handled either at the generation side or the distribution side. Focus on the transmission grid and its protection mechanisms are limited, though it forms the crucial link between the generation and distribution [6].

4 Frequency Monitoring

Frequency component is considered to be one of the most important parameter in understanding the operations of the power systems. As any fluctuations in this parameter can have adverse effects, it is imperative to determine the drop or increase in the frequency component of the power grid. This would thus aid the operator at the control center to initiate necessary action and help in better monitoring of the grid and restore normalcy onto the network [1].

However, the existing systems utilized to guard against these small signal instabilities are the offline tuning of power system stabilizers with mechanisms that have hard wired communication, and therefore the system values and status values monitored cannot be controlled.

In order to provide flexibility for the power grid management, the concept of variance monitoring is introduced.

Variance basically measures how far a set of numbers is spread out. It involves a two-pass algorithm, where in, the real-time frequency that enters the grid is monitored to be within the range of (49.2–50.7 Hz). If the range is faltered, the chances of system collapsing are high causing the entire system to collapse [7].

The two-pass algorithm is implemented as described below:

(1) Crossing over the threshold values, the mean of the values coming into the system are first calculated by

$$\bar{X} = \frac{\sum_{j=1}^{n} x_j}{n} \tag{1}$$

where, $\bar{X}$ represents the mean of the samples undertaken
n is the total samples taken
xj is the actual sample taken at an instant of time.

(2) From the mean represented by Eq. 1, the variance is determined as

$$\text{Variance} = S^2 = \frac{\sum_{i=1}^{n}(xi - \bar{x})^2}{n-1} \quad (2)$$

This variance value is treated as the cut-off point for determining the threshold. Once the threshold value is obtained, every frequency component coming into the system is compared with the variance threshold as shown [8], i.e., value $< \cdot >$ variance threshold. If this value is obtained to be greater than the threshold value represented by Eq. 2, the node triggers the alert phase leading to the individual monitoring of the specific node. This further aids to provide for the situational awareness with event localization to the operator for close monitoring where various parameters are defined as follows:

df is defined as the variance factor

Li is the initial load capacity which a node can handle

Li (*g*) is the intermediate node capacity

d is the average distance between the nodes.

Since the node has crossed its threshold, its impact on other nodes needs to be determined. To do this, the logic of Intentional Removal (IR) is applied where the node with the least value of [*Li* – *Li* (*g*)] is considered for removal [8, 9].

Once the node is identified, corresponding load of [*Li* – 2 *Li* (*g*)] is distributed to the remaining nodes. The initial capacity of node assumed to be $Li \leq Ci$ where *Ci* is the capacity of the node. *Ci* is further determined by

$$Ci = \lambda Li \ \ldots \quad (3)$$

where the capacity of a node is identified to be directly proportional to the initial load as represented by Eq. 3. λ defines the tolerance parameter and $\lambda \geq 1$ [10–12].

Further, the capacity of a line is defined as the maximum load that can be carried by the line and given as $Ci = (1 + \lambda\, Li)$.

Also, the load that can be given to a particular node needs to be decided based on not only the capacity of that node but also on the distance between the nodes which is given by

$$Lij = (D - 1) * (N - 1) \quad (4)$$

where, D = mean (Z)

Z = average distance between the nodes.

Thus, when a node i is removed from the network the total load on the remaining nodes increase by at least $Li - 2\,Li(g)$ and is based on the distance as represented by Eq. 4 [12–14].

This further prevents load to be given to a single node that could collapse the entire system [15].

5 Results and Evaluations

Since there are chances of false or short frequency fluctuations, the need for applying the concept of variance is required in order to reduce the impact of these fluctuations on the grid operations. Based on the two-pass algorithm described above, the values considered are as follows:

Since the frequency range is supposed to be in the range of 49.2–50.7 Hz, fluctuations with in this range can create problems too.

Consider two sets of frequency ranges namely $F1$ and $F2$ taking the values as: Let $F1 = [50, 50, 49, 49, 48, 48, 51, 51, 48, 48]$; $F2 = [50.5, 51, 50, 51, 49, 49, 51, 52, 50, 50]$;

This is then multiplied by random values,

$r = [r;\ F1(y) + (F2(y) - F1(y)) * \text{rand}\ (100, 1)];$

This value is determined for every node and represented as the output frequency for that node as shown in Fig. 1.

Where 1 s time frame represents 100 samples (Fig. 2).

As soon as the variance value crosses the threshold, it activates monitoring of that single node itself as shown in Fig. 3.

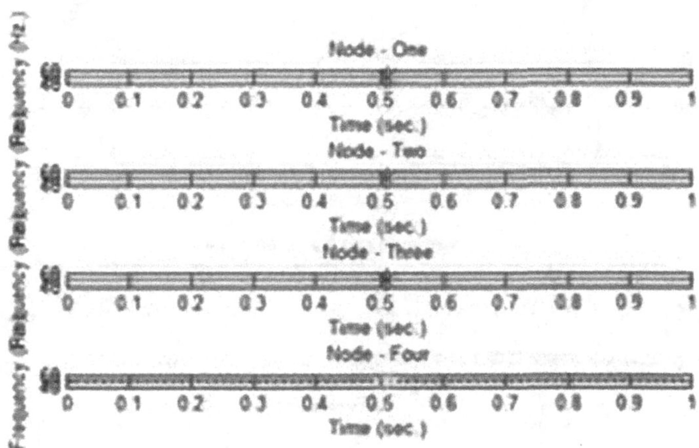

Fig. 1 Frequency monitoring of the nodes

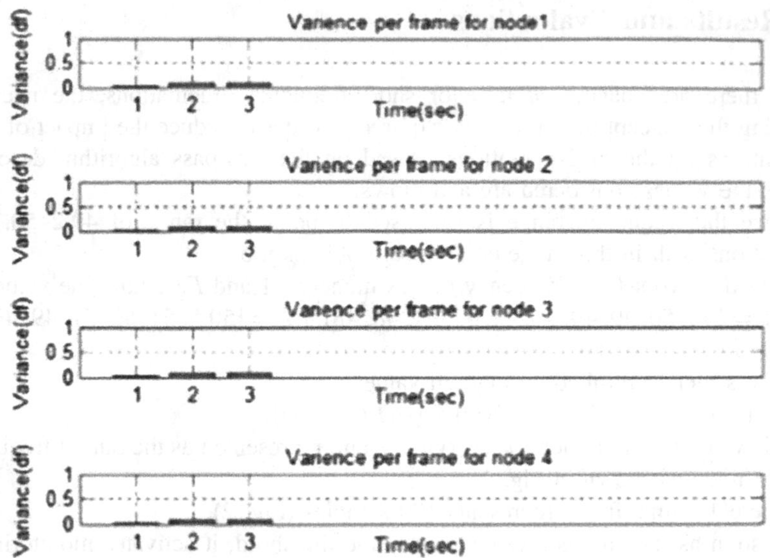

Fig. 2 Variance monitoring of the nodes

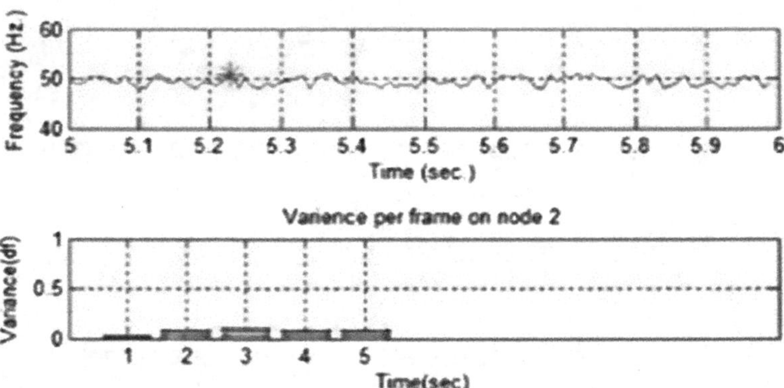

Fig. 3 Monitoring of the individual node

If there are a lot of frequency fluctuations, the node fails. This failure affects the other nodes which in turn increase their loads. This is represented by the Figs. 4 and 5. Figure 4 indicates the normal operation of the grid while Fig. 5 shows increase in the load of other nodes.

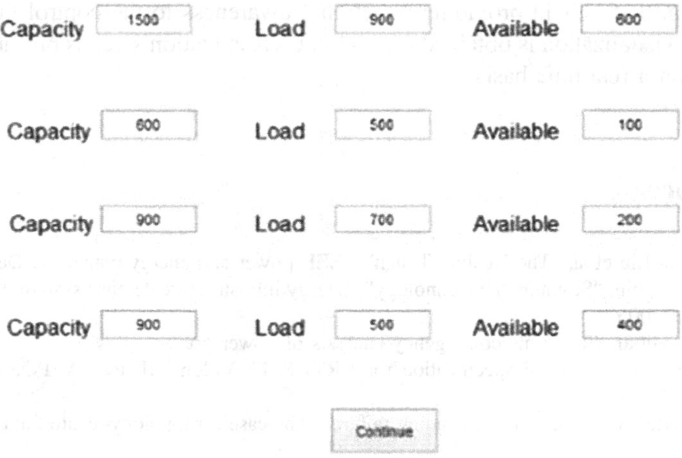

Fig. 4 Initial status of the grid

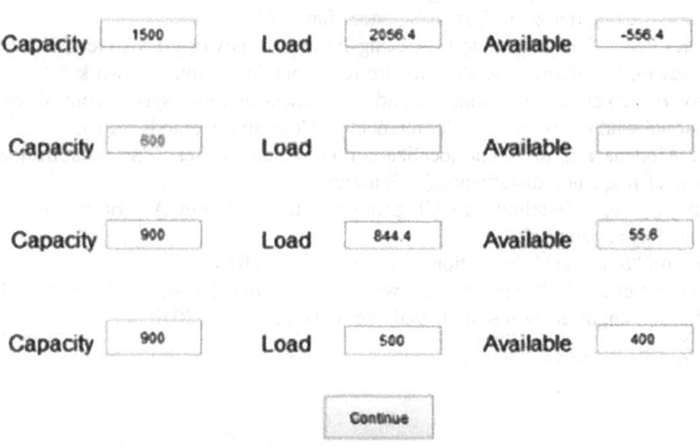

Fig. 5 Increase in the load of other nodes

6 Conclusion

Power grids are inevitable and the most crucial infrastructure that governs the economy on the whole. Fluctuations caused in the frequency parameter of the power grid causes low signal instability affects the smooth functioning of the grid and thereby causing outage and blackout condition. This paper proposes the methodology of frequency monitoring of the system through variance monitoring. Variance can be determined for over a 100 samples that help determine the threshold value to guard against small signal instability. Thus this provides a mechanism that aids in providing situational awareness to the control grid operators. This visualization is obtained through the visualization screens provided to the operator on a real-time basis.

References

1. Shanshan Liu et al, "The Healing Touch", IEEE power and energy magazine, Dec 2003.
2. Masood Amin, "Scanning the technology", Energy infrastructure defense system, Proceedings of IEEE 2005.
3. Anshul Mittal, "Real time contingency analysis of power grid".
4. P Penttayya, "Technical specification for URTDSM", Volume II, Part A, POSOCO report 2011.
5. Soila Pertet et al, "Handling cascading failures: The case for topology-aware fault tolerance", 2005.
6. Adilson E. Motter, "Cascade control and defense in complex networks", Phys. Rev. Lett 93, Aug 2004.
7. P. Pentayya et al, "Low frequency oscillations in the Indian grid", POSOCO paper.
8. B A Carreras et al, "Blackouts mitigation assessment in power transmission systems", Hawaii International Conference on System Science, Jan 2003.
9. Sachin Kadloor, "Understanding Cascading failures in power grids", Oct 2010.
10. David Newth, "Evolving cascading failure resilience in complex networks".
11. Mathaios Panteli et al, "Enhancing situation awareness in power system control centers", IEEE International Multi-Disciplinary Conference on Cognitive methods, 2013.
12. Deirche Alphenaar et al, "Event localization techniques for real time visualization of eastern interconnect frequency disturbances", Genscape.
13. R Kinney et al, "Modeling cascading failures in the North American power grid", The European Physics Journal.
14. R Podmore, "Smart grid restoration concepts", IEEE 2010.
15. Fazil Haneef et al, "Self healing framework for distributed system", International Journal of scientific and engineering research, Volume 4, Issue 7, July 2013.

Detection of Rogue Access Point Using Various Parameters

Sandeep Vanjale and P.B. Mane

Abstract The wireless local area network (WLAN) communication is a rapidly growing approach for data sharing. A wireless network provides network access to mobile devices. Benefits of WLAN are like flexibility, mobility, portability, imposes performance, and security requirements. Such communication brings new network security threats. Physical security of wireless networks is impossible because wireless network signals are unidirectional and can proceed out of intended coverage area. Intruder with an apt wireless receiver can snoop into the network still remaining virtually undetected. In a WLAN, the most important security apprehension is the presence of rogue access point (RAP). These RAPs can be definitely used by persons with inadequate security knowledge. Most of the security threats require an advanced technical knowledge or expensive intrusion devices. A RAP is a wireless AP, which is installed in a secure wireless network without network administrator permission. Such RAP allows intruder to do a man in the middle (MITM) attack. Existence of such RAP causes security threats in WLAN. The access point is very popular because of features like mobility, scalability, cost effectiveness, and ease of installation. Airtight report shows that lack of knowledge about secure wireless network causes number of security threats.

Keywords 802.11 · Rogue AP · Authorized AP · Network security

1 Introduction

The wireless local area network (WLAN) communication is a rapidly growing approach for data sharing. A wireless network provides network access to mobile devices. Benefits of WLAN are like flexibility, mobility, portability, imposes per-

Sandeep Vanjale (✉)
Department of Computer Engineering, BVDUCOE, Pune, India
e-mail: sbvanjale@bvucoep.edu.in

P.B. Mane
Department of Electronics Engineering, AISSMS, IOIT, Pune, India
e-mail: pbmane6829@rediffmail.com

© Springer Science+Business Media Singapore 2017

S.C. Satapathy et al. (eds.), *Proceedings of the International Conference on Data Engineering and Communication Technology*, Advances in Intelligent Systems and Computing 468, DOI 10.1007/978-981-10-1675-2_69

formance, and security requirements. Such communication brings new network security threats. Physical security of wireless networks is impossible because wireless network signals are unidirectional and can proceed out of intended coverage area. Intruder with an apt wireless receiver can snoop into the network still remaining virtually undetected [1].

In a WLAN, the most important security apprehension is the presence of RAPs. These RAPs are which can be definitely used by persons with inadequate security knowledge. Most of the security threats require an advanced technical knowledge or expensive intrusion devices [2]. A RAP is a wireless AP, which is installed in a secure wireless network without network administrator permission. Such RAP allows intruder to do a man in the middle (MITM) attack. Existence of such RAP causes security threats in WLAN [3].

The access point is very popular because of features like mobility, scalability, cost effectiveness, and ease of installation. Airtight [4] report shows that lack of knowledge about secure wireless network causes number of security threats. On the basis of Gartner research, we can say that 20 % of WLAN worldwide have unapproved access points. Intruder can use an AP with a high broadcast power to cover-up as a authentic AP [5]. Mobile agents were used for RAP detection but it has the limitation that client permission is required to run code [6].

2 Related Works

i. Jana and Kasara proposed a server side solution using clock skews of access point in a wireless network. In this approach, clock skews are used as a fingerprint to detect RAP in a network. Clock skews are calculated using the time stamp values from the beacon frames. This approach cannot detect MAC spoofing and has a lack of accuracy and speed in the calculation of clock skews in TCP/ICMP [7].

ii. S. Nikbakhsh et al. proposed client side approach for the detection of MITM attack and Evil Twin attack performed by RAP. It checks routes and the gateways through which packet travels as well as easily can be implemented without modifying a network. It is easy to implement on mobile devices. But attacker can easily break the security by using sniffing programs [8].

iii. Chao Yang et al. proposed to exploit fundamental communication structures and properties of evil twin attacks in wireless networks and to design new active, statistical, and anomaly detection algorithms. Their preliminary evaluation in real-world widely deployed 802.11b and 802.11g wireless networks shows very promising results. It can identify evil twins with a very high detection rate while maintaining a very low false positive rate [9].

iv. Kim et al. proposed a client side approach using the concept of received signal strength (RSS) for RAP detection. In this method, highly correlated RSS sequences are collected in the wireless devices. After that the received signal is normalized and classified whether the collected signal is multiple or not. For this, a sequential hypothesis technique is used. It is a lightweight solution to overcome the drawbacks of the client side approach. But in this technique, the distance between the client node and access points while calculating the signal strength was never considered. Distance affects the signal strength, hence reducing the effectiveness of this approach [10].

v. Han et al. proposed a timing-based scheme to detect RAP. It uses a client side approach, where the emphasis is on round trip time between the server and the user, to check whether the access point is authorized AP or not. The detection algorithm is effective and accurate, but only wireless traffic between AP and the station is considered to set the RAP. Large overhead due to the trade-off between the overhead and accuracy [11].

vi. Kao et al. proposed a client side RAP detection technique using bottleneck bandwidth analysis. It uses a passive packet analysis approach. It is based on bandwidth estimation using packet pair technology [12].

3 Limitations of Existing Methods

Following limitations were found in the existing methods by reviewing the above referred papers:

i. **Clock Skew Solution**: It is assumed that first, the authorized AP will be activated and then the malicious AP. But this assumption is weak, as one cannot control which AP will be activated first.

ii. **Inter Packet Arrival Time**: Can be used to detect RAPs, but it is not effective in case when Evil Twin is present [13].

iii. **Mobile Agent Code**: Mobile agent code is small, which is installed on a mobile device for the purpose of detecting RAP. But a mobile agent code cannot be installed without client permission, which results into a major drawback of this method [14, 15].

iv. **MAC Address & SSID**: SSID and MAC address is used to detect RAP. These properties can be spoofed by using many tools available on internet.

v. **RSS Level**: RSS of the access point is used by various methods to detect RAPs. But variations in RSS levels cause variation in results [16].

vi. **Wireless Traffic**: In wireless environment, network traffic can provide inaccurate results. Such inaccurate results create a suitable environment for RAP to perform attacks [17].

vii. **Workload of Access Point**: The effectiveness of detection of RAP is affected by the workload at the access point.

viii. **Server Side Approach**: The major drawback with the server side approach is that, if the central server is not available or compromised, then the system will not work properly. If client node is out of the reachability of a server then server cannot provide service to client. The server side approach is expensive, limited and cannot work for many real life scenarios [18].

Above vulnerabilities are observed in the existing methods, using which intruders perform various attacks on WLAN. These vulnerabilities can be eliminated by using multiple parameters for RAP detection.

4 Rogue Access Point Detection Parameters

4.1 SSID

SSID is a short form used for Service Set Identifier. SSID consist of 32 characters. In one network there can be multiple SSID's. There can be multiple access points having the same SSID in single network. Without SSID's it is difficult to communicate and interact with one another.

4.2 MAC Address

MAC address is a short form used for media access control. MAC address is used for communicated in between physical network segment and MAC address is assigned to network interfaces. It is a unique identifier.

4.3 RSSI

RSSI is known for received signal strength indicator. The quality of communication between the sensor unit and the access point is indicated by the RSSI value and it is expressed as decibels (dB). The RSSI values are always negative number because of low power levels and attenuation of free air. RSSI values can vary from 0 to −100. The value near to 0 signifies strong signal, whereas the value approaching −100 indicates weaker signal [19].

4.4 Channel and Frequency

Wireless channels are used to convey information signals from one network to another network. Channels can transmit the information signals from senders to receivers. The transmitting capacity of the channel is measured in bandwidth in Hz or its data rate in bits per second.

Wireless network consist of 13 channels which are unlicensed. Each channel has its own unique frequency from 2412 to 2484 MHz with difference of 5 MHz.

4.5 Authentication Type

User in any network wants security of its data being transferred from one source to destination. Transmission protocols and policies are known as authentication. Authentication types are given below:

4.5.1 WEP

WEP is a short from of wired equivalent policy. WEP is an older method of security used in case of older devices and it is easy to hack. So it is not used widely.

4.5.2 WPA

WPA is a term used in short form for Wi-Fi protected access. WPA provides guaranty to the user that only authorized people should have access to it. WPA is sub divided in two parts first one is WPA 1 and other is WPA 2.

4.5.3 WPA 2-PSK

PSK is term used for pre-shared key. This is the latest protocol used today for Wi-Fi security.

4.5.4 802.1X Authentication

It enhances security for 802.11 wireless network. It provides network access with validity [19].

4.6 Radio Type

IEEE has prepared different standards for wireless network with a suffix letter and it covers every standard including security aspects and quality service, e.g., 802.11a/b/g/n/ac.

5 System Architecture

In this system, Wi-Fi scanner scans all access points in the network. Access points in the network broadcast beacon frames after specific time. Capture beacon frame from each access point. From each beacon frame SSID, MAC address, Channel and frequency, RSSI of access point is extracted. This captured information is stored in whitelist.

Administrative login compares new AP parameter details with whitelist AP parameter values and sort authorized AP, unauthorized AP and Rogue AP in wireless network. Access point is scanned periodically and checks properties of access point from whitelist. We have used different parameters like SSID, MAC Address, Authentication type, Channel, Frequency, and Power (RSSI) for detection of RAP.

If SSID is same then check MAC address. If MAC address is also same then check authentication type. If authentication type is same then check channel & frequency of access point. If anyone parameter from all parameters is mismatch with whitelist content then declare that access point as rouge access point (Fig. 1).

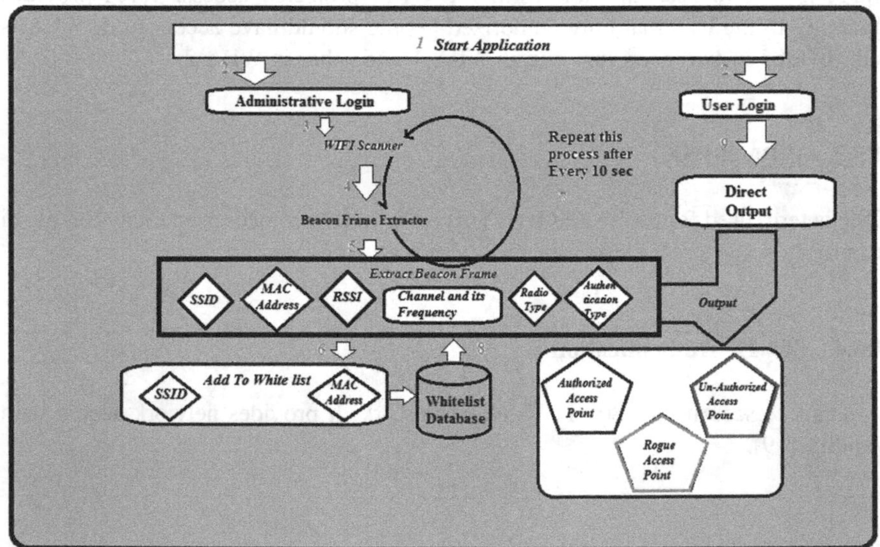

Fig. 1 Architecture of system

Table 1 List of all authorized and unauthorized access points present in network

Id	SSID	Authentication	MAC	RSSI	Channel	Frequency (MHz)	Radio_type	AP_type
1	Amit	Open	e8:de:27:50:45:d3	−30	6	2437	802.11n	Authorized
2	Paras123	WPA-Personal	5c:3c:27:o6:f6:67	−35	11	2462	802.11g	Unauthorized
3	Tataphtn3G	WPA-Personal	0a:1e:58:a0:c4:78	−26	6	2437	802.11n	Unauthorized
4	Sandeep123	WPA-Personal	5e:93:ac:b5:8f:44	−20	9	2452	802.11n	Authorized

Table 2 Rogue access point detection due to variation in RSSI

Id	SSID	Authentication	MAC	RSSI	Channel	Frequency (MHz)	Radio_type	AP_type
1	Amit	Open	e8:de:27:50:45:d3	−30	6	2437	802.11n	Authorized
2	Paras123	WPA-Personal	5c:3c:27:o6:f6:67	−35	11	2462	802.11g	Unauthorized
3	Amit	Open	e8:de:27:50:45:d3	−60	6	2437	802.11n	Rogue AP
4	Sandeep123	WPA-Personal	5e:93:ac:b5:8f:44	−20	9	2452	802.11n	Authorized

Table 3 Rogue access point detection using MAC address

Id	SSID	Authentication	MAC	RSSI	Channel	Frequency (MHz)	Radio_type	AP_type
1	Amit	Open	e8:de:27:50:45:d3	−30	6	2437	802.11n	Authorized
2	Paras123	WPA-Personal	5c:3c:27:o6:f6:67	−35	11	2462	802.11g	Unauthorized
3	Amit	Open	0a:1e:58:a0:c4:78	−40	6	2437	802.11n	Rogue AP
4	Sandeep123	WPA-Personal	5e:93:ac:b5:8f:44	−20	9	2452	802.11n	Authorized

Table 4 Rogue access point detection due to variation in RSSI

Id	SSID	Authentication	MAC	RSSI	Channel	Frequency (MHz)	Radio_type	AP_type
1	Amit	Open	e8:de:27:50:45:d3	−30	6	2437	802.11n	Authorized
2	Paras123	WPA-Personal	5c:3c:27:o6:f6:67	−35	11	2462	802.11g	Unauthorized
3	Amit	Open	e8:de:27:50:45:d3	−60	6	2437	802.11n	Rogue AP
4	Sandeep123	WPA-Personal	5e:93:ac:b5:8f:44	−20	9	2452	802.11n	Authorized

6 Results

Wi-Fi scanner captures the beacon frame. Following parameters values are extracted from the captured beacon frame (Table 1).

After giving authorization to access points, the result shows as they are authorized or unauthorized. If APs parameter value is changed by attacker then it becomes rogue access point (Tables 2, 3 and 4).

7 Conclusion

In this implemented solution, rogue access point is detected using various parameters. To detect RAP we have used various parameters like SSID, MAC address, RSSI value, channels and frequency, authentication, radio type, etc. It also detects MAC address and SSID spoofing attack with less false positive and false negative rates.

References

1. ia Sie Tung, Nurul Nadia Ahmad, Tan Kim Geok: Wireless LAN Security: Securing Your AP, IJCSNS International Journal of Computer Science and Network Security, VOL.6 No.5B, May 2006.
2. Beyah, R.; Venkataraman, A.: Rogue-Access-Point Detection: Challenges, Solutions, and Future Directions, IEEE Security & Privacy, vol.9, no.5, pp. 56,61, Sept.-Oct. 2011.
3. Gaogang XIE, Tingting HE, Guangxing ZHANG: Rogue Access Point Detection Using Segmental TCP Jitter, WWW 2008, April 21–25, 2008, Beijing, China. ACM 978–1-60558-085-2/08/04.
4. www.airmagnet.com.
5. www.netstumbler.com.
6. M. Asaka, S. Okazawa, A. Taguchi, and S. Goto: A Method of Tracing Intruders by Use of Mobile Agents, San Jose, USA, June 2005.
7. S. Jana and S. Kasera: On Fast and Accurate Detection of Unauthorized Wireless Access Points Using Clock Skews, IEEE Transactions on Mobile Computing, vol. 9, no. 3, March 2010.
8. Nikbakhsh, Somayeh, Azizah Bt Abdul Manaf, Mazdak Zamani, and Maziar Janbeglou: A novel approach for rogue access point detection on the client-side. In Advanced Information Networking and Applications Workshops (WAINA), 2012 26th International Conference on, pp. 684–687. IEEE, 2012.
9. Yang, Chao, Yimin Song, and Guofei Gu: Active user-side evil twin access point detection using statistical techniques. Information Forensics and Security, IEEE Transactions on 7, no. 5 (2012): 1638–1651.
10. Kim, Taebeom, Haemin Park, Hyunchul Jung, and Heejo Lee: Online Detection of Fake Access Points Using Received Signal Strengths. In Vehicular Technology Conference (VTC Spring), 2012 IEEE 75th, pp. 1–5. IEEE, 2012.
11. Hao Han and Sanglu Lu: A Timing-Based Scheme for Rogue AP Detection, IEEE Transactions on Parallel and Distributed Systems, vol. 99, no. 1, pp. 5555. ISSN: 1045–9219, 04 Apr. 2011. IEEE computer Society Digital Library.

12. Kao, Kuo-Fong, I-En Liao, and Yueh-Chia Li.: Detecting rogue access points client-side bottleneck bandwidth analysis, computers & security 28, no. 3 (2009): 144–152.
13. Shafiullah Khan, Noor Mast and KokKeong Loo, Ayesha Salahuddin: Cloned AP Detection and Prevention Mechanism in IEEE 802.11 Wireless Mesh Networks, Journal of Information Assurance and Security 3 (2008) 257–262.
14. Kannadiga, Pradeep, and Mohammad Zulkernine, DIDMA: A distributed intrusion detection system using mobile agents. First ACIS International Workshop on Self-Assembling Wireless Networks. SNPD/SAWN 2005. Sixth International Conference on, pp. 238–245. IEEE, 2005.
15. G. Helmer, J. Wong, Y. Wang, V. Honavar, and Les Miller: Lightweight Agents for Intrusion Detection, Journal of Systems and Software, Elsevier, vol. 67, pp. 109–122, 2010.
16. Liran Ma, Amin Y. Teymorian, Xiuzhen Cheng, Min Song: RAP: Protecting Commodity Wi-Fi Networks from Rogue Access Points. Supported by National Science Foundation grant CCF-0627322.
17. Bo Yan, Guanling Chen, JieWang, Hongda Yin: Robust Access Point Detection Using Segmental TCP Jitter, WWW 2008 Published online 1 November 2008 ©Springer Science + Business Media, LLC 2008.
18. S. Srilasak, K. Wongthavarawat and A. Phonphoem: Integrated Wireless Rogue Access Point Detection and Counterattack System, International Conference on Information Security and Assurance, (2008).
19. Chougule, S. B. Vanjale et al.: Detection and Prevention of Rogue Access Point in the 802.11 Using Various Parameters, International Journal of Advanced Research in Computer Science and Software Engineering 5(5), May-2015, pp. 1723–1727.

A Survey on Wireless Security: IP Security Concern

Garima Chopra, Rakesh Kumar Jha and Farida Lone

Abstract TCP/IP protocol suite is used to transfer the information among various networks/users located at remote sites. The transfer of information can be subjected to a lot of vulnerabilities, which can be overcome by providing authentication, confidentiality, integrity, etc. The TCP/IP has been implemented with IPSec to provide protection to packets that are transferred through internet. This paper discusses various protocols used for the protection from attacks by providing authentication check and confidentiality to data. The comparison has been made to identify the best suitable protocol and techniques used for the improvement in the performance of IPSec. This paper discusses various issues encountered when new techniques are implemented to enhance IPSec performance. The security flaws in the algorithms have also been summarized. VPN is also an example which describes the best suitable use of IPSec. This paper explores the VPN and various vulnerabilities present in it.

Keywords Security · IPSec · VPN · TCP/IP

Garima Chopra (✉) · Farida Lone
Department of Electronics and Communication, National Institute of Technology, Srinagar, India
e-mail: garimachopra100@gmail.com

Farida Lone
e-mail: fklone@rediffmail.com

R.K. Jha
Department of Electronics and Communication, Shri Mata Vaishno Devi University, Katra, India
e-mail: Jharakesh.45@gmail.com

© Springer Science+Business Media Singapore 2017

S.C. Satapathy et al. (eds.), *Proceedings of the International Conference on Data Engineering and Communication Technology*, Advances in Intelligent Systems and Computing 468, DOI 10.1007/978-981-10-1675-2_70

1 Introduction

There is increase in the use of Internet from the past few decades and the need for connecting them with best efforts arises and with better security. The security here is important because the requirement is to provide them with confidentiality in the data. For example, a company having branches located at distant locations and they are sharing the data. That data could be very confidential and anybody snooping in them is not acceptable at any cost. In this case, security to the data is required with integrity protection. The packets are transferred through internet using TCP/IP protocol and IPSec is required to provide confidentiality, integrity, non-repudiation, and authentication [1]. In IP security, two protocols are used: AH (authentication header) and ESP (encapsulating security payload). Using IP security, AH protocol has the feature to provide authentication by placing AH Header after data. It also provides data integrity depending upon the requirement and the usage area. ESP protocol encrypts the IP datagram and makes it invisible to the user and attaches IP header. ESP provides integrity of data but sometimes authentication and confidentiality but that is optional. Hence, the malicious person can see only the source and destination address, whereas required information remains hidden. The implementation of IPSec is done either by providing end-to-end security or by implementing security in the routers. In end-to-end security, end nodes are installed with IPSec software. The packets passing through it will only be protected. However, when packets are transferred outside the network, the confidentiality is not guaranteed.

2 IPSec Overview

In IPSec, the data are transferred through two modes: (1) tunnel mode or (2) transport mode. In tunnel mode, the entire datagram is encrypted and a new IP header is added at the end. Mostly, for data transfer between networks or within network tunnel mode is used commonly. The keys used for encryption and decryption will be shared only between sender and receiver gateways. In the transport mode, TCP header and datagram is encrypted and in this no new header is added at the end. However, tunnel mode is more secure than the transport as it provides protection to entire datagram in which only the routing details are only visible. For the combination of authentication and providing confidentiality, we need to use cryptographic algorithm. IKE (internet key exchange) is needed to exchange keys between nodes and establish a secured environment. Secured connections are needed for communications for that encryption is done either through public or private keys. The keys are managed manually or through automatic management. Before establishing a connection, both the communicating nodes need to agree upon a certain predefined rules called security association.

Many advancements have been made for the security in LAN's, VPN'S, etc., many papers describes the various types of attacks that could be possible while transferring data. It is also felt that sometimes TCP/IP is not sufficient for authentication, so stress on ICMP (internet control management protocol) is also used for controlling the transfer of information [2]. VoIP is introduced for voice communications and multimedia transfer, ANR conducted a survey and find out the probable attacks that could be present on it.

VPN is a network in which links are established among nodes virtually. I mean to say here that virtual means there is no direct connection between the communicating nodes. VPN is the common example in which IPsec is used because it isused for sharing information. The software implementation is done on public network but connection among nodes is purely private. Here the concept of tunneling is used in which private data is made to travel through the public network and VPN is sometimes also known as VPN tunnel. Although it suffers from vulnerabilities, which will be discussed further.

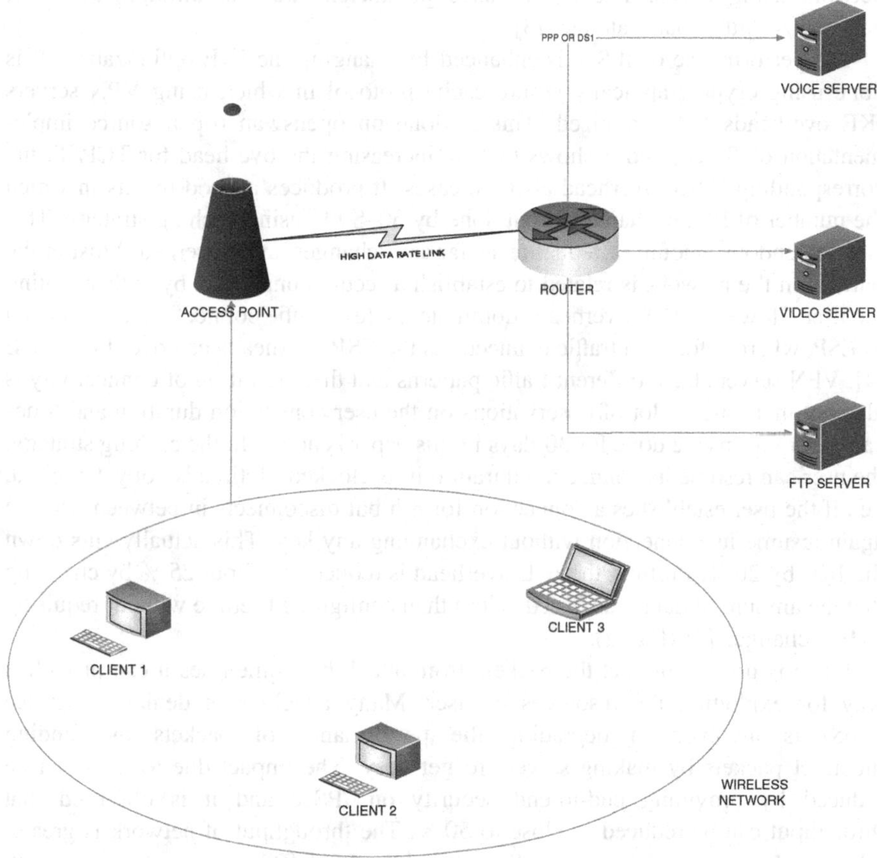

Fig. 1 Wireless network architecture

Internet protocol has two versions IPv4 and IPv6 but IPv6 only has the inbuilt feature of providing security, whereas in IPv4 security is needed and implementation is done either through software or by hardware. AH protocol is encrypting data received from upper layer and IP header is added at the end. All the information about source address, destination address, etc., will be visible. This only provides the authentication of the sender (Fig. 1).

3 Related Works on IPSec

For the better functionality of IPSec, we need to have better selection of configuration, policy, algorithm, and a key management. The ciphers that are used under IPSec for authentication or integrity may be weak either in theory or practical implementation. The intruder can take benefit about the weakness and can exploit the resources. Many works have been done so far to improve the performance in a network using IPSec. The performance parameters are like throughput, delay, packet size, jitter, data rate etc. [3].

The performance of IPSec is enhanced by changing the IKE optimization. It is done using cryptographically secure cache protocol in which using VPN servers IKE overheads are minimized. This is done on openswan (open source implementation of IPSec) and it shows that on increasing the overhead for TCP/IP, the corresponding IPSec overhead also increases. It produces a good results in which the number of IKE exchanges is cut done by 50–80 % using caching strategy. This is a tremendous amount of reduction as far as exchanges are concerned. Most of the burden on the network is related to establish a secure connection by authenticating the user. However, IKE overheads dominate the low traffic connection as compared to ESP, whereas for high traffic connections the ESP overheads can overdo the IKE [4]. VPN servers have different traffic patterns and there duration of connectivity is also not that long. A lot of observations on the user connection duration and times each user returns are done for 30 days in this paper is noted. In the caching strategy, the user can resume its connection duration if its clocked lifetime is not yet expired, i.e., if the user establishes a connection for 8 h but disconnects in between. He can again resume its connection without exchanging any key. This actually cuts down the IKE by 20 %. Further, the IKE overhead is reduced to about 25 % by checking that the amount of data transferred is less than configured lifetime without requiring IKE exchanges [5] (Fig. 2).

IPSec is used to protect the packets from attack but sometimes it can provide a way for exploiting the resources of user. Many attacks like denial of service (DoS) is involved in degrading the performance of packets by sending incorrect packets by making servers to get busy. The impact due to DoS can be reduced by providing end-to-end security on IPSec and it is observed that throughput can be reduced as close to 50 %. The throughput of network is greater when packets are transferred on the network without IPSec as compared to with IPSec. It is observed that throughput of packets reduced when different security

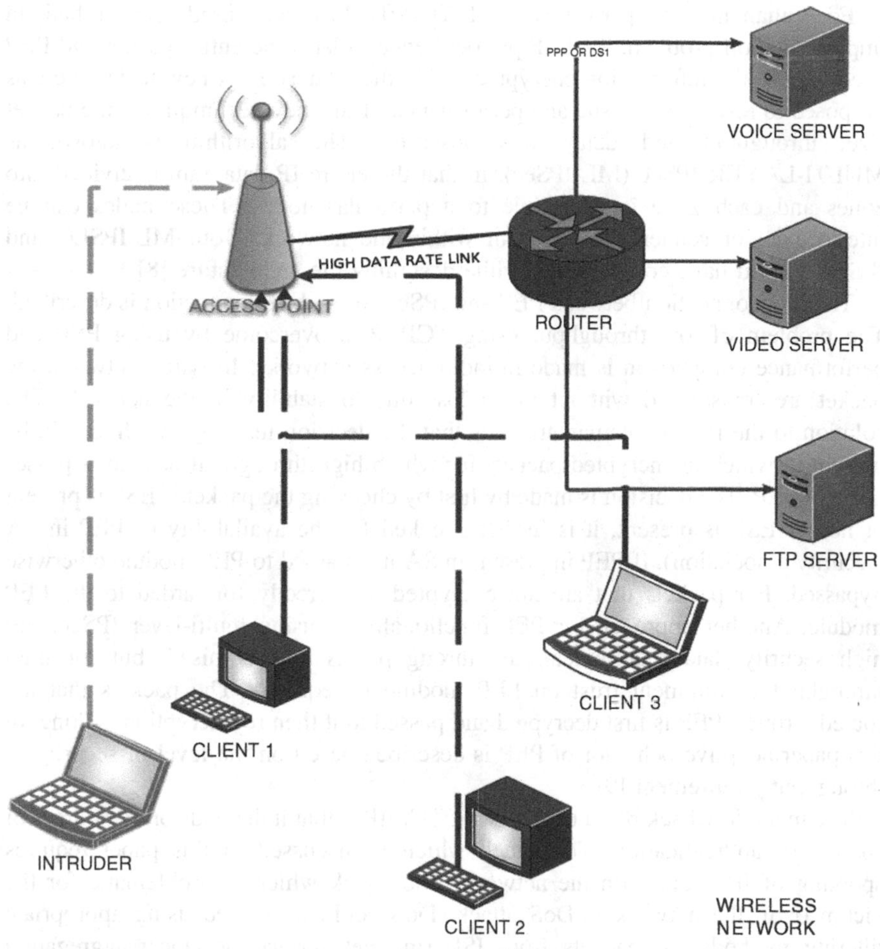

Fig. 2 Intruder accessing the access point

algorithm are use. The decrease is with solo (without IPSec) then md5 followed by sha1, des, and 3 des. However, it is observed that packets with larger size have lesser throughput than the smaller packets size. So, fragmentation is required for a better performance when IPSec is implemented. It is also noted that MD5 is comparatively faster than SHA1. But IPSec consumes a lot of CPU resources and providing opportunity for DoS. It is observed from the experiment in the paper that for smaller packet size high performance is not expected due to small block size used for encryption having initialization overhead. But the performance stabilizes even better for packet size of 512 bytes or even more. The efficiency due to UMAC algorithm speeds up for larger packet size but degrades in shorter messages. Reverse is the case with MD5, it works well for shorter message length. By using CPU intensive algorithm benefits more for nonce based security [6, 7].

For enhancing the performance of TCP/IP, PEP was used. But it had an implementation problem. As ESP protocol encapsulates the entire packet and PEP needs to use the information encrypted inside the IP header. A new technique was proposed to resolve this issue and performance of this new technique on the packet size, throughput and delay was observed. The algorithm is known as MULTI-LAYER IPSec (ML-IPSec). In that the entire IP datagram is divided into zones and each zone is accessible to a particular nodes. These nodes can be intermediate or sender/receiver from within the network. Both ML-IPSEC and IPSEC should have compatibility while designing the architecture [8].

The interconnection between PEP and IPSec for mobile connection is described. The problem of low throughput using TCP/IP is overcome by using PEP and performance comparison is made in radio access networks. In wired network, the packet are transmitted without much loss due to stability in the network. The solution to the PEP implementations is that the decision making switch should be present in which unencrypted packets for which high throughput desired is passes through PEP. The decision is made by first by checking the packet if ESP is present or not. If ESP is present, it is further checked for the availability of PEP in SA (security association). If PEP in present in SA it is passed to PEP module otherwise bypassed. For packets that are not encrypted are directly forwarded to the PEP module. Another approach for PEP functionality is using multi-layer IPSec. For high security data requirement, the throughput is compromised, but for high throughput requirement trust on PEP module is required. The packets that are routed through PEP is first decrypted and passed to it then re-encryption is done. In this paper adaptive behavior of PEP is described based on the level of security or throughput requirement [9].

The main drawback or security flaw of TCP/IP is that it depends on IP address of source for authentication. The attack which is discussed in this paper requires spoofing of IP address on the network. The attack which is problematic for the victim or in the network is DoS attack. DoS could be solved using appropriate filtering methods for packets from ISP (internet service provider) aggregation points. However, this scheme enables the ISP's to easily track the attackers source, because they (attacker) need to have the real/valid source address to reach. Routing and control protocols require stronger security mechanism rather than just filtering of IP packets. So, it is a viable solution that we need to rely on the IKE and some authentication methods which uses public cryptography for building a strong authentication, data integrity, and non-repudiation [10].

Due to tremendous growth of Internet in recent years, and outpacing growth of voice communication, IPSec is used with mobile IP without any other security like WEP. For this integration, only the mobile hosts and security gateways need to install IPSec in a wireless link. Using VPN, the nodes from offices, campus, or any other location can easily connect and form a network. The nodes which are connected can transfer their data through internet securely but with using IPSec. In IPSec [11], IKE is most required because the transferring medium is however very insecure and could be easily attacked if a little flaw is present in the designing phase. For IKE, first phase is the authentication phase used for shared secret key, digital signature creation and public key encryption. In second phase, pseudorandom function is used to combine the materials and creating the key. However, there is sometimes a restriction on the block size of plain text but the requirement of key is large. The VPN also suffers from issue which should be kept in mind, implementing are as follows [12, 13]: the issue in IKE is that first phase that is aggressive mode or main mode should not be used. In aggressive mode, both the communicating nodes need to exchange few packets and that too not encrypted. Through sniffing, anyone can capture this response using TCP dump, start dictionary, or brute force in IKE crack. IKE crack is an open source designed to brute force or dictionary attack using password with the pre-shared key [13]. The attacker needs to capture traffic for cracking IKE. Second, the issue of scalability of VPN is present. VPN is designed for the transfer of unicast packets. As VPN is used in tunnel mode, the server will be loaded for packet processing for multiple packages that need to be sent to end points. Thirdly, IPSec VPN does not support broadcast of messages. Another issue present which vendor's would not accept it as a problem is VPN fingerprinting. VPN fingerprinting provides the details about device that we are using and software version details [13].

Now the solutions to these problems can be sorted out using packet filtering and accepting ISAKMP negotiation from a trusted IP address. The use of aggressive mode should be avoided. The company should manage the operation and management of VPN gateway [14]. The most up-to-date antivirus should be used by computers which are connected to company's internal network. The dual tunneling is not permitted, only one way connection is allowed.

IPSec is designed with the way that it does not entertain NAT routers. So any packets destined to NAT router, in that case, would not be able to understand what to do with it. To resolve this issue, the NATT protocol is designed by IETF for IPSec in which packets are encapsulated in either TCP or UDP. Through this way, it is possible to integrate NAT with IPSec VPN [4]. Although there are also NAT routers available in market that can handle ESP packets more efficiently, NATT protocol cannot transmit when used in transport mode and also faces the problem of access control. For this reason IPSec over HTTP protocol based on IPSec tunnel and HTTP tunnel is designed for the conversion of communication to travel across the firewall which is responsible for controlling network communication. After doing experiments, it is observed that new system works worse in performance for heavy load.

The protocol adds 150 bytes header for encapsulation. It is suggested in this paper to use this protocol when your network environment is complex, while IPSec is successful in most of the implementations. The overall performance of the network will be reduced a little due to packet overhead. VPN technology can be implanted in wireless network communication in order to make it more secure [15]. However, it is observed that UDP is suitable for simple operation of request and reply which have a built in error detection capability. In TCP applications, the wireless VPN efficiency can be improved by sending data packets of optimal size.

Table 1 Detailed description of different security algorithms

S. no	Author	Algorithm	Description
1.	Yongguang Zhang, May 2004	MULTI-LAYER IPSec	Due to conflict between TCP and PEP implementation, ML-IPSec is added to IPSec. ML-IPSec divides the packet into zones and each zone has its own security association
2.	Sierra J.M., Hernindez J.C., Ribagorda A., and Jayaram N., 2002	ISAKMP protocol	ISAKMP is used to integrate different protocols together and establishing a negotiation on the parameters of each SA (security association) [14]
			Each SA is associated with a protocol that will be represented through Domain of Interpretation (DOI)
3.	Etzel Brower, LaTonya Jeffress, Jonah Pezeshki, Rohan Jasani, and Emre Ertekin	Header Compression	Large amount of packet overhead present in IP packets for ipv4, but this case further increases for ipv6. Due to this HC is used to solve the problem by compressing the header
4.	Mei Song, and Zhang Yun-he, 2009	IPSec over HTTP	To resolve the problem arises in VPN IPSec due to NAT routers. Throughput reduction due to header overhead and it needs to be compromised [17]
5.	Hao Han, Bo Sheng, Chiu C. Tan, Qun Li, and Sanglu Lu, November 2011	Timing based scheme	This algorithm is used for the detection of rogue access point (AP) in which it gives 100 % accuracy for light traffic but slightly less but more than 60 % for heavy traffic. This timing based algorithm is implemented on end users side [18]
			Rogue AP pretends to be a legitimate AP and attracts the node to connect through it. All the traffic will be passed through Rogue AP and it can snoop into the confidential information of the user like passwords, ids or their credit cards information

(continued)

Table 1 (continued)

S. no	Author	Algorithm	Description
6.	Nachiketh R. Potlapally, Srivaths Ravi, Anand Raghunathan, and Niraj K. Jha, February 2006	Energy consumption of cryptographic algorithms	The energy consumed by various cryptographic algorithms are analyzed based on key generation, verification of code, and signature generation
			The comparison among them is made based on all the three features and then the best alternative can be observed in place of others [19]
			Comprehensive analysis of various transport and network devices are done like SSL or TLs protocol
7.	Chao Yang, Yimin Song, and Guofei Gu, October 2012	ET-Sniffer is used using 2 protocols	In this approach, rogue AP is detected from user side but without the need for list of authorized users. Only the inter arrival time is used as detection statistics between two consecutive packets
			Also "immediate-ACK" is used in ET-Sniffer [20]
8.	Melbourne Barton, Derek Atkins, John Lee, Sanjai Narain, Deidra Ritcherson, Kemal E. Tepe, and K. Daniel Wong	Integration of IPSec in mobile IPv4	IPSec is implemented in the mobile nodes and security gateway. The home network needs not to trust on foreign network for security rules due to mobility of mobile host. However, the main drawback is of Routing inefficiency

4 Comparison

Various algorithm presented in papers are presented and concluded. However, it is observed that with time further improvements are done on IPSec. To implement PEP in TCP/IP and our purpose is to improve throughput and with minimum delay or jitter, ML-IPSec is used. However, PEP module can also be used in order to increase throughput for those packets which are unimportant. The data packets in which confidentiality is not of concern can be bypassed through PEP. Further, with the increase in header size using IPSec or IPSec over HTTP protocol, there is an algorithm called Header compression that will reduce the size header, so as to reduce the packet size and this could be one of the techniques which could result in throughput enhancement but results still need to be enhanced (Table 1).

Table 2 IPSec implementation and security threats in wireless networks

S. no	Network	IPSec implementation	Threats present
1.	1G	Not present	It was present in analog system that is highly insecure
			No support for encryption, poor sound quality and inefficient use of spectrum
2.	2G	Conversations are encrypted with temporary and randomly generated cipher keys. A5/1 stream cipher was used for security	
3.	3G	A5/3 or KASUMI was used earlier to provide security for 3GPP and UMTS standard. Today, encryption is provided from handset to base station to the RNC (radio network controller)	A5/3 was broken down over a decade ago
4.	4G	Security is present from mobile station to base station having authentication and encryption. Authentication but no encryption is present from base station to Mobility Management Entity	In 4G, threats can be made possible in physical layer by inserting intentionally man-made interference
			DoS attacks are initiated by flooding attacks from authentication management frames
			End users can also act as a source of worms, viruses, call or spam mails etc.
			VoIP leads to spoofing, modification of data, dropping of private communication etc.
5.	5G		5G connects many devices which introduces many threats like victims from ransomware are locked out of many connected devices
			The higher data rate leads to higher speed of malicious file transferring
			Remote survey or driverless cars are enabled by 5G and any security breach in them is highly intolerable

Table 2 describes the IPSec implementation from 1G to 5G. It gives the brief introduction about the related security issues which arises in the corresponding networks. It is observed that due to analog nature of 1G, it is prone to many attacks. The network is switched from analog to digital in 2G and further. In 2G, there is no security present. Day by day, modifications were performed in security in 3G architecture. However, there are little modifications in the security architecture from 3G to 4G.

5 Conclusion

In this paper, a review of all the papers, which are studied, have been merged with the best efforts to provide the knowledge about the various technological challenges encountered till now in IPSec and vulnerabilities during the integration of various new algorithms with IPSec for performance enhancement. It also describes how changing the parameters or the protocols used can enhance or affect the performance of IPSec. With the increasing need of security, improvement in designing of cryptographic algorithm is seen with the best alternatives among them and also seeing the energy issues that are today network as most of the systems have the capability of mobility. Some may be prone to excessive attacks but some have weak keys. By keeping all these attacks in mind and various cryptographic weaknesses, necessary improvements are expected [16]. So the security breaches should be less. It is also advised to purchase products for security from vendors which are most successful and trustful.

References

1. "Cryptographic algorithm for ESP & AH", RFC 4305 (2005).
2. Angelos D. Keromytis: A comprehensive survey of voice over IP security research, IEEE communication survey and tutorials (2012).
3. Craig A. Shue, and Steven A. Myers: IPSec: performance analysis and enhancements (2007).
4. Y.P. Kosta, Upena D. Dalal and Rakesh Kumar Jha: Security Comparison of Wired and Wireless Network with Firewall and Virtual Private Network (VPN), International Conference on Recent Trends in Information, Telecommunication and Computing (2010).
5. Joseph D. Touch, Yi-Hua Edward Yang: Reducing the impact of DoS attacks on endpoint IP security, IEEE (2006).
6. Joel Singh, and Ben Soh: A critical analysis of multilayer IP security protocol, IEEE ICITA (2005).
7. Yongguang Zhang: A Multilayer IP security protocol for TCP performance enhancement in wireless networks, IEEE journal on selected areas in communication, VOL. 22, NO. 4 (2004).
8. Nadim Assaf, Jijun Luo, Markus Dillinger, and Luis Menendez: Interworking between IP security and performance enhancing proxies for mobile networks, IEEE communication magazine (2002).
9. Madalina Baltatu, Antonio Lioy, Fabio Maino, and Daniele Mazzocchi: Security issues in control, management and routing protocols (2000).
10. Nadia Issa, and Chris Todd: IPSec implementation for a better security in IEEE 802.11 WLANs.
11. Radia Perlman, and Charlie Kaufman: Key exchange in IPSec, IEEE (2000).
12. Byeong-Ho Kang, and Maricel O. Balitanas: Vulnerabilities of VPN using IPSec and defensive measures, International journal of advanced science and technology (2009).
13. Thomas Berger: Analysis of current VPN Technologies, IEEE ARES (2006).
14. Sierra J.M., Hernindez J.C., Ribagorda A., and Jayaram N.: Migration of internet security protocols to the IPSec framework, IEEE (2002).
15. Bhavya Daya: Network security: history, importance, and future.
16. Kenneth G. Paterson, and Arnold K.L. Yau: Cryptography in theory and practice: The Case of Encryption in IPSec (2005).

17. Mei Song and Zhang Yun-he: One New Research About IPSec Communication Based on HTTP Tunnel (2009).
18. Hao Han, Bo Sheng, Chiu C. Tan, Qun Li, and Sanglu Lu: A Timing-Based Scheme for Rogue AP Detection, IEEE TRANSACTIONS ON PARALLEL AND DISTRIBUTED SYSTEMS, VOL. 22, NO. 11 (2011).
19. Nachiketh R. Potlapally, Srivaths Ravi, Anand Raghunathan, and Niraj K. Jha: A Study of the Energy Consumption Characteristics of Cryptographic Algorithms and Security Protocols", IEEE TRANSACTIONS ON MOBILE COMPUTING, VOL. 5, NO. 2 (2006).
20. Chao Yang, Yimin Song, and Guofei Gu: Active User-Side Twin Access Point detection Using Statistical Techniques, IEEE TRANSACTIONS ON INFORMATION FORENSICS AND SECURITY (2012).

An Enhanced Framework to Design Intelligent Course Advisory Systems Using Learning Analytics

V. Vaidhehi and R. Suchithra

Abstract Education for a person plays an anchor role in shaping an individual's career. In order to achieve success in the academic path, care should be taken in choosing an appropriate course for the learners. This research work is based on the framework to design a course advisory system in an efficient way. The design approach is based on overlapping of learning analytics, academic analytics, and personalized systems. This approach provides an efficient way to build course advisory system. Also, mapping of course advisory systems into the reference model of learning analytics is discussed in this paper. Course advisory system is considered as enhanced personalized system. The challenges involved in the implementation of course advisory system is also elaborated in this paper.

Keywords Learning analytics · Academic analytics · Personalized system · Reference model · Framework

1 Introduction

Higher education institutions across the globe offer a variety of courses in various disciplines. Learners find it difficult to choose the course because of its variety. Course selection in higher education is an important step in one's career. To help the students, in course selection, counsellors are available in higher education institutions. But, the amount of data with which the counsellors operate remains static. Due to dynamic changes occurring in the education system, the role of technology in the course advisory domain is highly recommended.

V. Vaidhehi (✉)
Department of Computer Science, Christ University, Bengaluru, Karnataka, India
e-mail: vaidhehi.v@christuniversity.in

R. Suchithra
Department of Computer Science, Jain University, Bengaluru, Karnataka, India
e-mail: suchithra.suriya@gmail.com

© Springer Science+Business Media Singapore 2017
S.C. Satapathy et al. (eds.), *Proceedings of the International Conference on Data Engineering and Communication Technology*, Advances in Intelligent Systems and Computing 468, DOI 10.1007/978-981-10-1675-2_71

In the past, course advisory systems were designed by using traditional database systems. Data in database remains static. Later on, databases were improved to store objects using object oriented database systems. Then the design of expert system using artificial intelligence approach started emerging. But the expert system, fails to capture all the new knowledge as more courses were introduced dynamically. Course advisory systems using fuzzy-and neural-based techniques were built, as these techniques have the capability to represent the complex real-world problems. The drawback of such systems is that they fail to analyze the learner's capability to the full potential. Analytics can be applied to every walk of life in today's world. Therefore, analytics can be applied on learner data and accordingly an efficient course can be suggested which enhances the performance of students.

This research work proposes a framework to build an intelligent course advisory system. Intelligent course advisory systems can be designed using learning analytics which analyzes the learner's data. From the learner's data, the caliber (potential) of the learner is evaluated by using learner's ability, skill and other parameters. According to the caliber (potential) of the learner, the courses could be recommended in a personalized way.

This paper is organized as Sect. 2 explains the various related work performed in this area. Section 3 describes about learning analytics. Section 4 explains how course advisory systems can be visualized as the overlapping of both academic as well as learning analytics system. Section 5 briefs about the mapping of learning analytics into course advisory systems. Section 6 describes course advisory system as an enhanced personalized system. Section 7 describes the enhanced framework that could be used in the design of course advisory systems.

2 Literature Review

The literature survey includes data mining algorithms, learning analytics, and technology-based educational systems. Forecasting methods for consumer products using data mining steps is discussed elaborately in [1]. Also real-world problem like this can be predicted using data mining techniques where demand rate is dynamic in nature. In [2], the implementation of random forest algorithm, Adaboost algorithm and ANN-based classifier MLPNN multilayer perceptron neural network-based classification algorithms and their performance based on the sound in the lung is discussed. In [3], integration of technology into the class room teaching based on the situation of the student is cited which will assist the teacher to plan the pedagogy of the subject to be delivered. Recommendation systems [4] and various algorithms used is discussed.

Genetic algorithms-based recommendation system for e-learning environment is discussed. Recommendation systems based on learning style using genetic algorithms is discussed in [5]. Assessment in collaborative learning is discussed in [6]. Understanding capabilities of learners in the science subject is measured and analyzed through content management [7]. Identification learning path using optimization procedures is discussed in [8]. The reference model of LA and its design challenges are elaborated in [9]. The technological requirements for integrating LA into HEI is discussed in [10]. The challenges that are to be addressed during the integration phase is presented in [10]. The framework for smart learning environment in a personalized approach is discussed in [11]. Various preprocessing strategies for effective data mining is discussed in [12]. The framework for smart learning environment in a context aware environment is discussed in [13].

3 Learning Analytics

Education data mining (EDM) mines the knowledge from educational data. There are two branches of EDM are Learning Analytics (LA) and Academic Analytics (AA). LA operates on learners data whereas AA operates on institutional data. AA is used for generating reports and assists in decision making. LA is used in a personalized environment for the students to improve their academic performance. Higher education institutions can use LA as well as AA into their functionality as it helps to enhance the performance of the educational system. Different users like teachers, students, institutions, parents, and industries can be benefitted by this approach. There are overlapping between LA and AA as the objective of these analysis is to improve and strengthen the educational system across the globe. This is shown in the following Fig. 1.

Fig. 1 Academic and learning analytics

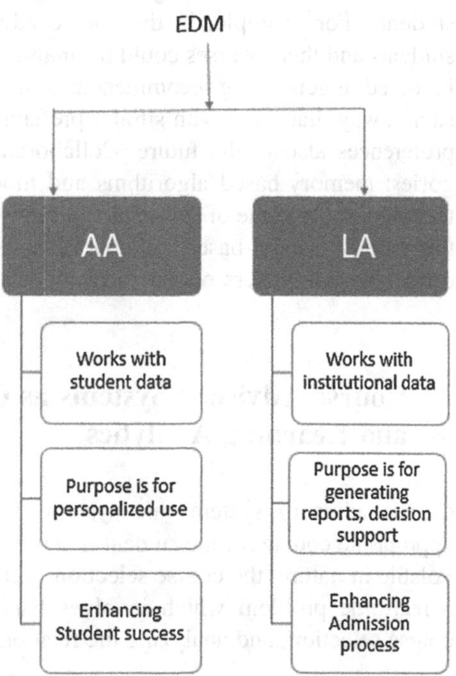

Similar to business sector, analytics can be applied to educational data. For example, the analytics can be applied to the student data (LA) which helps to predict the performance of the student, to monitor the progress of the student or to choose appropriate courses/subjects in a personalized way. Similarly, the analytics can be applied to institutional data (AA) which helps to monitor the performance of the various departments or to predict the number of students across each courses offered by the institution. Various steps involved in analytics are capture, report, predict, act, and refine. The first step is to collect data. Then the data selection for analytics is done by using patterns or queries. The third step is to predict the system by using statistical algorithms. Then, the next step is to take appropriate actions based on prediction. The final step is to refine system for improvements. Its cyclic and iterative process. This is continued until the final result is generated.

Various techniques are involved in analytics which include statistics, business intelligence, web analytics, artificial intelligence and data mining, social network analysis, and information visualization. The possible outcome of the analysis would be prediction, personalization and information visualization. Statistical operations like mean, median, and standard deviation help to generate reports in the course advisory system to provide basic information about the student. Information Visualization (IV) techniques help to visualize the Statistical reports (charts or graphs) effectively. Data representation becomes effective in charts or graphs compared to tables. Data mining (DM) methods includes supervised learning (or classification and prediction), unsupervised learning (or clustering), and association rule mining helps to provide appropriate recommendations, feedback for the students. Social Network Analysis (SNA) is used to analyze the similarity among the students. For example, in the course advisory system, similar characteristics of students and their courses could be analyzed by using SNA. Various techniques are involved in generating recommendations. Collaborative filtering approach works in such a way that users with similar preferences in the past will tend to have similar preferences also in the future. Collaborative filtering methods fall into two categories: memory-based algorithms and model-based algorithms. In memory-based techniques, the value of the unknown rating is computed as an aggregate function of the ratings. Model-based collaborative techniques provide recommendations by estimating parameters of statistical models for user rating.

4 Course Advisory Systems as Overlap of Academic and Learning Analytics

Course advisory systems are systems which will help the students to choose appropriate course for the student in a personalized way. As Course demand is very volatile in nature, the course selection becomes a complex task. Course selection is a real-time problem which involves handling variety of factors associated with course selection, and analyzing the relationships between the factors. Course which

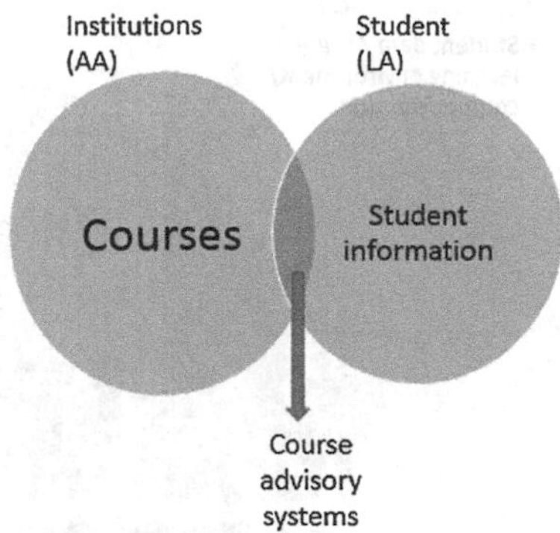

Fig. 2 Overlapping of academic and learning analytics

is in high demand, may not be there in the consecutive years. Therefore, sufficient knowledge about the courses plays a vital role in course selection. Also details about each courses across different institutions is essential for taking the decisions. Also, the student details and associated information should be made available so that recommendations/suggestions could be given for each student in a personalized way. Therefore, course advisory systems work in a overlapped environment by taking both the institutional as well as student data so that courses could be recommended to the student. The relationships between LA, AA, and course advisory system is shown in the Fig. 2.

5 Mapping the Reference Model of LA into Course Advisory System

Based on the reference of model of LA, the course advisory system is represented as four-dimensional model. As part of technology enhanced learning (TEL), LA plays a crucial role in enhancing the performance of the students. Other than the college admission management, prediction of student success, LA can be used effectively in the learning process as recommendation system or personalized learning system. Analytics process of EDM and LA are similar in nature as knowledge recovery is the primary concern. But, LA is focused on learners.

Course Advisory systems is visualized as four-dimensional system as what, who, why, and how. In what dimension of course advisory system using learning analytics, data/environment/context is considered. In who dimension of course advisory system using learning analytics, different stakeholders are considered. In why dimension of course advisory system using learning analytics, various objectives are considered. In how dimension of course advisory system using learning analytics, different methodologies are considered.

The present research work on course advisory system is focused on learner data in a regular learning environment for course selection. Different stakeholders like students, parents, teachers, industry, and institutions are considered in the

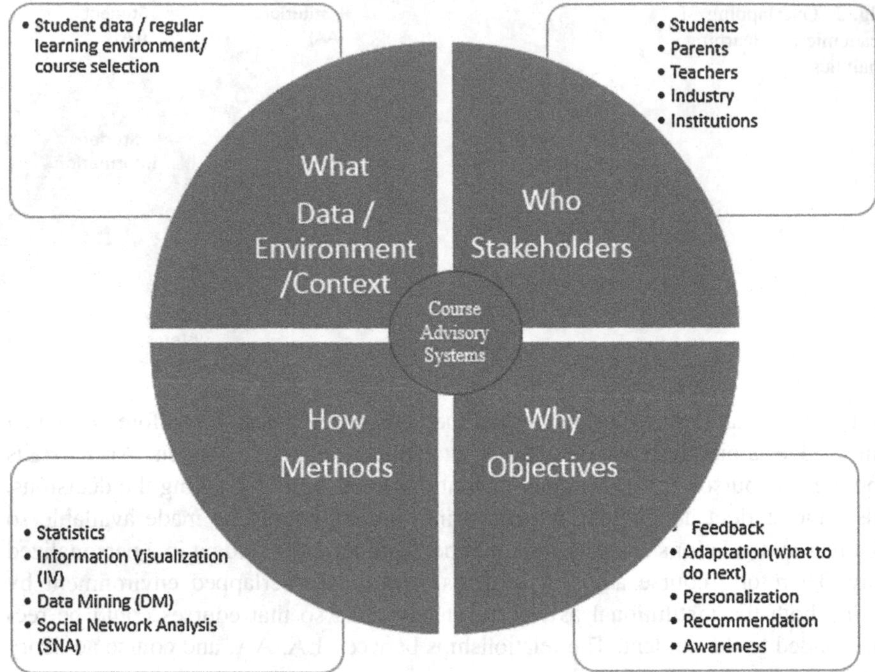

Fig. 3 Mapping learning analytics into course advisory system

implementation. Course advisory systems are designed with various objectives like feedback, adaptation, personalization, recommendation, and awareness. The various techniques like statistical methods, information visualization techniques, data mining, and social network analysis are involved in the design of course advisory system. The mapping of learning analytics in the course advisory system is shown in Fig. 3.

6 Course Advisory Systems as Enhanced Personalized Systems

Personalized recommendation system will help the learners to decide what to do in the next step of the academic journey. Personalized recommendation is the essential part of the personalized system. Course advisory systems are the part of recommendations of personalized systems. Intelligent Tutoring Systems (ITS) provides the learning sources depending upon the learning capability of the learners. For example, learner could be identified as fast or slow learner. Then accordingly, the learning sources are provided to the learners. Adaptive learning system (ALS) is to support the learners by providing appropriate learning interfaces according to their

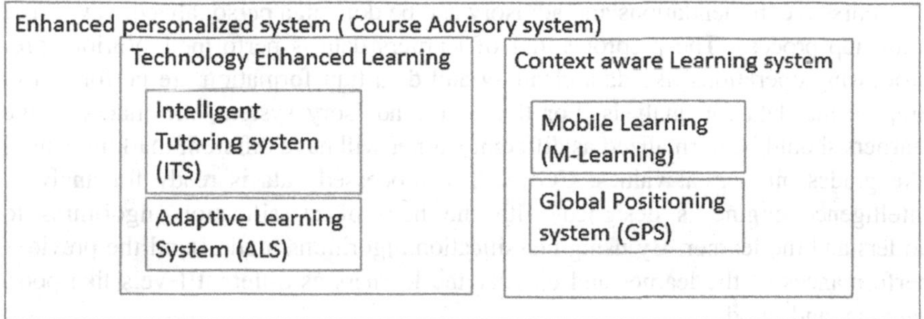

Fig. 4 Course Advisory system as enhanced personalized system

needs. Adaptive learning system is a part of ITS. Adaptability of the learner in the learning environment is the core attribute involved in the design of Adaptive learning systems. Such systems which helps the learners is referred as technology enhanced learning systems (TEL).

Due to the drift in the technology, various devices are involved in the learning environment. The usage of mobiles in e-learning environment has introduced a new learning environment as Mobile Learning (M-Learning). The mobility of the learner can be monitored by GPS. Therefore, the context in which the learner is participating in the learning environment has introduced a new branch of learning system called as context aware learning systems. Technology enhanced-learning systems and context aware learning systems are collective part of enhanced personalized system. Therefore, the course advisory systems could be designed by considering the learners capability on a personalized fashion is shown in the Fig. 4.

7 Framework

In the current scenario, universities collect huge data on attrition and retention. But it is not analyzed properly. Many times decisions are taken based on assumptions and intuitions. Therefore, the decisions taken at higher education institutions may not be effective. Also teaching and learning in higher education institutions become complicated. To enhance learning and help learners in an effective way, the course advisory system is essential in higher education institutions. Course advisory systems minimizes the retention rate and improves the performance of the learners. Learning Analytics is an overlapping of different fields like academic analytics, educational data mining, recommender systems, and personalized learning systems. Also, Learning Analytics blends different techniques like machine learning, artificial intelligence, information retrieval, statistics, and visualization. By using learning analytics, objectives of learners can be mapped into courses that are offered by the educational institutions.

Course recommendations and advisory can be done in a personalized way. It is a multistep process. The preprocessing of learners data is performed. Various preprocessing operations like data cleaning and data transformation are performed to prepare the data for analysis. For the course advisory system, the marks of the learners should be normalized as different learner will have different marking values like grades, numerical values, CGPA. The processed data is ready for analysis. Intelligence engine is designed with the help of classification algorithms to understand the learner. By using classification algorithms, understand the previous performances of the learner and classify the learners as different levels like poor, average, and good.

After, understanding the learner, the next step is to evaluate the learning style of the learner. Learning style helps to measure the ability of the individual learner like advanced, fast, and slow. Learning styles includes how a person learns, how a person acts in a learning group, participates in learning activities, relates to others, and solves problems, how a person gather and understand knowledge in a specific manner. Different learning style models are available. FSLSM model is best suited for evaluating learning style. Adaptive engine is designed to evaluate the learning style of the learner. The step is to design the academics engine. Academic analytics engine analyzes the learner profile with the courses of the institution. Context engine is also designed from the course details of the institution. Context engine is used to identify the learning environment of the learner. Recommender engine maps the learner, learning style, courses and the context of the learning environment, and suggests the appropriate courses to the learner.

Course advisory system in a personalized way can be designed by including preferences of learners, learning style, learning progress, knowledge level, higher order thinking level (cognitive style), learning content, and learning environment. Learning materials also play a vital role. Learning strategy, student profile, student participation in the learning process, and previous academic results are all considered for analytics. The framework is shown in the Fig. 5.

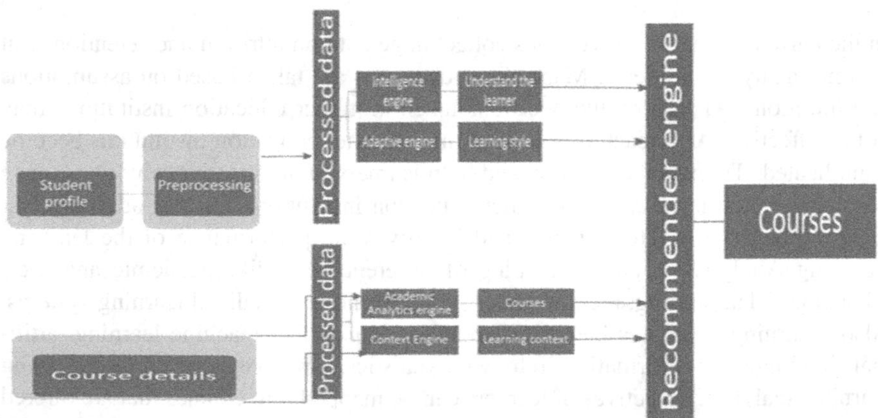

Fig. 5 Framework for course advisory system

8 Challenges

Open learning analytics is an upcoming research area in LA zone. It is challenging for all people involved in the education domain. Now a days, the curriculum designers have adopted open ended electives, choice-based credit system, inter-disciplinary courses, and certification courses. Therefore, course advisory systems should adopt to the changes incorporated in the education system. Different attributes of student profile is needed in different context namely personalization, adaptation, intelligent feedback, and recommendation. The different stakeholders should follow ethics in disclosing their details (transparency should be maintained). For example, the student should disclose his/her details in a truthful way for getting the appropriate feedback about the course selection. To define indicators for each analysis, prediction, feedback, adaptation, personalization, recommendation in the course selection process is a complex task. Finally as a tool, integration of learning analytics into system should be considered. Accordingly, appropriate visualization mechanisms for statistical, filtering, and mining tools should be used in designing the system.

9 Conclusion

This paper gives an outline to design an intelligent course advisory systems. Course advisory system is designed by using learning analytics, academic analytics, and personalized system. This paper provides an insight to the design of course advisory systems which could be used by the learners for their course selection. This system is helpful for the students before they join any course. The future work is to design the system according to the framework proposed in this paper and applying it in the Indian education system.

References

1. Dennis Maaß, Marco Spruit and Peter de Waal, “Improving short-term demand forecasting for short-lifecycle consumer products with data mining techniques”, Decision Analytics – A Springer open Access journal, vol: 1 Issue 4, 2014.
2. Nahit Emanet, Halil R Öz2, Nazan Bayram3 and Dursun Delen, “A comparative analysis of machine learning methods for classification type decision problems in healthcare’, Decision Analytics – A Springer open Access journal, vol: 1 Issue 6, 2014.
3. Pia Niemelä & Ville Isomöttönen & Lasse Lipponen, “Successful design of learning solutions being situation aware”, springer Science and Business Media, 2014.
4. Rasoul Karimi · Alexandros Nanopoulos · Lars Schmidt-Thieme, “A supervised active learning framework for recommender systems based on decision trees”, User Modeling and User-Adapted Interaction Volume 25 Issue 1, March 2015, 39–64.

5. Moushir M. El-Bishouty • Ting-Wen Chang •Sabine Graf • Kinshuk • Nian-Shing Chen, "Smart e-course recommender based on learning styles", journal of computers in education, 01, 2014, 99–111.
6. Lanqin Zheng • Nian-Shing Chen Ronghuai Huang • Kaicheng Yang, " A novel approach to assess collaborative learning processes and group performance through the knowledge convergence", journal of computers in education, 1, 2–3, 2014, 167-185.
7. Oliver Daems • Melanie Erkens • Nils Malzahn • H. Ulrich Hoppe, "Using content analysis and domain ontologies to check learners' understanding of science concepts", journal of computers in education, 1(2–3), 2014, 113–131.
8. Vincent Tam • Edmund Y. Lam • S. T. Fung," A new framework of concept clustering and learning path optimization to develop the next-generation e-learning systems", journal of computers in education, 01 (4), 2014, 335–352.
9. Mohamed Amine Chatti, Vlatko Lukarov, Hendrik Thüs Arham Muslim, Ahmed Mohamed Fahmy Yousef Usman Wahid, Christoph Greven, Arnab Chakrabarti, Ulrik Schroeder, "Learning Analytics: Challenges and Future Research Directions", eleed, Iss. 10.
10. Sérgio André Ferreira & António Andrade, 'Academic analytics: Anatomy of an exploratory essay", Education & Information Technologies, March 2014.
11. Vivekanandan Kumar, David Boulanger, Jeremie Seanosky, Kinshuk, Karthikeyan Panneerselvam, Thamarai Selvi Somasundaram, "Competence analytics", journal of computers in education, 01 (4), 2014, 251–270.
12. Syed Tanveer Jishan, Raisul Islam Rashu, Naheena Haque and Rashedur M Rahman, "Improving accuracy of students' final grade prediction model using optimal equal width binning and synthetic minority over-sampling technique", Decision Analytics – A Springer open Access journal, vol: 2 Issue 1, 2015.
13. Gwo-Jen Hwang, " Definition, framework and research issues of smart learning environments - a context-aware ubiquitous learning perspective", *Smart Learning Environments* 2014, **1**:4, 1–14.

Player Tracking in Sports Video Using Optical Flow Analysis

Chetan G. Kagalagomb and Sunanda Dixit

Abstract The core aspect of this work is a review for player detection and tracing. Coaches and players prepare widely by studying the opponents attacking and self-protecting formations, plays and metrics before every game. Moving object detection is one of the serious problems in the field of surveillance, traffic monitoring, computer vision, player tracking in sports video and so forth. Detecting the objects in the video and tracking its movement to recognise its qualities have been a demanding examination zone in the area of computer image and image processing. The core aspect of this work is a review for player detection and tracking. As coaches and players prepare widely by studying the opponents offensive and defensive formations, plays and metrics before every game, tracking of player becomes a major problem because of different problems in appearance, occlusion and so on. This work offers an analysis on detection of objects, segmentation of objects and tracking of objects. It also provides the study of comparison of diverse methods used for player tracking.

Keywords Occlusion · Surveillance · Segmentation

1 Introduction

In the field of computer image, detection of object and tracking of an object is an important job. In object tracking algorithms, the increase of computers, the convenience of low-cost and superior cameras have created an excessive interest in the tracking of objects. Three steps are involved in analysing the video: moving object detection, object tracking and analysing its behaviour.

C.G. Kagalagomb (✉)
Department of E&C, VSM Institute of Technology, Nipani, India
e-mail: chetangk006@hotmail.com

Sunanda Dixit
Department of ISE, Dayanand Sagar College of Engineering, Bengaluru, India
e-mail: sunanda.bms@gmail.com

© Springer Science+Business Media Singapore 2017
S.C. Satapathy et al. (eds.), *Proceedings of the International Conference on Data Engineering and Communication Technology*, Advances in Intelligent Systems and Computing 468, DOI 10.1007/978-981-10-1675-2_72

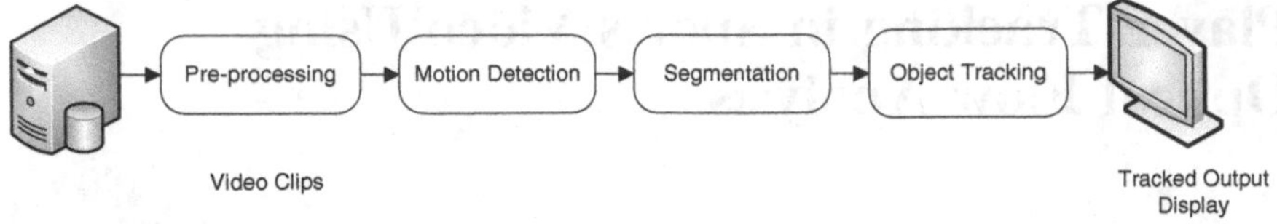

Fig. 1 Overview of player tracking in sports video

- ***Detection of Object***
  The process of finding the object in the video stream and to group the pixels of these is called detection of object. Detection of object is performed by different methods like background subtraction, frame extraction, optical flow and temporal differencing.
- ***Object Representation***
  Object representation can be done using number of techniques like motion based representation, colour representation, etc. Object may be people, animals, birds, vehicles, trees, etc.
- ***Tracking of Object***
  Finding the path of object in the image is defined as tracking of object. There are different types in tracking of objects such as kernel tracking, tracking of shadow, etc.

Automated player detection and tracking has got many advantage for both academic and professional athletics. Automated data provides information about the opposition's plays, approach and formation. Video broadcasting could also be improved by player tracking by automatically changing to the camera with the brilliant viewing angle or by aiming more on superstars. This is how sports broadcasting are done today by reducing the manual workload. This paper provides different techniques to detect each player and keep a track on each player.

The overview for player tracking is shown in Fig. 1, which contains three parts: video clip, detection of players, which consist of different motion detection algorithms and segmentation, tracking of players using tracking filters such as Kalman filter, particle filter, etc. Tracking players is done using tracking filter, which updates the current state of players. It first catches the current position of player using prediction equation and then updates the state of players. Video clip has number of frames. Frame extraction is done to get significant and complete information for motion detection and tracking of players. Different algorithms for all these blocks are studied and best of which is applied.

2 Literature Survey

Zhang et al. [1] and Olesova [2] offered a strong technique for spatiotemporal visual saliency. This method is based on the phase spectrum of the videos, which is computationally effective and easy to implement. Study on how the spatiotemporal saliency can be used in two important vision task, spatiotemporal interest point detection and deviation detection. This proposed technique is tested on different datasets with different state-of-the-art approaches. These tests reveal the efficiency of the proposed approach and its application to the above vision tasks giving an accuracy of about 93.78 %.

Khan et al. [3] proposed a method for background modelling technique. Spatiotemporal Gaussian mixture model (STGMM) is used for object detection. Here they took use of Kalman filter for partially occluded objects, whereas for high occlusion objects colour information and size attributes are used for tracking. Effective results were found by testing this system in real-time applications. Moderate accuracy is attained using this approach.

Liu et al. [4] showed a different idea from the traditional flow analysis. Based on apparent characteristic, this proposed system depends on motional graphic feature, which attains improved recognition properties with accuracy of about 69.4 %.

Mahmoudi et al. [5] Offered software that implements optical flow analysis using the lucas Kanade algorithm [6] for motion detection. Harris corner detector is also integrated with this system so as to perform spare tracking (tracking of the meaningful pixels only) which decreases the computational problem of the method. The software is greatly fast as a result of implementation of corner selection and tracking on GPU giving high accuracy.

Bialkowski et al. [7] presented a "role-based" illustration that gives each player's relative part at each frame dynamically and also shows how it captures thc short-term contcxt to permit both separate player and team analysis. Using a least entropy data partitioning technique, they discovered role directly from data. Also it is showed that how efficient is this approach to detect, visualise formations and analyse individual player behaviour with moderate accuracy.

Chen et al. [8] proposed an advanced particle filter tracking method based on moving edge information and colour, to offer more accurate outcomes in long-term tracking. To keep the accuracy of the target model, this method can be used to ensure that the target can be enclosed by the bounding box when it comes across different issues. The system could get the accuracy of about 91.8 % when matched to other existing methods.

Chauhan et al. [9] offered a new method for tracking uses Gaussian mixture model (GMM) along with optical flow analysis method. As these two methods are compatible with each other, they are used in many applications like video communication and compression, medical imaging, traffic control and video editing. This system could provide high accuracy.

Lu et al. [10] introduced a scheme that holds the facility to find and follow several players. Liu et al. [11] suggested a new method in which the complex inter-object relationships are implicitly merged by context-conditioned motion models. Demonstration of this process is done by taking 10 and 20 players, respectively, causing to give enough accuracy. Table 1 depicts the survey different segmentation algorithms.

Xing et al. [12] presented a new multiple objects tracking algorithm which contributes both in the tracking strategy level and observation modelling level.

Table 1 Survey of different segmentation algorithms

Year	Author	Segmentation methods	Pros	Cons
2015	Mohanty et al. (survey) [20]	Gaussian mixture model	Memory requirement is less	Does Not work in multimodal background
2014	Khan et al. [3]	Colour-based segmentation	Computational cost is low	Not always suitable because of accuracy
2014	Chen et al. [8]	Colour + Moving edge	Cost is less	Less accurate
2014	Deepjoy Das et al. [25]	Adaptive background mixture	It is able to handle bimodal backgrounds, long-term scene changes and repetitive motions	–
2013	Manikandan et al. [26]	Gaussian mixture model	Memory required is less	Will not support all kinds background
2013	Lu et al. [10]	Canny edge mapping	Works well in images with good contrast between object and background	• Sensitive to noise • Robust edge linking is not trivial
2013	Abhishek Kumar Chauhan et al. [9]	Foreground substraction	Implementation is easy	Needs a static background
2013	Manoranjan Paul et al. [16]	Silhouette segmentation	This method can support low quality video data and no need of video orientation	–
2012	Sreedevi et al. [27]	Segmentation using Edge + Colour	Cost of computation is less	Accuracy is less

Effectiveness of the offered technique is verified by experimenting this method on variety of sports videos like basketball and hockey.

Greggio et al. [13] proposed a method which is compared with state-of the-art alternatives and initialization efficacy and its improved segmentation performance are also shown. Nikhil et al. [14] compared different player tracking methods for analysis. The core aspects of this work are player detection and tracking. Similarly [15–24] depicts survey of different algorithms for object detection and tracking.

Table 1 gives the detailed survey of different segmentation algorithm along with its advantage and disadvantage. Table 2 gives the detailed survey of different tracking algorithms along with their advantages and disadvantages. Robust algorithms are chosen to get the efficient tracking system.

Table 2 Survey for different tracking algorithms

Year	Author	Motion detection algorithm	Pros	Cons
2015	Salehifar et al. [28]	Particle Filter	There is no complexity of distributions	The most efficient number of particles cannot be calculated
2015	Nandashri et al. [29]	Kalman Filter + Hungarian Algorithm	Occlusions are handled successfully	–
2015	Li et al. [30]	Cam-shift Algorithm	• It has the advantage of real-time output and framework. • Works well for human body tracking	–
2014	Khan et al. [3]	Kalman Filter	Find the points in blaring images	Variables Normally distributed (gaussian)
2013	MallikarjunaRao et al. [15]	Particle Filter	No Limitation for the complexity of distributions	The most effective number of elements cannot be calculated
2013	Shishido et al. [31]	Kalman Filter	Used to find the points in blaring images	Variables normally distributed (gaussian)

3 Conclusion

Tracking players in broadcast video is a challenging task as appearance, occlusion and so on cause poor tracking results. So many researches are going on at this context. Taking this to consideration, various object detection, segmentation and tracking methods have been explained in brief along with advantages and disadvantages. Advanced study is done to discover effective algorithm to decrease time and cost of computation needed to track variety of video with different features.

References

1. Qiang Zhang, Yilin Wang and Baoxin Li, "Unsupervised Video Analysis Based on a Spatiotemporal Saliency Detector", ACM, 2015.
2. Veronika Olesova, "Modified Methods of Generating Saliency Maps Based on Super pixels", Proceedings of CESCG: The 18th Central European Seminar on Computer Graphics, 2014.
3. Malik M. Khan, Tayyab W. Awan, Intaek Kim, and Youngsung Soh, "Tracking Occluded Objects Using Kalman Filter and Color Information", International Journal of Computer Theory and Engineering, Vol. 6, Issue 5, 2014.

4. Gang Liu, Deming Zhang and Hui Li, "Research on Action Recognition of Player in Broadcast Sports Video", International Journal of Multimedia and Ubiquitous Engineering, Vol. 9, No. 10, Pp. 297–306, 2014.
5. S.A. Mahmoudi, M. Kierzynka, P. Manneback and K. Kurowski, "Real-time motion tracking using optical flow on multiple GPUs", Bulletin of The Polish Academy of Sciences Technical Sciences, Vol. 62, Issue 1, 2014.
6. KausikBasak, Manjunatha M and Pranab Kumar Dutta, "Pyramidal Refinement of Lucas – Kanade Optical Flow based Tracking of Peripheral Air Embolism in OCT Contrast Imaging", International Journal of Computer Applications", Vol. 52, Issue 12, 2012.
7. Alina Bialkowski, Patrick Lucey, Peter Carr, Yisong Yue, Sridha Sridharan and Iain Matthews, "Large-Scale Analysis of Soccer Matches using Spatiotemporal Tracking Data", 2014.
8. Chao-Ju Chen, WernhuarTarng and Kuo-HuaLo, "An Improved Particle Filter TrackingSystem Based On Colour and Moving Edge Information", International Journal of Computer Science & Information Technology (IJCSIT), Vol. 6, Issue 4, 2014.
9. Abhishek Kumar Chauhan and Prashant Krishan, "Moving Object Tracking using Gaussian Mixture Model and Optical Flow", International Journal of Advanced Research in Computer Science and Software Engineering, Vol. 3, Issue 4, 2013.
10. Wei-Lwun Lu, Jo-Anne Ting, James J.Little and Kevin P. Murphy, "Learning to Track and Identify Players from Broadcast Sports Videos", IEEE, Vol. 35, Issue 7, Pp. 1704–1716, 2013.
11. Jingchen Liu, Peter Carr, Robert T. Collins and Yanxi Liu, "Tracking Sports Players with Context-Conditioned Motion Models", IEEE, Pp.1830–1837, 2013.
12. Junliang Xing, Haizhou Ai, LiweiLiu and ShihongLao, "Multiple Player Tracking in Sports Video: A Dual-Mode Two-Way Bayesian Inference Approach With Progressive Observation Modeling", Vol. 20, Issue 6, Pp. 1652–1667, IEEE, 2011.
13. Nicola Greggio, Alexandre Bernardino, Cecilia Laschi, Paolo Dario and Jose Santos-Victor, "Self-Adaptive Gaussian Mixture Models for Real-Time Video Segmentation and Background Subtraction", IEEE, Pp. 983–989, 2010.
14. Nikhil M and Sreejith S, "A Review Paper on Player Tracking and Automated Analysis in Sports Video", International Journal of Emerging Technology and Advanced Engineering, Vol. 5, Issue 6, 2015.
15. G. Mallikarjuna Rao and Dr.Ch. Satyanarayana, "Visual Object Target Tracking Using Particle Filter: A Survey", I.J. Image Graphics and Signal Processing, 2013.
16. Manoranjan Paul, Shah M E Haque and Subrata Chakraborty, "Human detection in surveillance videos and its applications - a review", Springer, 2013.
17. Ravi Prakash Singh, Syed Tariq Murtaza, Shamshad Ahmad, Arshad Hussain Bhat and Mohd Sharique, "Time Motion Analysis in Sports-A Review", Academic Sports Scholar, Vol. 3, Issue 10, 2014.
18. Mansi Manocha and Parminder Kaur, "Object Tracking Techniques for Video Tracking: A Survey", the International Journal of Engineering and Science (IJES), Vol. 3, Issue 6, 2014.
19. Payal Panchal, Gaurav Prajapati, Savan Patel, Hinal Shah and Jitendra Nasriwala, "A Review on Object Detection and Tracking Methods", International Journal for Research in Emerging Science and Technology, Vol. 2, Issue 1, 2015.
20. Aseema Mohanty and Sanjivani Shantaiya, "A Survey on Moving Object Detection using Background Subtraction Methods in Video", International Journal of Computer Applications, 2015.
21. Sarika Patela, Prof. Udesang K. Jaliyab and Prof. Mahashweta J. Joshib, "Review: Object Detection and Object Tracking Methods", International Journal of Innovative and Emerging Research in Engineering Vol. 2, Issue 1, 2015.
22. Dhara Trambadiya and ChintanVarnagar, "A Review on Moving Object Detection and Tracking Methods", National Conference on Emerging Trends in Computer, Electrical & Electronics (ETCEE-2015) International Journal of Advance Engineering and Research Development (IJAERD), 2015.

23. Gandhan Sindhuja and Dr Renuka Devi S M, "A Survey on Detection and Tracking of Objects in Video Sequence", International Journal of Engineering Research and General Science Vol. 3, Issue 2, 2015.
24. A. Hema and S. Vanathi, "A Survey on Moving Object Detection Using Image Processing Techniques", International Journal of Innovative Research in Computer and Communication Engineering, Vol. 3, Issue 7, 2015.
25. Mr. Deepjoy Das and Dr. SaratSaharia, "Implementation and Performance Evaluation of Background Subtraction Algorithms", International Journal on Computational Sciences & Applications (IJCSA), Vol. 4, No.2, 2014.
26. R.Manikandan and R.Ramakrishnan, "Human Object Detection and Tracking using Background Subtraction for Sports Applications", International Journal of Advanced Research in Computer and Communication Engineering, Vol. 2, Issue 10, 2013.
27. Sreedevi M, Yaswanth Kumar Avulapati, AnjanBabu G and Sendhil Kumar R, "Real Time Movement Detection for Human Recognition", World Congress on Engineering and Computer Science (WCECS),VolI, 2012.
28. Lili Zhang and Dazhi Wang, "Basketball player tracking model of hierarchical multiple feature fusion particle filter", Metallurgical and Mining Industry, No. 5, 2015.
29. Nandashri D and Smitha P, "An Efficient Tracking of Multi Object Visual Motion using Hungarian Method", International Journal of Engineering Research & Technology (IJERT), Vol. 4, Issue 4, 2015.
30. Jiude Li, "Research on Camera-Based Human Body Tracking Using Improved Cam-Shift Algorithm", International Journal on Smart Sensing and Intelligent Systems, Vol. 8, No. 2, 2015.
31. Hidehiko Shishido, Itaru Kitahara, Yoshinari Kameda and Yuichi Ohta, "A trajectory Estimation Method for Badminton shuttlecock Utilizing Motion Blur", 2013.

Integrating Hybrid FMEA Methodology with Path Planning Decisions in Autonomous Vehicles

Bishwajit Pal, Samitha Khaiyum and Y.S. Kumaraswamy

Abstract This paper demonstrates how to choose the best performing path planning algorithm for an autonomous vehicle to address different traffic conditions with the help of FMEA model. An autonomous vehicle uses path planning algorithms to navigate but no one algorithm is sufficient/address all kinds of traffic issues efficiently. This paper proposes an evaluation method. The quality and performance of each of the path planning algorithms are analyzed by a Hybrid FMEA framework which predicts the best path for a given set of traffic conditions. This decision can then be used by the vehicles master program to select the best path for execution instead of executing multiple algorithms. The Hybrid FMEA framework selects the appropriate algorithm for different road conditions/road curves/directions resulting in optimizing execution time and aids in maintaining real-time software conditions. This paper includes a case study of FMEA framework applied to autonomous driving vehicles to support decision-making in different traffic condition.

Keywords Autonomous vehicle · FMEA · Real time · Path planning

1 Introduction

The modern automobile industry is facing a new revolution in the form of autonomous vehicles. The concept of autonomous vehicle has been already proven by few of the modern automobile companies, but still the technology remains very costly and out of reach of the masses. Out of the entire biggest question on

Bishwajit Pal (✉) · Samitha Khaiyum · Y.S. Kumaraswamy
Dayanand Sagar College of Engineering, Kumaraswamy Layout, Bangalore 560078, Karnataka, India
e-mail: bishwajit.pal@gmail.com

Samitha Khaiyum
e-mail: samitha.athif@gmail.com

Y.S. Kumaraswamy
e-mail: yskldswamy2@yahoo.com

© Springer Science+Business Media Singapore 2017

S.C. Satapathy et al. (eds.), *Proceedings of the International Conference on Data Engineering and Communication Technology*, Advances in Intelligent Systems and Computing 468, DOI 10.1007/978-981-10-1675-2_73

autonomous vehicle are the safety and its quality of decision. Study has proven many benefits of these vehicles in terms of improved fuel economy, enhanced safety and reduced traffic congestion.

Many questions are raised on what type of path planning must be used by an autonomous vehicle. The path decided by a sport car will be different than of a large truck than of a tractor trailer. Also considering the safety of the vehicle, different path planned needs to be analyzed for its optimum output before being executed by the master program. These path planned also needs to take care of the dynamics of the vehicle on which it is going to get executed. Also not one single path planning algorithm will suffice all conditions. The algorithm needs to be mapped and measured against some index so that the quality of the planned path is derived. FMEA is an approach for identifying all possible failures in process, product or service and analyze its effects. Considering path planning as a service to the master program, failure indexes can be derived which can be used as quality attributes and analyzing the degree of failure.

2 Literature Survey

The present autonomous vehicle build by Google, GM, etc. uses very expensive LIDAR to map the external environment. Moreover these are electromechanical devices mounted on a spinning assembly. As long as the sensors are costly, autonomous vehicles will remain impractical to produce at industry scale.

Authors in [1] state that autonomous driving vehicles work on large-scale data collection of driver, vehicle and environment information in the real world. Safety critical events are currently identified using kinematic triggers, for instance searching for deceleration below a certain threshold signifying harsh braking. Due to the low sensitivity and specificity of this filtering procedure, manual review of video data is currently necessary to decide whether the events identified by the triggers are actually safety critical [2]. There are three interrelated spaces involved in fault diagnosis: fault space, observation space and diagnosis space. Each space contains the totality of all possibilities in their respective areas. Fault space contains all faults that can occur on the vehicle, observation space contains all observations from the point of view of the vehicle and diagnosis space contains all possible diagnoses available using the observation space [2, 3]. Authors, through their research show that usage of autonomous driven vehicles significantly increases the highway capacity by making of use sensors and having a vehicle to vehicle communication [4, 5]. Author demonstrates how LIDAR is used in real time to map all objects in external environment.

Table 1 Index levels of severity, occurrence and detection

Severity rating	Index level	Occurrence rating	Index level	Detection rating	Index level
7–10	Blocker	7–10	Probable	7–10	Delayed
4–6	Medium	4–6	Remote	4–6	Slow
1–3	Cosmetic	1–3	Improbable	1–3	Early

3 Autonomous Vehicle and FMEA

[3] The traffic condition is ever changing and the pattern of the traffic in front is always unexpected. The path selected by the master application must be selected based on some quality attribute and not only one path planning algorithm should be used for all condition. We are proposing a new hybrid FMEA to address the issue of selecting the best path from a set. In a typical FMEA model [6], the index is calculated on only three attributes. In this paper, we are proposing to make this attributes more flexible and open to new attributes based on the severity the attribute poses to the autonomous vehicle. A Hybrid FMEA will be pooling all the results and its detection algorithm for its effectiveness in various traffic situations. In an autonomous platform, the usage of FMEA is very likely as the environment is dynamic and the algorithms used for driving the vehicle is more challenged. An FMEA can give a potential prediction model to help the autonomous vehicle make quality decision and rule out risky decisions [7]. For example, it can keep many path planning algorithm, and execute only the best algorithm for a given condition. Table 1 shows the different index levels for factors like severity, occurrence, detection and the scales that determine the relative risk and then provide a mechanism to prioritize work. The impact or relative risk is measured using three different scales namely severity to measure consequence of failure if it occurs, occurrence which gives the frequency of failure and detection which is the ability to measure the potential failure before the consequences are observed.

4 Defining Quality Attributes of Planned Path

Whenever a path is generated by a path planning algorithm it should be measured against conditions or quality parameters that favors the vehicle. Quantifying the quality of a planned path is important for an autonomous vehicle. This can bring some quality attribute that the master program can count on. For example, for a [8, 9] heavy vehicle, the path planning algorithm needs to take special care that it does not make sudden curves at high speeds. A generic formula has been proposed in this paper for extending the FMEA index to any number of quality attribute or factors.

Fig. 1 The *Front Road view* of an autonomous vehicle and segmentation of the road into different regions, each region represent a distance from the vehicle

4.1 Nearest Distance from Each Obstacle

In Fig. 1, the autonomous vehicle maps the path ahead of it in the form of a grid. The selected path above start from the bottom of the grid, i.e. the start point of the path and goes till the top of the grid, which is the maximum range of the sensors used by the autonomous vehicle. An important factor in path planning is to keep the vehicle away from obstacle, as much as possible. Hence the nearest distance from the entire obstacle requires to be calculated. The path is a series of points starting from *Y* Axis as 0. Each point has an *X* and *Y* component on the 2D Grid. Generally a path is represented as a series of *x* and *y* axis coordinates as shown in Eq. 1.

$$\text{Path}x = (X_1, Y_1), (X_2, Y_2), \ldots, (X_n, Y_n). \tag{1}$$

For the planned path (pp), let's take PPj and PPk as the *x* and *y* coordinate respectively. Hence Eq. 1 can be rewritten as below (Eq. 2)

$$\text{Path planned} = (\text{PPj.1}, \text{PPk.1}), (\text{PPj.2}, \text{PPk.2}) \ldots (\text{PPj.n}, \text{PPk.n}). \tag{2}$$

Similarly we represent the obstacles in *x* and *y* coordinate by their position. For example, Obstacle Obs1 = (*X*1, *Y*1) or Obstacle Obs2 = (*X*2, *Y*2) and Obstacle Obsn = (*Xn*, *Yn*).

In this way, large objects can also be represented by series of 2D consecutive points. Now with the above Eqs. (1 and 2) the nearest distance for path is given by the following equation (Eq. 3):

$$
\begin{aligned}
\text{Near Distance for path1} = \min(&\ |\text{Obst\_1.x} - \text{PP\_1.X}| + |\text{Obst\_1.Y} - \text{PP\_1.y}|\ )/2,\\
&(|\text{Obst\_1.x} - \text{PP\_2.X}| + |\text{Obst\_1.Y} - \text{PP\_2.y}|\ )/2, \ldots\\
&(|\text{Obst\_1.x} - \text{PP\_n.X}| + |\text{Obst\_1.Y} - \text{PP\_n.y}|)/2 \ldots\\
&(\ |\text{Obst\_n.x} - \text{PP\_1.X}| + |\text{Obst\_n.Y} - \text{PP\_1.y}|\ )/2,\\
&(|\text{Obst\_n.x} - \text{PP\_2.X}| + |\text{Obst\_n.Y} - \text{PP\_2.y}|\ )/2 \ldots\\
&(|\text{Obst\_n.x} - |\text{PP\_n.X}| + |\text{Obst\_n.Y} - \text{PP\_n.y}|)/2.
\end{aligned} \tag{3}
$$

Here obs_n.x and obs_n.y is the x and y coordinate respectively for the nth obstacle. Similarly PP_n.x and PP_n.y is the x and y coordinate of the points on the path. The equation is like a cross join of every point in the planned path to the obstacle.

4.2 Average Distance from Obstacles

The average distance is calculated by summing up all the points in the path with all the points in the obstacle. This is a quality attribute for a path generated by a path planning algorithm. In this quality attribute, the average distance of the each point in the planned path calculated from the entire obstacle present in the context. For example, if the path planned by an algorithm is an array of x and y coordinate given by Eq. 2 and the obstacle in the context is given by 2D arrays representing their current position by (Eq. 4)

$$
\text{Obstacles} = (\text{Obst\_1.j},\ \text{Obst\_1.k}), (\text{Obst\_2.j},\ \text{Obst\_2.k}) \ldots (\ \text{Obst\_n.j},\ \text{Obst\_n.k}). \tag{4}
$$

The average distance for each point is calculated as given in Fig. 2:

4.3 Design and Execution of Hybrid FMEA

The AV will be executing tasks at the speed, nearly equivalent to human response (200 μs). The hybrid FMEA is calculated based on severity, occurrence and detection [10]. For the implementation in AV, we will modify the mapping of these

$$
\frac{\sum_{j=0}^{m}\left(\sum_{k=0}^{n}\sqrt[2]{(obst\_j.x - PP\_k.x)^2 + (obst\_j.y - PP\_k.y)^2}\right)}{j * k}
$$

Fig. 2 Average distance for each point

three variables to another three variables which represents the quality of the path generated by a path planning algorithm. The higher the number the better choice it will be.

The hybrid FMEA is implemented in the following four steps:

1. **Establish**: Establish the quality parameters (metrics like severity, detection and occurrence) to judge the path planning algorithms.
2. **Execute**: Run algorithm with classified/segmented sample data sets and record all the quality parameters of each of the path planning algorithm.
3. **Analysis**: Apply FMEA index formula to each of this segmented data to obtain the average FMEA index for each path planning algorithm.
4. **Decision**: Select the algorithm which has the highest FMEA index.

For the autonomous vehicle, the severity index is mapped into the product of the nearest point between path and obstacle and the number of time the same value is appearing in the entire path (Eq. 5).

$$\text{Severity} \propto \text{Nearest Point} * \text{Count of Nearest Point.} \tag{5}$$

Similarly, the occurrence is also mapped directly proportional to the number of obstacle that the autonomous vehicle's path planning algorithm has considered for planning/generating the path (Eq. 6). This is represented by the following equation:

$$\text{Occurrence} \propto \text{Path Length} / \text{Obstacle Count.} \tag{6}$$

The detection parameter is replaced by the total time taken by the path planning algorithm to generate the path in the given set of obstacle. The higher time it takes to generate the path the lesser preference it gets on the FMEA index (Eq. 7).

$$\text{Detection} \propto 1/\text{Time taken to generate the Path (in } \mu\text{s)}. \tag{7}$$

Combining the above Eqs. 5, 6 and 7 we get the FMEA index as (Eqs. 8 and 9)

$$\text{FMEA Index} = \text{Severity} * \text{Occurrence} * \text{detection.} \tag{8}$$

$$\begin{aligned} \text{FMEA Index} = {} & (\text{Nearest Point} * \text{Count of Nearest Points} * \text{Path Length}) \\ & /(\text{Time taken to generate the path} * \text{Count of Obstacle}). \end{aligned} \tag{9}$$

4.4 Generalizing the Hybrid FMEA Index

Let PP be the path planned for a given environment. The algorithms has many attributes like path length, search iteration, punish wrong move, etc. to make the FMEA index more generic in nature the following formula is proposed (Eq. 10).

For a given path planning algorithm which has A1, A2, A3…. An attributes which describes its behavior or some part of the quality attribute and is proportional or inversely proportional to the FMEA. W1, W2, W3 … Wn is their respective weight that needs to be considered for the index, then the FMEA index is given as

$$\text{FMEA Index} = (\text{A1} * \text{W1}) * (\text{A2} * \text{W2}) * (\text{A3} * \text{W3}) \ldots (\text{An} * \text{Wn}) \tag{10}$$

5 Data Analysis and Results

The new FMEA index was tested on different path with multiple iteration. The data was categorized into three different traffic conditions. First with low traffic (1–3 obstacles in front of the vehicle), second with medium traffic (4–6 obstacles in front of the vehicle) and finally with large traffic (7–9 obstacles in front of the vehicle). Five path planning algorithm was selected to run the test. These were Diagonal Short cut, Euclidean, Euclidean No Square, Manhattan and Maximum Dx-Dy. The data was generated from five iterations with every combination of obstacles over 10 times. The total data size was 5 * 3 * 10 = 150 records. The following is the outcome.

5.1 *Low Traffic Data*

For low traffic, Euclidean path planning has a much higher index than other path planning algorithm, as shown in Fig. 3. This is followed by the Manhattan algorithm. Clearly, for low traffic conditions the Euclidean algorithm must be used for

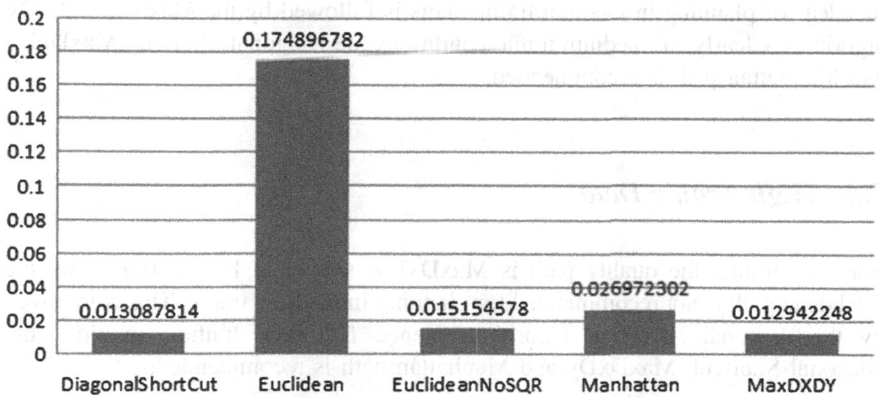

Fig. 3 Response time for low traffic condition

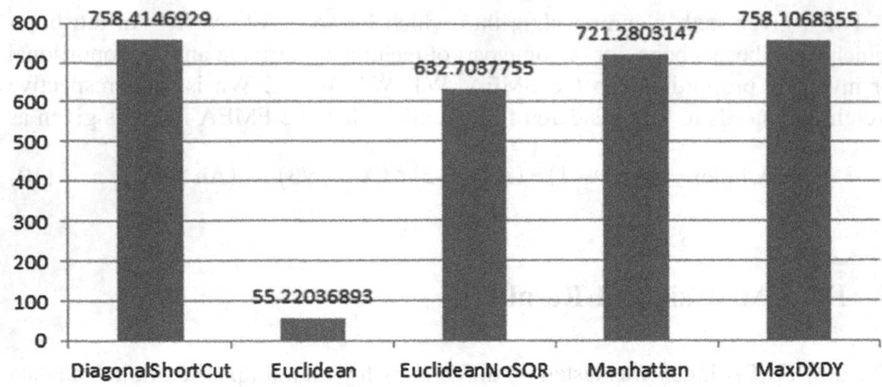

Fig. 4 Response time for medium traffic condition

low traffic condition. The Euclidean has lesser number of turns and has higher average distance from the obstacle. The Max DxDy also gives the same path but at higher time cost. From the data analyzed, we conclude that for low level of traffic conditions, i.e. 1–3 number of traffic the autonomous system can rely on the Euclidean algorithm for optimal path suited for real-time environment. The Euclidean algorithm takes less time to generate the path compared to all other path planning algorithm. Hence Euclidean's FMEA index is very high.

5.2 *Medium Traffic Data*

Conclusively for medium traffic, the scenario has changed and the best quality path generated is Diagonal Shot Cut, as shown in Fig. 4. Euclidean path is not recommended for planning in medium traffic. This is followed by the Maximum Dx-Dy algorithm. Clearly for medium traffic conditions the Diagonal-Shortcut, MaxDxDy and Manhattan path is recommended.

5.3 *High Traffic Data*

For high traffic, the quality path is MaxDxDy, shown in Fig. 5. Here also, the Euclidean path is not recommended for planning in medium traffic. This is followed by the diagonal short cut algorithm. Hence for High traffic conditions the Diagonal-Shortcut, MaxDxDy and Manhattan path is recommended.

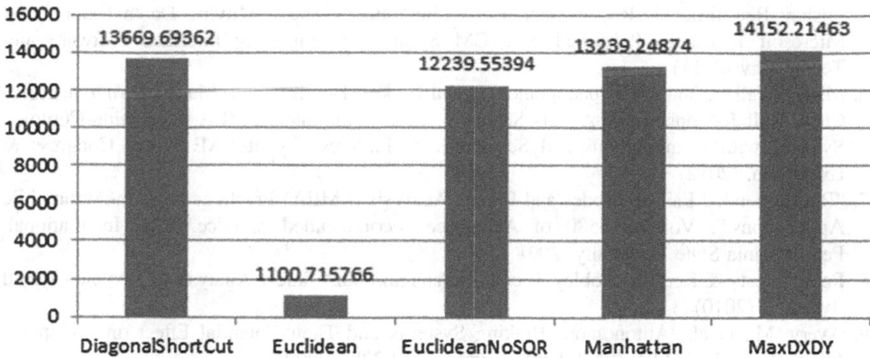

Fig. 5 Response time for medium traffic condition

6 Scope for Further Research and Conclusion

In the paper, we have considered only three dimensions to compare the algorithms. More attributes can be added to each algorithm to make the index stronger and safer [1]. A tie breaker attribute also can be added to the algorithm quality factors to decide on closely placed FMEA Index. This paper brings out the application of FMEA for autonomous vehicles to forecast decision outcomes. This model is most likely to be used on the AV platform. Because this model can also be executed over another machine the FMEA analysis is executed without interrupting the main program execution. With the response time of less than a microsecond the entire analysis table will be available to the main program to pool the right decision in real time. The main program can now rely on multiple path generating algorithms to achieve optimized decisions on which path to follow and also keep the AV from harms away.

References

1. Malta, L., Ljung Aust, M., Faber, F., Metz, B., Saint Pierre, G., Benmimoun, M., Schäfer, R., (2012): 'Final results: Impacts on traffic safety, EuroFOT (2012)
2. Kelvin Hamilton, Dave Lane, Nick Taylor and Keith Brown, "Fault Diagnosis on Autonomous Robotic Vehicles with recovery: An Integrated Heterogeneous-Knowledge Approach" at IEEE International Conference on Robotics & Automation. (2011)
3. Tientrakool, P.: 'Highway Capacity Benefits from Using Vehicle-to-Vehicle Communication and Sensors for Collision Avoidance', Vehicular Technology Conference VTC Fall), IEEE, 5–8 Sept., (2011)
4. Shahram Izadi, David Kim, Otmar Hilliges, David Molyneaux, Richard Newcombe, Pushmeet Kohli, Jamie Shotton, Steve Hodges, Dustin Freeman, Andrew Davison, Andrew Fitzgibbon," Kinect Fusion: Real-Time Dense Surface Mapping and Tracking", UIST'. (2011)
5. Richard A. Newcombe, Andrew J. Davison, Shahram Izadi, Pushmeet Kohli, Otmar Hilliges, Jamie Shotton, David Molyneaux, Steve Hodges, David Kim, Andrew Fitzgibbon, "Kinect

Fusion: Real-time 3D Reconstruction and Interaction Using a Moving Depth Camera", at Microsoft Research, Published by ACM Symposium on User Interface Software and Technology (2011)

6. Martin Walker, Yiannis Papadopoulos, David Parker, Henrik Lonn, Martin Torngren, DeJui Chen, Rolf Johannson and Anders Sandberg," Semi-Automatic FMEA Supporting Complex Systems with Combinations and Sequences of Failures" by at SAE World Congress & Exhibition. (2012)
7. "Recommended Failure Modes and Effects Analysis (FMEA) Practices for Non-Automobile Applications.". Volume 5580 of Aerospace recommended practice, SAE International, Pennsylvania State University, 2001
8. Peter H Jesty & Keith M Hobley, Richard Evans and Ian," Safety Analysis of Vehicle - Based Systems" (2010)
9. Avery, M., et al. 'Autonomous Braking Systems and Their Potential Effect on Whiplash Injury Reduction', ESV 2009. Paper Number 09-0328. (2009)
10. Samitha Khaiyum, Y. S. Kumaraswamy, K. Karibasappa, " Classification of failures in real time embedded software projects", International conference on Systemics, cybernetics and informatics proceedings, ISSN 0973-4864, SP-1.3 (2014)

An Evaluative Model for Discriminative Dictionary Learning with Hybrid Local Binary Patterns Based Feature Extraction for Bone Age Assessment in Content Based Image Retrieval

Ananthi Sheshasaayee and C. Jasmine

Abstract Skeletal maturity is evaluated visually by the comparison of hand radiographs against a standardized reference image atlas. Content-Based Image Retrieval (CBIR) yields a robust solution without the delineation and measurement of bones. This research work comprises of five major phases: Pre-processing, texture feature extraction, Relevance Score (RS) computation, bone age assessment and similarity matching. In the pre-processing phase, noise in the image samples is eliminated by making use of Kernel Fisher Discriminant Analysis (KFDA). Texture Feature extraction is performed using Hybrid Local binary Patterns (HLBP), Epiphysis—Metaphysis Region of interest (ROI) (EMROI) feature extraction is introduced for the pre-processed image. With the aim of evaluating the bone age the Tanner-Whitehouse scheme was utilized to inspect 20 bones. For the computation of the similarity matching between query and input image samples, Discriminative Dictionary Learning (DDL) is introduced in the research work; it shows that DDL performs better when compared to other state-of the-art approaches.

Keywords Similarity search · Pre-processing · Content based image retrieval (CBIR) · Kernel fisher discriminant analysis (KFDA) · Fuzzy neural network (FNN) · Bone age assessment (BAA)

Ananthi Sheshasaayee · C. Jasmine (✉)
Department of Computer Science, Quaid-E-Millath Govt. College for Women, Chennai 600 002, India
e-mail: jasminesamraj.research@gmail.com

Ananthi Sheshasaayee
e-mail: ananthi.research@gmail.com

© Springer Science+Business Media Singapore 2017

S.C. Satapathy et al. (eds.), *Proceedings of the International Conference on Data Engineering and Communication Technology*, Advances in Intelligent Systems and Computing 468, DOI 10.1007/978-981-10-1675-2_74

1 Introduction

Bone Age Assessment (BAA), which are based on hand radiographs is a task often performed and time consuming for radiologists. The two most generally employed techniques are on the basis of image comparison. In case of Tanner and Whitehouse (TW3) scheme [1], a specific subset of bones is examined. Different schemes have been taken to completely or moderately carry out the automation of bone age assessment [2–4]. Content-based image retrieval (CBIR)—supporting has gained attention for medical applications, specifically in radiology [5–7] for earlier cases. Not only images but even validated case information [8, 9] like chronological age, ethnic origin, gender, and multiple bone age readings are gathered, constructing a broad set of ground truth data. For the particular application to BAA, a specialized framework to recover Regions Of Interest (ROI) resembling those of a given hand radiograph is used instead of making use of the global similarity between entire images as used for example in the ImageCLEF contests [10]. Generally, the epiphyseal regions of interest (eROI) show the bone age reliably, particularly in the range of bone ages between 2 and 18 years [11]. CBIR yields a robust solution without any delineation and measurement of bones. This research work comprises of five major phases: Pre-processing, texture feature extraction, relevance score computation, bone age assessment and similarity matching which is explained as follows.

2 Proposed Pre-processing and Discriminative Dictionary Learning Methodology

Bone age assessment is a technique that is often utilized in forensic cases and by child development specialists. The goal of automated assessment with computers is renders the decision process with more objectivity, and also subsequently permitting more consistent results to be attained. In order to boost the result, the subsequent schemes are carried out. First, the noise in the image samples is eliminated by KFDA. Texture feature extraction is performed by using Hybrid Local Binary Pattern (HLBP) and SIFT feature. Epiphises–Metaphises ROI (EMROI) extraction (leave the Carpal ROI, ulna and radius out of focus based feature) is also extracted from images. Relevance Feedback (RF) has been generally used for assistance in getting the search purpose from the user and further improve the retrieval results. The FNN classifier is used from a large amount of labeled loosely medical images feature vectors and a fewer amount of BAA images that are accurately labeled. This is the first trial that makes use of the FNN learning for Relevance Score (RS). Similarity score computation is conducted for BAA. With the aim of assessing the bone age, the most effective scheme called Tanner-Whitehouse was employed. For this method, it is required to examine 20 bones. Also the DDL analysis scheme can learn a flexible nonlinear proximity function with local fisher discriminant in order to improve visual similarity search in CBIR for the cause of medical images.

2.1 Noise Removal Using Kernel Fisher Discriminant Analysis (KFDA)

Kernel Fisher Discriminant Analysis (KFDA) scheme [12] is presented in order to remove and reduce the noises from bone age assessment images by the definition of a kernel function. The efficiency of KFDA scheme is entirely based on the choice of the kernel. Consider $X = (x_1, \ldots x_n), x \in \mathcal{X}$ denotes the BAA samples, which is a random subset of $\mathbb{R}_n$, for every image samples pixels of the samples is denoted as $x \in P = \{ps_1, \ldots ps_m\}$ as the pixels 1 to m. A symmetric function $K: \mathcal{F} \times \mathcal{F} \rightarrow \mathbb{R}$ is referred to as kernel function among two BAA image pixels when it satisfies the finitely positive semi-definite characteristic: For any BAA image samples $ps_1, \ldots ps_m$ the gram matrix $G \in \mathbb{R}^{m \times m}$ is given as,

$$G_{ij} = K\left(ps_i, ps_j\right) \tag{1}$$

K in an implicit manner does the mapping of the input pixels with pixels P with a high-dimensional Hilbert space $\mathcal{H}$ equipped with the inner product $\langle ., . \rangle_{\mathcal{H}}$ by means of a mapping pixels $\phi: P \rightarrow \mathcal{H}$

$$K(o, p) = \langle \phi(o), \phi(p) \rangle_{\mathcal{H}}, \forall o, p \in P \tag{2}$$

Inner product $\langle \phi(o), \phi(p)_{\mathcal{H}} \rangle$ is referred to as Hilbert space. It is also recognized as noise removal space, it is completely dependent on the kernel function K and will be indicated as ϕ_k and $\mathcal{H}_k$. The noise removal phase of the BAA images is provided in the type of objective function h(p), to learn, pixel similarity the decision is conducted between two dissimilar pixels in the kernel function $\mathcal{H}_k$:

$$h(p) = \text{sgn}(w^t \phi_k(p) + b) \tag{3}$$

where $w \in \mathcal{H}_k$ denotes the pixels weight values $b \in \mathbb{R}$ be the bias value for pixels is the intercept, and

$$\text{sgn}(u) = \begin{cases} \text{positive pixel if } u > 0 \\ \text{negative pixel if } u < 0 \end{cases} \tag{4}$$

To carry out noise removal process for BAA images using KFDA,

$$\mu_K = \frac{1}{m}\sum_{i=1}^{m} \phi_k(p_i) \tag{5}$$

$$\Sigma_K = \frac{1}{m}\sum_{i=1}^{m} \phi_k(p_i) - \mu_K \tag{6}$$

As a result, KFDA maximizes the FDR,

$$FD_\lambda(w, K) = \frac{(w^T(\mu_K))^2}{w^T(\Sigma_K + \lambda I)w} \tag{7}$$

where λ denotes a positive regularization parameter and I represents the identity operator in HSI feature space $\mathcal{H}_k$. The weight value of every feature space which is determined according to the Gaussian Firefly Algorithm (GFA), which can show the weight vector for feature space,

$$w^* = (\Sigma_K + \lambda I)^{-1} \cdot (\mu_K) \tag{8}$$

increases the FDR. The maximum FDR is obtained by w^* is given by,

$$\begin{aligned} FD_\lambda^*(K) &= \max_{w \in \mathcal{H}_k\{0\}} FD_\lambda(w, K) \\ &= (\mu_K)^T(\Sigma_K + \lambda I)^{-1}(\mu_K) \end{aligned} \tag{9}$$

2.2 Feature Extraction

CBIR-based bone age assessment scheme works in accordance with the comparison of image content from a noise reduced images with previous cases in the database. All clinically utilized BAA techniques point to the epiphyseal area between the bones, such that the extraction of the features from the noise eliminated images becomes significant. Hence those areas are extracted as epiphyseal Regions Of Interest (eROI) of each image in a standard way. Rather than using a query on the entire image, every eROI extracted is employed for an individual query to the database. Subsequently, most important six categories of features are obtained in this phase. The five features are,

1. Color histogram and color moments (81 dimensions), 2. Gabor wavelets transform (120 dimensions), 3. Edge direction histogram (37 dimensions), 4. GIST features (512 dimensions), 5. Local binary pattern (59 dimensions), adaptive local binary pattern (72 dimensions) 6. eROI extraction.

Guo [13] exploited the oriented standard deviation details of the neighbourhood during the process of the matching step to increase rotation invariance. It is also called as adaptive LBP (ALBP) approach.

eROI Extraction: The initial stage, which is the process of extracting the eROIs, can be done either completely automatic or be semi-automated. The eROI size in case of larger images is chosen appropriately. The similarity comparison as given in the subsequent section is computed on 16×16 downscaled versions of the extracted eROIs.

2.3 Relevance Score (RS) Calculation Using Fuzzy Neural Network (FNN)

Through the assistance of Relevance Score (RS), it is easy to differentiate between the relevant and irrelevant group of medical images. On the other hand, the feature distributions of medical image from various domains may vary considerably and consequently and can have highly diverse statistical properties. To differentiate the medical images from these two domains, the labeled and unlabeled feature vector from last step $fvD_l^T = (fv_l^T, fvy_l^T)_{i=1}^{n_l}$ and $fvD_u^T = fv_l^T|_{i=n_l+1}^{n_l+n_u}$, are characterized correspondingly, in which $fvy_l^T \in \{\pm 1\}$ denotes the label of fv_l^T. Later FV^W is represented as the data set feature vector from the medical image database and $FV^T = FVD_l^T \cup FVD_u^T$ as the feature vector medical image samples data set from the input image with the size $n_T = n_l + n_u$. During the initial phase, the FNN scheme takes images features as input for the purpose of computing the relevance score results and subsequently carry out fuzzification operation to feature vectors along with fuzzy membership function. Here, a π-type [14] fuzzification parameter is employed for the fuzzification process. It yields the membership function results of individual feature vector results into two separate classes namely similar and non-similar images while measuring the score from Tanner-Whitehouse scheme. The results of FNN model generate relevance score results through the process of continuously executing Max operation during fuzzification step based on Feed Forward Multi-Layer Perceptron (FFMLP) classifier [15].

A fuzzy membership function $FV_{ij}(m)$ is produced for input features that correspond to relevance score results. As a result, the degree of fuzzy membership function relates to several features vectors of $(D)(FV_{ijD}, D \in (1, 15))$ together with two classes like (SF_1 and USF_1), where fv_{ijd} is the dth feature vector of pattern F_{ij}, with $F_{ijD} \in (1, 15)$ and $C \in (1, 2, \ldots, C)$. The depiction of this histogram in accordance with feature rh_{ij} is indicated as, $FV_{ij} = \left[FV_{11}, FV_{12}, \ldots FV_{ijd}, ..FV_{ijD}\right]^T$. Here, π-type fuzzification parameter is utilized during the fuzzification process, because of its simplicity and it is given as follows,

$$\pi(FV_{ij}; a, r, b) = \begin{cases} 0, & FV_{ij} \le a \\ 2^{m-1}[\frac{FV_{ij}-a}{r-a}]^m, & a < FV_{ij} \le p \\ 1 - 2^{m-1}[\frac{r-FV_{ij}}{r-a}]^m, & p < FV_{ij} \le r \\ 2^{m-1}[\frac{FV_{ij}-r}{(b-r)}]^m, & r < FV_{ij} \le q \\ 1 - 2^{m-1}[\frac{b-FV_{ij}}{b-r}]^m, & FV_{ij} \ge b \end{cases} \quad (10)$$

Here $m = 2$ is regarded as a fuzzification parameter for relevance score. Execute the π-type fuzzy membership function effectively, center r the values with two random parameters, for instance, p and q, with $r = \frac{p+q}{2}$. It is given as $r = \text{mean}(FV)$. In the same way, the crossover points p and q of the fuzzificaiton parameter *m* are approximated as, $p = \text{mean}(QF) - \frac{[\max(FV) - \min(FV)]}{2}$ and $q = \text{mean}(FV) + [\max(FV) - \min(FV)]/2$. Here, min and max respectively indicates the minimum and maximum value of feature vector with similar score to obtain the relevance feedback. Then, π-type fuzzification parameter is transformed as:

$$F(JSh_{ij}(m)) = \begin{bmatrix} f_{1,1}(FV_{11}) & f_{1,2}(FV_{12}(m)) & \dots & f_{1,1}(FV_{1C}(m)) \\ f_{2,1}(FV_{12}) & f_{2,2}(FV_{22}(m)) & \dots & f_{2,C}(FV_{2C}(m)) \\ \vdots & \vdots & \dots & \vdots \\ f_{D,1}(FV_{D1}(m)) & f_{D,2}(FV_{D2}(m)) & \dots & f_{D,C}(FV_D(m)) \end{bmatrix} \quad (11)$$

where $f_{D,C}(FV_D(m))$ stands for the membership of the *d*th feature to the *c*th class is taken as input to the neural network.

2.4 Tanner-Whitehouse Method

With the aim of evaluating the bone age for the extracted features, the Tanner-Whitehouse scheme was employed effectively [15]. In this scheme, it is essential that 20 bones have to be examined: distal radius, distal ulna, first, third and fifth metacarpals, proximal phalanges of the thumb, third and fifth fingers, middle phalanges of the third and fifth fingers, distal phalanges of the thumb, third and fifth fingers, the seventh carpal bones: capitate, hamate, triquetral, lunate, scaphoid, trapezium and trapezoid. Each bone, in compliance with the Tanner-Whitehouse scheme, was graded into 8 or 9 maturity stages. Staging were allocated in accordance with the rating system of the Tanner-Whitehouse scheme. When no sign of the bone was observed, the lowest rating was provided. The epiphyseal region is measured by identifying staging by difference of the size, shape, density, smoothness or thickening of the borders, thickness of epiphyseal line, extent of fusion and capping.

2.5 Discriminative Dictionary Learning (DDL) for Similarity Matching

Discriminative Dictionary Learning (DDL) is formulated which learns a flexible nonlinear proximity function to improve visual similarity search in CBIR. Let $Y = [y_1, y_2, \ldots, y_N] \in \mathbb{R}^{n \times N}$ indicates the feature vector RS learning results of query image and input medical image database, where $y_i \in \mathbb{R}^n$ indicates the ith input query together with input medical image database with n-dimensional feature description. Given a dictionary $A = [a_1, a_2, \ldots, a_N] \in \mathbb{R}^{n \times k}$, where a_i stands for the ith dictionary atom with l_1 regularization calculates the similarity score for input RS feature vector and it is provided as $[x_1, x_2, \ldots, x_N] \in \mathbb{R}^{k \times n}$, of the feature vector from RS, by means of solving the subsequent l_1 minimization complexity,

$$X^* = \arg\min_{X} \sum_{i=1}^{N} (||y_i - Ax_i||_2^2 + \gamma||x_i||_1) \quad (12)$$

where γ indicates a constant representing a sparsity constraint factor and the term $||y_i - Ax_i||_2^2$ represents the reconstruction error meant for matching results. In order to acquire discriminative feature vector x through the pairwise constrained dictionary A, the objective function for dictionary formation is given below:

$$\langle A^*, X^* \rangle = \arg\min_{A,X} \sum_{i=1}^{N} (||y_i - Ax_i||_2^2 + \gamma||x_i||_1) + \frac{\beta}{2} \sum_{i,j=1}^{N} ||x_i - x_j||_2^2 M_{ij} \quad (13)$$

$$\arg\min_{A,X} \sum_{i=1}^{N} (||y_i - Ax_i||_2^2 + \gamma||x_i||_1) + \beta(\mathrm{Tr}(X^T X D) - \mathrm{Tr}(X^T X M)) \quad (14)$$

where the constants γ and β take care of the virtual involvement of the resultant terms. The initial term $||y_i - Ax_i||_2^2$ indicates the reconstruction error term, which assesses the reconstruction error of the approximation to the query and input image matching results for RS feature vector. Second term $||x_i||_1$ stands for the regularization term. The final term, which is new and formulated at this instant, stands for the discrimination term recognized as 'pair-wise error feature vector' in accordance with the pairwise constraints which are encoded in matrix M. $D = diag\{d_1, \ldots, d_N\}$ indicates a diagonal matrix where diagonal components are the summations of the row elements of M,

$$d_i = \sum_{j=1}^{N} M_{ij} \cdot L = D - M \quad (15)$$

L indicates the Laplacian matrix. Matrix M has several forms in accordance with the setbacks that are taken into account. As a result, provided the sets of 'similar feature vectors from RS learning' and 'diverse feature vectors from RS learning' pairs *S* and *D*, matrix *M* is characterized for the purpose of encoding the (dis) similarity information as given below,

$$M_{ij} = \begin{cases} +1 \text{ if}(y_i, y_j) \in S \\ -1 \text{ if}(y_i, y_j) \in D \\ 0 \text{ else} \end{cases} \tag{16}$$

The objective function for the purpose of learning a pairwise constrained dictionary *A* for query and input image feature vectors from FNN with both reconstructive and discriminative power can be provided as below:

$$\begin{aligned} \langle A^*, X^*, W^* \rangle = \underset{A,X,W}{\arg\min} \sum_{i=1}^{N} (||y_i - Ax_i||_2^2 + \gamma||x_i||_1) \\ + \frac{\beta}{2} \sum_{i,j=1}^{N} ||x_i - x_j||_2^2 M_{ij} + \alpha \sum_{i,j=1}^{N} ||h_i - Wx_i||_2^2 + \lambda||W||_2^2 \end{aligned} \tag{17}$$

$||h_i - Wx_i||_2^2 + \lambda||W||_2^2$, $||h_i - Wx_i||_2^2$ stands for the classification query matching error, $||W||_2^2$ stands for the regularization penalty term, assists in learning an optimal linear predictive classifier $h_i = [0, 0, \ldots, 1, \ldots, 0, 0]^T \in \mathbb{R}^m$, (m stands for the amount of classes) is a label vector that is associated with a feature vectors resulting from KSVM learning, where non-zero position stands for the class label of y_i.

3 Experimentation Result

The evaluation for extended query based image retrieval system was implemented within the content based Image Retrieval with Medical Image (IRMA) framework. Generically, the epiphyseal regions of interest (eROI) reliably represent the bone age (Fig. 1), predominantly in the limit of bone ages in between 2 and 18 years [16].

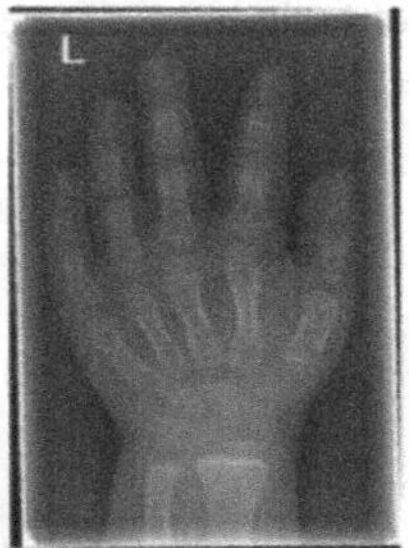

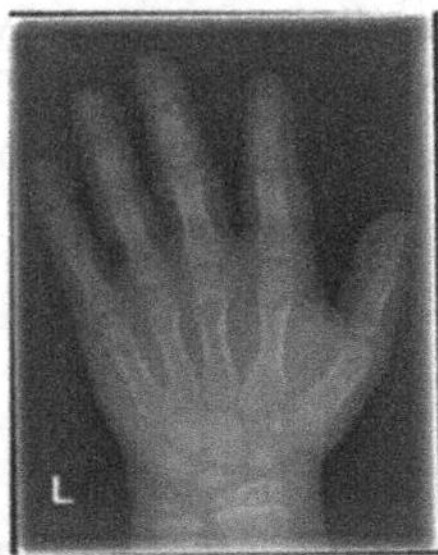

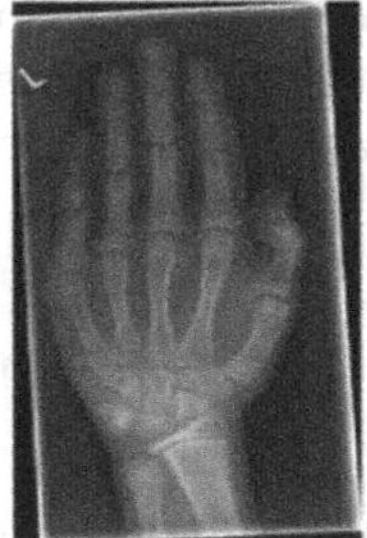
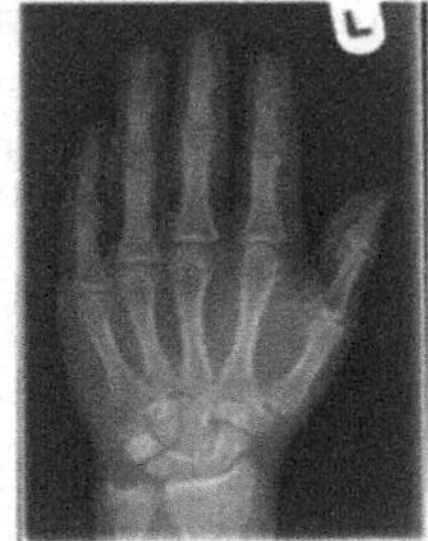

Fig. 1 Successive phases of skeletal progress of healthy male subjects. The radiographs have been obtained from the University Hospital Aachen, Germany

Fig. 2 Allocated labels for the eROI positions

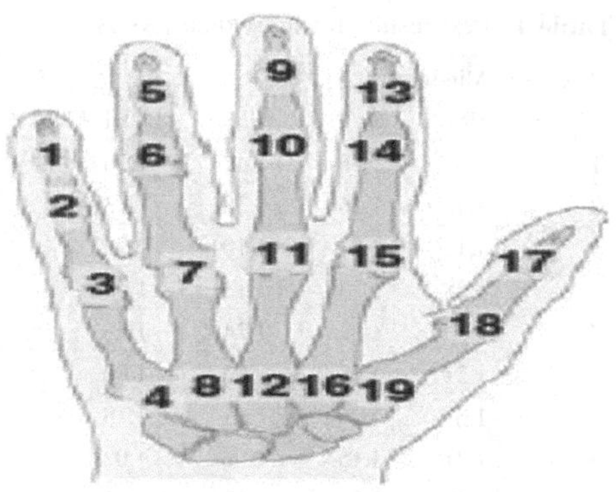

With the aim of acquiring similar image patches for CBIR, each eROI is extracted in a usual manner. The eROI centers can be computed either fully automatically or located in an interactive manner: Up to 19 eROIs may be utilized for BAA (Fig. 2). Based on the limitations of the automatic eROI extraction algorithm, only the 14 eROIs between metacarpals and upper phalanx are utilized for the age estimation. To be otherwise stated, eROIs 4, 8, 12, 16, and 19 are not considered in later processing. Mansourvar et al. [17] developed a Histogram based Matching (HM). Haak et al. [18] by SVM. The automatic localization of the eROIs will be improved by the Generalized Hough Transform (GHT) which is brought into use in [19]. The experimentation results of this work DDL with IRMA framework for Spine Pathology and Image Retrieval System (SPIRS) [20] was constructed at the U. S. National Library of Medicine for the retrieval of x-ray image and radiographs from clinical routine image is determined with SVM [21].

For each eROI, the observed minimum mean absolute prediction error and the minimum standard deviation are given with the relevant number of recovered eROIs, as K. The ranking yields an alternative ordering by the quality measures in place of the eROI number. In order to judge the prediction potential of individual eROIs, quality measures were got for each eROI using classification methods. The results indicate mean absolute error rates between 0.91 and 1.62 years and minimum standard deviations between 0.76 and 1.75 (Table 1).

Precision and Recall: It is to be noticed that CBIR is not especially being used for classification purpose, but also for discovery of similar images or cases. In order to assess CBIR, various existing performance evaluation measures [22], use the precision P and the recall R. The experimentation results precision and recall parameters of this work DDL with IRMA framework is shown in Figs. 3 and 4. It indicates that the precision and recall results of the research work DDL-FNN for BAA images are higher than the existing methods, since it removes noise from image samples.

Table 1 Best results for individual eROIs

Ages	Min-mean				Min-SD			
	HM	SVM	GHT	DDL-FNN	HM	SVM	GHT	DDL-FNN
1	1.62	1.56	1.42	1.36	1.75	1.62	1.541	1.428
2	1.43	1.36	1.28	1.138	1.61	1.56	1.438	1.3812
3	1.12	1.08	1.101	0.96	1.28	1.23	1.18	1.08
4	1.28	1.23	1.13	1.053	1.46	1.425	1.381	1.124
5	1.38	1.32	1.24	1.38	1.23	1.187	1.085	0.918
6	1.21	1.15	1.06	1.02	1.21	1.156	1.096	0.951
7	1.3	1.26	1.21	1.18	1.15	1.09	0.963	0.9218
8	1.21	1.15	1.06	1.02	1.48	1.398	1.238	1.18
9	1.09	1.04	1.02	0.91	1.32	1.256	1.181	1.015
10	1.43	1.38	1.32	1.264	1.24	1.212	1.163	1.017
11	1.41	1.32	1.241	1.185	0.962	0.921	0.865	0.8231
12	1.43	1.39	1.361	1.215	1.43	1.381	1.238	1.1381
13	1.62	1.58	1.46	1.416	1.27	1.23	1.168	0.963
14	1.58	1.46	1.43	1.382	1.102	1.042	1.0121	0.915
15	1.26	1.18	1.06	0.935	1.57	1.52	1.4231	1.218
16	1.63	1.52	1.463	1.381	1.24	1.138	1.0814	0.9812
17	1.68	1.43	1.381	1.2861	1.63	1.581	1.4238	1.3181
18	1.38	1.36	1.248	1.182	1.31	1.236	1.1821	1.0315

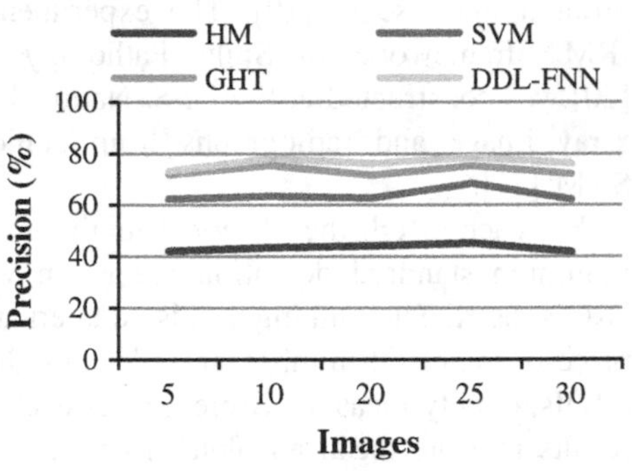

Fig. 3 Precision results comparison for X rays images in IRMA

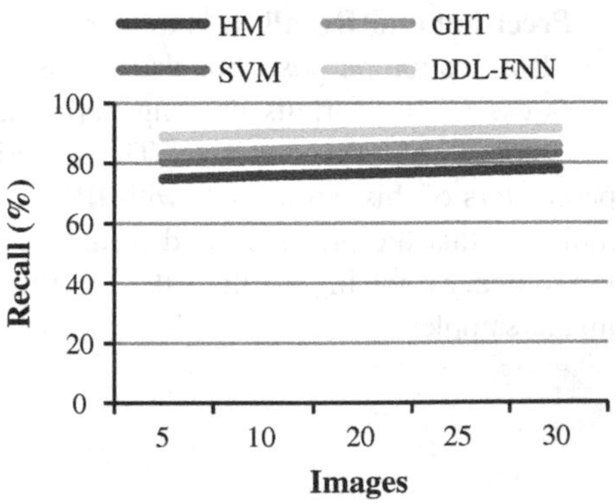

Fig. 4 Recall results comparison for X rays images in IRMA

4 Conclusion and Future Work

Assessment of the bone age is a subjective and meticulous examination process. Therefore, the study of automated assessment methods has been evolved for replacing the manual evaluation done for bone age assessment in medical applications. In this study, a novel fully automated method on the basis of the content-based image retrieval and making use of Discriminative Dictionary Learning (DDL) is modeled and adapted to evaluate skeletal maturity. The most significant uniqueness of this scheme is that it overcomes the bone age assessment complication as seen in existing systems. But, BAA image sets that include noisy images from every day routine can affect additional information. The noises existing in the fields of BAA and X-ray skull images noise are removed by making use of Kernel Fisher Discriminant Analysis (KFDA). All clinically employed BAA methods correspond to the epiphyseal area between the bones, so that the extraction of the features from the noise eliminated images becomes very significant. Hence those areas are extracted as epiphyseal Regions of Interest (eROIs) of each image in a standard way. FNN Classifiers used are trained by relevant and irrelevant regions being labeled and unlabeled as object and remaining partitions, respectively. The comparison of the estimation results of DDL with FNN models are done with those of HM, SVM and GHT models. Future work will focus on improvements of the similarity computation, web site usability, and comprehensive validation. The similarity computation will be improved by the comparison of eROIs only within the same gender and by extra image features like Tamura features and iterative distortion model (IDM).

References

1. Tann, J.M., Healy, M.R.J., Goldstein, H., Cameron, N.: Assessment of skeletal maturity and prediction of adult height (TW3). WB Saunders, London (2001).
2. Martin-Fernandez, M.A., Martin-Fernandez, M., Alberola-Lopez, C.: Automatic bone age assessment: A registration approach. Proc. SPIE, pp. 1765–76, (2003).
3. Park, K.H., Lee, J.M., Kim, W.Y.: Robust epiphyseal extraction method based on horizontal profile analysis of finger images. Proc ISITC. International Symposium on. IEEE. 278–82 (2007).
4. Gertych, A., Zhang, A., Sayre, J., Pospiech-Kurkowska, S., Huang, H. K.: Bone age assessment of children using a digital hand atlas. Computerized Medical Imaging and Graphics. 31(4-5):322–31 (2007).
5. Fischer, B., Brosig, A., Deserno, TM., Ott, B., Günther, R.W.: Structural scene analysis and content-based image retrieval applied to bone age assessment. Proc. SPIE.7260: 041–11 (2009).
6. Berner, E.S., McGowan, J.J.: Use of diagnostic decision support systems in medical education. Methods Inf Med. 49(4):412–417(2010).
7. Muller, H., Michoux, N., Bandon, D., Geissbuhler, A.: A review of content-based image retrieval systems in medical applications. Clinical benefits and future directions. Int J Med Inform. 73(1): 1–23(2004).

8. Lehmann, T.M., Güld, M.O., Thies, C., Fischer, B., Spitzer, K., Keysers, D., Ney, H., Kohnen, M., Schubert, H., Wein, B.B., Contentbased image retrieval in medical applications. Methods Inf Med.43(4):354–361, (2004).
9. Güld MO, Thies C, Fischer B, LehmannTM (2007) Ageneric concept for the implementation ofmedical image retrieval systems. Int J Med Inform. 76(2–3):252–259.
10. Güld, M.O., Thies, C., Fischer, B., Lehmann, T.M.: Content-based retrieval of medical images by combining global features, Lecturer notes in computer science, vol. 4022, pp. 702–711 (2006).
11. Fischer, B., Brosig, A., Welter P., Grouls, C., Guenther, R.W., Deserno, T.M Content-Based image retrieval applied to bone age assessment. In: Proceedings of SPIE, vol. 7624, pp 121–130(2010).
12. He, B., Tong, L., Han, X., Xu, W.: SAR image texture classification based on kernel fisher discriminant analysis. In Geosciences and Remote Sensing Symposium, IGARSS, pp. 3127–3129. IEEE International Conference (2006).
13. Guo, Z.: Rotation invariant texture classification using adaptive LBP with directional statistical features. In IEEE International Conference on Image Processing (ICIP). pp. 285–28, IEEE Piscataway (2010).
14. Ghosh, A., Meher, S. K., Shankar, B. U.: A novel fuzzy classifier based on product aggregation operator. Pattern Recognition, 41(3), 961–971(2008).
15. Marinai, S., Gori, M., Soda, G.: Artificial neural networks for document analysis and recognition. IEEE Transactions on Pattern Analysis and Machine Intelligence, vol.27(1), pp 23–35. IEEE (2005).
16. Gilsanz, V., Ratib, O.: Hand bone age. A digital atlas of skeletal maturity. Springer, Berlin (2005).
17. Mansourvara, M., Raj, R.G., Ismail, M.A., Kareem, S.A., Shanmugam, S., Wahid, S., Mahmud, R., Abdullah, R.H., Nasaruddin, F.H., Idris, N.: Automated Web Based System For Bone Age Assessment Using Histogram Technique. Malaysian Journal of Computer Science. 25(3), 107–121 (2012).
18. Haak, D., Simon, H., Yu, J., Harmsen, M., Deserno, T.M.: Bone Age Assessment Using Support Vector Machine Regression, Bildverarbeitung für die Medizin Informatik aktuell, pp 164–169. Springer Berlin Heidelberg (2013).
19. Brunk, M., Ruppertshofen, H., Schmidt, S., Beyerlein, P., Schramm, H.: Bone age classification using the discriminative generalized hough transform. In: Handels H, Ehrhardt J, Deserno TM, Meinzer HP, Tolxdorff T (eds) Bildverarbeitung für die Medizin., pp 284–288, Springer, Berlin (2011).
20. Hsu, W., Long, L.R., Antani, S.: SPIRS: a framework for content-based image retrieval from large biomedical databases. MEDINFO. pp. 188-192. Proceedings of the World Congress on Health (2007).
21. Igor, F., Amaral, Coelho, F., Pinto da Costa, J.F., Jaime Cardoso S.: Hierarchical Medical Image Annotation Using SVM-based Approaches, in Proceedings of the 10th IEEE International Conference on Information Technology and Applications in Biomedicine, (2010).
22. Muller, H., Muller, W., Squire, D.M., Maillent, S.M., Pun, T.: Performance Evaluation in Content-Based Image Retrieval: Overview and Proposals. Elsevier Pattern Rec. Lett. pp. 593–601. Elsevier (2001).

Evaluation of Prioritized Splay Tree Rotation Algorithm for Web Page Reorganization

Ananthi Sheshasaayee and V. Vidyapriya

Abstract The web usage mining (WUM) identifies user access patterns using knowledge acquired from web log file entries. As user interest changes, the static web site become outdated. Hence, the web site structure needs to be changed to meet user requirements. The core objective of the web site is to make possible for the user to retrieve required pages with minimum delay in page access. The splay trees are effective binary search trees whenever the average run time is important. This paper presents the preprocessing method to get the required data and a methodology for reorganization of web pages based on frequency count of accessed web pages and navigation link count in a web page using splay trees.

Keywords Web usage mining · Web log · Preprocessing · Web site reorganization · Splay trees

1 Introduction

Web mining is categorized into three areas based on which part of the web is mined [1]. The web usage mining (WUM) is a data mining technique used to determine interesting patterns from web log data file. Web log data captures the user's identity with their browsing behaviors at the web site.

Ananthi Sheshasaayee (✉) · V. Vidyapriya
Department of Computer Science, Quaid-E-Millath Government College for Women (A), Chennai 600002, India
e-mail: ananthi.research@gmail.com

V. Vidyapriya
e-mail: vidyapriya.research@gmail.com

© Springer Science+Business Media Singapore 2017
S.C. Satapathy et al. (eds.), *Proceedings of the International Conference on Data Engineering and Communication Technology*, Advances in Intelligent Systems and Computing 468, DOI 10.1007/978-981-10-1675-2_75

1.1 Web Log Format

The requests for the web pages in the server are stored in the log file of the web server in the same order. An extended log format (ELF) file is formed to maintain the requests occured for the web site.

The data is collected from a college web server log of Aug 2013. There are 6954 records collected from the college web server. The same log files' data and its attribute descriptions are explained in Table 1.

1.2 Web Log Data Preprocessing

The WUM process is accomplished in the following order: data collection, data preprocessing, pattern discovery, and pattern analysis [2–4].

To get a single file, the collected log files are merged, and from merged single log file, extract necessary elements [5–7]. The merged file has to be integrated and consistent to get a database that is easy to mine [8, 9]. The data pre-treatment works such as cleaning the data, identify the web users, identify the session, and finally path completion are done [10, 11].

Table 1 Log file attributes

Log file fields	Description	Example
date	Date of action occurred	2013-08-01
time	Time at which the action occurred	00:15:25
s-site name	The Internet domain name running on the client computer	W3SVC1401993659
s-ip (server IP)	The server IP address where the log entries were generated	67.218.96.180
cs-method	The action which was carried out by the user computer	GET
cs-uri-stem (URI stem)	The page (an HTML) read by client	route_map.htm
cs-uri-query	Client search string to seek match	80
s-port	The client port number	–
cs-username	User name who is trying to access the server	–
c-ip	Client IP address who access the server	157.55.33.27
cs (User-Agent)	The client browser name	Mozilla/5.0
sc-status	The HTTP status code sent to the client	200
sc-substatus	The HTTP sub status code sent to the client	0
sc-win32-status	The status of the action in Windows	0

2 Data Structure Technique—Splay Tree

Splay tree is a type of self-adjusting BST. It has some added access and update rule. Whenever a node of the tree is accessed for the following operations, insertion, deletion retrieval, and rotation, it is performed in amortized time O (log n) [12]. Above-mentioned operations are carried on the tree, to acquire recently updated node to come as root of amended tree.

When a node is accessed, either a single rotation is made or a series of rotations are made to shift the node to root of the tree. The operation of moving the node toward the root and rearranging the tree accordingly is called tree splaying. This can be completed through two ways namely top-down and bottom-up approaches. In bottom-up method, initially, a search for the binary tree is made for the required node. Then tree rotations are made on the access path to shift the node to the top. In the top-down method the search and the tree rotations are joined as a single phase.

The six types of rotations are feasible in a splay tree. They are zig rotation, zag rotation, zag-zag rotation, zig-zig rotation, zag-zig rotation, and zig-zag rotation. The detailed splay tree rotations are explained in the following Algorithm 1.

Algorithm 1- Splay Tree Rotations

Step 1. Let P_x = Node, P_p = Parent, P_g = Grandparent, R = Root.

Step 2. If P_x = R, then go to step10.

Step 3. Else if P_x = child of R, then go to step 4 else go to step 6.

Step 4. If P_x = left child of R, then right rotate about R, then go to step 10.

Step 5. Else left rotate about Root R, then go to step 10.

/ ZIG – ZIG ROTATION /

Step 6. If P_x = left – left child of P_g, then

(i) Right rotate about P_g

(ii) Righ rotate about P_p

(iii) Go to step 10

/ ZAG-ZAG ROTATION /

Step 7. Else if P_x = right – right child of P_g, then

(i) Left rotate about P_g

(ii) Left rotate about P_p

(iii) Go to step 10

/ ZIG-ZAG ROTATION /

Step 8. Else if P_x = right – left child of P_g, then

(i) Left rotate about P_p

(ii) Right rotate about $_{Pg}$

(iii) Go to step 10

/ ZAG-ZIG ROTATION /

Step 9. Else if P_x =left – right child of P_g, then

(i) Right rotate about P_p

(ii) Left rotate about P_g

Step 10. Stop.

---

3 Proposed Approach for Web Page Reorganization

The web site of 'n' pages with ID's PG1, PG2, PG3 … PGn is considered [13].

Let L: LF1, LF2, LF3 … LFn be the preprocessed log data files of above considered web site, and each LFi has the following attributes: page name, time spent, and user IP address.

Let the time spent by the users on each page is TP1, TP2, TP3 … TPn.

Let threshold for the least time is TPm.

Then for each page, the access frequency FQ1, FQ2, FQ3 … FQn is calculated by counting the respective requested page in the log data file for the time period TPi > = TPm.

Let FQm be the threshold for minimum frequency.

Then, the most frequently visited page is found out by the page with frequency of access FQi > = FQm.

Let the number of navigation links on each page be NL1, NL2, NL3 … NLn.

Let NLm be the minimum number of navigation links.

Then, the pages which are having more number of navigation links are found by the pages with navigation links NLi ≥ NLm.

Let the web pages PG[ ] be page with FQi ≥ FQm, NLi ≥ NLm, and TPi ≥ TPm in the given time period.

For the above approach for the reorganization of web pages, the target web site with 10 pages is considered and also this paper put forth the preprocessed the target web site log file data for the 1 month. Table 2 shows output of preprocessed log file.

The early binary search tree namely S is created as shown in Fig. 1 for the web site and depends on preliminary choice of the web site pages for the ID's.

To organize target web site, splay tree technique is developed which is one of the dynamic data structures [13]. Algorithm 2 enlightens the approach of the splay tree.

Table 2 Preprocessed log file with page ID

PGi	TPi	PGi	TPi	PGi	TPi	PGi	TPi
7	80	10	320	6	156	7	195
9	189	2	188	4	105	3	576
10	354	4	287	4	109	10	203
6	177	2	210	5	95	10	188
10	228	3	343	7	319	6	176
1	262	1	51	3	283	8	78
5	37	7	237	9	170	10	35
8	101	10	154	4	214	2	189
2	322	1	228	3	344	1	203
5	137	4	114	7	60	8	156
5	105	9	63	5	136	5	255
4	49	8	156	9	294	9	345
4	94	5	65	7	337	4	56

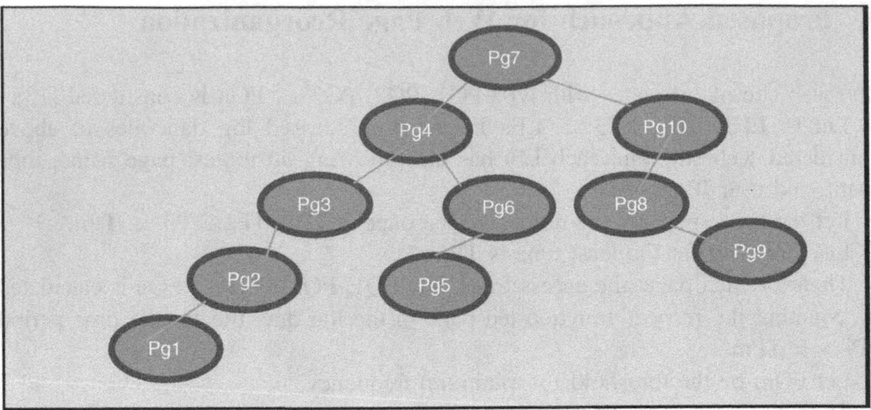

Fig. 1 Tree structure of initial BST R

Algorithm 2 – Proposed Technique - Splay Tree

Step 1: (Calculate the Access Frequency for each page of the website in the preprocessed log file [L])

For (each log entry LFi)

If (TPi > = TPm), then Increment FQi by 1.

Step 2: (Select the page ID's which is having at least minimal count for the threshold value)
Update the page list PG [] with FQi >=FQm.

Step 3: (Select the page ID's that have at least minimal count for the number of navigation links)
Update the page list PG [] with NLi>=NLm.

Step 4: Rearrange the page list PG[] based on access time so that the recently visited page as the last element of the page list.

Step 5: (Build the Splay tree based on PG[] in such a way that the last page accessed is at the root)

For (each page PGi in PG[])

Splay initial BST S based on PGi to create S_{Splay}.

Based on recently accessed pages which are having the number of navigation links more than or equal to NLm, the initial BST S is splayed using Table 3 as PG7, PG3, PG6, PG10, and PG9.

Figure 2 shows the splay tree S_{Splay} which is constructed based on Table 2 and Algorithms 1 and 2.

Table 3 Web page access delay

PG	Frequency	No of navigation links	Page access delay	
			Before reorganization of web site	After reorganization of web site
1	3	2	16	14
2	4	1	12	11
3	4	4	5	2
4	2	6	4	4
5	1	9	13	10
6	3	5	7	5
7	4	4	2	1
8	2	3	10	8
9	4	6	14	3
10	6	8	6	4

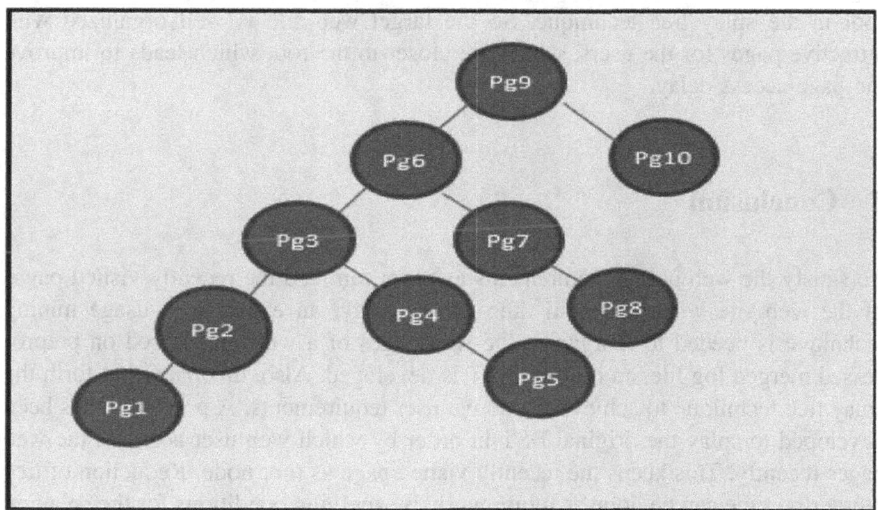

Fig. 2 Tree structure of R_{Splay}

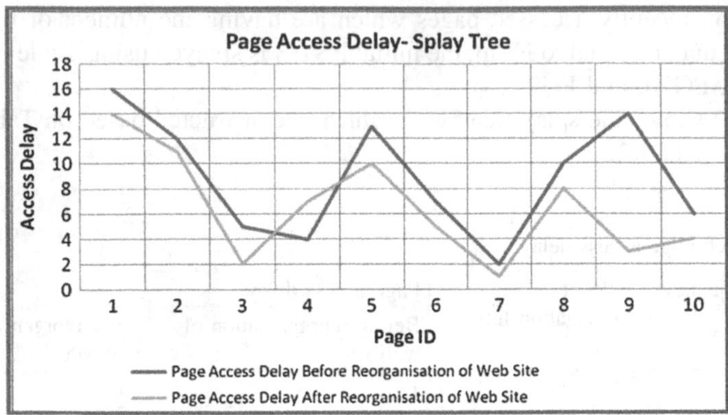

Fig. 3 Comparison of page access delay—splay tree

4 Experimental Results

For experimental analysis, a web site with 10 pages is considered. The log file for the 1 month has been considered. Table 3 shows the page access delay for the initial web site and after reorganization of web site using splay tree.

It is observed from Fig. 3 that web pages are having the following criteria: Web pages having more number of navigation links, frequently and recently accessed pages, are considered for reorganization. The recently visited pages are moved to root in the splay tree technique. So the target web site is well organized with attractive pages for the users, which are closer to the root which leads to improve the page access delay.

5 Conclusion

To satisfy the web user's requirements as users can read the recently visited pages of the web site with minimum time access delay, an expert web usage mining technique is needed to reorganize the web pages of a web site. Based on preprocessed merged log file, an original BST is developed. Also, this paper put forth the splay tree technique to achieve the above user requirements. A procedure has been developed to splay the original BST in order by which web user accesses the web pages recently. This keeps the recently visited page as root node. Reduction of tree reordering time can be done as future work, by applying conditions for the splaying operations on the proposed splay tree.

References

1. Rana C: A Study of Web Usage Mining Research Tools, Int. J. Advanced Networking and Applications, ISSN: 0975-0290, 3(06):1422–29 (2012).
2. Pamnani R, Chawan P: Web Usage Mining: A Research Area in Web Mining, 1–5 (2010).
3. Facca F M, Lanzi P L: Mining interesting knowledge from weblogs: a survey, Data & Knowledge Engineering, 53(3): 225–41 (2005).
4. Masseglia F, Poncelet P, Teisseire M, Marascu A: Web usage mining: extracting unexpected periods from web logs, Data Mining and Knowledge Discovery, 16(1):39–65 (2008).
5. Kumar B S, Rukmani K V: Implementation of Web Usage Mining Using APRIORI and FP Growth Algorithms, International Journal of Advanced Networking and applications, 1(06): pp. 400–404 (2010).
6. Srivastava M, Garg R, Mishra P K:Preprocessing Techniques in Web Usage Mining: A Survey, International Journal of Computer Applications, 97(18):1–9 (2014).
7. Rao M, Kumari M, Raju K:Understanding User Behavior using Web Usage Mining, International Journal of Computer Applications; 1(7):55–61 (2010).
8. Sheetal A. Raiyani and, Shailendra Jain:Efficient Preprocessing technique using Web log mining, International Journal of Advancements in Research & Technology, 1(6) ISSN 2278-7763 (2012).
9. J. Srivatsava, R.Cooley, M.Deshpande, and P.N. Tan: Web Usage Mining: Discovery and Applications of Usage Patterns from Web Data, ACM SIGKDD Explorat. Newsletter(2000).
10. Jiawai Han and Micheline Kamber:Data mining-Concepts and techniques, second edition, Elsevier(2010).
11. V. Chitraa, Dr. Antony Selvadoss Devamani: A Novel Technique for Sessions Identification in Web Usage Mining Preprocessing, International Journal of Computer Applications, Volume 34, No. 9 (2012).
12. Sleator D D, Tarjan R E: Self-adjusting binary trees, Journal of ACM 32(3):652–86 (1985).
13. Ananthi Sheshasaayee, V. Vidyapriya: A Framework for an Efficient Knowledge Mining Technique of Web Page Re-organisation using Splay Tree, Indian Journal of Science and Technology, Volume 8(29), IPL0647, November 2015.

A Prototype Model for Deriving Social Media Intelligence Using Opinion Mining from Microblog Data

Ananthi Sheshasaayee and R. Jayanthi

Abstract Online conversation has become very popular in recent years. Many people share their experience and opinions of an interesting topic on the social networking sites especially on microblogs. Since the opinions are preserved like historical data in the social networking sites, an individual's opinion has got the power to change trends and strategies of an organization. Organizations use those social media data to interpret and extract knowledge in their strategic decision-making process referred as social media intelligence. This article provides an efficient approach to extract opinions using text mining techniques integrated with supervised learning methods, classified and rated against an evaluation criteria ranging from excellent, good, average, and poor using fuzzy logic.

Keywords Opinion mining · Information extraction · Text mining · Social media intelligence

1 Introduction

Web 2.0 provides increasingly huge volume of different kinds of data available on the Internet in the form of reviews and opinions. Most of the internet users use microblogging sites such as Twitter, Facebook, etc., to share their information. Those information would be informal descriptions, mostly unstructured and does not follow any language grammar, but the potentially valuable intelligence hidden inside the unstructured data can be extracted from text-driven approaches [1, 2].

Social media content is enormous; in addition to that the user-generated data is unstructured making social media mining challenging. Social media mining is the

Ananthi Sheshasaayee (✉) · R. Jayanthi
PG and Research Department of Computer Science, Quaid-E-Millath Government College for Women (Autonomous), Chennai, Tamil Nadu, India
e-mail: ananthi.research@gmail.com

R. Jayanthi
e-mail: jayanthi.research@gmail.com

© Springer Science+Business Media Singapore 2017

S.C. Satapathy et al. (eds.), *Proceedings of the International Conference on Data Engineering and Communication Technology*, Advances in Intelligent Systems and Computing 468, DOI 10.1007/978-981-10-1675-2_76

process of representing, analyzing, and extracting actionable patterns from social media data. Since this is an emerging field, social media mining attracts the researchers to discover new patterns from human-generated content [3, 4].

The research directions of opinion mining are classified as document level, sentence level, and phrase level [5, 6]. The document level and sentence level uses two classification methods based on identifying opinion words/phrases. One of such classification methods is Lexicon-based which uses lexicon table with echo count to evaluate positive and negative opinions. The other method is rule-based which uses POS tagging with co-occurrence patterns of words/tags to find sentiments hidden in opinions. Opinion mining refers to the use of natural language processing, text mining, and computational linguistics to identify and extract subjective information in online sources [7, 8].

This research article makes use of social media data found in microblog. An approach ECO Miner is proposed to extract opinions from Twitter data and classify the opinions with the application of rule-based approach that leads to better decision making. The rest of the article is organized as follows: Sect. 2 deals with literature, Sect. 3 portrays the system architecture of ECO miner, and Sect. 2.5 presents the pseudocode for domain-specific feature extraction. Experimental results are discussed in Sect. 3 and finally Sect. 4 concludes the paper with the future direction of research in this area.

2 System Architecture

This article aims at collecting, extracting, classifying, and mining user-generated content in Twitter microblog. The general architecture of ECO Miner is shown in Fig. 1. The extract–classify–opinion miner (ECO Miner) consists of different phases like collecting tweets, preprocessing, generate parse tree using POS tagging with rule-based grammar, feature selection and feature extraction, classification of extracted features using Naive Bayes classifier, finding semantic word similarity and assign polarity using fuzzy logic, and finally evaluation and visualization of opinioned data.

2.1 Preprocessing

Initially, all informal user-generated content needs to be preprocessed before considering the data for the progress of the research. During preprocessing the following steps are followed for the better performance of the ECO Miner.

1. Data need to consider the structure of each tweet. Each Twitter post contains properties like *usernames, Hash tags,* and *Re-tweet*. The usernames contains '@' symbol which is a de facto standard must be followed. Hast tags (#) are

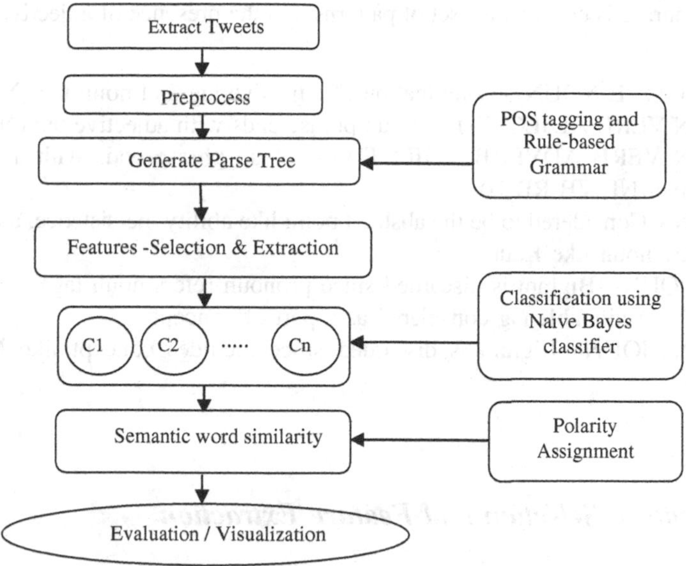

Fig. 1 System architecture of ECO miner

allowed in Twitter data to represent the content of the tweet. RT represents Re-tweet.

2. Eliminate the tweets that are not in English.
3. Remove the stop words like digits, special characters, smileys, etc.
4. WordNet dictionary is used to check whether the word is real word or not.
5. Repeated letters in a word are replaced with single letter, since the user may comment the post added with fun.
   For example, trueeeeeeeee and helloooooooo to true and hello.
6. Tweets that follow the grammar SVO are selected for processing.

This preprocessing phase is very important, because extra effort had to be applied to establish uniformity of syntax and interpretation of the expressed opinion.

2.2 Rule-Based Annotation

After extracting and preprocessing the tweets, the informal descriptions are converted into formal descriptions and now ready for the next phase. The sentences which follow the SVO pattern are extracted by applying part-of-speech tagging. Grammar chunks are framed for noun phrase and verb phrase using bigram rule-based grammar. In particular, the noun phrases are split into bigrams and then processed through part-of-speech patterns [9].

Noun phrase is checked for set of patterns and the presence of adjectives in POS tag:

- ADJECTIVE NOUN—Combination of adjective tag and noun tag (NP JJ NN)
- NOUN VERB ADJECTIVE—Noun phrase ends with adjective tag (NP VB JJ)
- NOUN VERB ADVERB ADJECTIVE—Noun phrase ends with adverb and adjective (NP VB RB JJ)
- NOUN—Considered to be the abstract noun like ability, persistence, and not the concrete noun like Kaur
- PRONOUN—Bigram is discarded since pronoun refers noun tag
- VERB—Actionable tag considered as a part of concept
- NOUN NOUN—Bigram is discarded since includes concept like Mahindar Kaur.

2.3 *Feature Selection and Feature Extraction*

This phase involves information extraction which combines feature selection and feature extraction. Features of interest are selected so as to improve the representation from redundant data. The proposed approach uses domain-specific feature knowledge base for feature extraction. Define a property F that finds the fine-grained keywords K from the selection list s.

A mathematical function *F(w)* is true, if and only if *w* is from the vocabulary of adjectives A or verb V or noun N (optional). The set of keywords K_s is represented as

$$K_s = \{w | w \in A \text{ OR } w \in V \text{ OR } w \in N\} \tag{1}$$

2.4 *Naive Bayes Classifier*

Naive Bayes classifier is a simple classification method in supervised learning. This model works well on text categorization [10]. Assume that the class attribute C contains the set of tweet value attribute T(t_1, t_2,…t_n}, then ς the class attribute value of instance T can be calculated by measuring

$$\varsigma = \arg\max_{ci} P(Ci|T) \tag{2}$$

By applying NBC with prior probability and current information, we can predict the future probabilities easier. The Naive Bayes classifier also finds the conditional independence. For the given class attribute value, other feature attributes are conditionally independent and that can be calculated using

$$P(Ci|T) = \frac{(\prod_{j=1}^{m} P(tj|ci))P(ci)}{P(T)} \quad (3)$$

This classifies the tweets into opinion class and non-opinion class which is independent on each other.

2.5 Semantic Similarity Between Lexicons

A very basic idea of this phase is that we know that a word may have different meanings in different situations. By finding the semantic similarity between words may contribute to make correct decision on the class attributes. WordNet dictionary is used to find the meanings of the given attribute which helps to enhance the evaluation of the proposed approach to obtain the better result [11]. The following is the pseudocode for finding the semantic similarity between words:

Pseudocode: Semantic similarity between words

```
If the phrases have the SVO format, then

        wordlist1: All valid objects in Noun Phrase(NP)
        wordlist2: All objects in Keywords List(KL)
    M1=Φ
    M2=Φ
    for all objects in NP do
        for all objects in KL do
            M1=M1∪{all semantic words matches from Wordnet}
            M2=M2∪{all semantic words matches from Wordnet}
            if {any one M1} and {any one M2} matches, then:
                wordsimvalue=M1.Word_Similarity(M2)
                WSList=Collection of all wordsimvalue
```

Display the maximum value of the Word_Similarity of M1 ∪ M2. Repeat this for all the objects of entire concept.

2.6 Training Classifiers in Discrete Polarity Values

The extracted word lexicons are difficult to process using computational linguistic methods. This prompted us to convert the lexicons into discrete values ranging from excellent, good, moderate, and poor. Discretization helps to convert continuous features into distinct values and this is achieved by changing the linguistic variables to numerical values using fuzzy logic [12–14]. Each lexicon is assigned a numeric value by finding word similarity with the set of polarity words (Excellent, Good,

Moderate, Poor). The maximum value obtained from the word similarity is assigned for the lexicon and thus assigned the relevant polarity to the linguistic variable. Opinions expressed in this manner are easy to understand and facilitate to interpret the result easily:

$$D(\text{opinion}) = \{0.2, 0.4, 0.6, 0.8\} \tag{4}$$

The above fuzzy set shows the discretization of opinions that relates 0.8 with *excellent*, 0.6 with *good*, 0.4 with *moderate*, and 0.2 with *poor*.

3 Experimental Results

This section of the article presents the evaluation of the extract–classify–opinion miner and comparison with other mining systems.

A. Dataset and Analysis

Sample data were collected from Twitter microblog about the cricket comments. These data were concerned with the performance ratings of the cricket players. Our focus of this experiment is to relate the performance of the cricket players with the relevance scoring and compare the result with the existing methods. Hence, the experimental setup is divided into the following phases and is developed using Python programming with Natural Language Toolkit.

Preprocessing—The input data were collected from user-generated content, and they were undoubtedly unstructured in nature. The unstructured data must be preprocessed to remove stop words, special characters, smilies, digits, repetition of letters in a word, and tweet containing only mail ids (id containing symbols like '@' and '#').

Rule-based Parse Tree Generation—Those twitter data which follow subject–verb–object pattern is considered for generation of parse tree. Grammar rules are written to create noun phrase chunks and verb phrase chunks with patterns which are already discussed in Sect. 2.2. This phase mainly applies the POS tagging with the focus to recognize nouns, verbs, and adjectives. The generated parse tree is now ready for the next phase.

Feature Extraction and Semantic Word Similarity—Features of interest in the cricket domain like batting, bowling, and all-rounder are selected and tweets are collected accordingly. From the parse tree, the important features are extracted pertaining to the performance of the cricket players. Noun, verb, and adjective tags are classified using NP and VB chunks and are considered as lexicons.

A set of keywords related to performance rating like excellent, good, average, and poor are compared by finding the word similarity semantically with the help of WordNet dictionary as shown in Fig. 2.

```
Python Shell
File Edit Shell Debug Options Windows Help
Python 2.7.6 (default, Nov 10 2013, 19:24:18) [MSC v.1500 32 bit (Intel)] on win32
Type "copyright", "credits" or "license()" for more information.
>>> ================================ RESTART ================================
>>>
                         Semantic Word Similarity
              Excellent        Good        Average        Poor        Max Value
unbeatable
               0.00            0.00        0.00           0.00        0.00
serious
               0.00            0.00        0.00           0.00        0.00
first
               0.00            0.43        0.27           0.33        0.43
make
               0.00            0.24        0.44           0.27        0.44
prove
               0.40            0.18        0.17           0.22        0.40
persistence
               0.00            0.38        0.24           0.29        0.38
perspiration
               0.00            0.25        0.24           0.29        0.29
combination
               0.00            0.33        0.31           0.60        0.60
maestro
               0.00            0.14        0.13           0.17        0.17
conversation
               0.00            0.31        0.29           0.36        0.36
competition
               0.00            0.33        0.31           0.40        0.40
true
               0.00            0.29        0.27           0.33        0.33
                                                                  Ln: 33 Col: 4
```

Fig. 2 Semantic word similarity between lexicons and keywords

B. Result Analysis

The proposed method focus on opinions with excellent, good, average, and poor to determine the semantic orientation of a tweet instead of positive, negative, and neutral. The accuracy of the proposed model is shown in Table 1.

Finally, the accuracy of the proposed model is compared with the existing opinion miner and is shown in Table 2.

The above table shows that the proposed model result is better than the existing two models of opinion mining.

Table 1 Result of the semantic similarity

	Excellent (4)	Good (36)	Average (20)	Poor (20)
Excellent	3	5	1	–
Good	1	26	4	2
Average	–	5	13	14
Poor	–	–	2	4

Table 2 Accuracy of ECO miner with other opinion miner

	Accuracy (%)	Reference
Opinion miner	70.39	[15]
Opinion mining system	83.19	[16]
ECO miner	85.23	–

4 Conclusion

Opinion mining or sentiment analysis is an emerging field of social media analysis. This article proposed an opinion mining approach to extract opinions from tweets and using classification methods, a training data set was built to improve the performance of the model and obtained good result compared with the existing models. The use of rule-based model is an added advantage of this model. The proposed model demonstrated with an experiment using sample data collected from microblog and accuracy is measured using precision, recall, and F score. Sentiment analysis is really a challenging area of research. The future work is to fine tune the model by considering the negative comments and make it generic opinion mining system so that when any opinion is given as input the influence of the opinion on decision making is measured and interpreted.

References

1. Zhou N, Cheng H, Chen H, Xiao S.: The Framework of Text-driven Business Intelligence. In: Wireless Communications, Networking and Mobile Computing WiCom IEEE, pp. 5468–5471 (2007).
2. Funk A, et al.: Opinion Analysis for Business Intelligence Applications. In: ACM OBI Proceedings of the first international workshop on Ontology-supported business intelligence, ISBN: 978-1-60558-219-1 doi:10.1145/1452567.1452570, pp. 1–7 (2008).
3. Kietzmann, J H, Hermkens K, et al.: Social media? Get serious! Understanding the functional building blocks of social media. In: Business Horizons, vol. 54(3), pp. 241–251 (2011).
4. Padmaja S, Fatima S S.: Opinion Mining and Sentiment Analysis –An Assessment of Peoples' Belief: A Survey. In: International Journal of Ad hoc, Sensor & Ubiquitous Computing (IJASUC), vol. 4(1), pp. 21–33 (2013).
5. Dave K., Lawrence S., and Pennock D.M.: Mining the peanut gallery: opinion extraction and semantic classification of product reviews. In: Proceedings of the 12th international conference on World Wide Web, Budapest, Hungary, ISBN: 1-58113-680-3, pp. 34–44, (2003).
6. Bo Pang, L. Lee, and S. Vaithyanathan.: Thumbs up?: sentiment classification using machine learning techniques. In: Proceedings of the ACL-02 conference on Empirical methods in natural language processing, doi: 10.3115/1118693.1118704, pp. 79–86 (2002).
7. Pang B, Lee L.: Opinion mining and sentiment analysis. In: Foundations and Trends in Information Retrieval, vol. 2(1-2), pp. 1–135 (2008).
8. Ritu M, Ajit S, Akash S.: Opinion Mining Techniques on Social Media Data. In: International Journal of Computer Applications(0975-8887), vol. 118(6), pp. 39–44 (2015).
9. Erik Cambria, Amir Hussain.: Sentic Computing: Techniques, Tools, and Applications. In: Springer (2012).
10. Pak. A and Paroubek. P.: Twitter as a Corpus for Sentiment Analysis and Opinion. In: Proceedings of Conference on International Language Resource and Evaluation LREC'10 (2010).
11. Yee W. LO1 and Vidyasagar POTDAR.: A Review of Opinion Mining and Sentiment Classification Framework in Social Networks. In: 3rd IEEE International Conference on Digital Ecosystems and Technologies (2009).
12. Shaidah Jusoh and Hejab M. Alfawareh.: Applying fuzzy sets for Opinion Mining. In: 978-1-4673-5285-7/13, IEEE (2013).

13. Jusoh S, Alfawareh H M.: Applying Information Extraction and Fuzzy Sets for Opinion Mining. In: ICICIE'2014, Kuala Lumpur, 15-16 Jan., pp. 91–94 (2014).
14. Haque A Md, Rahman T.: Sentiment Analysis by using Fuzzy Logic. In: International Journal of Computer Science, Engineering and Information Technology, vol. 4(1), pp. 33–48 (2014).
15. Po-Wei Liang and Bi-Ru Dai.: Opinion Mining on Social Media Data. In: IEEE 14th International Conference on Mobile Data Management (2013).
16. Poornima Singh, Gayatri S Pandi.: Opinion Mining on Social Media: Based on unstructured Data. In: International Journal of Computer Science and Mobile Computing, vol. 4(6), pp. 768–777 (2015).

Implementation for Securing the Cloud Data and Providing Secured Access Across the Network

Ananthi Sheshasaayee and K. Geetha

Abstract Storing and retrieving data in cloud are important in today's environment. It also adds insecurity as data sharing in cloud would be affected by hacking or modifying the original content of the data. For secure data transmission, encryption and decryption are the most followed methods. This existing method demands the authentication, security keys from a third party cannot be considered safe because the entire network becomes questionable when such party is not trustworthy. This paper proposes a method, where encrypting the public keys and decrypting the private keys would be generated from cloud node itself, not by a separate trusted authority. This paper also uses the attribute real-time parameters taken for building security keys for each requesting nodes in time dependent manner. Since attributes are real-time parameters and changes on random deployment add to security to the contributing nodes. Furthermore, some parameters like busy state help to route data and ensure good packet delivery and client data storage.

Keywords Cryptography · Public key · Request packet · Reply packet · Cloud data

1 Introduction

Cloud data storage is widely used in industries. It also helps in safe keeping data for individuals [1]. For knowledge sharing in open source and runtime compilation of various programming and for more soft data management, cloud computing is a base root [2]. Securing data in cloud is being a promising challenge today. Assuring

Ananthi Sheshasaayee (✉) · K. Geetha
Quaid-E-Millath College (Govt.), University of Madras,
Chennai, Tamil Nadu, India
e-mail: Ananthi.research@gmail.com

K. Geetha
e-mail: geethak.research@gmail.com

© Springer Science+Business Media Singapore 2017

S.C. Satapathy et al. (eds.), *Proceedings of the International Conference on Data Engineering and Communication Technology*, Advances in Intelligent Systems and Computing 468, DOI 10.1007/978-981-10-1675-2_77

users legal access and defending data from unauthorized used are the main focus in cloud data storage. In today's environment, accessing the content across the network without reference is common [3]. The content provider that is the central access unit responding to fake queries should be eradicated completely. The social response over securing own data and providing secured access to the end customer or user is today's universal need [4].

Data sharing between peer to peer is simply aimed to share content directly without any centralized authentication [5]. Resource aggregation and ad hoc communication have added security factors in data sharing. Integrated resource for computing is the basis for clustering. It consists of a large number of standalone computers interconnected together [6, 7]. Cloud is a parallel and distributed system mainly consisting of virtualized computers negotiated on certain service level agreements. Cloud adds features of both clustering and grid by its special capabilities and attributes like virtualization and third party assessment [8]. Irrespective of what network and configuration, the cloud server must be able to provide public, private keys to client. Sometimes cloud is applied as a single server computes multiple computers or virtualized computers (Figs. 1, 2, 3, 4, 5 and 6).

Fig. 1 Request packet format

01010010	Source Id	DestinationId

Fig. 2 Data packet format

SOF	SID	DID	Ch no	Pan ID	DATA	Bat pwr	No of sens	Busy Flag	EOF

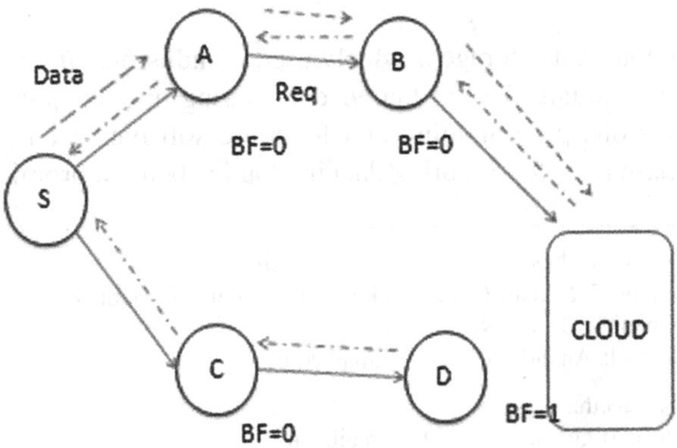

Fig. 3 Cloud data storage

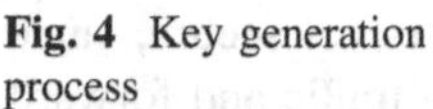

Fig. 4 Key generation process

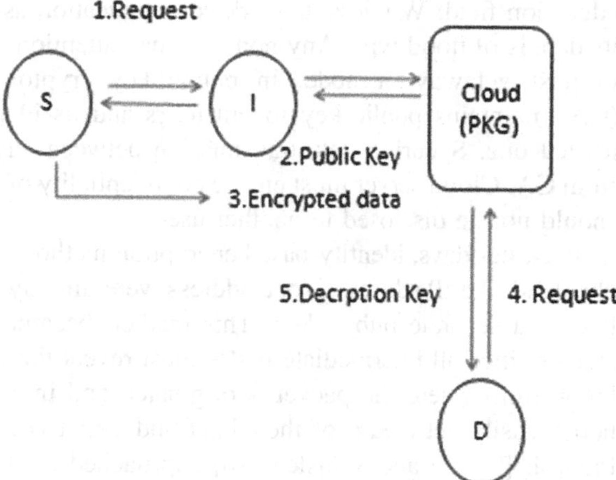

Fig. 5 Setup time versus no. of attribute

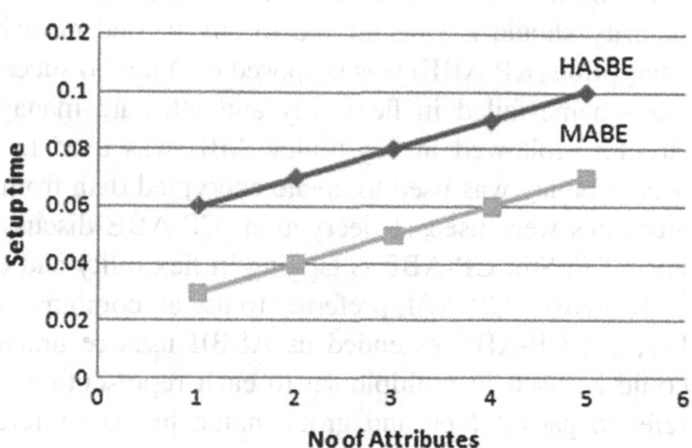

Fig. 6 Key generation time versus no. of attributes

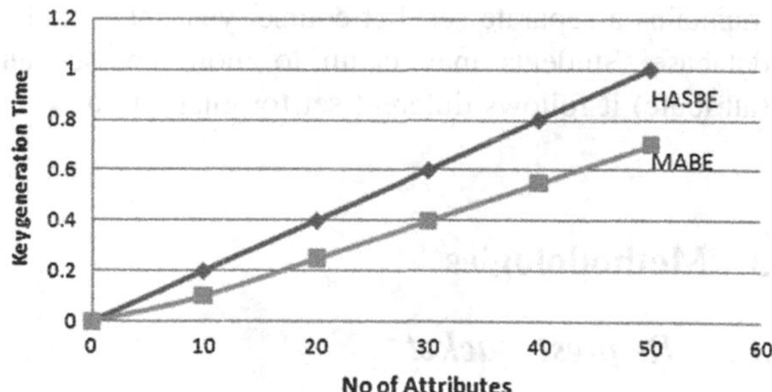

2 Related Works

It is mandatory to secure data in open network like cloud data sharing between N no of users. Sending plaintext directly in network is not the preferable way for data sharing. The impact of cloud computing is quit high in IT industries and in the education field. Wireless network communication as plaintext is not secure, since its data is of flood type. Any node can pay attention to ongoing traffic and forward to next getaway or node. In public key cryptography, certification authority (CA) maintains public key for all users and its identity. For that CA must be a trusted one. Securing data transmission between any two nodes needs assistance from CA. Cloud server must ensure confidentiality of data by individual users and it should not be disclosed to another user.

In earlier days, identity-based encryption method was used in that any devices or clients id like IP address, Node address were directly used as public key rather than having a separate public key. This method became easily vulnerable to security attacks since all intermediate nodes must reveal the IP address, identity number to know from where the packet is originated and in any kind of communication all nodes easily get aware of their inlet and outlet connecting nodes. McDaniel and Prakash [9], Yu and Winslett [10] approached earlier, which ensured safety only when both client and server reside at the same domain. In the case of cloud the security should not be limited to any trusted domain. Key-policy attribute-based encryption (KP-ABE) was proposed by Yu who succeed in access control; However the scheme failed in flexibility and attribute management. Common tree access structure followed in key-policy ABE was used to decrypt the cipher text. Tree access policy was used to create encrypted data from cipher text and similar set of attributes were used at decryption. CP-ABE discussed in paper is most likely to KP-ABE. Still CP-ABE is lagging in flexibility and efficiency.

Moreover, CP-ABE preferred to use all combinations of attributes in a single set. [11, 12] CP-ABE extended as ASBE uses recursive set, i.e., the same attribute could be used in multiple set to each representing unique behavior. Role id may refer to packet type and group name in two different sets. Two attributes from different sets cannot be combined freely, while it is possible when they belong to the same set. This solves the problem in ASBE for each different attribute which maintains a separate set. Let course, year, and grade be the attributes of students' database. Students may claim to more courses and for each different course (attribute) it follows different set for encryption.

3 Methodologies

3.1 Request Packet

Any node sending a request generally floods its request data to all neighbors. This request packet contains information such as source id, which initiates or originates the request and also includes the information as to which node it is sending, i.e., the destination to which it needs interaction. The initial 8 bit data 01010010 also indicates the packet as a request packet. The request data byte is followed by a 8 bit source id and destination id, respectively. 01010010 Source Id Destination Id.

3.2 Reply Packet

Upon receiving a request from any node, the intermediate node, which receives the request packet immediately, looks for the destination node. It assumes that the intended destination is availed. After acquiring the neighborhood availability of cloud destination node, intermediate node undergoes the attribute algorithm would be discussed later in this paper. On completing attribute analysis the node sets or clears its busy lag, which cleared initially. Every node senses the number of tasks scheduled in its queue. Depending on the number of tasks, it decides to contribute to other nodes network data transmission.

Since all nodes are wireless in nature their battery power impacts more in the performance of a node. In order to have a long living node its contribution to other nodes by carrying and forwarding data packets of others should be limited based on its residual energy. After going through all process as node replies in the packet format mentioned below.

3.3 Cloud Data Storage

Since this packet transmission follows asynchronous communication, no clock reference signal is required at individual nodes. While request and reply packets contain 24 bit packet frame, data packet carries 144 bit frame. All of them are identified.

Separately by start and end of the frame in data frame, start of frame is followed by 8 bit source identity, 8 bit destination and channel no, pan id, payload, battery power indication bits, and busy flag indication bits. Moreover this paper is implemented in 2.4 GHz bandwidth.

3.4 Security Factor

This method proposes a secured data transmission by incorporating private key generator in common cloud. Cloud data sharing becomes the a social necessity in all fields and impacts more in knowledge sharing and cloud computing like cloud compiler, cloud tools, etc., when the concepts come to a cloud, any node or any computer can demand the cloud to store and utilize data in it. The cloud alone would take decision on the reliability of the requester. So this paper implements and puts more security in cloud data with the help of a private key generator.

In public key encryption, all nodes aware of its public key and request for a private key from cloud. On obtaining a private key from cloud, it encrypts before sending as similar reverse process would be done in receiver side. That is, both encryption and decryption end client and node must seek the private key access from cloud. In wireless routing, a cloud may choose any node to be the next hop contributing to the conversation between cloud and end client. So, the cloud follows attribute-based routing. These attributes help in selecting routes and end to end delivery.

The attribute-based and cloud-based private key sharing and secured data transmission.

e = f{P,n(pk,pj)}

Encrypted data is a function of payload P, private key pk, and public key pj. All available networks are identified by its unique PAN id. A node is allowed to contribute in the conversation only if it is belonging to same network. If the source node S belongs to network group N1, it would like to store its data in cloud data and then can travel only through the node lying at similar group N1.

{S,I1,I2}€N1

Pvt Key = 3; Pub Key = 2; ch_id = 01; pan_id = 01; num_sen = 02; s_id = 01; d-id = 04;

The following is the MABE (multiple attribute-based encryption) algorithm used in the above methodology process.

Base station

Encryption Algorithm:

```
CT = msg + (pk*2) + node_id - attribute (ch) + panid
Pubkey = (PK[0] * 2) + (1) + (17);
Chid + pub key + Pvt key;
Pan_id + Pubkey + Pvt key;
Num_sen + pubkey +pvt key;
Busy + pubkey + Pvt key
```

Destination

S − id = 04; d − id = 01;

Decryption Algorithm:

```
attributed = Pvt key, pub key, pan id, ch no, residual energy, battery power, no of sensor connected
decrypted sk = child * node idreq - (3 * panid) + (Bat pwr) / num of nbr;
DK = 1 * 3 - (3 * 2) + (8) /3;
PK = req node id + randnum + child + panid.
```

3.5 Implementation

In the implementation each node is built with PIC16f877a microcontroller, Tarang p20, sensors, and power source. Tarang p20 is a Zigbee-based wireless transceiver module internally made of MC1322x 32 bit ARM-7 processor. It avails 16 channels in the range of 2.405–2.485 GHz. Point to Point, Peer to peer, and broadcast communication can be achieved in our proposed cryptography and attribute-based routing. In the same area, it is possible to have multiple numbers of nodes and there might be more number of network group existing within an area; some set of node together form a separate network. Likewise, more network group nodes can communicate. PAN id is used to find the network member of the belonging group to which it claims its network. All nodes transmit, receive, and carry forward packets only via this free scalar module. The data packets obtained at every module are sent to the master controller. The controller further analyzes and responds based on attribute-based algorithm. As explained earlier, each node identifies the request packet, reply packet and data packets, and also source id, destination id. After going through various steps o discovering routes, the master controller encrypts data and its attributes by public and private keys.

4 Result and Discussion

The cloud server node and other client node have been created and experimented in real time. The entire secured data storage and access process has four steps in it, such as setup, keygen, encryption, and decryption. The depth of key structure decides the creation of public key. It is the number of attributes and also includes how many of them are assigned with number o subsets. Setup process has been done by cloud server and results in the public key (PK) and master key (MK). Moreover the cloud computes the secret key (SK) using the key depth of user U. During encryption a node puts the data to the cloud server and encrypts its original message by using public key made available to the public. Data in the cloud is available in the form of cipher text (CT). A Node can perform decryption only using the secret key (SK) and depth of key structure of user U. Depending on the depth of the key structure, [5] i.e., the number of attributes, the setup time, and key generation time linearly increase.

Setup time is the time required for generating a public key (PK), while key generation time for generating secret key (SK). In comparison with the earlier method HASBE, this proposed method MABE would be able to generate a key faster with low time consumption. This paper concludes MABE as the better, faster, and more efficient method for data security yielding better and together security. The real-time behavior of node has been considered as a basic attribute frequently charging and dynamic in nature.

References

1. R. Buyya, C. ShinYeo, J. Broberg, and I. Brandic, "Cloud computing and emerging platforms: Future Generation Compute System" vol. 25, pp. 599–616, 2009.
2. B. Barbara, "Salesforce.com: Raising the level of networking", Inf. Today, vol. 27, pp. 45–45, 2010.
3. G. Wang, Q. Liu, and J. Wu, "Hierarchical attribute based encryption for fine-grained access control in cloud storage services", in Proc. ACM Conf. Computer and Communications Security (ACM CCS), Chicago, IL, 2010.

4. S. Yu, C. Wang, K. Ren, and W. Lou, "Achieving secure, scalable, and fine-grained data access control in cloud computing", in Proc. IEEE INFOCOM 2010, 2010, pp. 534–542.
5. R. Bobba, H. Khorana, and M. Prabhakaran, "Attribute-sets: A practically motivated enhancement to attribute-based encryption", in Proc. ESORICS, Saint Malo, France, 2009.
6. K. Barlow and J. Lane, "Like technology from an advanced alien culture: Google apps for education at ASU", in Proc. ACM SIGUCCS User Services Conf., Orland, FL, 2007.
7. K. J. Biba, Integrity Considerations for Secure Computer Sytems The MITRE Corporation, Tech. Rep., 1977.
8. J. Bethencourt, A. Sahai, and B. Waters, "Cipher textpolicy attribute- based encryption", in Proc. IEEE Symp. Security and Privacy, Oak- land, CA, 2007.
9. P. D. McDaniel and A. Prakash, "Methods and limitations of security policy reconciliation", in Proc. IEEE Symp. Security and Privacy, Berkeley, CA, 2002.
10. T. Yu and M. Winslett, "A unified scheme for resource protection in automated trust negotiation", in Proc. IEEE Symp. Security and Privacy, Berkeley, CA, 2003.
11. J. Bell, Hosting Enterprise Data in the Cloud Part9: Investment Value Zeta, Tech. Rep., 2010.
12. A. Ross, "Technical perspective: A chilly sense of security", Commun. ACM, vol. 52, pp. 90–90, 2009.

Integration of Mel-frequency Cepstral Coefficients with Log Energy and Temporal Derivatives for Text-Independent Speaker Identification

S.B. Dhonde, Amol Chaudhari and S.M. Jagade

Abstract This paper presents effect of possible integrations of delta derivatives and log energy with MFCC for text-independent speaker identification. MFCC features extracted from speech signal are used to create speaker model using vector quantization. First, the effect of varying MFCC filters and centroids of vector quantization is compared. Next, MFCC scheme is combined with delta derivatives and log energy. The effect of these possible combinations is compared by varying MFCC filters and centroids of vector quantization. Among all experiments carried out on 120 speakers of TIMIT database, average identification rate of 99.58 % is achieved for 29 MFCC filters and 32 centroids of vector quantization.

Keywords Speaker identification · Feature extraction · MFCC · Vector quantization

1 Introduction

Speaker identification refers to identifying a person from a fixed set of voices. Speaker identification is classified into text-dependent and text-independent approaches [1]. Unlike text-dependent speaker identification, text-independent system does not require any fixed phrase as a password [1, 2]. The utterances in

S.B. Dhonde (✉) · Amol Chaudhari
Department of Electronics Engineering, All India Shri Shivaji Memorial Society's, Institute of Information Technology, Pune 411001, India
e-mail: dhondesomnath@gmail.com

Amol Chaudhari
e-mail: amol191189@gmail.com

S.M. Jagade
Department of Electronics and Telecommunication Engineering, TPCT COE, Osmanabad 413501, India
e-mail: smjagade@yahoo.co.in

© Springer Science+Business Media Singapore 2017
S.C. Satapathy et al. (eds.), *Proceedings of the International Conference on Data Engineering and Communication Technology*, Advances in Intelligent Systems and Computing 468, DOI 10.1007/978-981-10-1675-2_78

training and testing phase are different in text-independent speaker identification. The important steps in speaker identification are feature extraction and feature matching. Feature extraction step represents a compact representation of raw speech signal [3]. Feature extraction represents speaker specific features by discarding the redundant information. Features extracted from speech signal are used to train speaker model in training mode. In testing mode, features computed from voice of unknown person are compared with the speaker model stored in database to compute a match score.

Identification rate mainly depends on extraction of effective and efficient features of the speech signal. MFCCs are popular for speech feature extraction scheme in speaker identification [4, 5]. MFCCs are obtained by taking Fourier transform of windowed speech frames [6]. MFCC feature extraction scheme is widely studied over recent years for speech feature extraction with different variations for robust features [4–11].

The integration of MFCC features with its delta derivatives is proposed in [12]. The different choices in creating feature vectors from MFCC are studied in [13]. The optimum number of MFCC features by incorporating delta derivatives is analyzed in [14] and authors proposed that these derivatives are not good substitute when computational burden is considered. The parameters related to vector quantization (number of clusters) and Gaussian Mixture Model (number of mixtures) are studied in [13, 14], respectively.

In this paper, we have compared the performance of speaker identification system for integration of MFCC features with log energy and delta derivatives by varying number of MFCCC filters as well as number of centroids of vector quantization. This paper is organized as follows. In Sect. 2, feature extraction scheme MFCC is discussed. Speaker modelling using vector quantization is discussed in Sect. 3. The experimental results are demonstrated in Sect. 4 followed by conclusion in Sect. 5.

2 MFCC Feature Extraction Scheme

In this section, the steps present in computation of MFCC are discussed. MFCC scheme starts with pre-processing which consists of pre-emphasis, framing and windowing.

(a) Pre-emphasis
In order to boost the higher frequencies of the signal, the first-order filter with transfer function given by the following equation is applied to speech signal [11]:

$$H(z) = 1 - az^{-1}, \text{ where } 0.9 \leq a \leq \tag{1}$$

(b) Framing
The speech signal is divided into small frames of duration 20–30 ms with 50 % overlap to avoid any loss of information. It is assumes that over this short duration, speech signal remains stationary.

(c) Windowing
Hamming window is used to make the signal smooth by tapering the signal at the beginning and end of the frame. It is multiplied with each frame. Hamming window is given by the following equation:

$$\mathrm{w(n)} = 0.54 - 0.46\cos\left(\frac{2\pi n}{M-1}\right) \tag{2}$$

(d) Fast Fourier transform
Fourier transform of each windowed frame is computed to estimate the spectrum of windowed signal. Here, DFT of length N = 256 is taken.

(e) Mel filter bank
Mel filter bank is based on non-linear frequency scale which shows the perception of sound by human ear. The scale is linear up to 1000 Hz and logarithmic above 1000 Hz and is given by following equation:

$$\mathrm{F_{mel}} = 2595 * \log_{10}\left(1 + \frac{F_{linear}}{700}\right) \tag{3}$$

Mel filter bank coefficients are multiplied with the windowed spectrum followed by logarithmic operation to separate the vocal tract response from the excitation.

(f) Discrete Cosine transforms
Discrete cosine transform (DCT) is used to compress the signal and to de-correlate the log energies.

(g) Log energy, delta and double delta
Delta derivatives are useful in speaker recognition for identifying speaking styles and duration [15]. The following equation is for delta coefficient from frame t in terms of static coefficients: $c_{t+p} - c_{t-p}$.

$$\overrightarrow{d_t} = \frac{\sum_{p=1}^{P} p\left(\overrightarrow{c_{t+p}^{c}} - \overrightarrow{c_{t-p}^{c}}\right)}{2\sum_{p=1}^{P} p^2} \tag{4}$$

Double delta coefficients computed from delta coefficients by replacing $\overrightarrow{c_t^c}$ by $\overrightarrow{d_t}$. Average energy per frame represented by log energy [16] is also combined to see whether it improves identification rate.

3 Speaker Modelling Using Vector Quantization

The features extracted in training mode are used to create speaker model. To create speaker model, vector quantization is used. Vector quantization is clustering procedure. Feature vectors are clustered into K separate clusters. These clusters are represented by average vector or code vector of that cluster which is also called as centroid. A set of resulting centroids or code vectors is called codebook. This codebook is then stored in database. Here, LBG algorithm is used for clustering feature vectors. In testing phase, feature vectors of an unknown speaker are compared with speaker models stored in the database and Euclidean distance is calculated. The speaker is identified on the basis of a selected codebook for which minimum Euclidean distance is calculated.

4 Experimental Results

The speaker identification experiments have been carried out on TIMIT [17] database which consists of 10 sentences spoken by each of 630 speakers. Here, a subset of TIMIT is used. Experimentation has been carried out on 120 speakers (78 male and 42 female). Out of ten sentences, eight sentences, five SX and three SI sentences are used to train speaker models. Remaining two SA sentences are used for testing and average identification rate is calculated.

The speech signal is first pre-emphasized with a = 0.95. The signal is then divided into number of frames of duration 16 ms with 50 % overlap. Hamming window is multiplied with each frame followed by FFT. The spectrum is multiplied with Mel filter bank coefficients followed by log and DCT. First, MFCC filters are varied as 13, 20 and 29 excluding 0th coefficient with centroids as 16, 32, 64 and 128 and average identification rate is observed as shown in Fig. 1. Next, possible integrations of MFCC features with delta derivatives and log energy are carried out by varying number of MFCC filters and centroids of vector quantization. The effect of these integrations is shown Figs. 2, 3, 4 and 5.

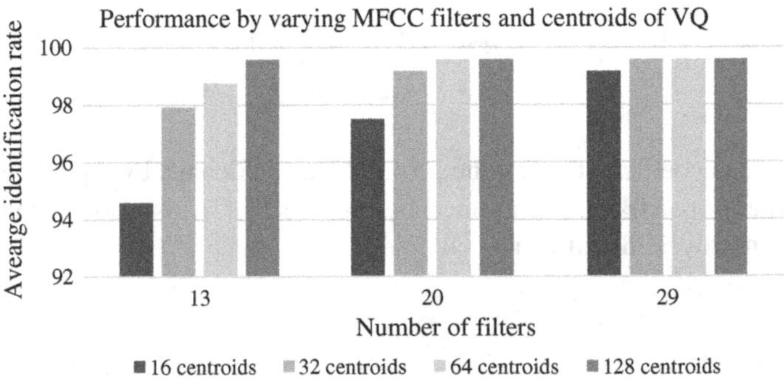

Fig. 1 Average identification rate of MFCC scheme by varying filters and centroids

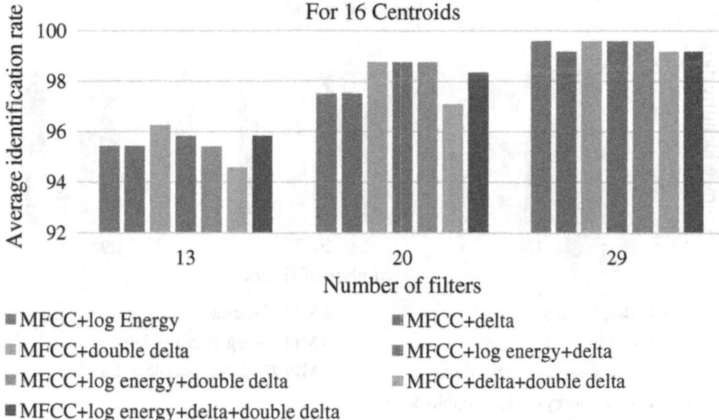

Fig. 2 Integration of MFCC with delta derivatives and log energy for 16 centroids

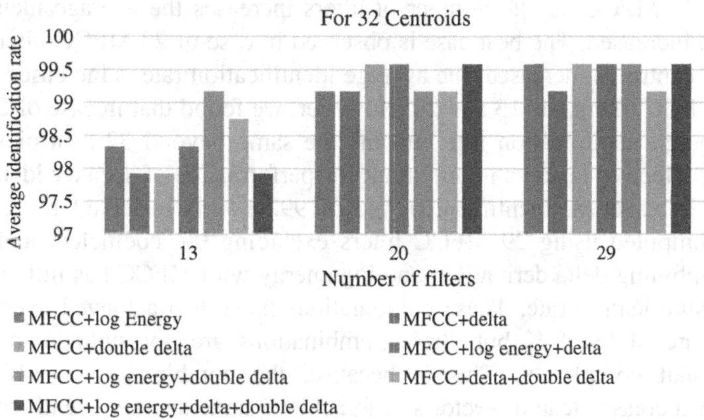

Fig. 3 Integration of MFCC with delta derivatives and log energy for 32 centroids

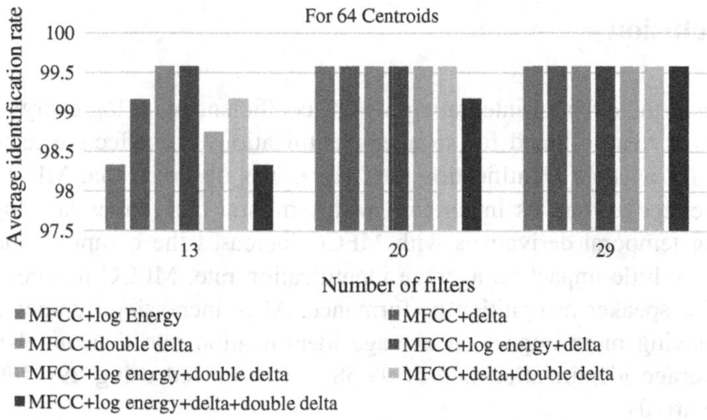

Fig. 4 Integration of MFCC with delta derivatives and log energy for 64 centroids

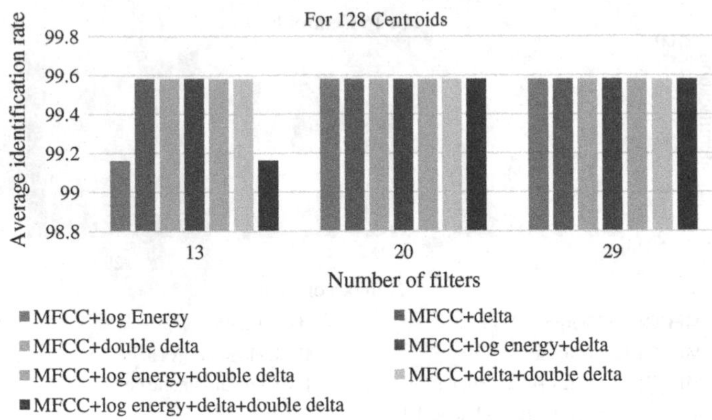

Fig. 5 Integration of MFCC with delta derivatives and log energy for 128 centroids

In case of MFCC, as the number of filters increases the average identification rate is also increased. The best case is observed in case of 29 MFCC filters. As the number of centroids increased, the average identification rate is increased for some cases of MFCC filters, i.e. 13 and 20. However, we found that in case of 29 MFCC filters, average identification rate remains the same beyond 32 centroids. Hence, number of effective features is important for performance of speaker identification system. Here, average identification rate of 99.58 % is observed for 28 MFCC features computed using 29 MFCC filters excluding 0th coefficient and 32 centroids. Combining delta derivatives and log energy with MFCC has little impact on average identification rate. These combinations have also achieved average identification rate of 99.58 % but, these combinations are not suitable in terms of computational complexity. This is because, the combination of MFCC with derivatives increases feature vector size to achieve same average identification rate. Identification rate mainly depends on effective feature vector size.

5 Conclusion

In this paper, the effect of integrating MFCC coefficients with log energy and delta derivatives is demonstrated for speaker identification. The effective feature set is important for average identification rate. Here, it is observed that MFCC features are more effective than its integration with temporal derivatives and log energy. Combining temporal derivatives with MFCC increases the computational burden and has very little impact on average identification rate. MFCC features are more effective for speaker recognition performance. Also, increasing number of MFCC filters is having more impact on average identification rate than number of centroids. Average identification rate of 99.58 % is achieved using 29 MFCC filters and 32 centroids.

References

1. Douglas A. Reynolds, "An over view of automatic speaker recognition technology", Acoustics, Speech, and Signal Processing (ICASSP), IEEE International Conference, vol.4 (2002).
2. Tomi Kinnunen, Haizhou Li, "An overview of text-independent speaker recognition: From features to supervectors", Journal on Speech Communication, Elsevier, vol.52, issue 1, pp. 12–40 (2010).
3. Md Jahangir Alam, Tomi Kinnunen, Patrick Kenny, Pierre Ouellet, Douglas O'Shaughnessy, "Multitaper MFCC and PLP features for speaker verification using i-vectors", Journal on Speech Communication, Elsevier, vol.55, issue 2, pp. 237–251 (2013).
4. WU Zunjing, CAO Zhigang, "Improved MFCC-Based Feature for Robust Speaker Identification", TUP Journals & Magazines, vol.10, no. 2, pp. 158—161 (2005).
5. Tomi Kinnunen, Rahim Saeidi, FilipSedlák, Kong Aik Lee, Johan Sandberg, Maria Hansson-Sandsten, and Haizhou Li, "Low-Variance Multitaper MFCC Features: A Case Study in Robust Speaker Verification", IEEE Transaction on Audio, Speech, and Language Processing, vol.20, no.7 pp. 1990–2001 (2012).
6. CemalHanilçi, Tomi Kinnunen, FigenErtaş, Rahim Saeidi, JouniPohjalainen, and PaavoAlku, "Regularized All-Pole Models for Speaker Verification Under Noisy Environments", IEEE Signal Processing Letters, vol.19, no.3, pp. 163–166 (2012).
7. K. Sri Rama Murty and B. Yegnanarayana, "Combining Evidence From Residual Phase and MFCC Features for Speaker Recognition", IEEE Signal Processing Letters, vol.13, no.1, pp. 52—55 (2006).
8. Seiichi Nakagawa, Longbiao Wang, and Shinji Ohtsuka, "Speaker Identification and Verification by Combining MFCC and Phase Information", IEEE Transaction on Audio, Speech, and Language Processing, vol.20, no.4, pp. 1085–1095 (2012).
9. MdSahidullah, Student Member, IEEE, and GoutamSaha, Member, IEEE, "A Novel Windowing Technique for Efficient Computation of MFCC for Speaker Recognition", IEEE Signal Processing Letters, vol.20, no. 2, pp. 149–152 (2013).
10. Pawan K. Ajmera, Dattatray V. Jadhav, Ragunath S. Holambe, "Text-independent speaker identification using Radon and discrete cosine transforms based features from speech spectrogram", Journal on Pattern Recognition, Elsevier, vol.44, issue 10–11, pp. 2749–2759 (2011).
11. Pawan K. Ajmera, Raghunath S. Holambe, "Fractional Fourier transform based features for speaker recognition using support vector machine", Journal on Computers and Electrical Engineering, Elsevier, vol. 39, issue 2, pp. 550–557 (2013).
12. Ahmad, K.S.; Thosar, A.S.; Nirmal, J.H.; Pande, V.S., "A unique approach in text independent speaker recognition using MFCC feature sets and probabilistic neural network," Advances in Pattern Recognition (ICAPR), 2015, pp. 1–6 (2015).
13. M.HassanShirali-Shahreza, SajadShirali-Shahreza, "Effect of MFCC Normalization on Vector Quantization Based Speaker Identification", Signal Processing and Information Technology (ISSPIT), 2010, pp. 250–253, (2010).
14. Almaadeed, N.; Aggoun, A.; Amira, A., "Speaker identification using multimodal neural networks and wavelet analysis," Biometrics, IET, vol.4, no.1, pp. 18–28, (2015).
15. Togneri, R.; Pullella, D., "An Overview of Speaker Identification: Accuracy and Robustness Issues," Circuits and Systems Magazine, IEEE, vol.11, no.2, pp. 23–61, Second quarter (2011).
16. FrédéricBimbot, Jean-François Bonastre, Corinne Fredouille, Guillaume Gravier, Ivan Magrin-Chagnolleau, Sylvain Meignier, Teva Merlin, Javier Ortega-García, DijanaPetrovska-Delacrétaz, Douglas A. Reynolds, "A tutorial on text-independent speaker verification", EURASIP Journal on Applied Signal Processing, 2004, pp. 430–451 (2004).
17. J.S. Garofolo, L.F. Lamel, W.M. Fisher, J.G. Fiscus, D.S. Pallett, N.L. Dahlgren, V. Zue, TIMIT acoustic-phonetic continuous speech corpus, http://catalog.ldc.upenn.edu/ldc93s1, (1993).

Traffic Violation Detection and Penalty Generation System at a Street Intersection

Pratik Chaudhari, Rajesh Yawle and Pratiksha Chaudhari

Abstract The world is rapidly urbanizing. This has resulted in a manifold increase in the number of vehicles plying on city roads. Random lane changing and red signal violation has become a major cause of traffic jams and accidents at street intersections. This tendency is increasing due to lack of proper penalty system. Also it is not possible to assign traffic police at each and every junction. This paper proposes an automated traffic violation detection system depending purely on video processing techniques. It evaluates red signal as well as lane change violation with the help of vehicle movement trend, approach lane, stop line, and traffic signal status without taking any signal from other systems such as traffic signal control box, inductive loop detectors, etc. On detecting the violation, system captures snapshot of violating vehicle and identifies the vehicle license plate number. As a part of penalty system, an android application is developed which provides all the details of violation on just entering the vehicle license plate number by authorized officer.

Keywords Violation detection · Lane change · Traffic penalty

1 Introduction

The number of accidents on roads has increased in recent times. A majority of these accidents occur at road junctions [1], when the drivers do not obey red light rules. There is a tendency among the drivers to change lanes before the stop line to avoid

Pratik Chaudhari (✉) · Rajesh Yawle
Department of Electronics and Telecommunication, Sinhgad Academy of Engineering, Pune, India
e-mail: pratik.pvc@gmail.com

Rajesh Yawle
e-mail: rajeshyawle@yahoo.co.in

Pratiksha Chaudhari
Department of Electronics and Telecommunication, Datta Meghe Institute of Engineering Technology and Research, Wardha, India
e-mail: pratiksha.pvc@gmail.com

© Springer Science+Business Media Singapore 2017

S.C. Satapathy et al. (eds.), *Proceedings of the International Conference on Data Engineering and Communication Technology*, Advances in Intelligent Systems and Computing 468, DOI 10.1007/978-981-10-1675-2_79

delays. With advances in signal processing and computer vision technology many image-based solutions have been proposed to monitor the traffic junctions.

Chen and Yang [2] have proposed a system for red-light violation detection of vehicles in the detection region, which is based on analysis of vehicle movement. Lim et al. [3] have proposed a system that detects almost all violations at road junctions such as red light violation, speed violation, stop line violation, and lane change violation by tracking individual vehicles. Both the systems depend on inputs from the traffic signal box. If there is any error in functioning of the traffic signal box, then the systems may give wrong outputs. Lai et al. [4] have proposed a system to detect color image sequence by using a stationary background estimation algorithm. The algorithm implies running mode and running average algorithms that are commonly used for background estimation. It is very important to store the true position of vehicles for red light violation detection. Sochor [5] has proposed a technique for tracking the vehicles in real time with the help of lanes analysis using kalman filter. All the systems mentioned above have not assigned any technique for generation of penalty. A proper penalty system will help in improving discipline among the drivers.

This paper proposes a system, totally independent of interconnections from the traffic signal box. This independent system works only on video monitoring at junctions. The violations detected are used to generate penalties, assisted by an android application. The paper is organized as follows: Section 2 describes the methodology for traffic violation detection using video processing. Section 3 describes penalty system based on android application. Section 4 shows the results and in Sect. 5, we conclude the paper.

2 Traffic Violation Detection

The main causes of traffic jams and accidents at a signalized junction are lane change and red signal violation. To avoid these incidents many systems have been deployed in the past, ranging from simple devices like inductive loops, laser-based devices to traffic light queue control system [6, 7]. These systems heavily depend on the signals from the traffic signal box for their functioning. For example, the inductive loop system is activated, when it receives an input from the signal box upon red signal. If any vehicle crosses the magnetic flux generated by the loops, then the video camera is activated for recording the violation. A major loophole of this system is its excessive dependence on the traffic signal box. If any interconnection between the various components fails, then the entire system is bound to fail.

An overview of the system is illustrated in Fig. 1. The propose system works on video processing techniques and is independent of the inputs from the traffic signal box. This system also includes a smart penalty ticket system based on an android application. In India, proposed penalty system can be easily targeted without making extensive changes in the current functioning of traffic management

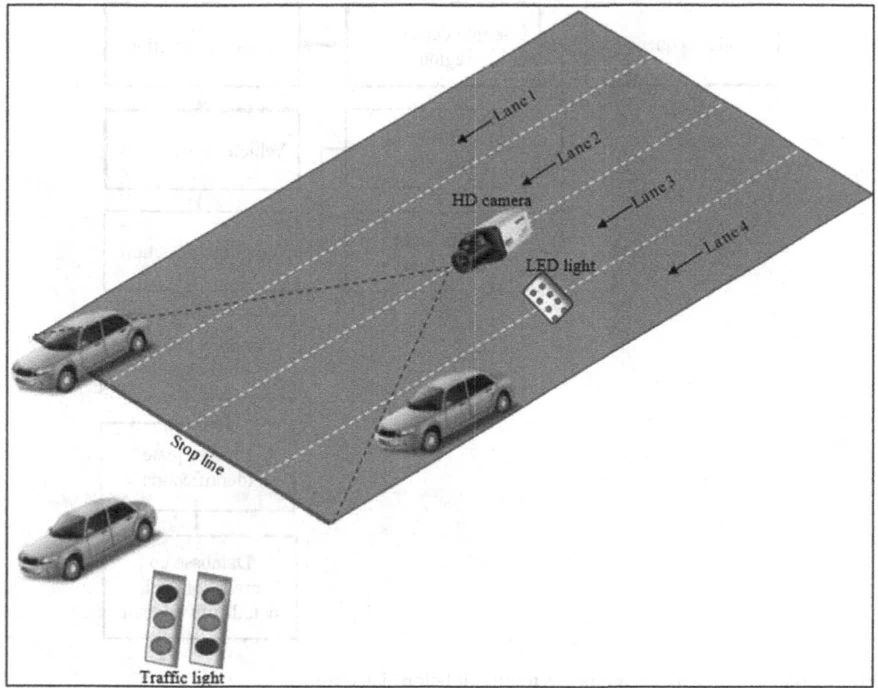

Fig. 1 An overview of purely video processing violation detection system

system. The system involves digital camera and flashlight installed at convenient location and height to get precise rear view of the vehicles and license plate. The system also takes into consideration the current status of the signal. Figure 2 shows block schematic diagram for violation detection system.

2.1 Video Capture and Setting Detection Region

Upon initialization, the cameras start basic monitoring of the area. Setting up of the detection region as shown in Fig. 3 includes marking of stop lines, different lanes and traffic signal location, using a simple graphical user interface (GUI). The pixels in region within the lines are considered and others will be ignored.

The detection region are all of pixels inside the yellow line, thus the pixels outside the detection region are zero. The detection region should have a certain length for computation of vehicle speed and vehicles movement behavior. The detection region helps reducing the computation complexity of the algorithm for vehicles movement due to only pixels interior of the yellow line are computed.

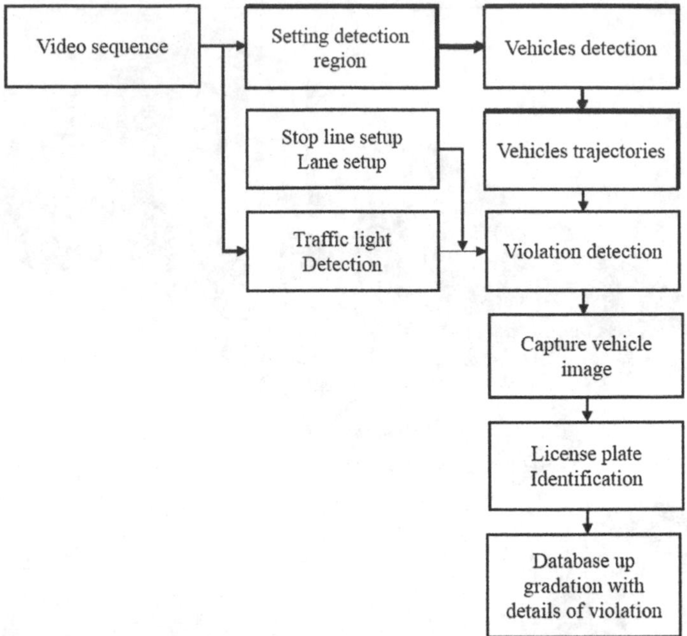

Fig. 2 Block schematic diagram for traffic violation detection

Fig. 3 Detection region setting

2.2 *Identifying the Red Signal*

The basic requirement to detect red-signal violation is to check whether the signal status is red or not. The detection of the traffic light is done from the detection region with the green lines shown in Fig. 3.

Algorithm used for detection of traffic signal: Acquire single frame and extract the red layer matrix from the RGB frame. Get the gray image of RGB frame. Now subtracting the gray frame from the red frame and filtering out unwanted noise using median filter. Convert filtered frame into corresponding binary image. From the binary image, get the centroids and bounding boxes of the blobs. Centroids are converted into integer for comparing it with the marked detection region (green box) co-ordinates. If centroid co-ordinates found inside the marked detection region, pixels takes value as 1. So, the red signal is detected in the region of interest.

2.3 Vehicle Detection

Once the red signal is detected, the system looks for any vehicle inside the yellow box region. For detection of vehicles, most of the methods [6, 8] assume that the camera is static and then desired vehicles can be detected by image differencing. In Frame differencing [9], pixel-wise differencing between two or three consecutive frames in an image sequence is used to detect the regions corresponding to moving object such as human and vehicles. In this technique an absolute difference is taken between every current image $It(x, y)$ and the reference background image $B(x, y)$ to find out the motion detection mask $D(x, y)$. The reference image of the background is generally the first frame of a video, without containing foreground object.

$$\begin{aligned} D(x, y) = 1 \text{ if } |It(x, y) - B(x, y)| \geq \tau \\ 0 \text{ otherwise} \end{aligned} \tag{1}$$

2.4 Violation Detection

Red signal and lane change violations contributes to trouble at road cross section.
Red Signal Violation—After vehicle detection on red light, the system scans for vehicles that have crossed the stop line in spite of the red signal. This is done by tracking the vehicles that have crossed the first line marked by red color. Vehicles are detected and their positions are identified by assigning bounding box and detecting centroid. If those co-ordinates cross the region marked for stop line, system detects red light violation.
Lane Change Violation—The algorithm uses the centroid co-ordinates and compares it with the co-ordinates of box marked blue lines to check lane change violation. If any vehicle moves from one blue box to other, then a violation is detected.

When a vehicle violates any one of the above rules then the camera will capture two images. One image is at current position of violation and second image after fraction of seconds to detect the direction in which the vehicle has moved. These images are accompanied by the details such as location (which is already feed in system while the system installation), time and date at which the violation happened in text file. All these details are used for penalizing the vehicle owners.

2.5 Vehicle License Plate Number Detection and Database up Gradation

On capturing the images of violation, the license plate number of the vehicles is extracted by optical character recognition (OCR) technique. Based upon this number, the fields in the MySQL database such as license plate number date, place, time and type of violation and the image of actual violation are stored. In case the system fails to detect the license plate number, then the database can be manually updated by identifying the license plate number from image by the operator. This database is used to generate the proper penalty using the system explained in Sect. 3.

3 Penalty System

The proposed penalty system has many advantages than the existing method. As a penalty system, android application is developed using software tools i.e. Eclipse Indigo (v 3.7) and Android software development kit. Proposed application is targeted for all the currently operating android versions. Android application is used to search all previously committed violation by a specific vehicle. The working of android application is explained in Fig. 4 with the help of flowchart.

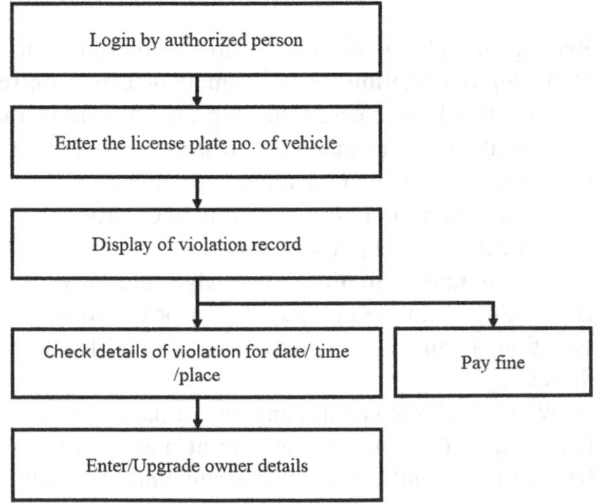

Fig. 4 Algorithm flowchart for working of penalty system android application

3.1 Login Page

For security reasons and to restrict the misuse of the content, penalty system application is provided with login module. Log in details will be provided to traffic police by respective higher authorities.

3.2 Violation Search

After providing correct log in details, application will enter to home screen as shown in Fig. 5a. Search field is used to search the violations by entering the license plate number of vehicle. 'Control room' button is provided to contact the control room through call. On clicking 'control room' button, automatically call will be connected to control room to provide information like accidents, robberies, etc. Local map button will help to display the local city map with traffic report to solve the traffic jam problems by diverting traffic to alternate routes.

Application will fetch all the previously done violation recorded details from the database and display it as shown in Fig. 5b. For collecting fine, traffic police will collect the cash from a violator and click on 'Pay Fine' button to update the fine details in database for respective violation.

3.3 Details of Violation Page

Single click on 'details of violation' button shows new page where violation image and details of violation such as place, date and time is displayed. Also on the same page, pay fine button and button to 'Enter or Update the owner details' is shown in Fig. 5c.

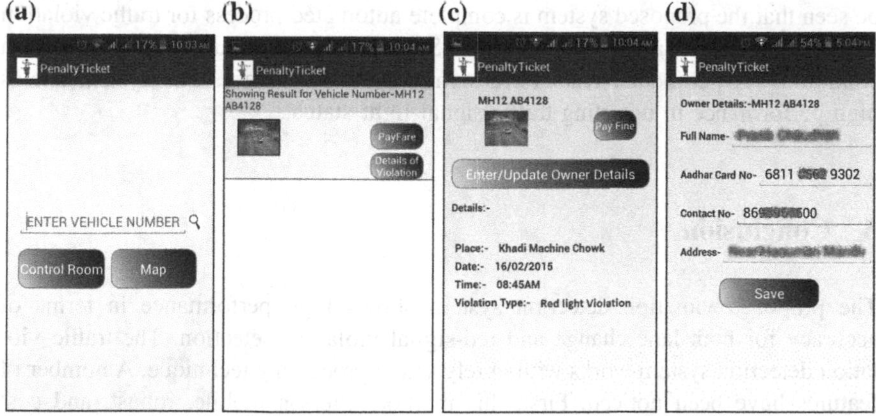

Fig. 5 **a** Home screen; **b** Violation results; **c** Details of violation page; **d** Owner details page

Table 1 Experimental results

Performance metric	The no. of vehicles				Total	Success	Results (%)
	14:00–15:00	Success	18:00–19:00	Success			
Traffic signal detection	–	–	–	–	–	–	100
Red light violation detection	69	56	46	32	115	88	76.5
Lane change violation detection	81	68	73	62	154	130	84.4
Classification of violation type	–	–	–	–	–	–	100

3.4 Owner Details Page

Owner details such as full name, Aadhar card number, contact number, and address if updated earlier will be shown otherwise blank field to fill information are provided as shown in Fig. 5d. Aadhar card number is issued by Unique Identification Authority of India as a proof of identity. Aadhar card number will be utilized for further schemes and advancements. After filling all the details, 'Done' button will progress the updating of information to the database.

4 Experimental Results

The proposed system has been tested for both of the violations such as red signal violation and lane change violation using algorithms for different modules, i.e., setting detection region, traffic signal detection, vehicle detection, violation detection, license plate identification, database up gradation, and penalty system. It can be seen that the proposed system is complete automated process for traffic violation detection system at a street intersection. System is tested at different illumination conditions. Experimental results are summarized in Table 1. The system shows high performance in detecting traffic signal light status.

5 Conclusion

The proposed violation detection system shows high performance in terms of accuracy for both lane change and red-signal violation detection. The traffic violation detection system works with purely video processing technique. A number of features have been noticed. First, this method offers a mobile, robust, and cost effective solution as it works efficiently without needing any signal from traffic signal box or buried loop detectors. The set up process is simple and fast. The violation type classification rate is 100 %.

This paper proposes a penalty system using android application which can be used for multiple approaches such as to charge violator with fine, to update vehicle owner details in database. This idea will reduce small amount of bribery percentage in India.

References

1. C. Yang, K.M. Lum. "An overview of red-light surveillance cameras in Singapore". ITE Journal, pages 87–91, (1998).
2. Y. Chen and C. Yang, "Vehicle red-light detection base on region", IEEE transection on intelligence Transportation System, vol. 9, 170–187, (2010).
3. Joon-Suk Jun, Dae-Woon Lim, Sung-Hoon Choi, "Automated detection of all kinds of violations at a street intersection using real time individual vehicle tracking". Fifth IEEE Southwest Symposium on Image Analysis and Interpretation (SSIAI02), pages 126–129, (2002).
4. Nelson H.C. Yung and Andrew H.S. Lai. "An effective video analysis method for detecting red-light runners". IEEE transaction on vehicular technology, 50(4):1074–1084, (2001).
5. Jakub Sochor. "Fully automated real-time vehicles detection and tracking with lanes analysis". In Proceedings of the 18th Central European Seminar on Computer Graphics, (2014).
6. Ali, S.S.M.; George, B.; Vanajakshi, L., "A magnetically coupled inductive loop sensing system for less-lane disciplined traffic," Instrumentation and Measurement Technology Conference (I2MTC), 2012 IEEE International, 827–832, 13–16 May (2012).
7. Cheng, H.H., Shaw, B.D., Palen, J.; Larson, J.E.; Xudong Hu; Van katwyk, K., "A real-time laser-based detection system for measurement of delineations of moving vehicles," Mechatronics, IEEE/ASME Transactions on, vol. 6, no. 2, 170–187, Jun (2001).
8. Andrew H. S. lai and Nelson H. C. Yung. "A fast and accurate scoreboard algorithm for estimating stationary backgrounds in an image sequence." IEEE International Symposium on Circuits and Systems Proceedings. 4:241–244, (1998).
9. Manisha Chaple and S.S. Paygude. Vehicle detection and tracking from video frame sequence. International Journal of Scientific and Engineering Research. 4(3):2773–2781, March (2013).

Novel Speech Processing Algorithm for Perception Improvement and Needed Research for Hearing Impaired

Bhagyashree M. Magdum and Pravin A. Dhulekar

Abstract Now-a-days, huge amount of people face the problems of hearing impairment. Hearing defines the capacity to recognize sound. People facing problems of hearing impairment has obscurity in classification of sound due to hearing defects. It is not easy to mimic the performance of human auditory system in its whole and thus reimburse for the hearing impairment. Due to the current technological advancement in signal processing area, a quite simple artificial hearing implant can be designed, thus we can achieve improvement in perception of the impaireds. For the design of artificial hearing implant, human auditory system is the top model to begin with. This paper focuses on the mechanism of human auditory system, types of hearing impairments, etc.

Keywords Hearing mechanism · Hearing loss · Hearing implants · Dyadic filters

1 Introduction

1.1 Hearing Mechanism

Outer ear, middle ear and inner ear are the three main divisions of human ear as shown in Fig. 1 [1]. Pinna and the ear canal together form the outer ear. Pinna helps in transforming the incoming signal at higher frequencies and thus helps in finding out the origin, direction and distance of the sound signal. The ear canal defends the ear from distinct bodies such as dust, dirt particles, etc. and keep the canal moist. Ear canal behaves as a filter and boosts the intensity of sound [1]. Eardrum splits the outer and middle ear.

B.M. Magdum (✉) · P.A. Dhulekar
Sandip Institute of Technology and Research Center, Nasik, India
e-mail: bhagyamagdum@gmail.com

P.A. Dhulekar
e-mail: pravin.dhulekar@sitrc.org

© Springer Science+Business Media Singapore 2017

S.C. Satapathy et al. (eds.), *Proceedings of the International Conference on Data Engineering and Communication Technology*, Advances in Intelligent Systems and Computing 468, DOI 10.1007/978-981-10-1675-2_80

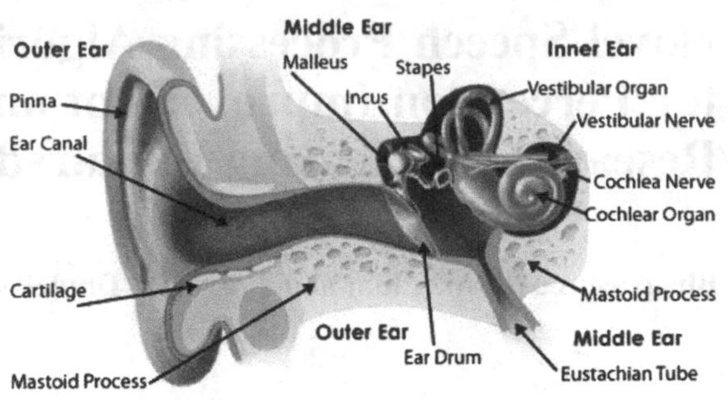

Fig. 1 Human auditory system

Middle ear is an air-filled cavity which comprises of ear drum, three ossicles and middle ear muscles. Ossicles are further subdivided as the malleus, the incus and the stapes. Ear drum is attached to Malleus at one end and the incuses attached at the other end. Incus makes contact with the stapes, the footplates of which are attached to the membrane of oval window (beginning point of the inner ear). The middle ear provides impedance matching and decreases the quantity of reflecting sound [1]. Due to impedance matching intensity of sound is about 25 dB in the middle ear. The increase in sound intensity is uneven over all the frequencies. For intense sound contraction of the muscles takes place to which the ossicles are attached and decreases transmission of sound, this is known as acoustic reflex, which plays an crucial role in defending the inner ear from extremely intense sound. Due to slow activation of acoustic reflex, impulsive signals are likely to reach the inner ear and cause damage. Reflex reduces the self-generated sounds that are emitted during chewing and before vocalization [1]. Inner ear is also called as cochlea which is a snail shaped cavity having spherical arrangement, filled with fluid [2]. The beginning of cochlea is called as oval window. Due to the incoming sound signal the oval window is set in motion due to which a pressure difference is applied athwart basilar membrane. The movement that take place depends on the involuntary properties of the basilar membrane and the rate of the incoming signal fluctuates from bottom to peak. The high-rate signals generate highest displacement at the bottom while low-rate signals generate highest displacement at the peak [1]. The organ of corti contains large number of sensory hair cells, which is situated in between basilar and tectorial membrane. The inner hair cells contains 3000–3500 hairs cells, each cell contains about 40 hairs and the outer hair cells contains 20000–25000 hairs cells, each cell contains about 140 hairs, respectively. [1]. The sound information is conveyed by means of internal hair cells which excite the tactile neurons. Outer hair cells helps in improving responses of the basilar membrane which generate high acuity, and fine-tuning to the basilar membrane [1].

2 Possible Impairment in Auditory System and Causes

Hearing impairment is generally categorized as conductive and sensorineural depending upon the location of damage in the auditory system [1]. Conductive impairment occurs due to damage in the outer ear, ear drum or middle ear, due to which the transmission of sound to the inner ear gets blocked. It generates attenuation of the stimulus and enhance both the hearing threshold levels and discomfortable loudness levels. Conductive impairment can be treated medically or surgically in most cases [1]. When the transduction mechanism of the inner ear gets damaged then such a impairment is known as Sensorineural hearing impairment. The injury caused due to deficiency in the hearing part is known as cochlear impairment [1] and the injury taking place due to defects in the auditory nerve is retrocochlear hearing impairment. The cochlear hearing impairment is associated with absence of cochlear hair cells. Impairment caused due to ageing causes malfunctions in auditory system such as stiffening of basilar membrane, of the cochlea which causes loss of hair cells in the organ of Corti and deterioration of neurons in the intact auditory system. Impairment is bilaterally symmetrical, with greater hearing impairment at high frequencies, and increased difficulty in understanding speech. Initially, the hearing loss may be mild with a loss of sensitivity in the high frequencies that continues to increase with increasing age. When hearing loss occurs due to both conductive and sensorineural loss, in other words there may be any defect in the outer/middle or inner ear or auditory nerve is called as mixed hearing loss.

3 Related Work on Hearing Aids

Hearing impairment can be partially compensated using hearing aid which amplifies acoustic signals with a frequency/gain characteristic which best reimburses for the deficiency in hearing [3]. Conventional hearing aids are usually used for addressing the problem of elevated thresholds. The hearing aids comprises of a microphone, electronic filter, control for the fine-tuning and frequency response, an earphone, a battery for supplying the power source, and elastic tubing and an ear mold, for merging the response of the earphone to the peripheral ear channel. Conventional hearing aids are categorized by their size and the way of wearing, as body-worn hearing aid, eyeglass hearing aid, etc. [2–4]. The BTE Hearing aids are the electronic components which are snowed under a tiny oval case that is fitted at the back of the ear. Receiver distributes the responses to the ear canal through a flexible tube terminating in an ear mold. These are the hearing aids most commonly used severe to profound hearing impaired. ITE hearing aid is made up of a small plastic case containing all the components, which is worn in the external section of the peripheral ear and the cochlea. Such aid is normally used by mild to moderate and moderately severe hearing impaired. ITC aid is the smallest in size, among all hearing aids that fits entirely in the ear canal. Because of small size, available power output is low. Hence mild or moderate, flat, and gradual sloping hearings impaired generally use these aids [4]. The compression techniques are used in hearing aids to overcome the problem of acoustic amplification, which compose weak signals audible to intense signals but makes uncomfortably loud due to loudness recruitment in sensorineural hearing impairment. In these techniques the amplification decreases with intensity, such that the wide dynamic range of input signal acquired compression in a smaller dynamic range of the output [3]. An epochal method is projected where the critical bands were compacted inorder to regulate the outline of the acoustic filters of hearing impaired persons. The compression using Fourier transform-based approach get better quality and intelligibility of speech for hearing impaired persons. The compression rates vary between 20 to 40 % depending individual shaped auditory filter [5]. Frequency transposition technique is a method in which energy in the high frequency region is transposed to low frequency region. The effects of the simultaneous and non simultaneous masking have been compensated by various speech enhancing schemes in case of the impairments in the inner ear. Some methods which use alteration of energy for certain frequency components are used in order to reduce the spectral masking. Schemes for enhancing the speech perception for the persons with reduced temporal resolution are based on the use of clear speech [2–4]. Masking effects plays an vital role in our day to day life. When two people are having conversion in silent environment very small amount of speech power is required for both the persons to hear and understand each other. However, the speech in the crowded shopping mall is completely impossible to hear. A signal is most likely to get masked by another signal with frequency components which are near to, or the same as, that of the signal. When masking event occurs among any two signals which appear at the same time is called simultaneous masking or spectral masking, on the other hand, if the signals are comparatively postponed in time it is called temporal masking or non simultaneous masking [6].

4 Proposed System

The proposed work is divided into two sections, first Decomposition by Dyadic Analysis Filter Bank represented in Fig. 2. And second Reconstruction by Synthesis of Dyadic Analysis Filter Bank represented in Fig. 3.

4.1 Decomposition

Figure 2 represents the decomposition part of the proposed system. First, the own recorded VCV speech signals are acquired. The sampling rate of the signal is lowered in the process of decimation. Then the signals are downsampled by factor

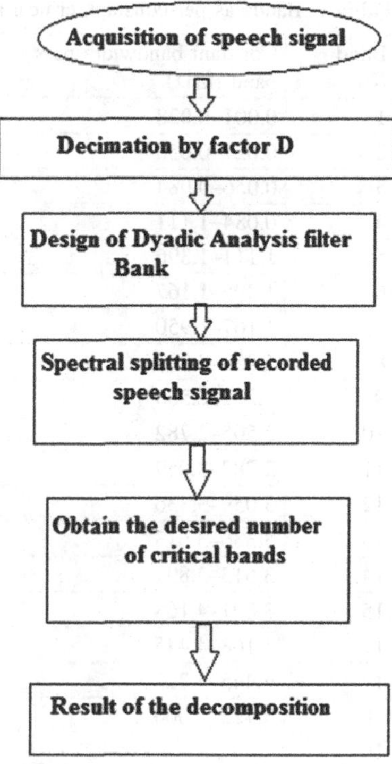

Fig. 2 Flowchart for decomposition

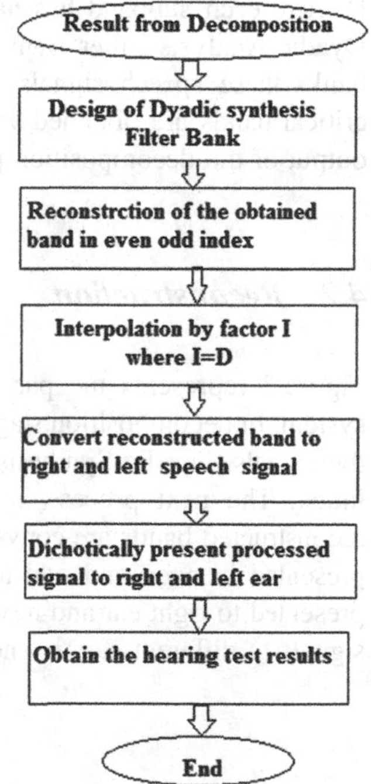

Fig. 3 Flowchart for reconstruction

Table 1 Bands as per constant, critical and 1/3 octave bandwidth

Band No.	Constant bandwidth pass band (kHz)	Critical bandwidth pass band (kHz)	1/3 Octave bandwidth
1	0.001–0.028	0.01–0.20	0.0708–0.089
2	0.027–0.056	0.20–0.30	0.089–0.112
3	0.056–0.084	0.30–0.40	0.112–0.141
4	0.084–1.111	0.40–0.51	0.141–0.178
5	1.111–1.396	0.51–0.63	0.178–0.224
6	1.396–1.167	0.63–0.77	0.224–0.282
7	1.167–1.950	0.77–0.92	0.282–0.355
8	1.950–2.227	0.92–1.08	0.355–0.447
9	2.227–2.505	1.08–1.27	0.447–0.562
10	2.505–2.782	1.27–1.48	0.562–0.708
11	2.782–3.059	1.48–1.72	0.708–0.891
12	3.059–3.336	1.72–2.00	0.891–1.120
13	3.336–3.613	2.00–2.32	1.120–1.410
14	3.613–3.891	2.32–2.70	1.410–1.780
15	3.891–4.168	2.70–3.15	1.780–2.240
16	4.168–4.445	3.15–3.70	2.240–2.820
17	4.445–4.722	3.70–4.40	2.820–3.550
18	4.722–5.000	4.40–5.00	3.550–4.470
19			4.470–5.000

D, here each subband has half bandwidth and half sample rate. Next Design of Dyadic Analysis Filter Bank is done. By filtering the signals by using the filter banks these speech signals are spectrally splitted. Then the desired number of critical bands are obtained which are listed below in Table 1. At last the desired output of the decomposition part is obtained.

4.2 Reconstruction

Figure 3 represents the part B that is the Reconstruction part of the proposed system. In decomposition stage we decimate the speech signals here we reconstruct the signals. Dyadic Synthesis Filter bank reconstruct the signals in even and odd index. The next process is upsampling of the signal where I = D. Next, the reconstructed bands are converted to right and left speech signal. These signals are presented in even and odd terms i.e. if there are 16 bands then eight bands are presented to right ear and next eight bands to left ear. This presentation of different signals to different ears is known as dichotic presentation of the speech signals..

5 Results and Discussion

For achieving the above mentioned objectives, the algorithm is developed to obtain a desired number of bands of selected speech material for binaural dichotic presentation. These bands are obtained using comb filters which gives sharp transitions between selected bands. For implementing this strategy MATLAB software is used along with its Simulink expansion. In the first stage, 18 bands of Constant Bandwidth, nineteen bands of 1/3 Octave bandwidths and eighteen bands of critical bandwidth are obtained by using a couple of corresponding comb filters. There are nine pass bands analogous to auditory critical bandwidth filters in the corresponding comb filters. These bands are shown in the following tables as:

Simulation results obtained after applying above scheme of analysis on sample speech signal is shown below.

The first quadrant of Fig. 4 shows the unprocessed sample speech signal, 2nd and 3rd quadrant shows the result of splitted speech having 9 bands each with constant bandwidth.

Figure 5 represents the unprocessed sample speech signal in 1st quadrant while 2nd and 3rd quadrant shows splitted speech having 9 bands each based on critical bandwidth.

Similarly, Fig. 6 represents unprocessed speech signal and 19 bands of processed speech based on 1/3 octave bandwidth.

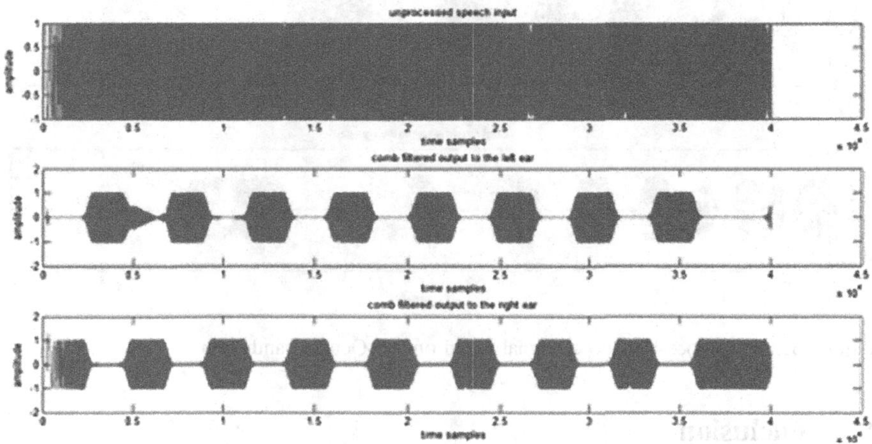

Fig. 4 Bands of processed speech signal based on constant bandwidth

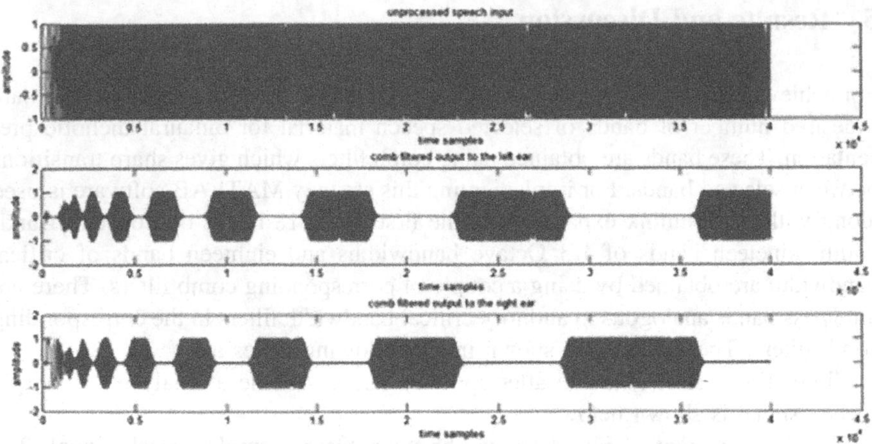

Fig. 5 Bands of processed speech signal based on critical bandwidth

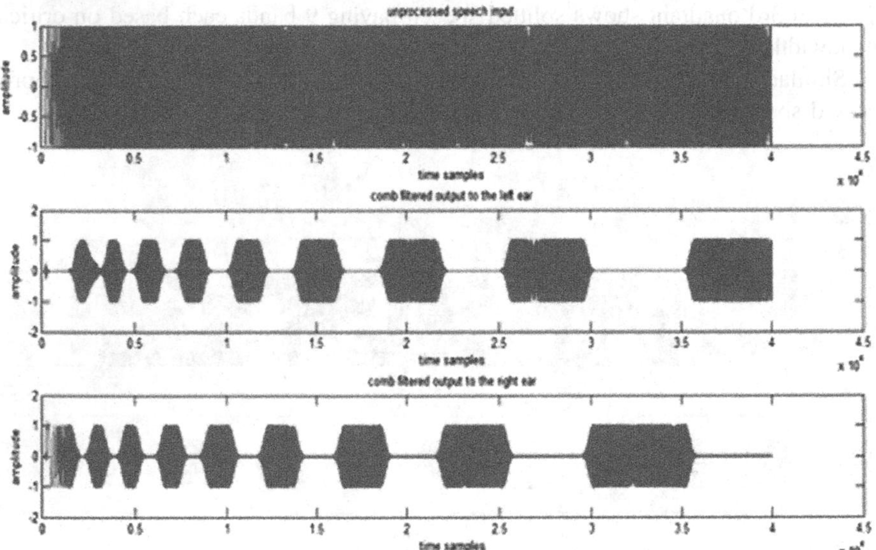

Fig. 6 Bands of processed speech signal based on 1/3 Octave bandwidth

6 Conclusion

This paper gives a brief study of the Human Auditory system, Types of hearing losses, Different hearing aids and measures to overcome the hearing losses. The proposed algorithm is useful in removing the masking effect caused due to the improper frequency selection of the people suffering from hearing loss. To obtain the desired result, a novel algorithm is designed. Eighteen bands of Constant Bandwidth, 19 bands of 1/3 Octave bandwidths and 18 bands of critical bandwidth using pair of complementary comb filters from the filter sets are obtained successfully.

References

1. Moore, B. C. J.: An Introduction to Psychology of Hearing, 4th Ed. (Academic, London), 1997.
2. Loizou, P. C., Mani, Arunvijay, and Dorman, M. F.: Dichotic speech recognition in noise using reduced spectral cues, J. Acoust. Soc. Am. vol. 114(1), 475–483, 2003.
3. GL Ward, MD MacAllister.: Apparatus and method for conveying amplified sound to ear, Google Patent, 1993.
4. CHABA.: Speech-perception aids for hearing-impaired people: Current status and needed research, J. Acoust. Soc. Am. vol. 90, 637–683, 1991.
5. Yasu, K., Kobayashi, K., Hishitani, M., Arai, T.: and Murahara Y. "Critical band-based frequency compression for digital hearing aids, J. Acoust. Sci. and Tech. vol. 25, 61–63, 2004.
6. Peng Dai, Frank Rudzicz, Ing Yann Soon, Alex Mihailidis, Huijun Ding.: 2D Psychoacoustic modeling of equivalent masking for automatic speech recognition, ELSEVIER, 2015.

Author Index

© Springer Science+Business Media Singapore 2017

S.C. Satapathy et al. (eds.), *Proceedings of the International Conference on Data Engineering and Communication Technology*, Advances in Intelligent Systems and Computing 468, DOI 10.1007/978-981-10-1675-2

Printed in the United States
By Bookmasters